Pro/ENGINEER® Wildfire 2.0™

... James Gee

De Anza College

ISBN 1-58503-194-1

Schroff Development Corporation

www.schroff.com
www.schroff-europe.com

Preface

Pro/ENGINEER® is one of the most widely used CAD/CAM software programs in the world today. Any aspiring engineer will benefit from the knowledge contained herein, while in school or upon graduation as a newly employed engineer.

The text involves creating a new part, an assembly, or a drawing, using a set of Pro/E commands that walk you through the process systematically. Instructors will find the text short enough for a one-term Engineering Graphics class at the university level and comprehensive enough for a full in-depth class on Pro/ENGINEER at a technical school or community college.

Projects are not included in the text to keep the length and cost to the user down. For instructors and students wanting more material, compressive supplemental lessons (Sweeps, Blends, Patterns, Shells, Sheetmetal, NC, Expert Machinist, and Surfaces) can be downloaded at www.cad-resources.com.

If you wish to contact the author concerning orders, questions, changes, additions, suggestions, comments, or to get on our email list, please send an email to one of the following:

- **CAD-Resources.com** (Louis Gary Lamit and Associates)
- Web Site: **www.cad-resources.com**
- Email: **cad@cad-resources.com**

For a small handling and shipping fee (10.00 US), a CD (with Pro/E files) is available from the author for **instructors** (not individuals) who adopt this text. The Pro/E files will open only on Academic and Commercial versions *(not SE or TE Editions)* of Pro/ENGINEER software. *CD's will only be sent to instructors at their college or school department.*

CADTRAIN's ***COAch for Pro/ENGINEER*** www.cadtrain.com is a computer-based training (CBT) product designed to provide a comprehensive training program for Pro/ENGINEER users in their actual CAD environment. COAch for Pro/ENGINEER®, has been referenced in the book's figures with the authorized use of illustrations. COAch is one of the best ways available for expanding your knowledge of Pro/E software. *A free sampler CD can be requested from CADTRAIN, or downloaded from their website.*

Dedication

This book is dedicated to my first grandchild, Madison Elizabeth Champagne (MEC)

Om Mani Padme Hum

Acknowledgments

I want to thank the following people and organizations for the support and materials granted the author:

Ken Page	Parametric Technology Corporation
Leslie Minasian	Parametric Technology Corporation
Larry Fire	Parametric Technology Corporation
Thuy Dao Lamit	Lamit and Associates
Dennis Stajic	CADTRAIN
Stephen Schroff	SDC

About the Authors

Louis Gary Lamit is currently a full time instructor at De Anza College in Cupertino, Ca., where he teaches Pro/ENGINEER, Pro/SURFACE, Pro/SHEETMETAL, Pro/NC, and Expert Machinist.

Mr. Lamit has worked as a drafter, designer, numerical control (NC) programmer, technical illustrator, and engineer in the automotive, aircraft, and piping industries. A majority of his work experience is in the area of mechanical and piping design. He started as a drafter in Detroit (as a job shopper) in the automobile industry, doing tooling, dies, jigs and fixture layout, and detailing at Koltanbar Engineering, Tool Engineering, Time Engineering, and Premier Engineering for Chrysler, Ford, AMC, and Fisher Body. Mr. Lamit has worked at Remington Arms and Pratt & Whitney Aircraft as a designer, and at Boeing Aircraft and Kollmorgan Optics as an NC programmer and aircraft engineer. He also owns and operates his own consulting firm (CAD-Resources.com- Lamit and Associates), and has been involved with advertising, and patent illustration. He is the author of over 30 textbooks, workbooks, tutorials, and handbooks.

Mr. Lamit received a BS degree from Western Michigan University in 1970 and did Masters' work at Wayne State University and Michigan State University. He has also done graduate work at the University of California at Berkeley and holds an NC programming certificate from Boeing Aircraft.

Since leaving industry, Mr. Lamit has taught at all levels (Melby Junior High School, Warren, Mi.; Carroll County Vocational Technical School, Carrollton, Ga.; Heald Engineering College, San Francisco, Ca.; Cogswell Polytechnical College, San Francisco and Cupertino, Ca.; Mission College, Santa Clara, Ca.; Santa Rosa Junior College, Santa Rosa, Ca.; Northern Kentucky University, Highland Heights, Ky.; and De Anza College, Cupertino, Ca.). His textbooks include:

- ***Industrial Model Building***, with Engineering Model Associates, Inc. (1981),
- ***Piping Drafting and Design*** (1981),
- ***Piping Drafting and Design Workbook*** (1981),
- ***Descriptive Geometry*** (1983),
- ***Descriptive Geometry Workbook*** (1983), and
- ***Pipe Fitting and Piping Handbook*** (1984), published by Prentice-Hall.
- ***Drafting for Electronics*** (3rd edition, 1998),
- ***Drafting for Electronics Workbook*** (2nd edition 1992), and
- ***CADD*** (1987), published by Charles Merrill (Macmillan-Prentice-Hall Publishing).
- ***Technical Drawing and Design*** (1994),
- ***Technical Drawing and Design Worksheets and Problem Sheets*** (1994),
- ***Principles of Engineering Drawing*** (1994),
- ***Fundamentals of Engineering Graphics and Design*** (1997),
- ***Engineering Graphics and Design with Graphical Analysis*** (1997), and
- ***Engineering Graphics and Design Worksheets and Problem Sheets*** (1997), published by West Publishing (ITP/Delmar).

James Gee is currently a part time instructor at De Anza College in Cupertino, Ca., where he teaches Pro/ENGINEER, and Pro/MECHANICA. Mr. Gee graduated from the University of Nevada- Reno with a BSME. He has worked in the Aerospace industry for Lockheed Missiles and Space Company, Sunnyvale, Ca. and Space Systems/Loral in Palo Alto, Ca. Mr. Gee has assisted in checking and writing a number of articles and textbooks with Mr. Lamit including:

- ***Basic Pro/ENGINEER in 20 Lessons*** (1998) (Revision 18) and
- ***Basic Pro/ENGINEER (with references to PT/Modeler)*** (1999) (Revision 19 and PT/Modeler), published by PWS Publishing (ITP).
- ***Pro/ENGINEER 2000i*** (1999) (Revision 2000i), and
- ***Pro/E 2000i² (includes Pro/NC and Pro/SHEETMETAL)*** (2000) (Revision 2000i²), published by Brooks/Cole Publishing (ITP).
- ***Pro/ENGINEER Wildfire*** (2003) (Revision Wildfire) published by Brooks/Cole Publishing (ITP).

Table of Contents

Downloads

Extra material can be downloaded from **www.cad-resources.com** ⇒ click: **Downloads**

- Part, Assembly, and Drawing Projects
- Extra Lessons
- Pro/E Files
- Pro/E related articles and information

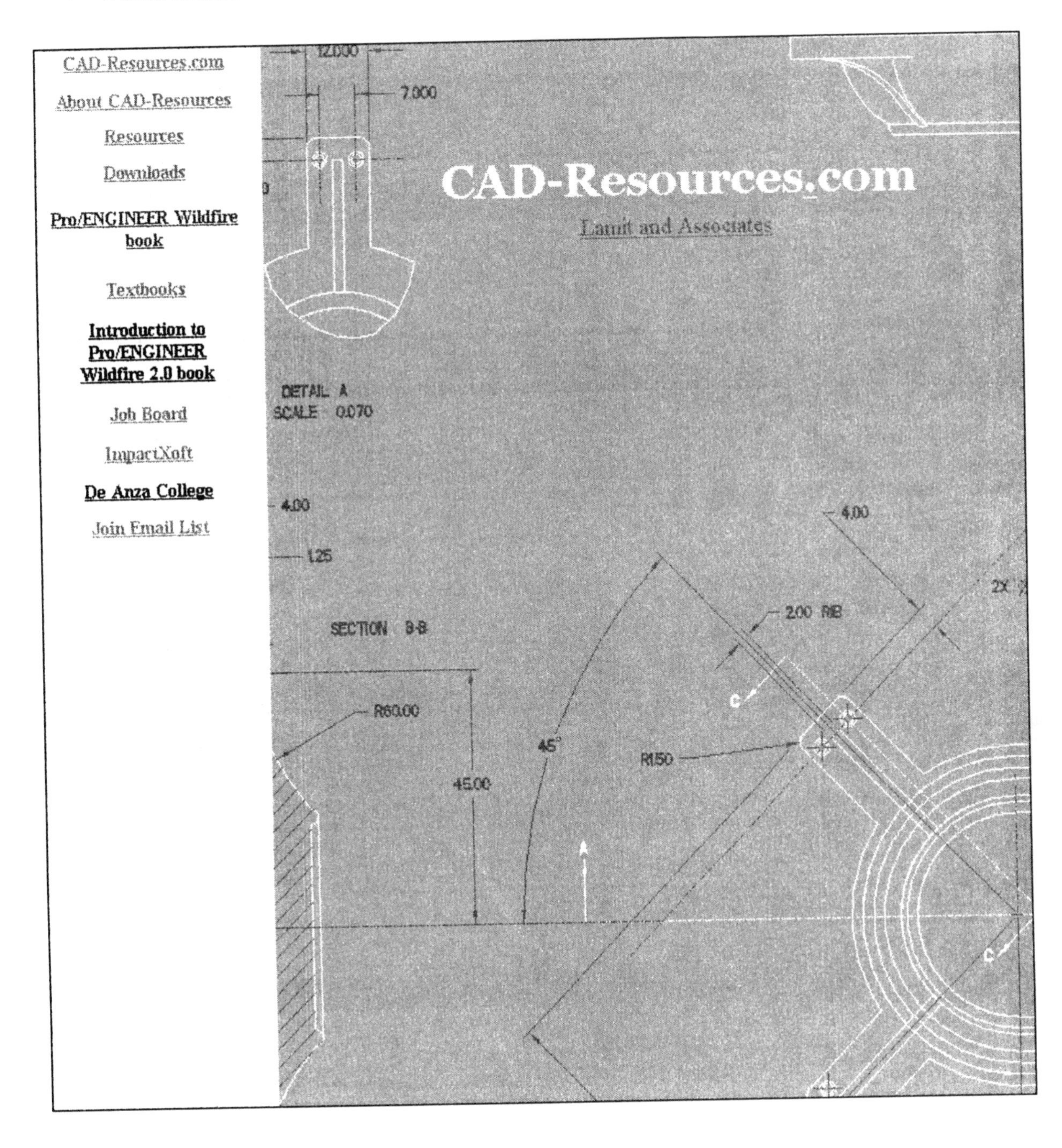

Projects from Website

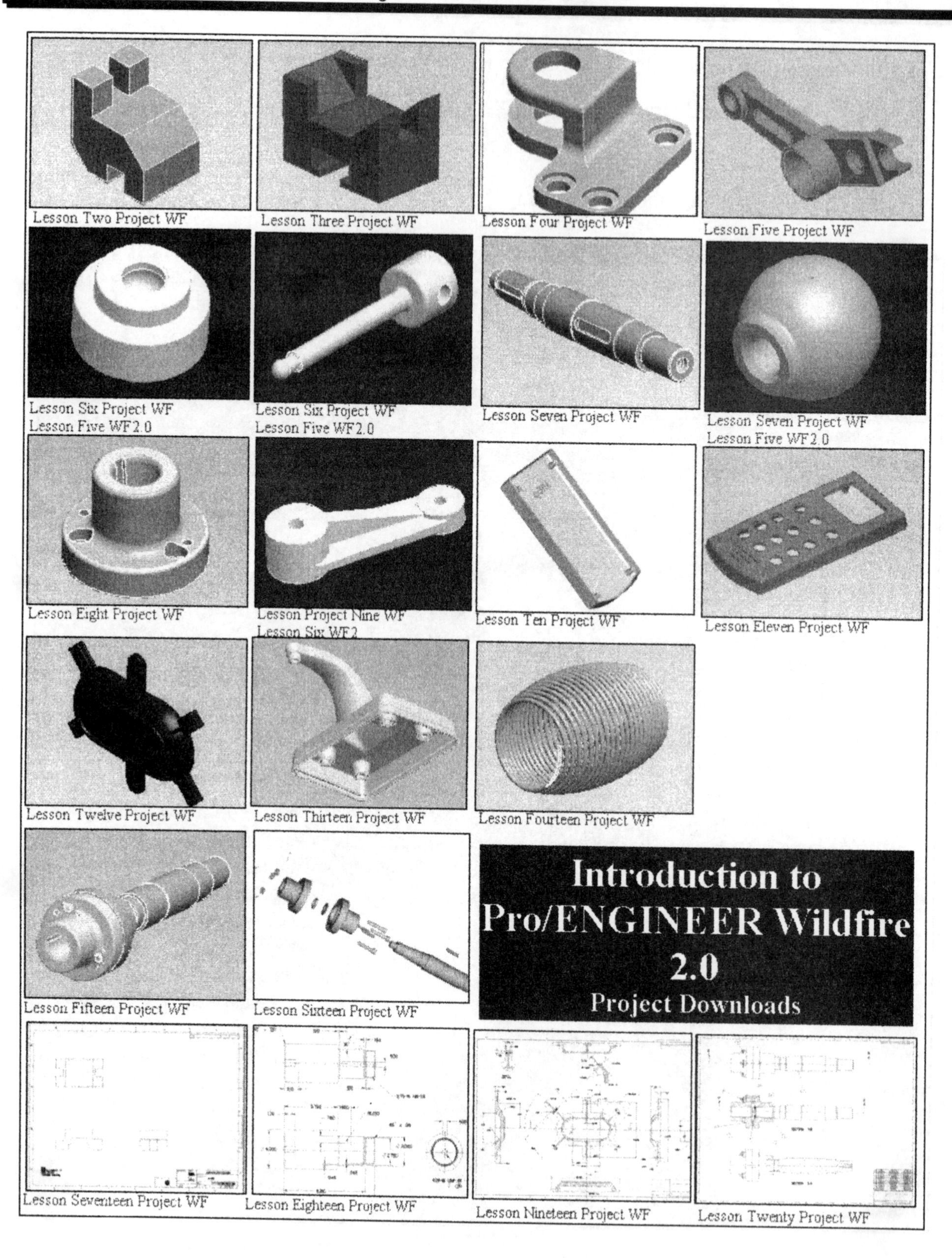
Lesson Two Project WF
Lesson Three Project WF
Lesson Four Project WF
Lesson Five Project WF
Lesson Six Project WF
Lesson Five WF2.0
Lesson Six Project WF
Lesson Five WF2.0
Lesson Seven Project WF
Lesson Seven Project WF
Lesson Five WF2.0
Lesson Eight Project WF
Lesson Project Nine WF
Lesson Six WF2
Lesson Ten Project WF
Lesson Eleven Project WF
Lesson Twelve Project WF
Lesson Thirteen Project WF
Lesson Fourteen Project WF
Lesson Fifteen Project WF
Lesson Sixteen Project WF
Introduction to
Pro/ENGINEER Wildfire
2.0
Project Downloads
Lesson Seventeen Project WF
Lesson Eighteen Project WF
Lesson Nineteen Project WF
Lesson Twenty Project WF

Introduction to Pro/ENGINEER Wildfire 2.0

This text introduces the basic concepts of parametric design using Pro/ENGINEER® Wildfire 2.0™. While using this text, you will create individual parts, assemblies, and drawings.

Parametric can be defined as *any set of physical properties whose values determine the characteristics or behavior of an object.* **Parametric design** enables you to generate a variety of information about your design: its mass properties, a drawing, or a base model. To get this information, you must first model your part design.

Parametric modeling philosophies used in Pro/E include the following:

Feature-Based Modeling Parametric design represents solid models as combinations of engineering features (Fig. 1).

Creation of Assemblies Just as features are combined into parts, parts may be combined into assemblies (Fig. 1).

Capturing Design Intent The ability to incorporate engineering knowledge successfully into the solid model is an essential aspect of parametric modeling.

Figure 1 Parts and Assembly Design

Parametric Design

Parametric design models are not drawn so much as they are *sculpted* from solid volumes of materials. To begin the design process, analyze your design. Before any work is started, take the time to ***tap*** into your own knowledge bank and others that are available. Think, Analyze, and Plan. These three steps are essential to any well-formulated engineering design process.

Break down your overall design into its basic components, building blocks, or primary features. Identify the most fundamental feature of the object to sketch as the first, or base, feature. Varieties of **base features** can be modeled using extrude, revolve, sweep, and blend tools.

Sketched features (*extrusions, sweeps, etc.)* and pick-and-place features called **referenced features** (*holes, rounds, chamfers, etc.*) are normally required to complete the design. With the SKETCHER, you use familiar 2D entities (points, lines, rectangles, circles, arcs, splines, and conics) (Fig. 2). There is no need to be concerned with the accuracy of the sketch. Lines can be at differing angles, arcs and circles can have unequal radii, and features can be sketched with no regard for the actual objects' dimensions. In fact, exaggerating the difference between entities that are similar but not exactly the same is actually a far better practice when using the SKETCHER.

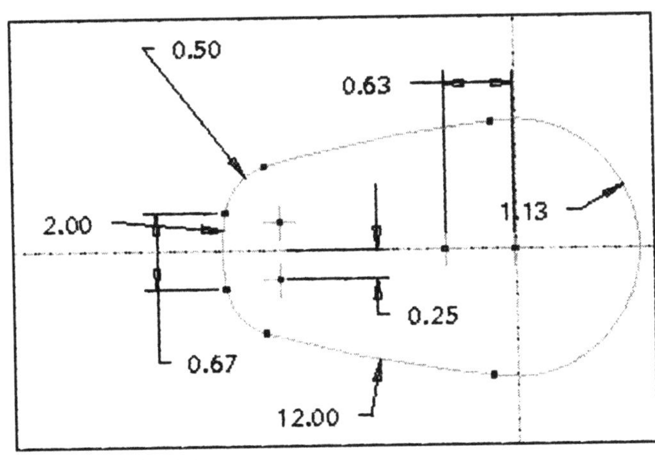

Figure 2 Sketching

Geometry assumptions and constraints will close ends of connected lines, align parallel lines, and snap sketched lines to horizontal and vertical (orthogonal) orientations. Additional constraints are added by means of **parametric dimensions** to control the size and shape of the sketch.

Features are the basic building blocks you use to create an object (Fig. 3). Features "understand" their fit and function as though "smarts" were built into the features themselves. For example, a hole or cut feature "knows" its shape and location and the fact that it has a negative volume. As you modify a feature, the entire object automatically updates after regeneration. The idea behind feature-based modeling is that the designer constructs an object so that it is composed of individual features that describe the way the geometry is supposed to behave if its dimensions change. This happens quite often in industry, as in the case of a design change. Feature-based modeling is diagramed in Figure 4.

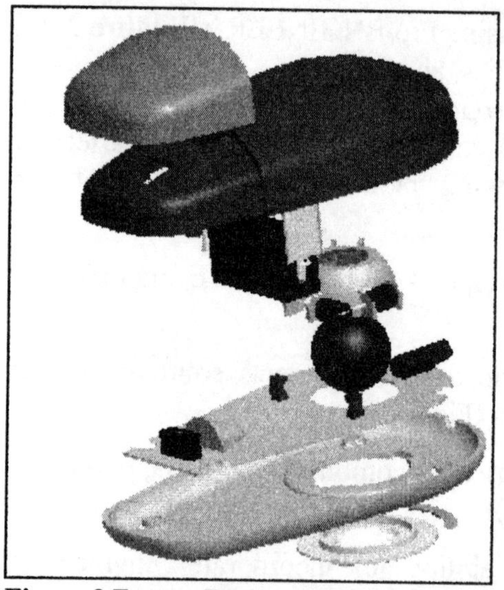

Figure 3 Feature Design (Courtesy CADTRAIN)

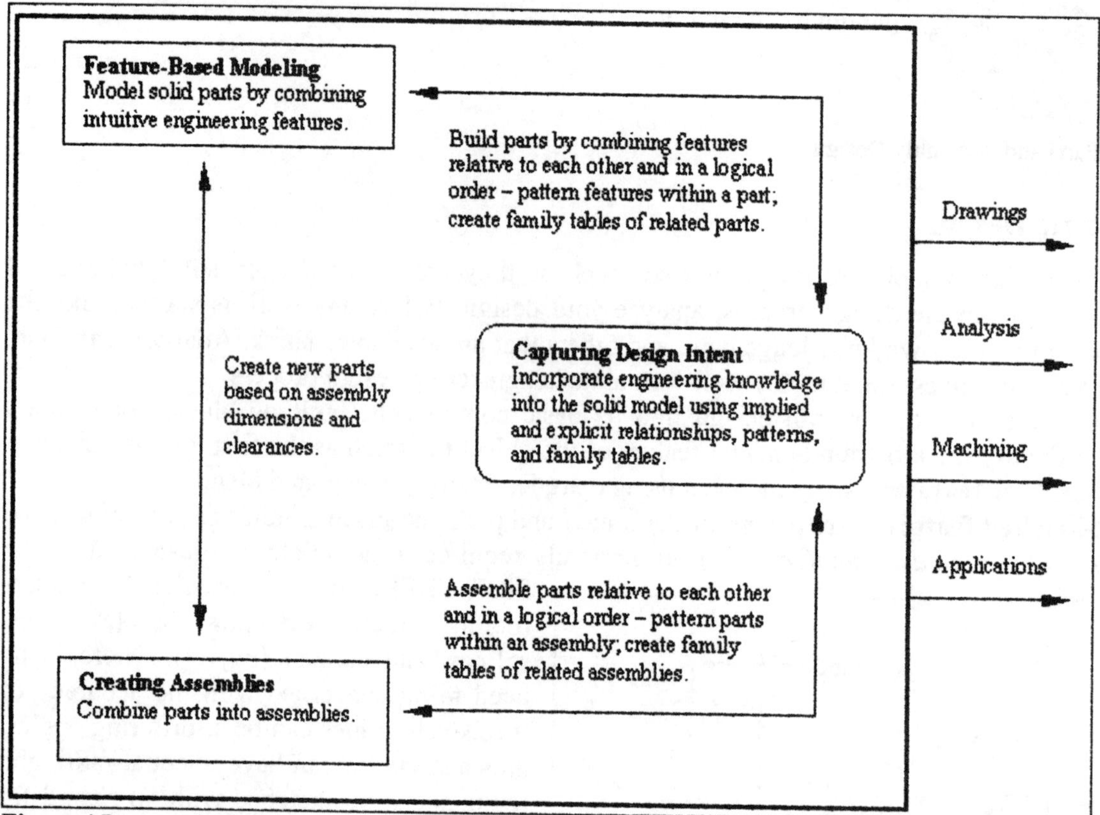
Feature-Based Modeling
Model solid parts by combining intuitive engineering features.
Build parts by combining features relative to each other and in a logical order – pattern features within a part; create family tables of related parts.
Drawings
Analysis
Capturing Design Intent
Incorporate engineering knowledge into the solid model using implied and explicit relationships, patterns, and family tables.
Create new parts based on assembly dimensions and clearances.
Machining
Applications
Assemble parts relative to each other and in a logical order – pattern parts within an assembly; create family tables of related assemblies.
Creating Assemblies
Combine parts into assemblies.

Figure 4 Parametric Design

Parametric modeling is the term used to describe the capturing of design operations as they take place, as well as future modifications and editing of the design. The order of the design operations is significant. Suppose a designer specifies that two surfaces be parallel, such that surface two is parallel to surface one. Therefore, if surface one moves, surface two moves along with it to maintain the specified design relationship. The second surface is a **child** of surface one in this example. Parametric modelers allow the designer to **reorder** the steps in the object's creation.

Various types of features are used as building blocks in the progressive creation of solid objects. Figure 5 shows base features, datum features, sketched features, and referenced features. The "chunks" of solid material from which parametric design models are constructed are called **features**.

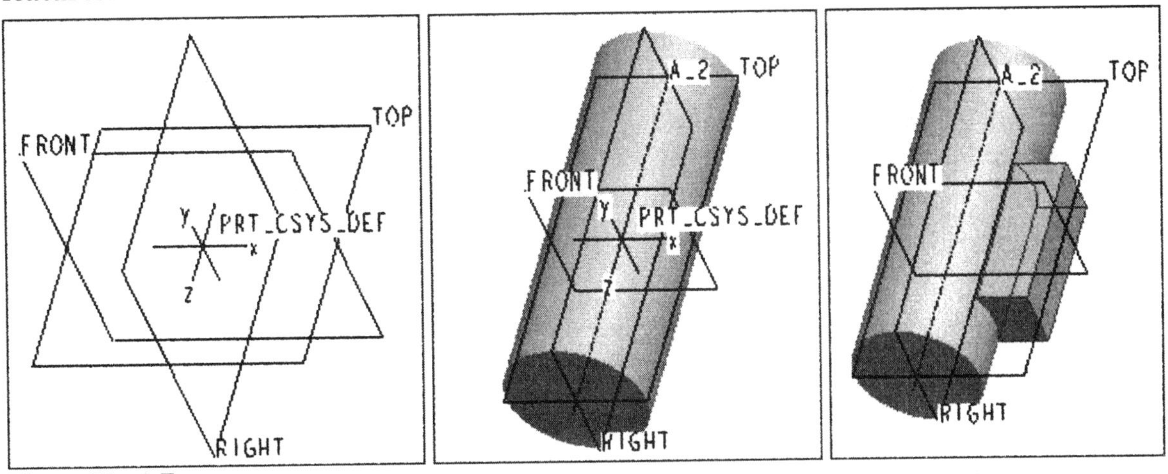

Figure 5(a-c) Features

Features generally fall into one of the following categories:

Base Feature The base feature is normally a set of datum planes referencing the default coordinate system. The base feature is important because all future model geometry will reference this feature directly or indirectly; it becomes the root feature. Changes to the base feature will affect the geometry of the entire model (Fig. 6).

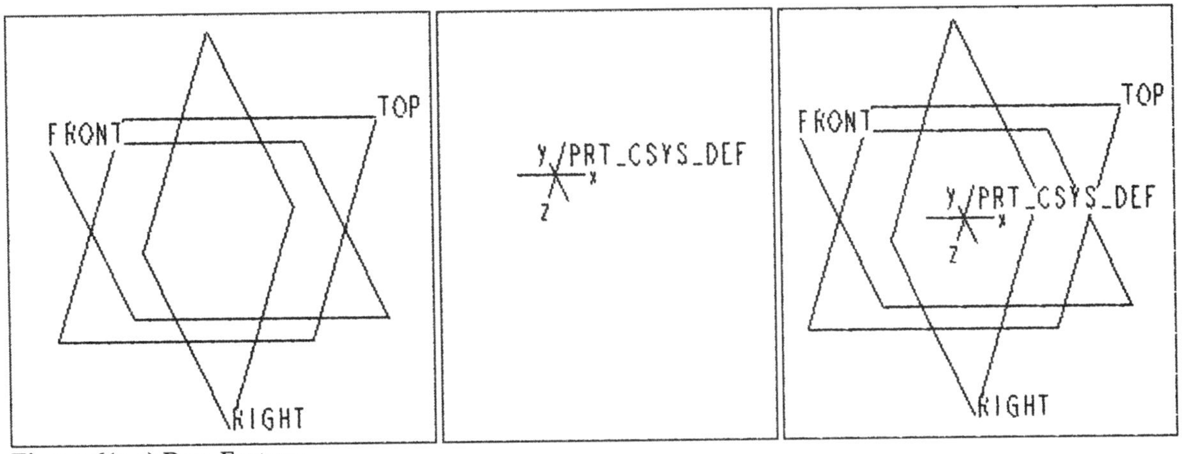

Figure 6(a-c) Base Features

Datum Features Datum features (lines, axes, curves, and points) are generally used to provide sketching planes and contour references for sketched and referenced features. Datum features do not have volume or mass and may be visually hidden without affecting solid geometry (Fig. 7).

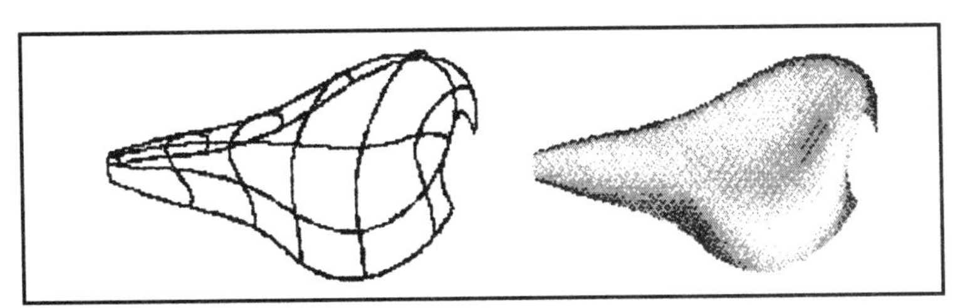

Figure 7 Datum Features

Sketched Features Sketched features are created by extruding, revolving, blending, or sweeping a sketched cross section. Material may be added or removed by protruding or cutting the feature from the existing model (Fig. 8).

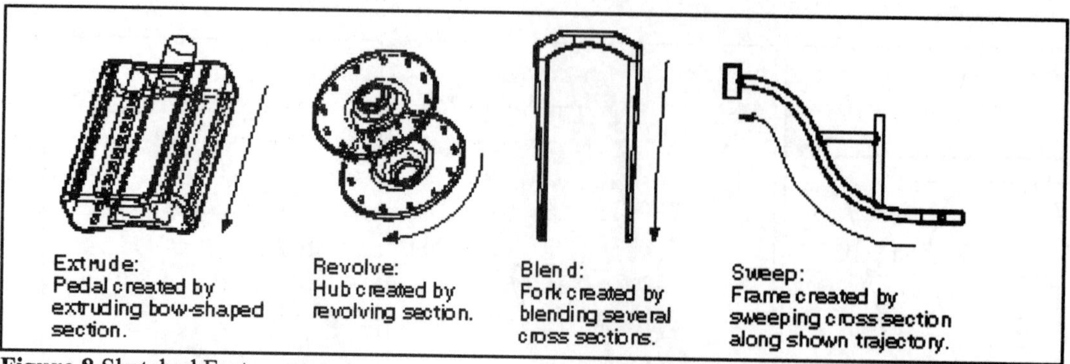

Figure 8 Sketched Features

Referenced Features Referenced features (rounds, holes, shells, and so on) utilize existing geometry for positioning and employ an inherent form; they do not need to be sketched (Fig. 9).

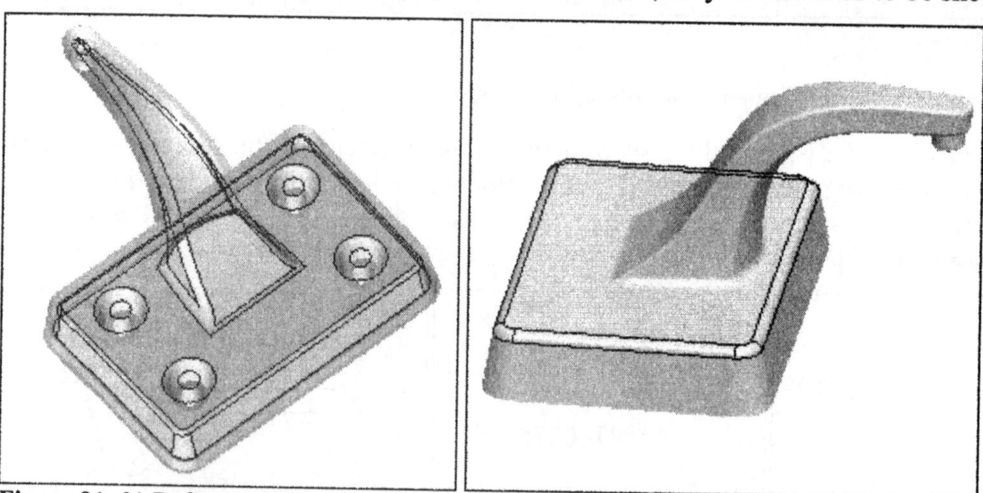

Figure 9(a-b) Referenced Features- Shell and Round (Spout is a Swept Blend Feature)

A wide variety of features are available. These tools enable the designer to make far fewer changes by capturing the engineer's design intent early in the development stage (Fig.10).

Figure 10 Parametric Designed Part

Fundamentals

The design of parts and assemblies, and the creation of related drawings, forms the foundation of engineering graphics. When designing with Pro/ENGINEER, many of the previous steps in the design process have been eliminated, streamlined, altered, refined, or expanded. The model you create as a part forms the basis for all engineering and design functions.

The part model contains the geometric data describing the part's features, but it also includes non-graphical information embedded in the design itself. The part, its associated assembly, and the graphical documentation (drawings) are parametric. The physical properties described in the part drive (determine the characteristics and behavior of) the assembly and drawing. Any data established in the assembly mode, in turn, determines that aspect of the part and, subsequently, the drawings of the part and the assembly. In other words, all the information contained in the part, the assembly, and the drawing is interrelated, interconnected, and parametric (Fig. 11).

Part Design

In many cases, the part will be the first component of this interconnected process. Therefore, in this text, the first set of Lessons (1-6) cover part design. The *part* function in Pro/E is used to design components.

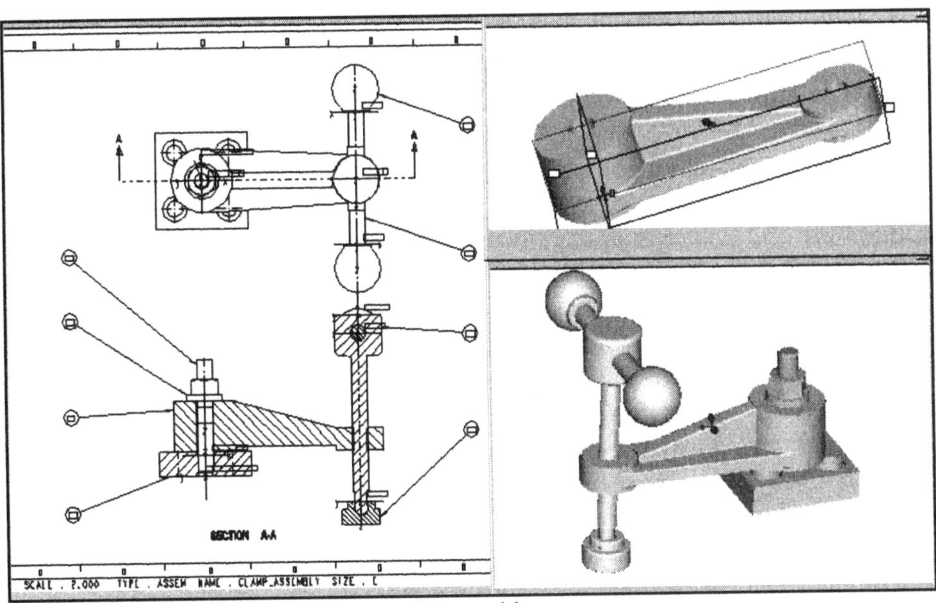

Figure 11 Assembly Drawing, Part, and Assembly

During part design (Fig. 12), you can accomplish the following:

- Define the base feature
- Define and redefine construction features to the base feature
- Modify the dimensional values of part features (Fig. 13)
- Embed design intent into the model using tolerance specifications and dimensioning schemes
- Create pictorial and shaded views of the component
- Create part families (family tables)
- Perform mass properties analysis and clearance checks
- List part, feature, layer, and other model information
- Measure and calculate model features
- Create detail drawings of the part

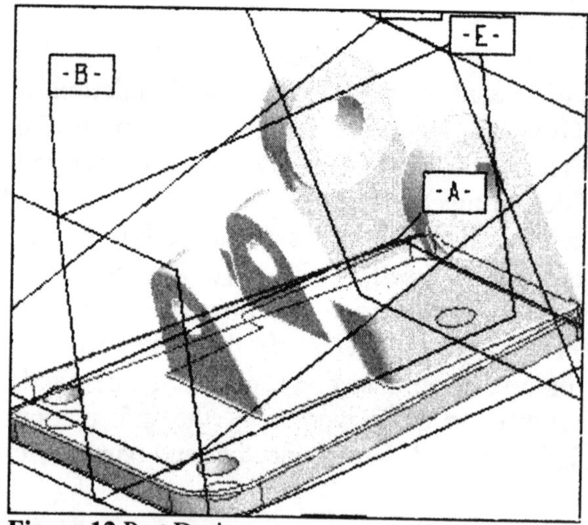

Figure 12 Part Design

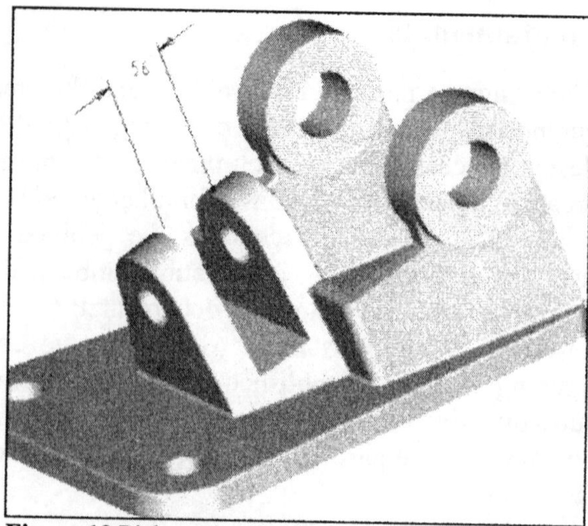
Figure 13 Pick on the **.56** dimension and type in a new value

Establishing Features

The design of any part requires that the part be *confined*, *restricted*, *constrained*, and *referenced*. In parametric design, the easiest method to establish and control the geometry of your part design is to use three datum planes. Pro/E automatically creates the three **primary datum planes**. The default datum planes (**RIGHT**, **TOP**, and **FRONT**) constrain your design in all three directions.

Datum planes are infinite planes located in 3D model mode and are associated with the object that was active at the time of their creation. To select a datum plane, you can pick on its name or anywhere on the perimeter edge. Datum planes are *parametric*--geometrically associated with the part. Parametric datum planes are associated with and dependent on the edges, surfaces, vertices, and axes of a part.

Datum planes are used to create a reference on a part that does not already exist. For example, you can sketch or place features on a datum plane when there is no appropriate planar surface. You can also dimension to a datum plane as though it were an edge. In Figure 14, three **default datum planes** and a **default coordinate system** were created when a NEW part was started using the default template. Note that in the **Model Tree** window they are the first four features of the part, which means that they will be the *parents* of the features that follow. The three *default datum planes* and the *default coordinate system* appear in the Model Tree as the first four features of a new part (**PRT0003.PRT**).

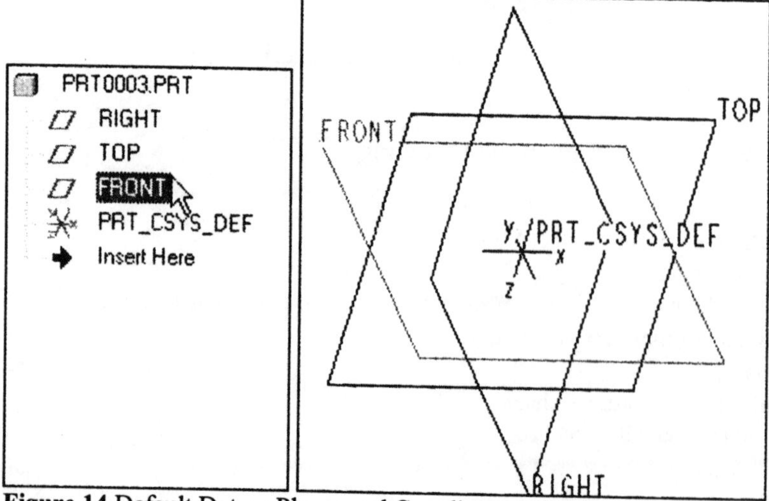

Figure 14 Default Datum Planes and Coordinate System

Datum Features

Datum features are planes, axes, and points you use to place geometric features on the active part. Datums other than defaults can be created at any time during the design process.

As we have discussed, there are three (primary) types of datum features (Fig. 14): **datum planes, datum axes**, and **datum points** (there are also ***datum curves*** and ***datum coordinate systems***). You can display all types of datum features, but they do not define the surfaces or edges of the part or add to its mass properties. In Figure 15, a variety of datum planes are used in the creation of the cell phone.

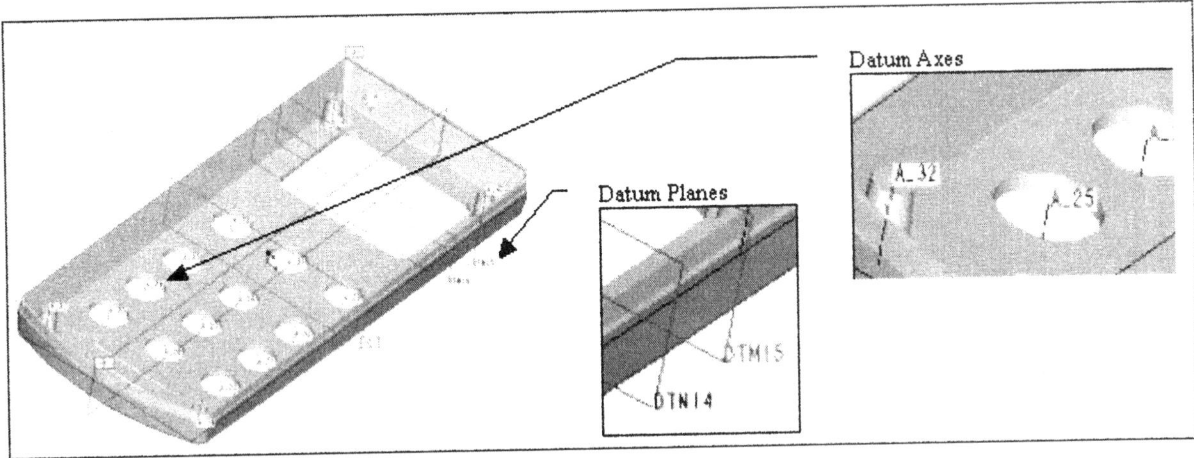

Figure 15 Datums in Part Design

Specifying constraints that locate it with respect to existing geometry creates a datum. For example, a datum plane might be made to pass through the axis of a hole and parallel to a planar surface. Chosen constraints must locate the datum plane relative to the model without ambiguity. You can also use and create datums in assembly mode.

Besides datum planes, datum axes and datum points can be created to assist in the design process. You can also automatically create datum axes through cylindrical features such as holes and solid round features by setting this as a default in your Pro/E configuration file. The part in Figure 16 shows **A_1** through the hole. **A_1** is the default axis of the circular cut.

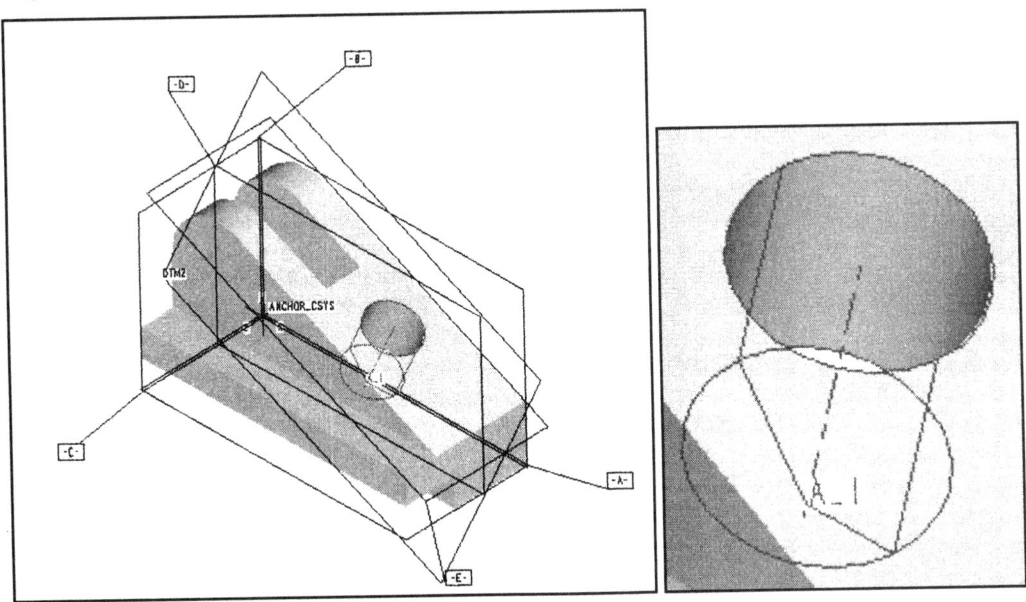

Figure 16(a-b) Feature Default Datum Axis

Parent-Child Relationships

Because solid modeling is a cumulative process, certain features must, of necessity precede others. Those that follow must rely on previously defined features for dimensional and geometric references. The relationships between features and those that reference them are termed ***parent-child relationships***. Because children reference parents, parent features can exist without children, but children cannot exist without their parents. This type of CAD modeler is called a history-based system. Using Pro/E's information command will list the models' information as shown in Figure 17.

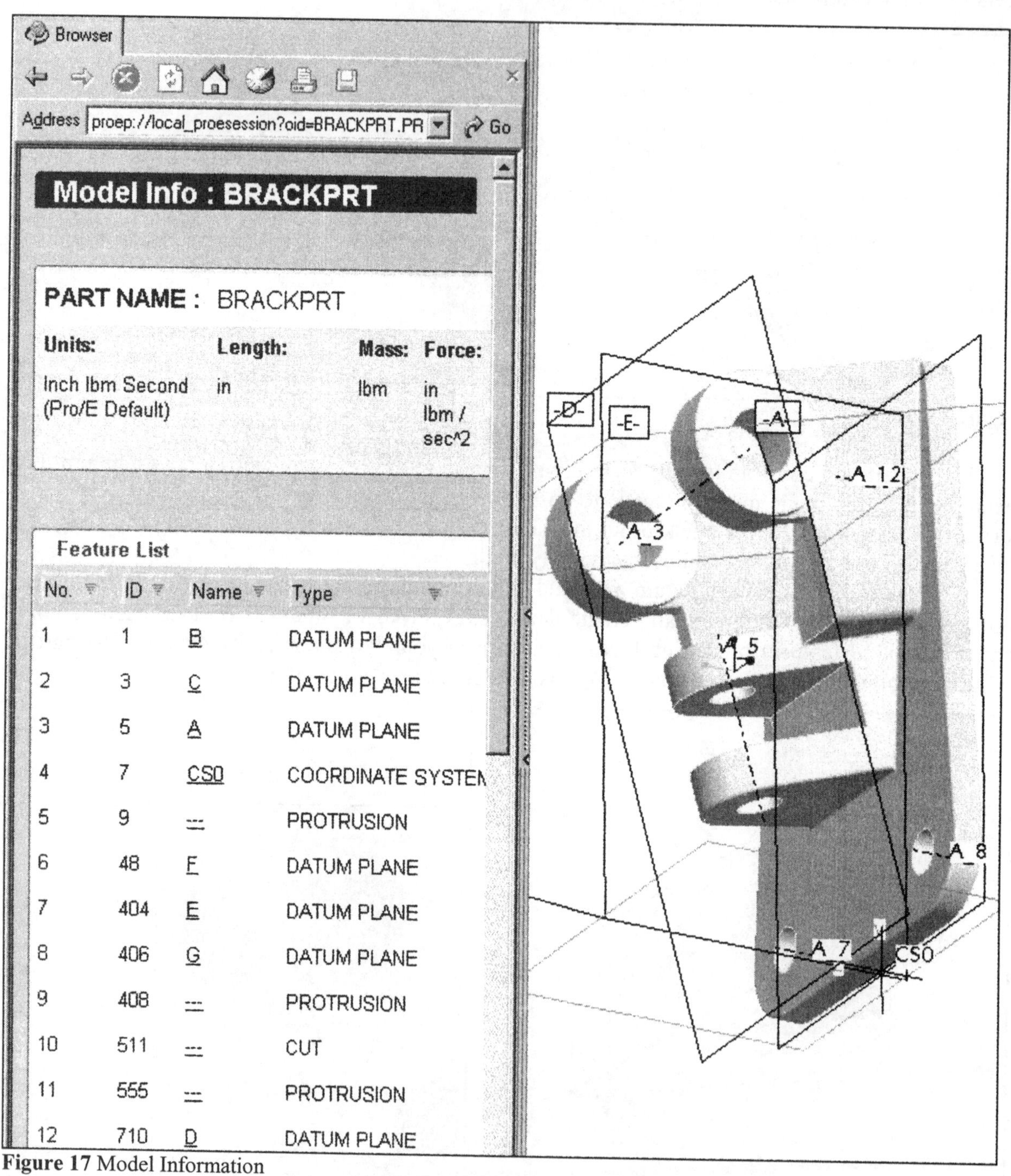

Figure 17 Model Information

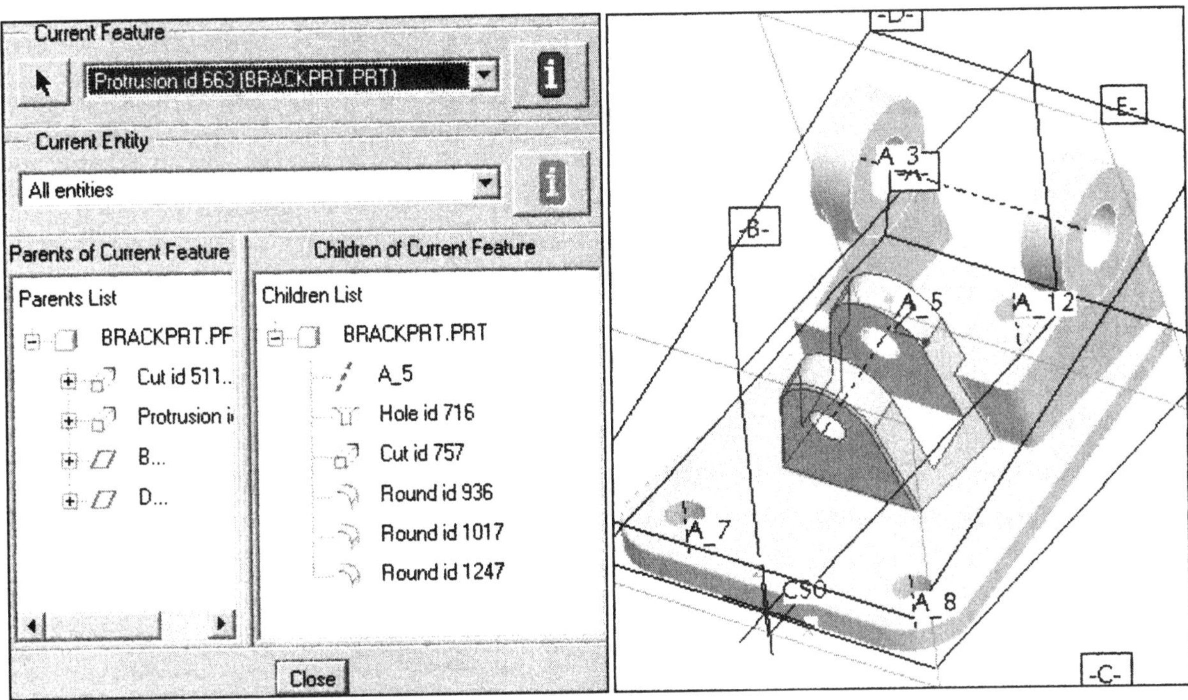

Figure 18 Parent-Child Information

The parent-child relationship (Fig. 18) is one of the most powerful aspects of parametric design. When a parent feature is modified, its children are automatically recreated to reflect the changes in the parent feature's geometry. It is essential to reference feature dimensions so that design modifications are correctly propagated through the model/part. Any modification to the part is automatically propagated throughout the model (Fig. 19) and will affect all children of the modified feature.

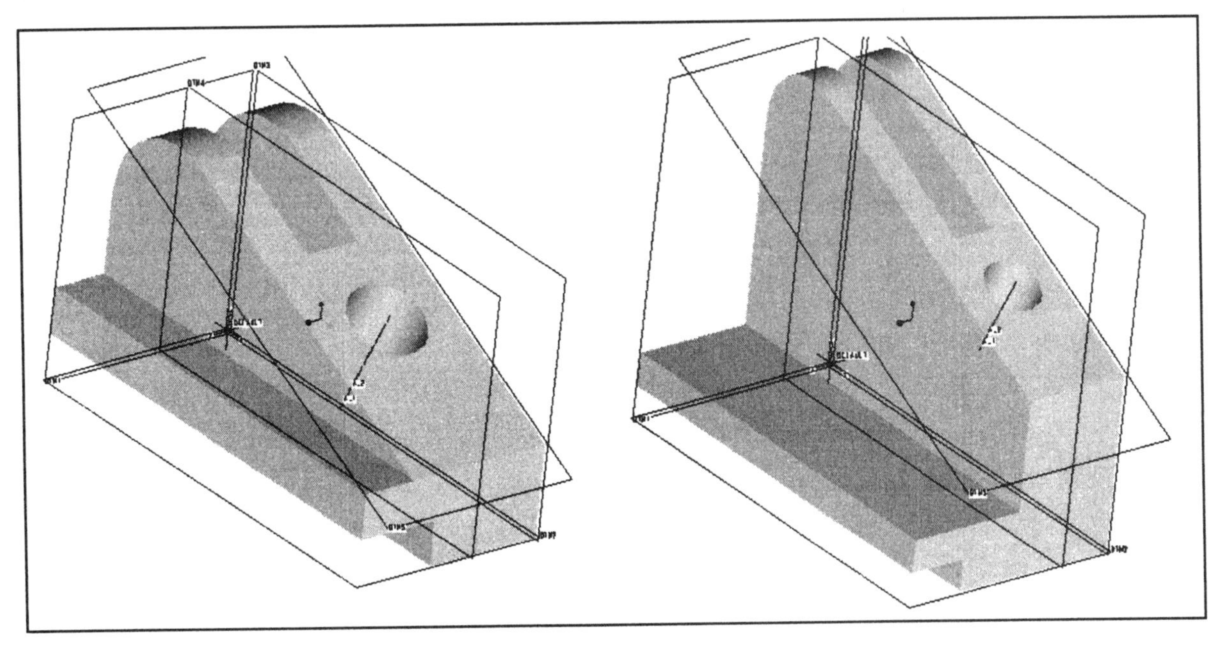

Figure 19 Original Design and Modification

Capturing Design Intent

A valuable characteristic of any design tool is its ability to *render* the design and at the same time capture its *intent* (Fig. 20). Parametric methods depend on the sequence of operations used to construct the design. The software maintains a *history of changes* the designer makes to specific parameters. The point of capturing this history is to keep track of operations that depend on each other. Whenever Pro/E is told to change a specific dimension, it can update all operations that are referenced to that dimension.

For example, a circle representing a bolthole circle may be constructed so that it is always concentric to a circular slot. If the slot moves, so does the bolt circle. Parameters are usually displayed in terms of dimensions or labels and serve as the mechanism by which geometry is changed. The designer can change parameters manually by changing a dimension or can reference them to a variable in an equation (**relation**) that is solved either by the modeling program itself or by external programs such as spreadsheets.

Features can also store non-graphical information. This information can be used in activities such as drafting, numerical control (NC), finite-element analysis (FEA), and kinematics analysis.

Capturing design intent is based on incorporating engineering knowledge into a model by establishing and preserving certain geometric relationships. The wall thickness of a pressure vessel, for example, should be proportional to its surface area and should remain so, even as its size changes.

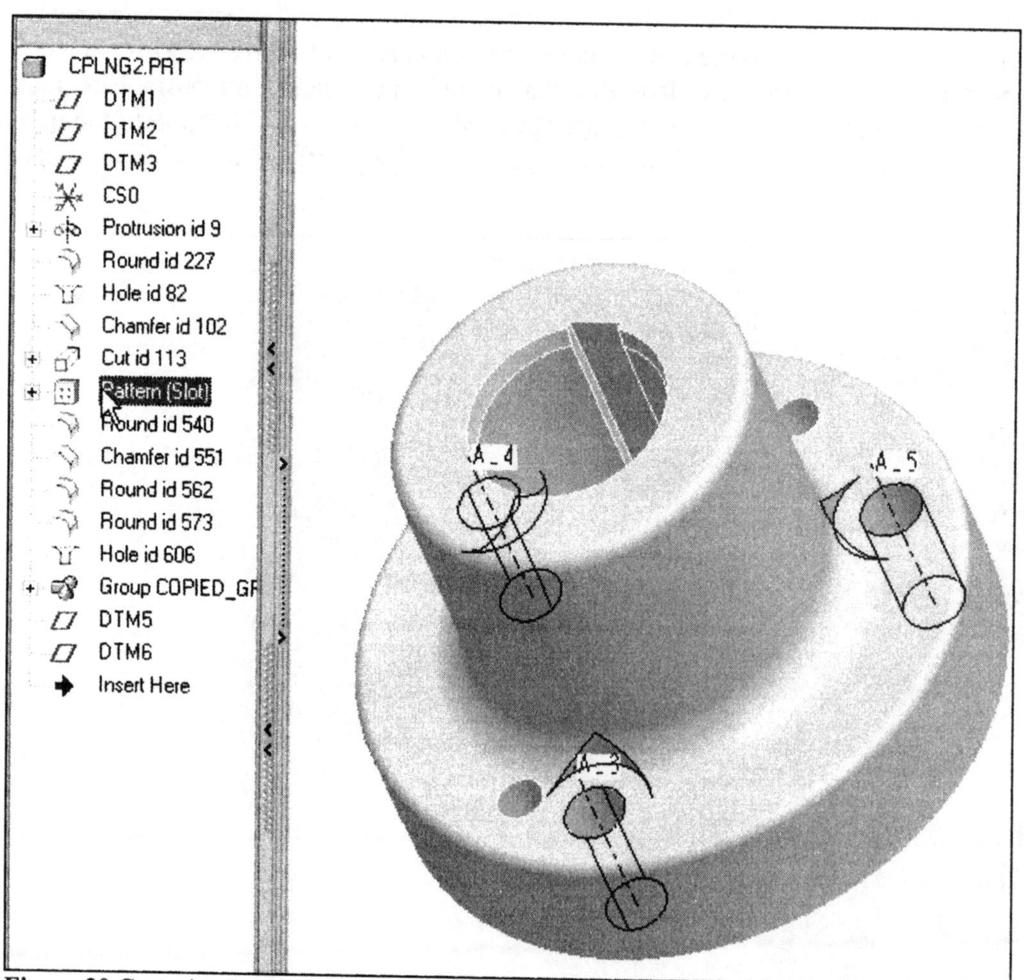

Figure 20 Capturing Design Intent

Parametric designs capture relationships in several ways:

Implicit Relationships Implicit relationships occur when new model geometry is sketched and dimensioned relative to existing features and parts. An implicit relationship is established, for instance, when the section sketch of a tire (Fig. 21) uses rim edges as a reference.

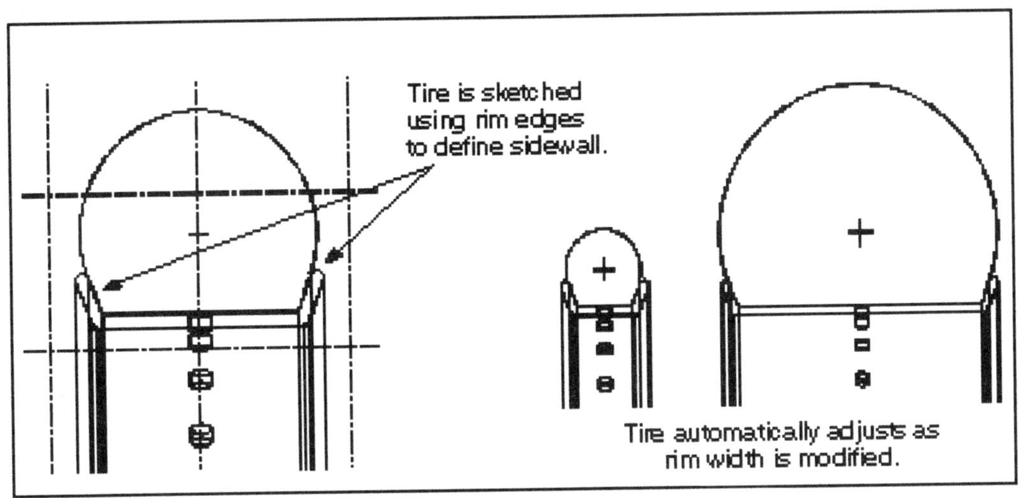

Figure 21 Tire and Rim

Patterns Design features often follow a geometrically predictable pattern. Features and parts are patterned in parametric design by referencing either construction dimensions or existing patterns. One example of patterning is a wheel hub with spokes (Fig. 22). First, the spoke holes are radially patterned. The spokes can then be strung by referencing this pattern.

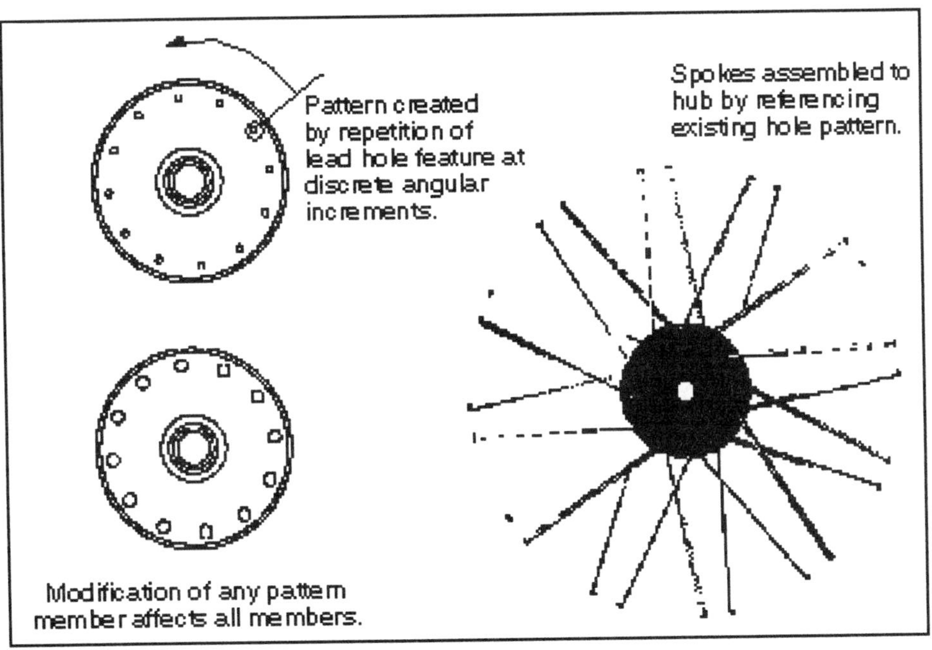

Figure 22 Patterns

Modification to a pattern member affects all members of that pattern. This helps capture design intent by preserving the duplicate geometry of pattern members.

The modeling task is to incorporate the features and parts of a complex design while properly capturing design intent to provide flexibility in modification. Parametric design modeling is a synthesis of physical and intellectual design (Fig. 23).

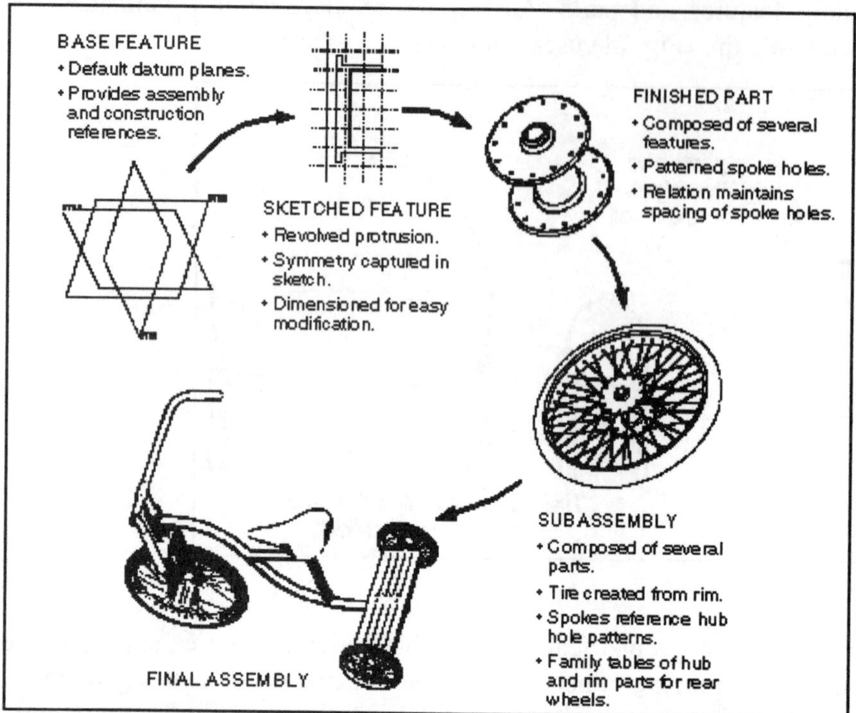

Figure 23 Relations

Explicit Relations Whereas implicit relationships are implied by the feature creation method, the user mathematically enters an explicit relation. This equation is used to relate feature and part dimensions in the desired manner. An explicit relation (Fig. 24) might be used, for example, to control sizes on a model.

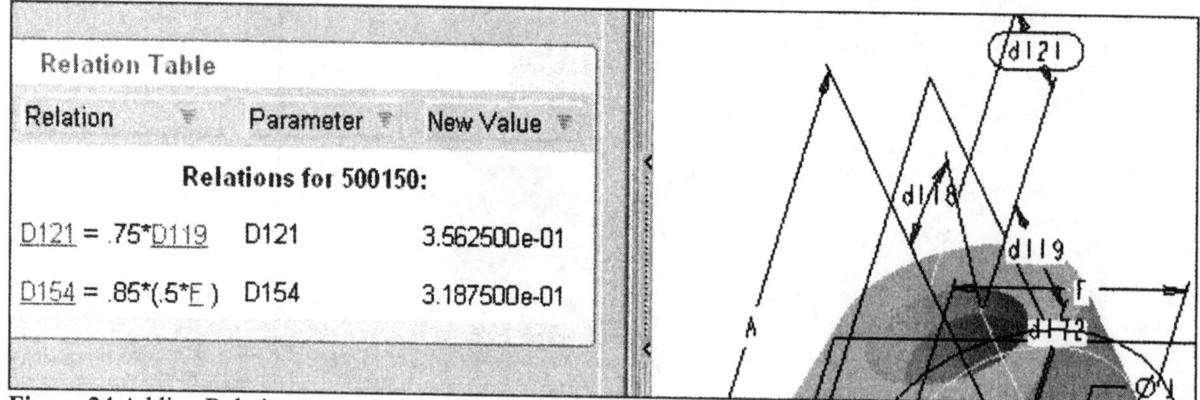

Relation Table

Relation	Parameter	New Value
Relations for 500150:		
D121 = .75*D119	D121	3.562500e-01
D154 = .85*(.5*E)	D154	3.187500e-01

Figure 24 Adding Relations

Family Tables Family tables are used to create part families (Fig. 25) from generic models by tabulating dimensions or the presence of certain features or parts. A family table might be used, for example, to catalog a series of couplings with varying width and diameter as shown in Figure 26.

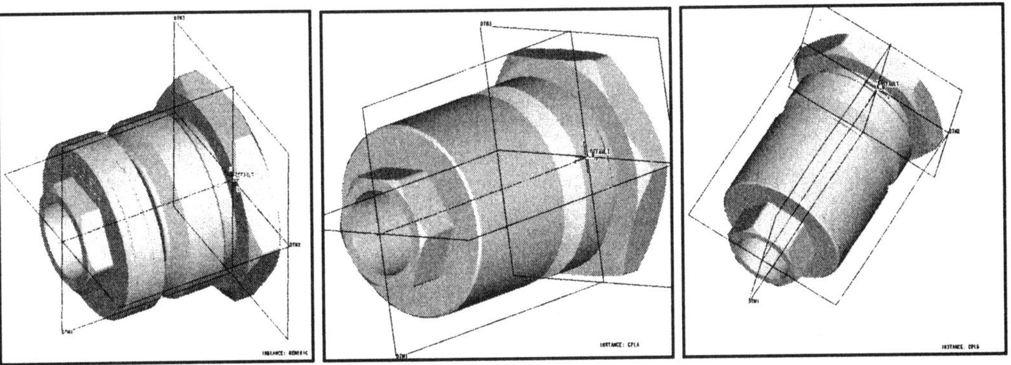

Figure 25(a-c) Family of Parts- Coupling

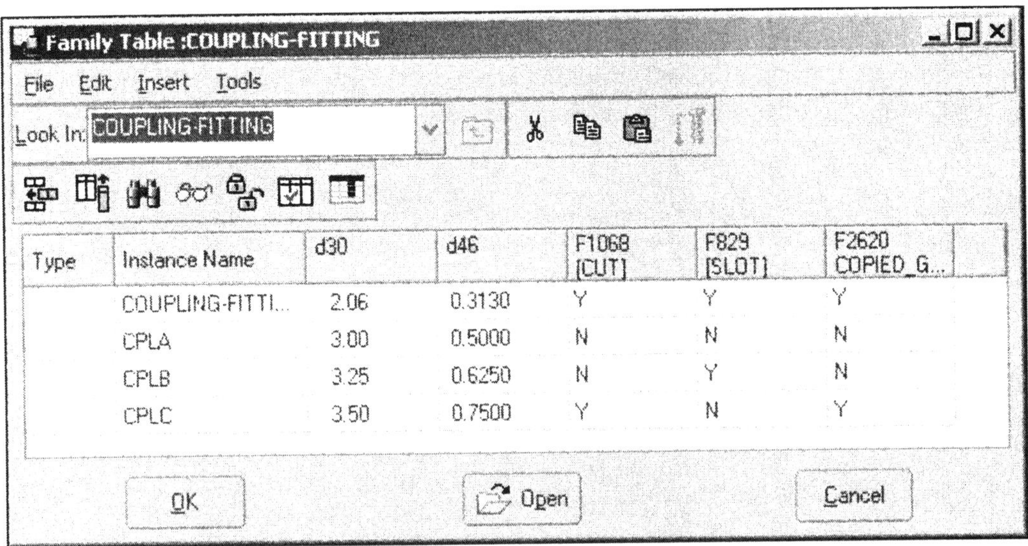

Type	Instance Name	d30	d46	F1068 [CUT]	F829 [SLOT]	F2620 COPIED_G...
	COUPLING-FITTI...	2.06	0.3130	Y	Y	Y
	CPLA	3.00	0.5000	N	N	N
	CPLB	3.25	0.6250	N	Y	N
	CPLC	3.50	0.7500	Y	N	Y

Figure 26 Family Table for Coupling

Assemblies

Just as parts are created from related features, **assemblies** are created from related parts. The progressive combination of subassemblies, parts, and features into an assembly creates parent-child relationships based on the references used to assemble each component (Fig. 27).

The *Assembly* functionality is used to assemble existing parts and subassemblies.

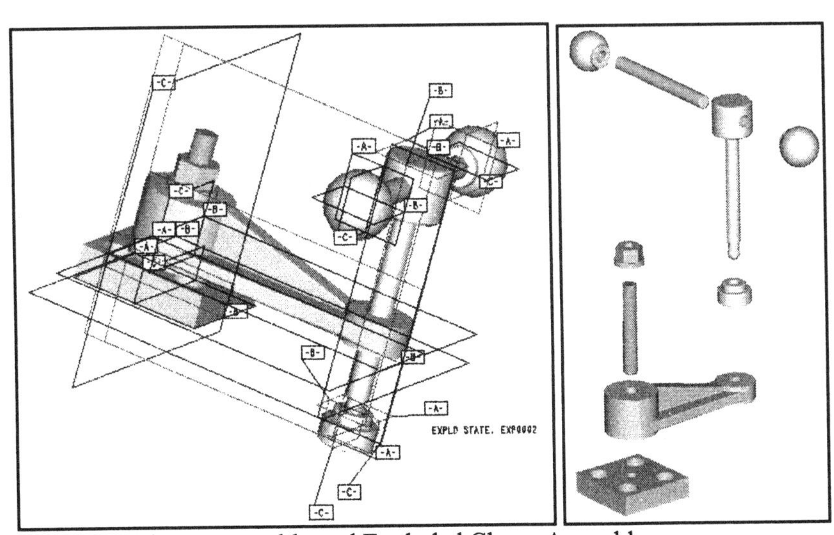

Figure 27 Clamp Assembly and Exploded Clamp Assembly

During assembly creation, you can:

- Simplify a view of a large assembly by creating a simplified representation
- Perform automatic or manual placement of component parts
- Create an exploded view of the component parts
- Perform analysis, such as mass properties and clearance checks
- Modify the dimensional values of component parts
- Define assembly relations between component parts
- Create assembly features
- Perform automatic interchange of component parts
- Create parts in Assembly mode
- Create documentation drawings of the assembly

Just as features can reference part geometry, parametric design also permits the creation of parts referencing assembly geometry. **Assembly mode** allows the designer both to fit parts together and to design parts based on how they should fit together.

In Figure 28, an assembly *Bill of Materials* report is generated.

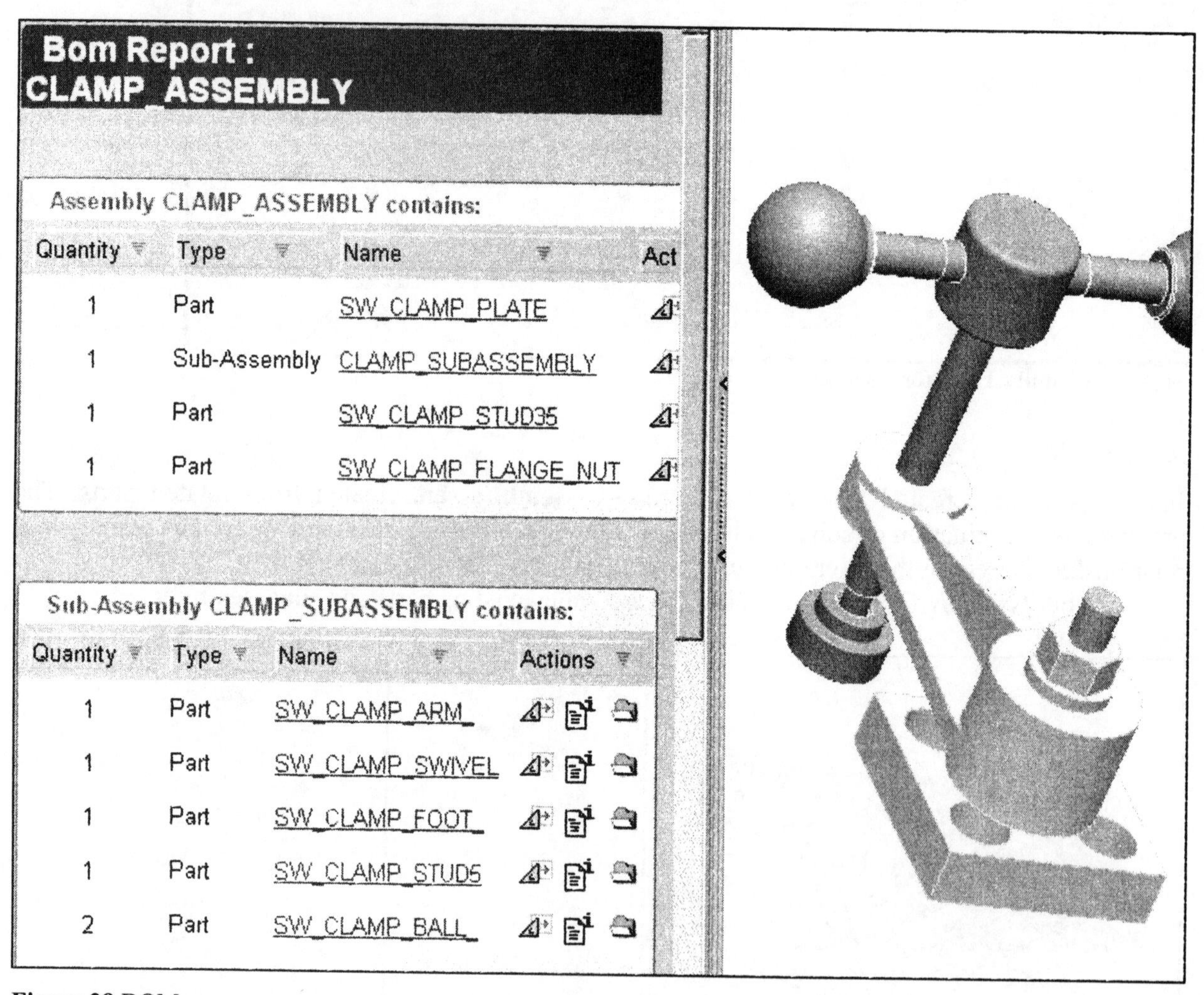

Bom Report :
CLAMP_ASSEMBLY

Assembly CLAMP_ASSEMBLY contains:

Quantity	Type	Name	Act
1	Part	SW_CLAMP_PLATE	
1	Sub-Assembly	CLAMP_SUBASSEMBLY	
1	Part	SW_CLAMP_STUD35	
1	Part	SW_CLAMP_FLANGE_NUT	

Sub-Assembly CLAMP_SUBASSEMBLY contains:

Quantity	Type	Name	Actions
1	Part	SW_CLAMP_ARM	
1	Part	SW_CLAMP_SWIVEL	
1	Part	SW_CLAMP_FOOT	
1	Part	SW_CLAMP_STUD5	
2	Part	SW_CLAMP_BALL	

Figure 28 BOM

Drawings

You can create drawings of all parametric design models (Fig. 29). All model views in the drawing are *associative:* if you change a dimensional value in one view, other drawing views update accordingly. Moreover, drawings are associated with their parent models. Any dimensional changes made to a drawing are automatically reflected in the model. Any changes made to the model (e.g., addition of features, deletion of features, dimensional changes, and so on) in Part, Sheet Metal, Assembly, or Manufacturing modes are also automatically reflected in their corresponding drawings.

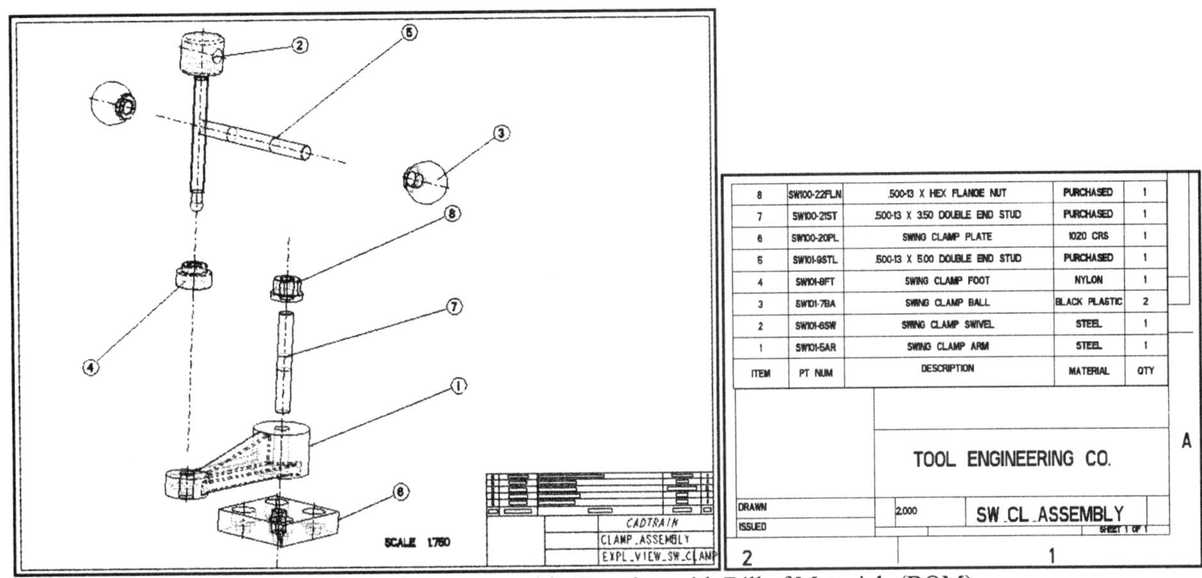

ITEM	PT NUM	DESCRIPTION	MATERIAL	QTY
8	SW100-22FLN	.500-13 X HEX FLANGE NUT	PURCHASED	1
7	SW100-21ST	.500-13 X 3.50 DOUBLE END STUD	PURCHASED	1
6	SW100-20PL	SWING CLAMP PLATE	1020 CRS	1
5	SW101-9STL	.500-13 X 5.00 DOUBLE END STUD	PURCHASED	1
4	SW101-8FT	SWING CLAMP FOOT	NYLON	1
3	SW101-7BA	SWING CLAMP BALL	BLACK PLASTIC	2
2	SW101-6SW	SWING CLAMP SWIVEL	STEEL	1
1	SW101-5AR	SWING CLAMP ARM	STEEL	1

Figure 29(a-b) Ballooned Exploded View Assembly Drawing with Bill of Materials (BOM)

The **Drawing** functionality is used to create annotated drawings of parts and assemblies. During drawing creation, you can:

- Add views of the part or assembly
- Show existing dimensions
- Incorporate additional driven or reference dimensions
- Create notes on the drawing
- Display views of additional parts or assemblies
- Add sheets to the drawing
- Create draft entities on the drawing
- Balloon components on an assembly drawing (Fig. 30)
- Create an associative BOM

You can annotate the drawing with notes, manipulate the dimensions, and use layers to manage the display of different items on the drawing. The module **Pro/DETAIL** can be used to extend the drawing capability or as a stand-alone module allowing you to create, view, and annotate models and drawings.

Pro/DETAIL supports additional view types and multi-sheets and offers commands for manipulating items in the drawing and for adding and modifying different kinds of textural and symbolic information. In addition, the abilities to customize engineering drawings with sketched geometry, create custom drawing formats, and make numerous cosmetic changes to the drawing are available.

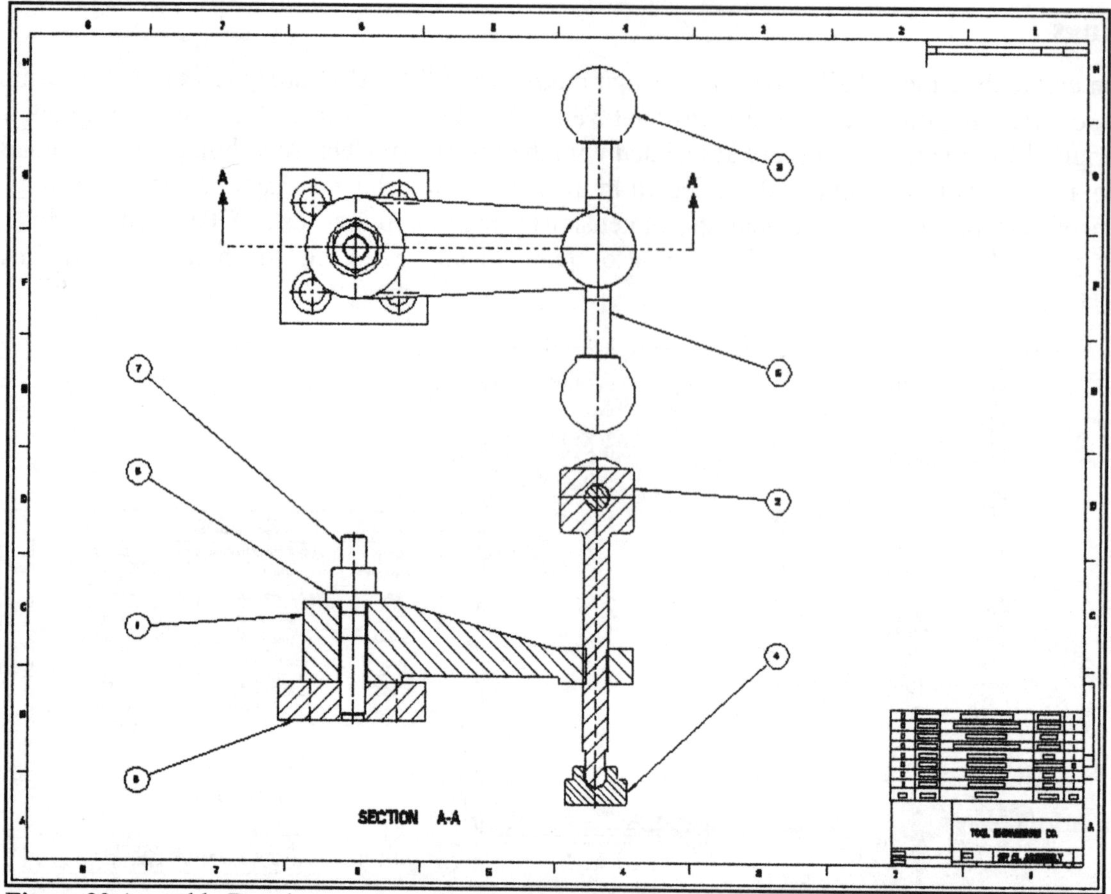

Figure 30 Assembly Drawing

Drawing mode in parametric design provides you with the basic ability to document solid models in drawings that share a two-way associativity (Fig. 31).

Changes that are made to the model in Part mode or Assembly mode will cause the drawing to update automatically and reflect the changes. Any changes made to the model in Drawing mode will be immediately visible on the model in Part and Assembly modes. The part shown in Figure 31 has been detailed in Figure 32. Basic Pro/E allows you to create drawing views of one or more models in a number of standard views with dimensions.

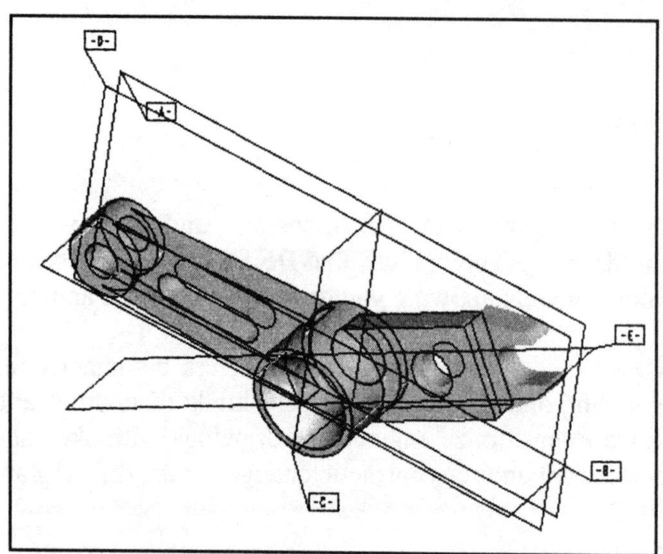

Figure 31 Angle Frame Model

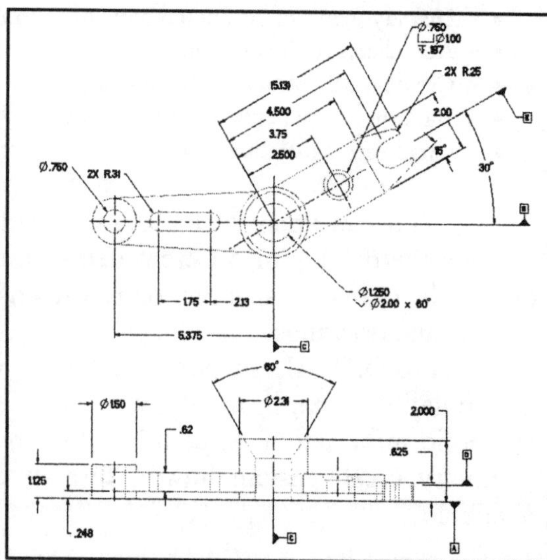

Figure 32 Angle Frame Drawing

Using the Text

The text utilizes a variety of command boxes and descriptions to lead you through construction sequences. Also, see the Appendix for Pro/ENGINEER WILDFIRE 2.0 Quick Reference Cards.

The following icons, symbols and conventions will be used throughout the text *(command sequences are always in a box)*:

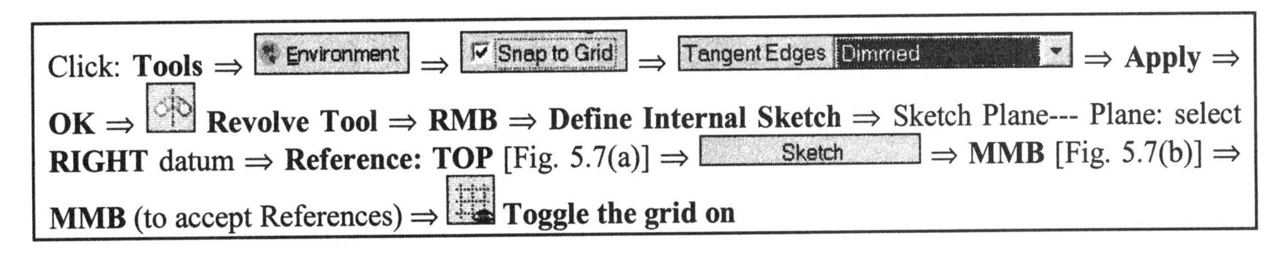

Click: **Tools** ⇒ Environment ⇒ Snap to Grid ⇒ Tangent Edges Dimmed ⇒ **Apply** ⇒ **OK** ⇒ **Revolve Tool** ⇒ **RMB** ⇒ **Define Internal Sketch** ⇒ Sketch Plane--- Plane: select **RIGHT** datum ⇒ **Reference: TOP** [Fig. 5.7(a)] ⇒ Sketch ⇒ **MMB** [Fig. 5.7(b)] ⇒ **MMB** (to accept References) ⇒ **Toggle the grid on**

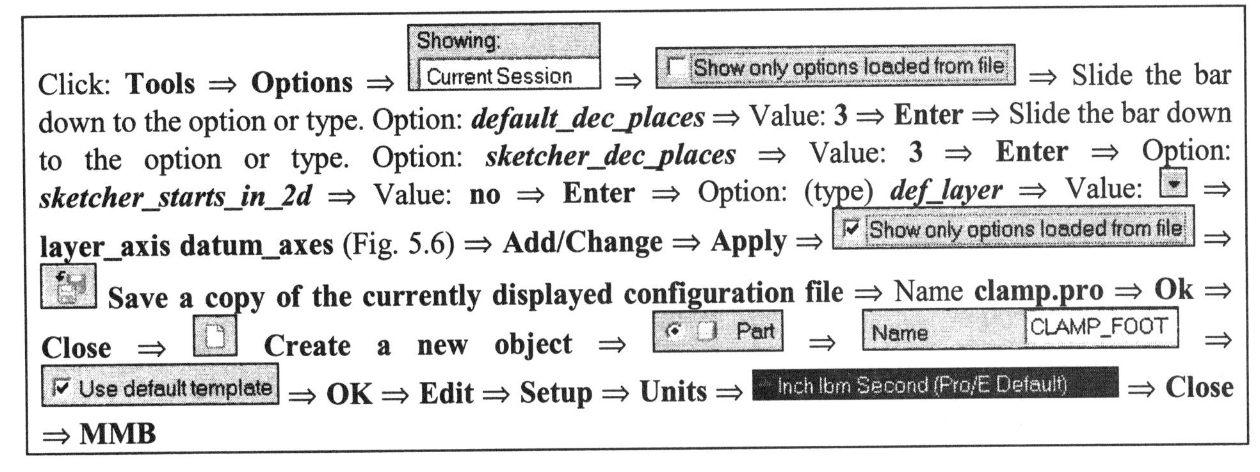

Click: **Tools** ⇒ **Options** ⇒ Showing: Current Session ⇒ Show only options loaded from file ⇒ Slide the bar down to the option or type. Option: ***default_dec_places*** ⇒ Value: **3** ⇒ **Enter** ⇒ Slide the bar down to the option or type. Option: ***sketcher_dec_places*** ⇒ Value: **3** ⇒ **Enter** ⇒ Option: ***sketcher_starts_in_2d*** ⇒ Value: **no** ⇒ **Enter** ⇒ Option: (type) ***def_layer*** ⇒ Value: ⇒ **layer_axis datum_axes** (Fig. 5.6) ⇒ **Add/Change** ⇒ **Apply** ⇒ Show only options loaded from file ⇒ **Save a copy of the currently displayed configuration file** ⇒ Name **clamp.pro** ⇒ **Ok** ⇒ **Close** ⇒ **Create a new object** ⇒ Part ⇒ Name CLAMP_FOOT ⇒ Use default template ⇒ **OK** ⇒ **Edit** ⇒ **Setup** ⇒ **Units** ⇒ Inch lbm Second (Pro/E Default) ⇒ **Close** ⇒ **MMB**

Commands:

- ⇒ — Continue with command sequence or screen picks using **LMB**
- **Create lines** — icon (with description) indicates command to pick using **LMB**

Mouse or keyboard terms used in this text:

- **LMB** — **L**eft **M**ouse **B**utton
  - or "**Pick**" — term used to direct an action (i.e., "Pick the surface")
  - or "**Click**" — term used to direct an action (i.e., "Click on the icon")
  - or "**Select**" — term used to direct an action (i.e., "Select the feature")
- **MMB** — **M**iddle **M**ouse **B**utton (accept the current selection or value)
  - or **Enter** — press **Enter** key to accept entry
  - or ✔ — Click on this icon to accept entry
- **RMB** — **R**ight **M**ouse **B**utton (toggles to next selection, or provides a list of available commands)

(Refer to the Pro/ENGINEER Wildfire 2.0 Quick Reference Cards in the Appendix)

Text Organization

Text Lessons- Parts, Assemblies, and Drawings

- **Lesson 1** introduces Pro/ENGINEER's Wildfire 2.0 interface and embedded Browser.
- **Lesson 2** provides uncomplicated instructions to model a variety of simple-shaped parts.
- **Lessons 3** and **4** involve part modeling, using a variety of commands and tools.
- **Lessons 5** through **9** involve modeling the parts, creating the assembly [Fig. 33(a)], and documenting the design using detail and assembly drawings [Fig. 33(b)].
- **Lesson 10** introduces more features, capabilities, and options (same methodology as Lesson 2).
- **Appendix** includes Screen Customizing, Mapkeys, and ***Pro/E Quick Reference Cards***.

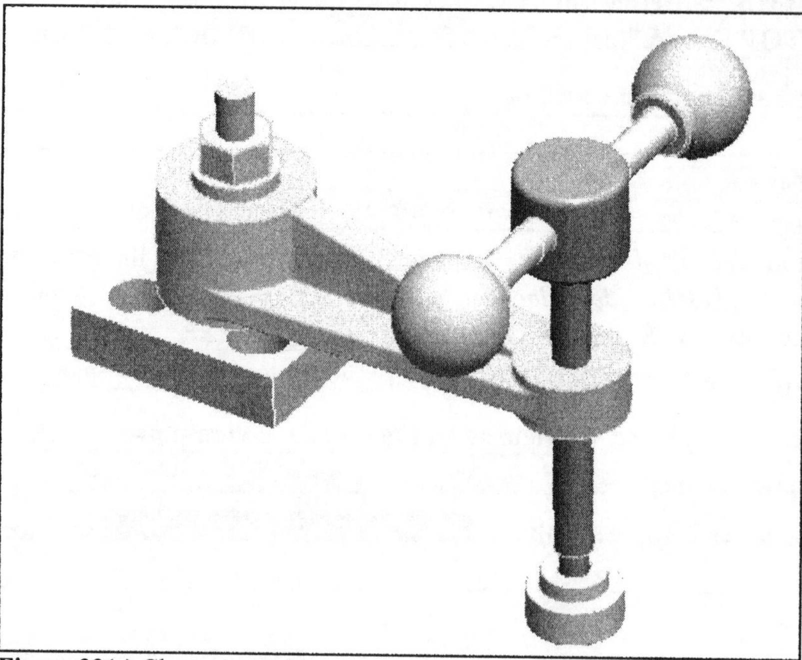

Figure 33(a) Clamp Assembly

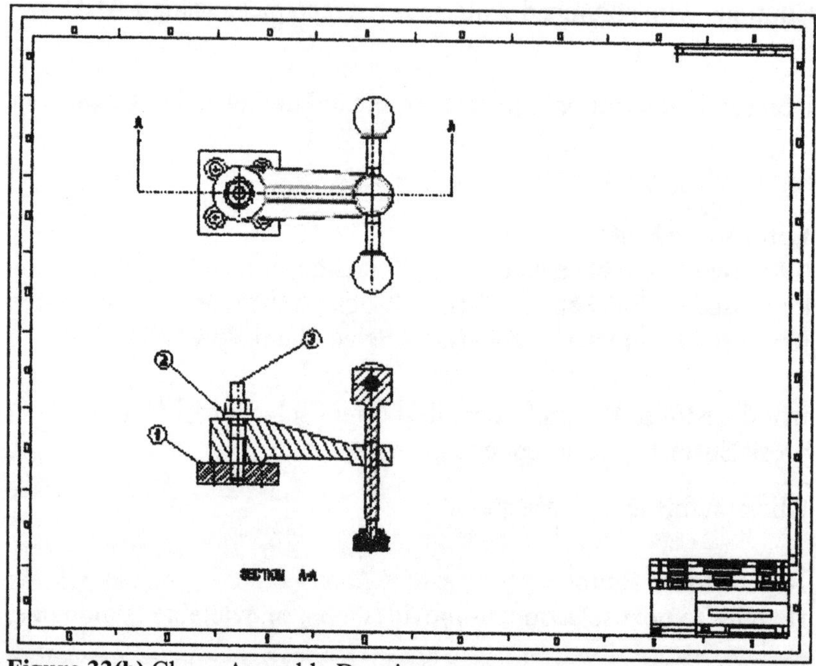

Figure 33(b) Clamp Assembly Drawing

Lesson 1 Introduction to Pro/ENGINEER Wildfire 2.0

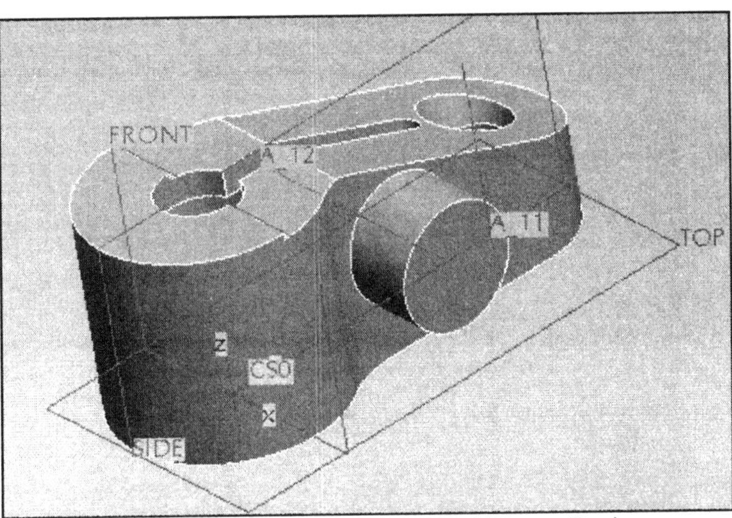

Figure 1.1 Swing/Pull Clamp Arm (SPX Fluid Power Part)

OBJECTIVES

- Understand the **User Interface (UI)**
- Download **Catalog Parts** using the **Browser**
- Master the **File Functions**
- Learn how to **Email** active Pro/ENGINEER objects
- Become familiar with the **Help** facility
- Be introduced to the **Display** and **View** capabilities
- Use **Mouse Buttons** to **Pan**, **Zoom**, and **Rotate** the object
- Change **Display Settings**
- Investigate an object with **Information Tools**
- Experience the **Model Tree** functionality

Pro/ENGINEER Wildfire 2.0

The first lesson will introduce you to Pro/ENGINEER's working environment. An existing part model (Fig. 1.1) will be downloaded from the Catalog and used to demonstrate the UI (user interface) and the general interaction required to master Pro/E.

You will be using a part available thru the **Catalog** using the **Browser**. If you are not connected to the Internet, your instructor will provide you with a simple start part. In addition, if you have Pro/Library installed, you may use any library part that you wish.

Pro/ENGINEER's Main Window

The Pro/ENGINEER user interface consists of a navigation window, an embedded Web browser, the menu bar, toolbars, information areas, and the graphics window (Fig. 1.2).

- **Navigation Window** Located on the left side of the Pro/E main window, it includes tabs for the Model Tree and Layer Tree, Folder Browser, Favorites, History and Connections.
- **Browser Window** An embedded Web browser is located to the right of the navigation window. This browser provides you access to internal or external Web sites.

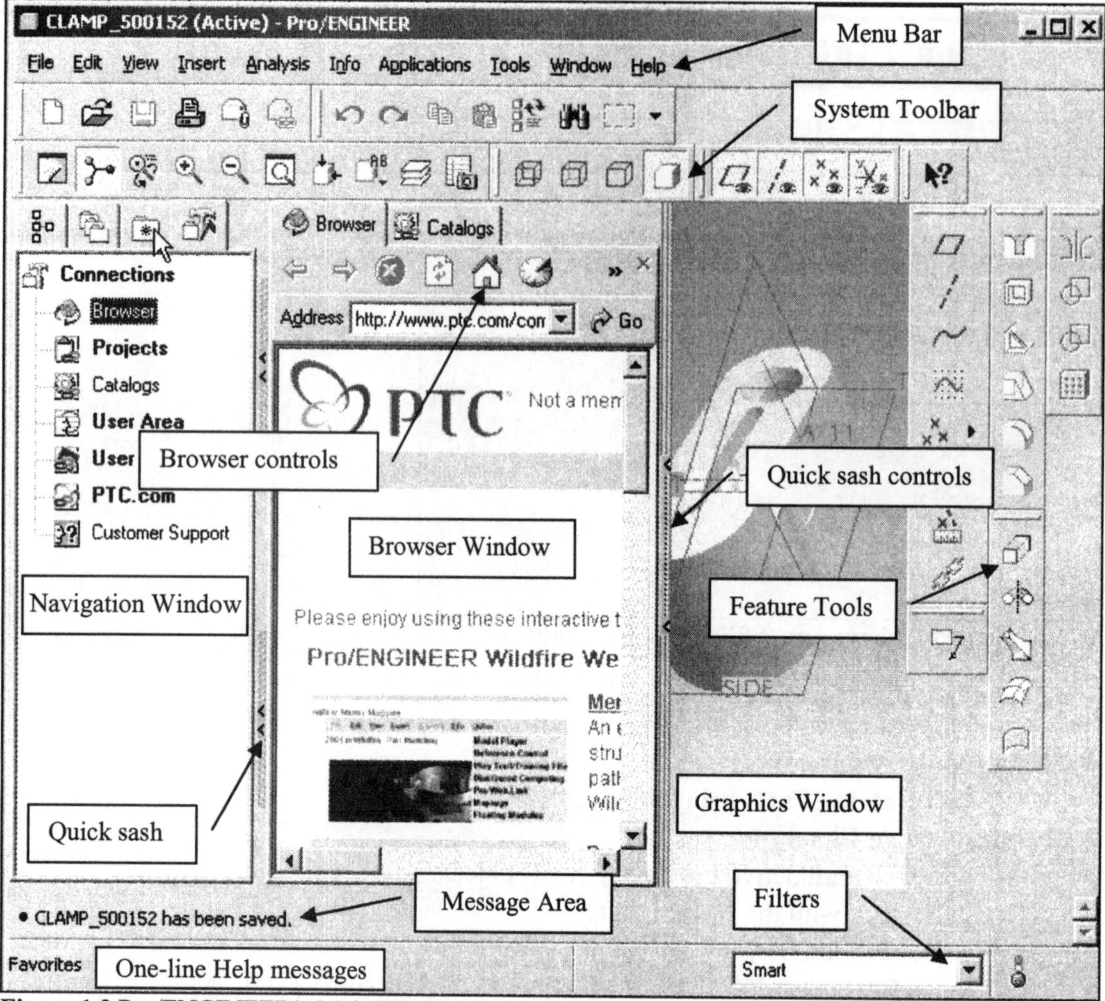

Figure 1.2 Pro/ENGINEER's Main Window

- **Menu Bar** The menu bar contains menus with options for creating, saving, and modifying models, and also contains menus with options for setting environment and configuration options.
- **Information Areas** Each Pro/E window has a message area near the bottom of the window for displaying one-line Help messages.
    - **Message Area** Messages related to work performed are displayed here.
    - **Status Bar** At the bottom of the window is a one-line Help area that dynamically displays one-line context-sensitive Help messages. If you move your mouse pointer over a menu command or dialog box option, a one-line description appears in this area.
    - **Screen Tips** The status bar messages also appear in small yellow boxes near the menu option or dialog box item or toolbar button that the mouse pointer is passing over.
- **Toolbars** The toolbars contain icons to speed up access to commonly used menu commands. By default, the toolbars consist of a row of buttons located directly under the main menu bar. Toolbar buttons can be positioned on the top, left, and right of the window. Toolbar buttons can be added or removed from the Toolchest by customizing the layout.
- **Graphics Window** The graphics window is the main working space (main window) for modeling and is to the right of the embedded Browser. Normally the Browser is collapsed when modeling.

Lesson 1 STEPS

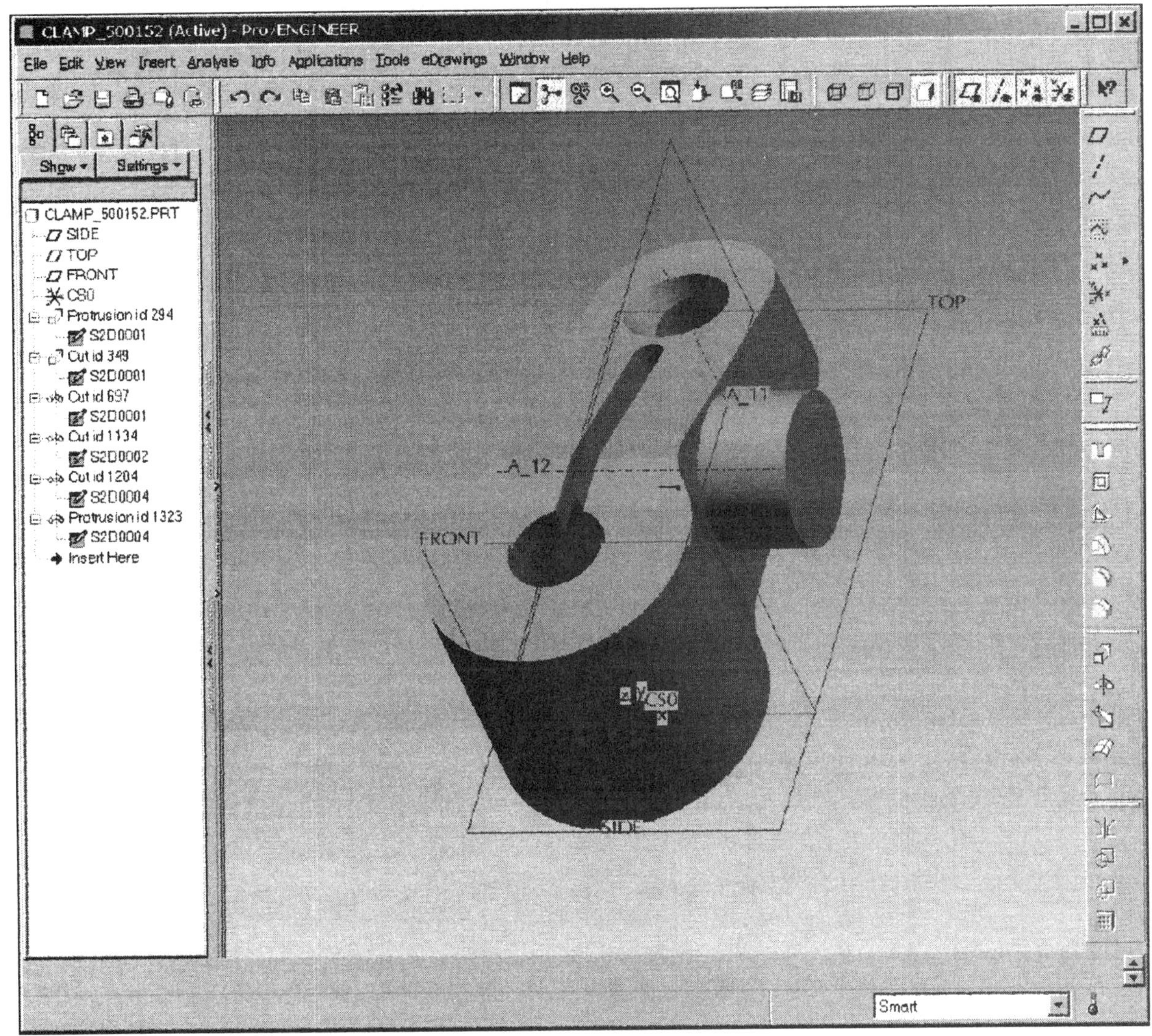

Figure 1.3 Swing/Pull Clamp Arm

Catalog Parts Swing/Pull Clamp

In order to see and use Pro/ENGINEER's UI, we must have an active object (Fig. 1.3). Since you will not be modeling the part, you must download an existing model from the Catalog of parts from PTC. If you are using an Academic or Commercial version of Pro/ENGINEER Wildfire 2.0, you can download catalog parts directly. The Student Edition (SE) and Tryout Edition (TE) do not have access to these selections.

*If you do not have access to the catalog, go to www.cad-resources.com ⇒ **Downloads** ⇒ **Lesson 1 Clamp** ⇒ drag and drop the part into the Graphics Window as per the instructions.*

Throughout the text, a box surrounds all commands and menu selections.

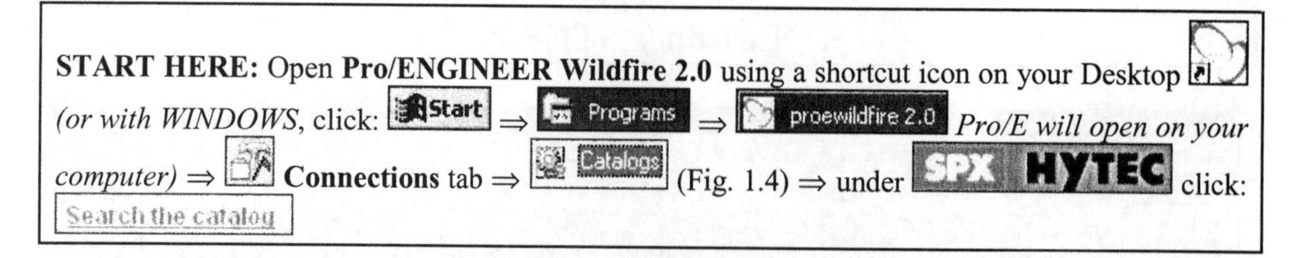

START HERE: Open **Pro/ENGINEER Wildfire 2.0** using a shortcut icon on your Desktop *(or with WINDOWS*, click: Start ⇒ Programs ⇒ proewildfire 2.0 *Pro/E will open on your computer)* ⇒ **Connections** tab ⇒ Catalogs (Fig. 1.4) ⇒ under SPX HYTEC click: Search the catalog

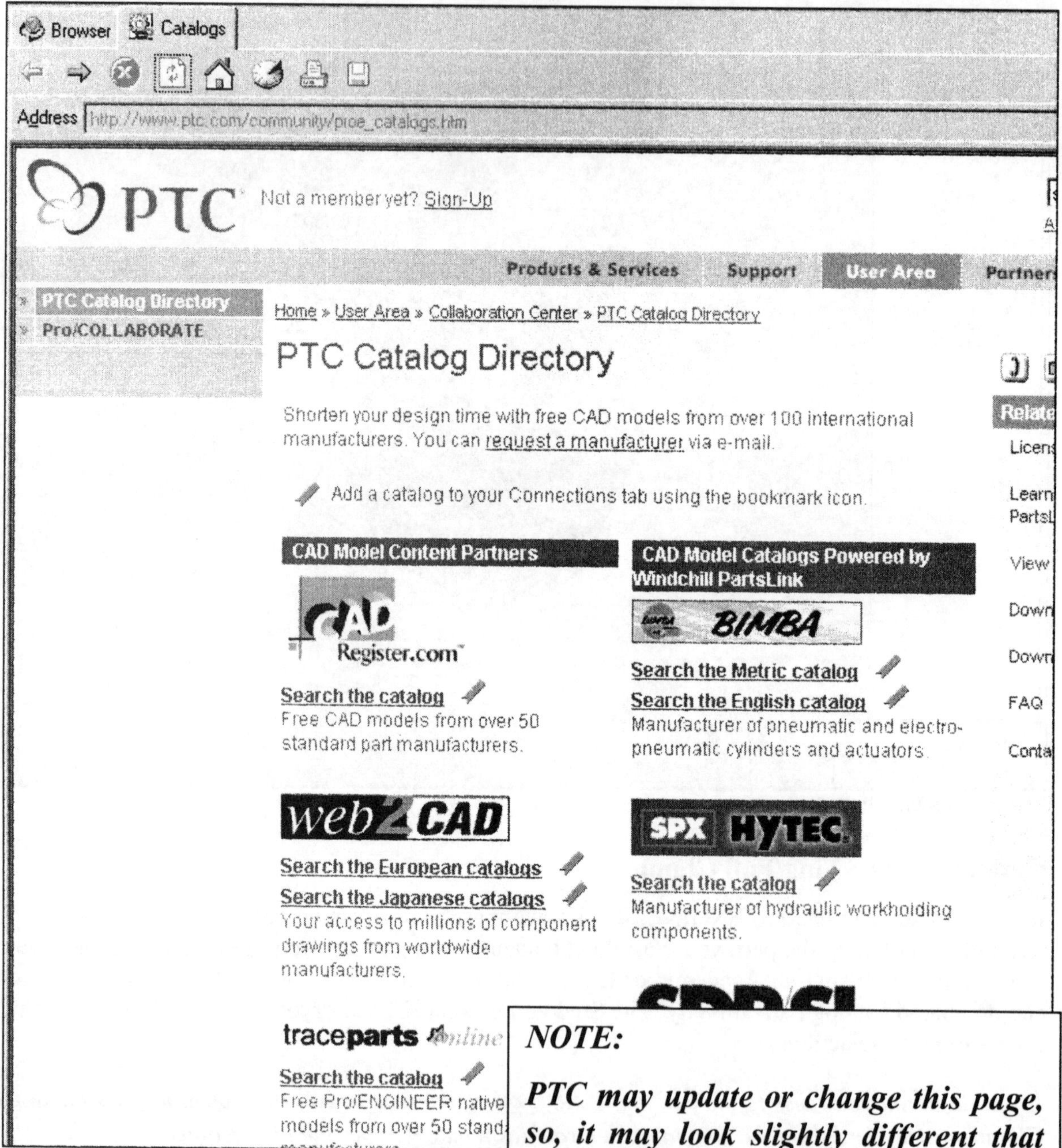

Figure 1.4 Connections and Catalog

NOTE:

PTC may update or change this page, so, it may look slightly different that what is displayed here.

Click: 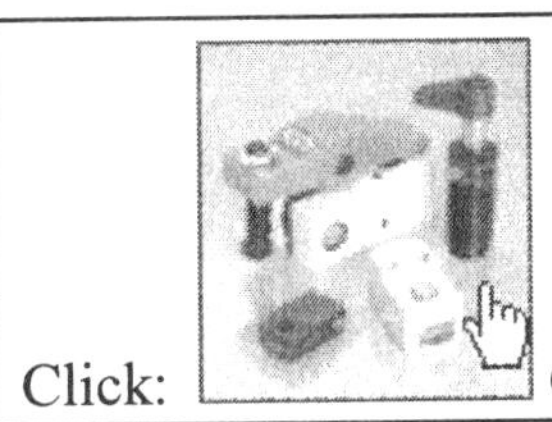 **Clamping** [Fig. 1.5(a)]

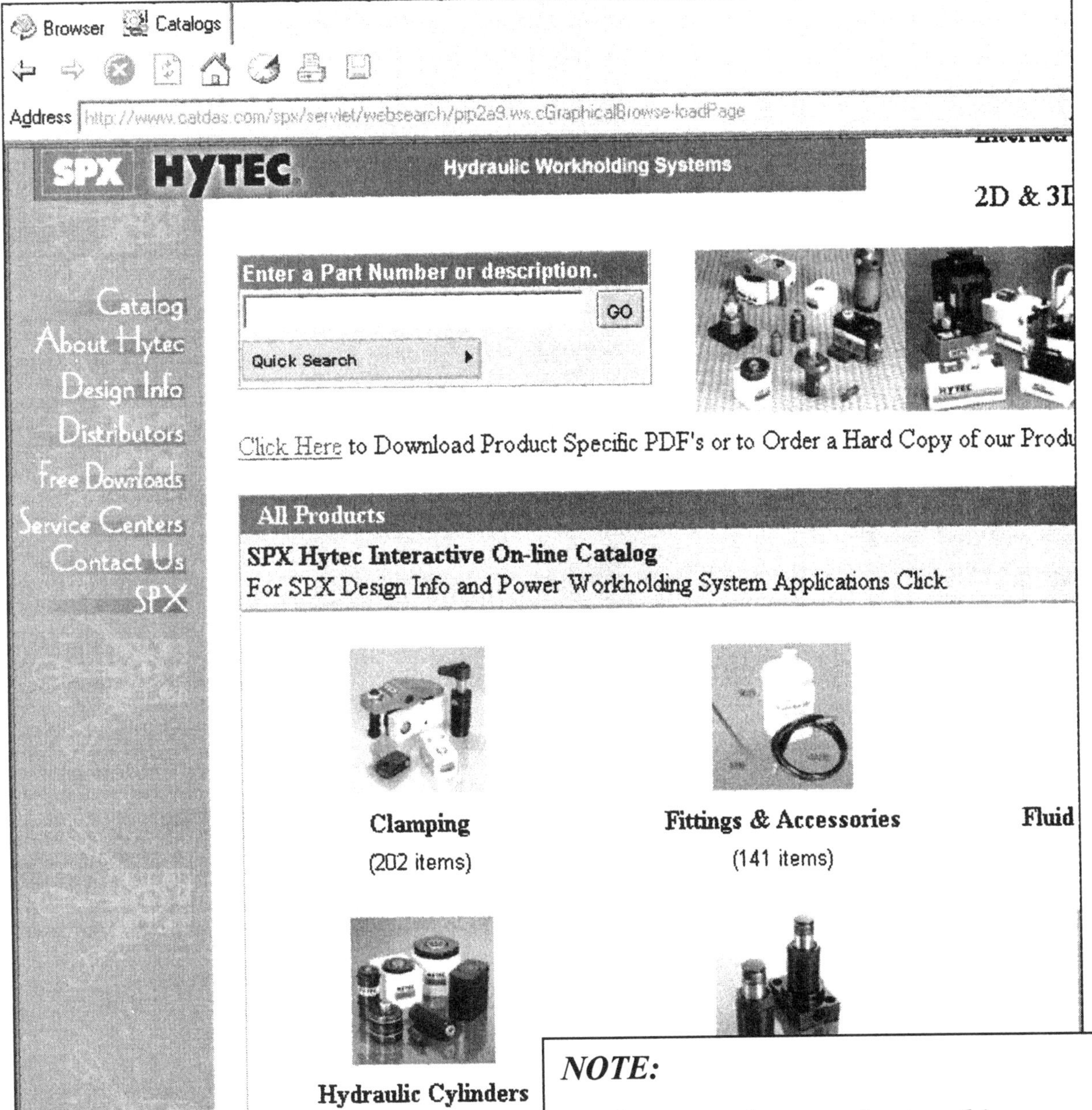

Figure 1.5(a) SPX Hytec Interactive On-Line Catalog

> ***NOTE:***
>
> ***PTC may update or change this page, so, it may look slightly different that what is displayed here.***

Click: 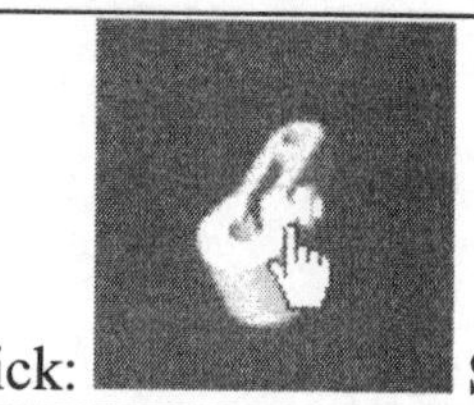 **Swing/Pull Clamp Arms** [Fig. 1.5(b)]

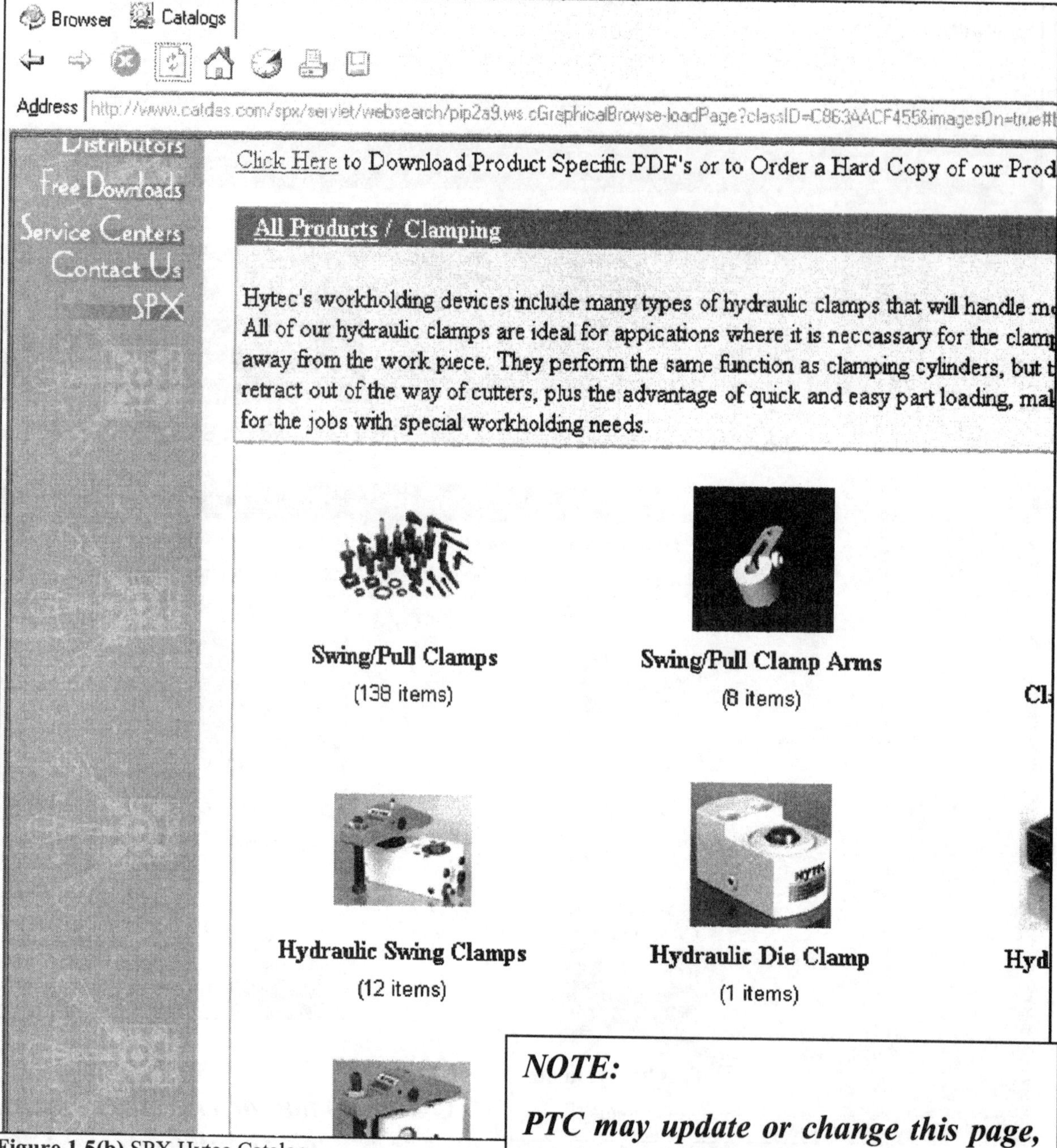

Figure 1.5(b) SPX Hytec Catalog

NOTE:

PTC may update or change this page, so, it may look slightly different that what is displayed here.

Double-click on the entry for Product Number **500152** (use the slide bar if needed) [Fig. 1.5(c)]:

500152 4500 Lbs. 4500 Lbs. 5000.0 PSI 250 in.**3/min. 0.5 sec. 2.5 inches

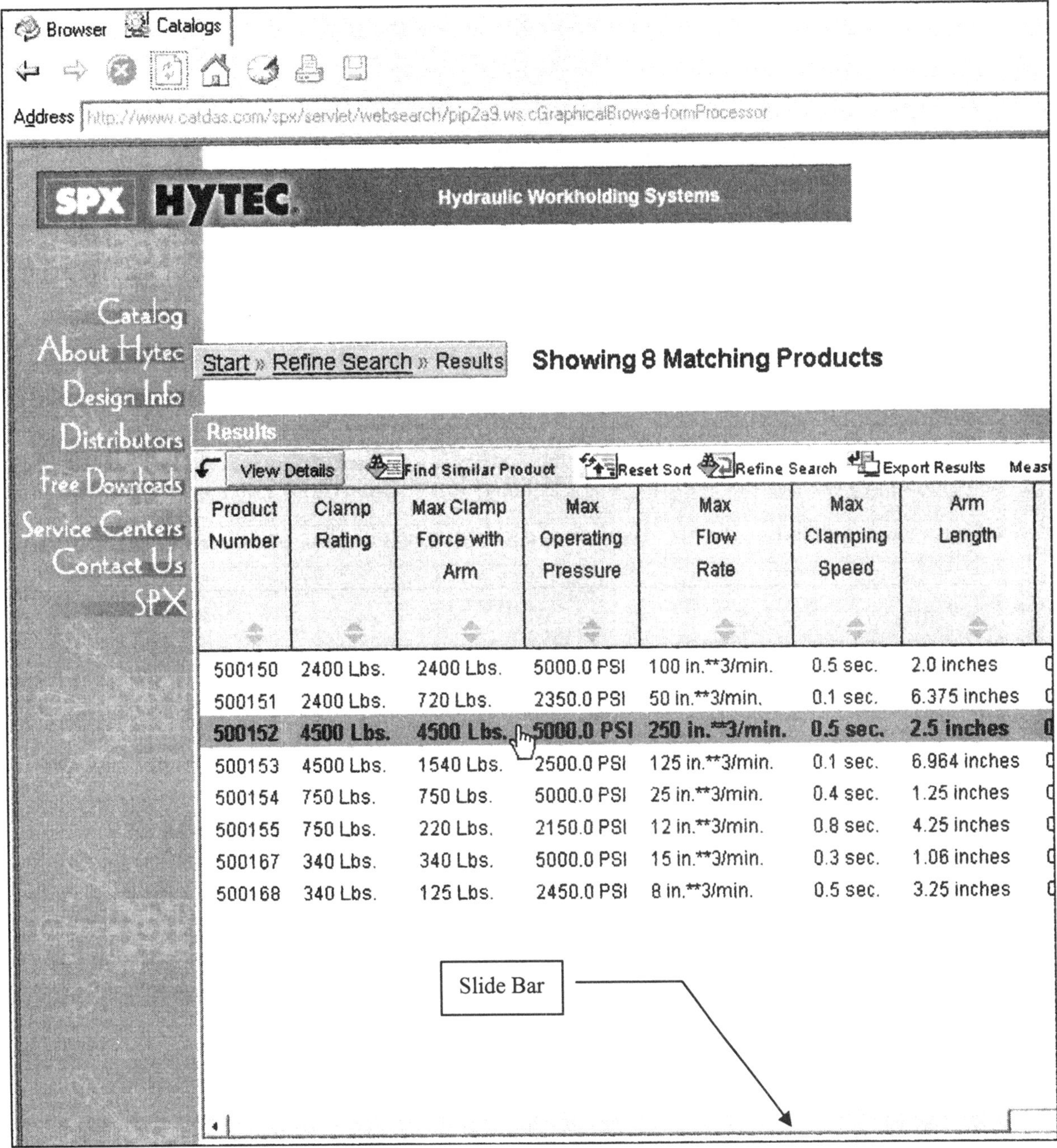

Figure 1.5(c) Swing Clamp Arms 500152

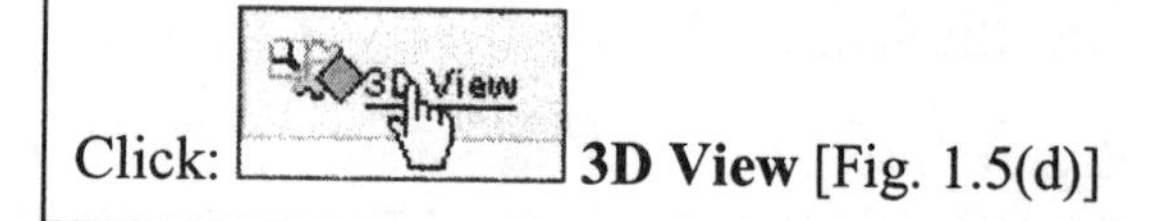

Click: **3D View** [Fig. 1.5(d)]

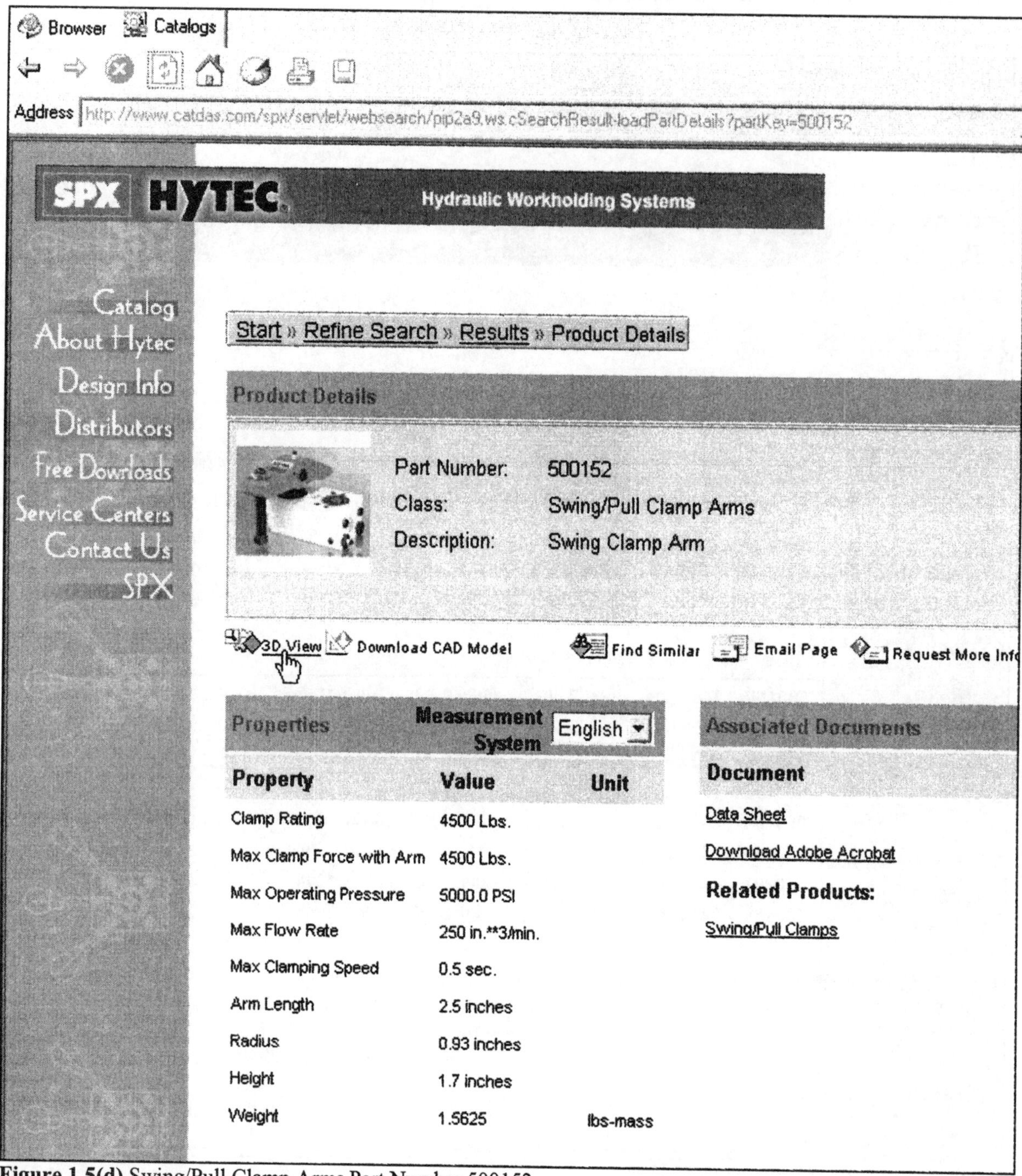

Figure 1.5(d) Swing/Pull Clamp Arms Part Number 500152

ProductView Lite Edition

ProductView Lite is a data visualization tool that provides basic viewing capabilities for 3D models, including measurement and annotation. The Model Viewer lets you perform navigation tasks such as zoom, pan, rotate, fly-through, and exploding components of 3D models. You can also display models in wireframe, HLR (hidden lines removed), or shaded render modes. ProductView Lite provides the ability to view annotations that are associated with drawings or models. The annotations must be stored along with the file that you are viewing. Annotation sets appear along the left side of the ProductView Lite window as thumbnails. Click a thumbnail to view that annotation set.

You can use this tool to view annotations only; you cannot create annotations in ProductView Lite. In addition, 2D markups currently cannot be viewed with ProductView Lite; as a result, markups do not appear in the list of annotations.

The View Part window opens with the clamp displayed, click: Viewer Commands [Figs. 1.6(a-b)]

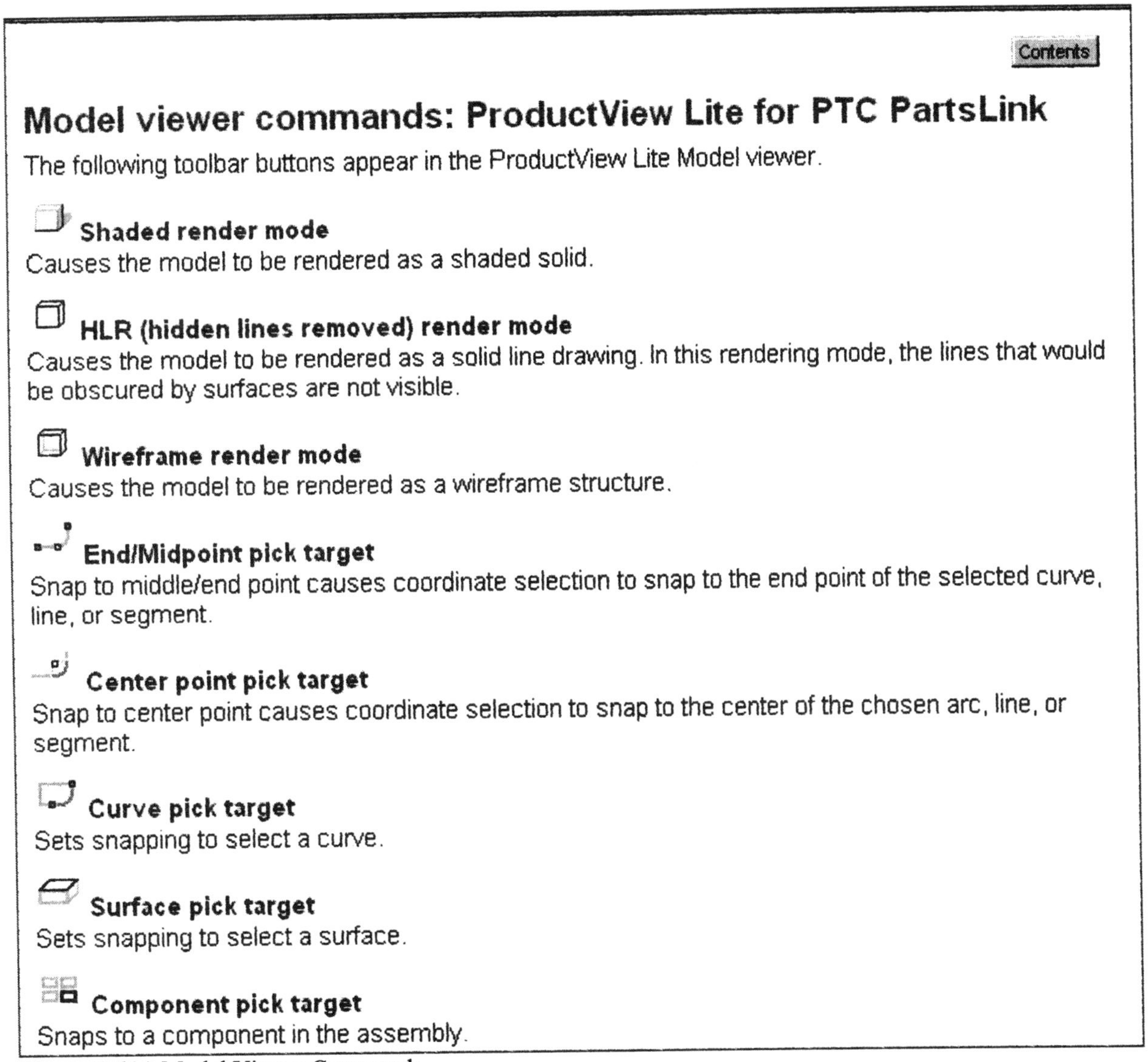

Contents

Model viewer commands: ProductView Lite for PTC PartsLink

The following toolbar buttons appear in the ProductView Lite Model viewer.

Shaded render mode
Causes the model to be rendered as a shaded solid.

HLR (hidden lines removed) render mode
Causes the model to be rendered as a solid line drawing. In this rendering mode, the lines that would be obscured by surfaces are not visible.

Wireframe render mode
Causes the model to be rendered as a wireframe structure.

End/Midpoint pick target
Snap to middle/end point causes coordinate selection to snap to the end point of the selected curve, line, or segment.

Center point pick target
Snap to center point causes coordinate selection to snap to the center of the chosen arc, line, or segment.

Curve pick target
Sets snapping to select a curve.

Surface pick target
Sets snapping to select a surface.

Component pick target
Snaps to a component in the assembly.

Figure 1.6(a) Model Viewer Commands

Zoom all
Adjusts the image size in the display to show the entire drawing within the window.

Zoom window
Magnifies the window to fill the screen. To use the Zoom Window tool, click and drag with the mouse to define a window. When you release the mouse button, the window is magnified to the maximum allowable size, while still showing all the areas within the zoom window borders.

Restore location
Allows you to move all parts back to their original positions. This command restores the location of selected components. If no components are selected, it restores all components.

Select
The Select command allows you to select a component. The selected component is highlighted in red in the graphical display. The selected component(s) become the target of certain features. To select a component, click the Select button. Press **Ctrl** to select multiple objects. Press **Shift** to indicate when selection is available for your current pick target (such as point, curve, or surface).

Dimension measurement
Select **Dimension measurement** to display measurements for a selected pick target. To display dimensions of a component, click on that component. The dimensions appear in a label callout. For example, you can click on a cylinder to display the height, radius, and area. Or, you can click on an arc to display the length and radius. For more information, see Taking measurements .

Distance measurement
Select **Distance measurement** to click and measure the distance between two pick targets. To measure distance between items, specify the pick target (snap) for item 1, click on it to select it, then specify the pick target for item 2, and click to select that item. In other words, you can change pick targets between clicks for distance measurement. For more information, see Taking measurements .

Clear measurements
Causes all currently displayed measurements and annotations to be cleared from the view.

Figure 1.6(b) Model Viewer Commands (continued)

From the Model viewer commands window, click: **File** ⇒ **Close** ⇒ click on the model with **MMB** and "fly the shuttle icon" down to zoom out from the model until it is centered in the window [Fig. 1.6(c)] ⇒ click: **RMB** to rotate the model (Fig. 1.7) ⇒ use both capabilities to position the model similar to that shown in Figure 1.7 ⇒ click on the model with **LMB** to turn on the enclosing box and coordinate system (Fig. 1.8) ⇒ reposition the model as desired.

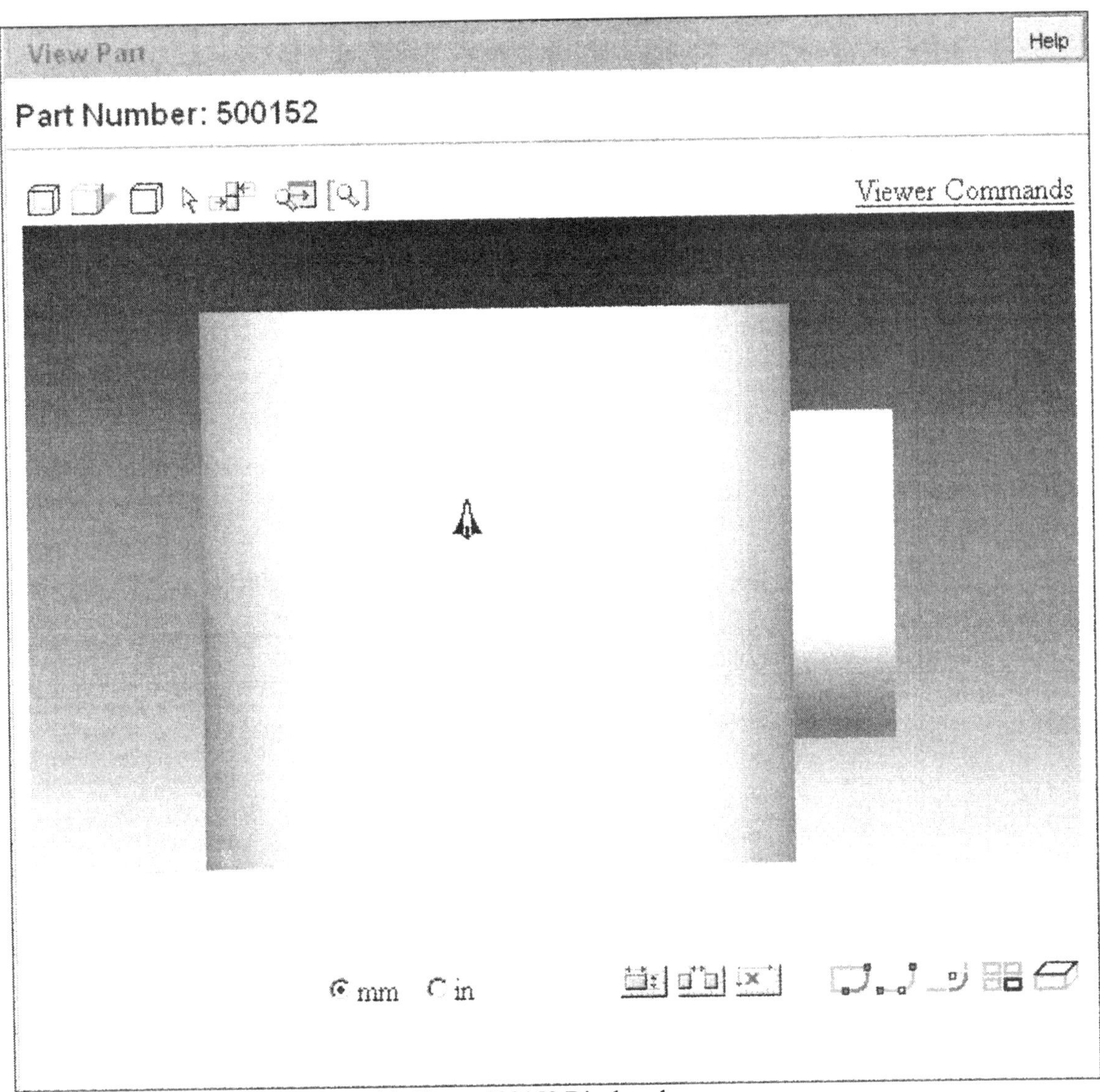

Figure 1.6(c) View Part Window with Clamp 500152 Displayed

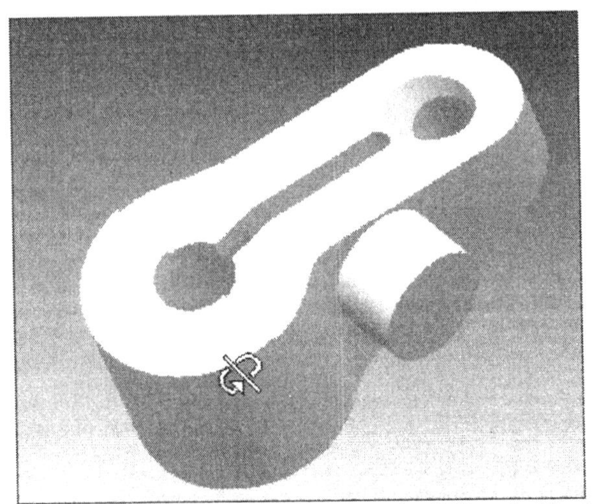

Figure 1.7 Fly and Rotate until Desired View is Obtained

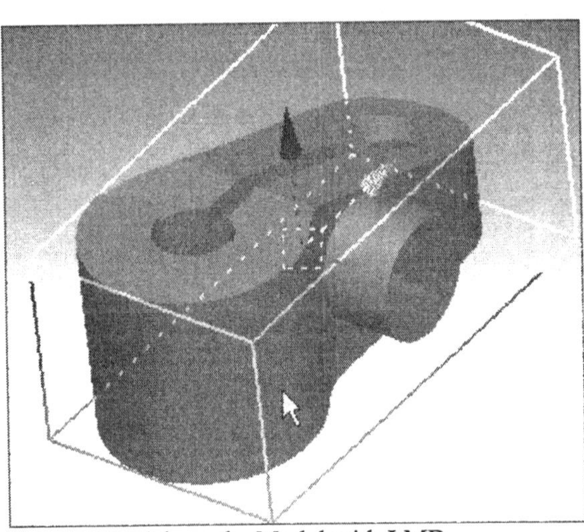

Figure 1.8 Pick on the Model with LMB

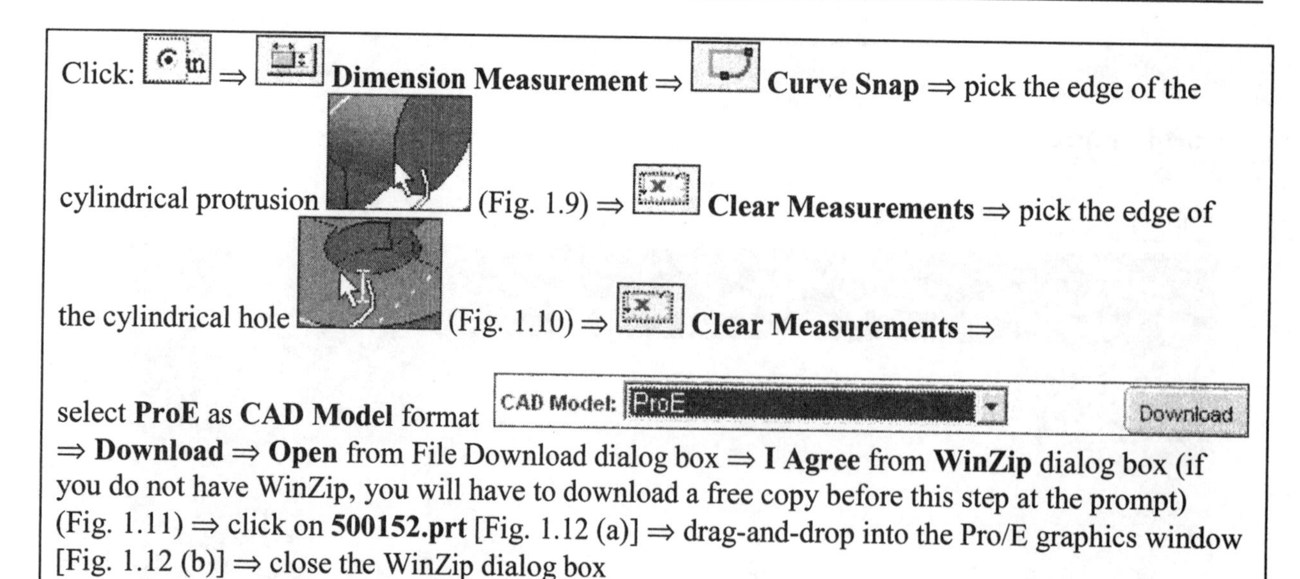

Click: ⇒ **Dimension Measurement** ⇒ **Curve Snap** ⇒ pick the edge of the cylindrical protrusion (Fig. 1.9) ⇒ **Clear Measurements** ⇒ pick the edge of the cylindrical hole (Fig. 1.10) ⇒ **Clear Measurements** ⇒ select **ProE** as **CAD Model** format

⇒ **Download** ⇒ **Open** from File Download dialog box ⇒ **I Agree** from **WinZip** dialog box (if you do not have WinZip, you will have to download a free copy before this step at the prompt) (Fig. 1.11) ⇒ click on **500152.prt** [Fig. 1.12 (a)] ⇒ drag-and-drop into the Pro/E graphics window [Fig. 1.12 (b)] ⇒ close the WinZip dialog box

*(If you are using the **Tryout Edition** or the **Student Edition**, type www.cad-resources.com in the Browser Address Bar or in your Internet Browser's Address Bar ⇒ **Downloads** ⇒ **Clamp** ⇒ **TE & SE Version** ⇒ drag and drop the part from the WinZip dialog box into the Graphics Window)*

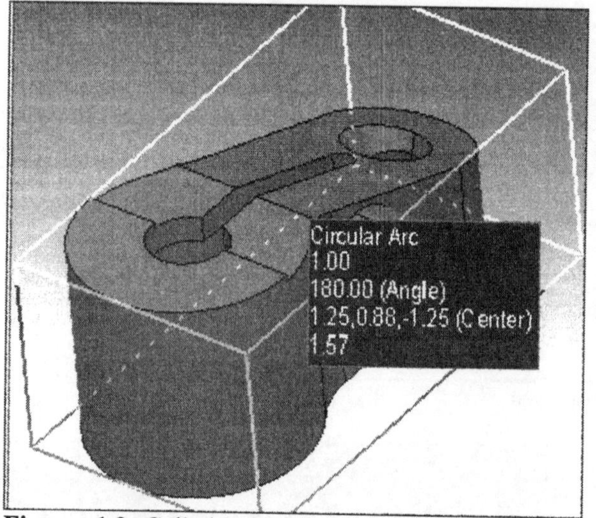

Figure 1.9 Cylindrical Protrusion Measurement

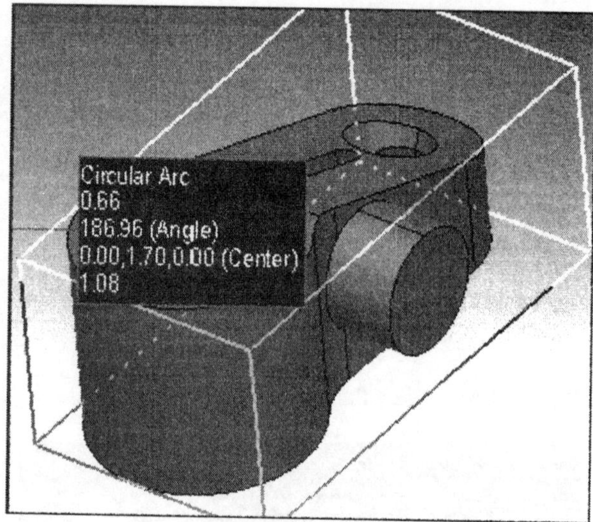

Figure 1.10 Hole Measurement

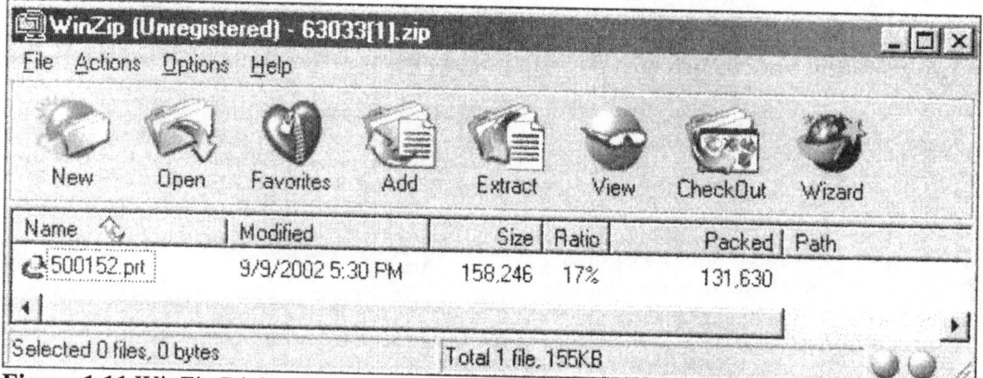

Figure 1.11 WinZip Dialog Box Displaying 500152.prt

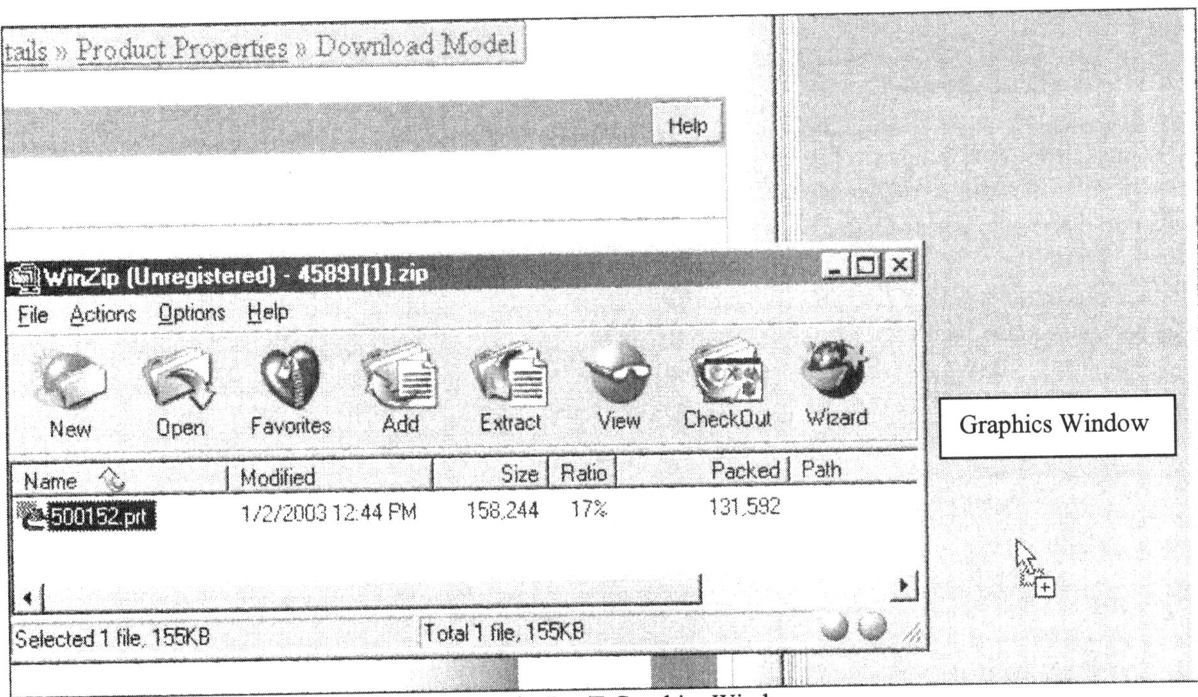

Figure 1.12(a) Drag-and-Drop 500152.prt into the Pro/E Graphics Window

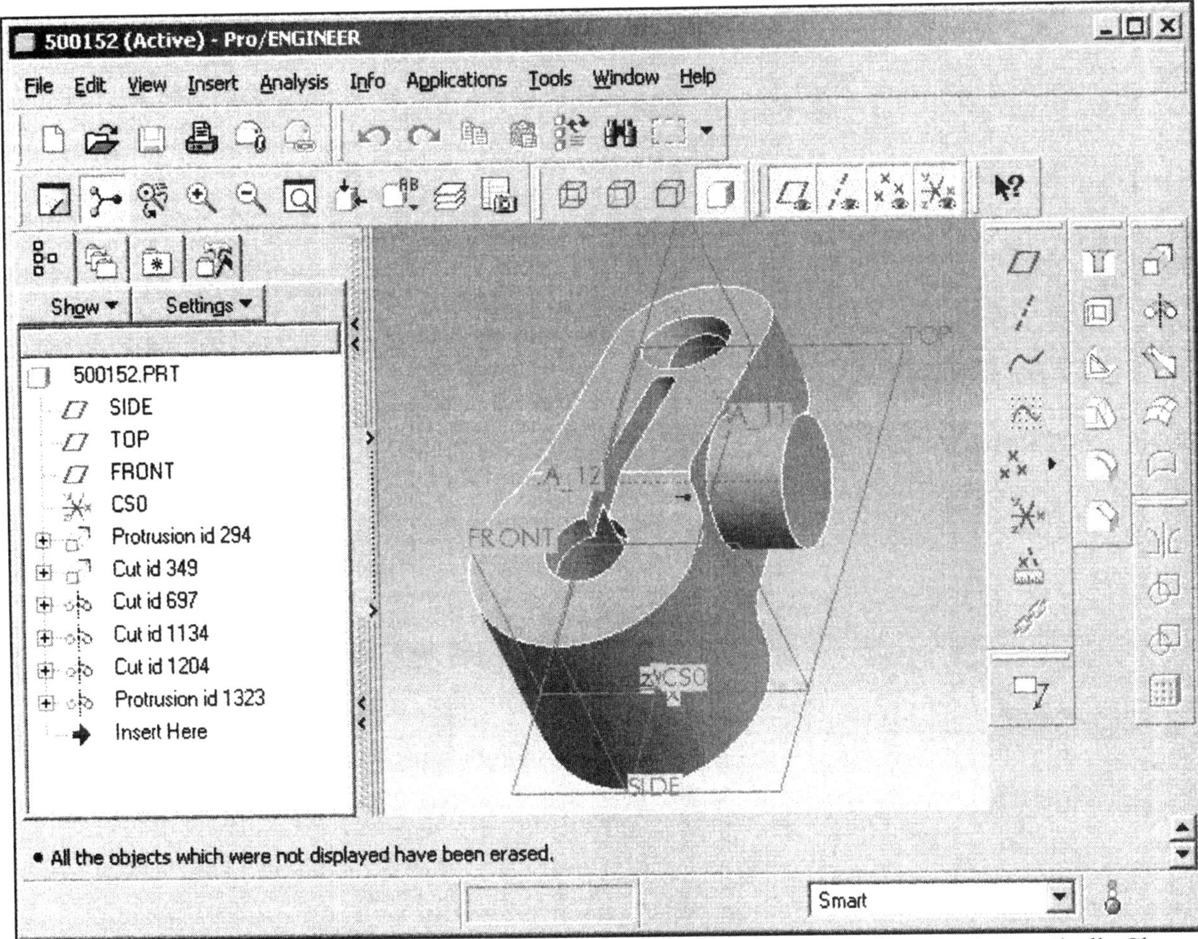

Figure 1.12(b) Active Part 500152.prt Displayed in the Graphics Window and the Browser Automatically Closes

File Functions

The **File** menu on the Pro/E main window provides options for manipulating files (such as opening, creating, saving, renaming, backing up files, and printing). File functions include options for importing files from and exporting files to external formats, setting your working directory and performing operations on instances. Before using any File tool, make sure you have set your working directory to the folder where you wish to save objects for the project on which you are working. The Working Directory is a directory that you set up to contain Pro/E files.

You can save Pro/E files using Save or Save a Copy commands on the File menu. The Save a Copy dialog box also allows you to export Pro/E files to different formats, and to save files as images. Since the name already exists in session, you cannot save or Rename a file using the same name as the original file name, even if you save the file in a different directory. Pro/E forces you to enter a unique file name by displaying the message: *An object with that name exists in session. Choose a different name.*

Names for all Pro/E files are restricted to a maximum of 31 characters and must contain no spaces. A File can be a part, assembly, or drawing. Each is considered an "object".

Click: **File** ⇒ **Set Working Directory** (Fig. 1.13) ⇒ select the directory you wish all objects to be saved to for this Pro/E project ⇒ **OK** ⇒ **File** ⇒ **Save a Copy** ⇒ type in the New Name field: **CLAMP_500152** ⇒ **OK** (Fig. 1.14) ⇒ **Window** ⇒ **Close** ⇒ **File** ⇒ **Open** ⇒ **clamp_500152** ⇒ **Open**

[Depending on the computer you are using (yours or an organizations) the following may not work. If you have your mail server configured as the default then proceed, otherwise skip this step]

Send email with object in active window ⇒ email the object to yourself (or a friend with Pro/E software) ⇒ check on **Create a ZIP file** ⇒ **OK** (Fig. 1.15) ⇒ follow your email procedure as required

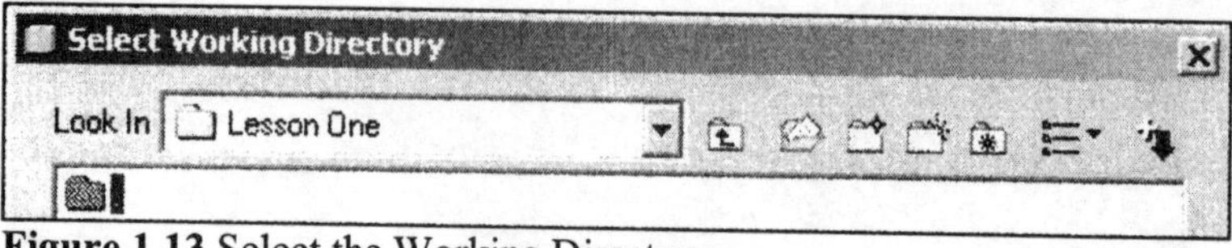

Figure 1.13 Select the Working Directory

Model Name 500152.PRT
New Name CLAMP_500152
Type Part (*.prt)
OK Cancel

Figure 1.14 Save a Copy

Send As Attachment
Model Name 500152.PRT
New Name 500152
Type Part (*.prt)
Create a ZIP file containing the attachment
OK Cancel

Figure 1.15 Create a Zip File and Send as Attachment

Help

Accessing the Help function is one of the best ways to learn CAD software. Use the Help tool as often as possible to understand the tool or command you are using at the time and to expand your knowledge of the other capabilities provided by Pro/ENGINEER. Use the **Help** menu to gain access to online information, Pro/E release information, and customer service information. The following commands are available on the Pro/E **Help** menu in standard Part and Assembly modes.

- **Help Center** Displays the context-sensitive online help system. When you select this, your supported network browser opens to display a navigation tree and search tools to aid you in finding specific help topics. You can also access these topics by clicking for context-sensitive Help from windows, menu commands, and dialog boxes.
- **What's This?** Enables context-sensitive Help mode.
- **Release Notes** Displays what is new in this release of Pro/E software.
- **Technical Support Info** Displays product information, including the release level, license information, installation date, and customer support contact information.
- **About Pro/ENGINEER** Displays Pro/E copyright and release information.

Click: **Help ⇒ Help Center** (Fig. 1.16) ⇒ **Fundamentals ⇒ Pro/ENGINEER Fundamentals**

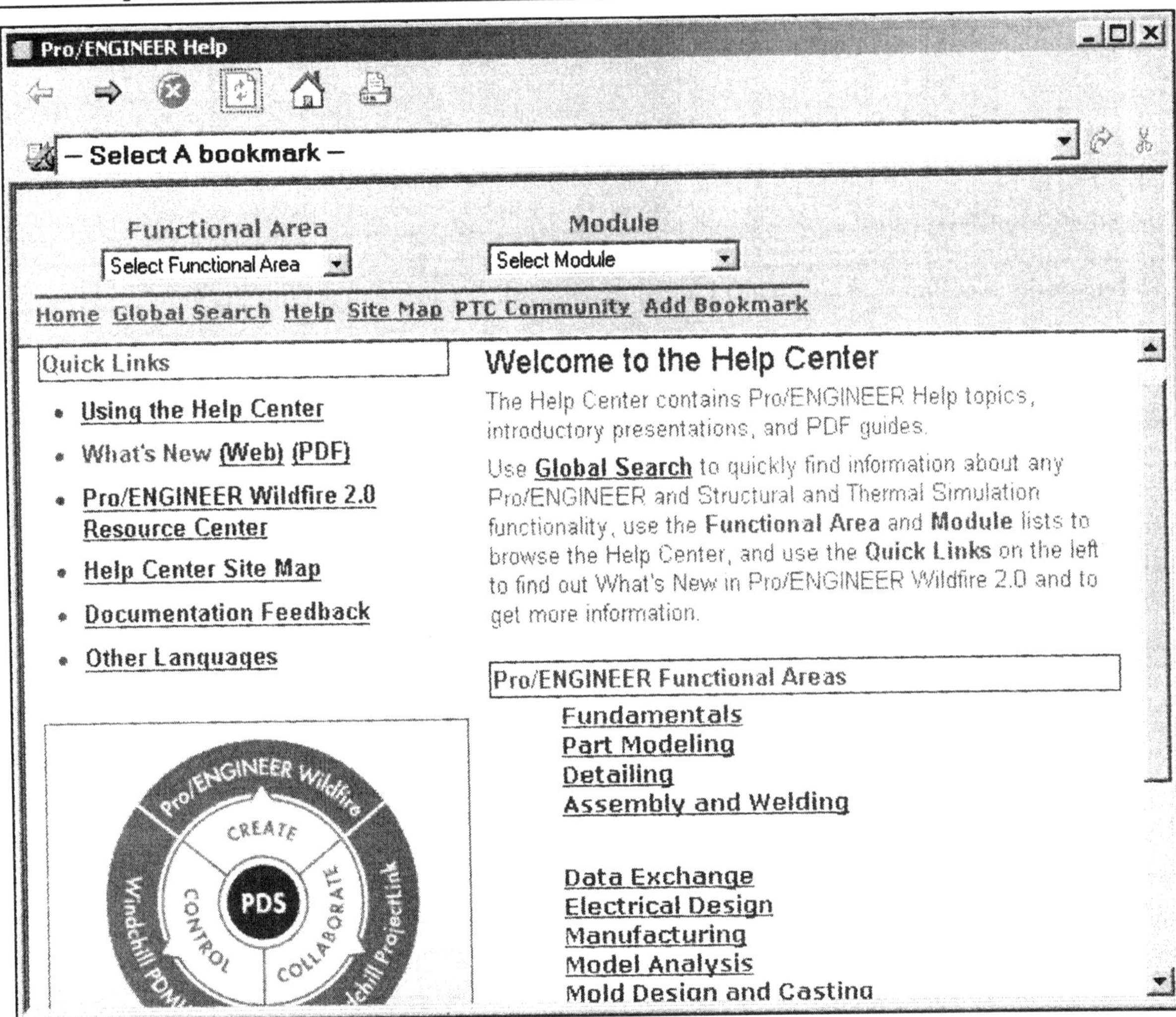

Figure 1.16 Pro/ENGINEER Help Center

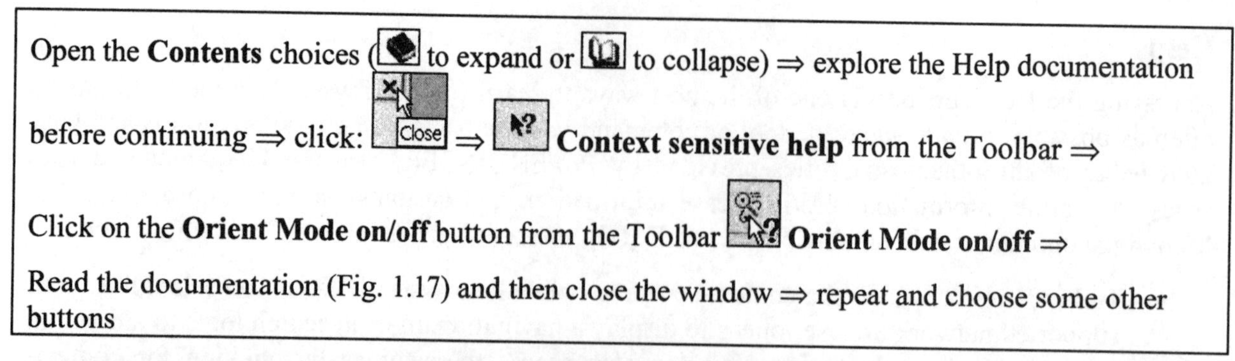

Open the **Contents** choices (to expand or to collapse) ⇒ explore the Help documentation before continuing ⇒ click: Close ⇒ **Context sensitive help** from the Toolbar ⇒

Click on the **Orient Mode on/off** button from the Toolbar **Orient Mode on/off** ⇒

Read the documentation (Fig. 1.17) and then close the window ⇒ repeat and choose some other buttons

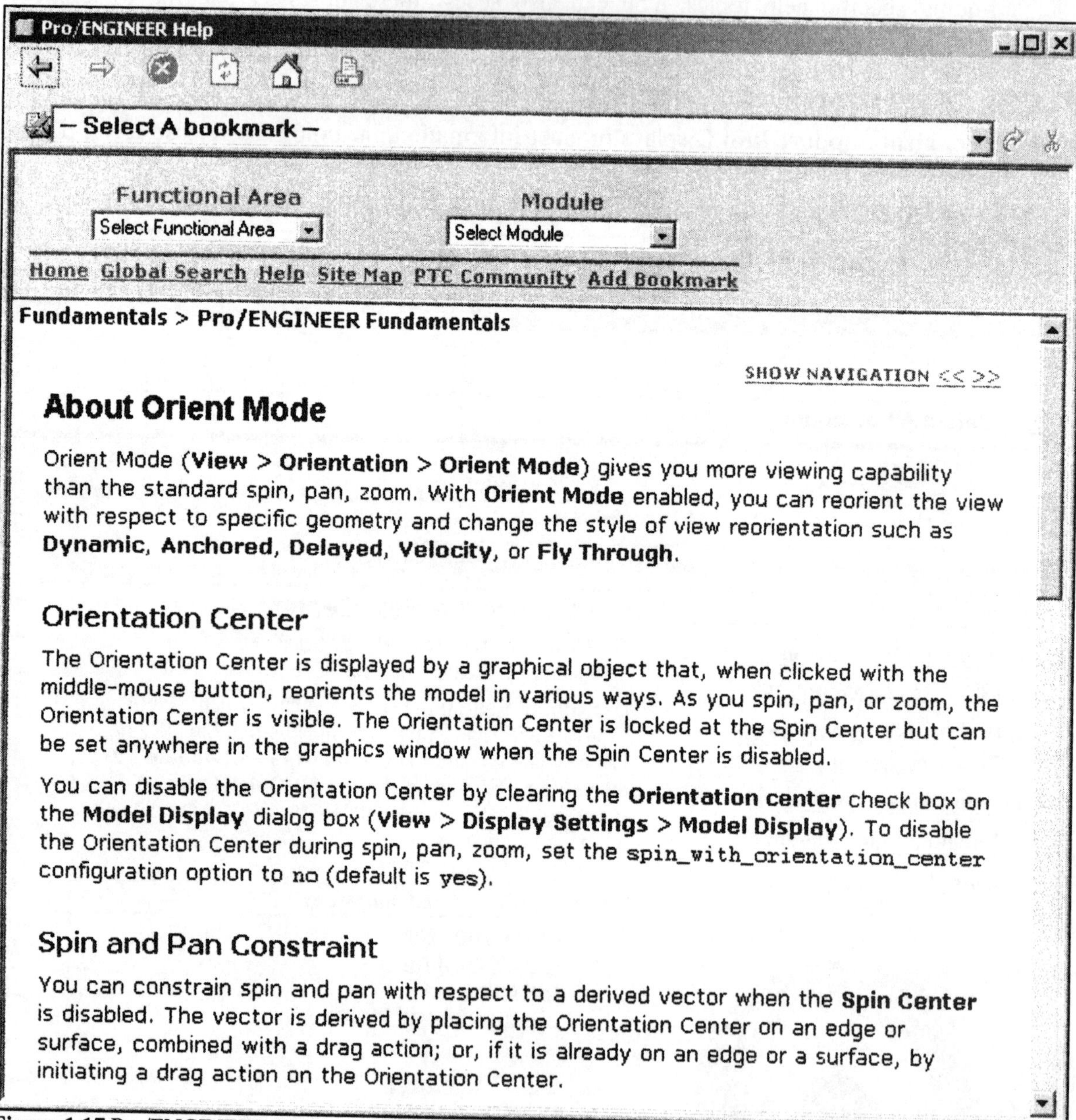

Figure 1.17 Pro/ENGINEER Help

View and Display Functions

Using the Pro/E **View** menu, you can adjust the model view, orient the view, hide and show entities, create and use advanced views, and set various model display options. The following list includes some of the View operations you can perform:

- Orient the model view in the following ways, using the **Orientation** dialog box: spin, pan, and zoom models and drawings, display the default orientation, revert to the previously displayed orientation, change the position or size of the model view, change the orientation (including changing the view angle in a drawing), and create new orientations.
- Temporarily shade a model by using cosmetic shading
- Show, dim, or remove hidden lines
- Highlight items in the graphics window when you select them in the Model Tree
- Explode or unexplode an assembly view
- Repaint the Pro/E graphics window
- Refit the model to the Pro/E window after zooming in or out on the model
- Update drawings of model geometry
- Hide and show entities, and hide or show items during spin or animation
- Use advanced views
- Add perspective to the model view

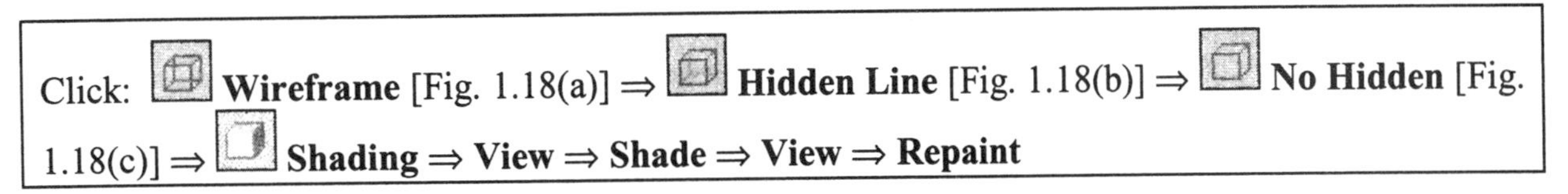
Click: **Wireframe** [Fig. 1.18(a)] ⇒ **Hidden Line** [Fig. 1.18(b)] ⇒ **No Hidden** [Fig. 1.18(c)] ⇒ **Shading ⇒ View ⇒ Shade ⇒ View ⇒ Repaint**

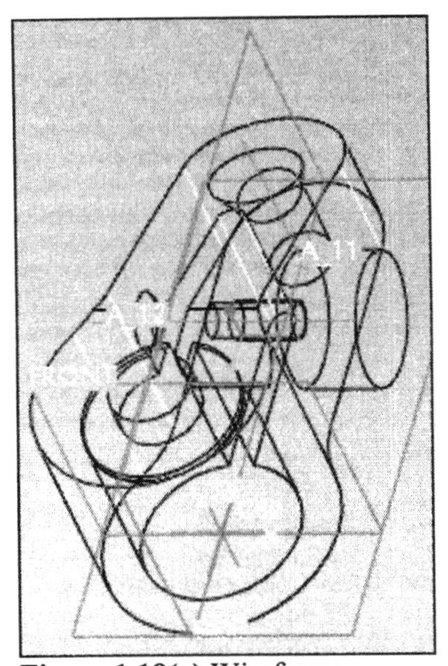

Figure 1.18(a) Wireframe

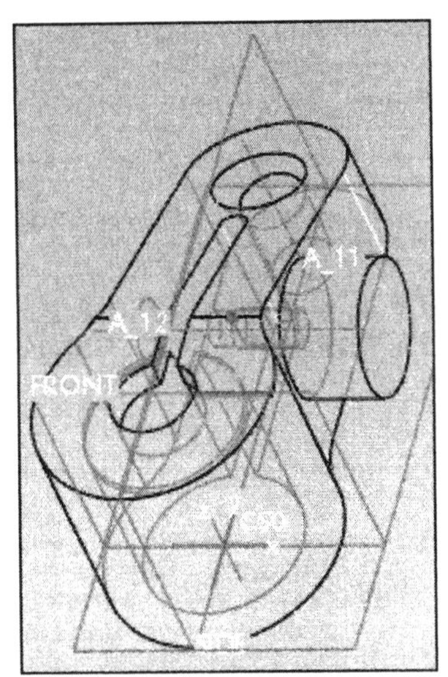

Figure 1.18(b) Hidden Line

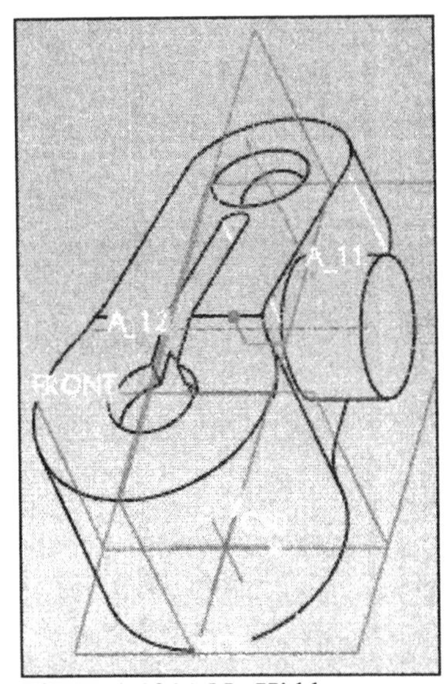

Figure 1.18(c) No Hidden

To see the standard views provided, click: ⇒ **Front** ⇒ ⇒ **Right** [Figs. 1.19(a-b)] ⇒ try all the variations ⇒ ⇒ **Standard Orientation**

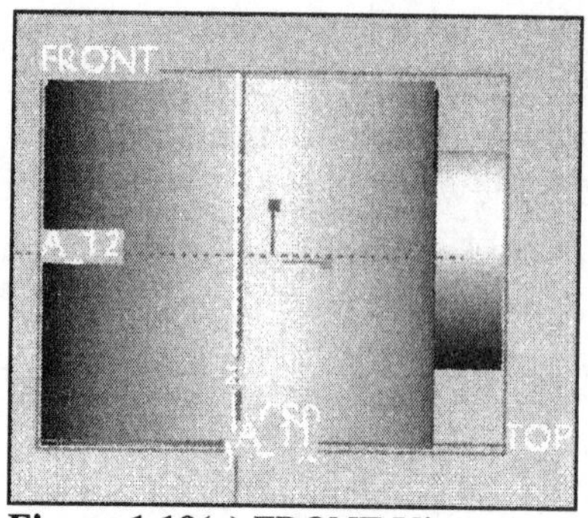

Figure 1.19(a) FRONT View

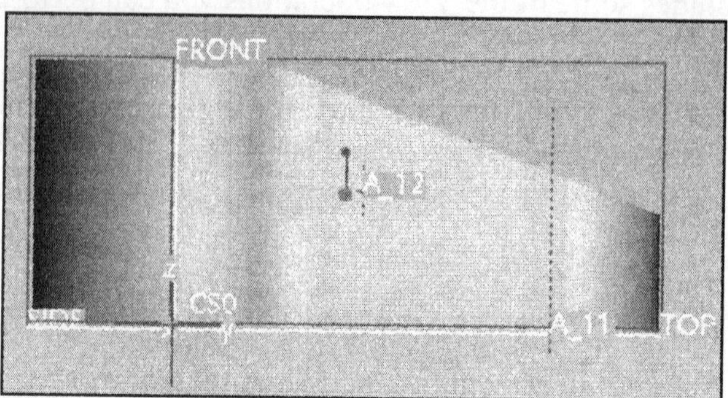

Figure 1.19(b) RIGHT View

View Tools

As with all CAD systems, Pro/E provides the typical view tools associated with CAD:

- **Zoom In** Use this tool to zoom in on a specific portion of the model. Pick two positions of a rectangular zoom box.
- **Zoom Out** Use this tool to reduce the view size of the model on the screen by 50%.
- **Refit object to fully display it on the screen** Use this tool to refit the model to the screen so that you can view the entire model. A refitted model fills 80% of the graphics window.

Click: **Zoom In** [Fig. 1.20(a)] ⇒ pick two positions about an area you wish to enlarge [Fig. 1.20(b)] ⇒ **Zoom Out** ⇒ **Refit** [Fig. 1.20(c)] ⇒ **Redraw the current view**

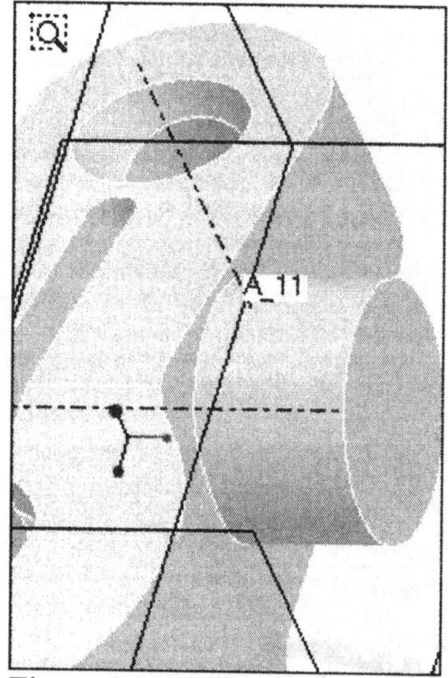

Figure 1.20(a) View

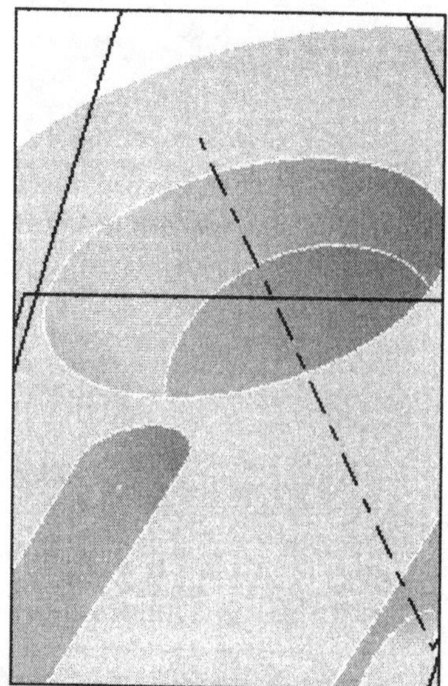
Figure 1.20(b) Zoom In

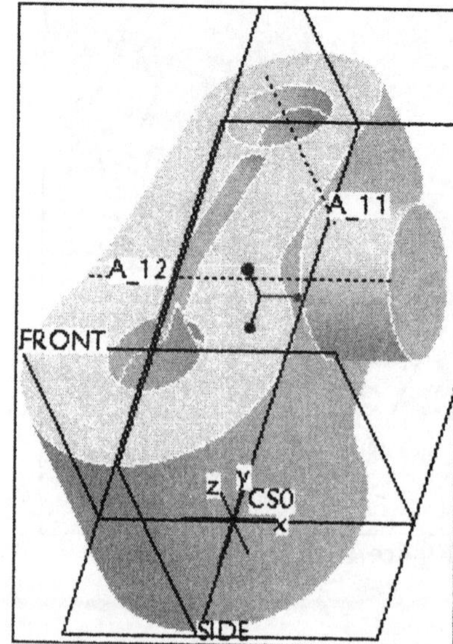

Figure 1.20(c) Refit

Using Mouse Buttons to Manipulate the Model

You can also dynamically reorient the model using the **MMB** by itself (**Spin**) or in conjunction with the **Shift** key (**Pan**) or **Ctrl** key (**Zoom, Turn**).

Hold down **Ctrl** key and **MMB** in the graphics area near the model and move the cursor up (zoom out) [Fig. 1.21(a)] ⇒ hold down **Shift** key and **MMB** in the graphics area near the model and move the cursor about the screen (pan) [Fig. 1.21(b)] ⇒ hold down **MMB** in the graphics area near the model and move the cursor around (spin) [Figs. 1.22(a-b)] ⇒ ⇒ **Standard Orientation**

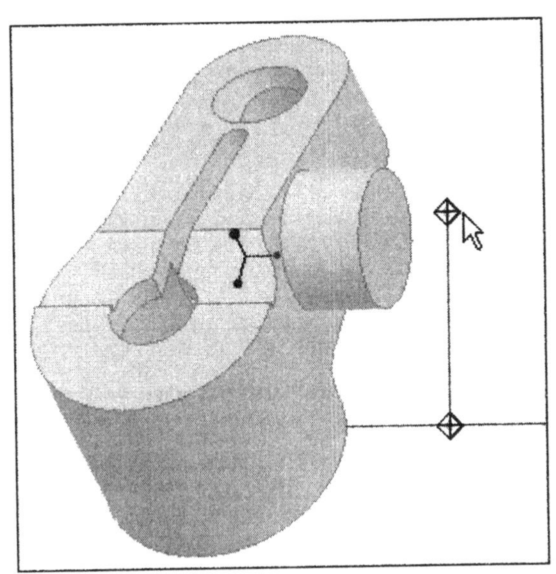

Figure 1.21(a) Zoom

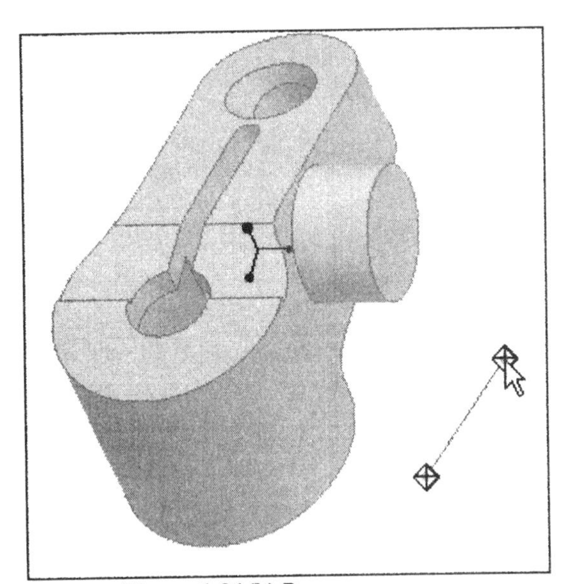

Figure 1.21(b) Pan

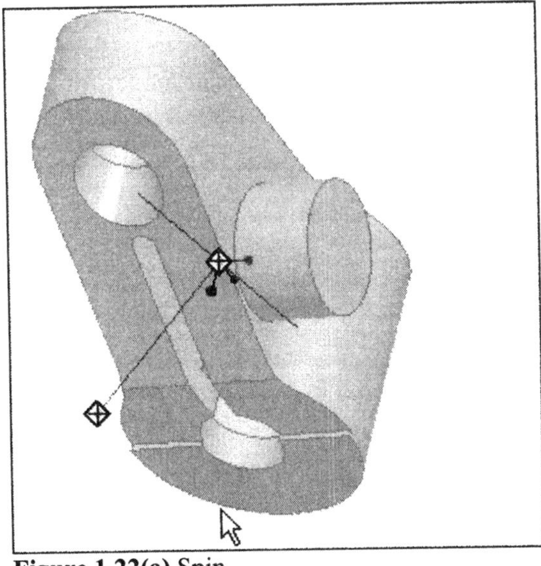

Figure 1.22(a) Spin

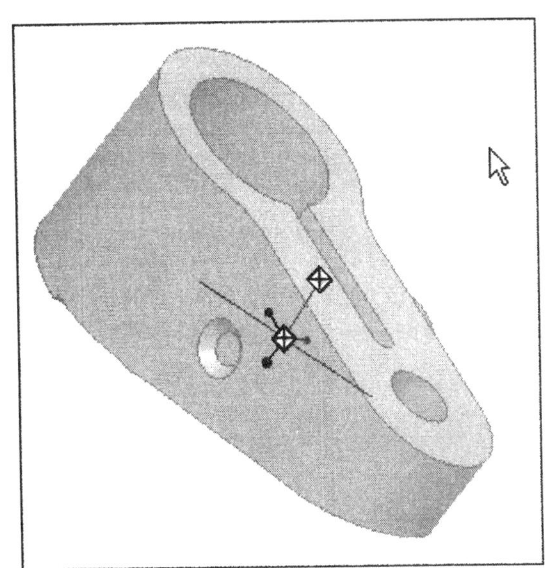

Figure 1.22(b) Spin Again

You may have noticed that the illustrations of the text have changed since Figure 1.20. The default for the background and geometry colors has been changed so that the illustrations will capture and print clearer. In the next section, you will learn how to change the system display settings.

System Display Settings

You can make a number of changes to the default colors furnished by Pro/E, customizing them for your own use:

- Define, save, and open color schemes
- Customize colors used in the user interface
- Change your entire color scheme to a predefined color scheme (such as black on white)
- Change the top or bottom background colors
- Redefine basic colors used in models
- Assign colors to be used by an entity
- Store a color scheme so you can reuse it
- Open a previously used color scheme

The **Scheme** menu includes the following color schemes (text uses the **Black on White** selection):

- **Black on White** Black entities shown on a white background
- **White on Black** White entities shown on a black background
- **White on Green** White entities shown on a dark-green background
- **Initial** Reset the color scheme to the one defined by the configuration file settings
- **Default** Reset the color scheme to the system default

Click: View ⇒ Display Settings ⇒ System Colors [Figs. 1.23(a-c)] ⇒ Blended Background ⇒ Scheme ⇒ Black on White ⇒ experiment with different color schemes and system colors ⇒ Scheme ⇒ Default ⇒ OK

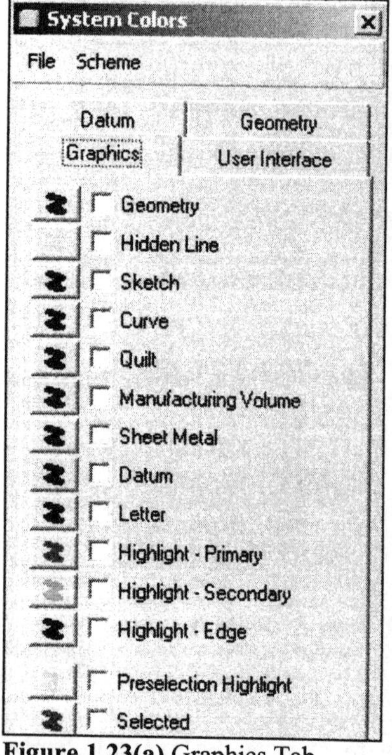

Figure 1.23(a) Graphics Tab

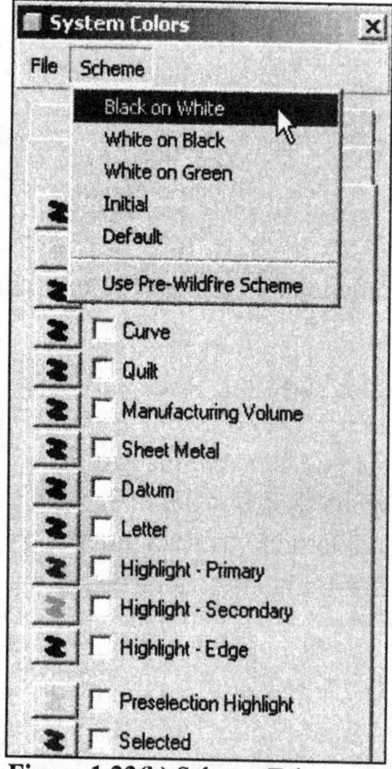

Figure 1.23(b) Scheme Tab

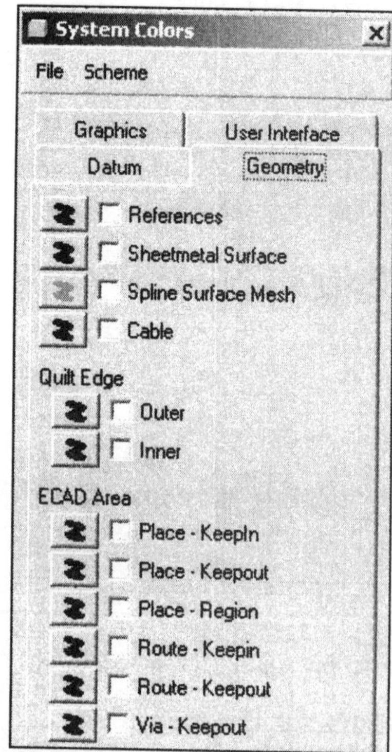

Figure 1.23(c) Geometry Tab

Information Tools

At any time during the design process you can request model, feature, or other information. Picking on a feature in the graphics window and then **RMB** ⇒ **Info** (Fig. 1.24) will provide information about that feature in the Browser [Figs. 1.25(a-b)]. This can also be accomplished by clicking on the feature name in the Model Tree and then **RMB**. Both feature and model information can be obtained using this method (Fig. 1.26). A variety of information can also be extracted using the **Info** tool from the menu bar.

Pick once on the revolved protrusion (Fig. 1.24) ⇒ **RMB** ⇒ **Info** ⇒ **Feature** [Figs. 1.25(a-b)] ⇒ click on the "quick sash" to collapse the Browser ⇒ in the graphics window, **RMB** (Fig. 1.26) ⇒ **Info** ⇒ **Model** (Fig. 1.27) ⇒ collapse the Browser

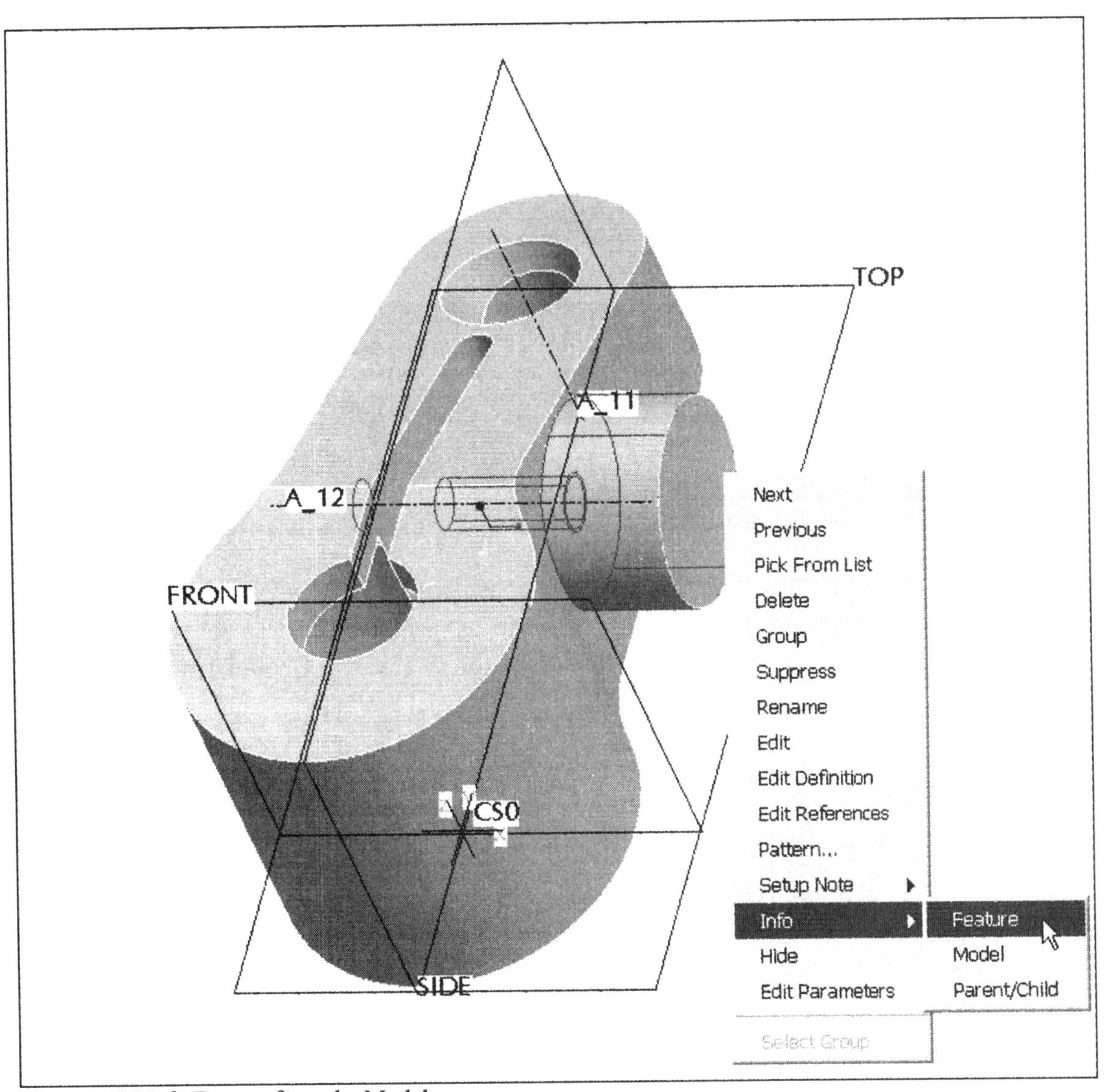

Figure 1.24 Info Feature from the Model

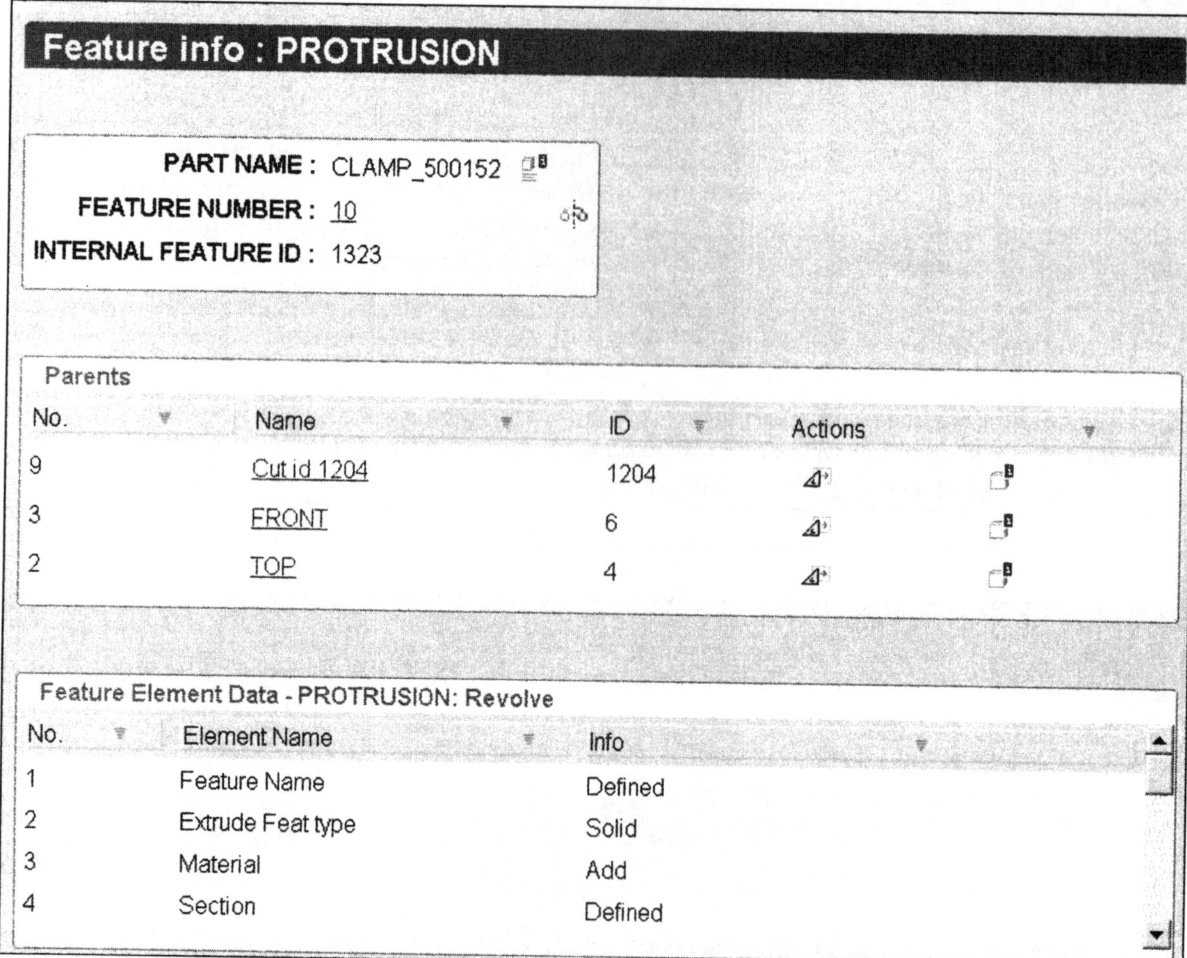

Figure 1.25(a) Feature Info Displayed in the Browser

FEATURE'S DIMENSIONS:

Dimension ID	Dimension Value
HD	1.0000 Dia
DIMENSION IS IN LAYER(S) :	
LAYER_DIM	OPERATION = SHOWN
LAYER_PARAMETER_DIM	OPERATION = SHOWN
HT	.656
DIMENSION IS IN LAYER(S) :	
LAYER_DIM	OPERATION = SHOWN
LAYER_PARAMETER_DIM	OPERATION = SHOWN

Figure 1.25(b) Feature Dimension Information

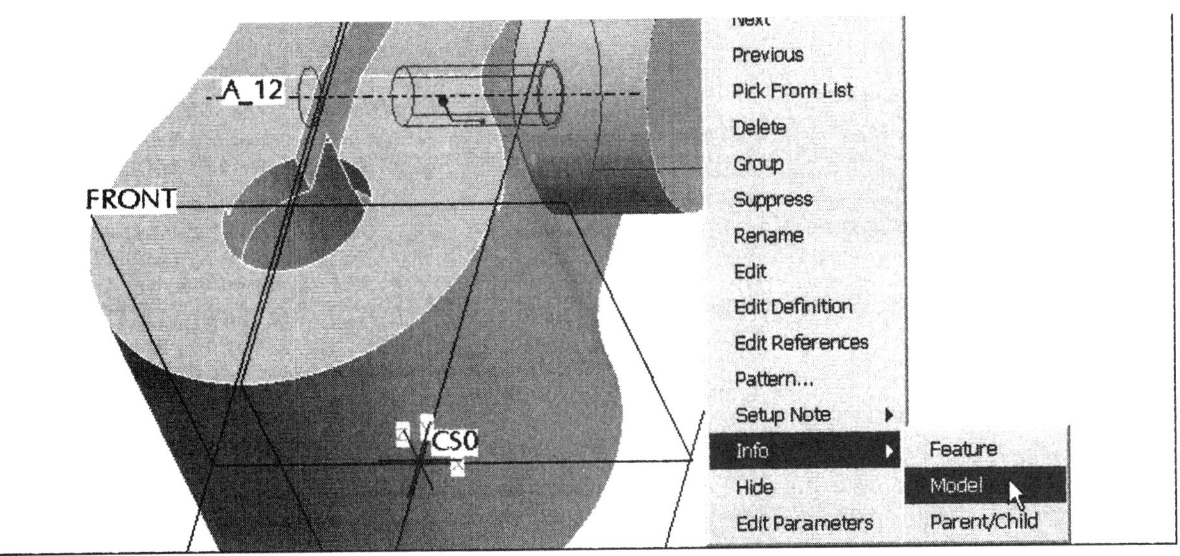

Figure 1.26 Info Model

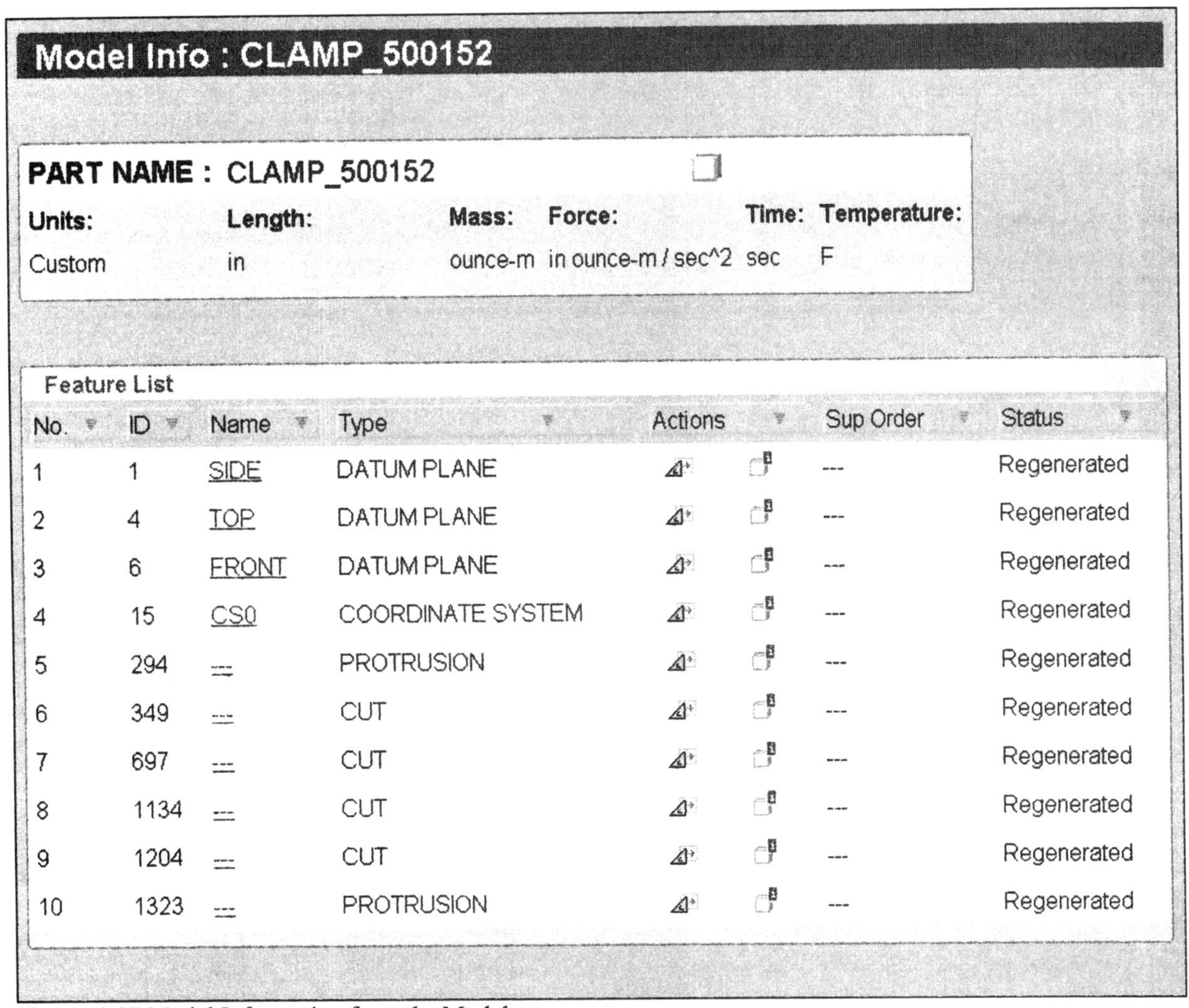

Model Info : CLAMP_500152

PART NAME : CLAMP_500152

Units:	Length:	Mass:	Force:	Time:	Temperature:
Custom	in	ounce-m	in ounce-m / sec^2	sec	F

Feature List

No.	ID	Name	Type	Actions	Sup Order	Status
1	1	SIDE	DATUM PLANE		---	Regenerated
2	4	TOP	DATUM PLANE		---	Regenerated
3	6	FRONT	DATUM PLANE		---	Regenerated
4	15	CS0	COORDINATE SYSTEM		---	Regenerated
5	294	---	PROTRUSION		---	Regenerated
6	349	---	CUT		---	Regenerated
7	697	---	CUT		---	Regenerated
8	1134	---	CUT		---	Regenerated
9	1204	---	CUT		---	Regenerated
10	1323	---	PROTRUSION		---	Regenerated

Figure 1.27 Model Information from the Model

The Model Tree

The **Model Tree** is a tabbed feature on the Pro/E navigator that displays a list of every feature or part in the current part, assembly, or drawing.

The model structure is displayed in hierarchical (tree) format with the root object (the current part or assembly) at the top of its tree and the subordinate objects (parts or features) below. If you have multiple Pro/E windows open, the Model Tree contents reflect the file in the current active window.

The Model Tree lists only the related feature and part level items in a current object and does not list the entities (such as edges, surfaces, curves, and so forth) that comprise the features.

Each Model Tree item contains an icon that reflects its object type, for example, hidden, assembly, part, feature, or datum plane (also a feature). The icon can also show the display status for a feature, part, or assembly, for example, suppressed.

Selection in the Model Tree is object-action oriented; you select objects in the Model Tree without first specifying what you intend to do with them. You can select components, parts, or features using the Model Tree. You cannot select the individual geometry that makes up a feature (entities). To select an entity, you must select it in the graphics window.

With the **Settings** tab you can control what is displayed in the Model Tree.

You can add informational columns to the Model Tree window, such as **Tree Columns** containing parameters and values, assigned layers, or feature name for each item. You can use the cells in the columns to perform context-sensitive edits and deletions. These options will be covered elsewhere in the text, as they are needed in the design process.

Click: **Settings** tab ⇒ Tree Filters... [Figs. 1.28(a-b)] ⇒ Display (if not selected, check all active options) ⇒ **Apply** ⇒ **OK** ⇒ in the graphics window click **LMB** to deselect

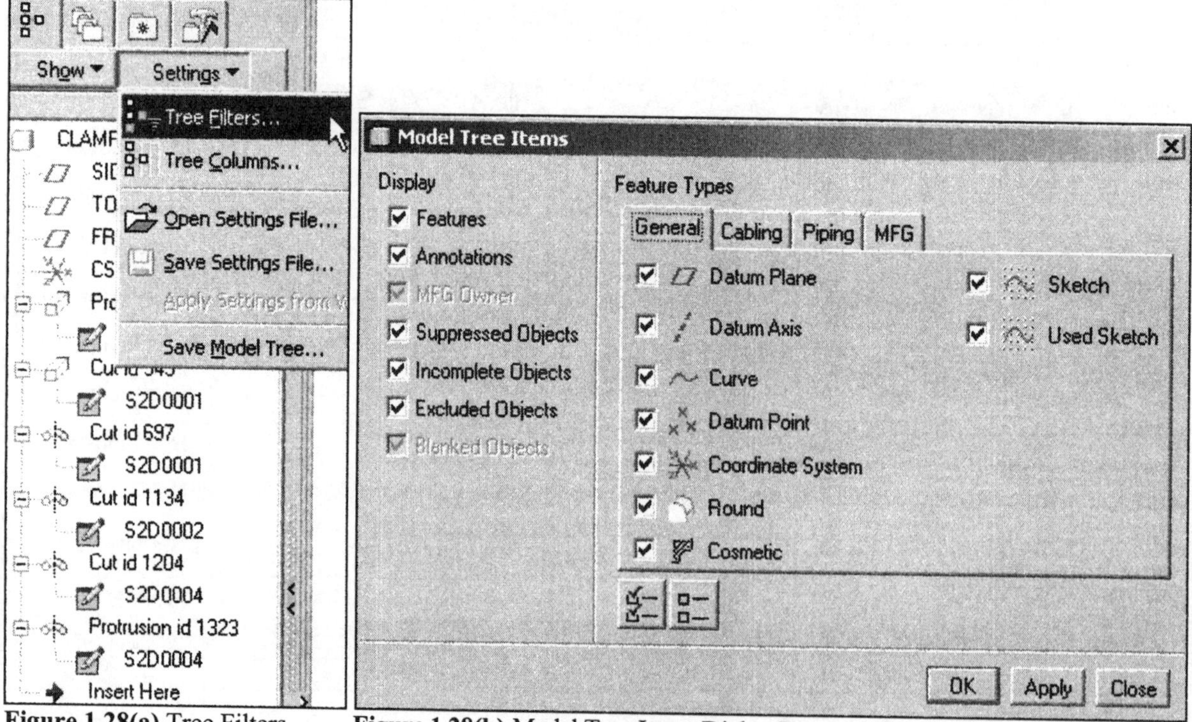

Figure 1.28(a) Tree Filters **Figure 1.28(b)** Model Tree Items Dialog Box

Working on the Model

In Pro/E, you can select objects to work on from within the graphics window or in the Model Tree by using the mouse or the keyboard. The object types that are available for selection vary depending on whether you select an object from within the graphics window or in the Model Tree. You can select any type of object, including features, 3-D notes, parts, datum objects (planes, axes, curves, points, and coordinate systems), and geometry (edges and surfaces) from within the graphics window. Additionally, since the Model Tree displays only parts, components, and features, you can select only those object types from within the Model Tree.

Selection in both the graphics window and the Model Tree can be action-object or object-action oriented, depending on the process you choose within Pro/E to build your model. You can specify the action you want to perform on an object before you select the object, or you can select the object before you specify the action.

You can *dynamically* modify certain features from within the graphics window as you work. Features can be edited by selecting them from the graphics window directly, or from the Model Tree. Dimensions of the following features can be modified dynamically:

- **Protrusions** Extruded protrusions with variable depth, and revolved protrusions of variable angle
- **Cuts** Extruded cuts with variable depth, and revolved cuts with variable angle
- **Surfaces** Extruded surfaces with variable depth, and revolved surfaces with variable angle
- **Rounds** Simple, constant, and edge chain

Pick on the cylindrical protrusion ⇒ **RMB** [Fig. 1.29(a)] ⇒ **Delete** ⇒ **OK** [Fig. 1.29(b)]

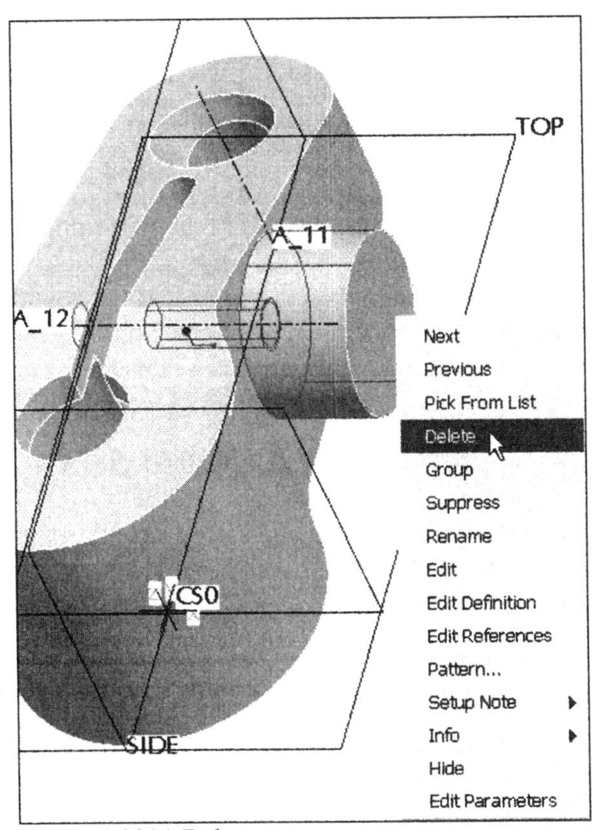

Figure 1.29(a) Delete

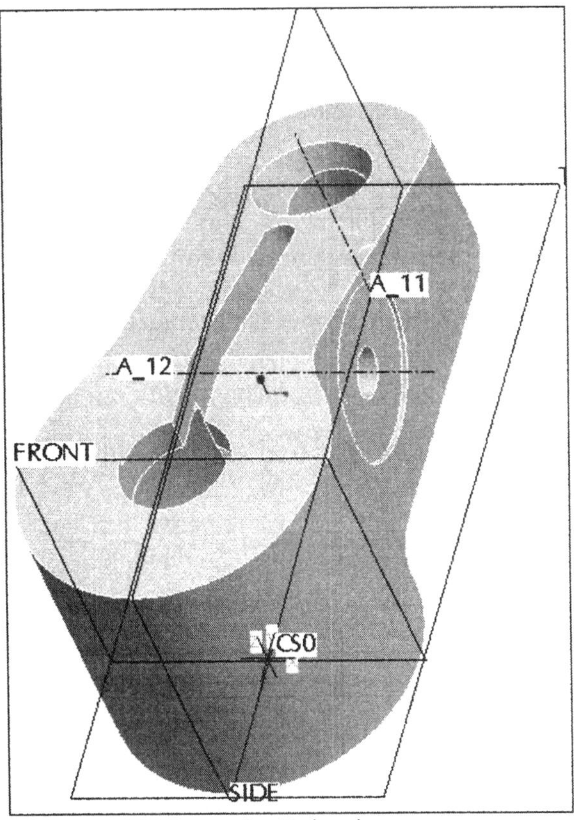

Figure 1.29(b) Protrusion Deleted

Click on the protrusion in the Model Tree (Fig. 1.30) ⇒ **RMB** ⇒ **Edit Definition** (Fig. 1.31)

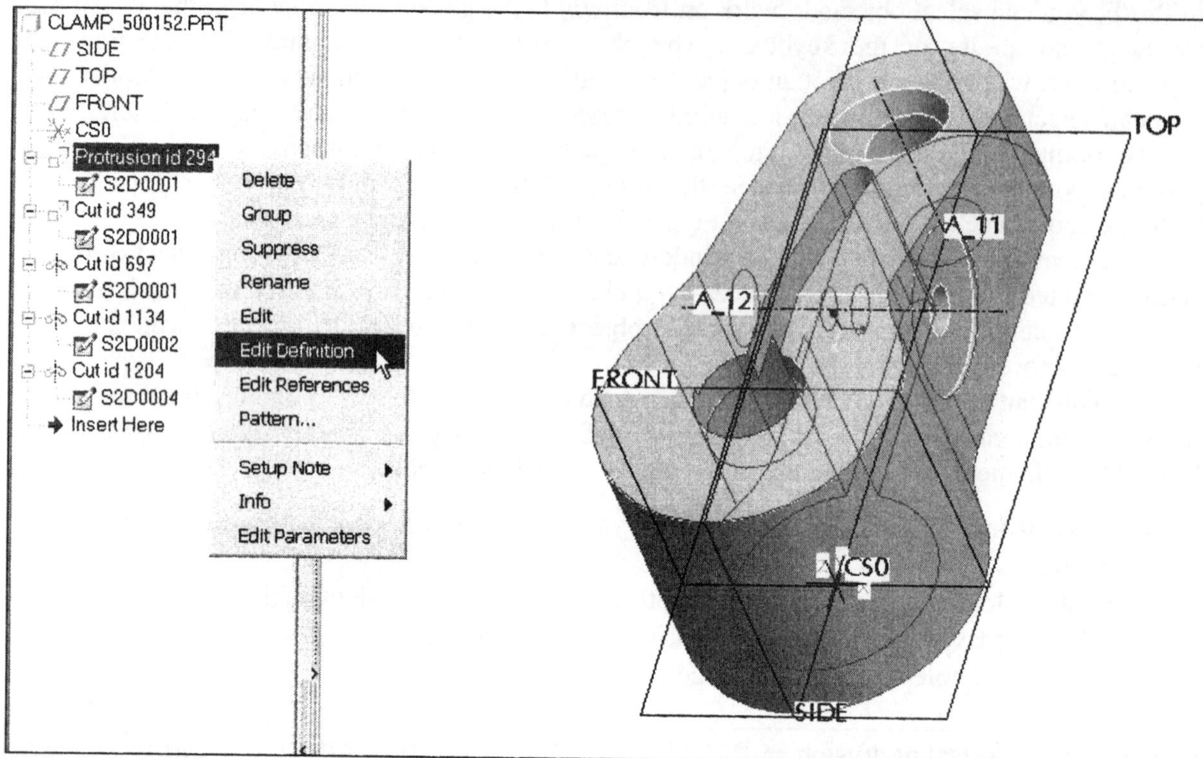

Figure 1.30 Redefining a Feature

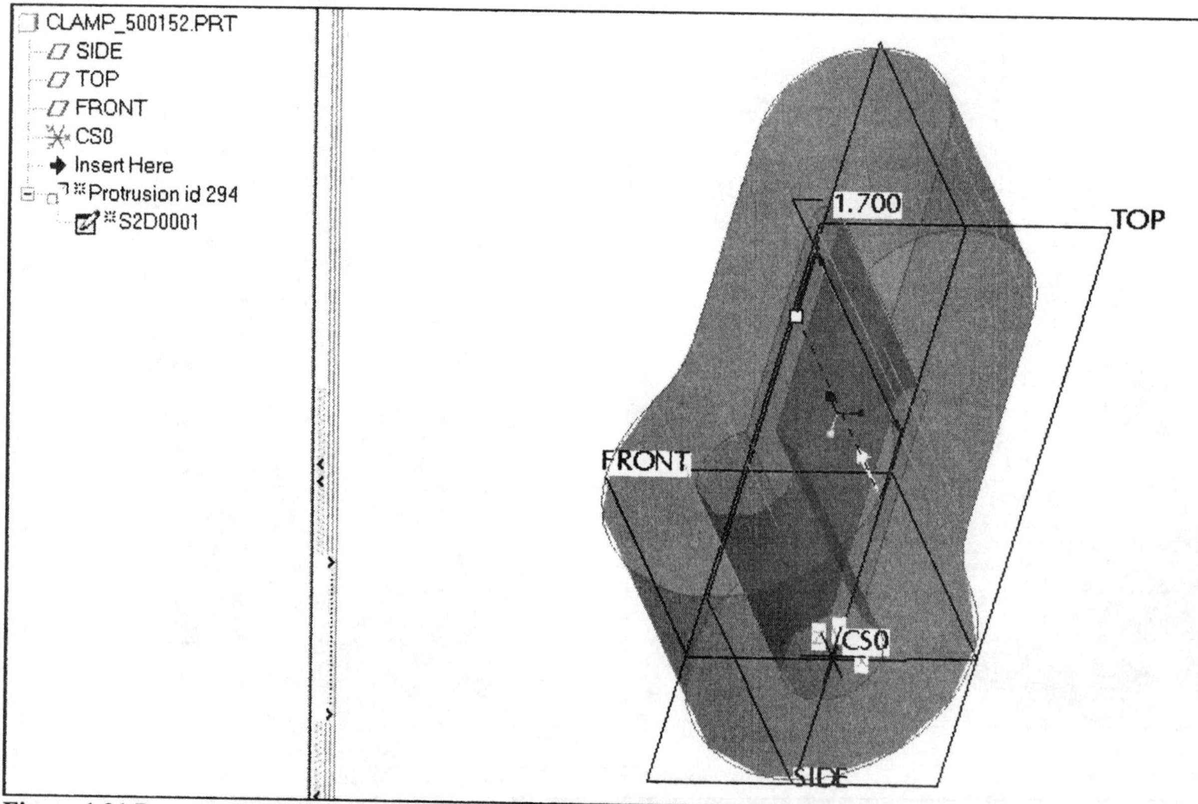

Figure 1.31 Dynamically Redefining a Feature

Pick on and drag the white nodal handle to change the protrusion's height from **1.700** (Fig. 1.32) to **2.000** (Fig. 1.33) ⇒ **MMB** to regenerate the model ⇒ **LMB** to deselect

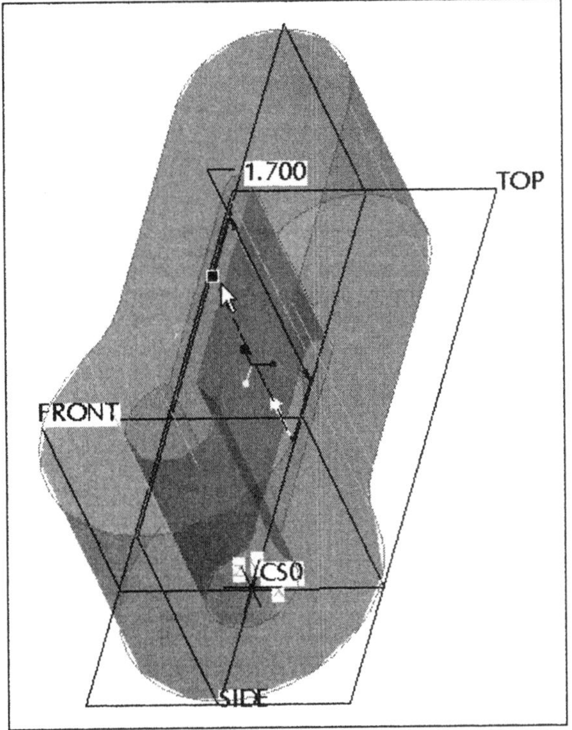

Figure 1.32 Dynamically Redefining a Feature

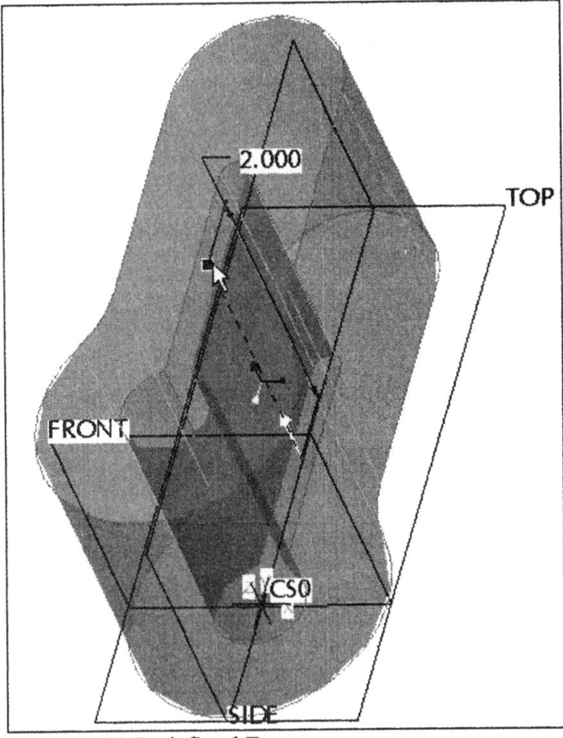

Figure 1.33 Redefined Feature

Although you may have not noticed it, there was other information and editing available during the redefine process. Repeat the command: Click on the protrusion in the Model Tree ⇒ **RMB** ⇒ **Edit Definition** ⇒ look at the "dashboard" on the lower part of the screen [Fig. 1.34(a)] ⇒ **Options** tab [Fig. 1.34(b)] ⇒ change the **2.000** to **3.000** ⇒ **Enter** ⇒ **MMB** (or ✓) [Figs. 1.35(a-b)] ⇒ **Window** ⇒ **Close**

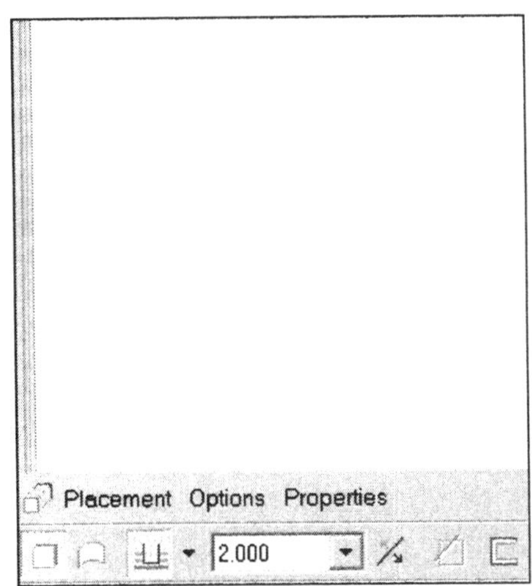

Figure 1.34(a) Dashboard

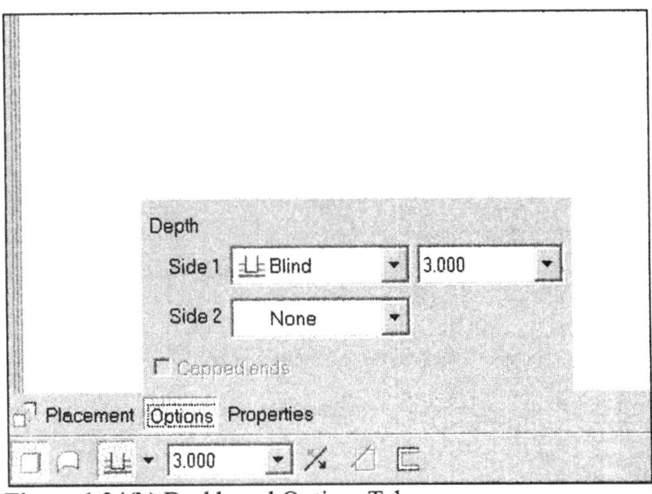

Figure 1.34(b) Dashboard Options Tab

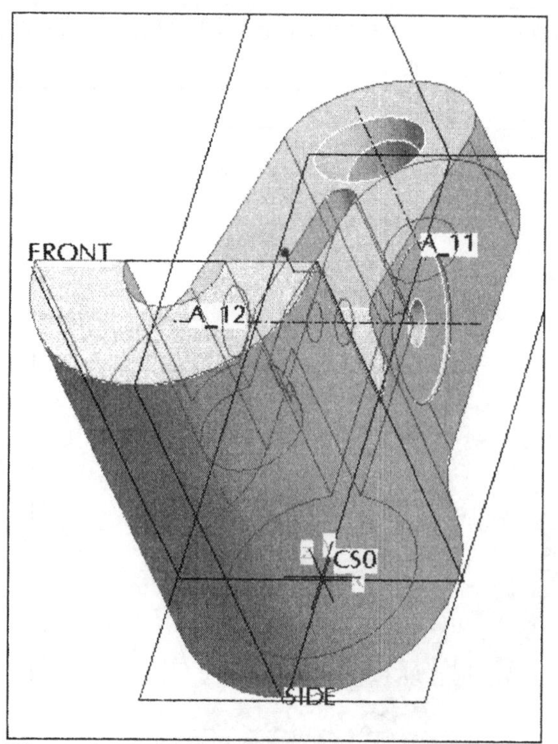

Figure 1.35(a) Edited Model

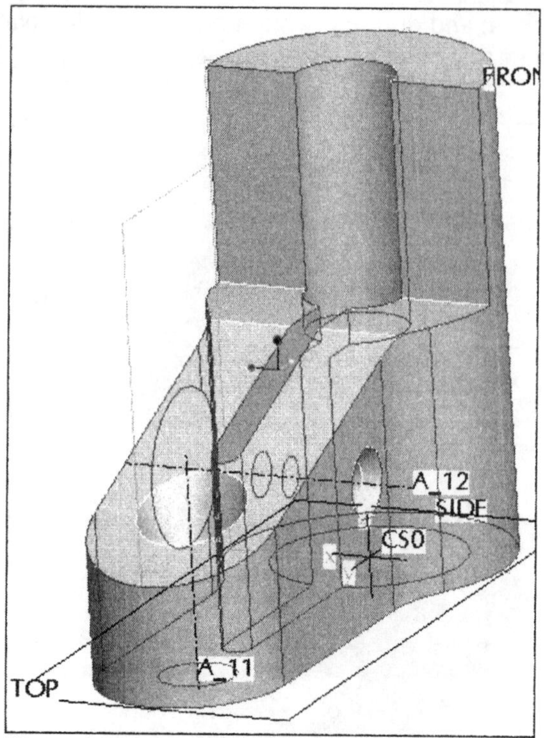

Figure 1.35(b) Edited Model Reoriented

About the Dashboard

As you create and modify your models using direct graphical manipulation in the graphics window, the **Dashboard** guides you throughout the modeling process. This context sensitive interface monitors your actions in the current tool and provides you with basic design requirements that need to be satisfied to complete your feature. As you select individual geometry, the Dashboard narrows the available options enabling you to make only targeted modeling decisions. For advanced modeling, separate slide-up panels provide all relevant advanced options for your current modeling action. These Dashboard panels remain hidden until needed. This enables you to remain focused on successfully capturing the design intent of your model. The Dashboard for a revolve feature is different from an extrude feature [Figs. 1.36(a-c)].

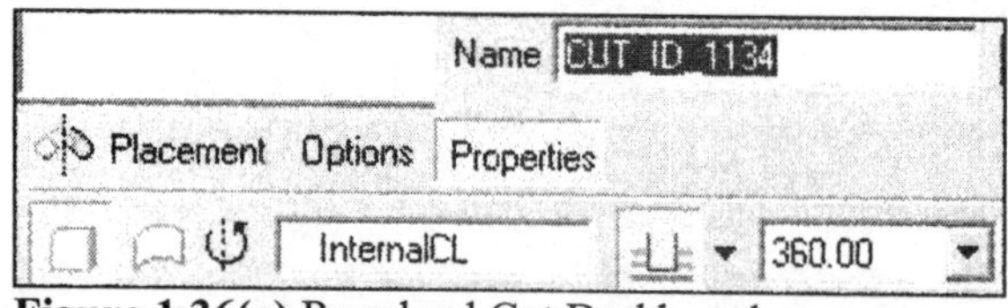

Figure 1.36(a) Revolved Cut Dashboard

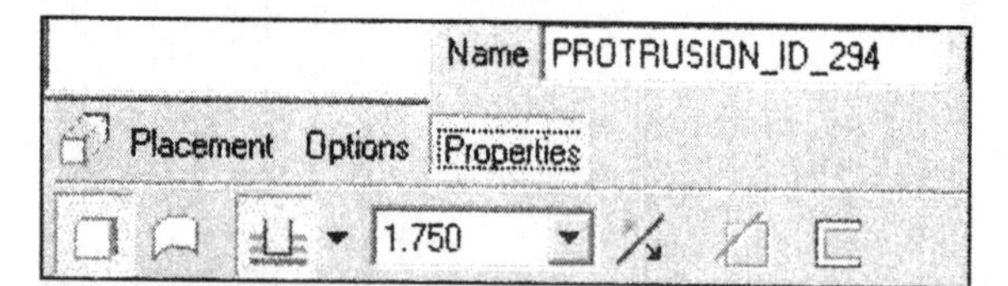

Figure 1.36(b) Extruded Protrusion Dashboard

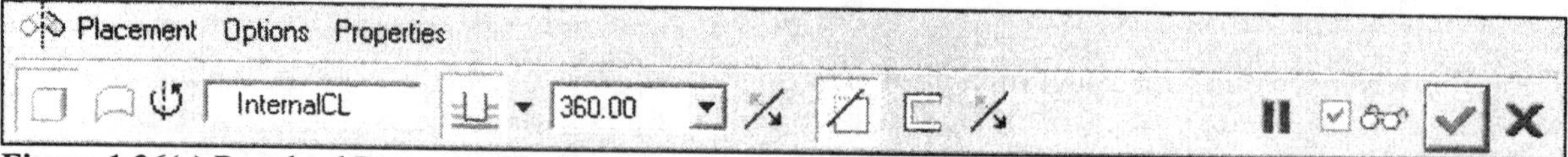

Figure 1.36(c) Revolved Protrusion Dashboard

This concludes the basic tour of Pro/E, Lesson 1. A Lesson Project is not provided here, at the end of Lesson 1. Instead, you may download another model from the PTC Catalog and practice navigating the user interface and using commands and tools introduced previously. If you do not have Internet access, ask your instructor to provide another model. If you are using the TE or SE Edition, email cad@cad-resources.com or see the website www.cad-resources.com.

Lesson 2 Direct Modeling

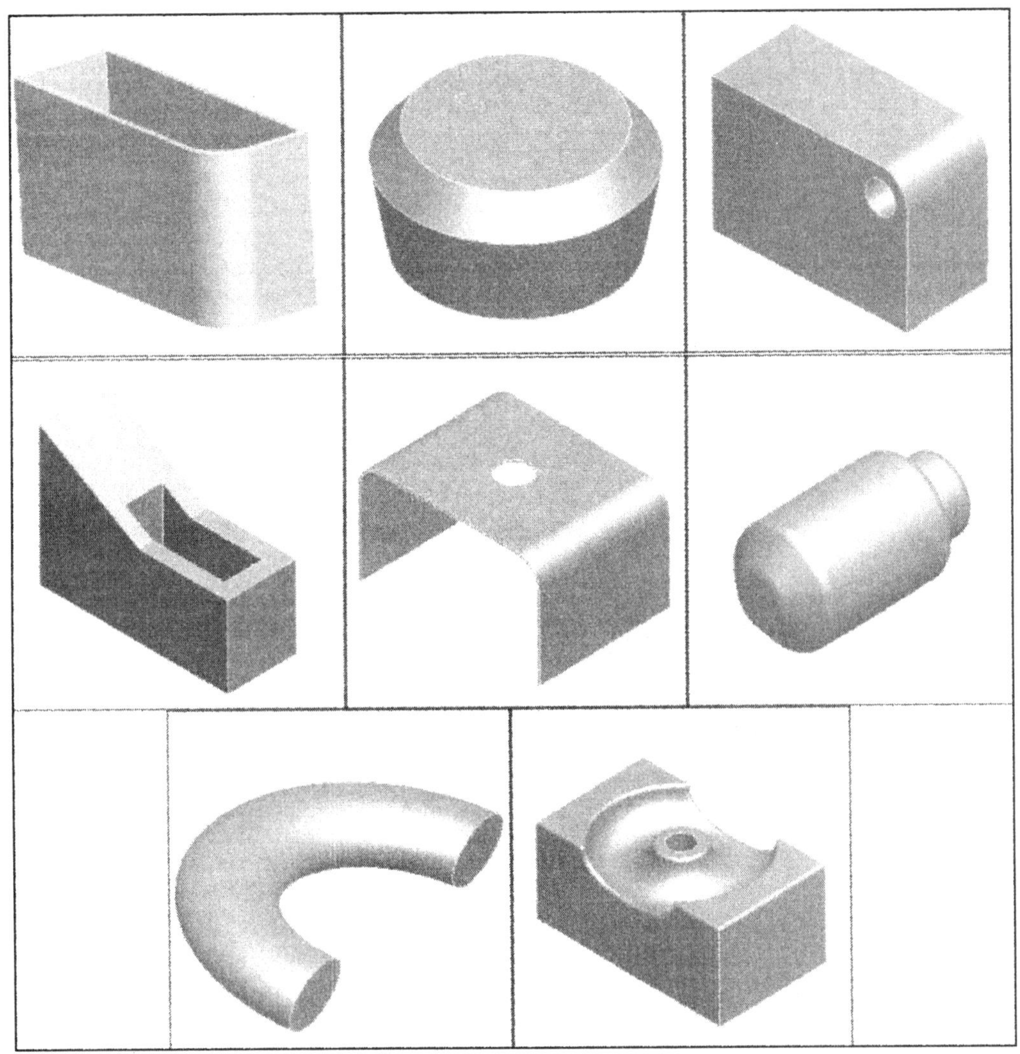

Figure 2.1 Eight Quick Modeling Parts

OBJECTIVES

- Modeling simple parts quickly using **default selections**
- Sample the **Extrude Tool** and the **Revolve Tool**
- Try out a variety of **Engineering Tools** including: **Hole**, **Shell**, **Round**, **Chamfer**, and **Draft**
- Sketch simple **sections**

Modeling

The purpose of this lesson is to quickly introduce you to a variety of **Feature Tools** and **Engineering Tools**. You will model a variety of very simple parts (Fig. 2.1). Little or no explanation of the methodology or theory of the Tool or process will accompany the instructions. By using almost all default selections; you will create models that will display the power and capability of Pro/E Wildfire 2.0.

In Lesson 3, a detailed step-by-step systematic description will accompany all commands. Here, we hope to get you up and running on Pro/E without any belabored explanations.

Lesson 2 STEPS

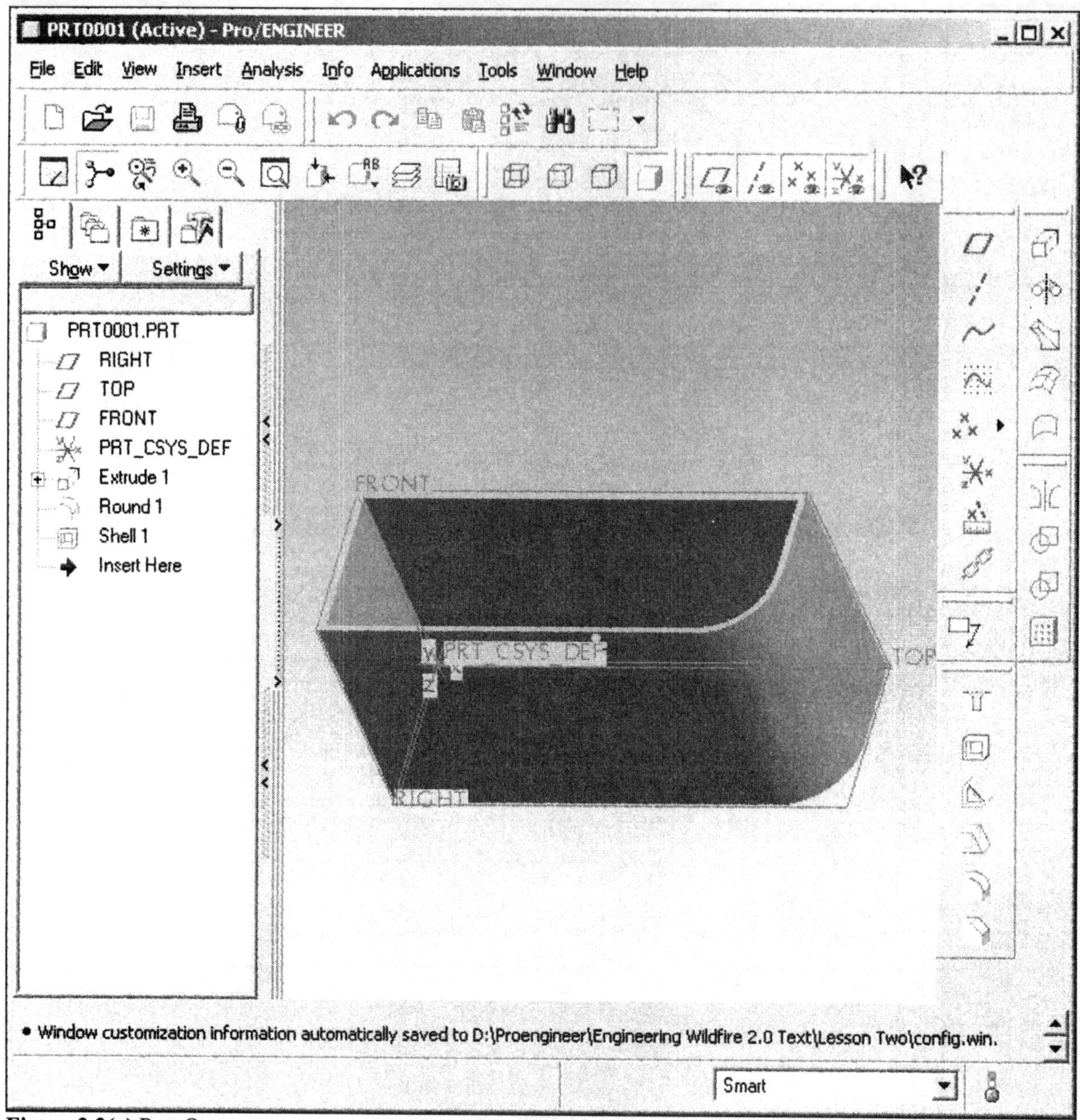

Figure 2.2(a) Part One

Part Model One (PRT0001.PRT) (Extrude)

The first model [Fig. 2.2(a)] will introduce the **Extrude Tool** to create a simple box-shape, the **Round Tool** to add a round to one edge, and the **Shell Tool** to remove one surface and make the part walls a consistent thickness. In this first quick modeling part we will provide step-by-step illustrations. For subsequent parts, only important steps or illustrations displaying aspects of the command sequence that represent new material will be provided. The same applies to ***Tool Tips*** that appear as you pass your cursor over a button or icon. All parts have been created with out-of-the-box system settings for Pro/E Wildfire 2.0 including default templates, grid settings, and so forth.

Click: **Launch Pro/E** ⇒ **File** ⇒ **Set Working Directory** ⇒ select the working directory ⇒ **OK** ⇒ **Create a new object** ⇒ ●**Part** ⇒ **OK** ⇒ click the **FRONT** datum plane in the Model Tree [Fig. 2.2(b)]

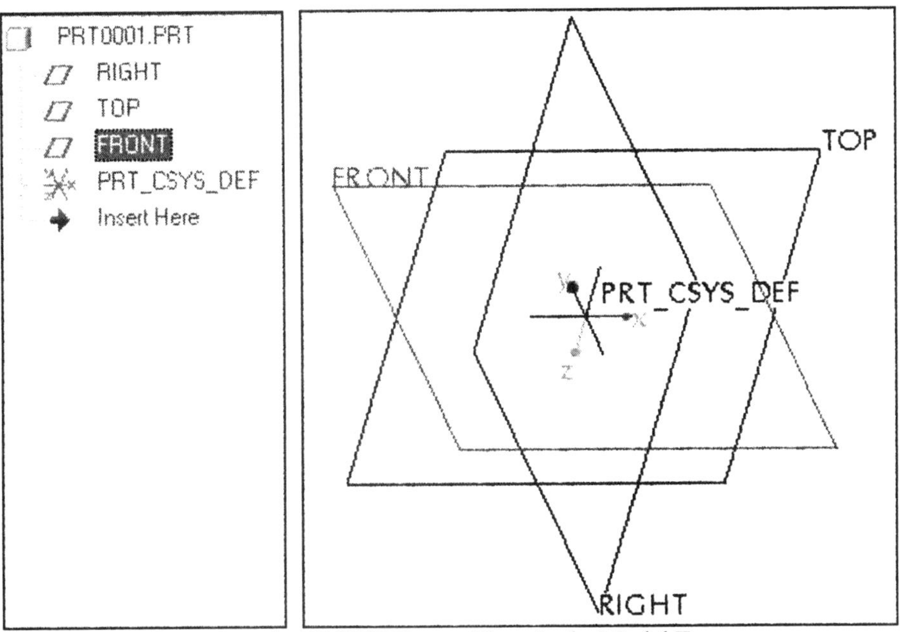

Figure 2.2(b) Pre-select the FRONT Datum Plane in the Model Tree

Click: **Extrude Tool** ⇒ Placement ⇒ Define.. Sketch dialog box opens with Sketch Plane and Sketch Orientation selected [Fig. 2.2(c)]

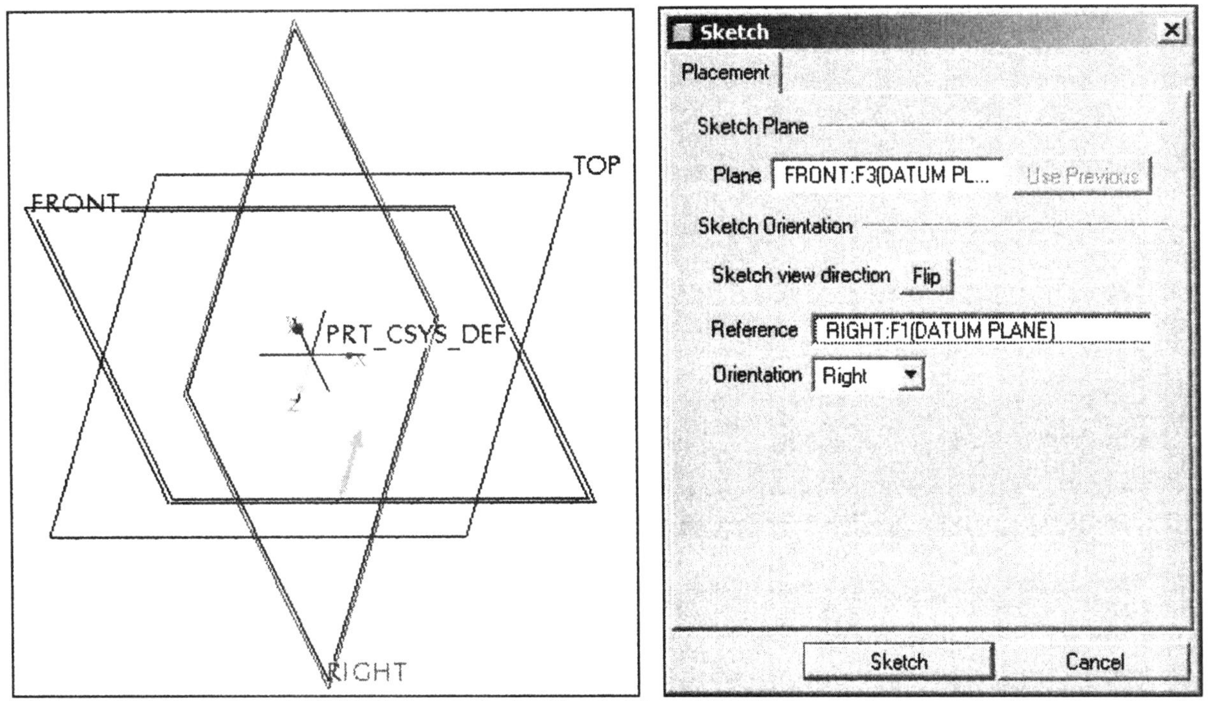

Figure 2.2(c) Pre-selected FRONT Datum is the Sketch Plane, RIGHT Datum is Automatically Selected as the Sketch Orientation Reference

With your mouse pointer in the Pro/E graphics window, click: **MMB** References dialog box displays [Fig. 2.2(d)] ⇒ **MMB** accepts selection ⇒ **MMB** closes dialog box

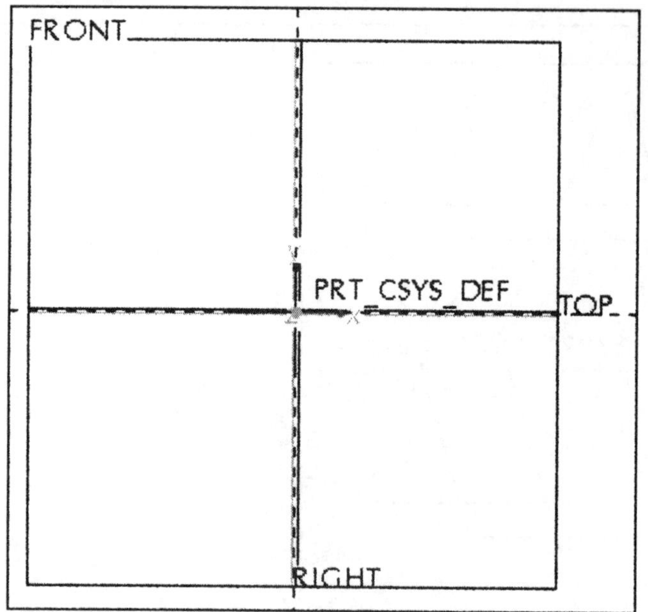

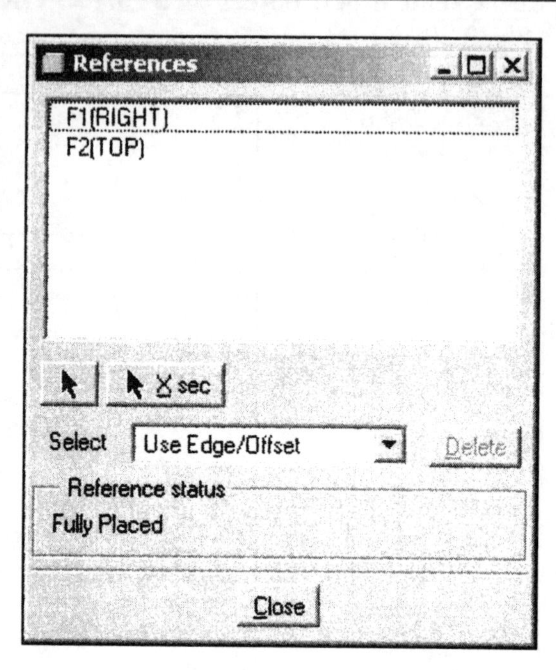

Figure 2.2(d) References Dialog Box Displays

Click: **Create rectangle** ⇒ sketch a rectangle by picking two corners [Fig. 2.2(e)] ⇒ **Continue with the current section**

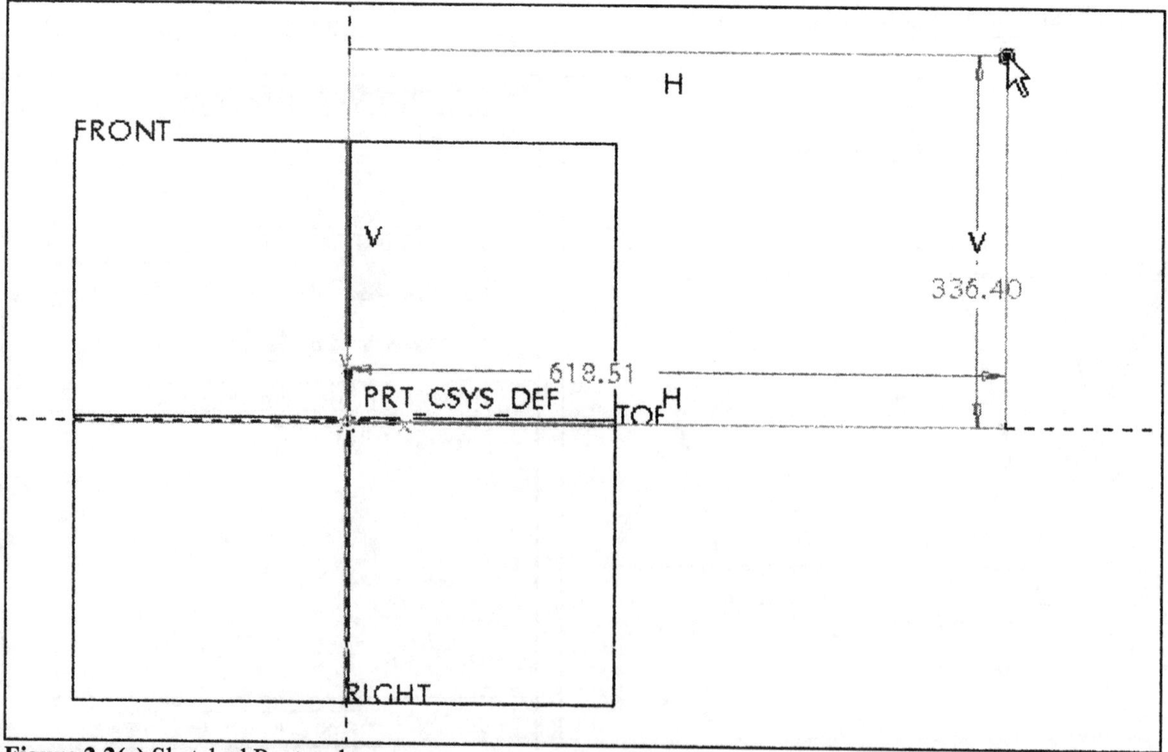

Figure 2.2(e) Sketched Rectangle

Click: ⇒ **Standard Orientation** [Fig. 2.2(f)] ⇒ **Shading** ⇒ **MMB** [Fig. 2.2(g)]

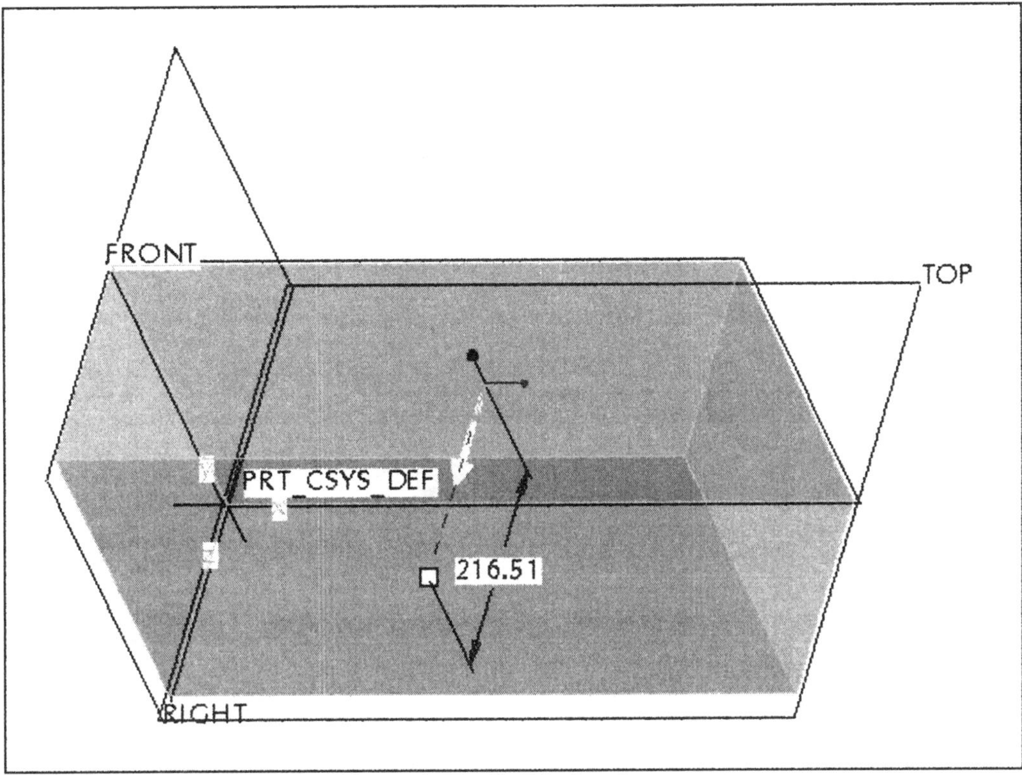

Figure 2.2(f) Trimetric View with Depth Handle Displayed

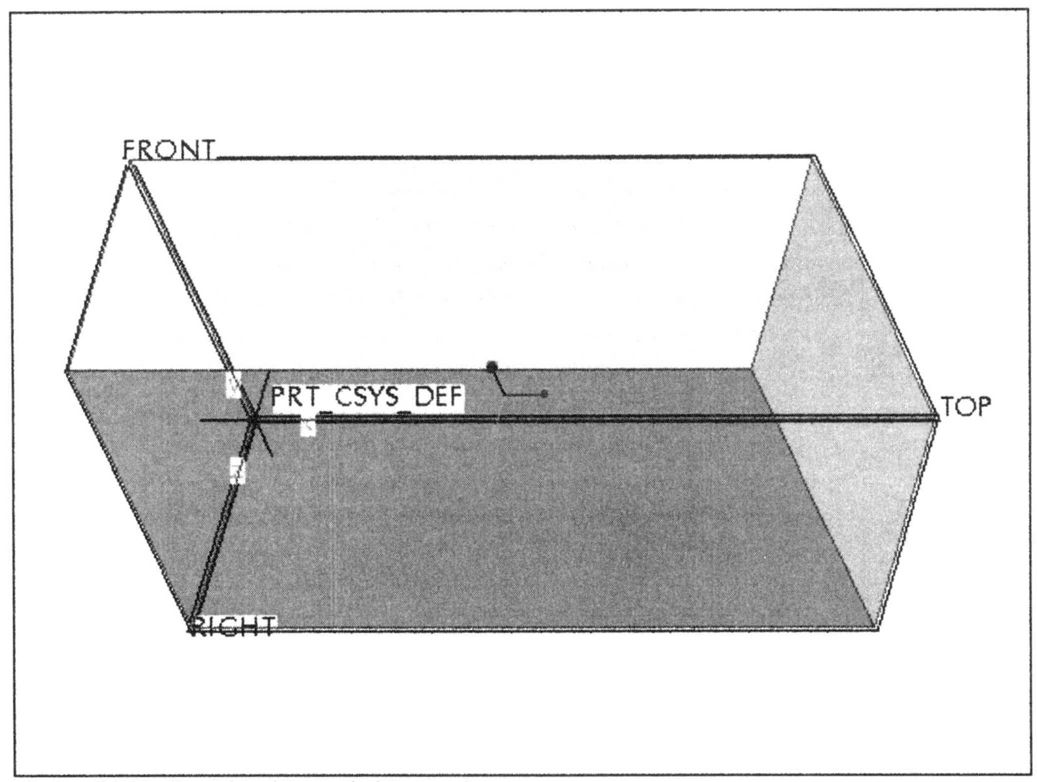

Figure 2.2(g) Completed Extruded Protrusion

Slowly pick on the right front edge of the part until it highlights [Fig. 2.2(h)] ⇒ **RMB** ⇒ **Round Edges** [Fig. 2.2(i)] ⇒ pick on and move the drag handle [Fig. 2.2(j)] ⇒ **MMB** [Fig. 2.2(k)]

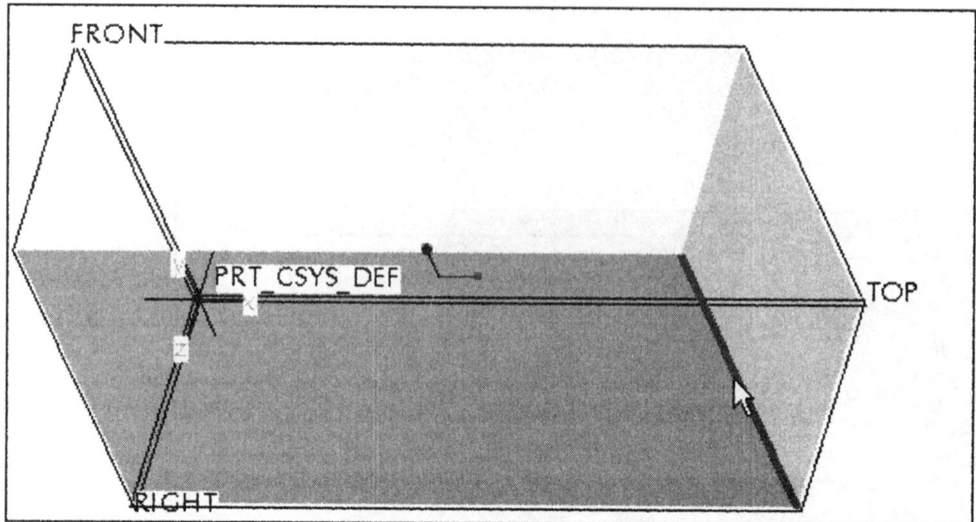

Figure 2.2(h) Pick on Edge to Highlight

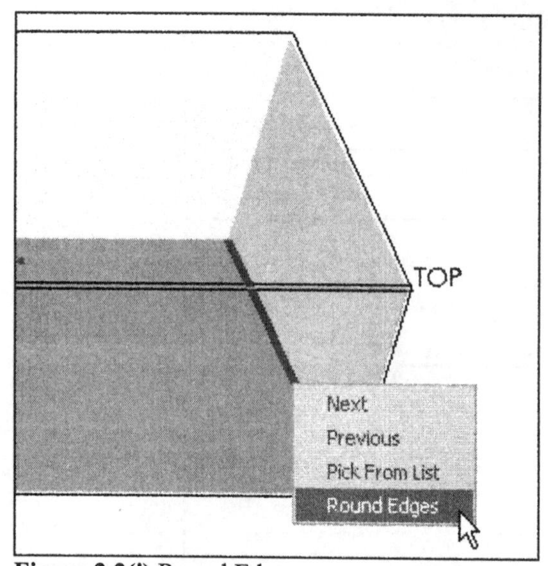

Figure 2.2(i) Round Edges

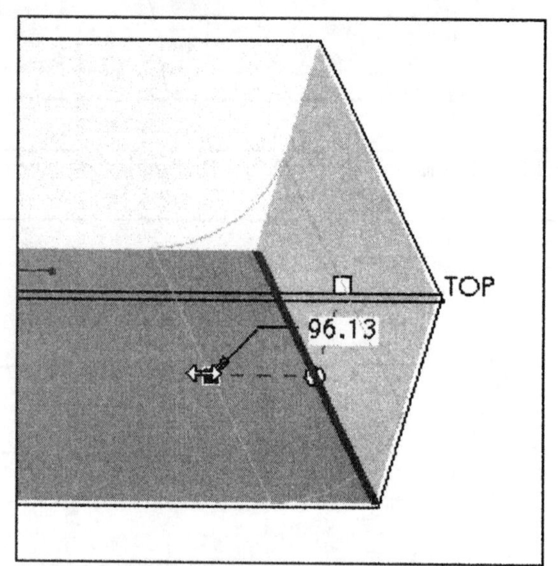

Figure 2.2(j) Move the Drag Handle

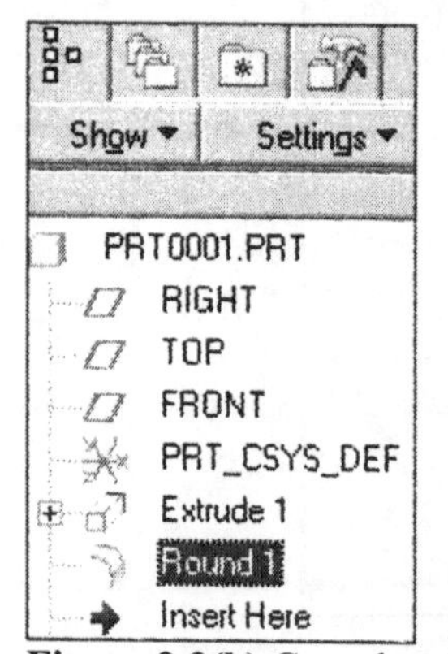

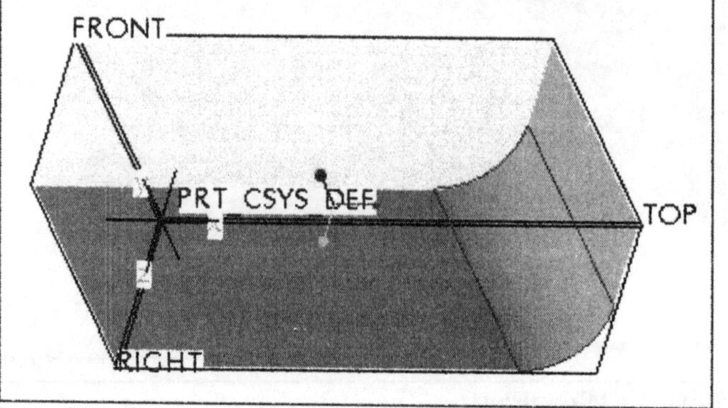

Figure 2.2(k) Completed Round

Slowly pick on the top surface of the part until it highlights [Fig. 2.2(l)] ⇒ **Shell Tool** [Fig. 2.2(m)] ⇒ **MMB** [Fig. 2.2(n)] ⇒ **Save the active object** ⇒ **MMB** ⇒ **File** ⇒ **Close Window**

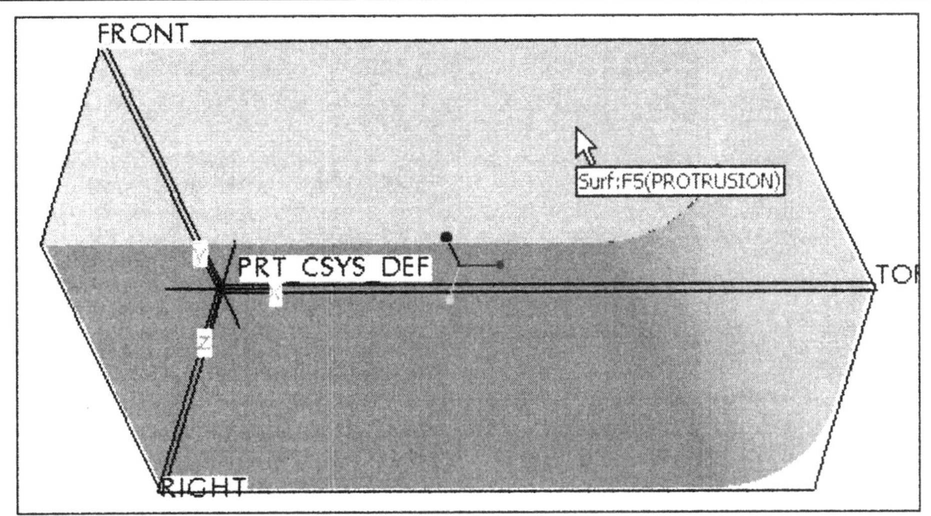

Figure 2.2(l) Selected Top Surface is Highlighted

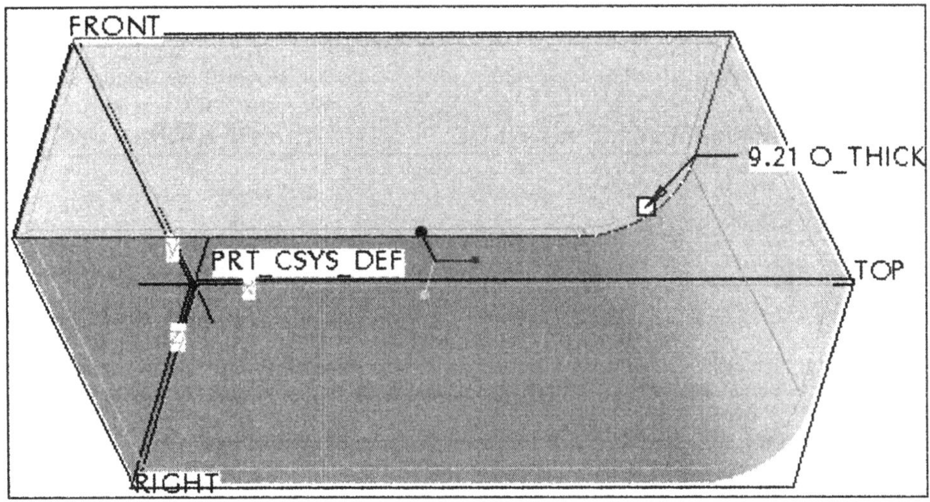

Figure 2.2(m) Shell Tool Applied

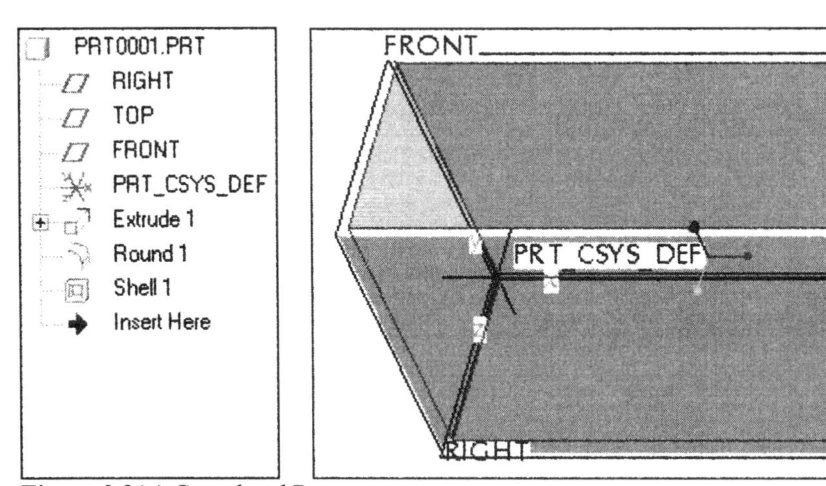

Figure 2.2(n) Completed Part

Part Model Two (PRT0002.PRT) (Draft)

Click: ⇒ ●**Part (PRT0002.PRT)** ⇒ **MMB** ⇒ click on the **TOP** datum plane in the Model Tree ⇒ **Extrude Tool** ⇒ Placement ⇒ Define.. Section dialog box opens with Sketch Plane and Sketch Orientation selected [Fig. 2.3(a)] ⇒ **MMB** in the graphics window ⇒ **MMB** ⇒ **MMB**

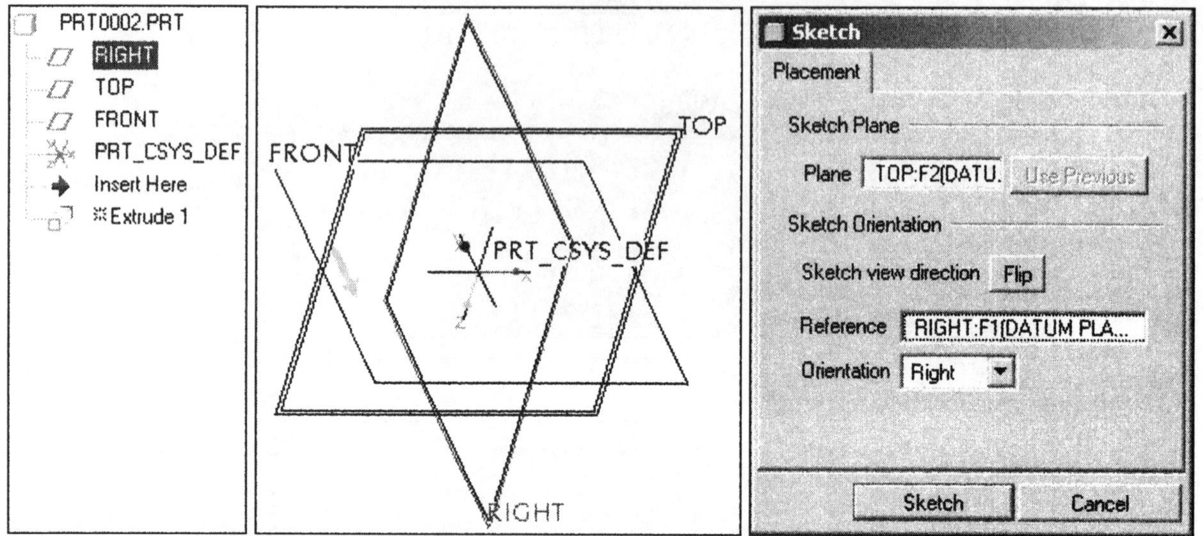

Figure 2.3(a) Pre-selected TOP Datum is the Sketch Plane, RIGHT Datum is Automatically Selected as the Sketch Orientation Reference

Click: **Create circle** ⇒ sketch a circle by picking at the origin and then stretching the circle's diameter [Fig. 2.3(b)] to a convenient size ⇒ pick again to establish the circle's diameter [Fig. 2.3(c)]

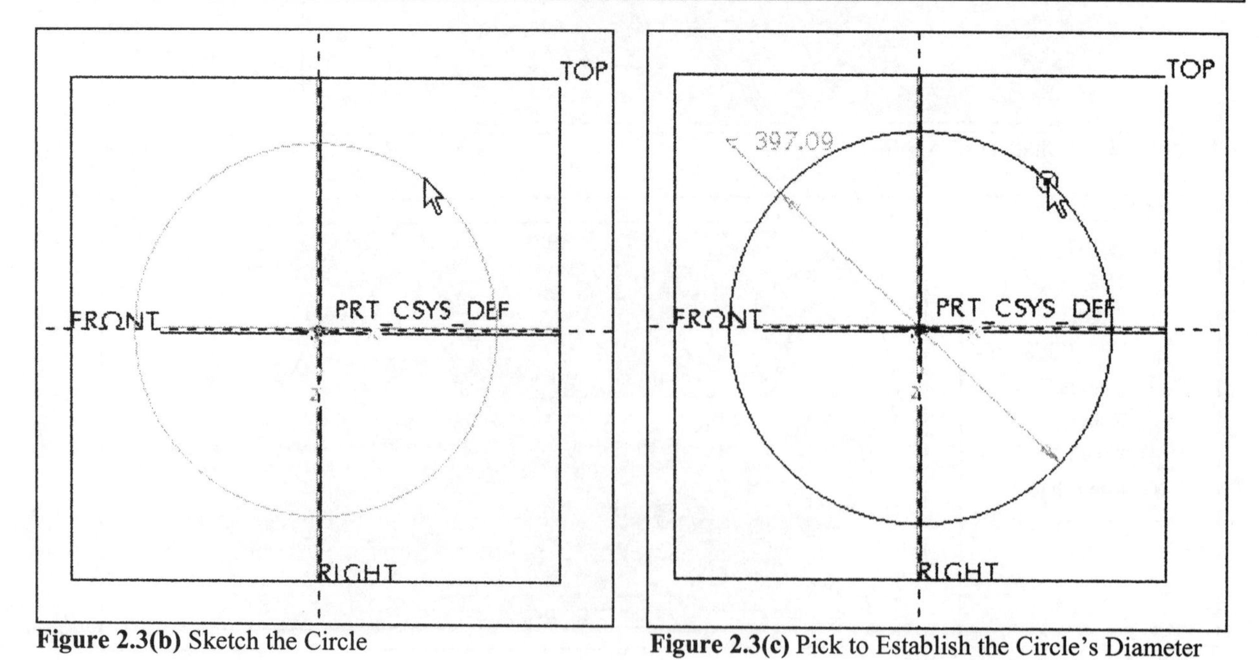

Figure 2.3(b) Sketch the Circle

Figure 2.3(c) Pick to Establish the Circle's Diameter

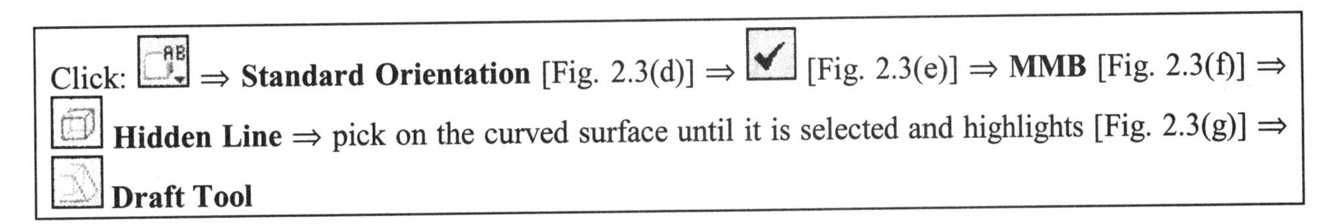

Click: ⇒ **Standard Orientation** [Fig. 2.3(d)] ⇒ [Fig. 2.3(e)] ⇒ **MMB** [Fig. 2.3(f)] ⇒ **Hidden Line** ⇒ pick on the curved surface until it is selected and highlights [Fig. 2.3(g)] ⇒ **Draft Tool**

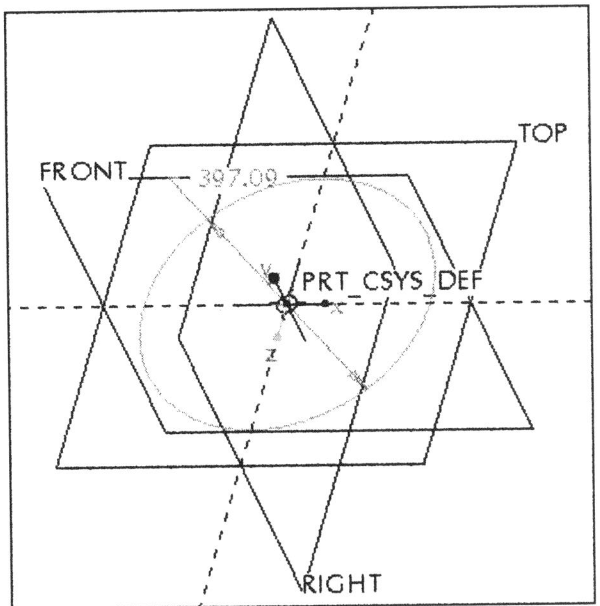

Figure 2.3(d) Sketch in Standard Orientation

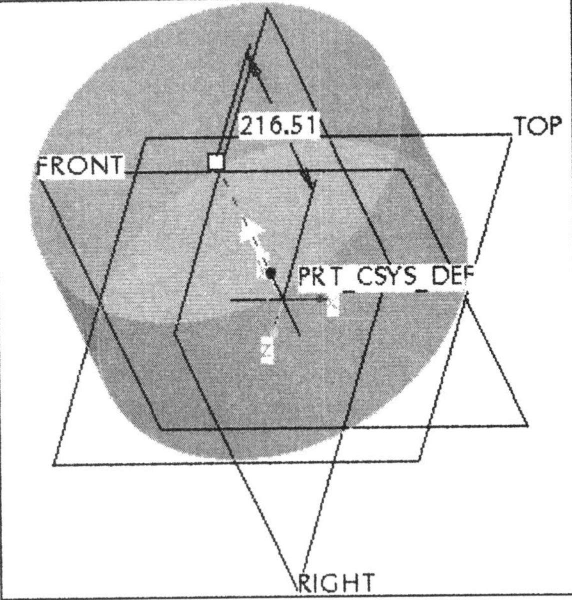

Figure 2.3(e) Depth Preview of Extruded Circle

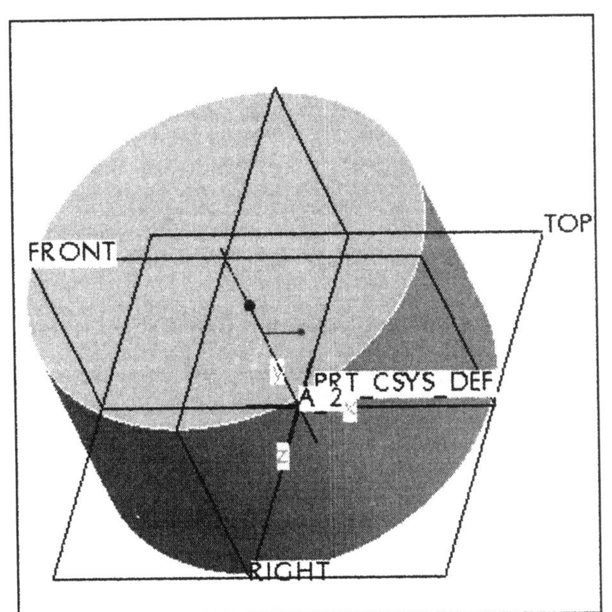

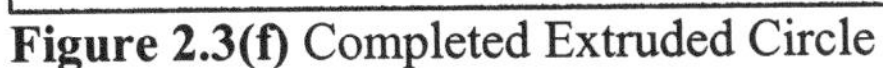

Figure 2.3(f) Completed Extruded Circle

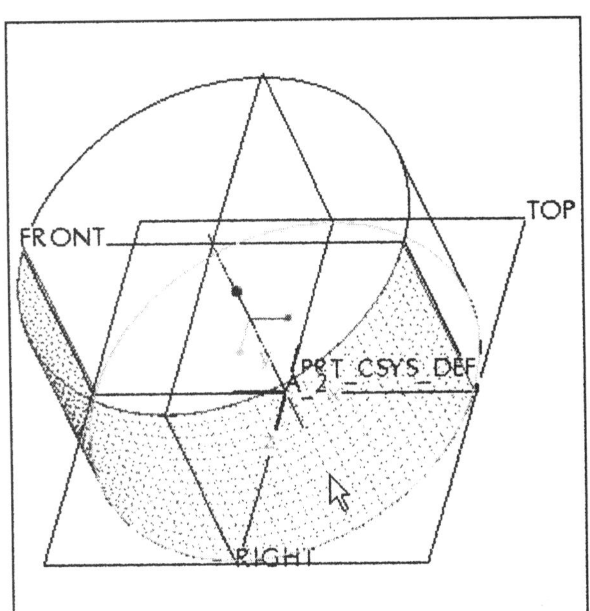

Figure 2.3(g) Select the Curved Surface

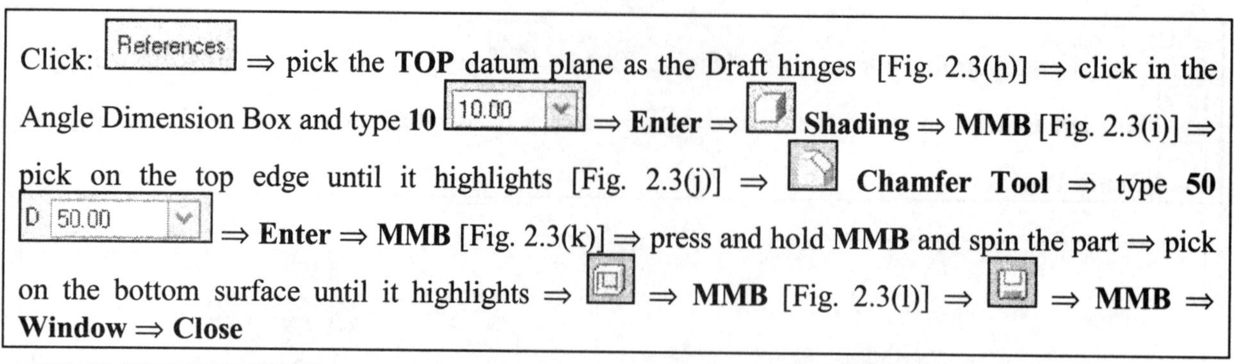

Click: References ⇒ pick the **TOP** datum plane as the Draft hinges [Fig. 2.3(h)] ⇒ click in the Angle Dimension Box and type **10** 10.00 ⇒ **Enter** ⇒ **Shading** ⇒ **MMB** [Fig. 2.3(i)] ⇒ pick on the top edge until it highlights [Fig. 2.3(j)] ⇒ **Chamfer Tool** ⇒ type **50** D 50.00 ⇒ **Enter** ⇒ **MMB** [Fig. 2.3(k)] ⇒ press and hold **MMB** and spin the part ⇒ pick on the bottom surface until it highlights ⇒ ⇒ **MMB** [Fig. 2.3(l)] ⇒ ⇒ **MMB** ⇒ **Window** ⇒ **Close**

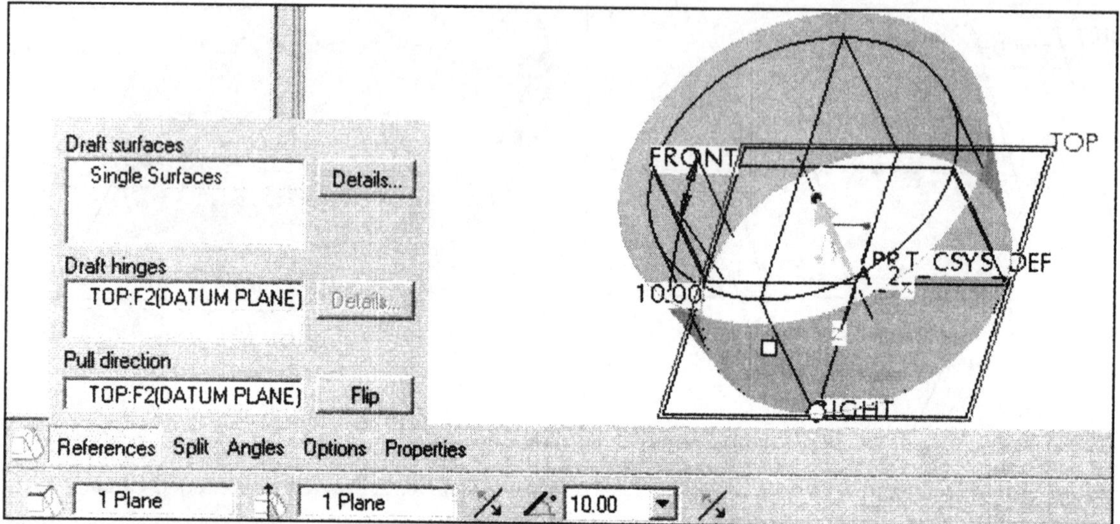

Figure 2.3(h) TOP Datum Plane Selected as the Draft Hinges

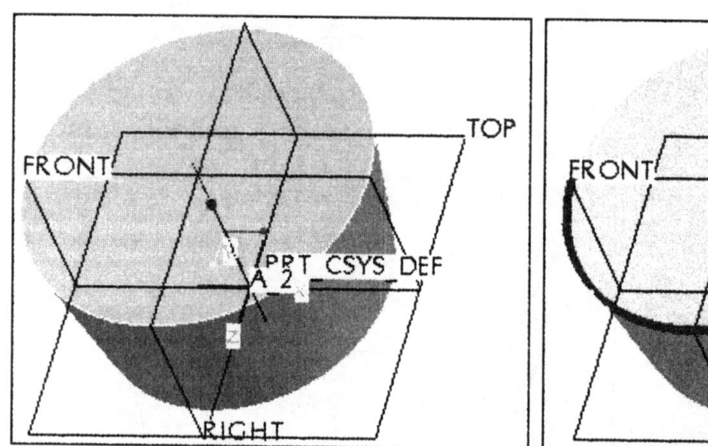

Figure 2.3(i) Completed Draft

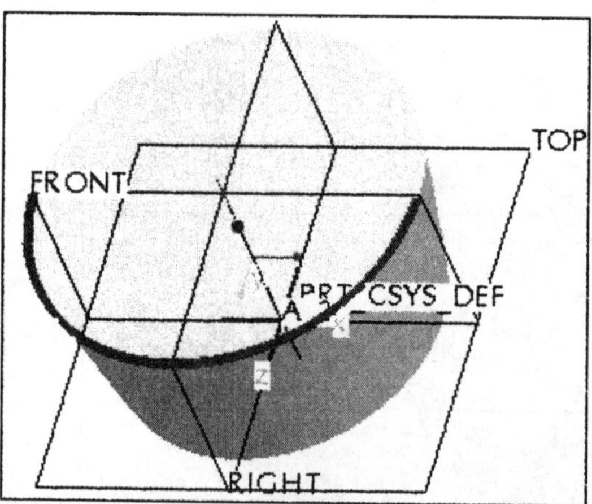

Figure 2.3(j) Select Front Edge

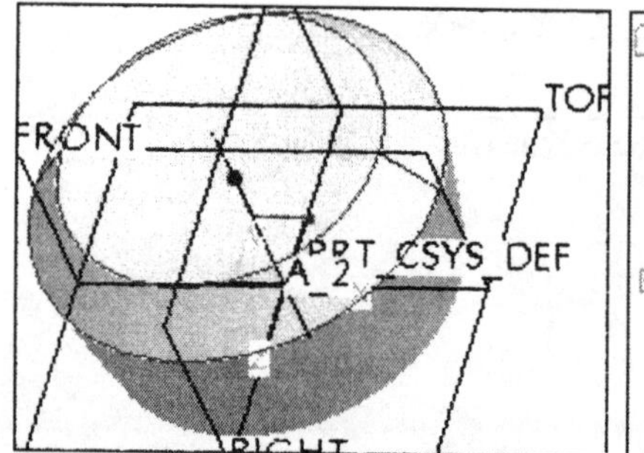

Figure 2.3(k) Completed Chamfer

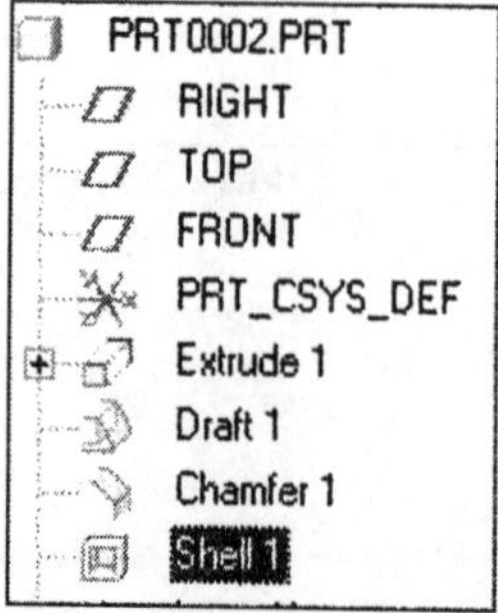

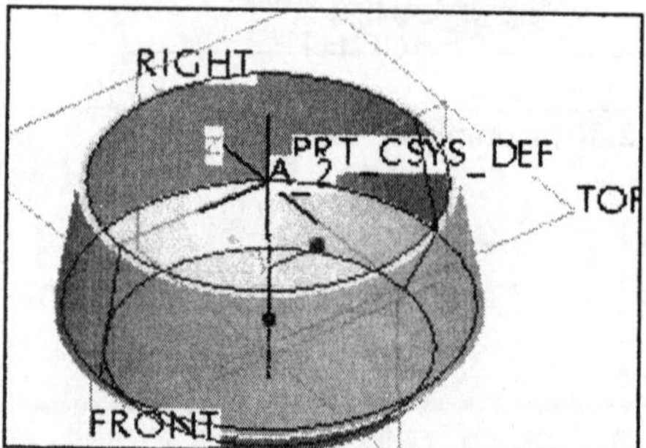

Figure 2.3(l) Completed Part

Part Model Three (PRT0003.PRT) (Hole)

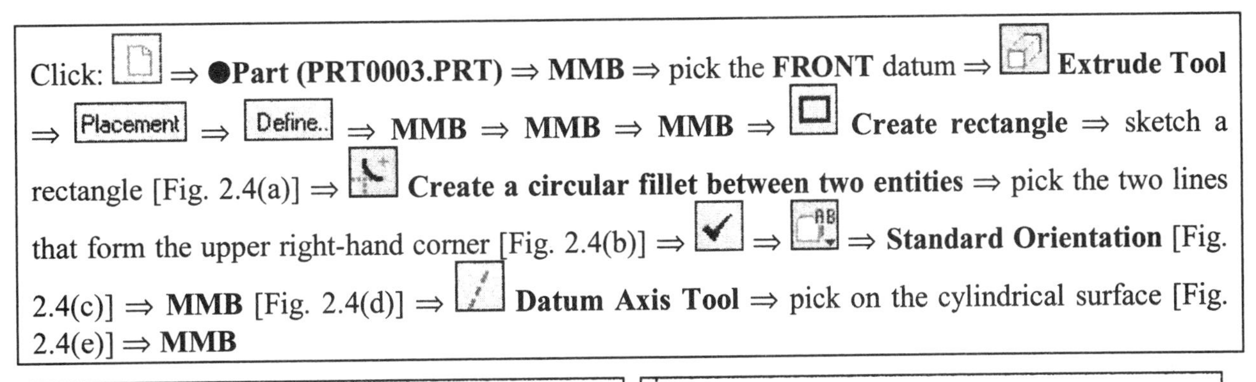

Click: [icon] ⇒ ●**Part (PRT0003.PRT)** ⇒ **MMB** ⇒ pick the **FRONT** datum ⇒ [icon] **Extrude Tool** ⇒ [Placement] ⇒ [Define..] ⇒ **MMB** ⇒ **MMB** ⇒ **MMB** ⇒ [icon] **Create rectangle** ⇒ sketch a rectangle [Fig. 2.4(a)] ⇒ [icon] **Create a circular fillet between two entities** ⇒ pick the two lines that form the upper right-hand corner [Fig. 2.4(b)] ⇒ [icon] ⇒ [icon] ⇒ **Standard Orientation** [Fig. 2.4(c)] ⇒ **MMB** [Fig. 2.4(d)] ⇒ [icon] **Datum Axis Tool** ⇒ pick on the cylindrical surface [Fig. 2.4(e)] ⇒ **MMB**

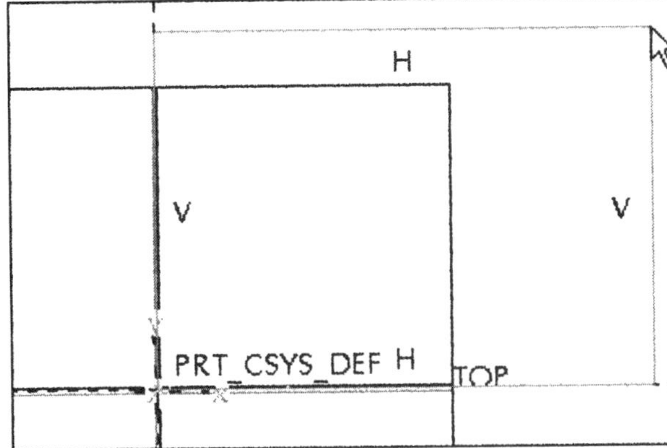

Figure 2.4(a) Sketch a Rectangle

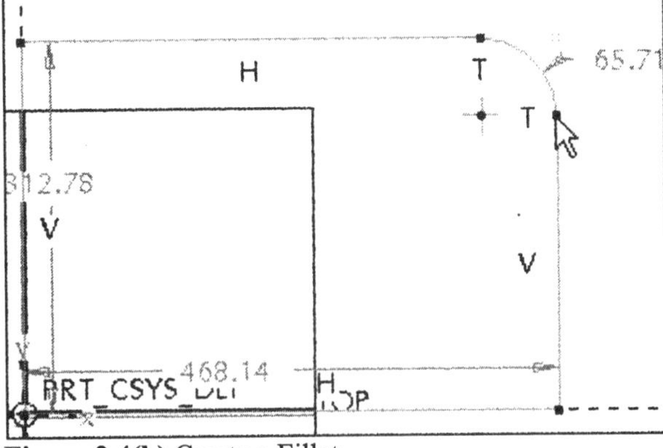

Figure 2.4(b) Create a Fillet

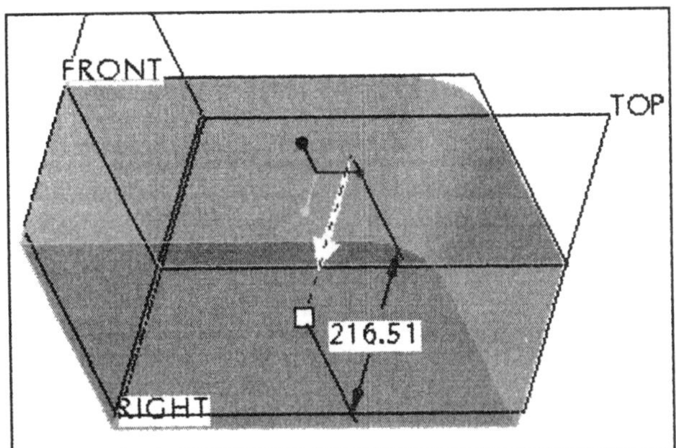

Figure 2.4(c) Depth Preview

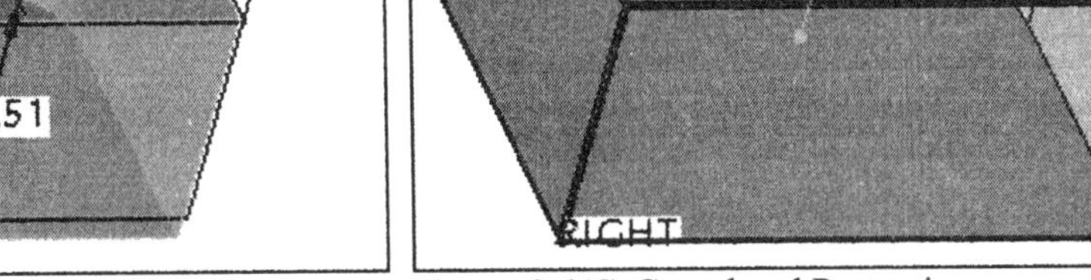

Figure 2.4(d) Completed Protrusion

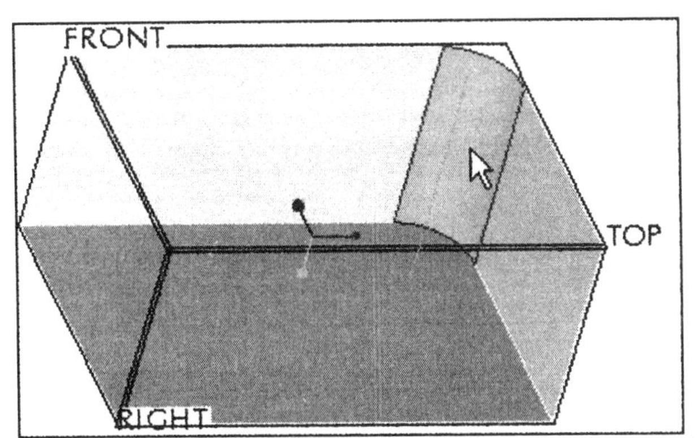

Figure 2.4(e) Pick on the Cylindrical Surface to Establish the References for the Datum Axis

With the datum axis still highlighted, click: **Hole Tool** [Fig. 2.4(f)] ⇒ Placement ⇒ click inside the Secondary references *collector* (turns yellow) [Fig. 2.4(g)] ⇒ pick on the front face [Fig. 2.4(h)] ⇒ expand depth options by opening slide-up panel ⇒ **Drill to intersect with all surfaces** [Fig. 2.4(i)] ⇒ **MMB** [Fig. 2.4(j)] ⇒ ⇒ **MMB** ⇒ **Window** ⇒ **Close**

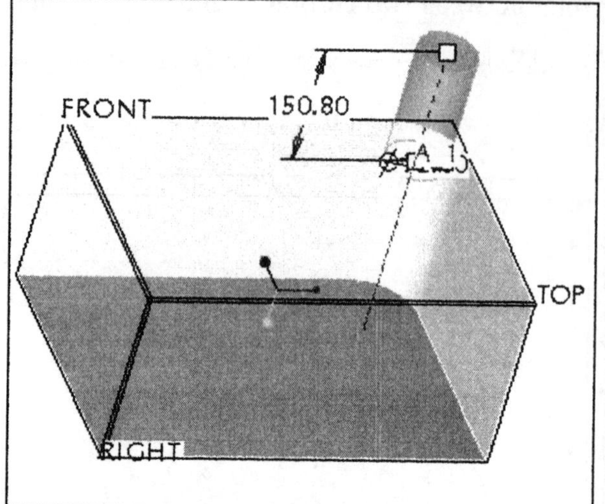

Figure 2.4(f) Hole Tool Display Preview

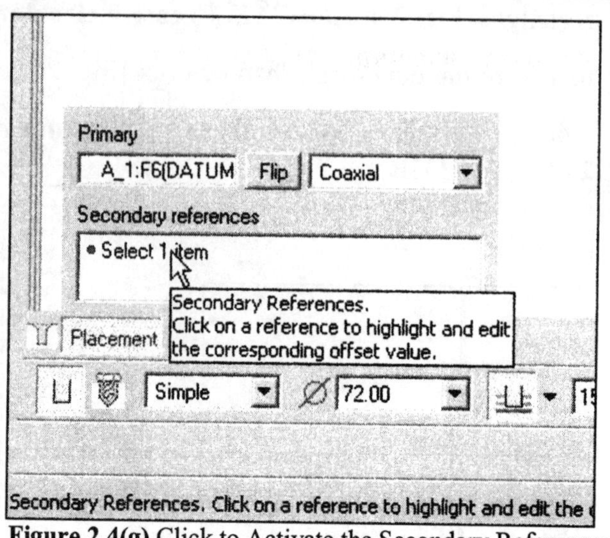

Figure 2.4(g) Click to Activate the Secondary References

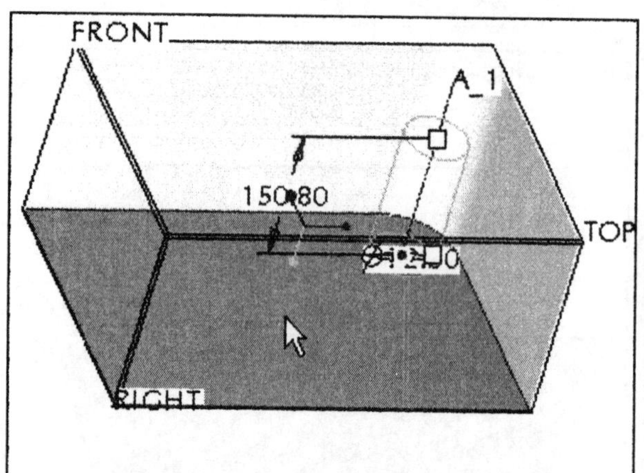

Figure 2.4(h) Pick on the Front Face

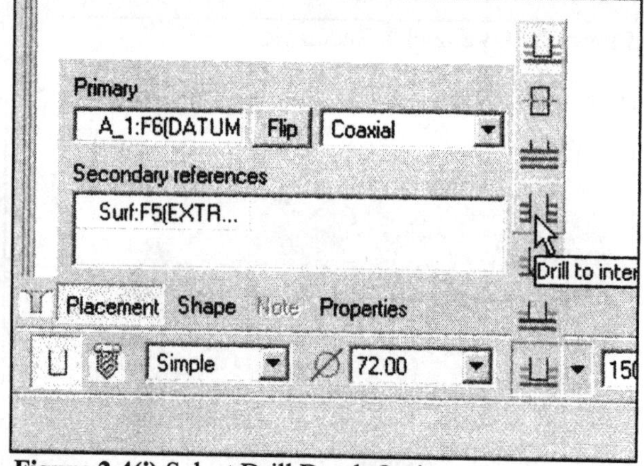

Figure 2.4(i) Select Drill Depth Option

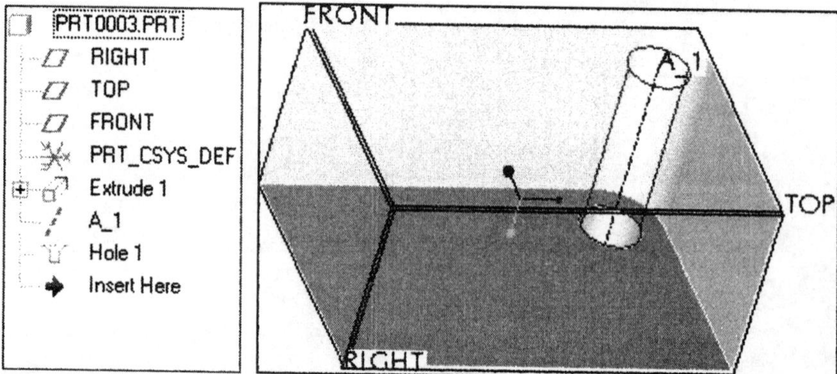

Figure 2.4(j) Completed Part

Part Model Four (PRT0004.PRT) (Cut)

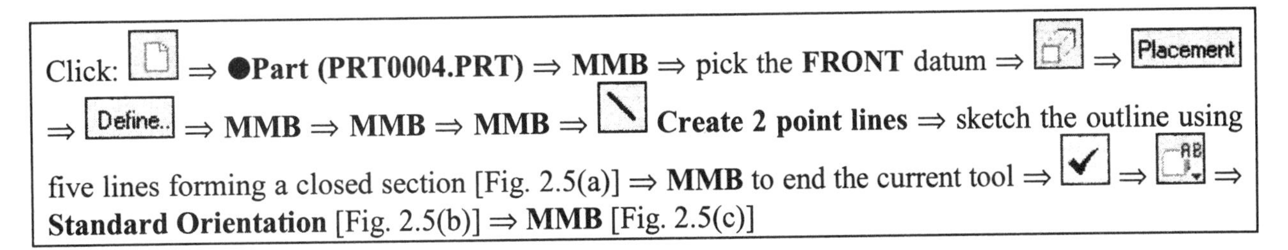

Click: [icon] ⇒ ●**Part (PRT0004.PRT)** ⇒ **MMB** ⇒ pick the **FRONT** datum ⇒ [icon] ⇒ Placement ⇒ Define.. ⇒ **MMB** ⇒ **MMB** ⇒ **MMB** ⇒ [icon] **Create 2 point lines** ⇒ sketch the outline using five lines forming a closed section [Fig. 2.5(a)] ⇒ **MMB** to end the current tool ⇒ [icon] ⇒ [icon] ⇒ **Standard Orientation** [Fig. 2.5(b)] ⇒ **MMB** [Fig. 2.5(c)]

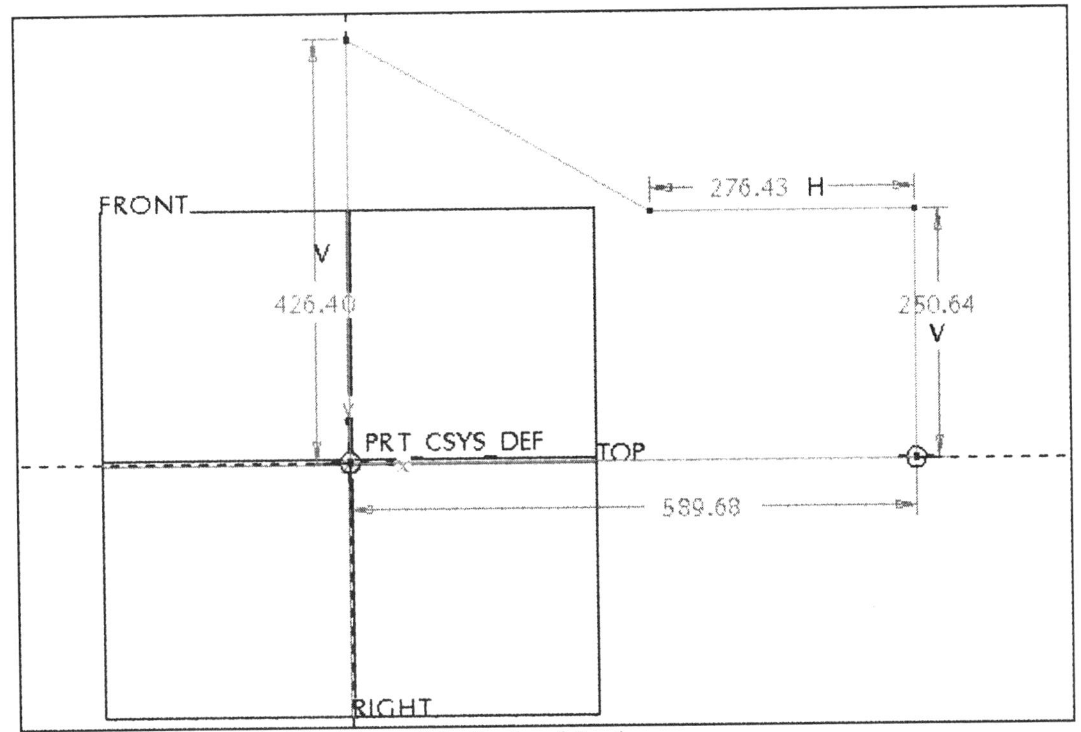

Figure 2.5(a) Sketch the Five Lines of the Enclosed Section

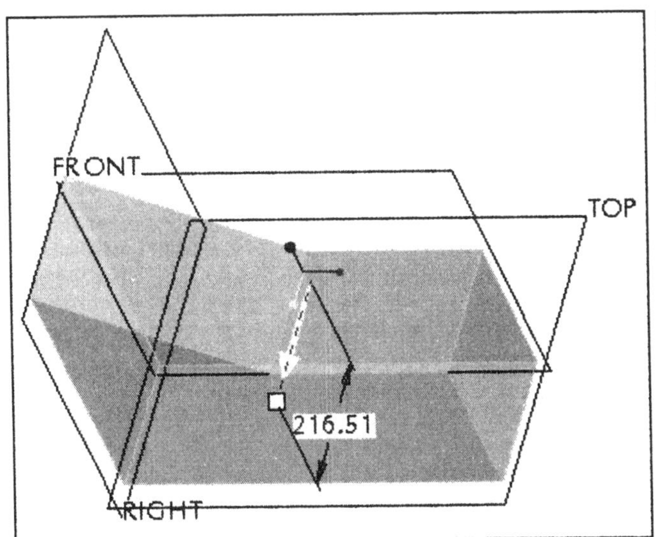

Figure 2.5(b) Depth Preview

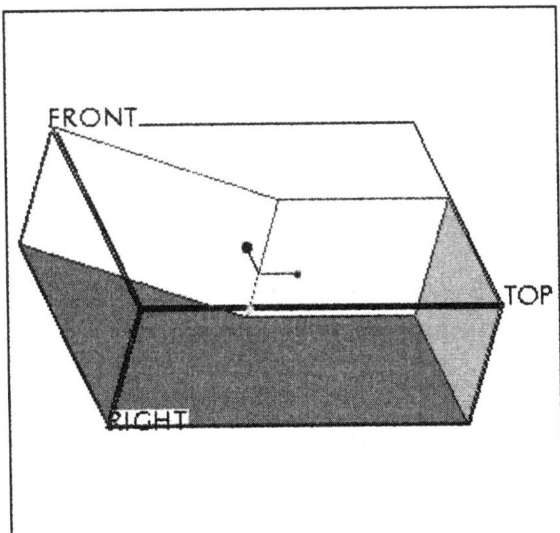

Figure 2.5(c) Completed Protrusion

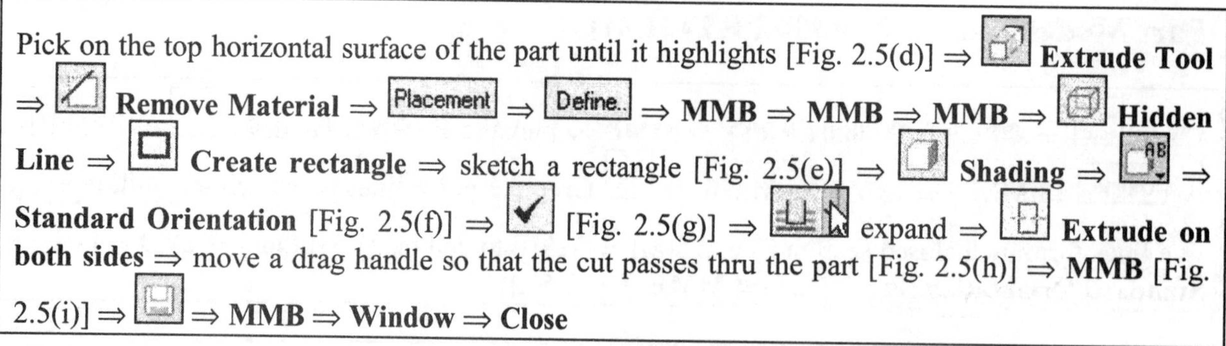
Pick on the top horizontal surface of the part until it highlights [Fig. 2.5(d)] ⇒ **Extrude Tool** ⇒ **Remove Material** ⇒ Placement ⇒ Define.. ⇒ **MMB** ⇒ **MMB** ⇒ **MMB** ⇒ **Hidden Line** ⇒ **Create rectangle** ⇒ sketch a rectangle [Fig. 2.5(e)] ⇒ **Shading** ⇒ ⇒ **Standard Orientation** [Fig. 2.5(f)] ⇒ [Fig. 2.5(g)] ⇒ expand ⇒ **Extrude on both sides** ⇒ move a drag handle so that the cut passes thru the part [Fig. 2.5(h)] ⇒ **MMB** [Fig. 2.5(i)] ⇒ ⇒ **MMB** ⇒ **Window** ⇒ **Close**

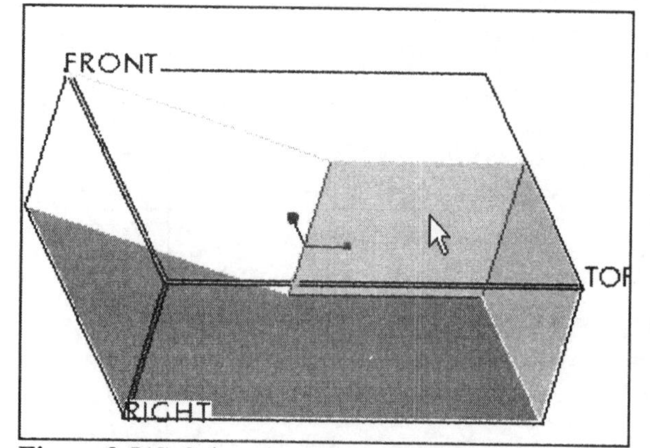

Figure 2.5(d) Select the Horizontal Surface

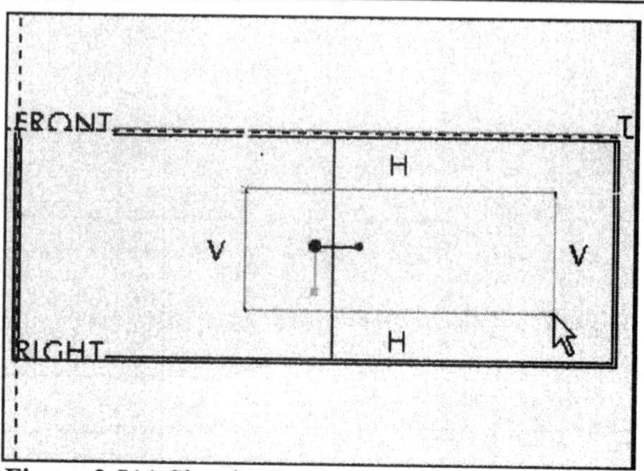

Figure 2.5(e) Sketch a Rectangle

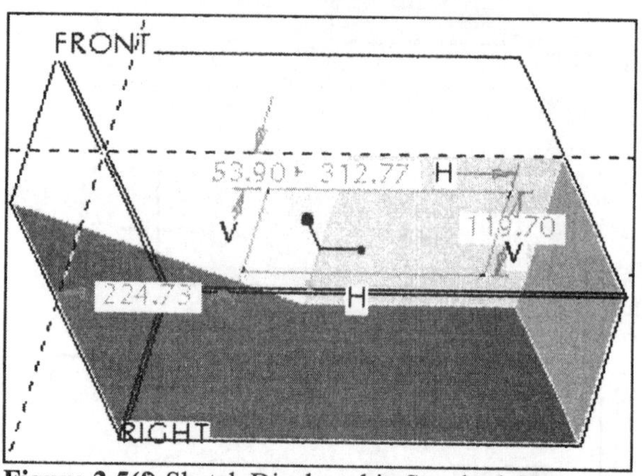

Figure 2.5(f) Sketch Displayed in Standard Orientation

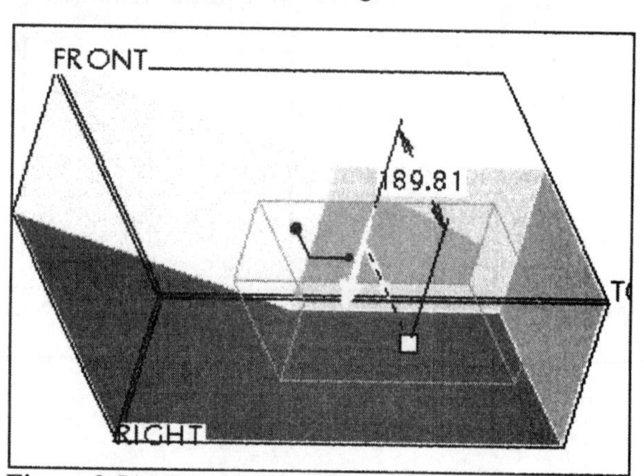

Figure 2.5(g) Previewed Cut

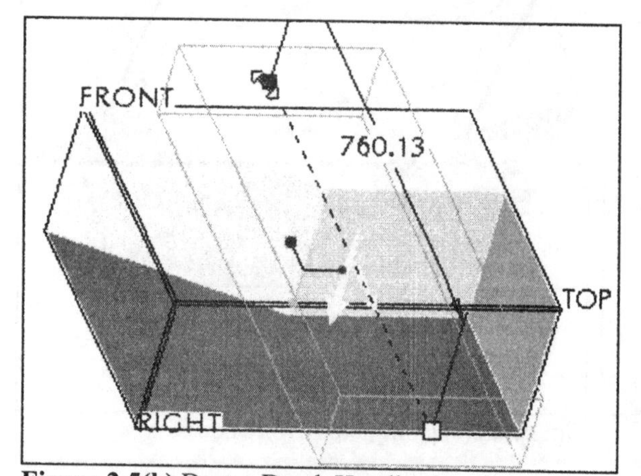

Figure 2.5(h) Drag a Depth Handle

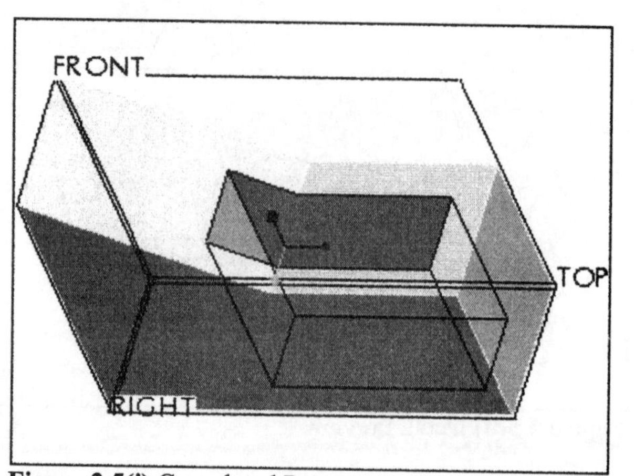

Figure 2.5(i) Completed Part

Part Model Five (PRT0005.PRT) (Mirror)

Click: [icon] ⇒ ●**Part (PRT0005.PRT)** ⇒ **MMB** ⇒ pick the **FRONT** datum ⇒ [icon] ⇒ [icon] **Thicken Sketch** ⇒ [Placement] ⇒ [Define..] ⇒ **MMB** ⇒ **MMB** ⇒ **MMB** ⇒ **RMB** ⇒ **Centerline** ⇒ pick two points vertically on the edge of the RIGHT datum plane [Figs. 2.6(a-b)] ⇒ **MMB** to end the current tool ⇒ **RMB** ⇒ **Line** [Fig. 2.6(c)] ⇒ sketch the vertical and horizontal lines [Fig. 2.6(d)] ⇒ **MMB** to end the current line

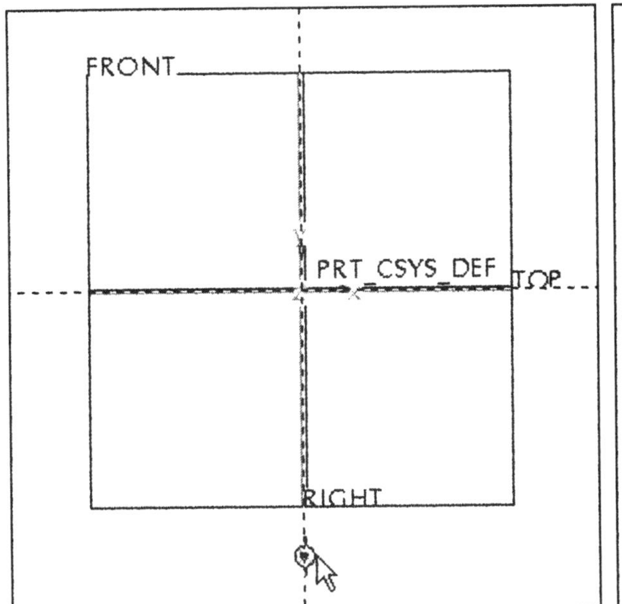

Figure 2.6(a) Pick the First Point of the Vertical Vertical Centerline

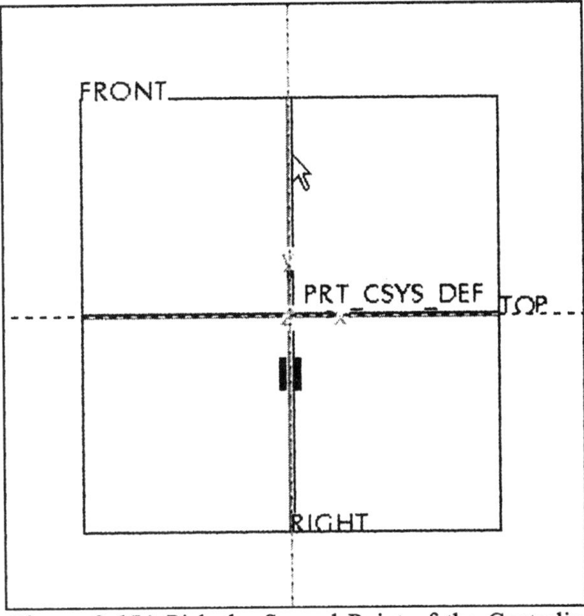

Figure 2.6(b) Pick the Second Point of the Centerline

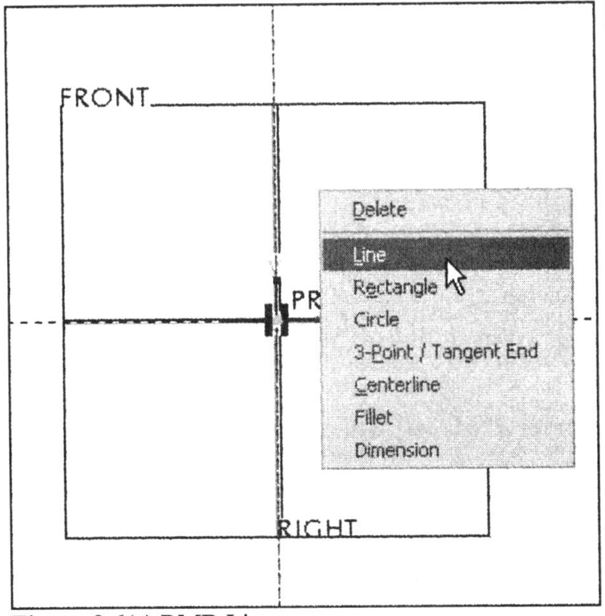

Figure 2.6(c) RMB Line

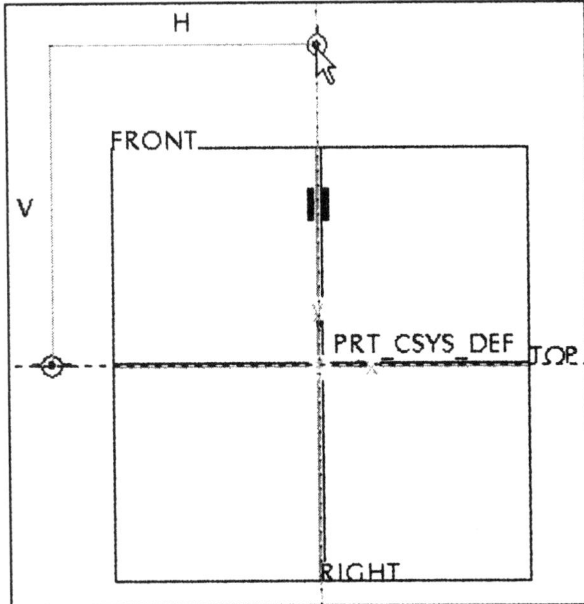

Figure 2.6(d) Sketch Two Lines

Click: **RMB** ⇒ **Fillet** [Fig. 2.6(e)] ⇒ pick the lines near the corner ⇒ **MMB** ⇒ press **LMB** and hold while dragging a window until it incorporates the two lines and fillet [Fig. 2.6(f)] ⇒ **Mirror selected entities** ⇒ pick the centerline [Fig. 2.6(g)] ⇒ ⇒ **Standard Orientation** ⇒ ✓ [Fig. 2.6(h)]

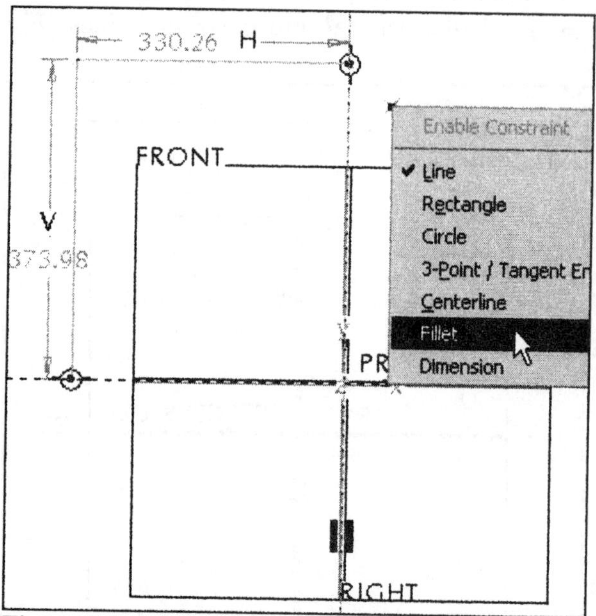

Figure 2.6(e) RMB Fillet

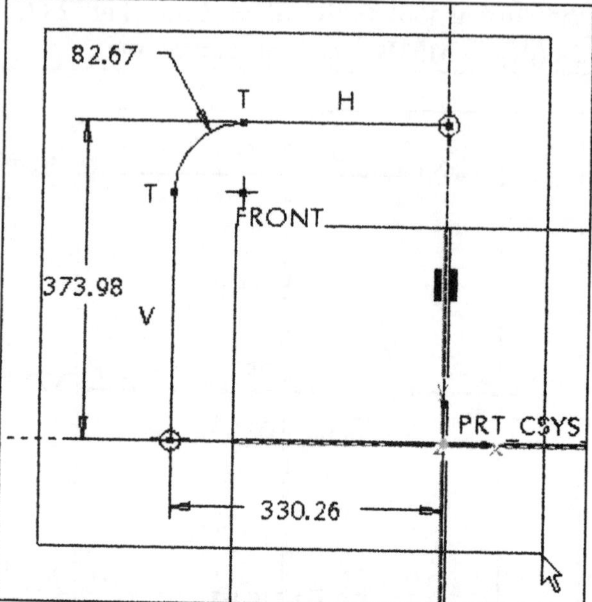

Figure 2.6(f) Select Sketch Entities by Windowing

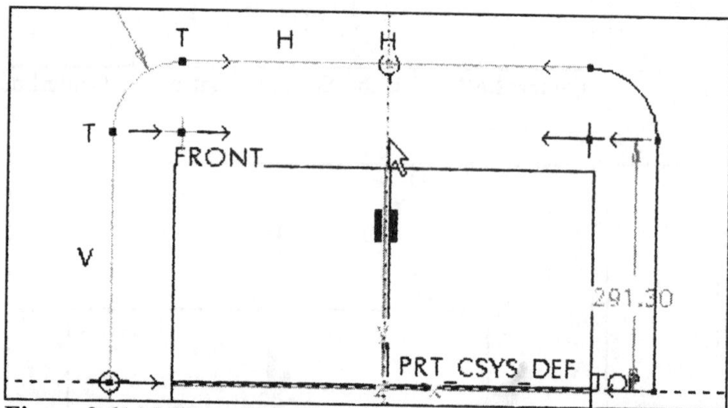

Figure 2.6(g) Mirrored Sketch

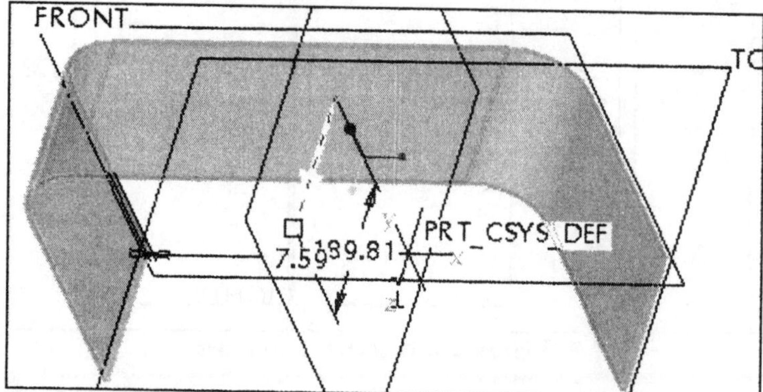

Figure 2.6(h) Depth Preview

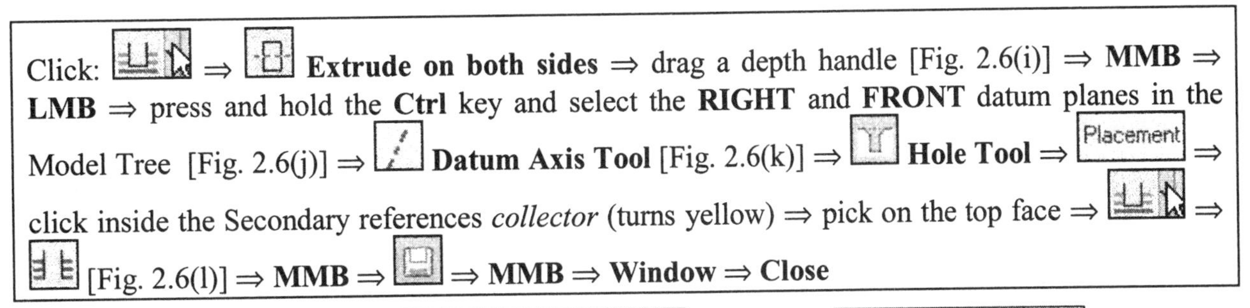

Click: ⇒ **Extrude on both sides** ⇒ drag a depth handle [Fig. 2.6(i)] ⇒ **MMB** ⇒ **LMB** ⇒ press and hold the **Ctrl** key and select the **RIGHT** and **FRONT** datum planes in the Model Tree [Fig. 2.6(j)] ⇒ **Datum Axis Tool** [Fig. 2.6(k)] ⇒ **Hole Tool** ⇒ Placement ⇒ click inside the Secondary references *collector* (turns yellow) ⇒ pick on the top face ⇒ ⇒ [Fig. 2.6(l)] ⇒ **MMB** ⇒ ⇒ **MMB** ⇒ **Window** ⇒ **Close**

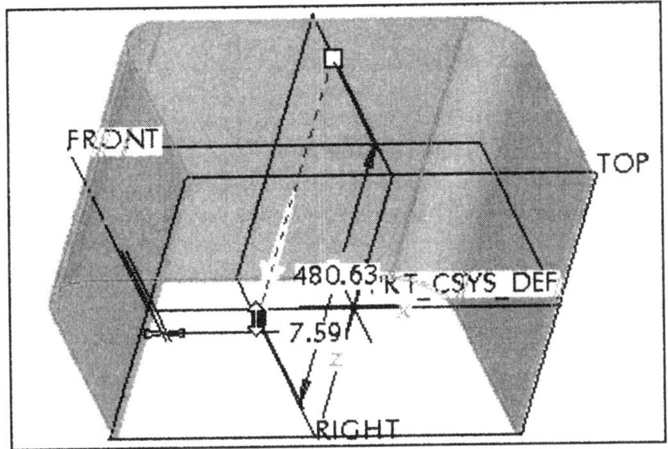

Figure 2.6(i) Drag Depth Handle

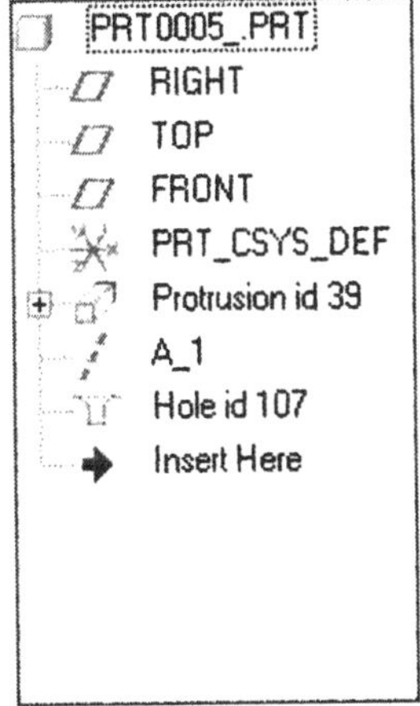

Figure 2.6(j) Select Datums

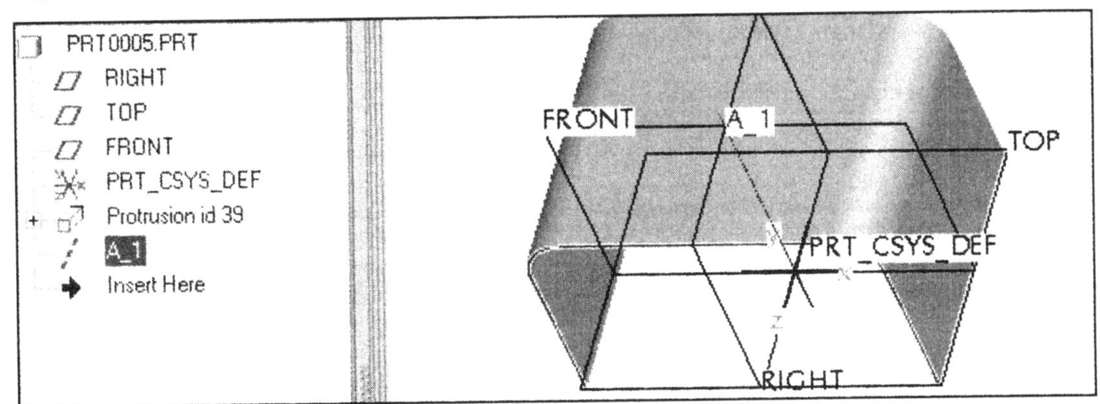

Figure 2.6(k) A_1 Axis Created

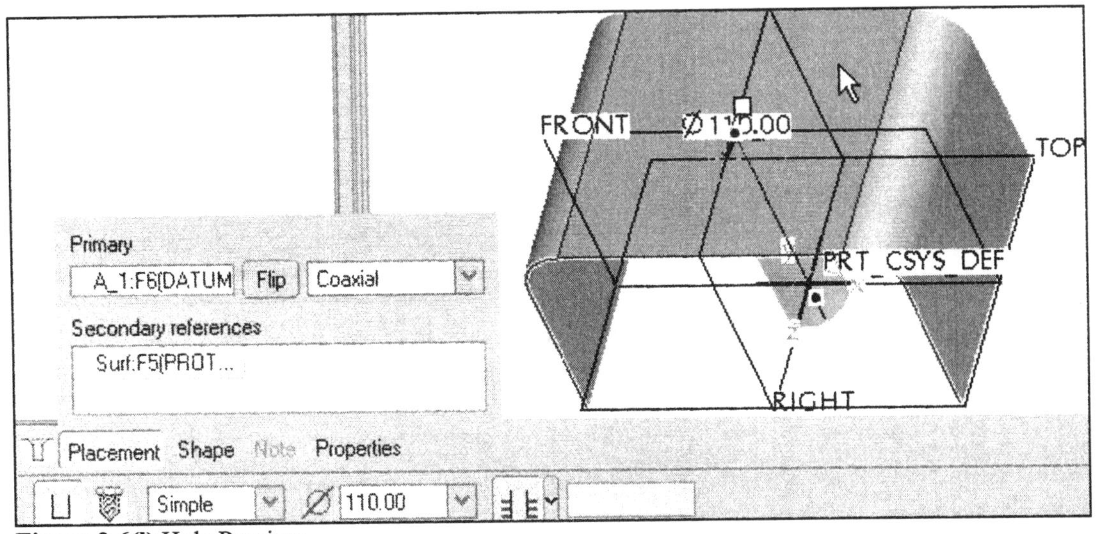

Figure 2.6(l) Hole Preview

Part Model Six (PRT0006.PRT) (Revolve)

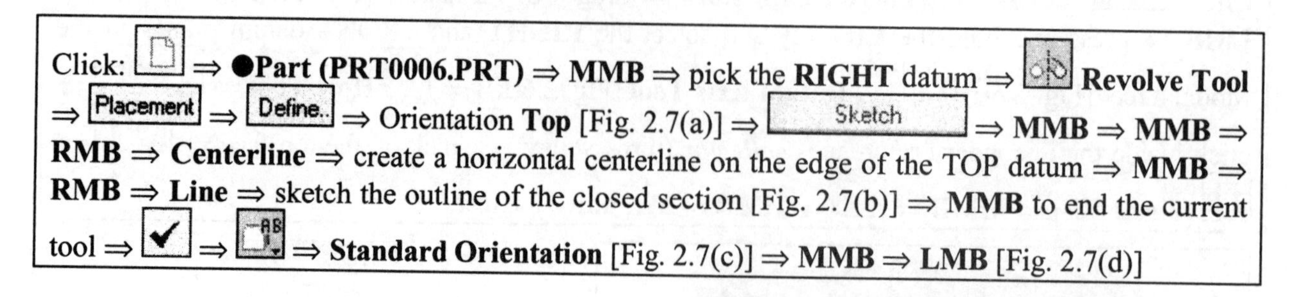

Click: [New] ⇒ ●**Part (PRT0006.PRT)** ⇒ **MMB** ⇒ pick the **RIGHT** datum ⇒ [Revolve] **Revolve Tool** ⇒ Placement ⇒ Define.. ⇒ Orientation **Top** [Fig. 2.7(a)] ⇒ Sketch ⇒ **MMB** ⇒ **MMB** ⇒ **RMB** ⇒ **Centerline** ⇒ create a horizontal centerline on the edge of the TOP datum ⇒ **MMB** ⇒ **RMB** ⇒ **Line** ⇒ sketch the outline of the closed section [Fig. 2.7(b)] ⇒ **MMB** to end the current tool ⇒ ✔ ⇒ [Saved View List] ⇒ **Standard Orientation** [Fig. 2.7(c)] ⇒ **MMB** ⇒ **LMB** [Fig. 2.7(d)]

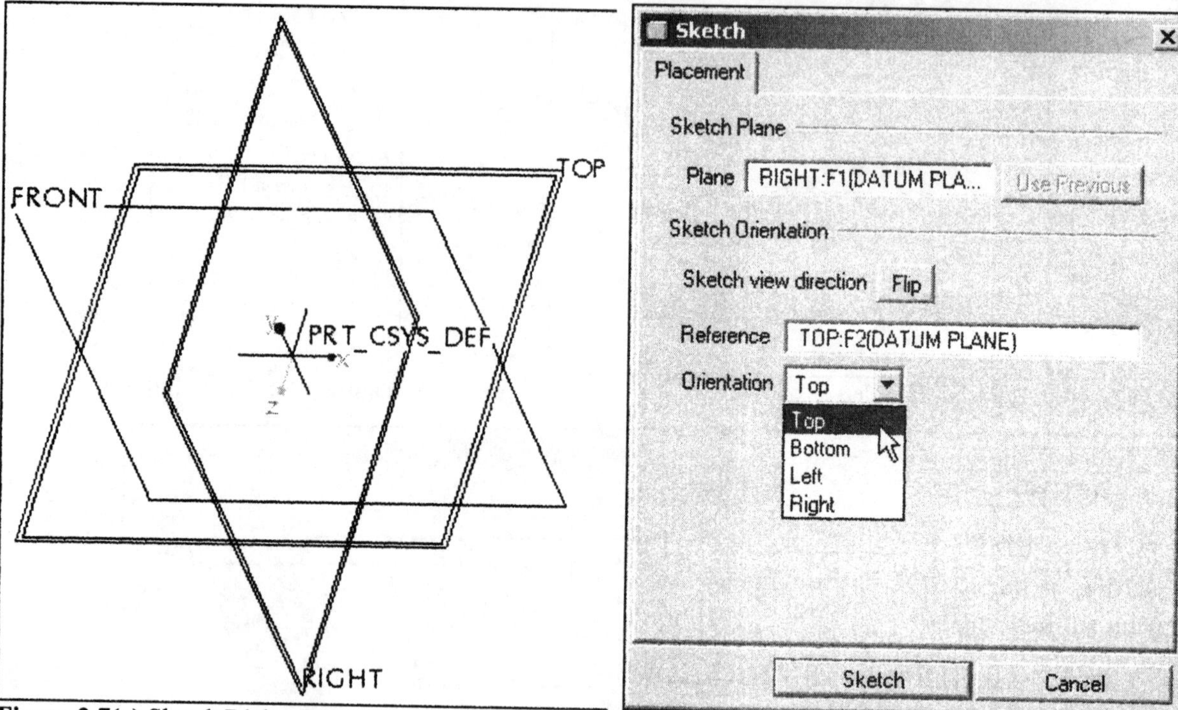

Figure 2.7(a) Sketch Dialog Box

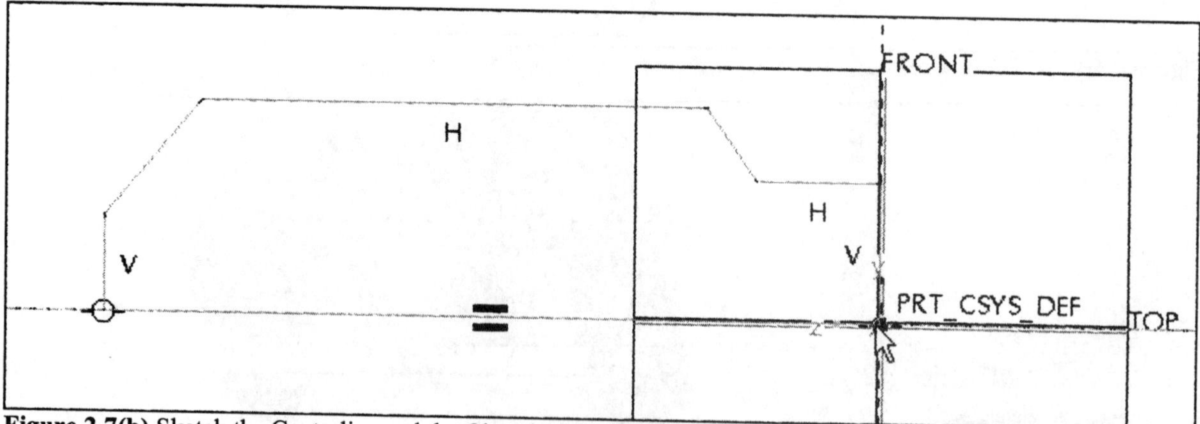

Figure 2.7(b) Sketch the Centerline and the Closed Section Outline

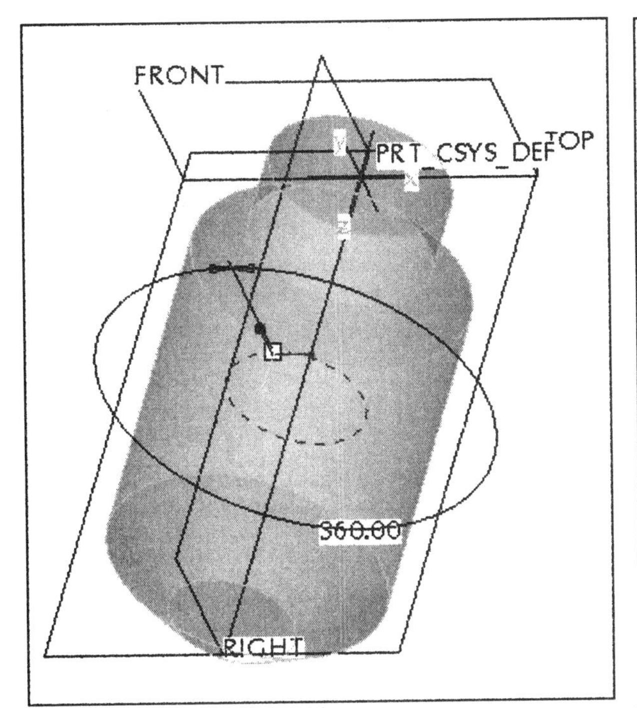

Figure 2.7(c) Revolve Preview

Figure 2.7(d) Completed Revolved Protrusion

Slowly pick on the front edge of the part until it highlights ⇒ press and hold the **Ctrl** key ⇒ pick the other visible edges ⇒ **RMB** ⇒ **Round Edges** [Fig. 2.7(e)] ⇒ drag the handle [Fig. 2.7(f)] ⇒ **MMB** ⇒ **LMB**

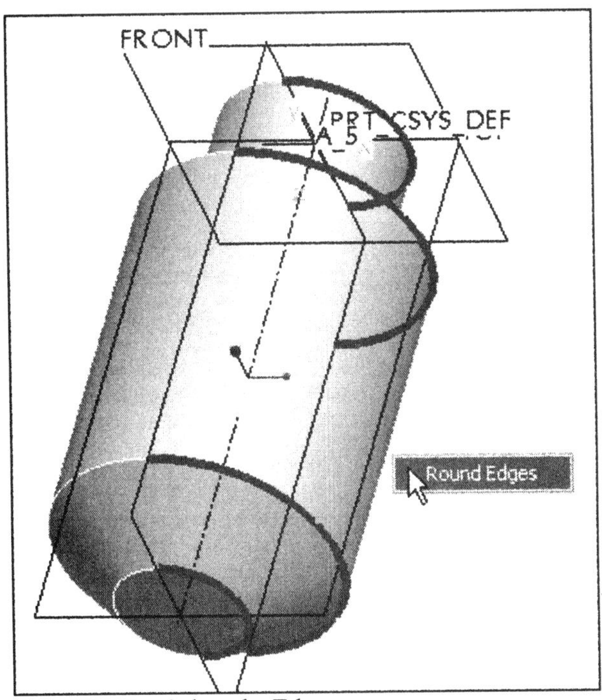

Figure 2.7(e) Select the Edges

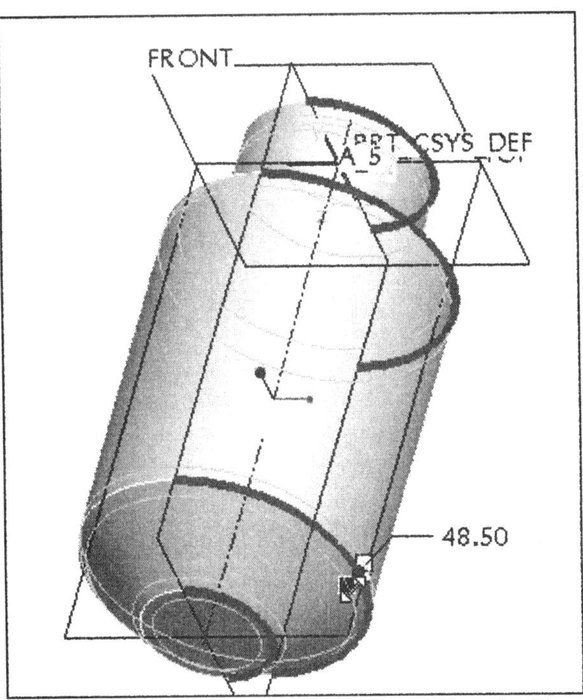

Figure 2.7(f) Move the Drag Handle to Adjust the Size

Click: **View** from the Menu Bar ⇒ View Manager [Fig. 2.7(g)] ⇒ Xsec [Fig. 2.7(h)] ⇒ New [Fig. 2.7(i)] ⇒ type **A** ⇒ **Enter** [Fig. 2.7(j)] ⇒ **MMB** ⇒ select **RIGHT** from the Model Tree [Fig. 2.7(k)] ⇒ Display ⇒ Show X-Hatching [Fig. 2.7(l)] ⇒ Close ⇒ **LMB** ⇒ (Save) ⇒ **MMB** ⇒ **Window** ⇒ **Close**

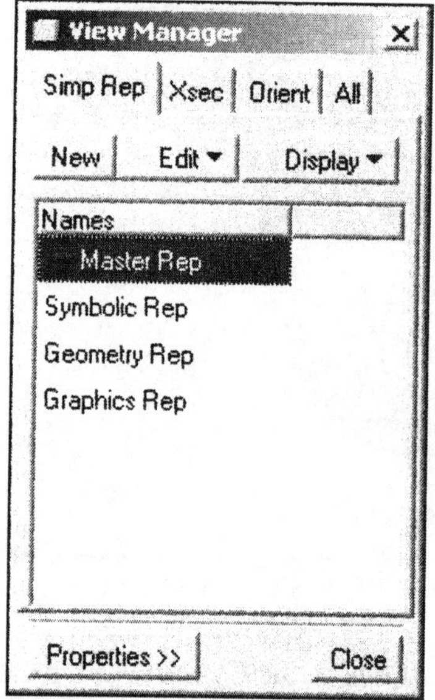

Figure 2.7(g) View Manager

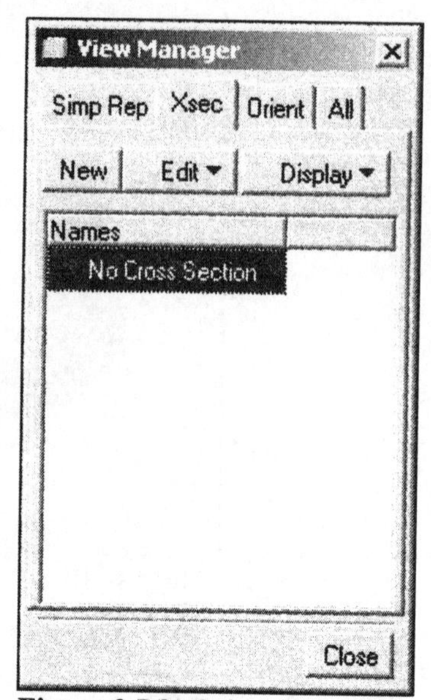

Figure 2.7(h) Select Xsec Tab

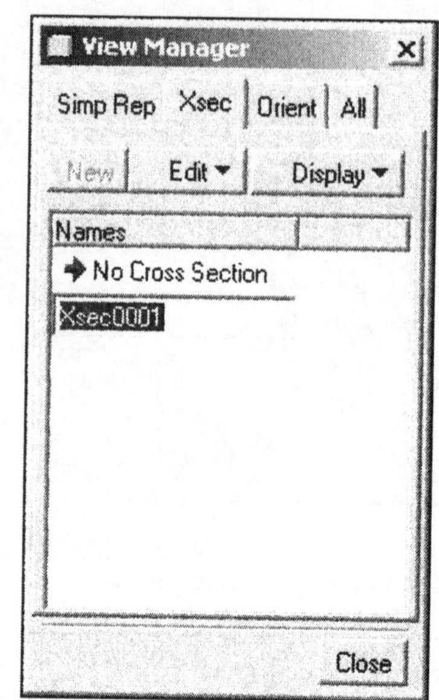

Figure 2.7(i) New Xsec

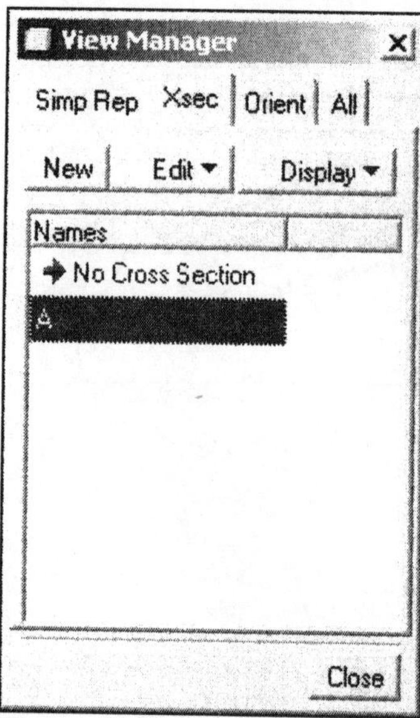

Figure 2.7(j) Section A

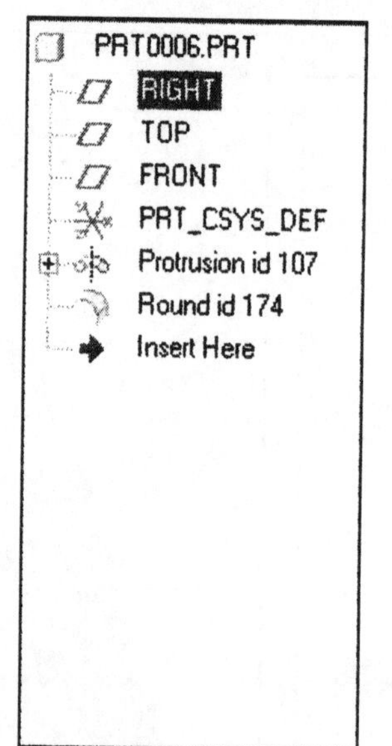

Figure 2.7(k) Select RIGHT

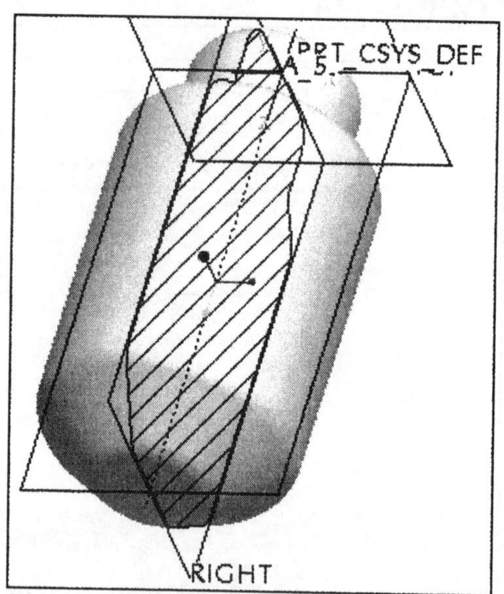

Figure 2.7(l) Sectioned Part

Part Model Seven (PRT0007.PRT) (Revolve Ellipse)

Click: ⇒ ●**Part (PRT0007.PRT)** ⇒ **MMB** ⇒ pick the **FRONT** datum ⇒ **Revolve Tool** ⇒ **RMB** ⇒ **Define Internal Sketch** ⇒ **MMB** ⇒ **MMB** ⇒ **MMB** ⇒ **RMB** ⇒ **Centerline** ⇒ create a vertical centerline on the edge of the RIGHT datum ⇒ ⇒ **Create a full ellipse** ⇒ pick one point to locate the center and the second to determine the shape of the ellipse [Fig. 2.8(a)] ⇒ **MMB** to end the current tool ⇒ ⇒ ⇒ **Standard Orientation** ⇒ Placement ⇒ 180.00 [Fig. 2.8(b)] ⇒ **MMB** [Fig. 2.8(c)] ⇒ **LMB**

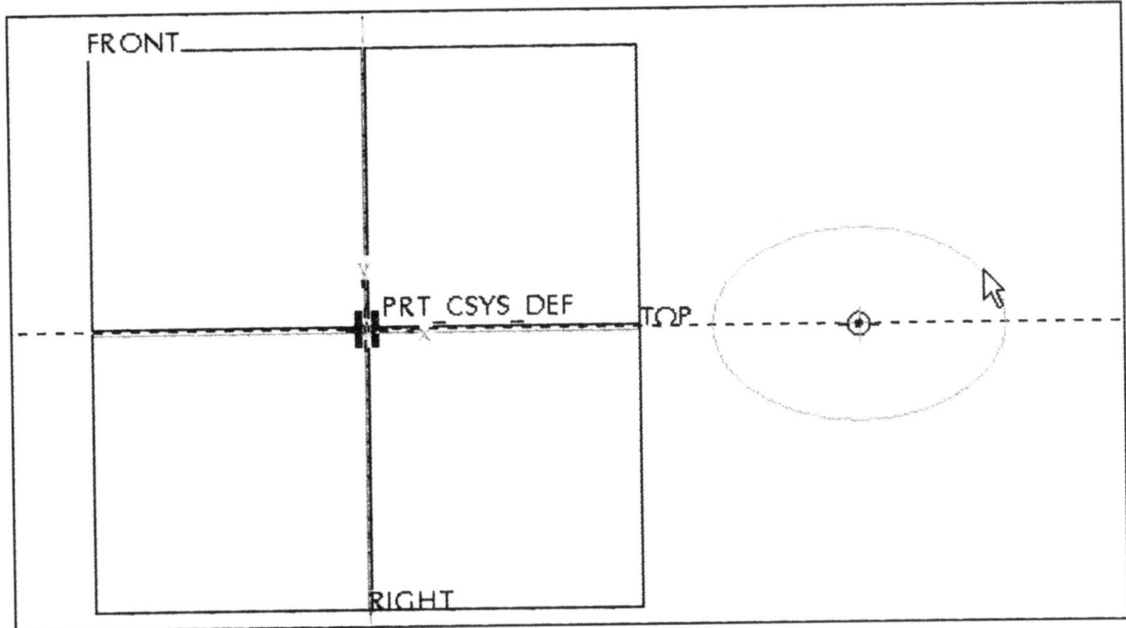

Figure 2.8(a) Sketch a Horizontal Centerline and an Ellipse

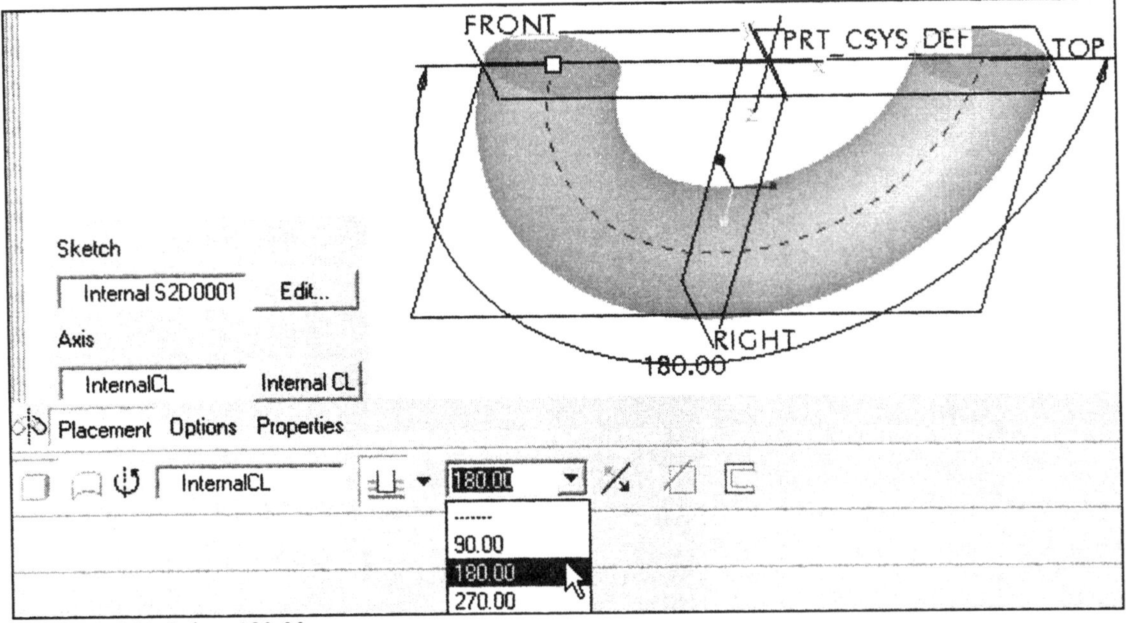

Figure 2.8(b) Select **180.00**

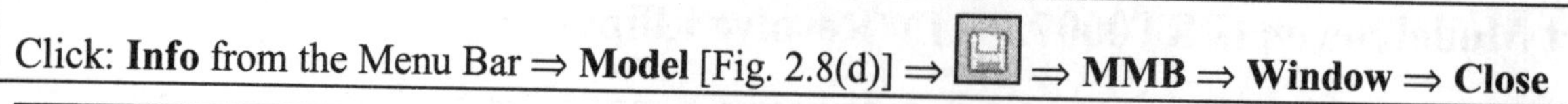

Click: **Info** from the Menu Bar ⇒ **Model** [Fig. 2.8(d)] ⇒ ⇒ **MMB** ⇒ **Window** ⇒ **Close**

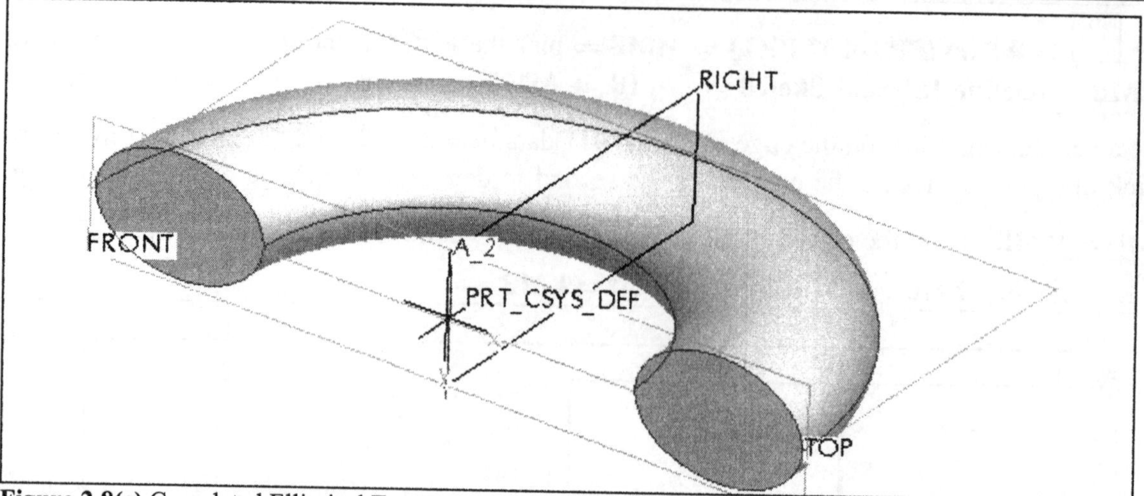

Figure 2.8(c) Completed Elliptical Torus

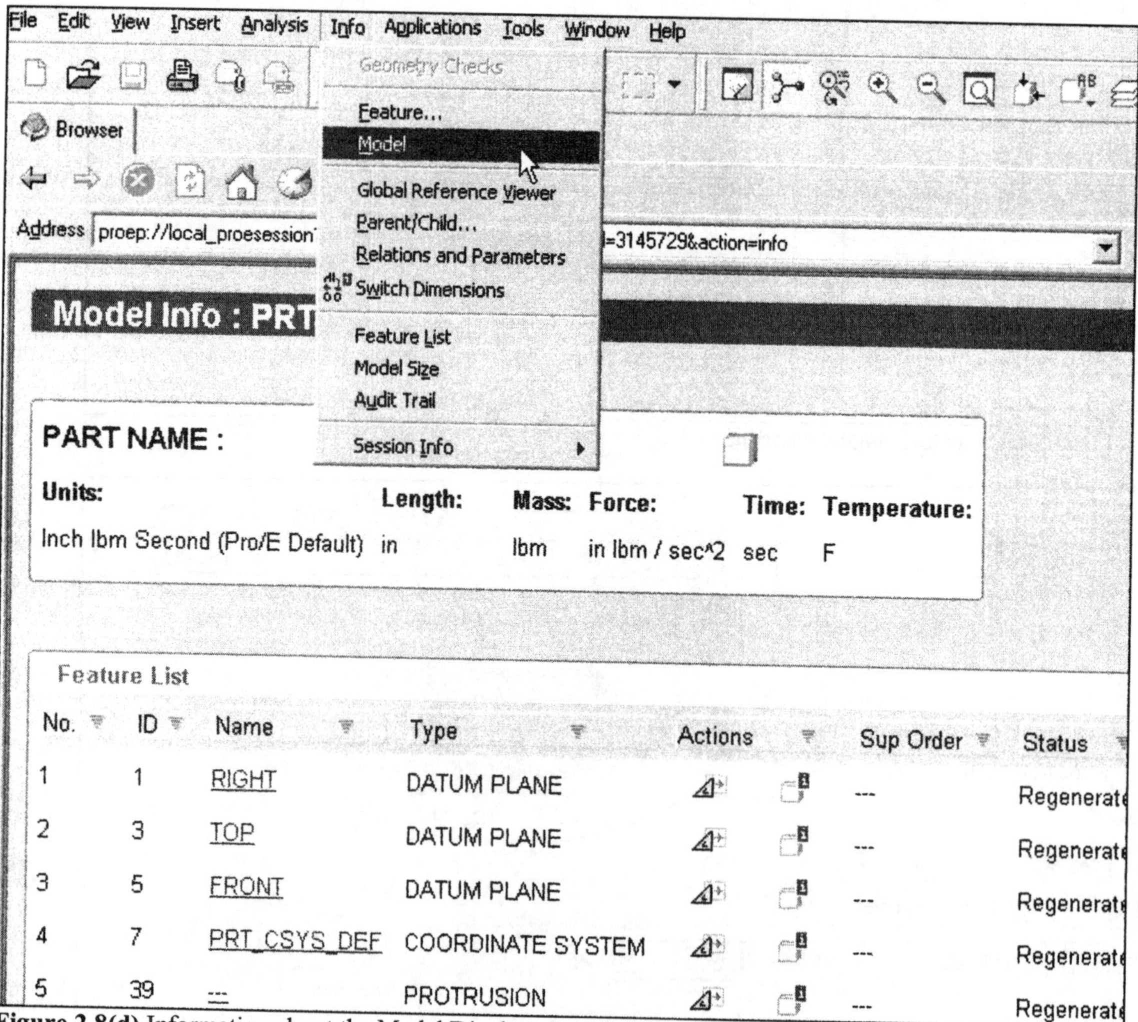

Figure 2.8(d) Information about the Model Displayed in Web Browser

Part Model Eight (PRT0008.PRT) (Revolve Cut)

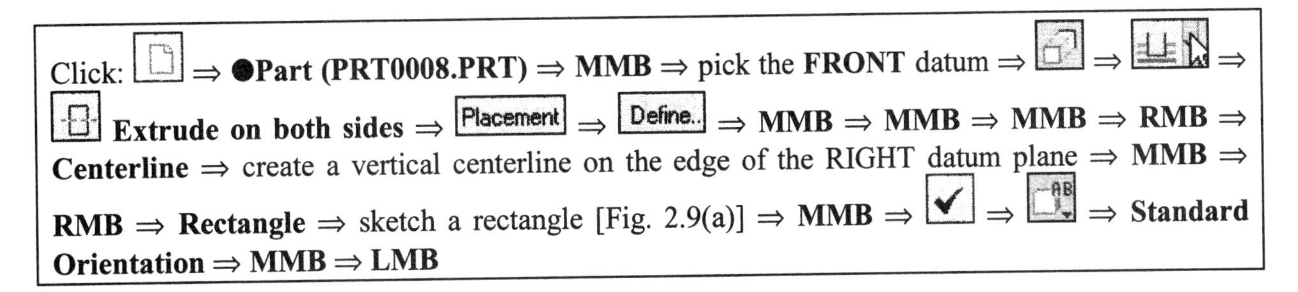
Click: ⇒ ●**Part (PRT0008.PRT)** ⇒ **MMB** ⇒ pick the **FRONT** datum ⇒ ⇒ ⇒ **Extrude on both sides** ⇒ Placement ⇒ Define.. ⇒ **MMB** ⇒ **MMB** ⇒ **MMB** ⇒ **RMB** ⇒ **Centerline** ⇒ create a vertical centerline on the edge of the RIGHT datum plane ⇒ **MMB** ⇒ **RMB** ⇒ **Rectangle** ⇒ sketch a rectangle [Fig. 2.9(a)] ⇒ **MMB** ⇒ ⇒ ⇒ **Standard Orientation** ⇒ **MMB** ⇒ **LMB**

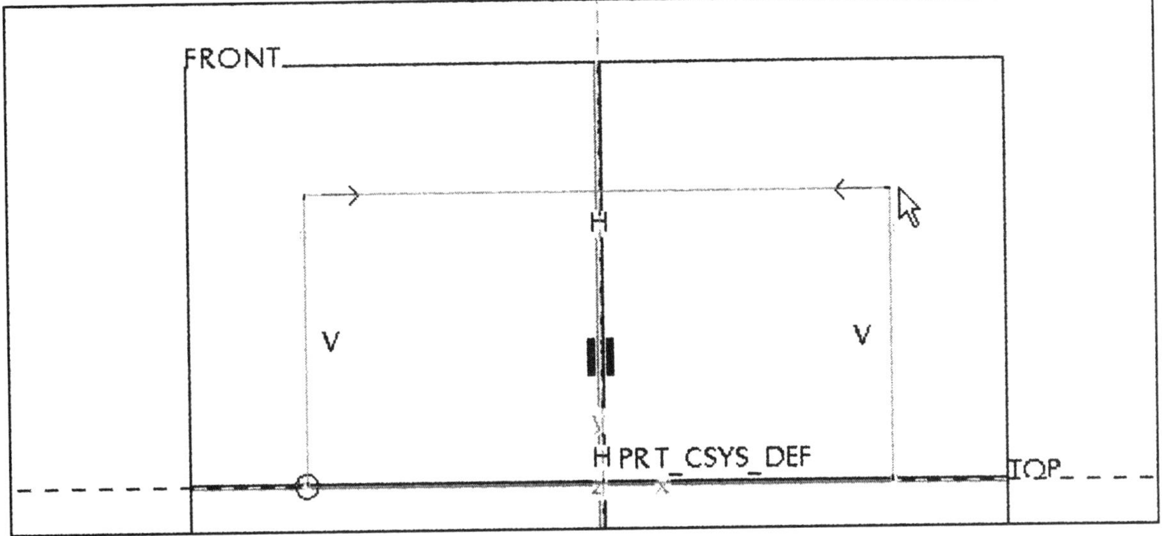

Figure 2.9(a) Sketch a Rectangle

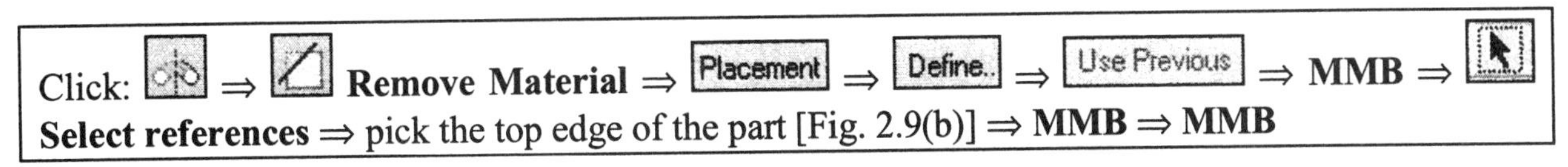
Click: ⇒ **Remove Material** ⇒ Placement ⇒ Define.. ⇒ Use Previous ⇒ **MMB** ⇒ **Select references** ⇒ pick the top edge of the part [Fig. 2.9(b)] ⇒ **MMB** ⇒ **MMB**

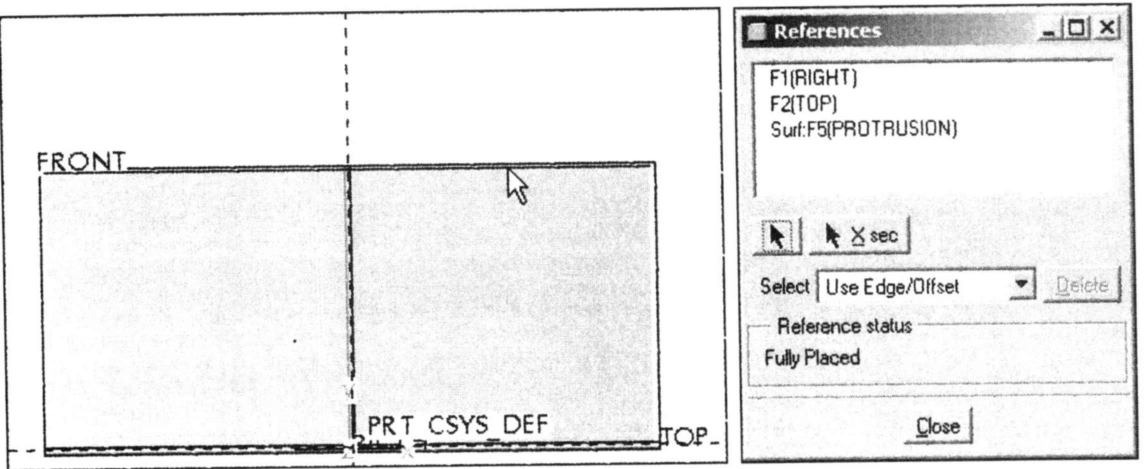

Figure 2.9(b) Select the Top Edge as a Reference

Click: **RMB** ⇒ **Centerline** ⇒ create a vertical centerline through the middle of the part ⇒ **RMB** ⇒ **Circle** ⇒ sketch a circle [Figs. 2.9(c-d)] ⇒ **MMB** ⇒ ✔ ⇒ [icon] ⇒ **Standard Orientation** [Fig. 2.9(e)] ⇒ **MMB** [Fig. 2.9(f)] ⇒ **LMB**

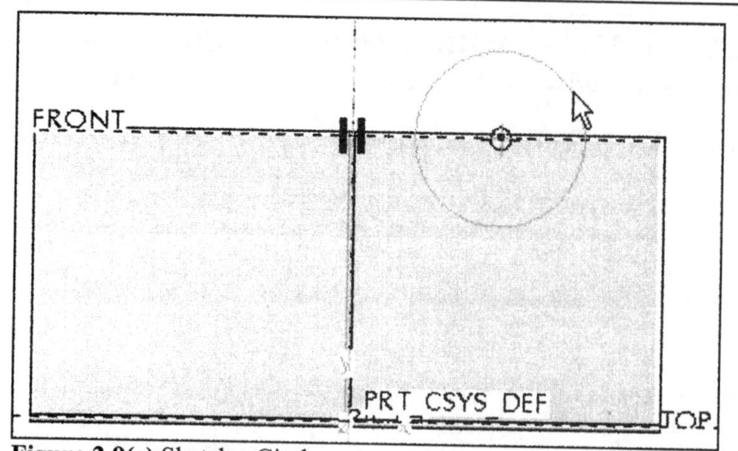

Figure 2.9(c) Sketch a Circle

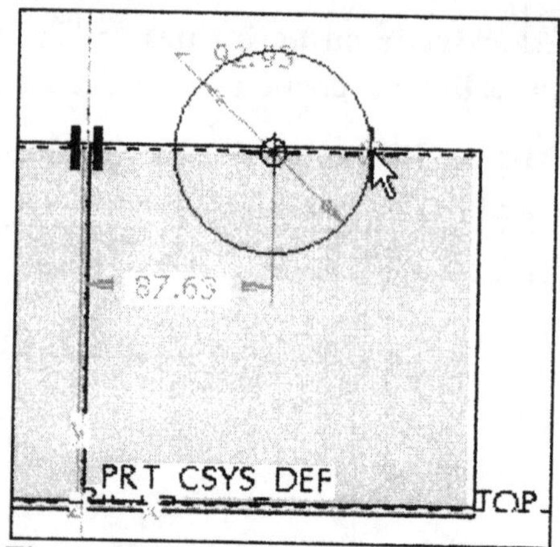

Figure 2.9(d) Completed Sketched Circle

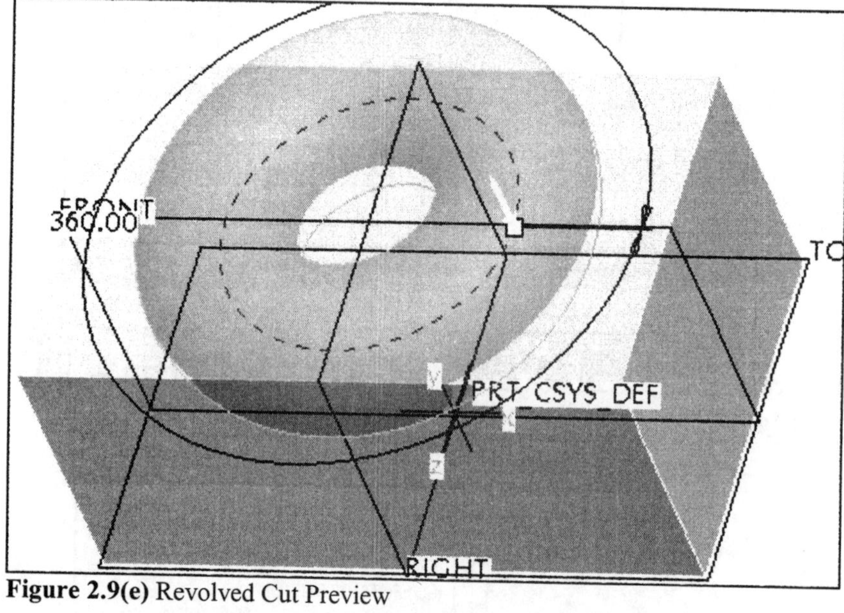

Figure 2.9(e) Revolved Cut Preview

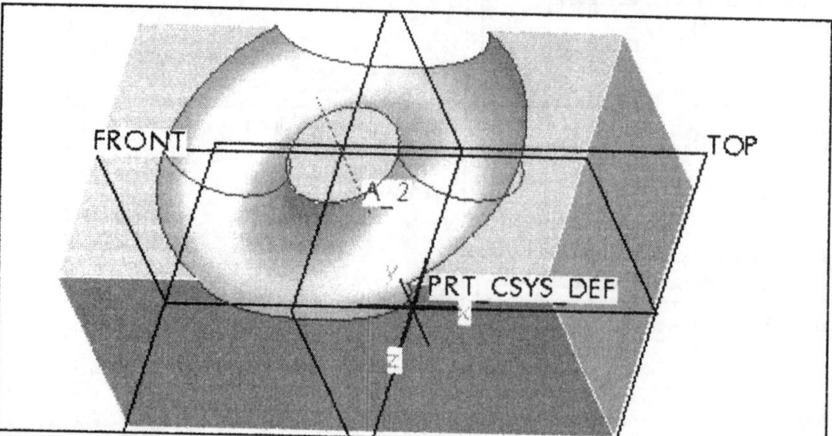

Figure 2.9(f) Completed Cut

Slowly pick on the edge of the part until it highlights ⇒ press and hold the **Ctrl** key ⇒ pick the other edges ⇒ **RMB** ⇒ Round Edges [Figs. 2.9(g-h)] ⇒ **MMB** [Fig. 2.9(i)] ⇒ **LMB**

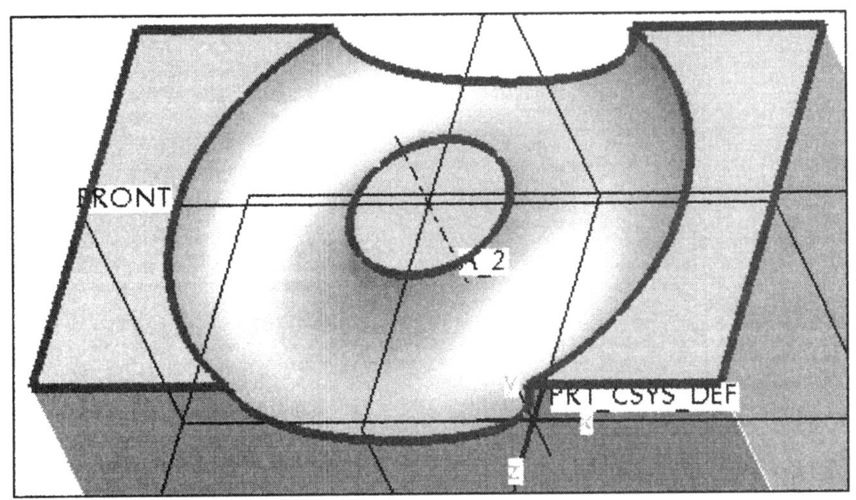

Figure 2.9(g) Select the Edges

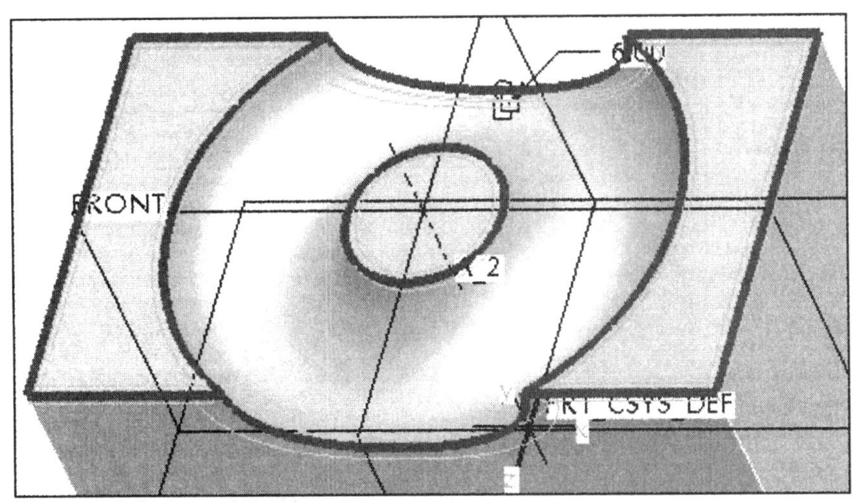

Figure 2.9(h) Round Preview

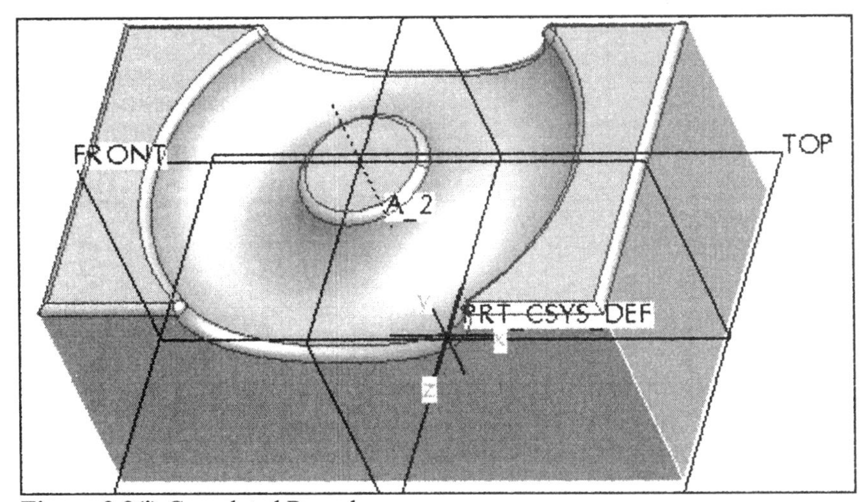

Figure 2.9(i) Completed Round

Place your cursor on Axis **A_2** ⇒ **LMB** until **A_2** is highlighted [Fig. 2.9(j)] ⇒ **Hole Tool** ⇒ Placement ⇒ click inside the Secondary references *collector* (turns yellow) ⇒ pick on the top face of the part [Fig. 2.9(k)] ⇒ ⇒ ⇒ **MMB** [Fig. 2.9(l)] ⇒ **LMB** ⇒ ⇒ **MMB** ⇒ **Window** ⇒ **Close**

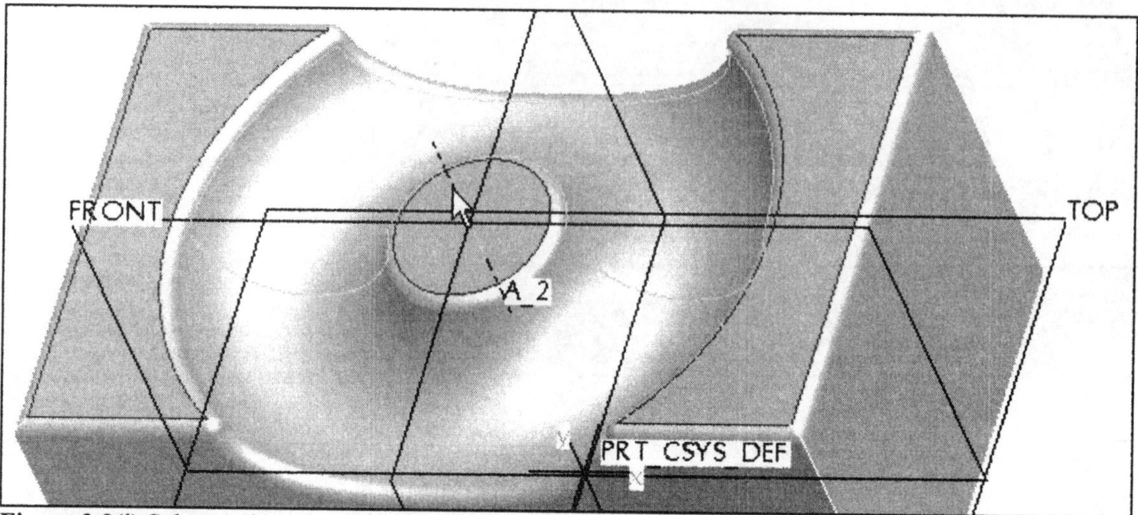

Figure 2.9(j) Select Axis **A_2**

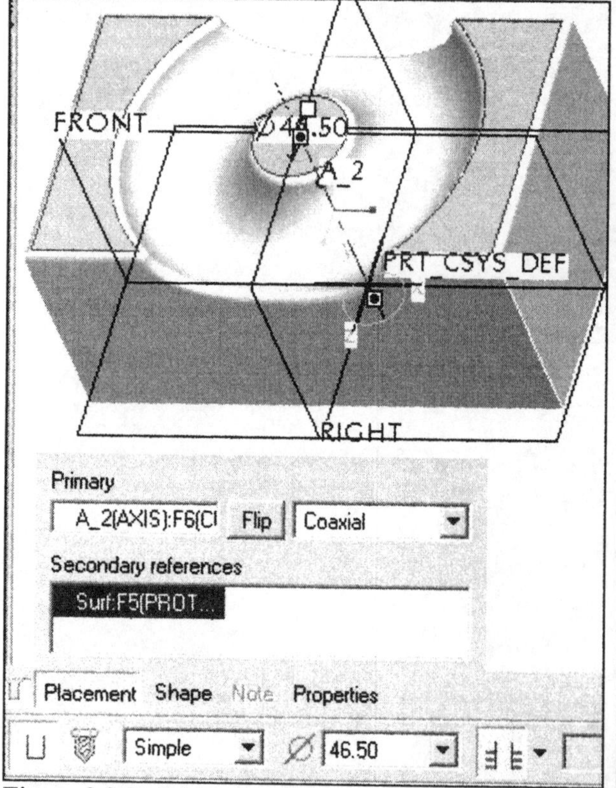

Figure 2.9(k) Select the Top Face of the Part

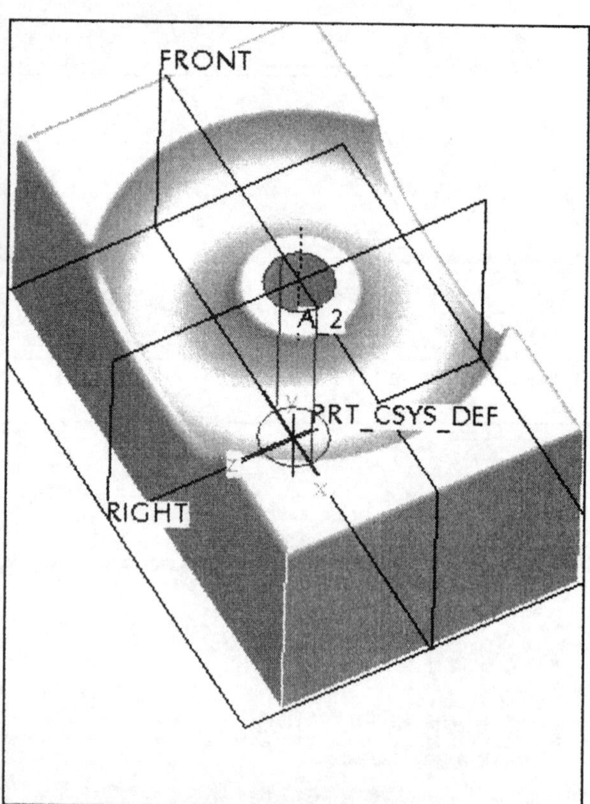

Figure 2.9(l) Completed Part

See *www.cad-resources.com* ⇒ ***Downloads***, for extra Lessons and projects.

Lesson 3 Extrusions

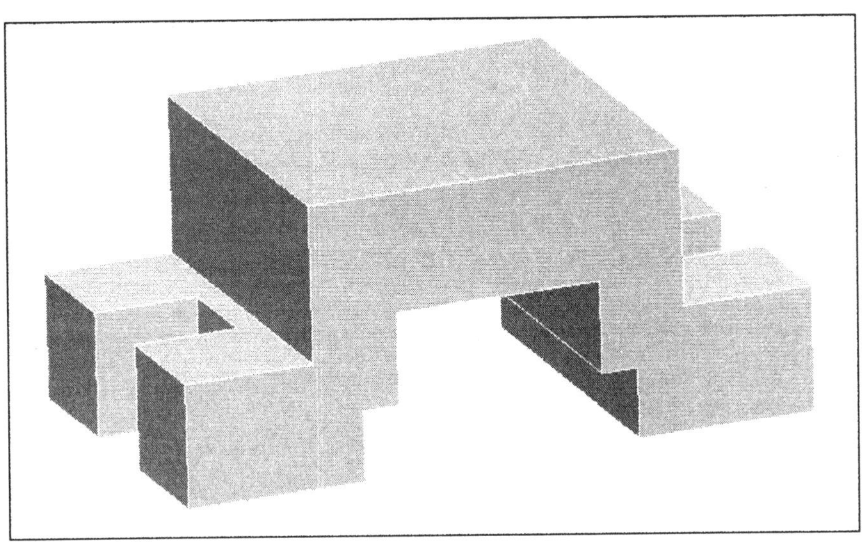

Figure 3.1 Clamp

OBJECTIVES

- Create a feature using an **Extruded** protrusion
- Understand **Setup** and **Environment** settings
- Define and set a **Material** type
- Create and use **Datum** features
- Sketch protrusion and cut feature geometry using the **Sketcher**
- Understand the feature **Dashboard**
- **Copy** a feature
- **Save** and **Delete Old Versions** of an object

Extrusions

The design of a part using Pro/E starts with the creation of base features (normally datum planes), and a solid protrusion. Other protrusions and cuts are then added in sequence as required by the design. You can use various types of Pro/E features as building blocks in the progressive creation of solid parts (Fig. 3.1). Certain features, by necessity, precede other more dependent features in the design process. Those dependent features rely on the previously defined features for dimensional and geometric references.

The progressive design of features creates these dependent feature relationships known as *parent-child relationships*. The actual sequential history of the design is displayed in the Model Tree. The parent-child relationship is one of the most powerful aspects of Pro/E and parametric modeling in general. It is also very important after you modify a part. After a parent feature in a part is modified, all children are automatically modified to reflect the changes in the parent feature. It is therefore essential to reference feature dimensions so that Pro/E can correctly propagate design modifications throughout the model.

An **extrusion** is a part feature that adds or removes material. A protrusion is *always the first solid feature created.* This is usually the first feature created after a base feature of datum planes. The **Extrude Tool** is used to create both protrusions and cuts. A toolchest button is available for this command or it can be initiated using Insert ⇒ Extrude from the menu bar. Figure 3.2 shows four different types of basic protrusions.

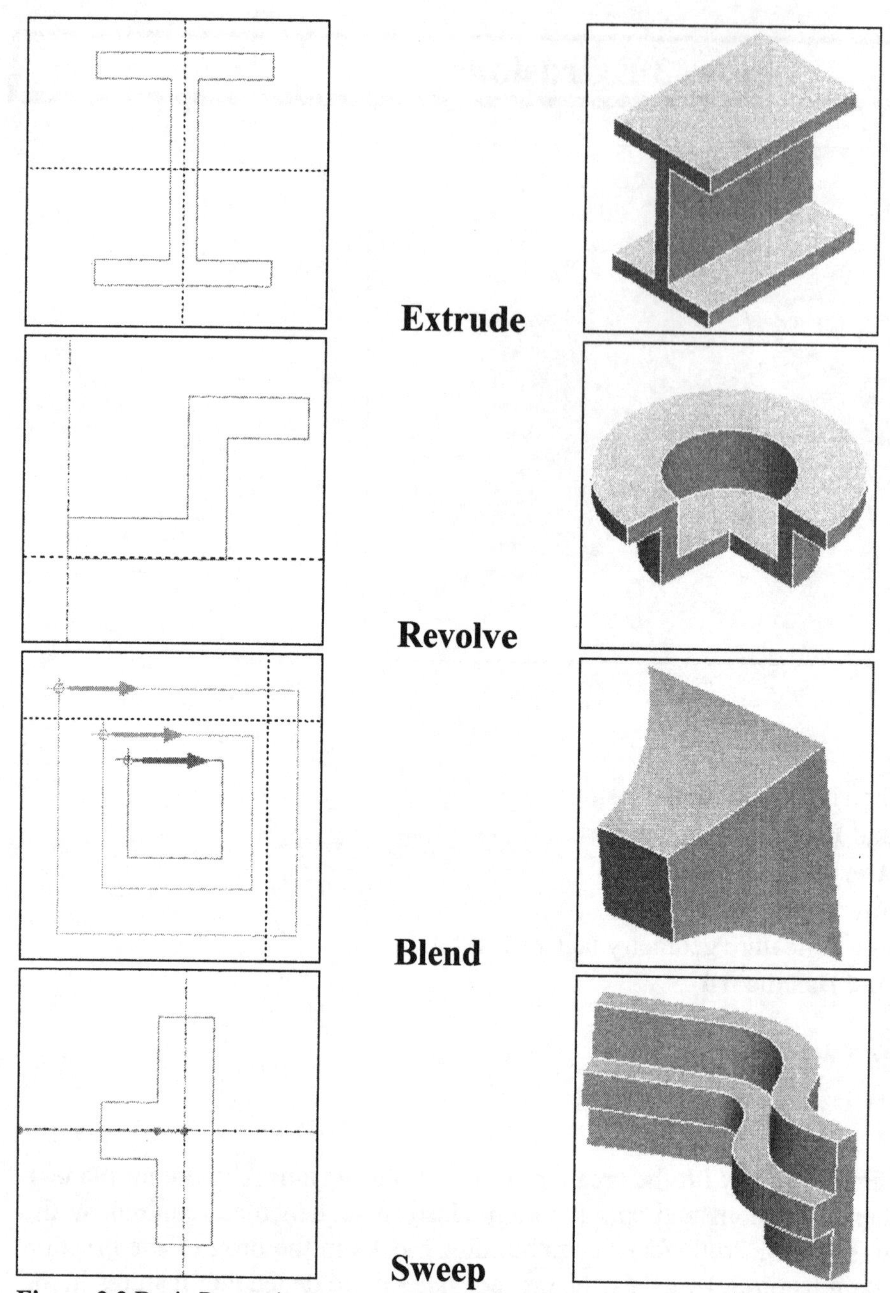

Figure 3.2 Basic Protrusions

The Design Process

It is tempting to directly start creating models. Nevertheless, in order to build value into a design, you need to create a product that can keep up with the constant design changes associated with the design-through-manufacturing process. Flexibility must be "built in" to the design. Flexibility is the key to a friendly and robust product design while maintaining design intent, and you can accomplish it through planning. To plan a design, you need understand the overall function, form, and fit of the product. This understanding includes the following points:

- Overall size of the part
- Basic part characteristics
- The way in which the part can be assembled
- Approximate number of assembly components
- The manufacturing processes required to produce the part

Lesson 3 STEPS

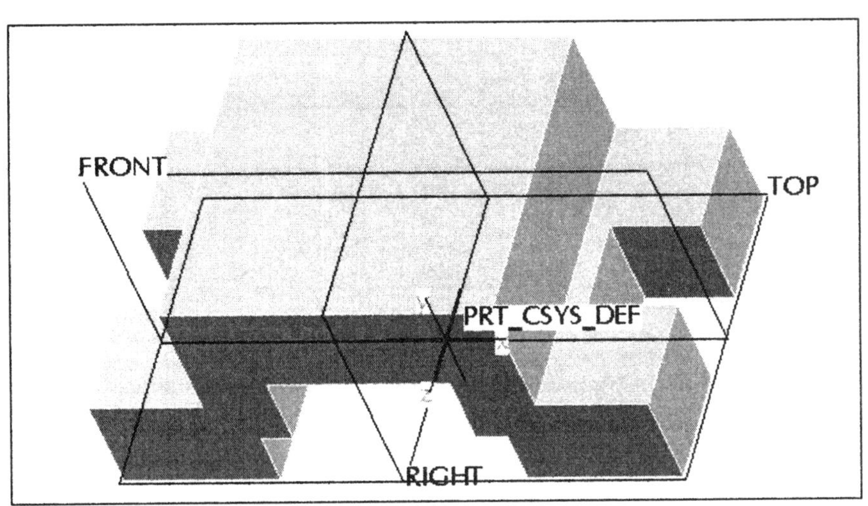

Figure 3.3 Clamp and Datum Planes

Clamp

The clamp in Figure 3.3 is composed of a protrusion and two cuts. A number of things need to be established before you actually start modeling. These include setting up the *environment*, selecting the *units*, and establishing the *material* for the part.

Before you begin any part using Pro/E, you must plan the design. The **design intent** will depend on a number of things that are out of your control and on a number that you can establish. Asking yourself a few questions will clear up the design intent you will follow: Is the part a component of an assembly? If so, what surfaces or features are used to connect one part to another? Will geometric tolerancing be used on the part and assembly? What units are being used in the design, SI or decimal inch? What is the part's material? What is the primary part feature? How should I model the part, and what features are best used for the primary protrusion (the first solid mass)? On what datum plane should I sketch to model the first protrusion? These and many other questions will be answered as you follow the systematic lesson part. However, you must answer many of the questions on your own when completing the *lesson project*, which does not come with systematic instructions.

Launch **Pro/ENGINEER WILDFIRE 2.0** ⇒ **File** ⇒ **Set Working Directory** ⇒ select the working directory ⇒ **OK** ⇒ **Create a new object** ⇒ ●**Part** ⇒ Name **CLAMP** ⇒ Use default template ⇒ **OK** ⇒ **Edit** ⇒ **Setup** ⇒ **Units** ⇒ Units Manager **millimeter Newton Second (mmNs)** ⇒ **Set** ⇒ ●**Convert dimensions** [Figs. 3.4(a-b)] ⇒ **OK** ⇒ **Close** ⇒ **Material** ⇒ **Define** ⇒ type **STEEL** Enter material name STEEL ⇒ [Fig. 3.4(c)] ⇒ **File** from the material table ⇒ **Save** ⇒ **File** ⇒ **Exit** ⇒ **Assign** ⇒ pick **STEEL** ⇒ **Accept** ⇒ **MMB** ⇒ ⇒ **MMB** *(or **Enter** or **OK**)*

The material file, STEEL, is without any file information [Fig. 3.4(c)]. As an option, if your instructor provides you with the specifications, or you are familiar with setting up material specs, you can edit the file using: *Edit ⇒ Setup ⇒ Material ⇒ Edit ⇒ Steel ⇒ Accept ⇒ fill in the information ⇒ File ⇒ Save ⇒ File ⇒ Exit ⇒ Done.*

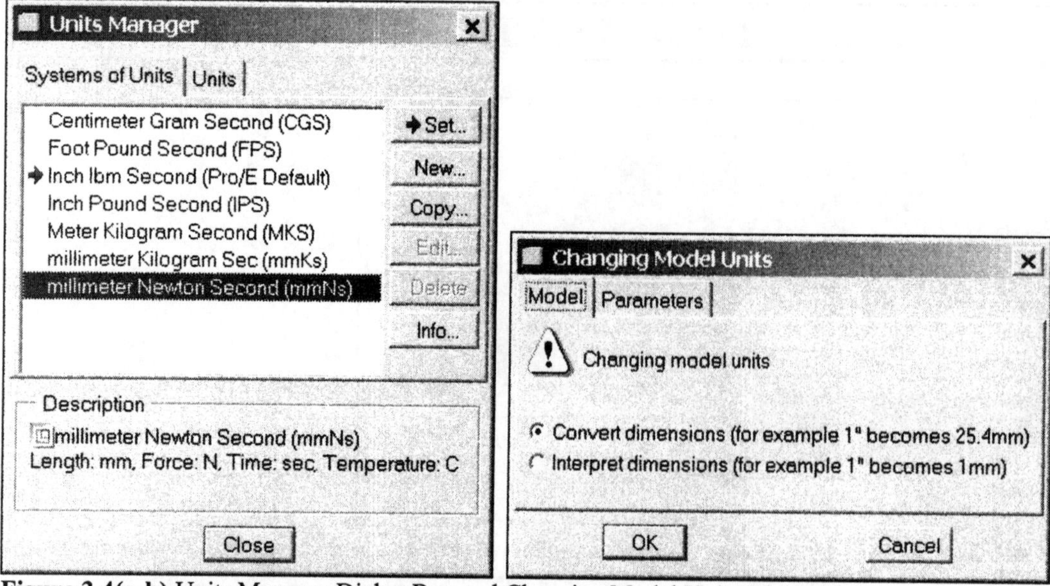

Figure 3.4(a-b) Units Manager Dialog Box and Changing Model Units Dialog Box

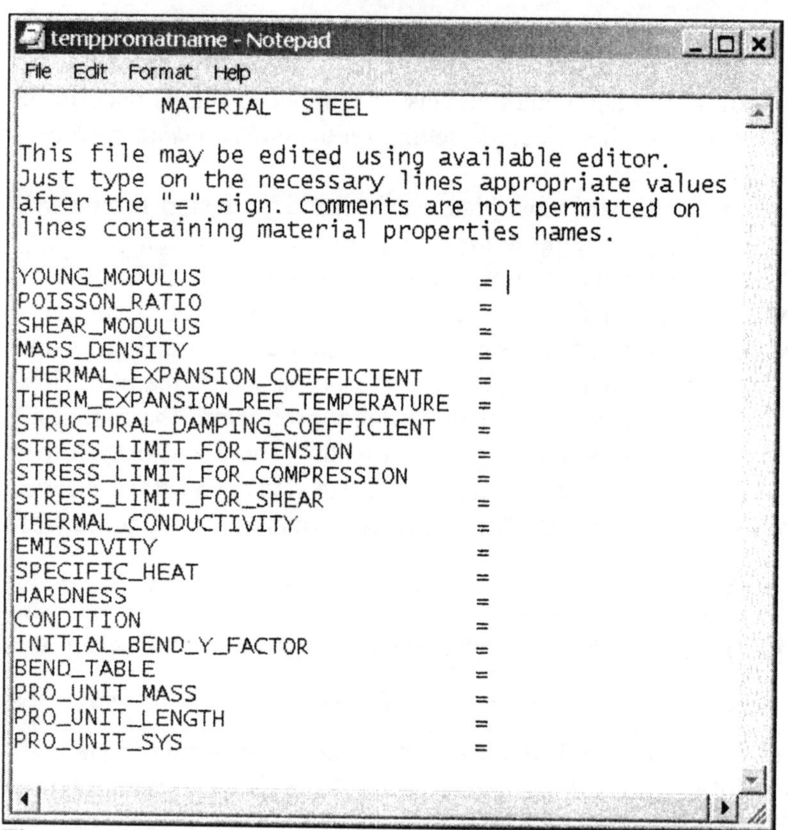

Figure 3.4(c) Material File

Since [✓ Use default template] was selected, the default datum planes and the default coordinate system are displayed in the graphics window and in the Model Tree (Fig. 3.5). *The **default datum planes** and the **default coordinate system** will be the first features on all parts and assemblies.* The datum planes are used to sketch on and to orient the part's features. Having datum planes as the first features of a part, instead of the first protrusion, gives the designer more flexibility during the design process. Picking on items in the Model Tree will highlight that item on the model (Fig. 3.5).

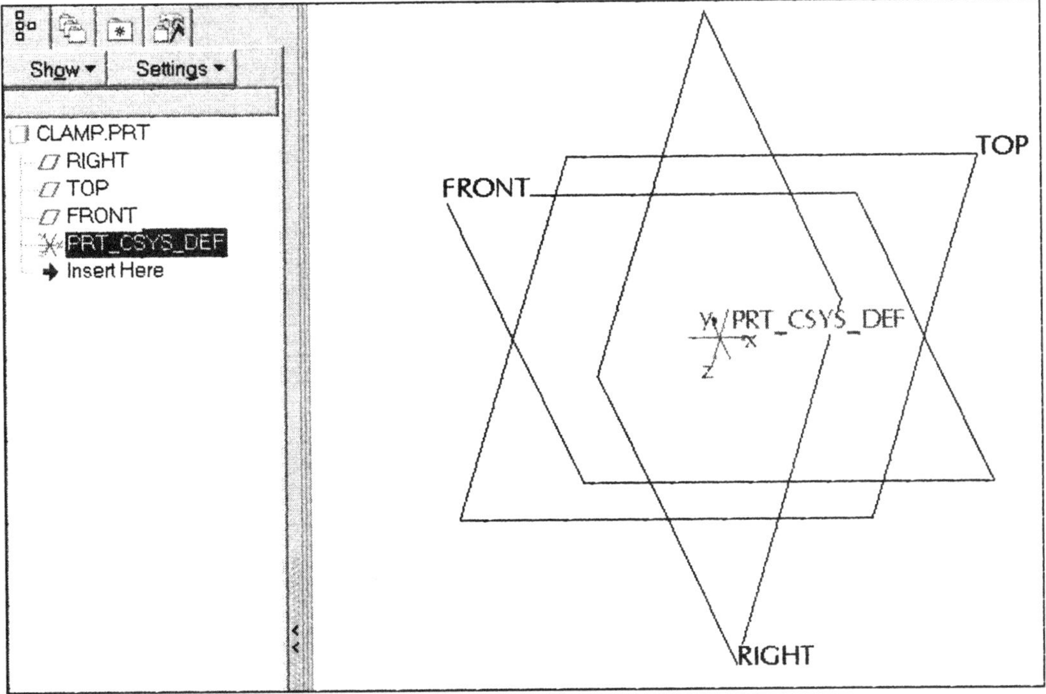

Figure 3.5 Default Datum Planes and Default Coordinate System

Pick the **FRONT** datum plane in the Model Tree ⇒ **Sketch Tool** from the right Toolbar ⇒ Sketch dialog box opens [Fig. 3.6(a)], click: Sketch

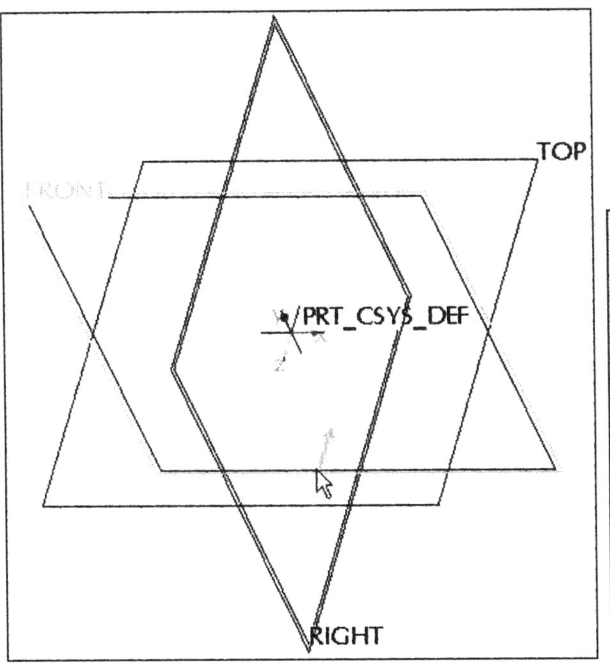

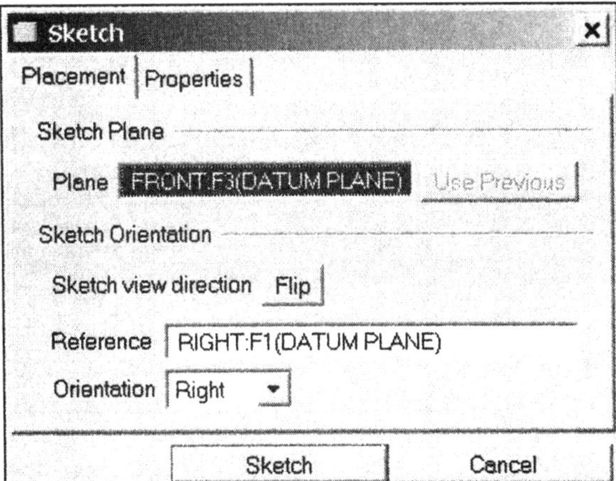

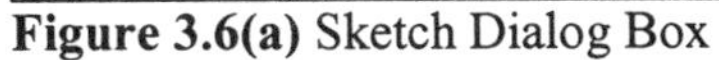

Figure 3.6(a) Sketch Dialog Box

Click: **Close** to accept the References [Fig. 3.6(b)] ⇒ **Toggle the grid on** from the top toolbar [Fig. 3.6(c)]

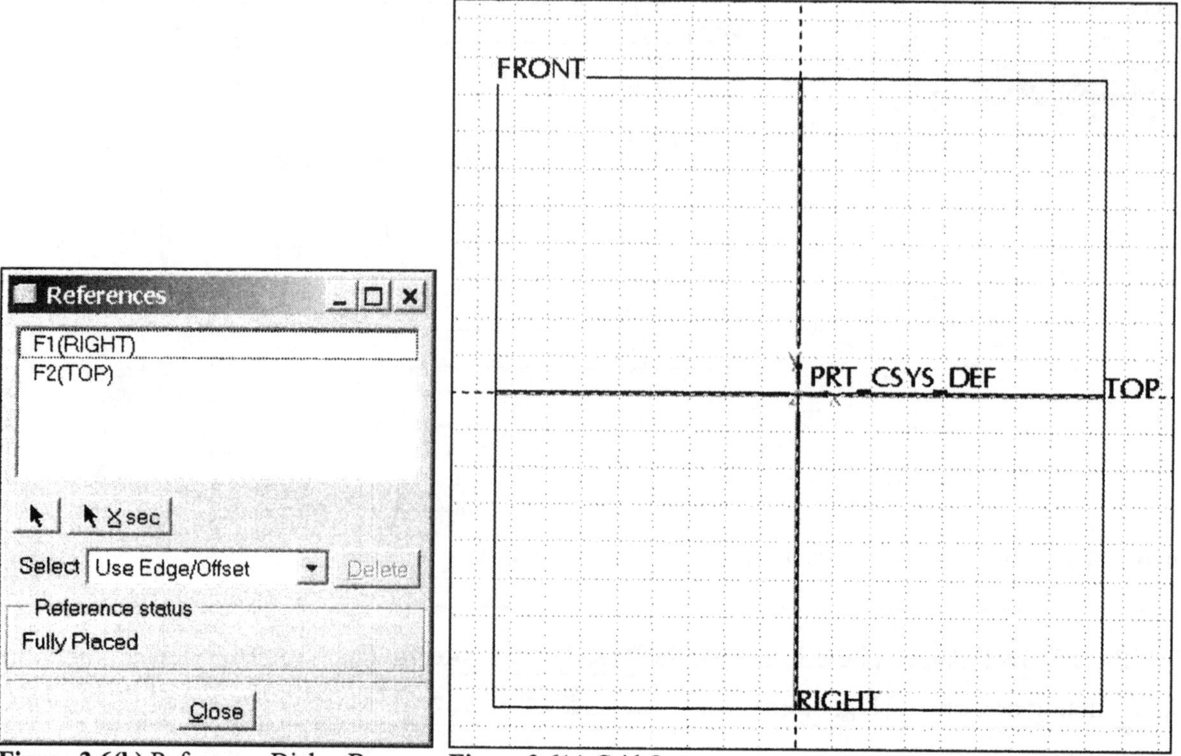

Figure 3.6(b) References Dialog Box **Figure 3.6(c)** Grid On

The sketch is now displayed and oriented in 2D [Fig. 3.6(c)]. The coordinate system is at the middle of the sketch, where datum RIGHT and datum TOP intersect. The X coordinate arrow points to the right and the Y coordinate arrow points up. The Z arrow is pointing toward you (out from the screen). The square box you see is the limited display of datum FRONT. This is similar to sketching on a piece of graph paper. Pro/E is not coordinate-based software, so you need not enter geometry with X, Y, and Z coordinates as with many other CAD systems.

Use **Shift MMB** and **Ctrl MMB** to reposition and resize the sketch as needed. Since you now have a visible grid, it is a good idea to have your sketch picks snap to the grid. Click: **Tools** from the menu bar ⇒ **Environment** ⇒ Snap to Grid [Fig. 3.6(d)] ⇒ **Apply** ⇒ **OK**

You can control many aspects of the environment in which Pro/E runs with the Environment dialog box. To open the Environment dialog box, click Tools ⇒ Environment on the menu bar or click the appropriate icon in the toolbar. When you make a change in the Environment dialog box, it takes effect for the current session only. When you start Pro/E, the environment settings are defined by your configuration file, if any; otherwise, by Pro/E configuration defaults.

Depending on which Pro/E Mode is active, some or all of the following options may be available in the **Environment** dialog box:

Display:

Dimension Tolerances Display model dimensions with tolerances
Datum Planes Display the datum planes and their names
Datum Axes Display the datum axes and their names
Point Symbols Display the datum points and their names
Coordinate Systems Display the coordinate systems and their names
Spin Center Display the spin center for the model
3D Notes Display model notes

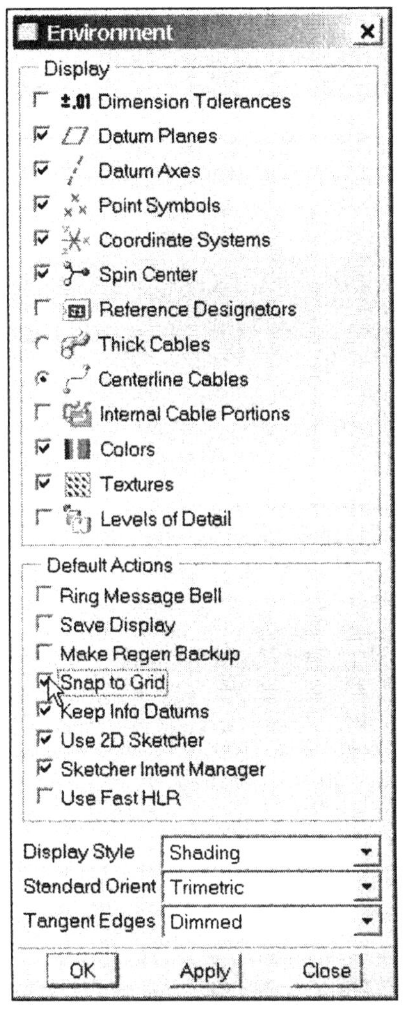

Figure 3.6(d) Environment Dialog Box

Notes as Names Display the note as a name, not the full note
Reference Designators Display reference designation of cabling, ECAD, and Piping components
Thick Cables Display a cable with 3-D thickness
Centerline Cables Display the centerline of a cable with location points
Internal Cable Portions Display cable portions that are hidden from view
Colors Display colors assigned to model surfaces
Textures Display textures on shaded models
Levels of Detail Controls levels of detail available in a shaded model during dynamic orientation

Default Actions:

Ring Message Bell Ring bell (beep) after each prompt or system message
Save Display Save objects with their most recent screen display
Make Regen Backup Backs up the current model before every regeneration
Snap to Grid Make points you select on the Sketcher screen snap to a grid
Keep Info Datums Control how Pro/E treats datum planes, datum points, datum axes, and coordinate systems created on the fly under the Info functionality
Use 2D Sketcher Control the initial model orientation in Sketcher mode
Sketcher Intent Manager Use the Intent Manager when in Sketcher
Use Fast HLR Make possible the hardware acceleration of dynamic spinning with hidden lines, datums, and axes

Display Style:

- **Wireframe** Model is displayed with no distinction between visible and hidden lines
- **Hidden Line** Hidden lines are shown in gray
- **No Hidden** Hidden lines are not shown
- **Shading** All surfaces and solids are displayed as shaded

Standard Orient:

- **Isometric** Standard isometric orientation
- **Trimetric** Standard trimetric orientation
- **User Defined** User-defined orientation

Tangent Edges:

- **Solid** Display tangent edges as solid lines
- **No Display** Blank tangent edges
- **Phantom** Display tangent edges in phantom font
- **Centerline** Display tangent edges in centerline font
- **Dimmed** Display tangent edges in the Dimmed Menu system

Because you checked [✓ Snap to Grid], you can now sketch by simply picking grid points representing the part's geometry (outline). Because this is a sketch in the true sense of the word, you need only create geometry that *approximates* the shape of the feature; the sketch does not have to be accurate as far as size or dimensions are concerned. No two sketches will be the same between those using these steps, unless you count grid spaces (which is not necessary). Even with the grid snap off Pro/E, constrains the geometry according to rules, which include but are not limited to the following:

- **RULE:** Symmetry
  DESCRIPTION: Entities sketched symmetrically about a centerline are assigned equal values with respect to the centerline
- **RULE:** Horizontal and vertical lines
  DESCRIPTION: Lines that are approximately horizontal or vertical are considered exactly horizontal or vertical
- **RULE:** Parallel and perpendicular lines
  DESCRIPTION: Lines that are sketched approximately parallel or perpendicular are considered exactly parallel or perpendicular
- **RULE:** Tangency
  DESCRIPTION: Entities sketched approximately tangent to arcs or circles are assumed to be exactly tangent

The outline of the part's primary feature is sketched using a set of connected lines. The part's dimensions and general shape are provided in Figure 3.6(e). The cut on the front and sides will be the created with separate sketched features. Sketch only one series of lines (8 lines in this sketch). ***Do not sketch lines on top of lines.***

It is important not to create any unintended constraints while sketching. Therefore, remember to exaggerate the sketch geometry and not to align geometric items that have no relationship. Pro/E is very smart: If you draw two lines at the same horizontal level, Pro/E thinks they are horizontally aligned. Two lines the same length will be constrained as so.

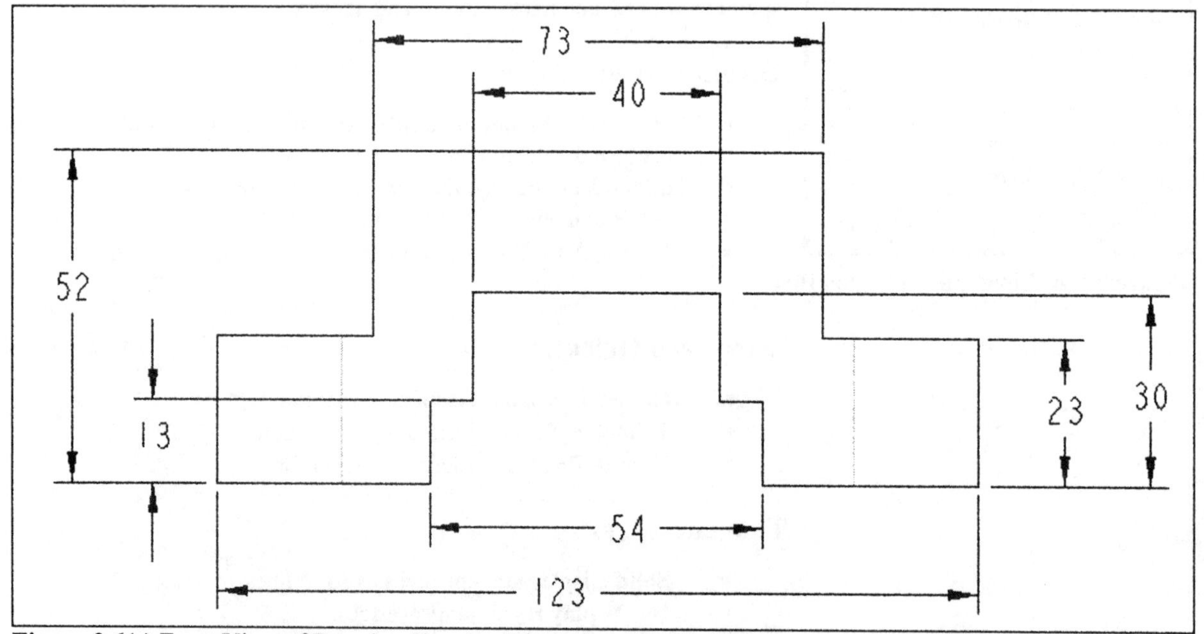

Figure 3.6(e) Front View of Drawing Showing Dimensions for the Clamp

Click: **RMB** [Fig. 3.6(f)] ⇒ **Centerline** [Fig. 3.6(g)] ⇒ pick two positions to create the vertical centerline

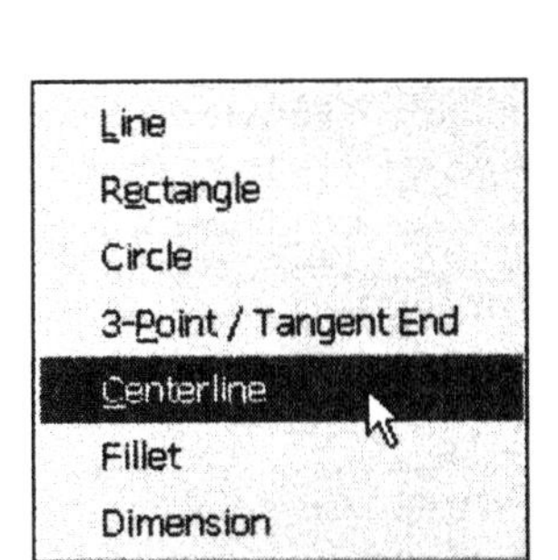

Figure 3.6(f) RMB Options

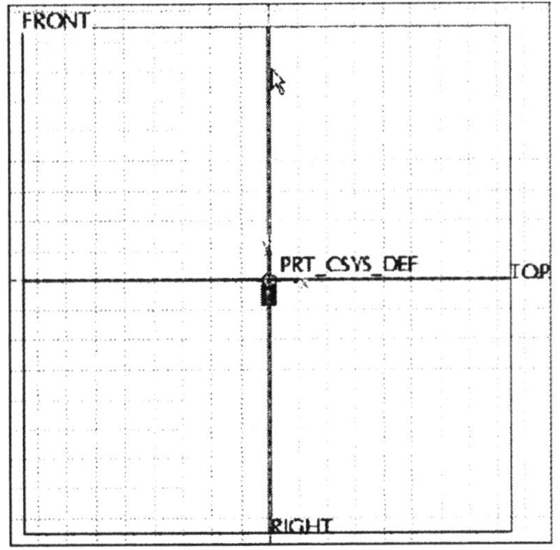

Figure 3.6(g) Create the Centerline

Click: ⇒ **RMB** ⇒ **Line** ⇒ sketch the eight lines of the outline [Fig. 3.6(h)] ⇒ **MMB** to end the line sequence [Fig. 3.6(i)] ⇒ **MMB**

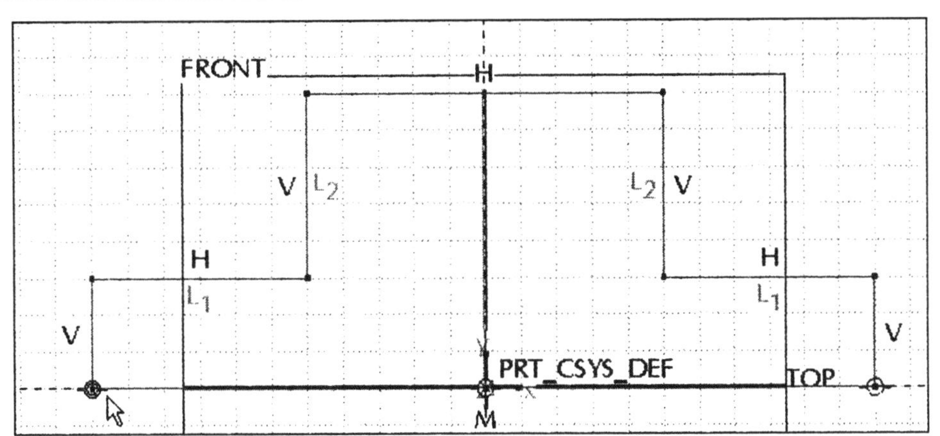

Figure 3.6(h) Sketching the Outline

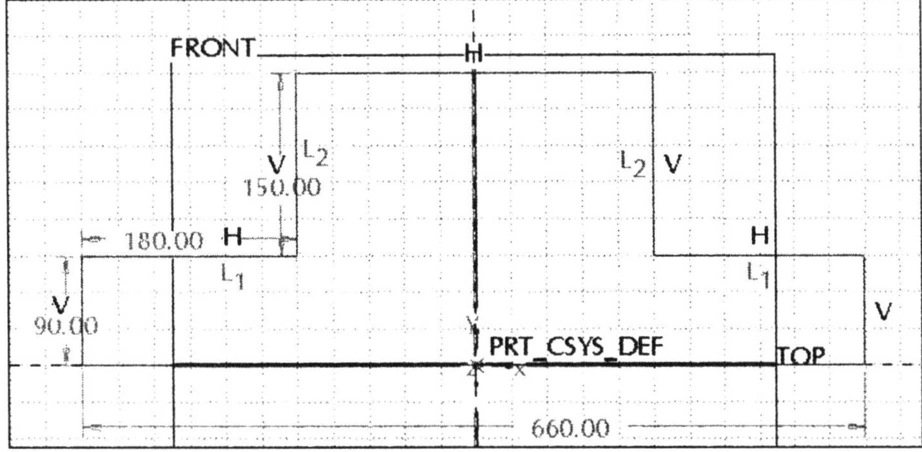

Figure 3.6(i) Default Dimensions Display

Dimensions, Constraints, Grid, and Vertices can be toggled on and off, as needed using the toolbar buttons . A sketcher *constraint symbol* appears next to the entity that is controlled by that constraint. Sketcher constraints can be turned on or off (enabled or disabled) while sketching. An **H** next to a line means horizontal; a **T** means tangent. Dimensions display, as they are needed according to the references selected and the constraints. Seldom are they the same as the required dimensioning scheme needed to manufacture the part. You can add, delete, and move dimensions as required. *The dimensioning scheme is important, **not the dimension value**, which can be modified now or later.*

Place and create the dimensions as required [Fig. 3.6(j)]. Do not be concerned with the perfect positioning of the dimensions, but try to, in general, follow the spacing and positioning standards found in the **ASME Geometric Tolerancing and Dimensioning** standards. This saves you time when you create a drawing of the part. Dimensions placed at this stage of the design process are displayed on the drawing document by simply showing all the dimensions.

To dimension between two lines, simply pick the lines with the left mouse button (LMB) and place the dimension value with the middle mouse button (MMB). To dimension a single line, pick on the line (LMB), and then place the dimension with MMB.

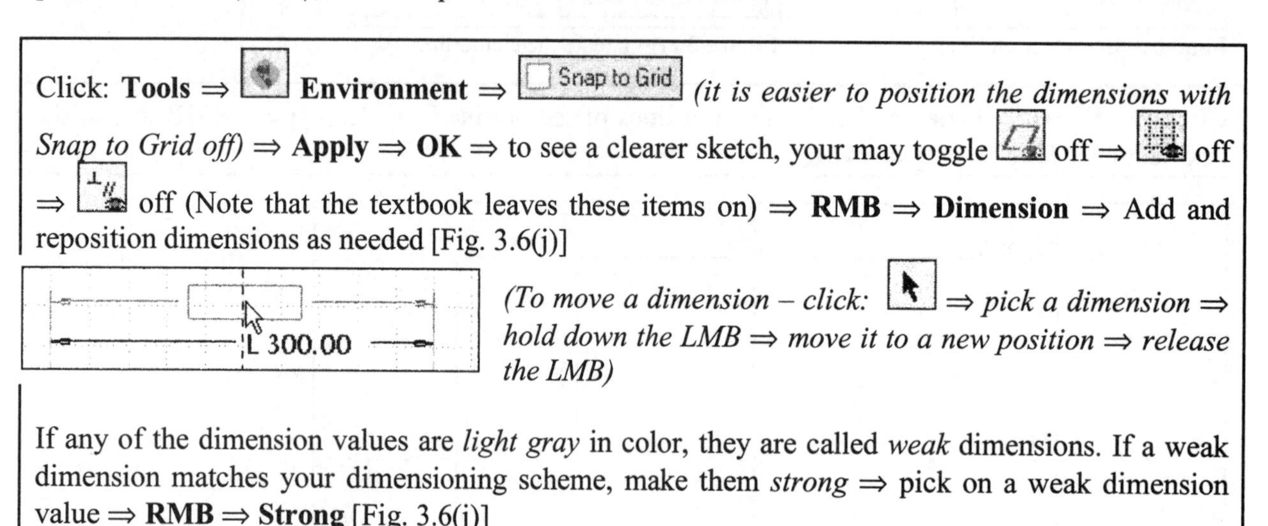

Click: **Tools** ⇒ **Environment** ⇒ *(it is easier to position the dimensions with Snap to Grid off)* ⇒ **Apply** ⇒ **OK** ⇒ to see a clearer sketch, your may toggle off ⇒ off ⇒ off (Note that the textbook leaves these items on) ⇒ **RMB** ⇒ **Dimension** ⇒ Add and reposition dimensions as needed [Fig. 3.6(j)]

(To move a dimension – click: ⇒ pick a dimension ⇒ hold down the LMB ⇒ move it to a new position ⇒ release the LMB)

If any of the dimension values are *light gray* in color, they are called *weak* dimensions. If a weak dimension matches your dimensioning scheme, make them *strong* ⇒ pick on a weak dimension value ⇒ **RMB** ⇒ **Strong** [Fig. 3.6(j)]

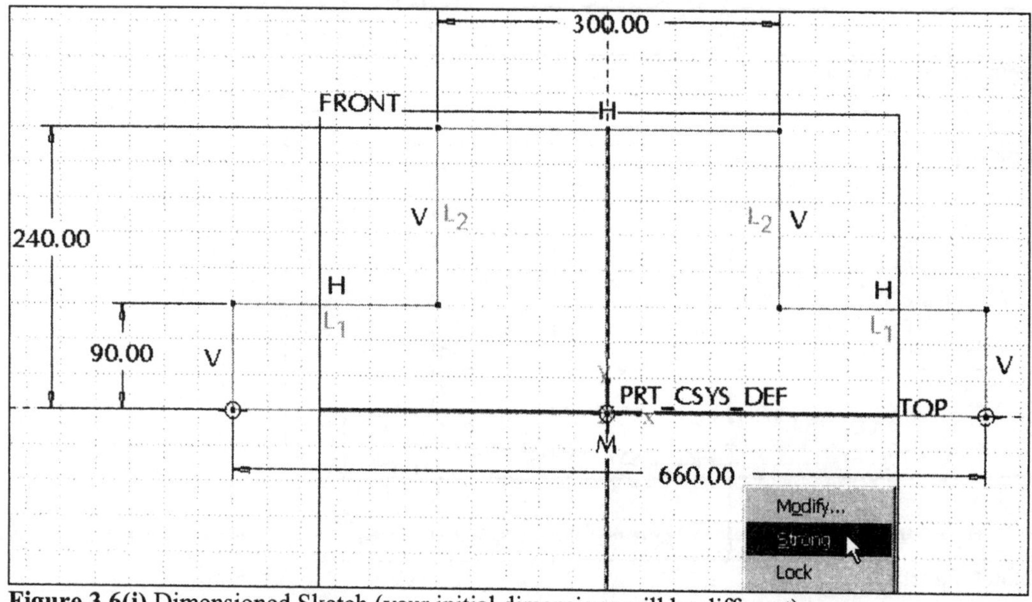

Figure 3.6(j) Dimensioned Sketch (your initial dimensions will be different)

Next, control the sketch by adding symmetry constraints, click: **Impose sketcher constraints on the section** ⇒ **Make two points or vertices symmetric about a centerline** [Fig. 3.6(k)] ⇒ pick the centerline and then pick two vertices to be symmetric [Fig. 3.6(l)] ⇒ repeat the process and make the sketch symmetrical [Fig. 3.6(m)] ⇒ **Close**

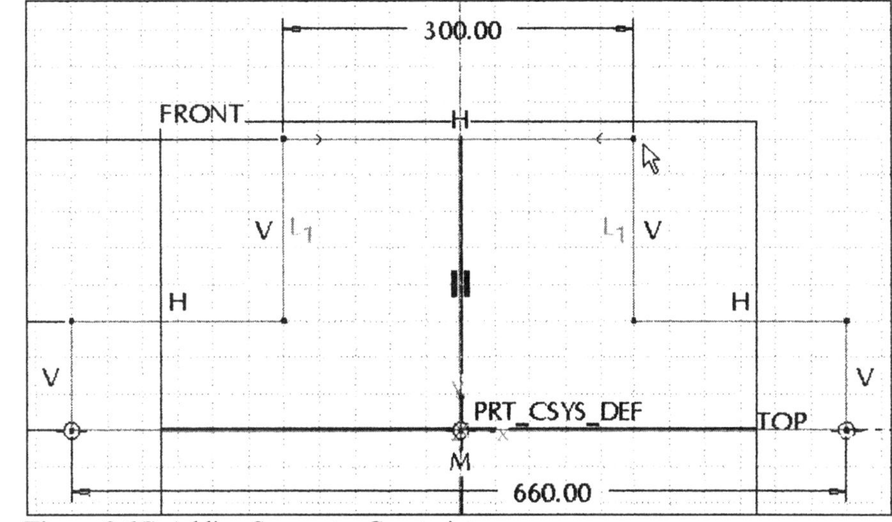

Figure 3.6(k) Constraint Dialog

Figure 3.6(l) Adding Symmetry Constraint

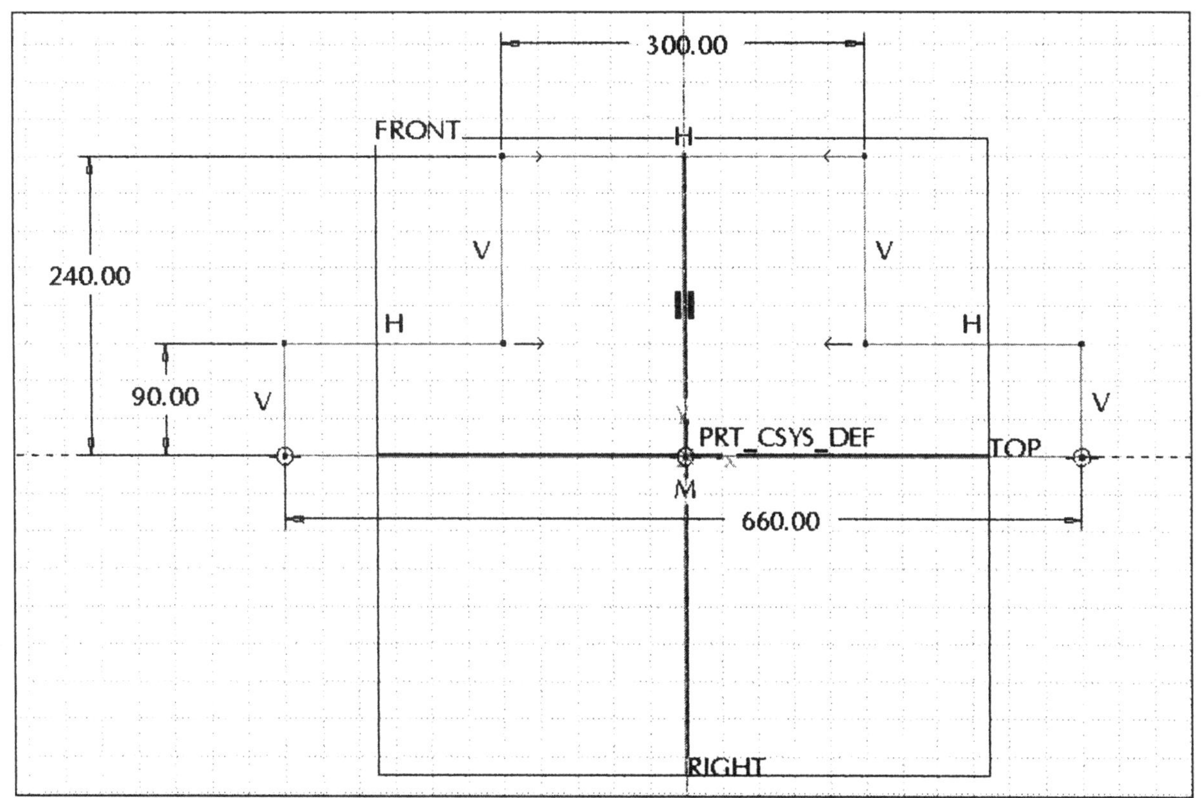

Figure 3.6(m) Sketch is Symmetrical

You can now modify the dimensions to the *design sizes*. Your original sketch values will be different from the example, but the final design values will be the same.

Click: [arrow icon] ⇒ Window-in the sketch (place the cursor at one corner of the window with the **LMB** depressed, drag the cursor to the opposite corner of the window and release the **LMB**) to capture all four dimensions. They will turn red. ⇒ **RMB** ⇒ **Modify** ⇒ [Lock Scale] ⇒ [Regenerate] [Fig. 3.6(n)] ⇒ double-click on length dimension (*here it is 660, but your dimension will be different*) in the Modify Dimensions dialog box and type the design value at the prompt (**123**) [Fig. 3.6(o)] ⇒ **Enter** ⇒ [✓] **Regenerate the section and close the dialog** ⇒ double-click on another dimension and modify the value [Fig. 3.6(p)] ⇒ **Enter** ⇒ continue until all of the values are changed to the design sizes [Fig. 3.6(q)]

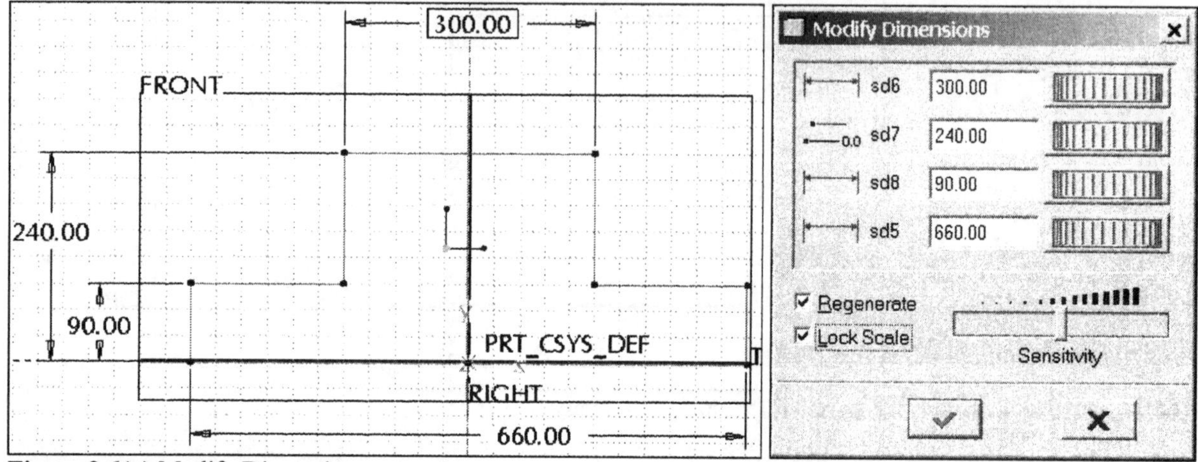

Figure 3.6(n) Modify Dimensions

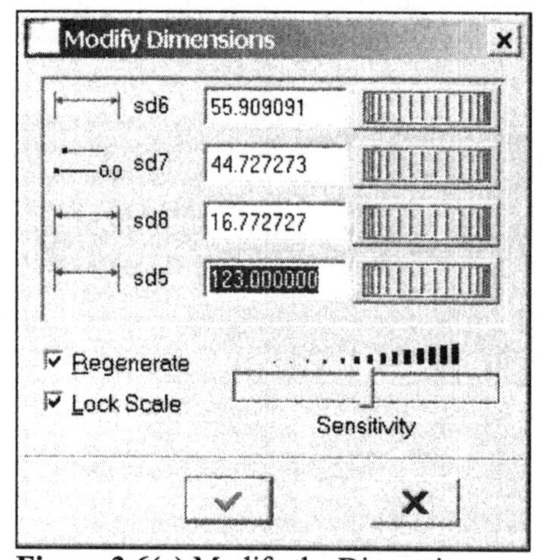

Figure 3.6(o) Modify the Dimension

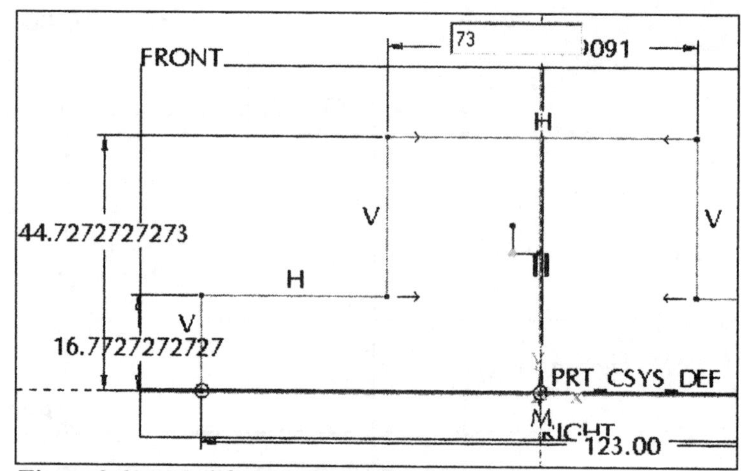

Figure 3.6(p) Modify each Dimension Individually

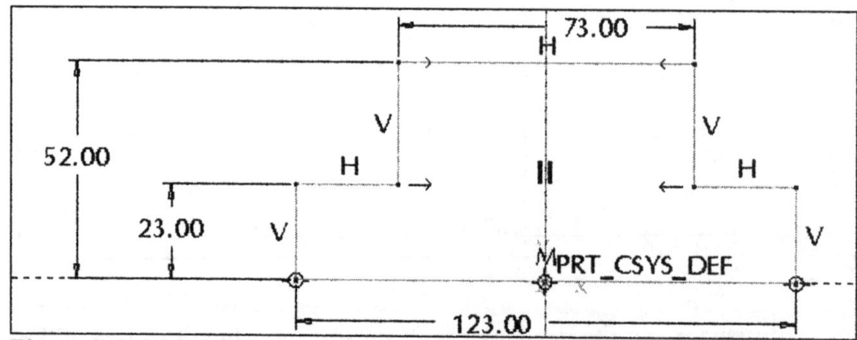

Figure 3.6(q) Modified Sketch showing the Design Values

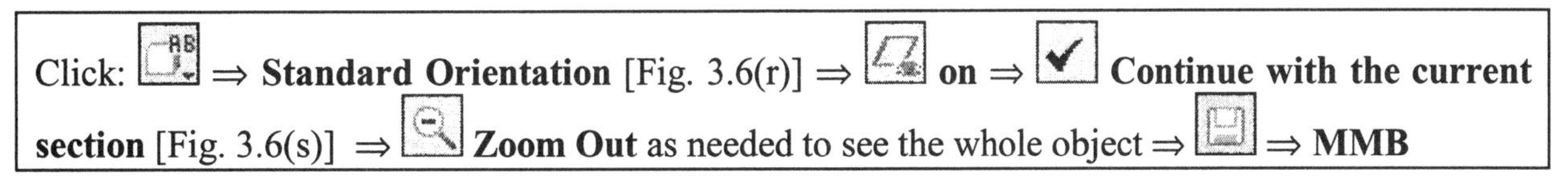

Click: ⇒ **Standard Orientation** [Fig. 3.6(r)] ⇒ **on** ⇒ **Continue with the current section** [Fig. 3.6(s)] ⇒ **Zoom Out** as needed to see the whole object ⇒ ⇒ **MMB**

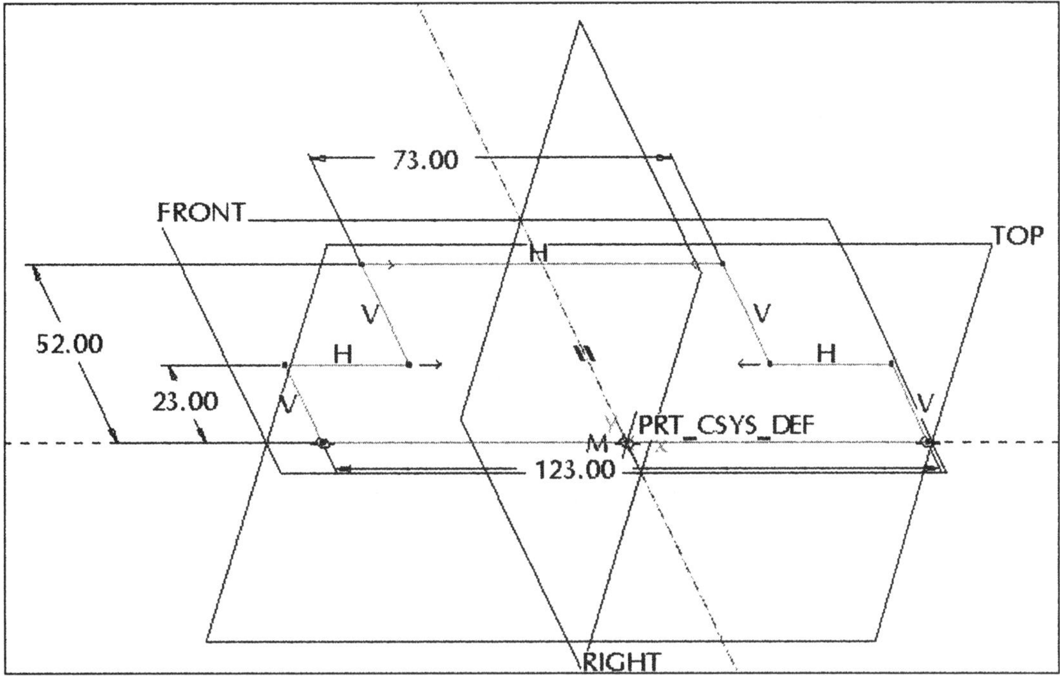

Figure 3.6(r) Regenerated Dimensions

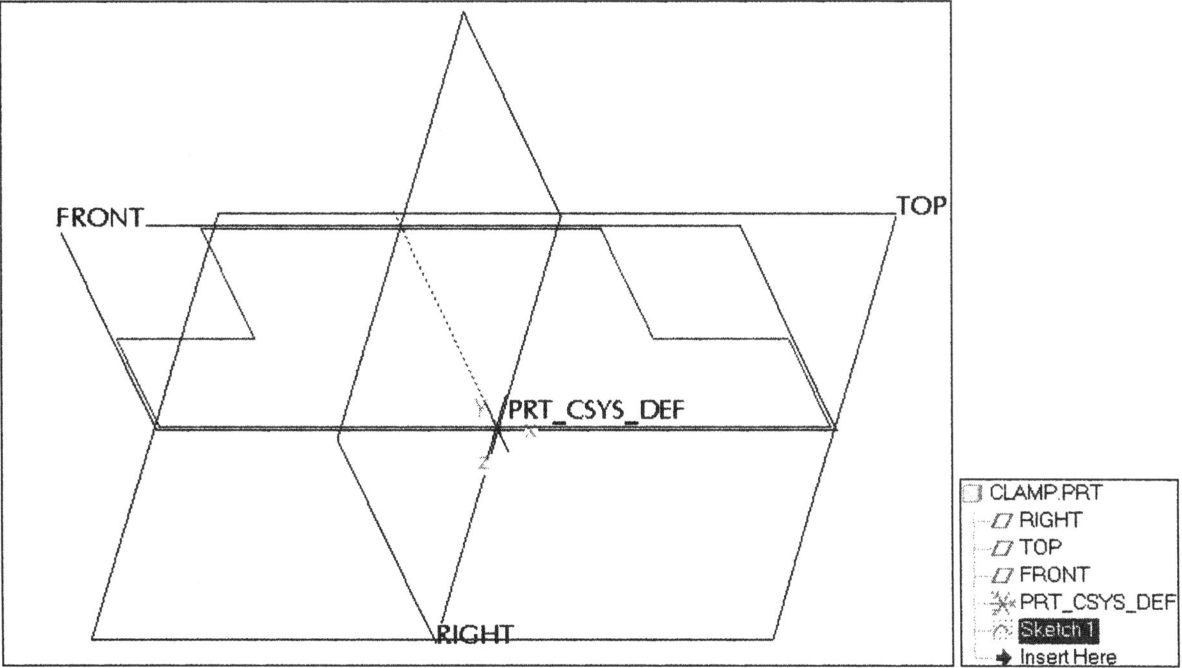

Figure 3.6(s) Completed Sketched Curve (Datum Curve)

The datum curve (Sketch1) will remain red, active and therefore selected.

Click: **Extrude Tool** [Fig. 3.7(a)] ⇒ double-click on the depth value on the model ⇒ type **70** [Fig. 3.7(b)] ⇒ **Enter** ⇒ place your pointer over the square drag handle (it will turn black) ⇒ **RMB** ⇒ **Symmetric** [Fig. 3.7(c)] ⇒ **MMB** [Fig. 3.7(d)] ⇒ ⇒ **MMB**

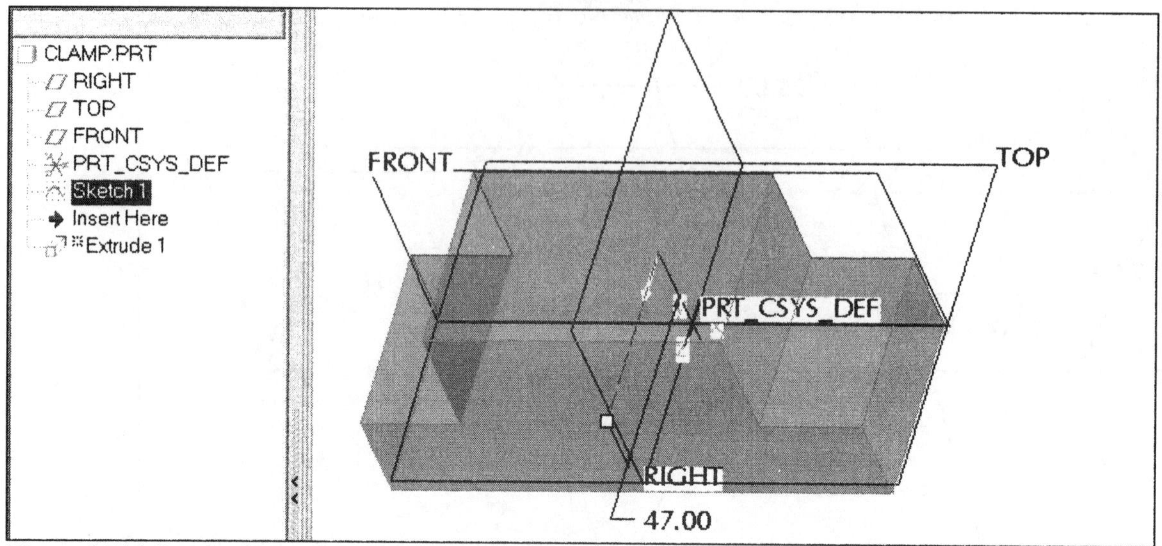

Figure 3.7(a) Depth of Extrusion Previewed

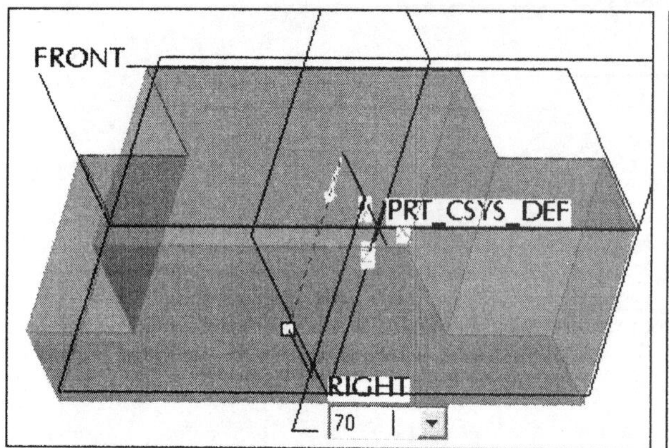

Figure 3.7(b) Modify the Depth Value

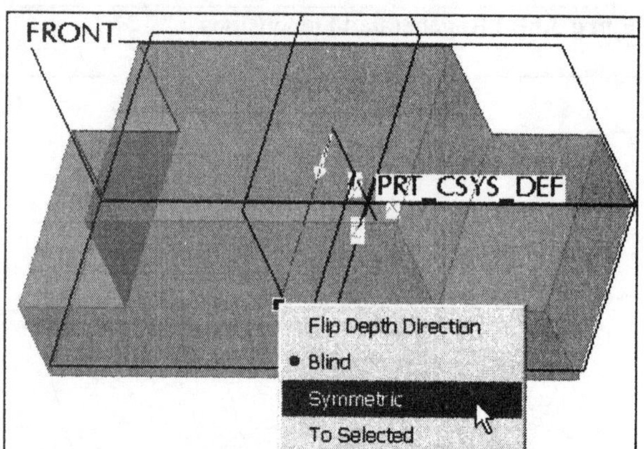

Figure 3.7(c) Symmetric

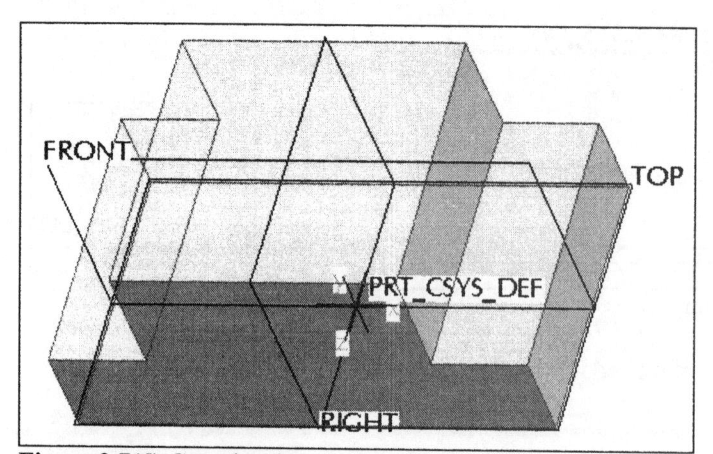

Figure 3.7(d) Completed Extrusion

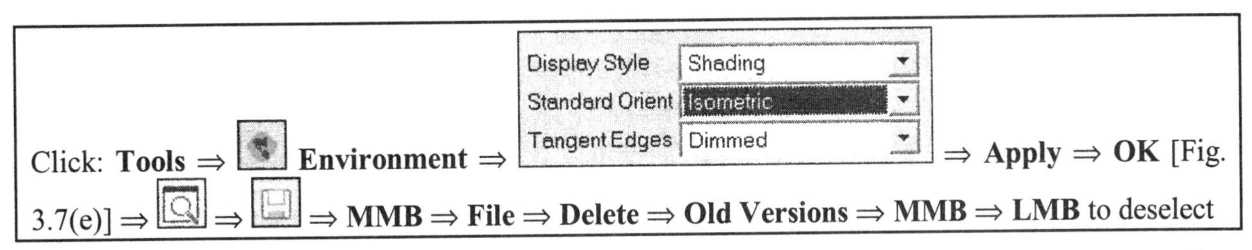

Click: **Tools** ⇒ **Environment** ⇒ ⇒ **Apply** ⇒ **OK** [Fig. 3.7(e)] ⇒ ⇒ ⇒ **MMB** ⇒ **File** ⇒ **Delete** ⇒ **Old Versions** ⇒ **MMB** ⇒ **LMB** to deselect

Storing an object on the disk does not overwrite an existing object file. To preserve earlier versions, Pro/E saves the object to a new file with the same object name but with an updated version number. Every time you store an object using Save, you create a new version of the object in memory, and write the previous version to disk. Pro/E numbers each version of an object storage file consecutively (for example, box.sec.1, box.sec.2, box.sec.3). If you save 25 times, you have 25 versions of the object, all at different stages of completion. You can use *File ⇒ Delete ⇒ Old Versions* after the *Save* command to eliminate previous versions of the object that may have been stored.

When opening an existing object file, you can open any version that is saved. Although Pro/E automatically retrieves the latest saved version of an object, you can retrieve any previous version by entering the full file name with extension and version number (for example, **partname.prt.5**). If you do not know the specific version number, you can enter a number relative to the latest version. For example, to retrieve a part from two versions ago, enter **partname.prt.3** *(or **partname.prt.-2**)*.

You use *File ⇒ Erase* to remove the object and its associated objects from memory. If you close a window before erasing it, the object is still in memory. In this case, you use *File ⇒ Erase ⇒ Not Displayed* to remove the object and its associated objects from memory. This does not delete the object. It just removes it from active memory. *File ⇒ Delete ⇒ All Versions* removes the file from memory and from disk completely. You are prompted with a Delete All Confirm dialog box when choosing this command. Be careful not to delete needed files.

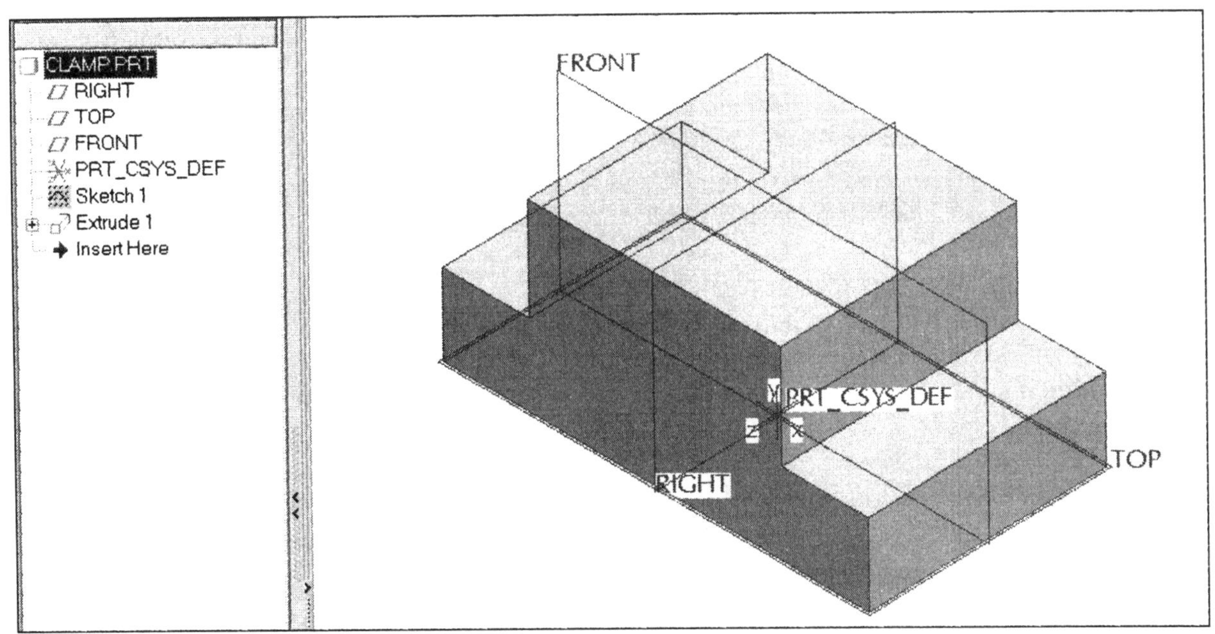

Figure 3.7(e) Isometric Orientation

Next, the cut through the middle of the part will be modeled.

Click: Extrude Tool ⇒ **RMB** ⇒ **Remove Material** ⇒ **RMB** ⇒ **Define Internal Sketch** [Fig. 3.8(a)] ⇒ **Use Previous** from the Sketch dialog box Use Previous [Fig. 3.8(b)] ⇒ **Sketch** ⇒ **RMB** ⇒ **Centerline** [Fig. 3.8(c)] ⇒ create a vertical centerline ⇒ **RMB** ⇒ **Line** ⇒ sketch the seven lines of the open outline [Fig. 3.8(d)] ⇒ **MMB** ⇒ **Impose sketcher constraints on the section** ⇒ [Fig. 3.6(k)] ⇒ pick the centerline and then pick two vertices to be symmetric ⇒ repeat the process and make the sketch symmetrical ⇒ If you attempt to create too many constraints, Pro/E will open the Resolve Sketch dialog box [Fig. 3.8(e)]. **Delete** the extra symmetric constraint if this happens. ⇒ **Close**

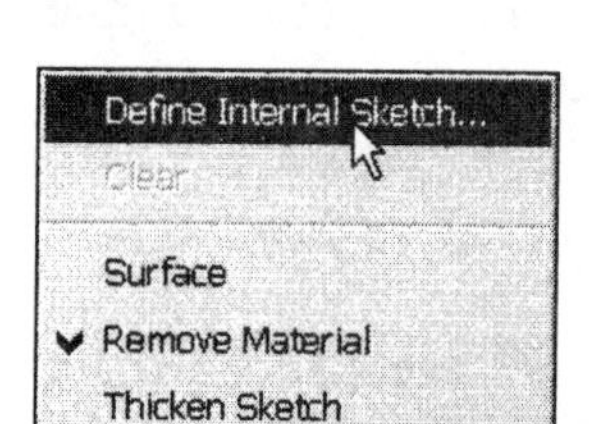

Figure 3.8(a) RMB Options

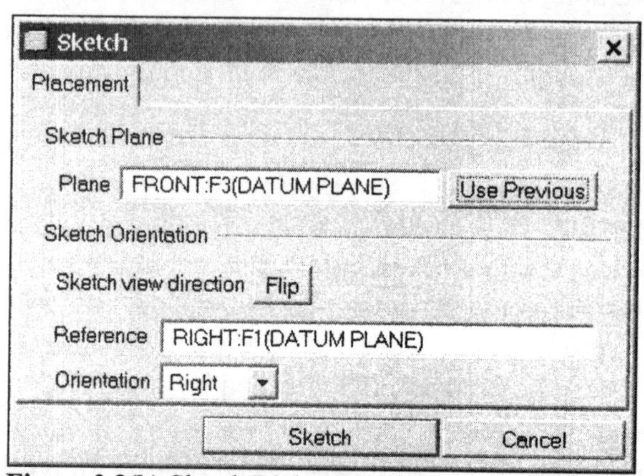

Figure 3.8(b) Sketch Dialog Box

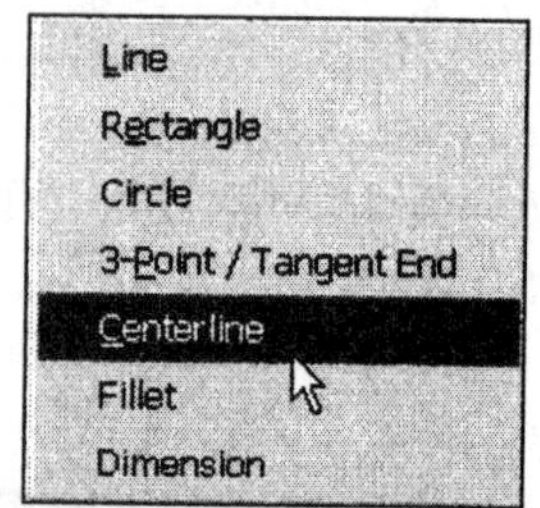

Figure 3.8(c) Centerline

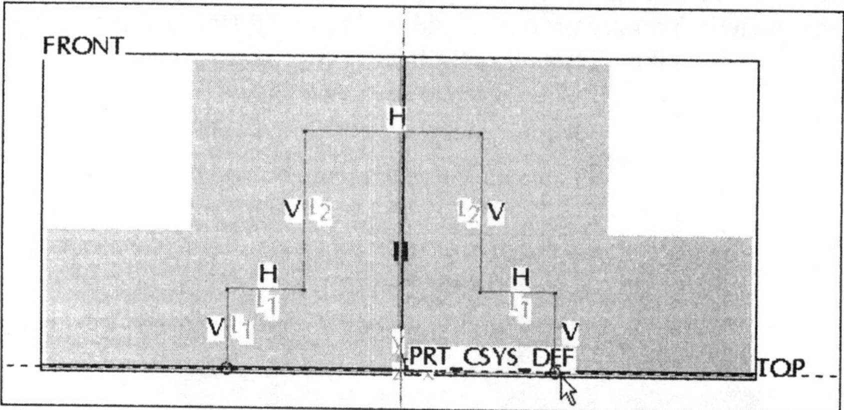

Figure 3.8(d) Sketch

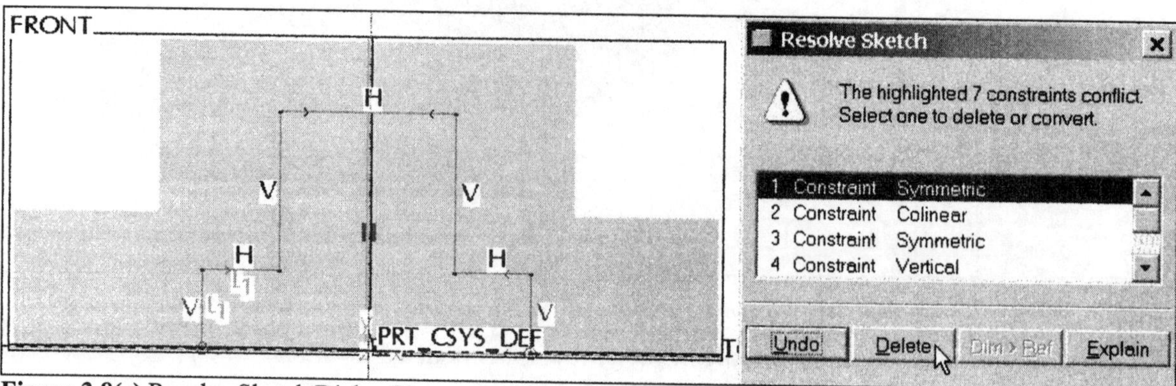

Figure 3.8(e) Resolve Sketch Dialog Box

Click: **RMB** ⇒ **Dimension** ⇒ add and reposition dimensions, your values will be different ⇒ Hidden Line [Fig. 3.8(f)] ⇒ **MMB** to deselect dimension tool and activate **Select items**

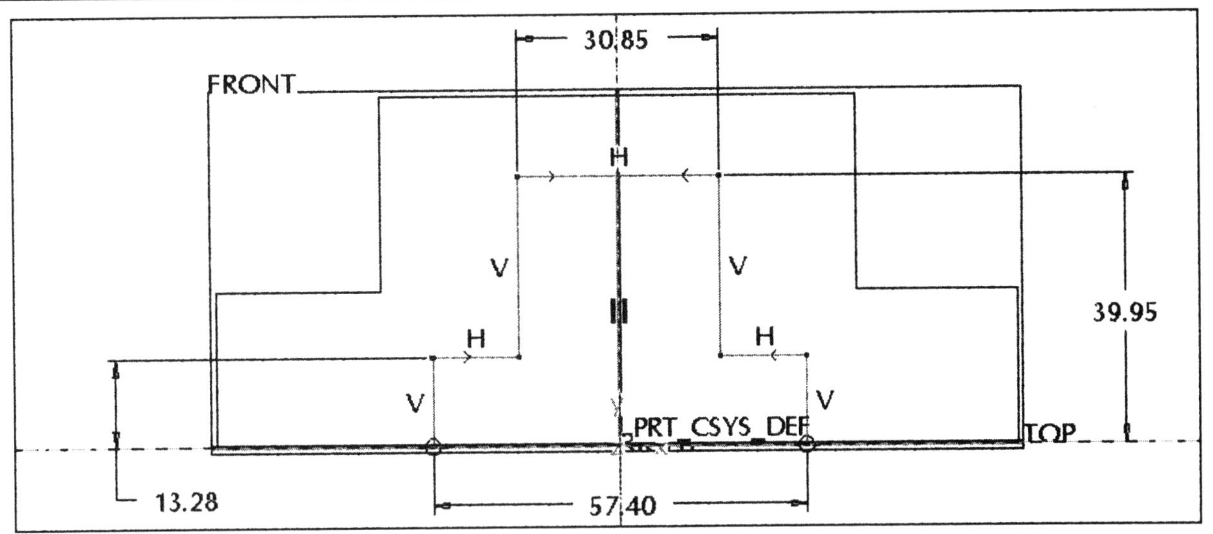

Figure 3.8(f) Dimensioned Sketch

Window-in the sketch to capture all four dimensions. They will turn red. ⇒ **RMB** ⇒ **Modify** ⇒ Regenerate Lock Scale ⇒ modify the values [Fig. 3.8(g)] ⇒ [Fig. 3.8(h)]

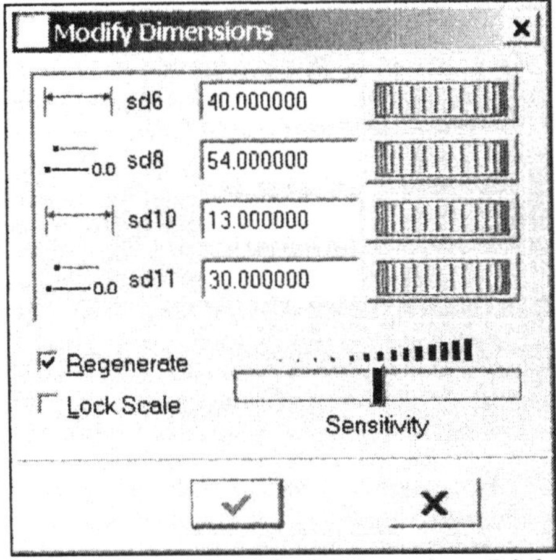

Figure 3.8(g) Modified Dimensions Dialog Box

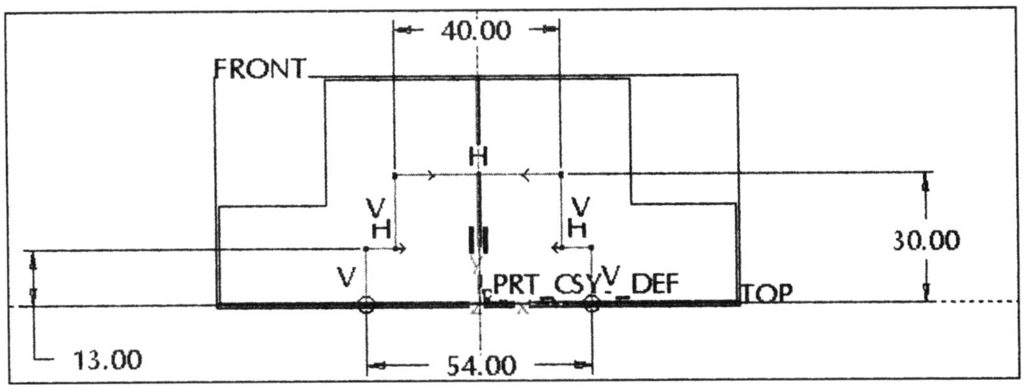

Figure 3.8(h) Modified Sketch Dimensions

Click: **Shading** ⇒ ⇒ **Standard Orientation** ⇒ ⇒ note the yellow direction arrow ⇒ **Zoom Out** ⇒ **Options** from the Dashboard ⇒ **Side 1** ⇒ ⇒ **Through All** [Fig. 3.8(i)] ⇒ **Side 2** ⇒ ⇒ **Through All** [Fig. 3.8(j)] ⇒ **MMB** [Fig. 3.8(k)] ⇒ ⇒ ⇒ **MMB** ⇒ **LMB**

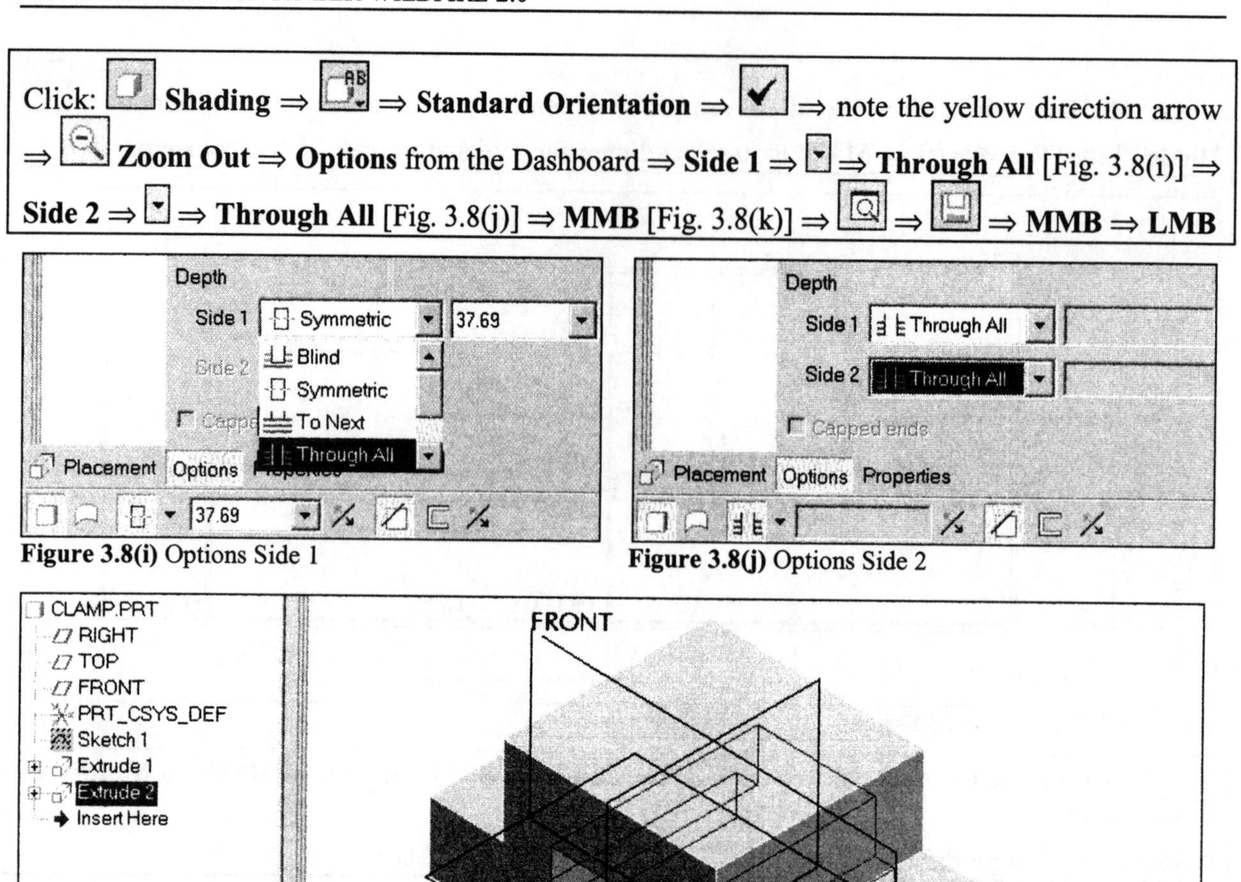

Figure 3.8(i) Options Side 1

Figure 3.8(j) Options Side 2

Figure 3.8(k) Completed Cut

The next feature will be a **20** × **20** centered cut (Fig. 3.9). Because the cut feature is identical on both sides of the part, you can mirror and copy the cut after it has been created.

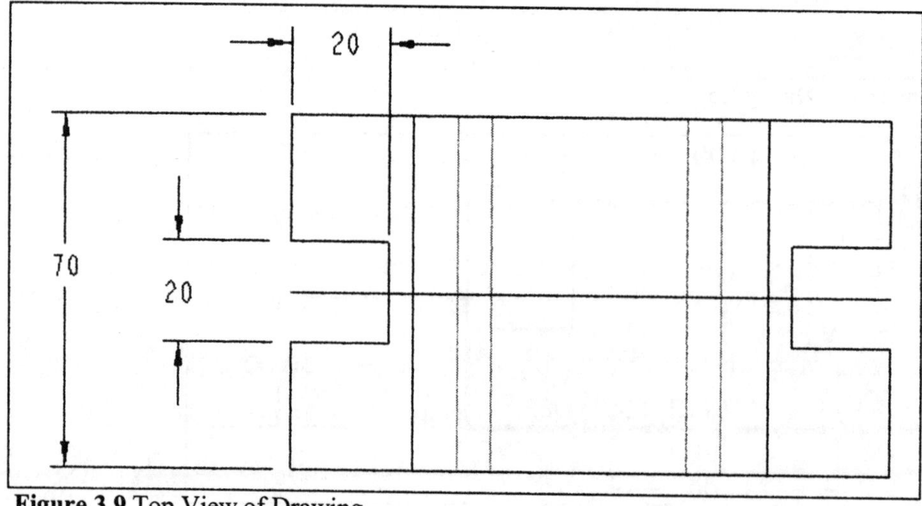

Figure 3.9 Top View of Drawing

Create the cut, click: **Extrude Tool** ⇒ Placement from the Dashboard ⇒ Define... Sketch dialog box opens ⇒ Sketch Plane--- Plane: select **TOP** datum from the model as the sketch plane [Fig. 3.10(a)] ⇒ Sketch ⇒ pick the left edge of the part to add it to the References dialog box [Fig. 3.10(b)] ⇒ click **Close** to accept the References ⇒ **Tools** ⇒ **Environment** ⇒ Snap to Grid ⇒ **OK**

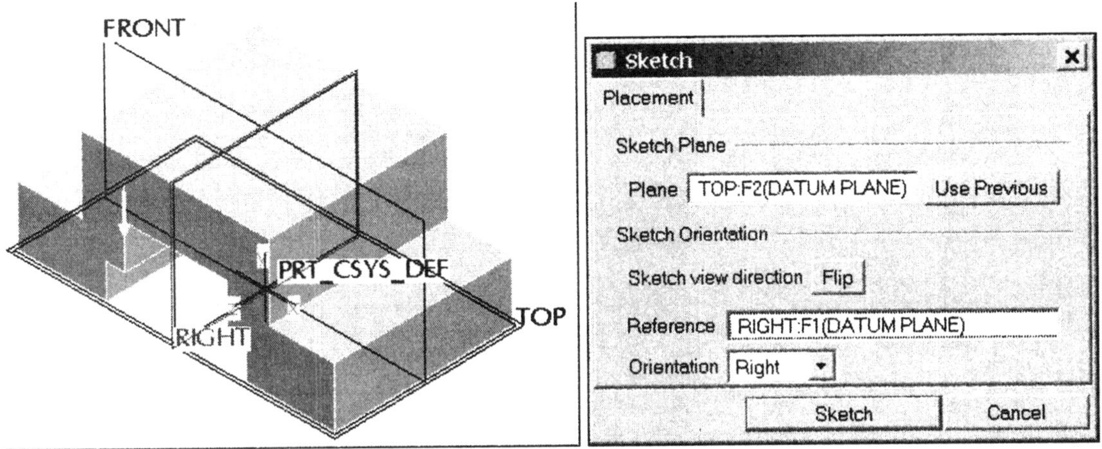

Figure 3.10(a) Sketch Plane Selection and Sketch Dialog Box

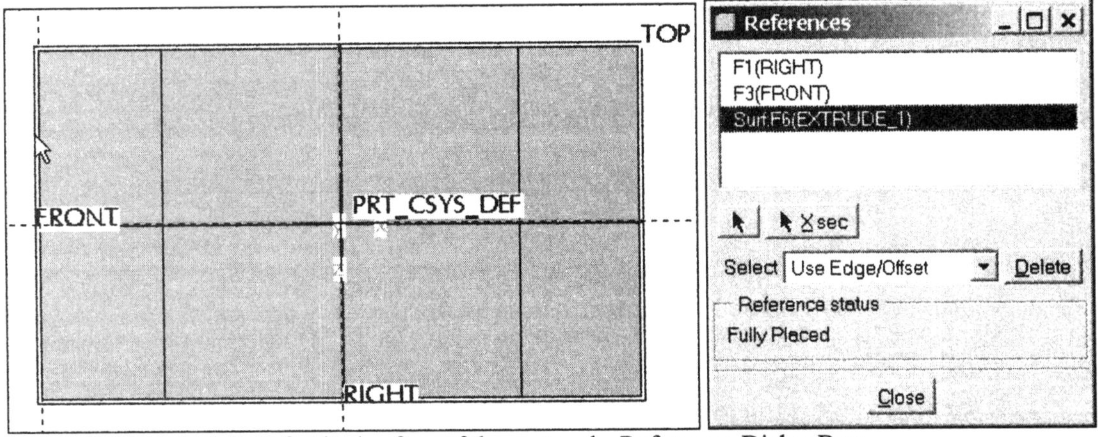

Figure 3.10(b) Add the left edge/surface of the part to the References Dialog Box

Click: Hidden Line ⇒ **RMB** ⇒ **Centerline** ⇒ create a horizontal centerline through the center of the part ⇒ **RMB** ⇒ **Line** ⇒ place the mouse on the left edge and create three lines [Fig. 3.10(c)] ⇒ **MMB** to end the line sequence [Fig. 3.10(d)] ⇒ **Impose sketcher constraints on the section** ⇒ ⇒ pick the centerline and then pick two vertices to be symmetric add the required dimensions ⇒ **MMB** ⇒ **MMB** ⇒ move and modify the values for the two dimensions **(20 X 20)** [Fig. 3.10(e)] ⇒ ⇒ ⇒ **Standard Orientation** ⇒ from the dashboard **Remove Material** ⇒ **Options** tab ⇒ ⇒ **Through All (Extrude in first direction to interest with all surfaces)** [Fig. 3.10(f)] ⇒ ⇒ ⇒ **Shading** ⇒ rotate your model using **MMB** to see the cut clearly [Fig. 3.10(g)] ⇒ ⇒ **MMB**

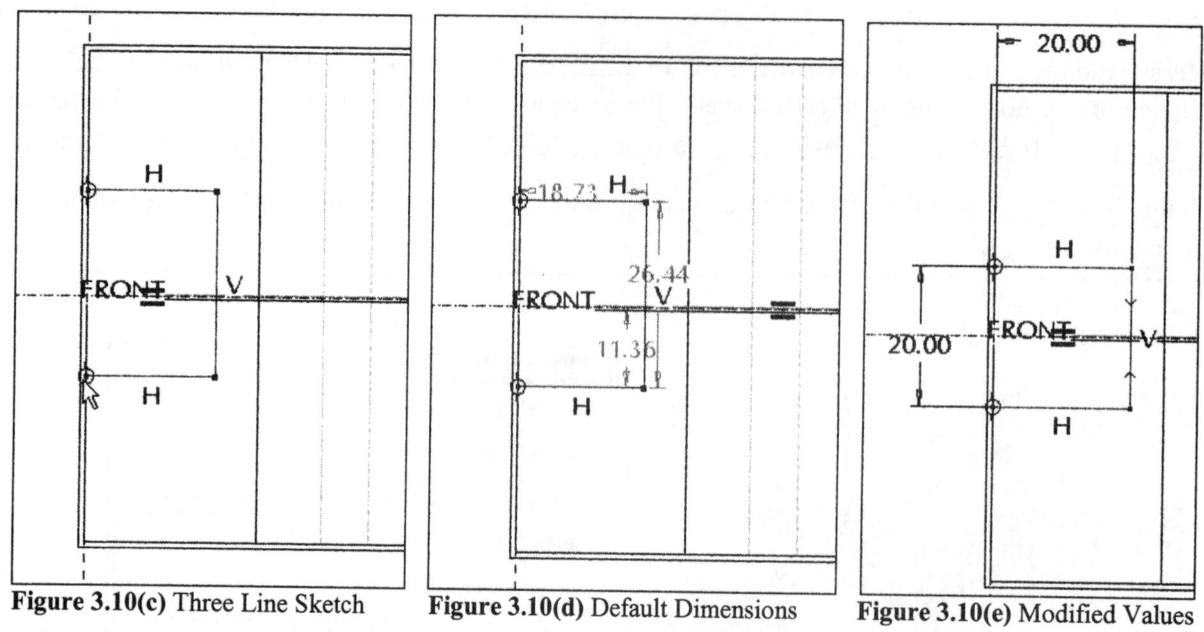

Figure 3.10(c) Three Line Sketch **Figure 3.10(d)** Default Dimensions **Figure 3.10(e)** Modified Values

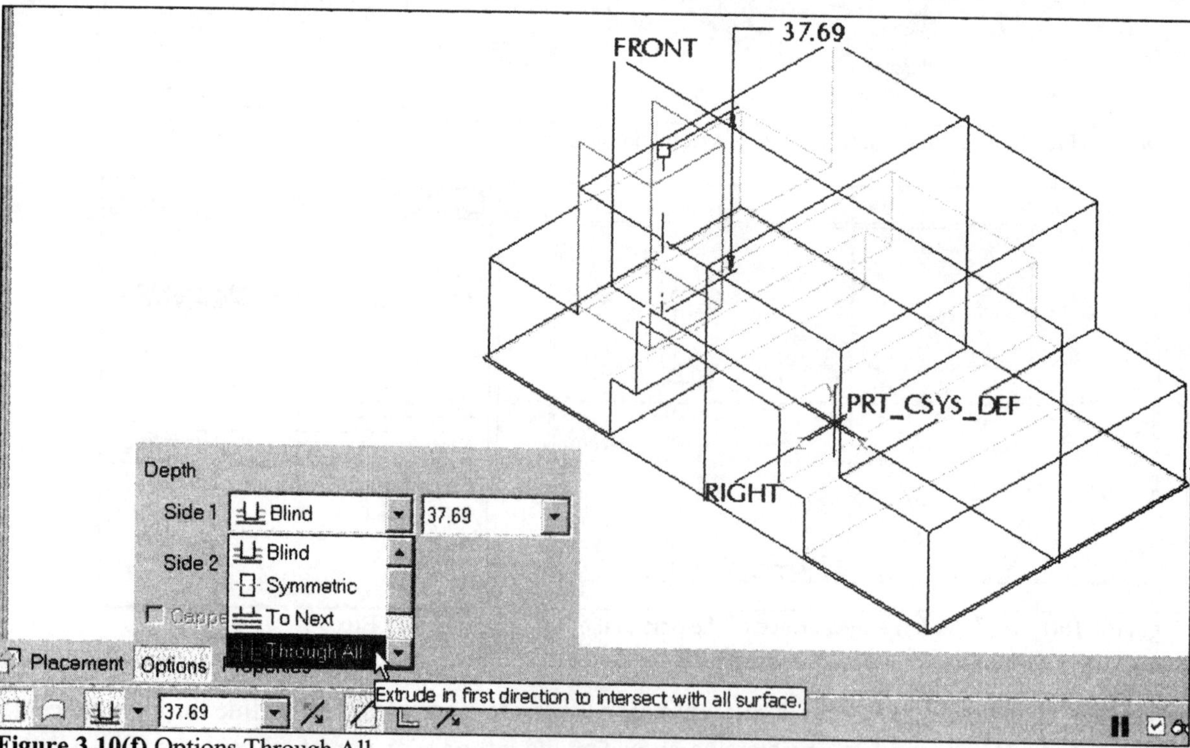

Figure 3.10(f) Options Through All

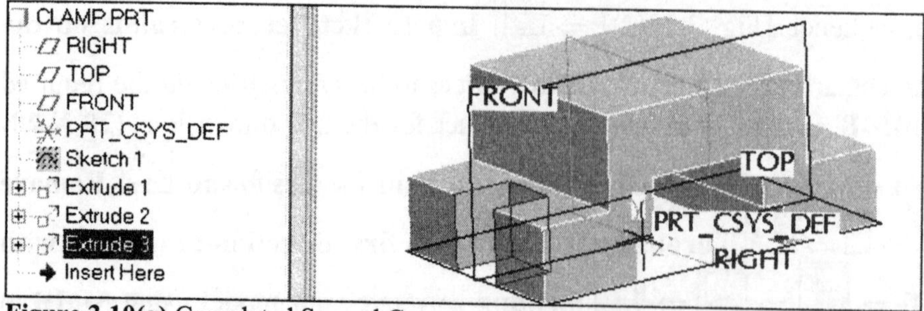

Figure 3.10(g) Completed Second Cut

Click: ⇒ **Standard Orientation** ⇒ with the new cut still highlighted [Fig. 3.11(a)], click: **Mirror Tool** ⇒ select the **RIGHT** datum plane from the Model Tree [Fig. 3.11(b)] ⇒ **MMB** [Fig. 3.11(c)] ⇒ **File** ⇒ **Save** ⇒ **MMB** ⇒ **MMB** rotate part [Fig. 3.11(d)] ⇒ **File** ⇒ **Close Window**

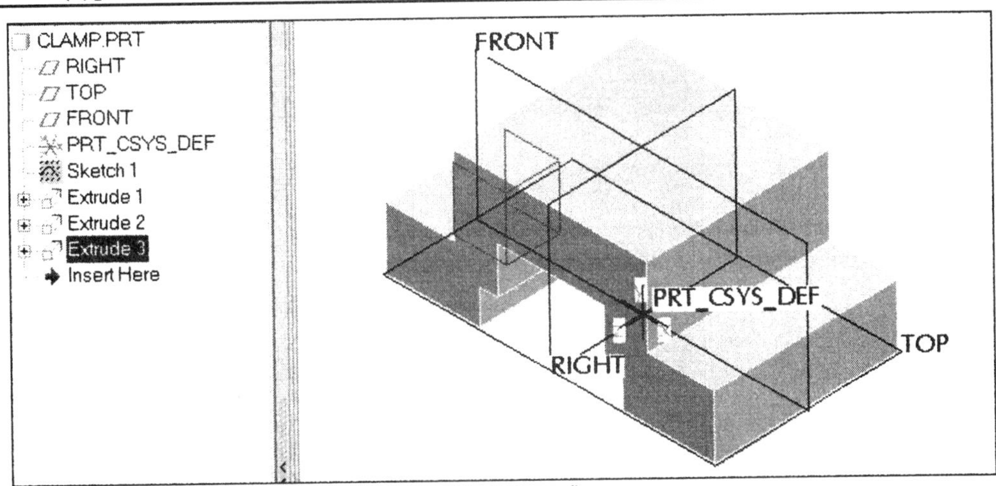

Figure 3.11(a) Extruded Cut is Highlighted (Selected)

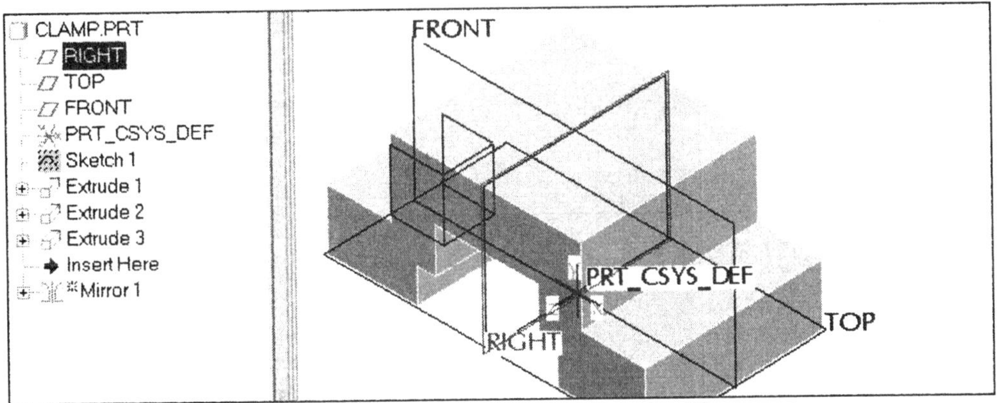

Figure 3.11(b) Select the RIGHT Datum Plane as the Mirroring Plane

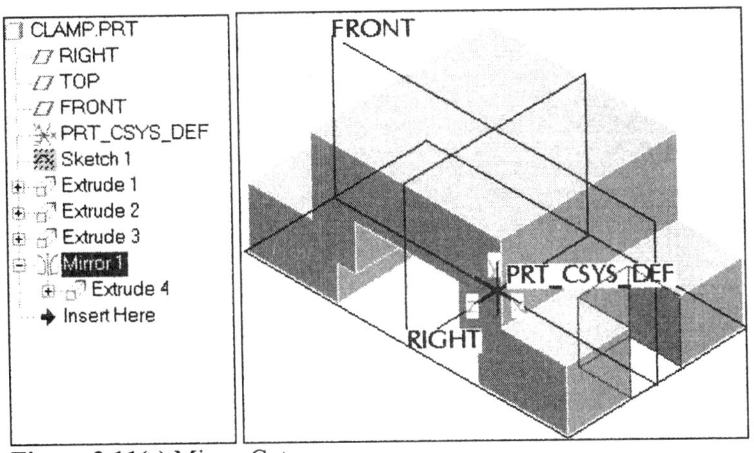

Figure 3.11(c) Mirror Cut

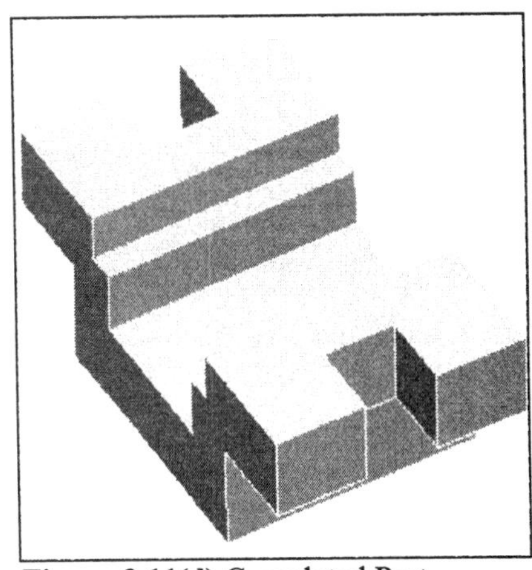

Figure 3.11(d) Completed Part

Lesson 3 is now complete. If you wish to model a project without instructions, a complete set of projects and illustrations are available at ***www.cad-resources.com*** ⇒ ***Downloads***.

NOTES:

Lesson 4 Datums, Layers, and Sections

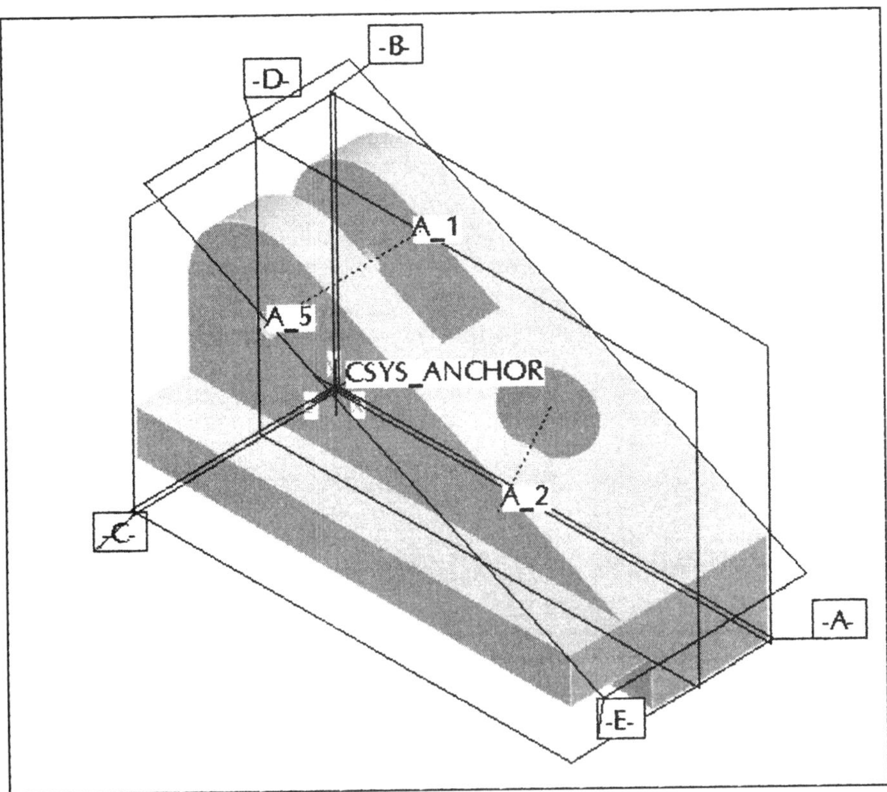

Figure 4.1(a) Anchor Model with Datum Features

OBJECTIVES

- Create **Datums** to locate features
- Set datum planes for **geometric tolerancing**
- Learn how to change the **Color** and **Shading** of models
- Use **Layers** to organize features
- Use datum planes to establish **Model Sectioning**
- Add a simple **Relation** to control a feature
- Use **Info** command to extract **Relations** information

DATUMS AND LAYERS

Datums and **layers** are two of the most useful mechanisms for creating and organizing your design [Fig. 4.1(a)]. Features such as *datum planes* and *datum axes* are essential for the creation of all parts, assemblies, and drawings using Pro/E.

Layers are an essential tool for grouping items and performing operations on them, such as selecting, hiding or unhiding, plotting, and suppressing. Any number of layers can be created. User-defined names are available, so layer names can be easily recognized.

Most companies have a layering scheme that serves as a *default standard* so that all projects follow the same naming conventions and objects/items are easily located by anyone with access. Layer information, such as display status, is stored with each individual part, assembly, or drawing.

Lesson 4 STEPS

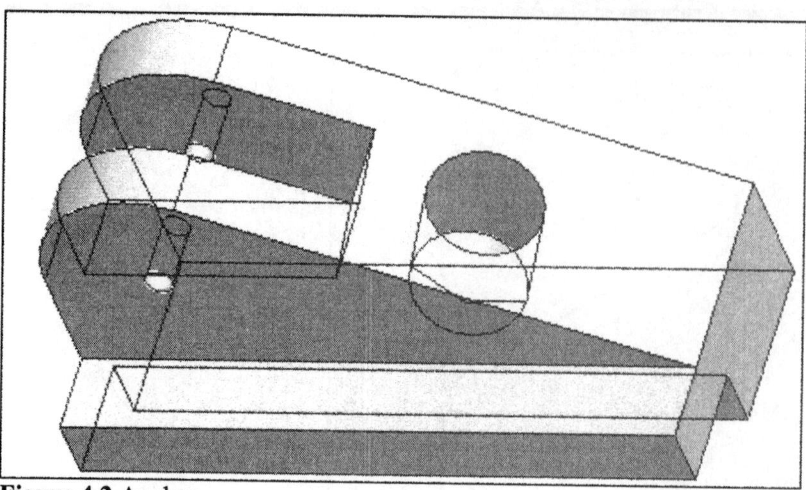

Figure 4.2 Anchor

Anchor

Though default datum planes have been sufficient in previous lessons, the Anchor (Fig. 4.2) incorporates the creation of user-defined datums and the assignment of datums to layers. The datum planes will be set as geometric tolerance features and put on a separate layer.

Launch **Pro/ENGINEER WILDFIRE 2.0** ⇒ Click: **File** ⇒ **Set Working Directory** ⇒ **OK** ⇒ **Create a new object** ⇒ ●**Part** ⇒ Name **Anchor** ⇒ Use default template ⇒ **OK** ⇒ **Edit** ⇒ **Setup** ⇒ **Units** ⇒ Units Manager **Inch lbm Second** ⇒ **Close** ⇒ **Material** ⇒ **Define** *(or if you have previously saved a material file you may simply **Assign** it **From File**)* ⇒ type **Steel** ⇒ **Enter** ⇒ **File** ⇒ **Save** ⇒ **File** ⇒ **Exit** ⇒ **Assign** ⇒ pick **STEEL** ⇒ **Accept** ⇒ **Done** ⇒ click on the default coordinate system name on the model-- **PRT_CSYS_DEF** ⇒ **RMB** [Fig. 4.3(a)] ⇒ **Rename** ⇒ type **CSYS_ANCHOR** in the **Model Tree** [Fig. 4.3(b)] ⇒ **Enter**

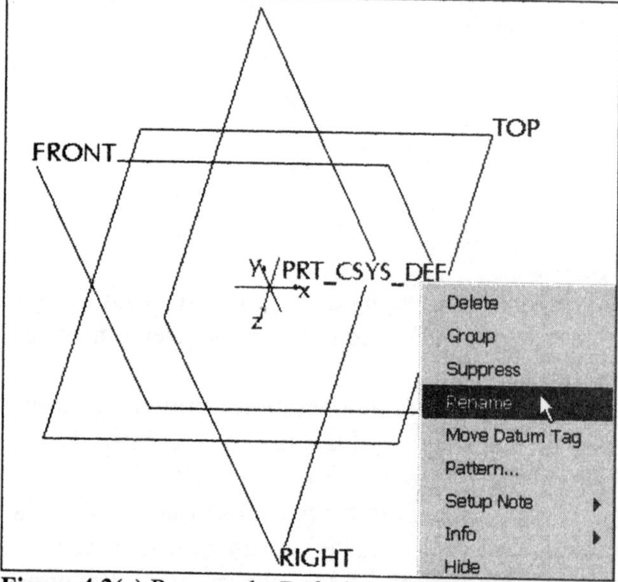

Figure 4.3(a) Rename the Default Coordinate System

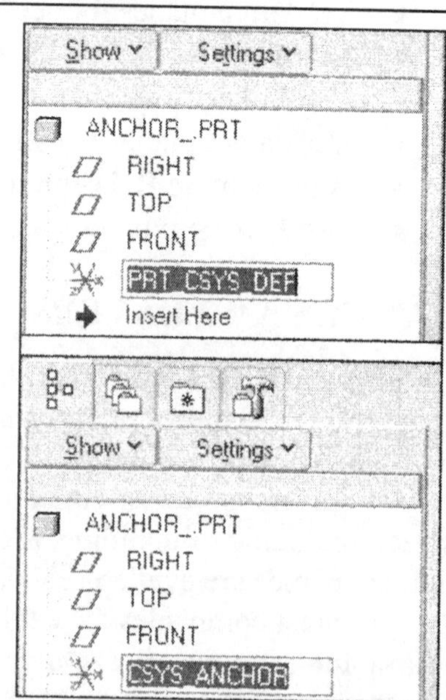

Figure 4.3(b) Rename the CSYS

It is considered good practice to rename the default coordinate system to something similar to the components name. If an assembly has 25 components, it will have 25 default coordinate systems. Renaming the coordinate system to the component name will make it easier to identify.

In the next set of steps, you will show the Layer Tree and see the default layering that automatically takes place as you model. The default datum planes and the default coordinate system are automatically placed on two layers each. For the datum planes; the *part default datum plane layer* and the *part all datum plane layer.*

The coordinate system will also be layered in a similar fashion. In Figure 4.3(d), the *part all datum layer* for the coordinate system displays the new name created for the coordinate system.

Click: **Show** in the **Navigator** ⇒ **Layer Tree** Layer Tree displays in place of the Model Tree [Figs. 4.3(c-d)] ⇒ **Save** ⇒ **MMB** ⇒ **Show** in the **Navigator** ⇒ **Model Tree**

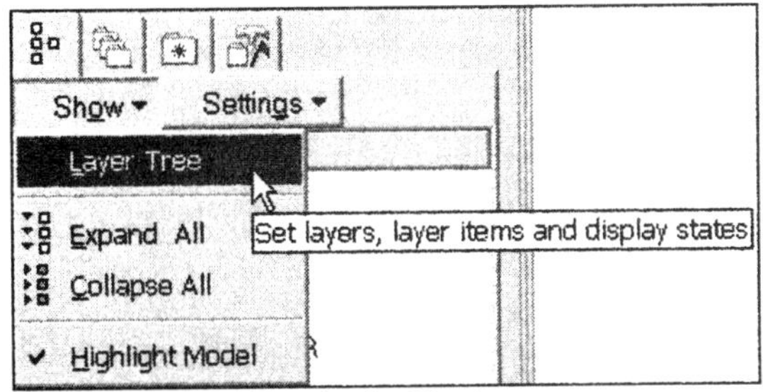

Figure 4.3(c) Show in the Navigator

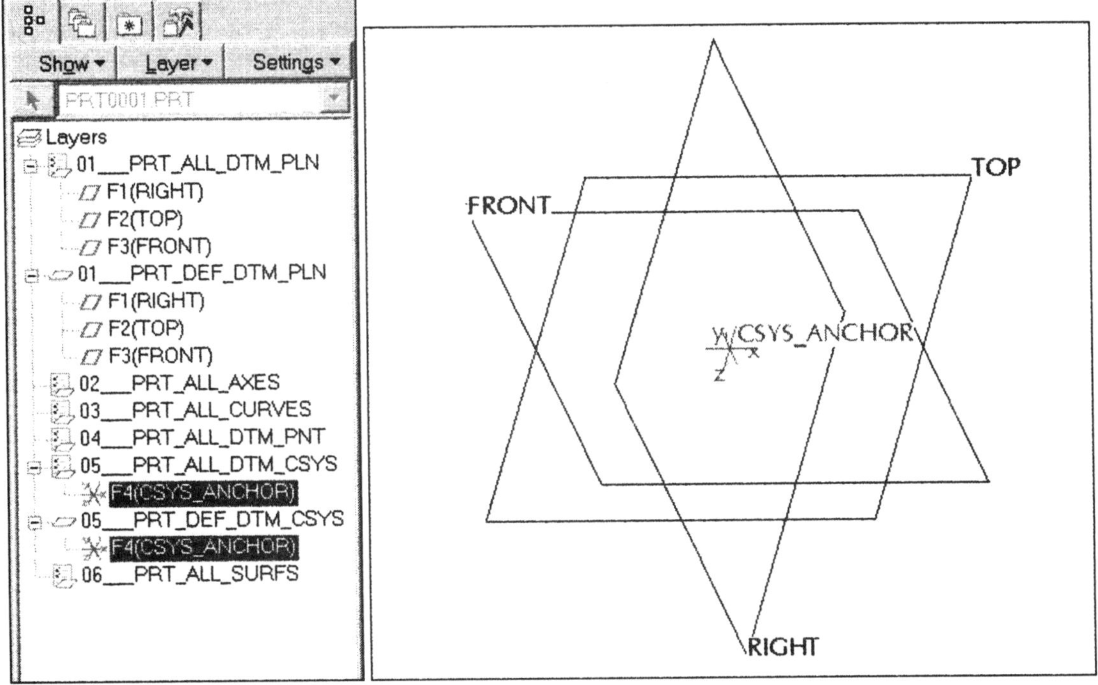

Figure 4.3(d) Layers Displayed in Layer Tree

As in previous lessons, the first protrusion will be sketched on **FRONT**. Use Figure 4.4 for the protrusion dimensions. Use only the **4.50, R1.00, 1.125**, and **25°** dimensions.

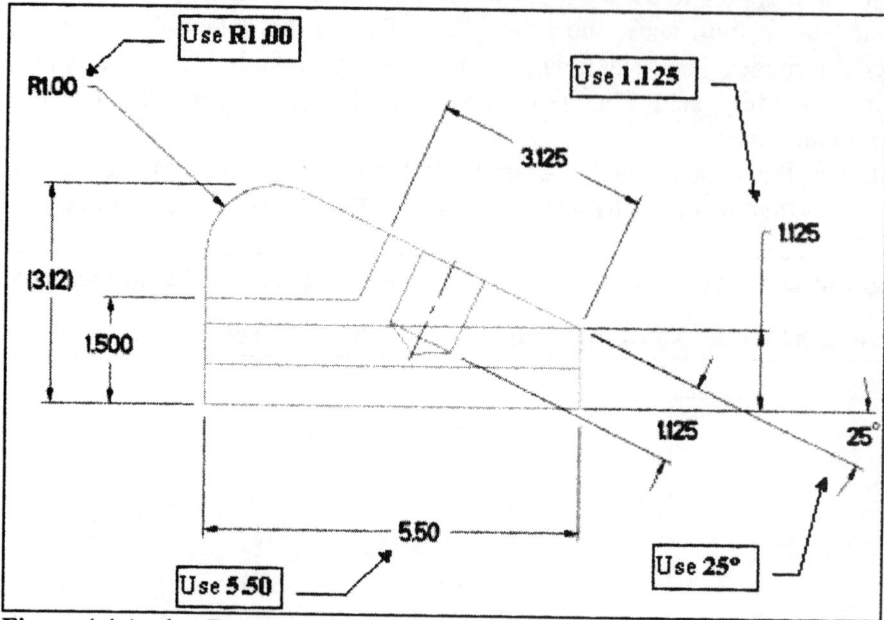

Figure 4.4 Anchor Drawing Front View

Create the first protrusion, click: ⇒ **Extrude Tool** ⇒ **Placement** ⇒ **Define** [Fig. 4.5(a)] ⇒ Sketch Plane--- Plane: select **FRONT** datum ⇒ **Sketch** from Sketch dialog box ⇒ **Tools** ⇒ **Environment** ⇒ Snap to Grid ⇒ **OK** ⇒ **Toggle the grid on** ⇒ **RMB** anywhere in the graphics window ⇒ **Line** [Fig. 4.5(b)]

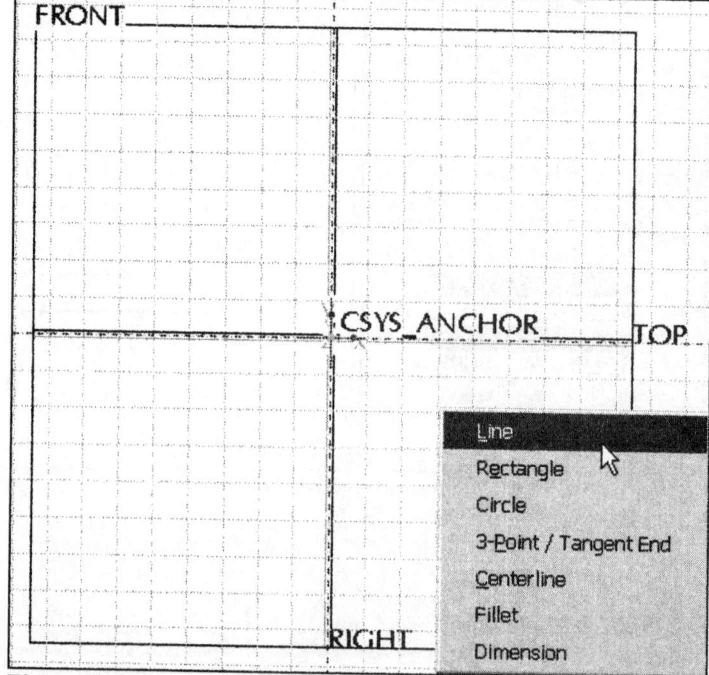

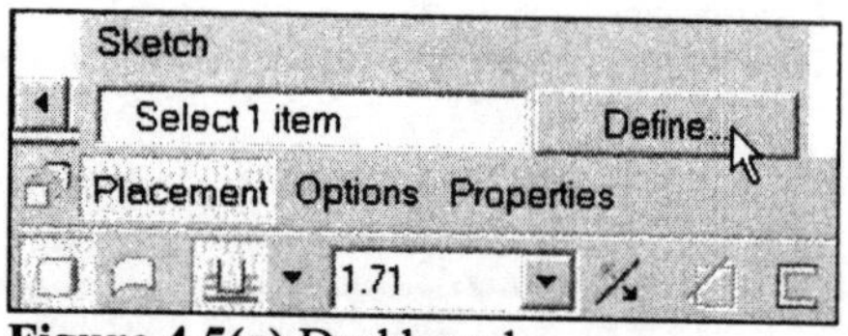

Sketch
Select 1 item
Define...
Placement Options Properties
1.71

Figure 4.5(a) Dashboard

Figure 4.5(b) Selecting the Line Tool with the Pop-up Menu

Sketch the four lines ⇒ **MMB** [Fig. 4.5(c)] ⇒ **Create a circular fillet between two entities** ⇒ pick the vertical left line and the angled line to create a fillet [Fig. 4.5(d)] ⇒ ⇒ pick on the horizontal line and then the angled line and use **MMB** to position the dimension value [Fig. 4.5(e)] ⇒ **Tools** from menu bar ⇒ **Environment** ⇒ Snap to Grid ⇒ **OK** ⇒ **Toggle the grid off**

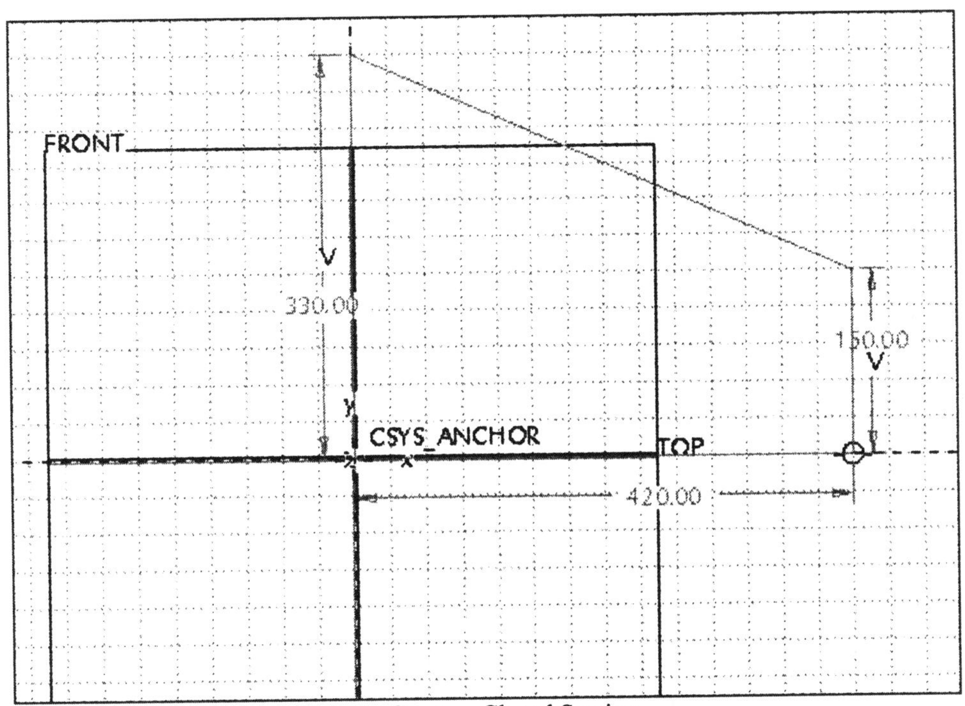

Figure 4.5(c) Sketch Four Lines to Create a Closed Section

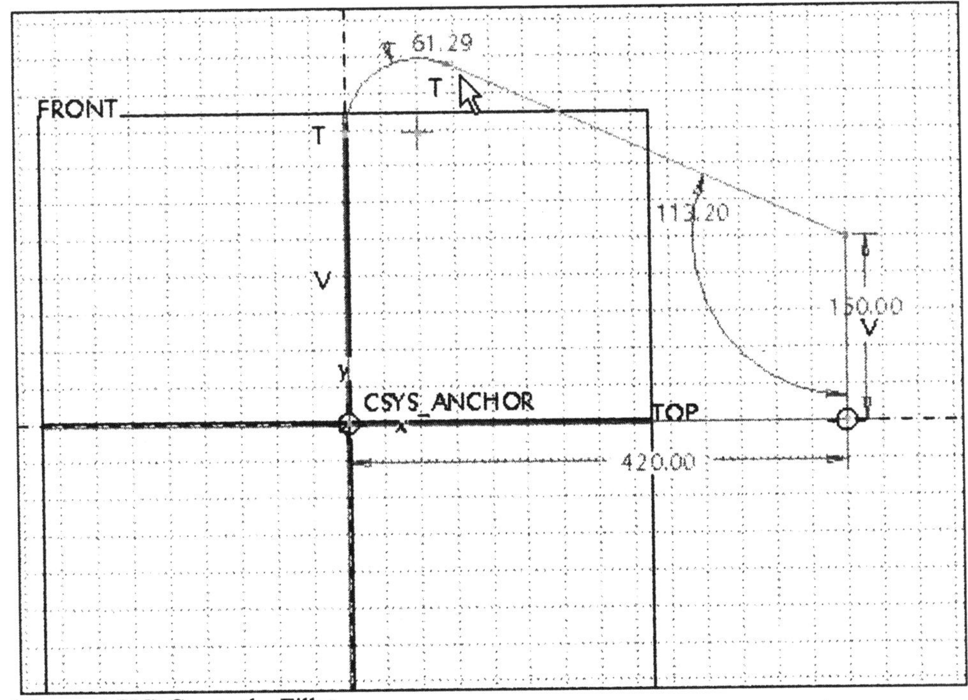

Figure 4.5(d) Create the Fillet

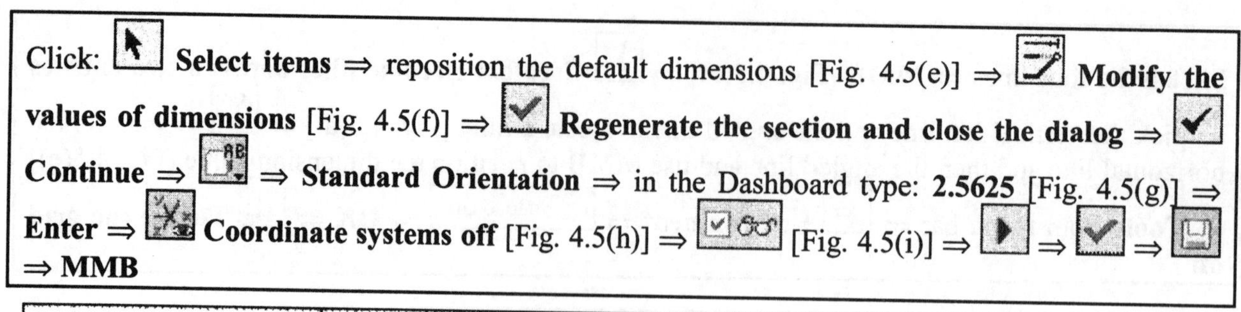

Click: **Select items** ⇒ reposition the default dimensions [Fig. 4.5(e)] ⇒ **Modify the values of dimensions** [Fig. 4.5(f)] ⇒ **Regenerate the section and close the dialog** ⇒ **Continue** ⇒ ⇒ **Standard Orientation** ⇒ in the Dashboard type: **2.5625** [Fig. 4.5(g)] ⇒ **Enter** ⇒ **Coordinate systems off** [Fig. 4.5(h)] ⇒ [Fig. 4.5(i)] ⇒ ⇒ ⇒ ⇒ **MMB**

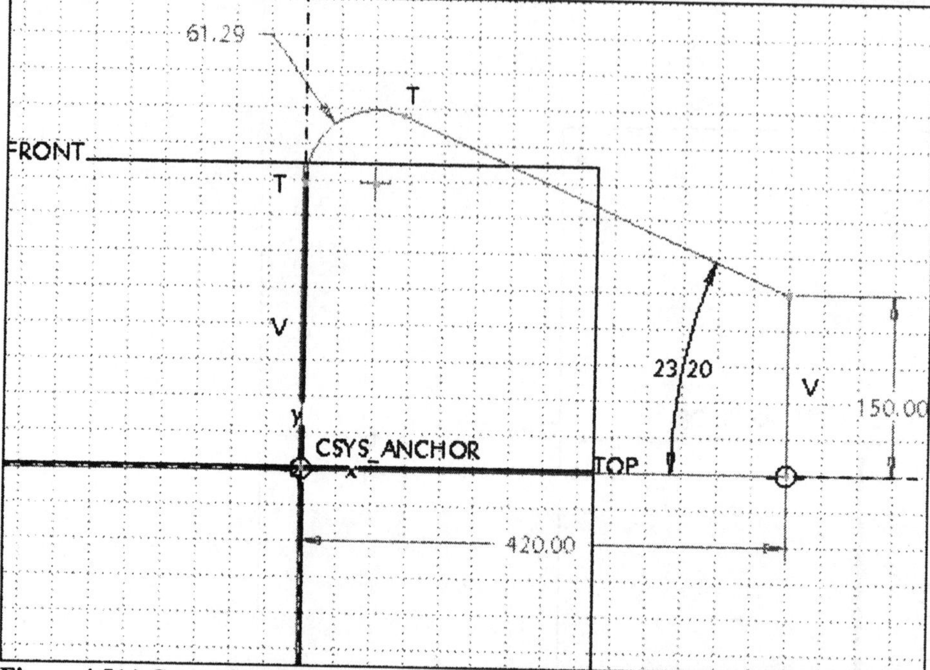

Figure 4.5(e) Create the Angle Dimension and Reposition the other Dimensions

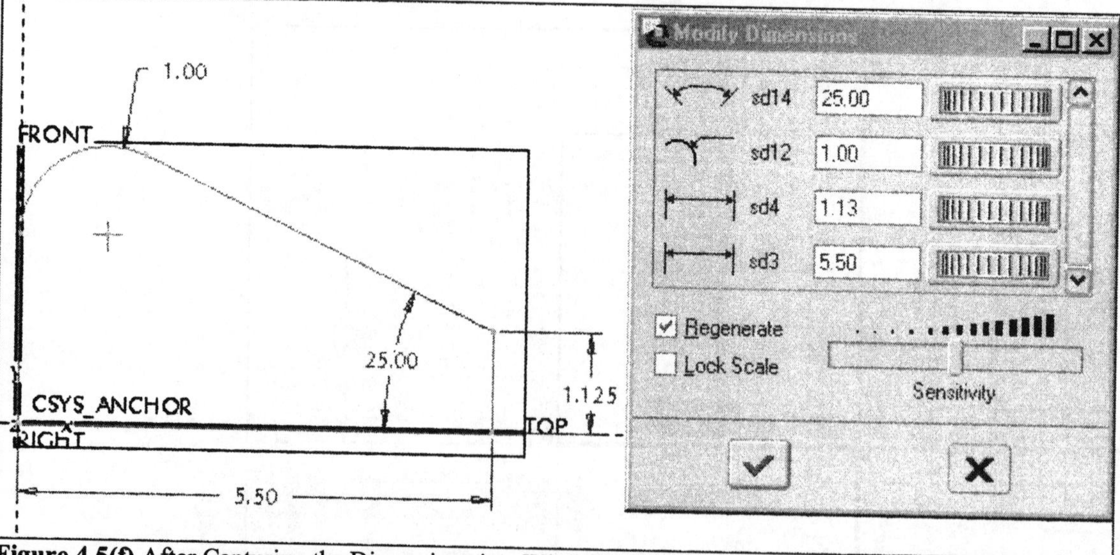

Figure 4.5(f) After Capturing the Dimensions in a Window, Modify the Dimension Values

Figure 4.5(g) Modify the Depth to **2.5625** in the Dashboard

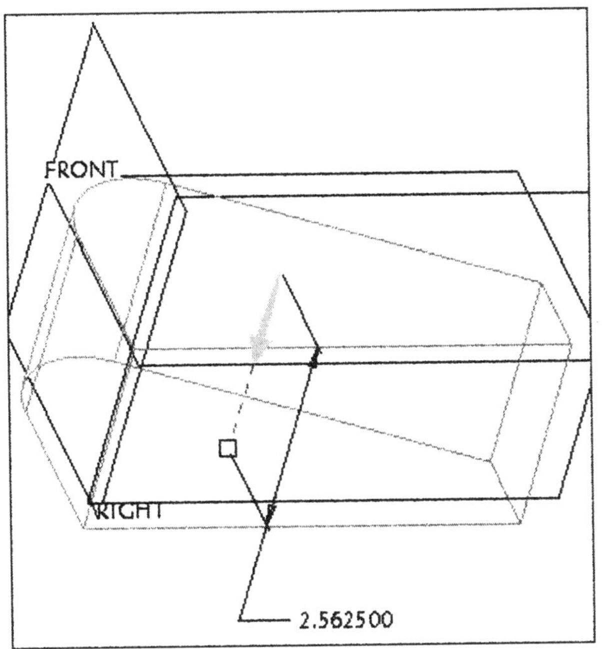

Figure 4.5(h) Feature Depth

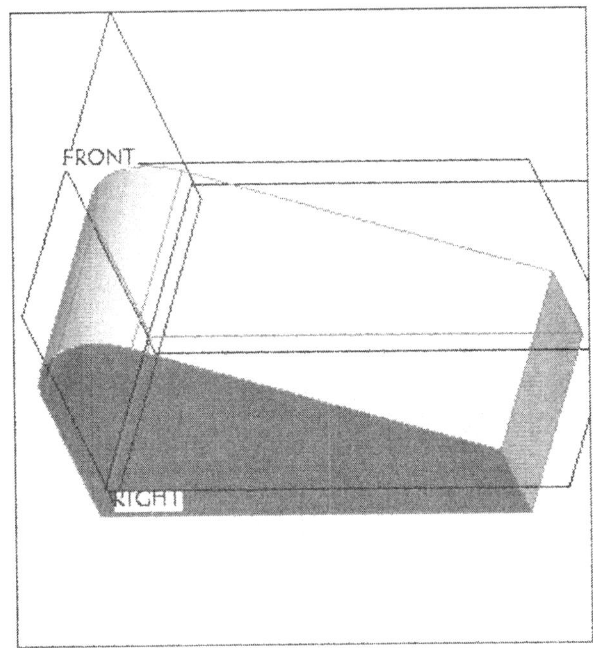

Figure 4.5(i) Preview (with Shading on)

Colors

Colors are used to define material and light properties. Use the **Color Editor** to create and modify colors and to customize the display of geometric items such as curves and surfaces and interface items such as fonts and the background. You can access the Color Editor by clicking a color swatch to set the **Appearance Color** or the **Highlight Color**.

Next, you will change the color of the model from the default. In general, try to avoid the colors that are used as Pro/E defaults. Because feature and entity highlighting is defaulted to *red*, colors similar to *red* should be avoided. Datum planes have *tan* and *gray* sides; therefore, they should be avoided. Shades of *blue* and *green* work well. It is really up to you to choose colors that are pleasant to look at and work well with the type of project you are modeling. Remember, if you will be using the parts in an assembly, each component should have a unique color scheme.

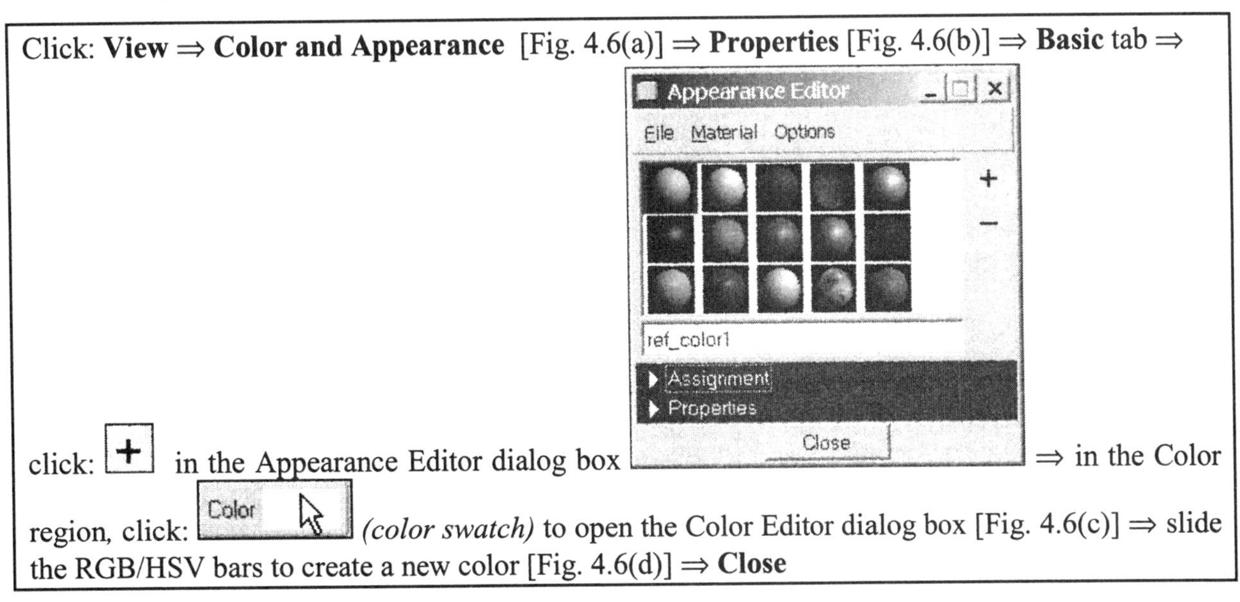

Click: **View** ⇒ **Color and Appearance** [Fig. 4.6(a)] ⇒ **Properties** [Fig. 4.6(b)] ⇒ **Basic** tab ⇒ click: + in the Appearance Editor dialog box ⇒ in the Color region, click: Color *(color swatch)* to open the Color Editor dialog box [Fig. 4.6(c)] ⇒ slide the RGB/HSV bars to create a new color [Fig. 4.6(d)] ⇒ **Close**

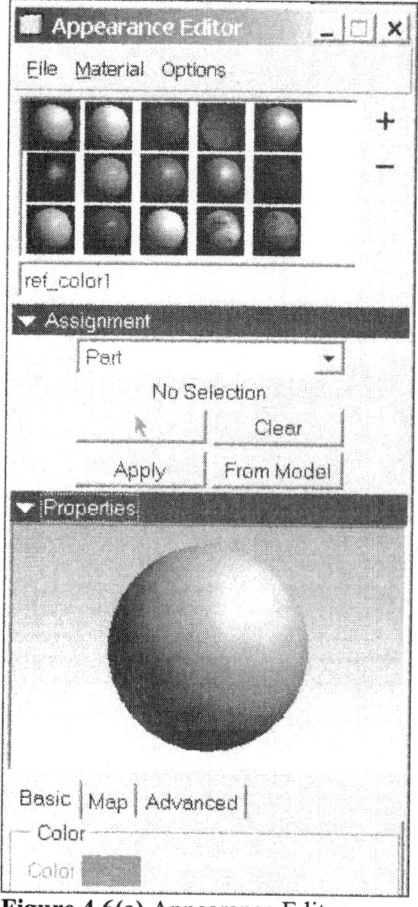

Figure 4.6(a) Appearance Editor

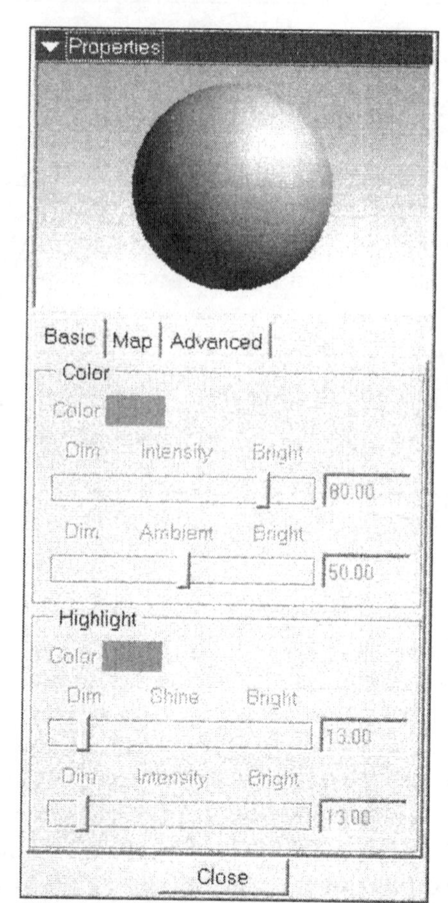

Figure 4.6(b) Properties

Repeat to make a few more colors, adding each to the palette [Fig. 4.6(h)] ⇒ also, add some Highlights to the colors using the same basic method [Fig. 4.6(e)] ⇒ also, try creating colors with the **Color Wheel** or the **Blending Palette** [Figs. 4.6(f-g)] ⇒ **File** ⇒ **Save** ⇒ type a name for your color file *(you can open and use this file for other projects instead of recreating new colors)* ⇒ **OK** [Fig. 4.6(h)] ⇒ **Apply** ⇒ **Close** from the Appearance Editor dialog box

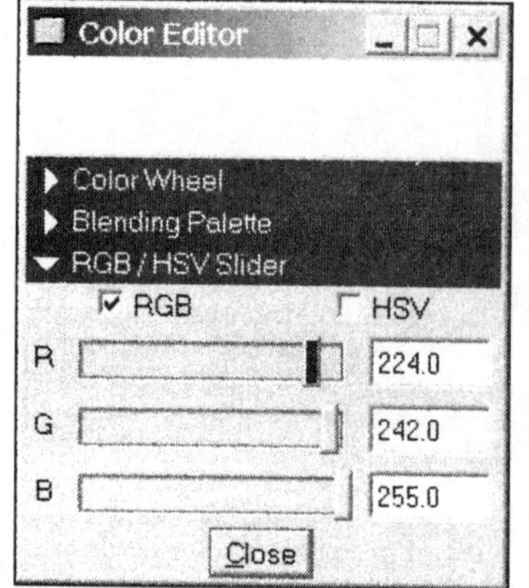

Figure 4.6(c) Color Editor Default

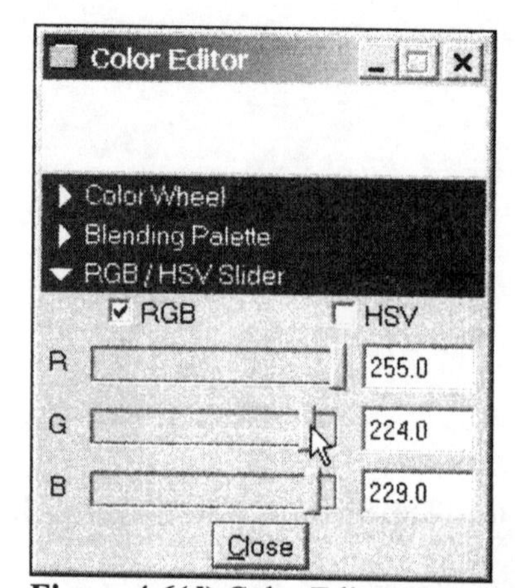

Figure 4.6(d) Color Editor New Color

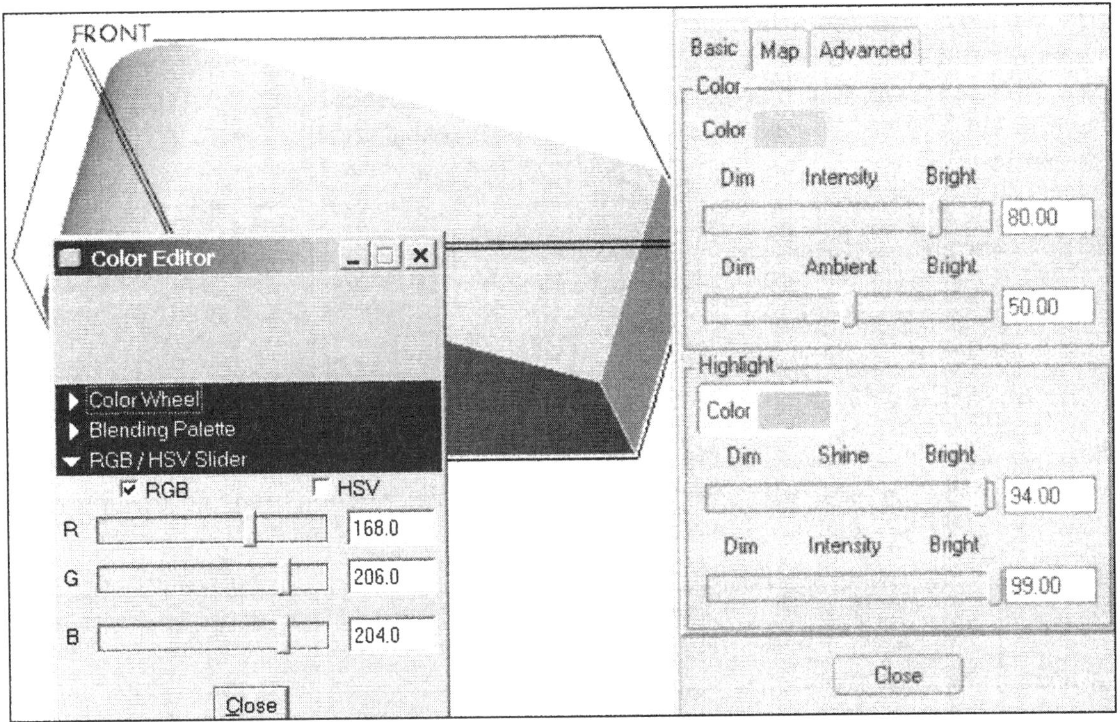

Figure 4.6(e) Highlight Color Editor

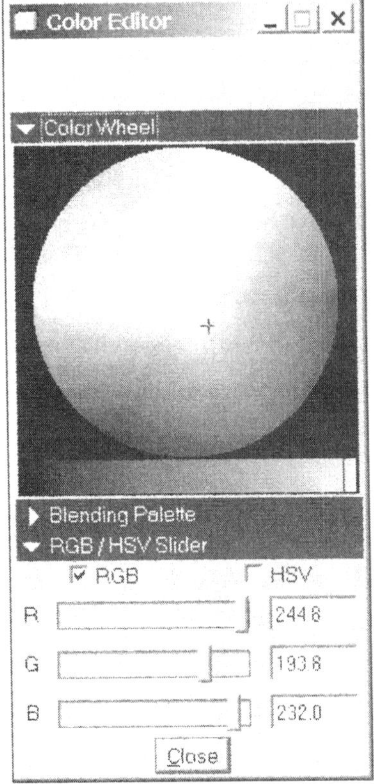

Figure 4.6(f) Color Wheel

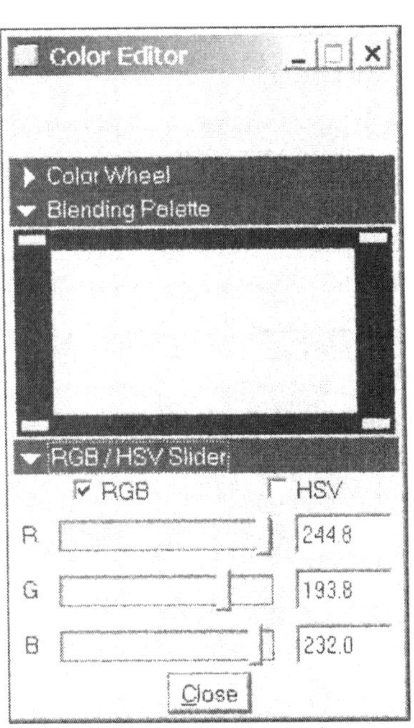

Figure 4.6(g) Blending Palette

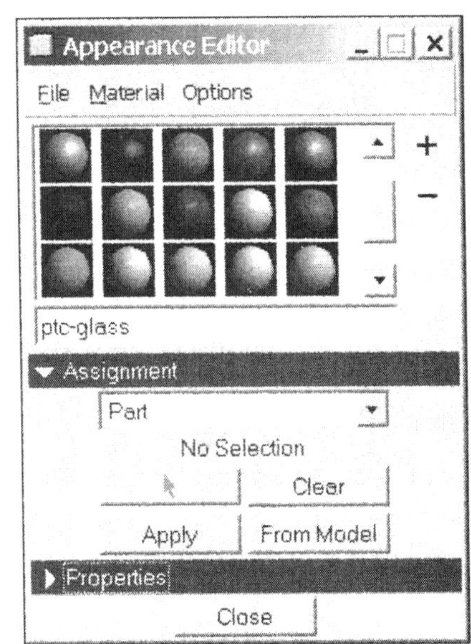

Figure 4.6(h) Completed Color Palette

The next two features will remove material. For both of the cuts, use RIGHT as the sketching plane, use TOP as the reference plane. Each cut requires just two lines, and two dimensions. Use the first cut's sketching/placement plane and reference/orientation **(Use Previous)** for the second cut. Create the cuts separately, each as an open section. Figure 4.7 shows the dimensions for each cut.

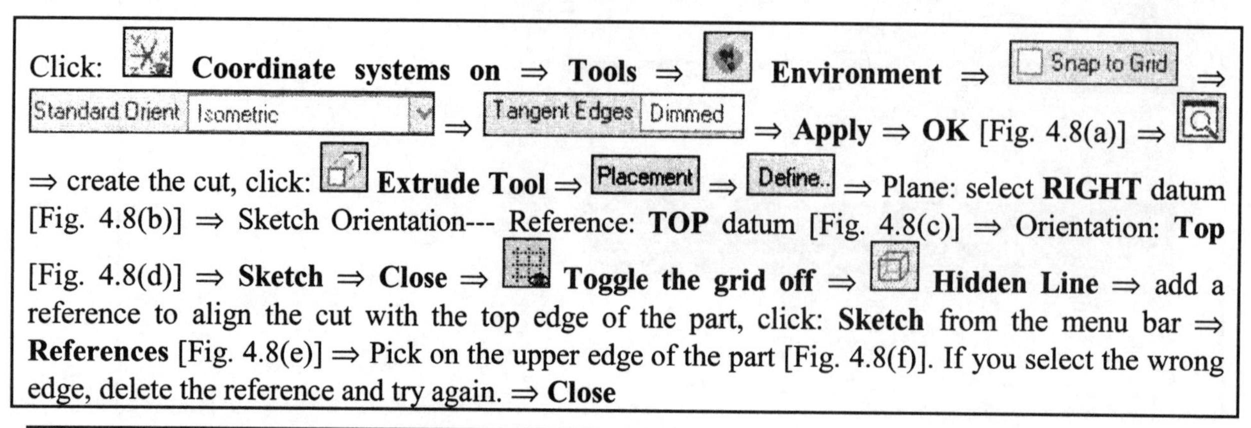
Click: **Coordinate systems on** ⇒ **Tools** ⇒ **Environment** ⇒ Snap to Grid ⇒ Standard Orient Isometric ⇒ Tangent Edges Dimmed ⇒ **Apply** ⇒ **OK** [Fig. 4.8(a)] ⇒ ⇒ create the cut, click: **Extrude Tool** ⇒ Placement ⇒ Define.. ⇒ Plane: select **RIGHT** datum [Fig. 4.8(b)] ⇒ Sketch Orientation--- Reference: **TOP** datum [Fig. 4.8(c)] ⇒ Orientation: **Top** [Fig. 4.8(d)] ⇒ **Sketch** ⇒ **Close** ⇒ **Toggle the grid off** ⇒ **Hidden Line** ⇒ add a reference to align the cut with the top edge of the part, click: **Sketch** from the menu bar ⇒ **References** [Fig. 4.8(e)] ⇒ Pick on the upper edge of the part [Fig. 4.8(f)]. If you select the wrong edge, delete the reference and try again. ⇒ **Close**

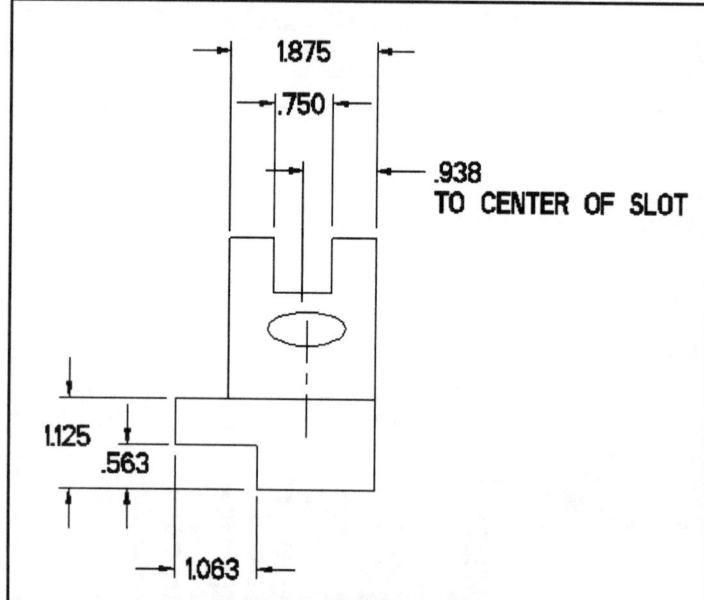

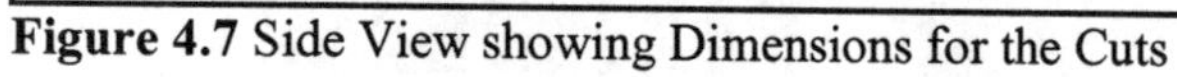
Figure 4.7 Side View showing Dimensions for the Cuts

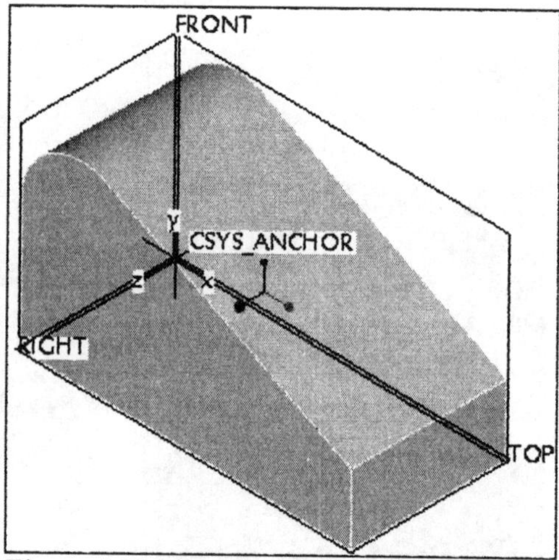

Figure 4.8(a) Standard Orientation Isometric

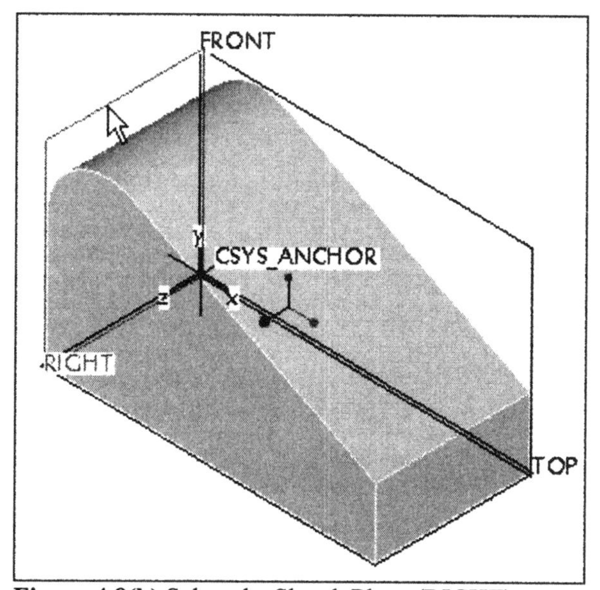

Figure 4.8(b) Select the Sketch Plane (RIGHT)

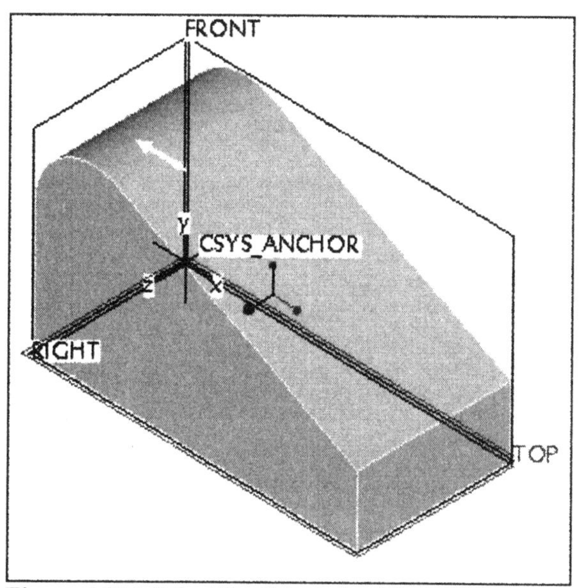

Figure 4.8(c) Select the Orientation Plane (TOP)

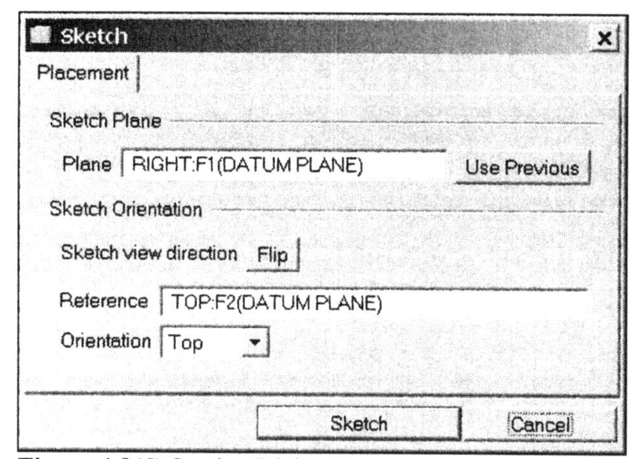

Figure 4.8(d) Section Dialog Box

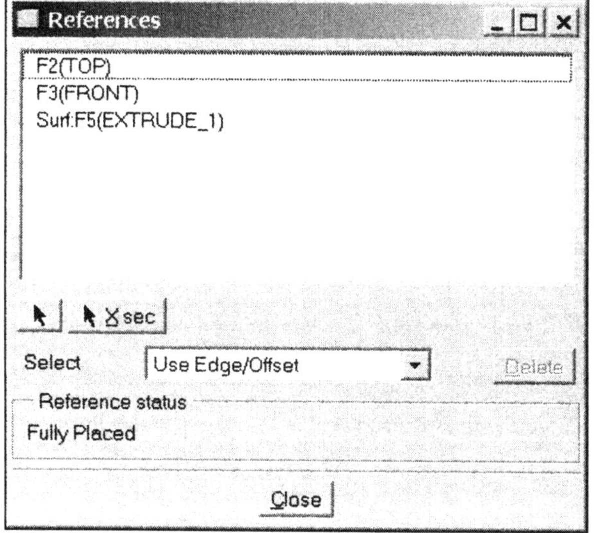

Figure 4.8(e) References Dialog Box

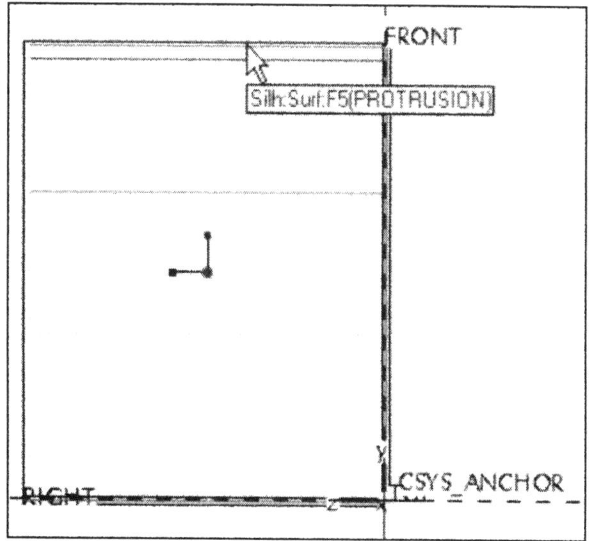

Figure 4.8(f) Add the upper Surface/Edge as a Reference

Click: **Create 2 point lines** ⇒ sketch the lines- start from the top with a vertical line [Fig. 4.8(g)] ⇒ **MMB** to end the line sequence ⇒ the Confirm dialog box will display asking if you want the end point of the line to be aligned with the edge of the part [Fig. 4.8(h)] ⇒ **Yes** [Fig. 4.8(i)] ⇒ ⇒ reposition the default dimensions [Fig. 4.8(j)]

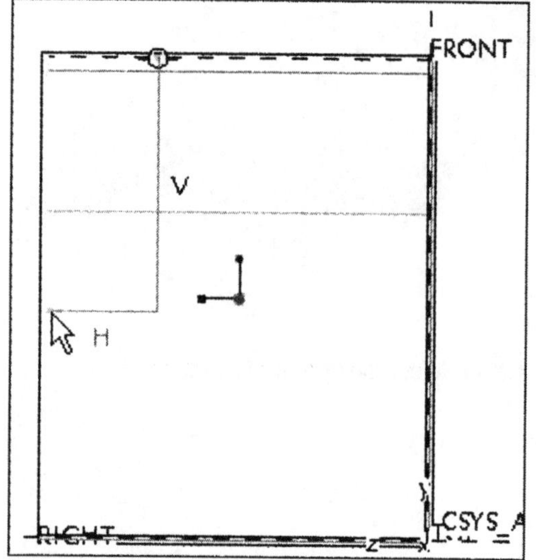

Figure 4.8(g) Sketch a Vertical and a Horizontal Line

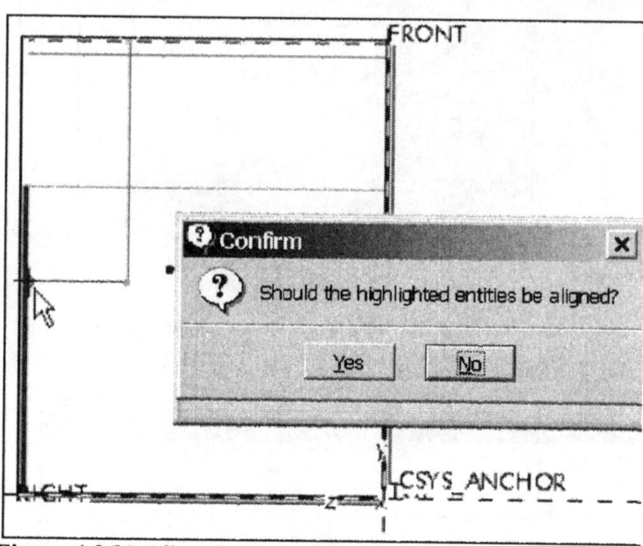

Figure 4.8(h) Align the End of the Horizontal Line

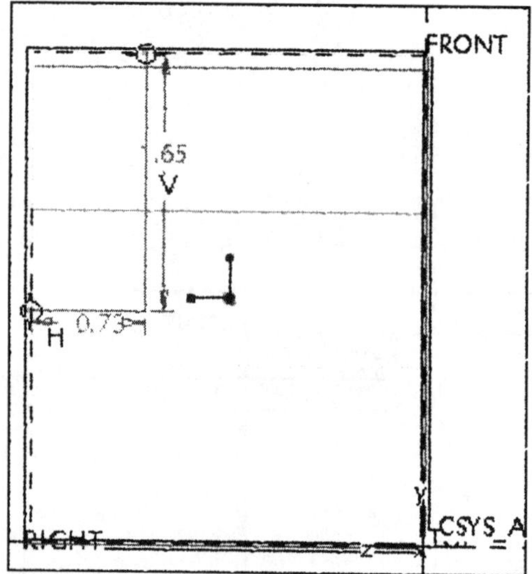

Figure 4.8(i) Original position of Default Dimensions

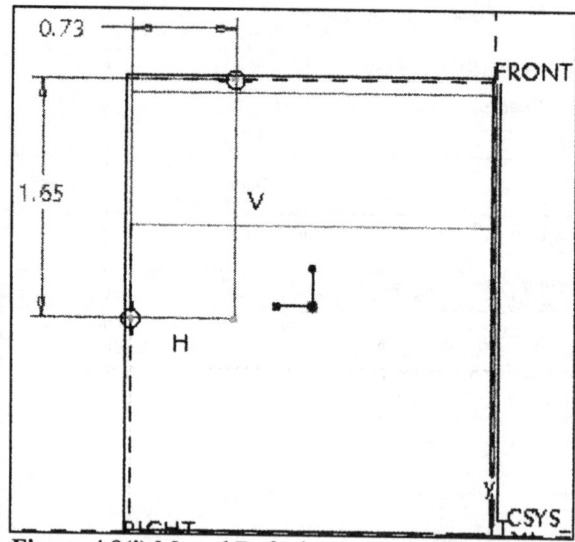

Figure 4.8(j) Moved Default Dimensions

Add the two dimensions [Fig. 4.8(k)] ⇒ ⇒ modify the dimension values [Figs. 4.8(l-m)] [your sd (sketcher dimension) numbers will be different] ⇒ ⇒ ⇒ ⇒ **Standard Orientation** ⇒ **Shading** [Fig. 4.8(n)]

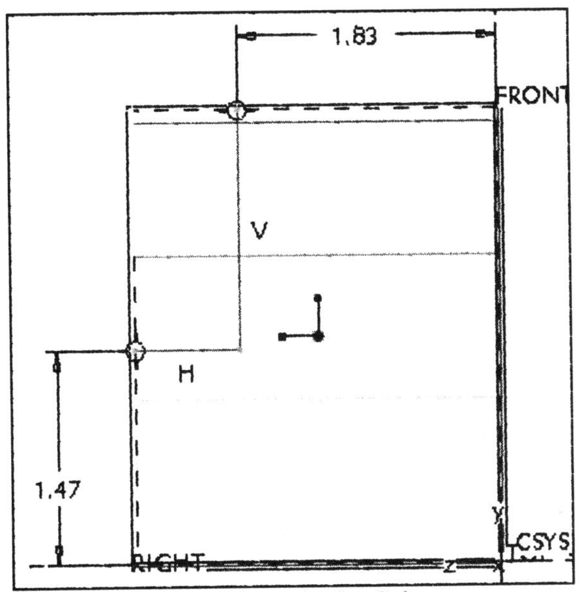

Figure 4.8(k) New Dimensioning Scheme

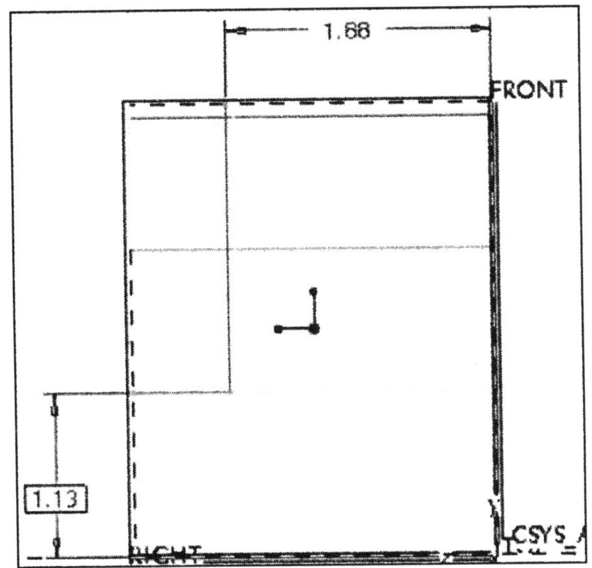

Figure 4.8(l) Modified Dimensions **1.875** and **1.125**

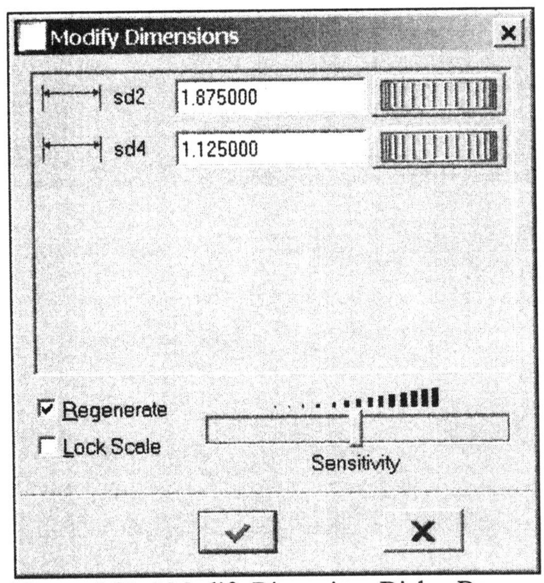

Figure 4.8(m) Modify Dimensions Dialog Box

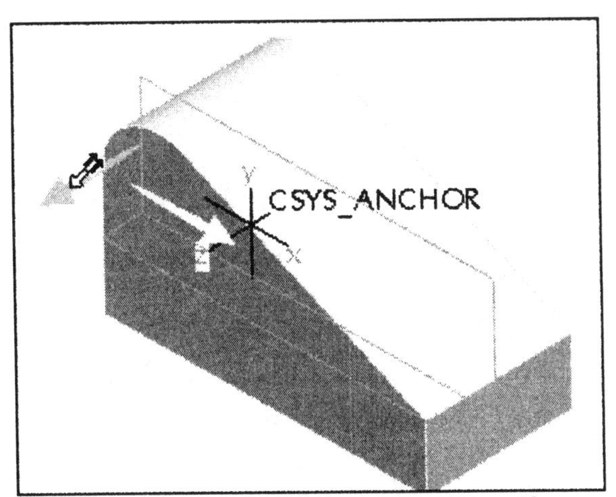

Figure 4.8(n) Extrusion Preview (flip arrow if necessary)

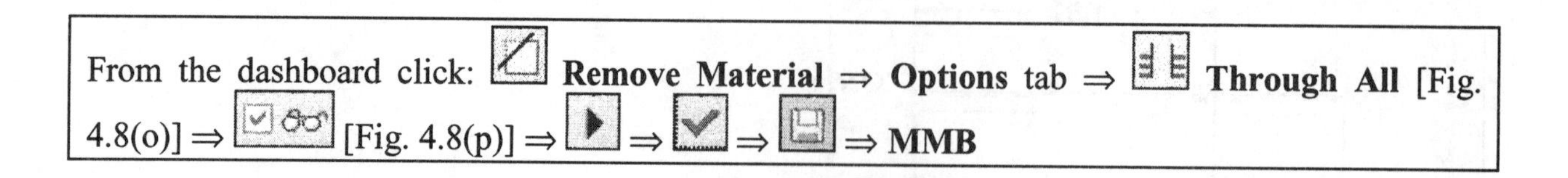

From the dashboard click: **Remove Material** ⇒ **Options** tab ⇒ **Through All** [Fig. 4.8(o)] ⇒ [Fig. 4.8(p)] ⇒ ⇒ ⇒ ⇒ **MMB**

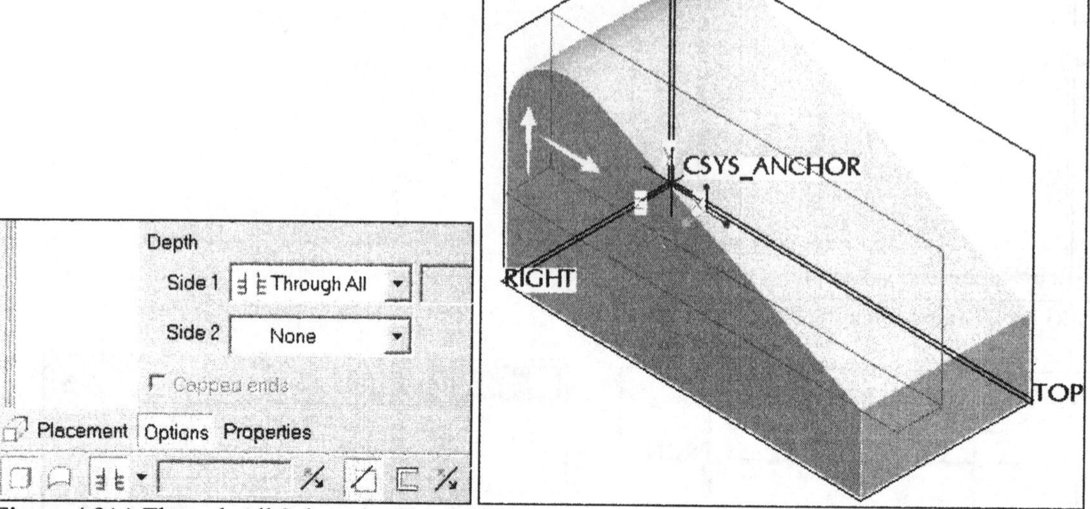

Figure 4.8(o) Through All Selected as Depth

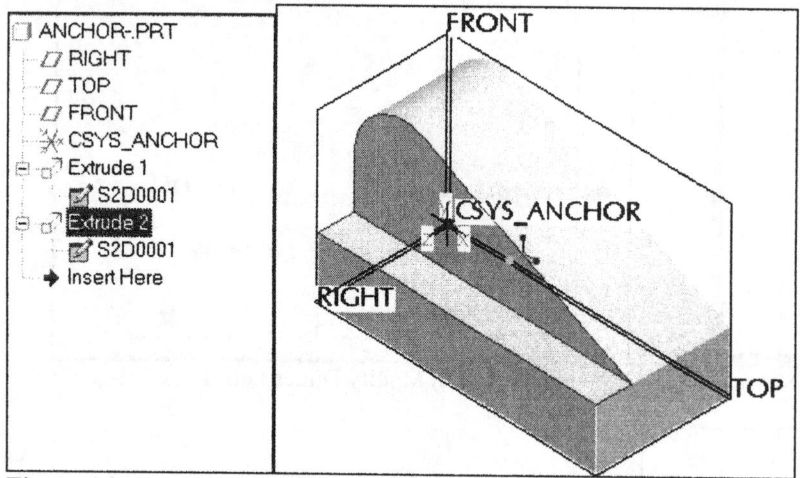

Figure 4.8(p) Extruded Cut

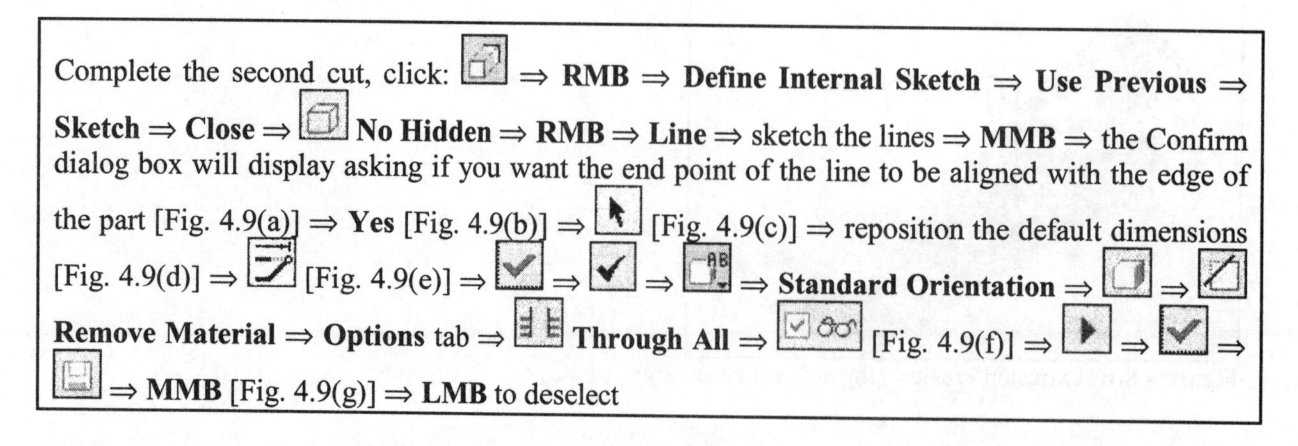

Complete the second cut, click: ⇒ **RMB** ⇒ **Define Internal Sketch** ⇒ **Use Previous** ⇒ **Sketch** ⇒ **Close** ⇒ **No Hidden** ⇒ **RMB** ⇒ **Line** ⇒ sketch the lines ⇒ **MMB** ⇒ the Confirm dialog box will display asking if you want the end point of the line to be aligned with the edge of the part [Fig. 4.9(a)] ⇒ **Yes** [Fig. 4.9(b)] ⇒ [Fig. 4.9(c)] ⇒ reposition the default dimensions [Fig. 4.9(d)] ⇒ [Fig. 4.9(e)] ⇒ ⇒ ⇒ ⇒ **Standard Orientation** ⇒ ⇒ **Remove Material** ⇒ **Options** tab ⇒ **Through All** ⇒ [Fig. 4.9(f)] ⇒ ⇒ ⇒ ⇒ **MMB** [Fig. 4.9(g)] ⇒ **LMB** to deselect

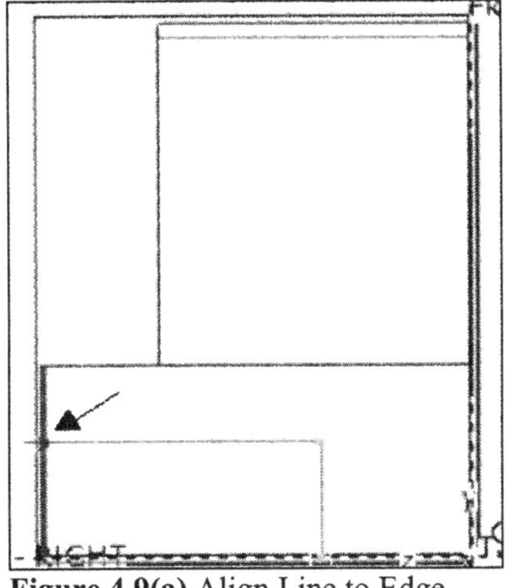

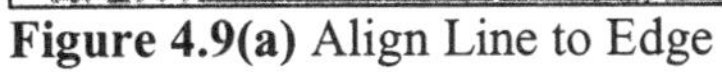

Figure 4.9(a) Align Line to Edge

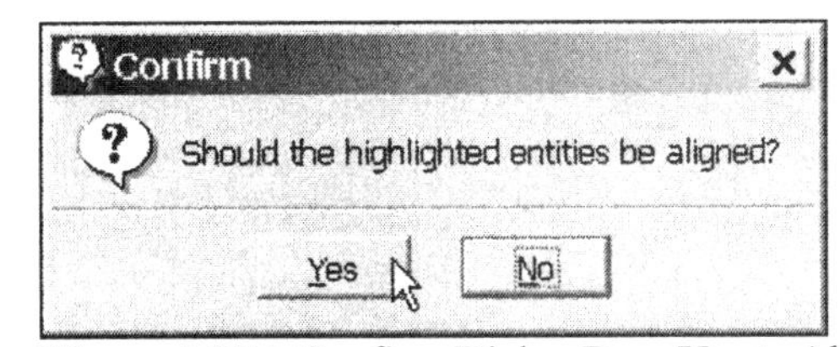

Figure 4.9(b) Confirm Dialog Box, Yes to Alignment

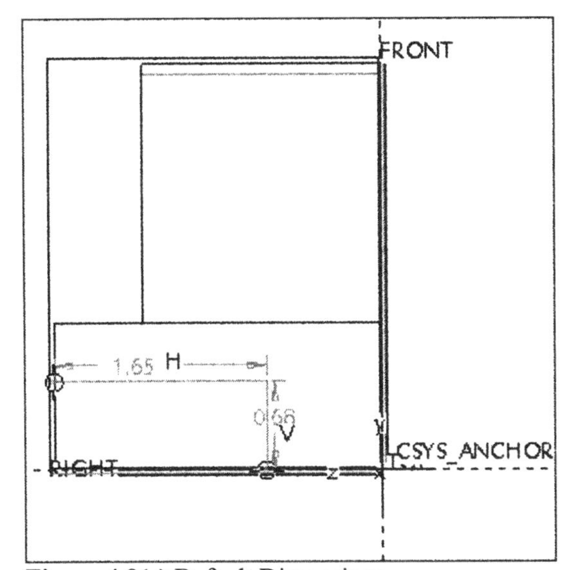

Figure 4.9(c) Default Dimensions

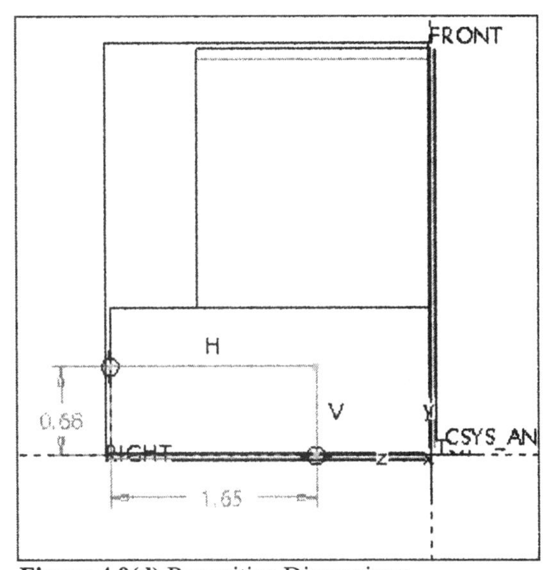

Figure 4.9(d) Reposition Dimensions

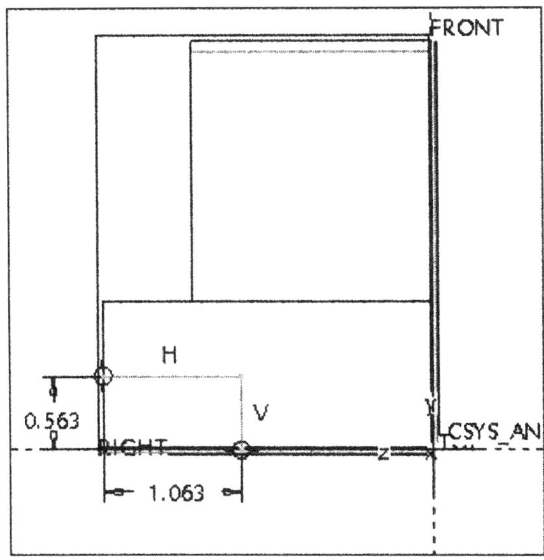

Figure 4.9(e) Modify Dimensions

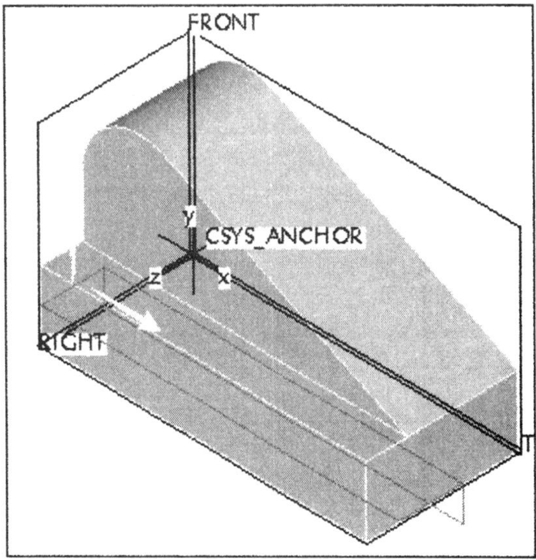

Figure 4.9(f) Cut Preview

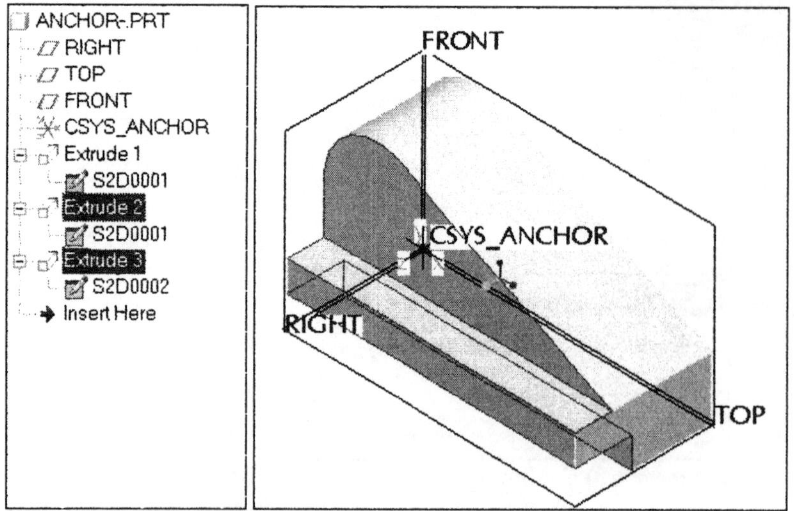

Figure 4.9(g) Completed Cuts

For the next feature, you will need to create a new datum plane on which to sketch. New part datum planes are by default numbered sequentially, DTM1, DTM2, and so on.

Also, a datum axis (A-1) will be created through the curved top of the part and used later for the axial location of a small hole. A datum axis can be inserted through any curved feature. Holes and circular features automatically have axes when they are created. Features created with arcs, fillets, and so on need to have axes added, unless set in the *configuration file* as the default.

Click: **Datum Plane Tool** ⇒ References: pick datum **FRONT** as the plane to offset from [Fig. 4.10(a)] ⇒ in the DATUM PLANE dialog box, Offset: Translation, type the distance **1.875/2** ***(1.875/2 = .9375)*** ⇒ **Enter** ⇒ **OK** ⇒ **LMB** to deselect ⇒ **Datum Axis Tool** ⇒ References: pick the curved surface [Fig. 4.10(b)] ⇒ **OK** ⇒ **LMB** to deselect ⇒ **Datum Plane Tool** ⇒ References: pick the angled surface [Fig. 4.10(c)] ⇒ click **Offset** [Fig. 4.10(d)] ⇒ click **Offset** again to see options [Fig. 4.10(e)] ⇒ click **Through** [Figs. 4.10(f-h)] ⇒ **OK** ⇒ ⇒ **MMB** ⇒ **LMB** to deselect

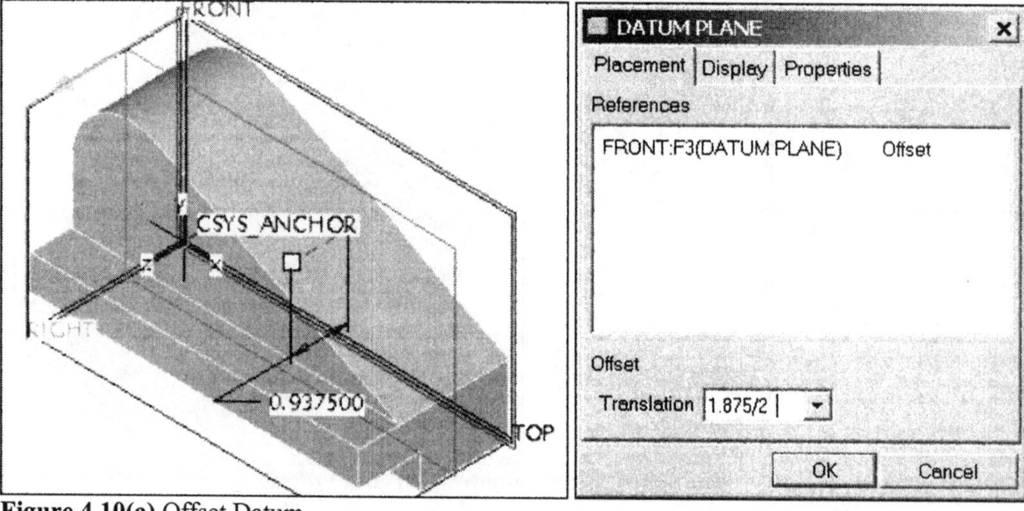

Figure 4.10(a) Offset Datum

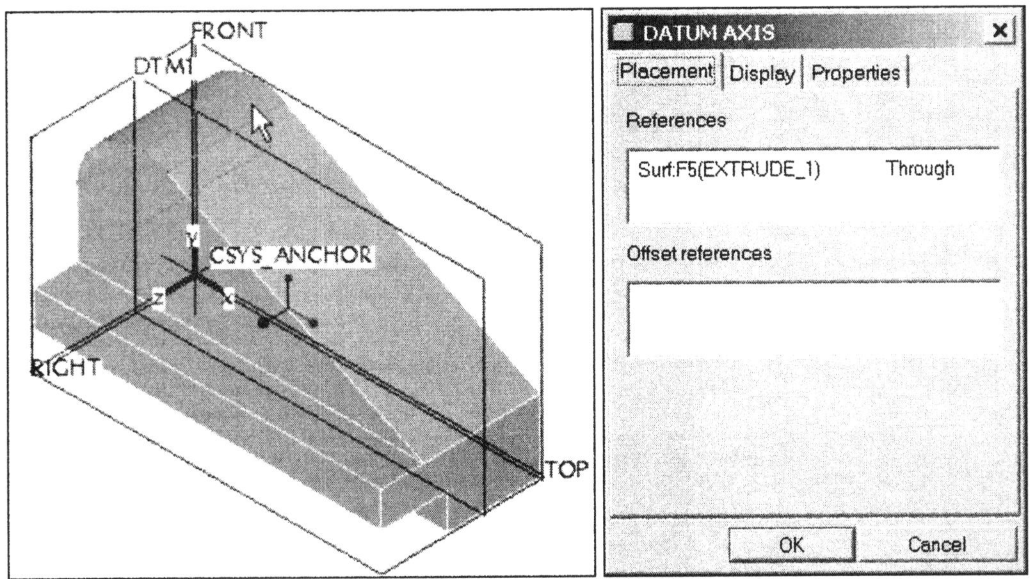

Figure 4.10(b) Datum Axis A_1 Through the Center of the Curved Surface

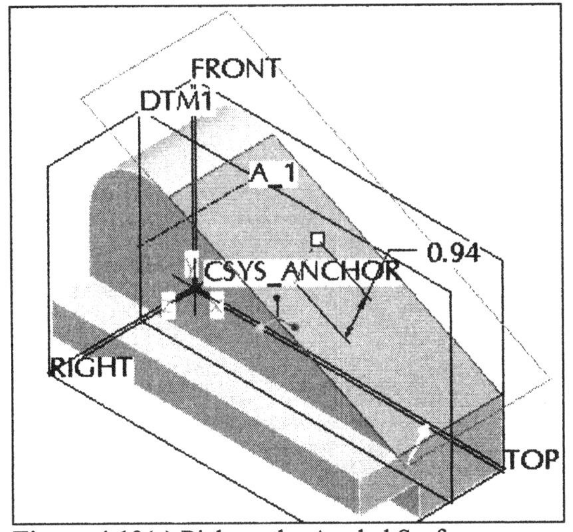

Figure 4.10(c) Pick on the Angled Surface

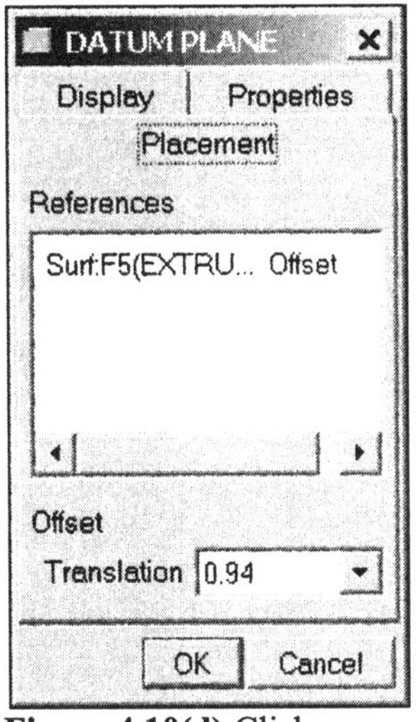

Figure 4.10(d) Click on Offset

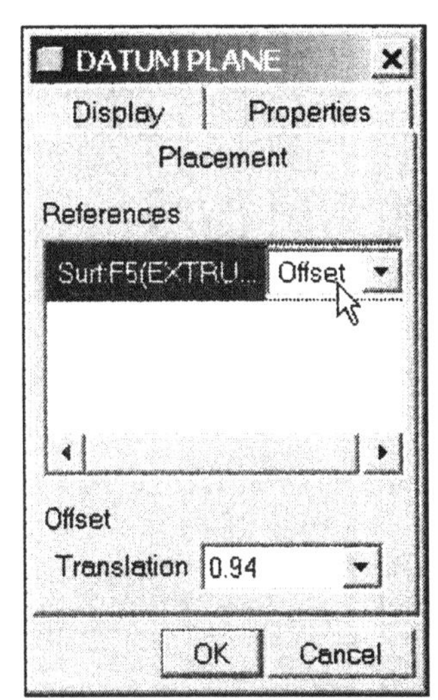

Figure 4.10(e) Click Offset Again

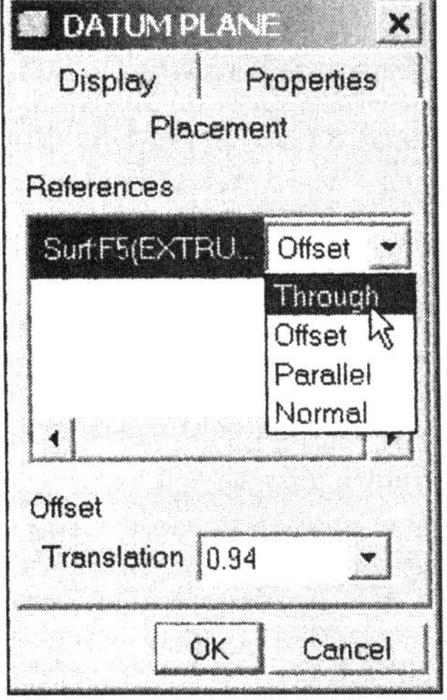

Figure 4.10(f) Click Through

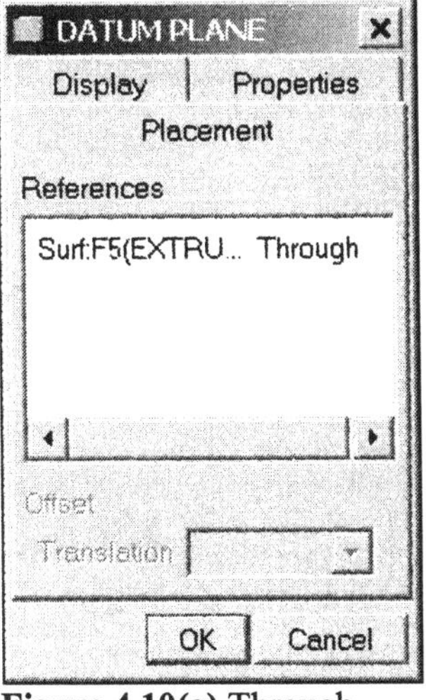

Figure 4.10(g) Through

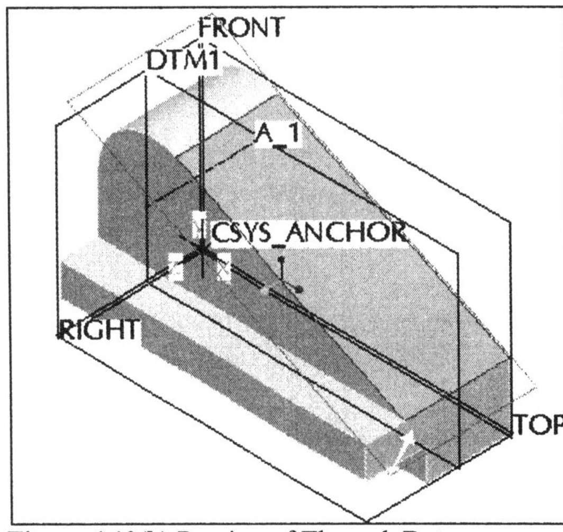

Figure 4.10(h) Preview of Through Datum

Create a new layer and add the new datum planes and axis to it, click: **Show** in the **Navigator** ⇒ **Layer Tree** --Layer Tree displays in place of the Model Tree [Fig. 4.11(a)] ⇒ click on and highlight **Layers** in the Layer Tree ⇒ **RMB** ⇒ **New Layer** [Fig. 4.11(b)] ⇒ Name: type **DATUM_FEATURES** *(do not press Enter)* as the name for new layer ⇒ select the two new datum planes from the model ⇒ select the axis from the model, use **RMB** ⇒ **Next** to toggle to the correct selection [Fig. 4.11(c)] ⇒ select the axis with **LMB** ⇒ **OK** [Fig. 4.11(d)] ⇒ expand layer to see items [Fig. 4.11(e)] ⇒ **Show** in the **Navigator** ⇒ **Model Tree** ⇒ **Ctrl+R**

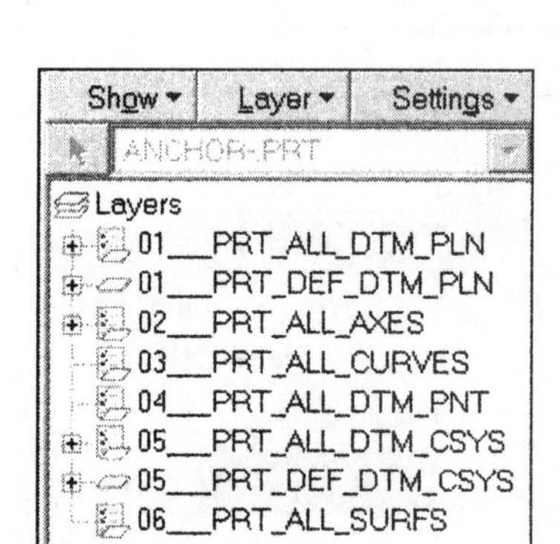

Figure 4.11(a) Layer Tree

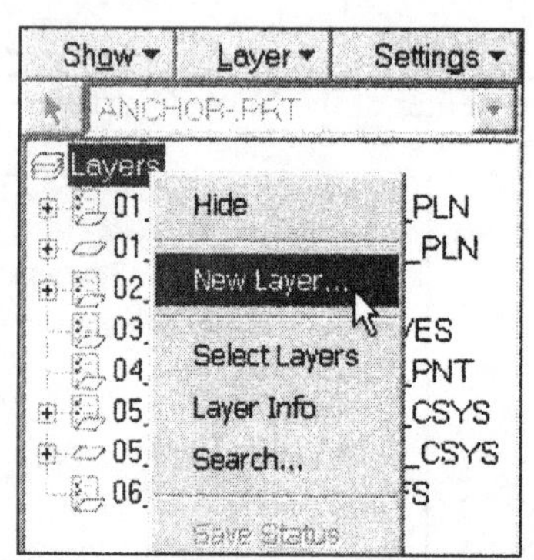

Figure 4.11(b) New Layer

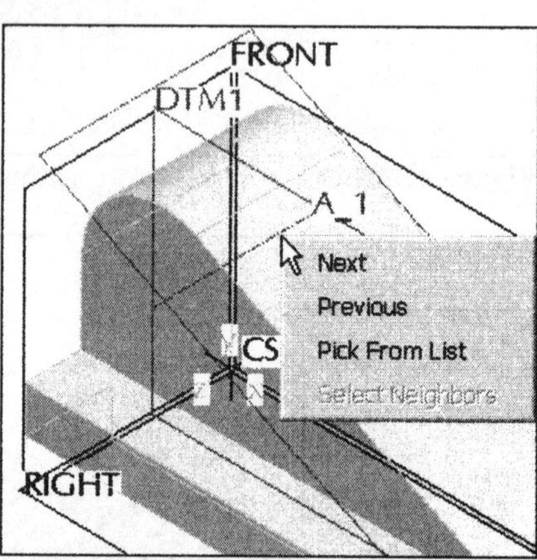

Figure 4.11(c) Highlight Axis using RMB Next

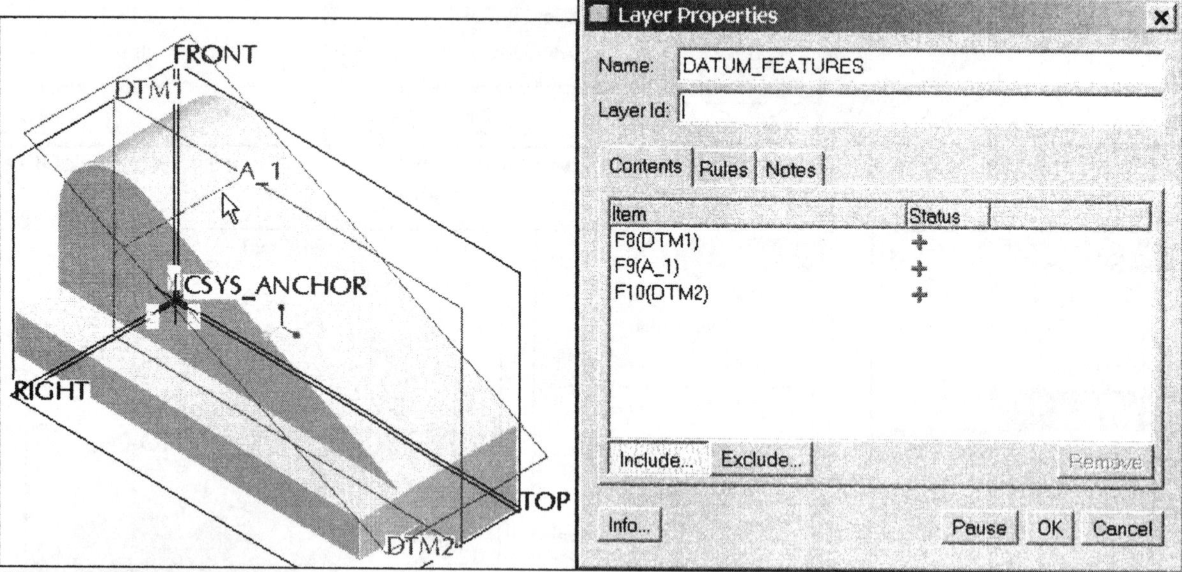

Figure 4.11(d) Adding Items to a Layer

Figure 4.11(e) DATUM_FEATURES Layer in Layer Tree

Geometric Tolerances

Before continuing with the modeling of the part, the datums used and created thus far will be "set" for geometric tolerancing. Geometric tolerances (GTOLs) provide a comprehensive method of specifying where on a part the critical surfaces are, how they relate to one another, and how the part must be inspected to determine if it is acceptable. They provide a method for controlling the location, form, profile, orientation, and run out of features. When you store a Pro/E GTOL in a solid model, it contains parametric references to the geometry or feature it controls—its reference entity—and parametric references to referenced datums and axes. As a result, Pro/E updates a GTOL's display when you rename a referenced datum.

In Assembly mode, you can create a GTOL in a subassembly or a part. A GTOL that you create in Part or Assembly mode automatically belongs to the part or assembly that occupies the window; however, it can refer only to set datums belonging to that model itself, or to components within it. It cannot refer to datums outside of its model in some encompassing assembly, unlike assembly created features. You can add GTOLs in Part or Drawing mode, but they are reflected in all other modes. Pro/E treats them as annotations, and they are always associated with the model. Unlike dimensional tolerances, though, GTOLs do not affect part geometry.

Before you can reference a datum plane or axis in a GTOL, you must set it as a reference. Pro/E encloses its name using the set datum symbol. After you have set a datum, you can use it in the usual way to create features and assemble parts. You enter the set datum command by clicking Edit ⇒ Setup ⇒ Geom Tol ⇒ Set Datum, and then select the datum plane or axis. Pro/E encloses the datum name in a feature control frame. If needed, type a new name in the Name field of the Datum dialog box. Most datums will follow the alphabet, A, B, C, D, and so on. You can hide (not display) a reference datum plane by placing it on a layer and then hiding the layer *(or using View ⇒ Visibility ⇒ Hide)*.

Click: **Edit** ⇒ **Setup** ⇒ **Geom Tol** ⇒ **Set Datum** ⇒ pick **TOP** from the model [Fig. 4.12(a)] ⇒ Name- type **A** ⇒ **OK** ⇒ pick **FRONT** ⇒ Name- **B** ⇒ **OK** ⇒ pick **RIGHT** ⇒ Name- type **C** ⇒ **OK** ⇒ pick **DTM1** ⇒ Name- type **D** ⇒ **OK** ⇒ pick **DTM2** ⇒ Name- type **E** ⇒ **OK** ⇒ **MMB** ⇒ **MMB** ⇒ **MMB** [Fig. 4.12(b)] ⇒ ⇒ ⇒ **MMB** ⇒ **File** ⇒ **Delete** ⇒ **Old Versions** ⇒ **MMB** ⇒ **LMB** to deselect

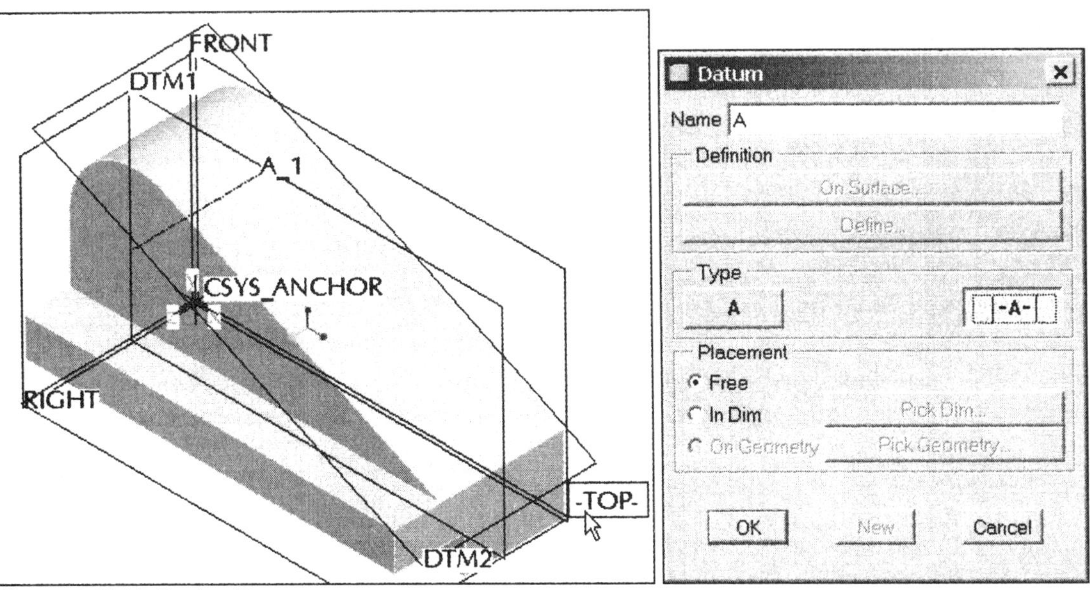

Figure 4.12(a) Setting Datums

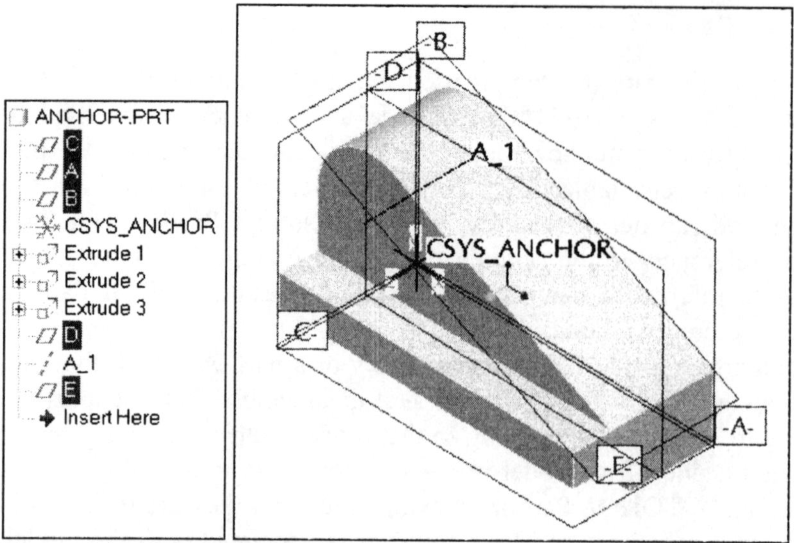

Figure 4.12(b) Set Datums

Create the slot on the top of the part by sketching on datum D and projecting the cut toward both sides. Use the Model Tree to select the appropriate datum planes.

Click: ⇒ **RMB** ⇒ **Define Internal Sketch** ⇒ Plane: select datum **D** [Fig. 4.13(a)] ⇒ Sketch Orientation--- Reference: datum **C** ⇒ Orientation: **Right** ⇒ **Sketch** ⇒ **Close** ⇒ ⇒ **Standard Orientation** ⇒ add a reference to align the cut with the angled edge of the part, click: **Sketch** from the menu bar ⇒ **References** ⇒ pick on datum **E** ⇒ pick the small vertical surface on the right side of the part as the fourth reference [Fig. 4.13(b)] ⇒ **Close**

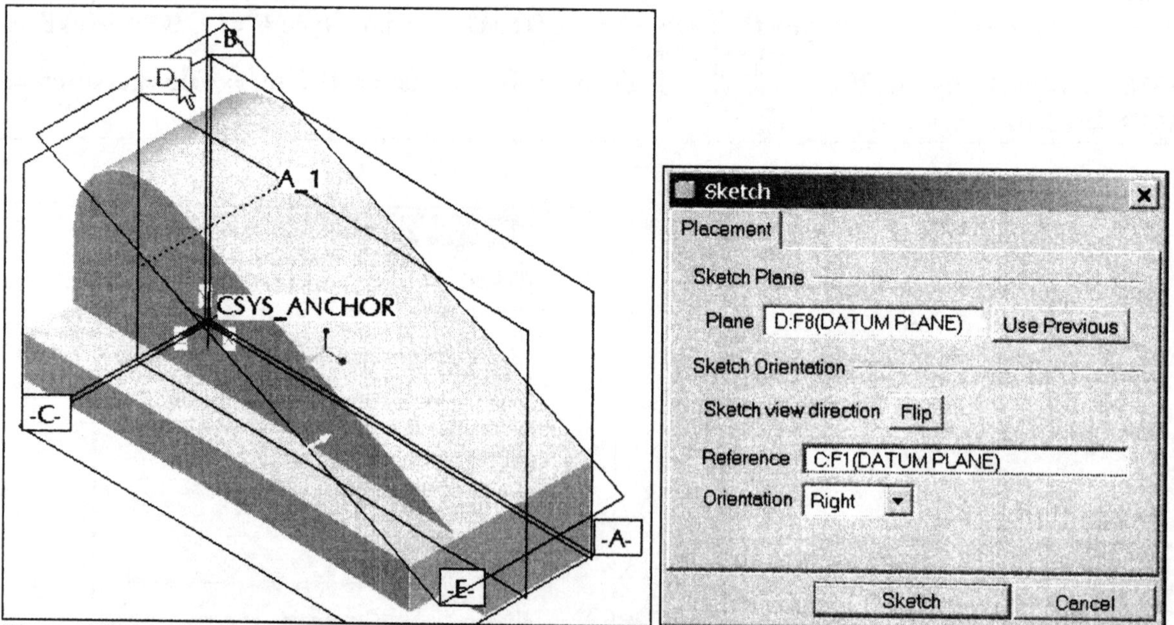

Figure 4.13(a) Sketch Plane and Sketch Orientation Reference

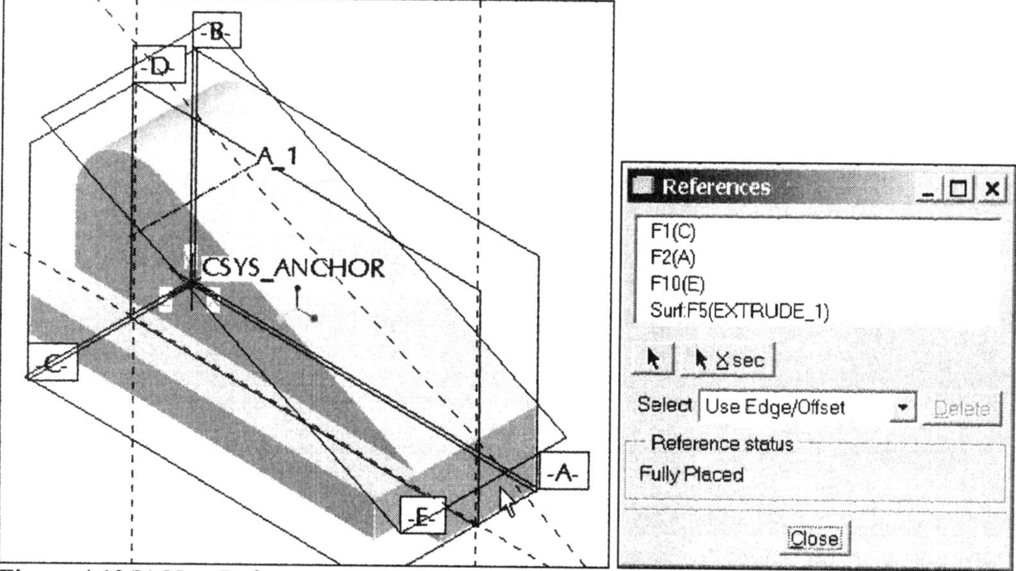

Figure 4.13(b) New References (Datum E and Small Right Vertical Surface)

Click: **Hidden Line** ⇒ **Orient the sketching plane parallel to the screen** ⇒ **Create points** pick at the corner of the angled datum **E** and the right vertical reference [Fig. 4.13(c)] ⇒ **MMB**

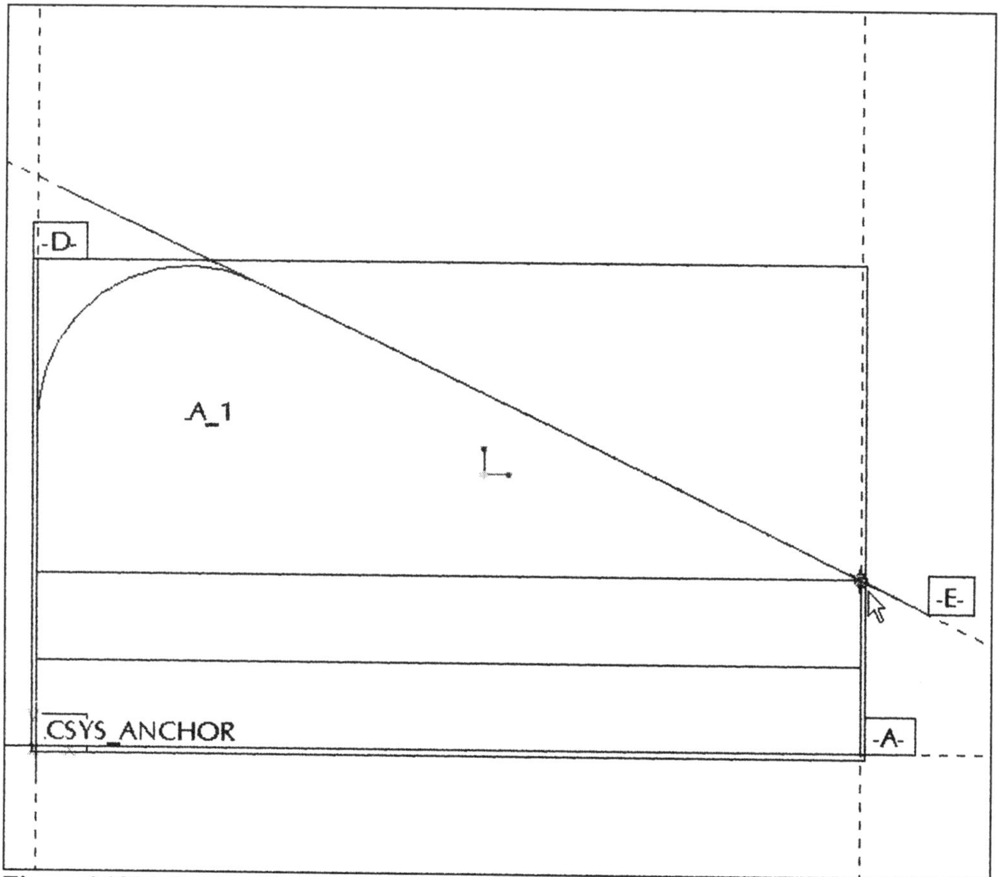

Figure 4.13(c) Create a Sketcher Point

Click: **Create 2 point lines** ⇒ sketch the first line from and perpendicular to the angled edge [Fig. 4.13(d)] ⇒ continue to sketch the second line horizontal [Fig. 4.13(e)] ⇒ **MMB** [Fig. 4.13(f)]

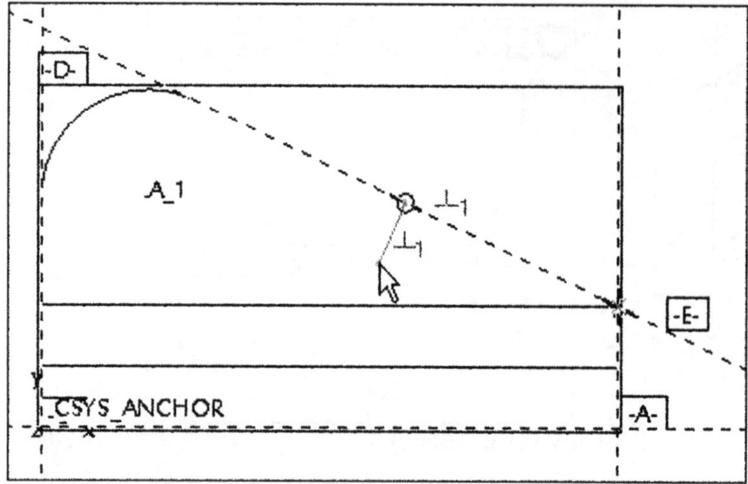

Figure 4.13(d) Create the first Line Perpendicular to Datum E

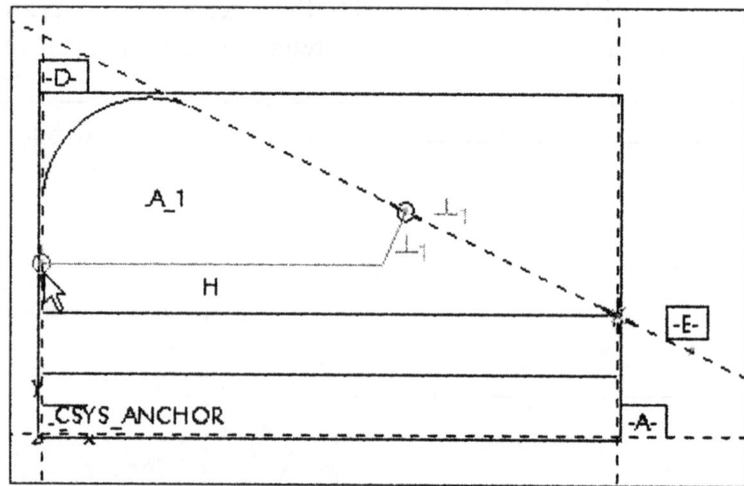

Figure 4.13(e) Create the Second Line Horizontal

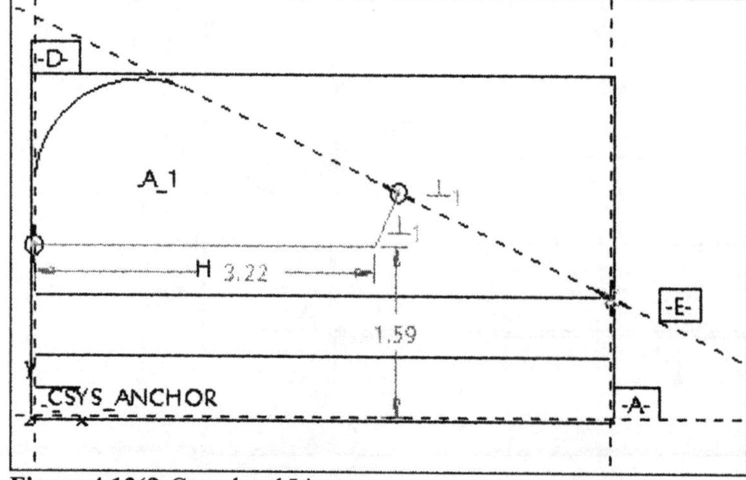

Figure 4.13(f) Completed Lines

Click: [dimension icon] ⇒ add a dimension by picking the first line [Fig. 4.13(g)] and then the point *(do not pick endpoint to point)* ⇒ **MMB** to place the dimension [Fig. 4.13(h)] ⇒ add a vertical dimension [Fig. 4.13(i)] ⇒ **Ctrl+R**

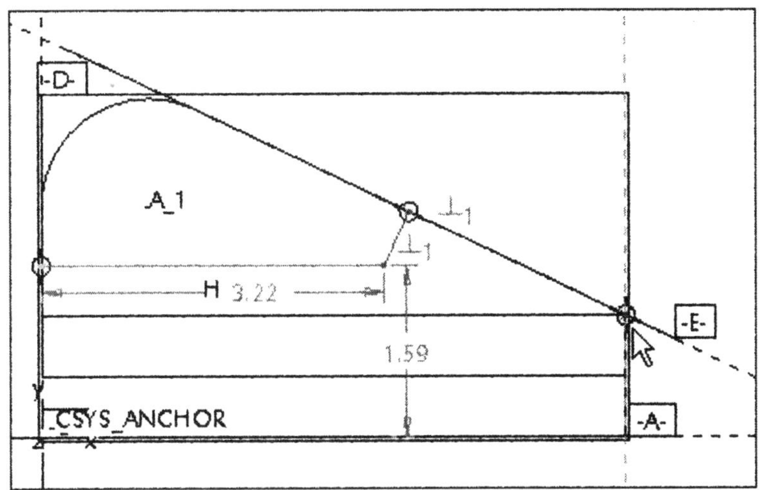

Figure 4.13(g) Create Dimension from First Line to Point (not point to point!)

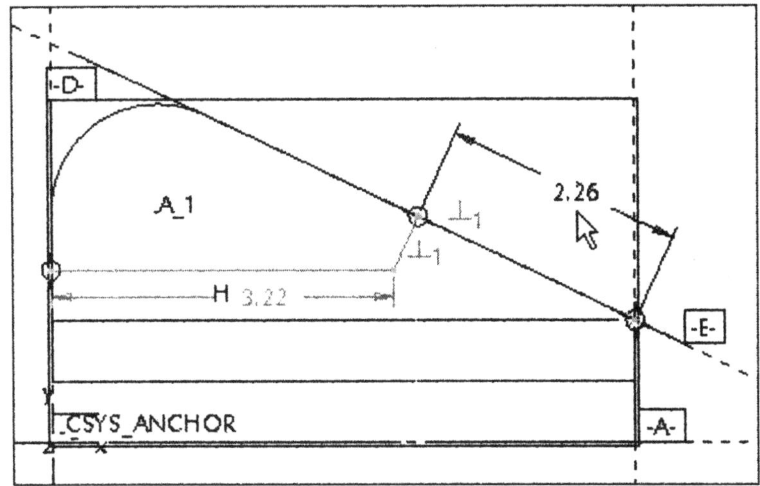

Figure 4.13(h) Place Dimension

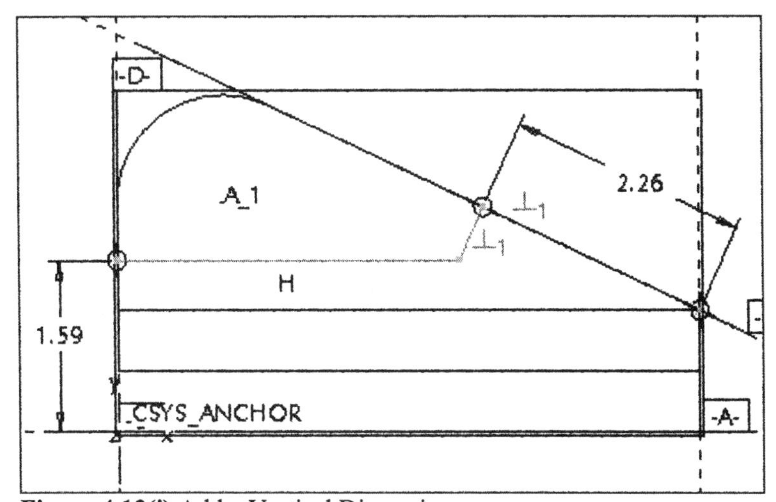

Figure 4.13(i) Add a Vertical Dimension

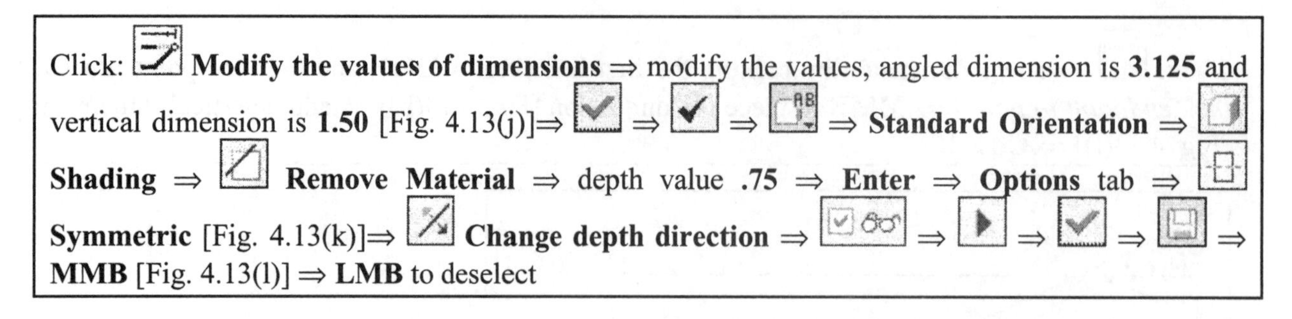

Click: **Modify the values of dimensions** ⇒ modify the values, angled dimension is **3.125** and vertical dimension is **1.50** [Fig. 4.13(j)]⇒ ⇒ ⇒ ⇒ **Standard Orientation** ⇒ **Shading** ⇒ **Remove Material** ⇒ depth value **.75** ⇒ **Enter** ⇒ **Options** tab ⇒ **Symmetric** [Fig. 4.13(k)]⇒ **Change depth direction** ⇒ ⇒ ⇒ ⇒ ⇒ **MMB** [Fig. 4.13(l)] ⇒ **LMB** to deselect

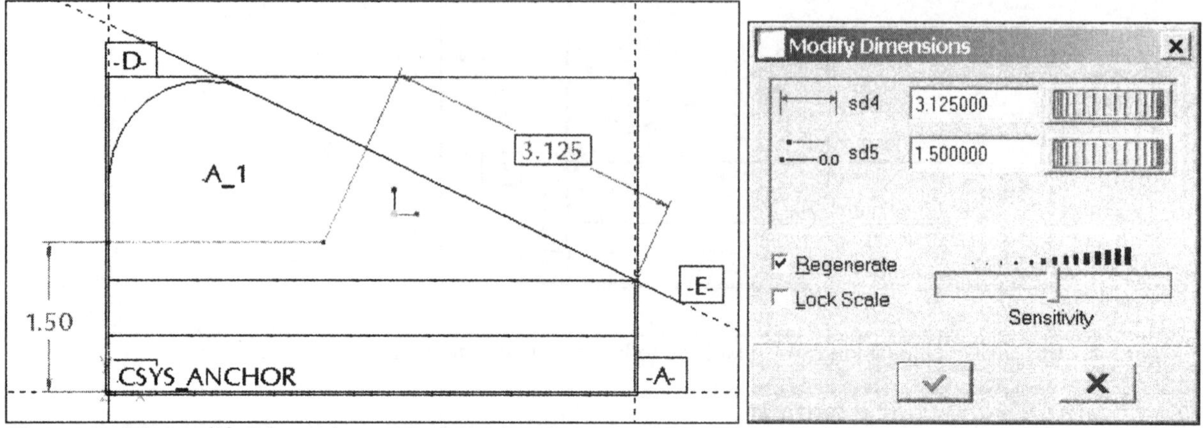

Figure 4.13(j) Modify Dimensions to **3.125** and **1.50**

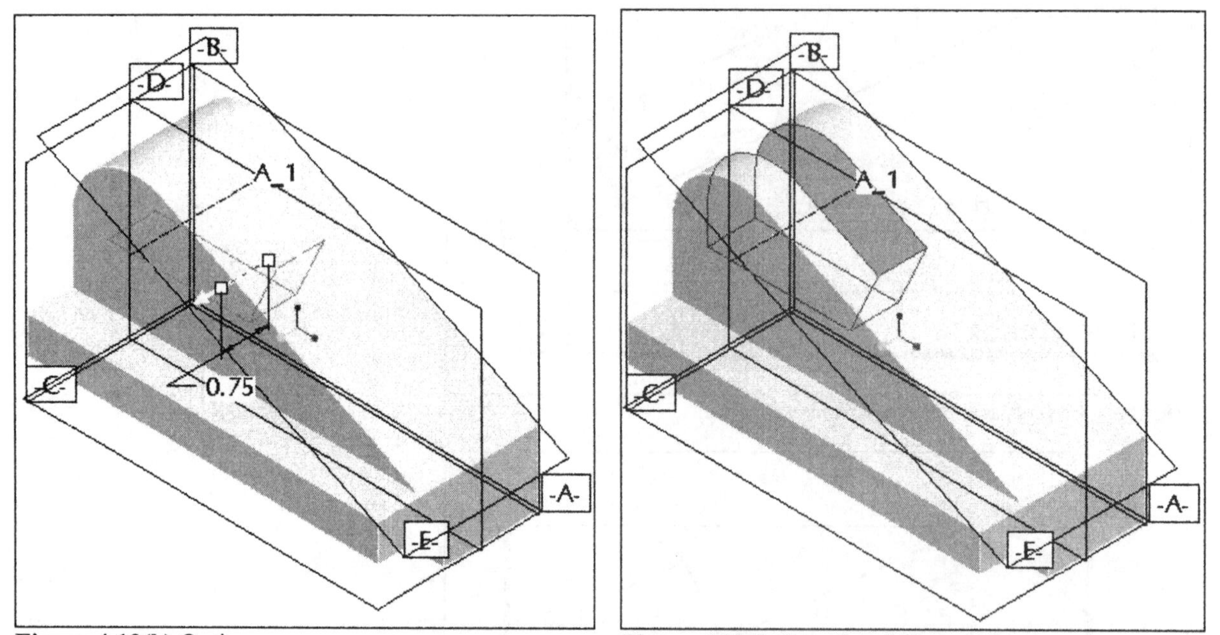

Figure 4.13(k) Options

Figure 4.13(l) Completed Slot

The hole drilled in the angled surface appears to be aligned with datum D. Upon closer inspection (Fig. 4.7), it can be seen that the hole is at a different distance *(.875 from B)* and is not in line with the slot and datum plane *(D was offset .9375 from B)*. Create the feature using a sketched hole. The drill tip (**118** degrees) at the bottom of the hole needs to be modeled so the hole is created as a sketched hole.

Click: **Hole Tool** ⇒ Simple ⇒ Sketched ⇒ pick on the angled face and a hole will display with handles ⇒ drag one handle to datum **B** and the other to the edge between the angled surface and the right vertical face ⇒ double click on the dimension from the hole's center to datum **B** and modify the value to **.875** ⇒ **Enter** ⇒ double click on the dimension from the hole's center to the edge and modify the value to **2.0625** ⇒ **Enter** ⇒ **Placement** tab [Figs. 4.14(a-b)] ⇒ ⇒ **Tools** ⇒ **Environment** ⇒ Snap to Grid ⇒ **Apply** ⇒ **OK** ⇒ **Toggle the grid on** ⇒ ⇒ sketch a vertical centerline ⇒ ⇒ sketch the four lines required to describe half of the hole's shape [Figs. 4.14(c-d)] ⇒ **MMB** ⇒ ⇒ Create a diameter dimension by picking the centerline with the **LMB**, then the edge to be dimensioned with the **LMB**, and then the centerline a second time with the **LMB**. Place the dimension by picking a position with the **MMB**. ⇒ add the angle and the depth dimension [Fig. 4.14(e)] ⇒ add centerlines, point, and symmetric constraint [Fig. 4.14(f)] ⇒ add the full angle dimension ⇒ delete the half-angle dimension ⇒ **Modify** to the design values [Fig. 4.14(g)] ⇒ ⇒ ⇒ ⇒ **Standard Orientation** ⇒ [Fig. 4.14(h)] ⇒ ⇒ **MMB** ⇒ **LMB** to deselect

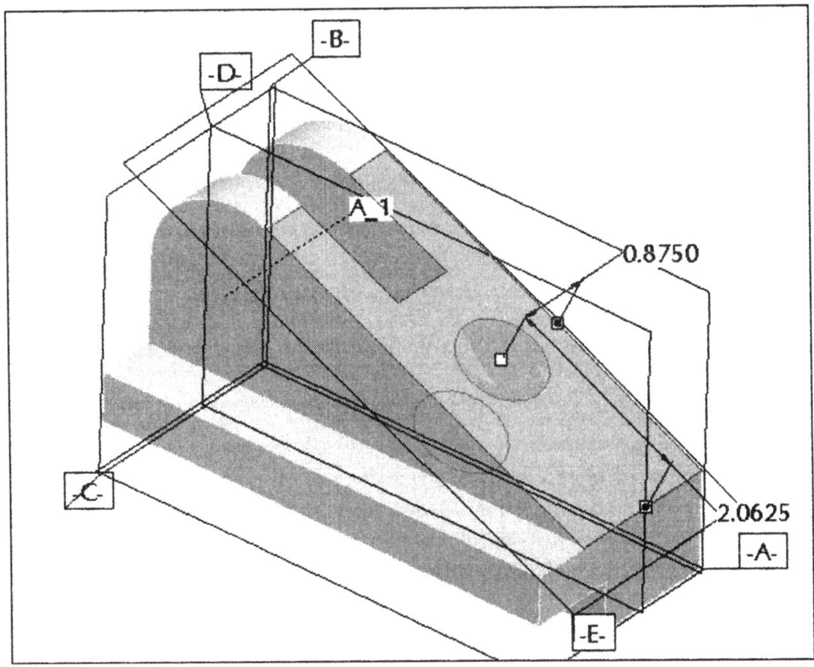

Figure 4.14(a) Hole Placement

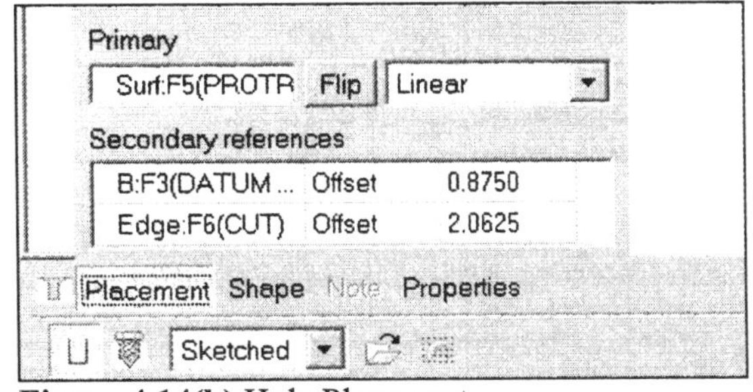

Figure 4.14(b) Hole Placement

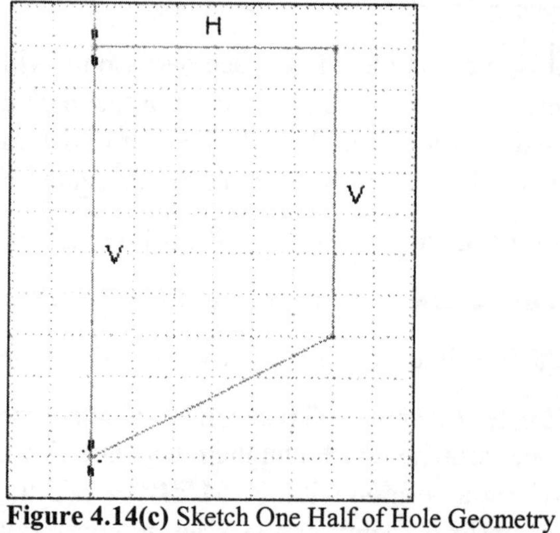

Figure 4.14(c) Sketch One Half of Hole Geometry

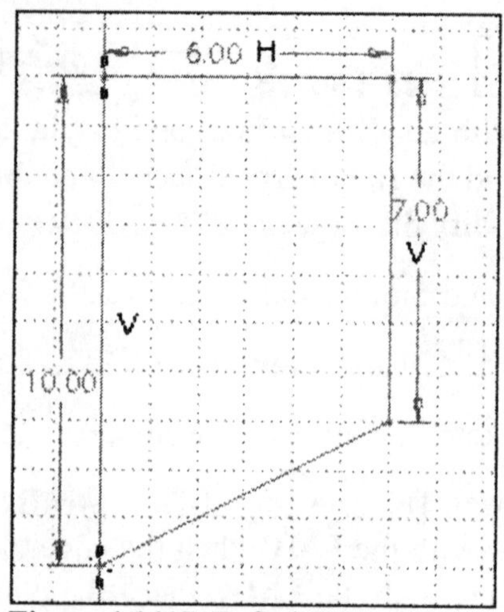

Figure 4.14(d) Default Dimensions

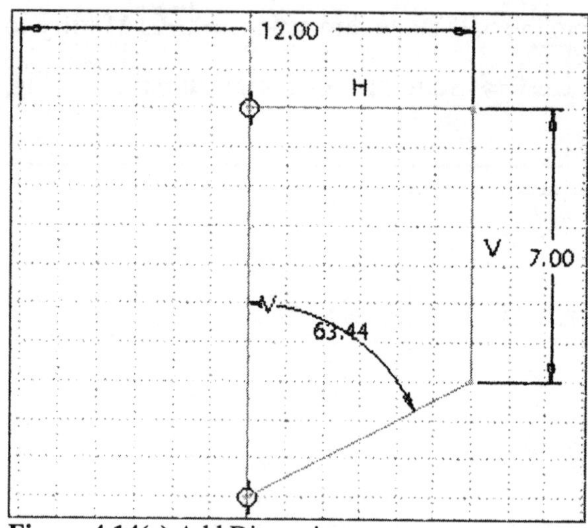

Figure 4.14(e) Add Dimensions

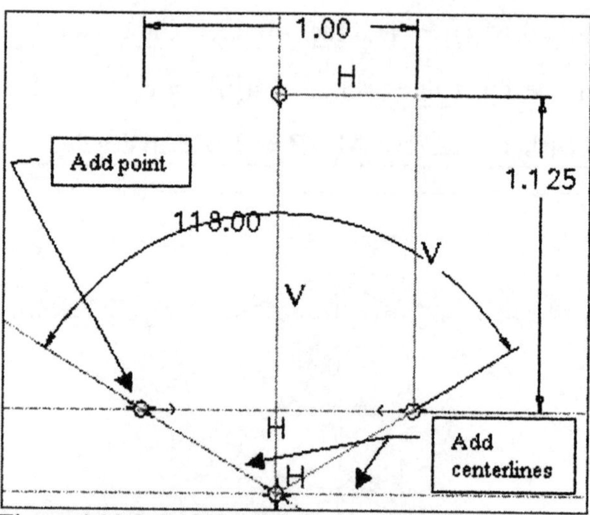

Figure 4.14(f) Add Centerlines, Point, and Symmetry Constraint

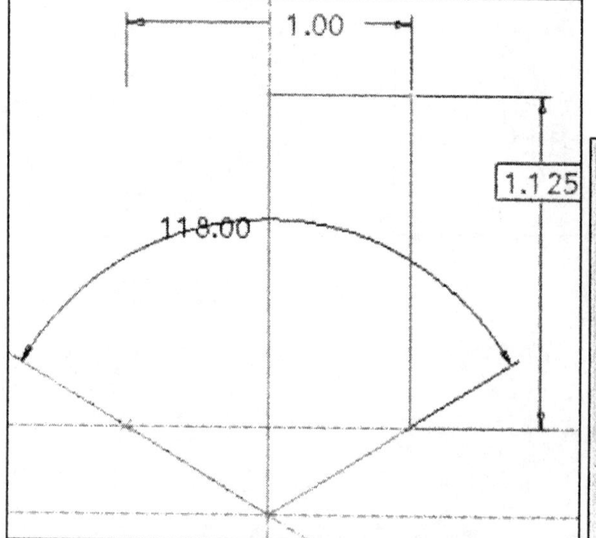

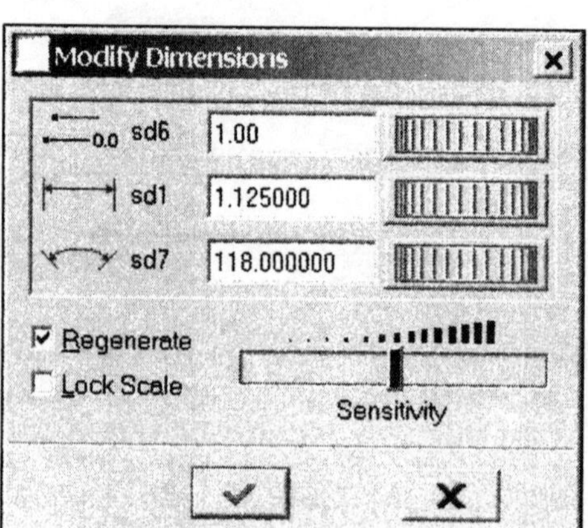

Figure 4.14(g) Add and Modify Dimensions

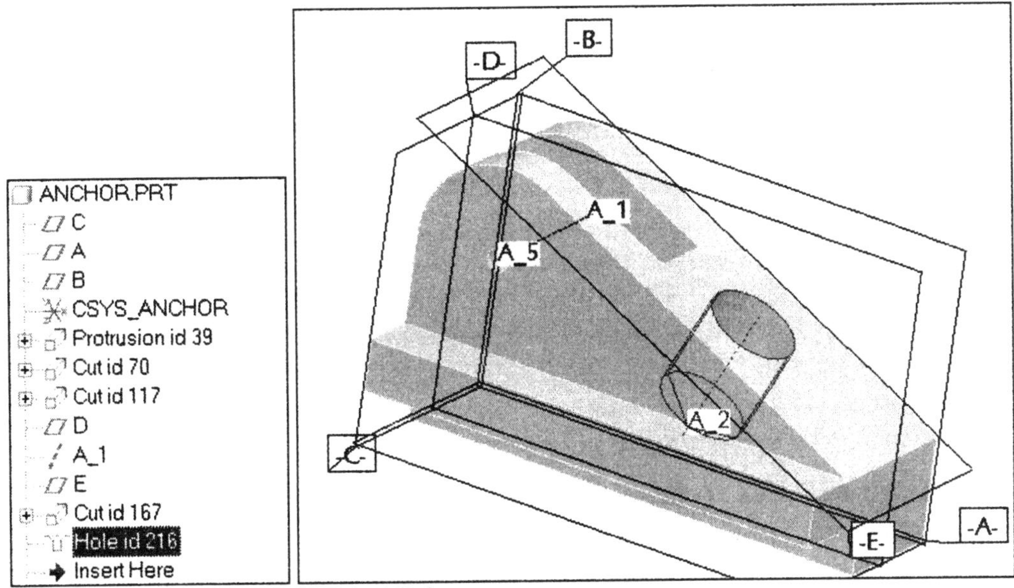

Figure 4.14(h) Completed Sketched Hole

The last feature to create is a ∅**.250** hole. Click: **Hole Tool** ⇒ pick **A_1** from the model ⇒ **RMB Secondary References Collector** ⇒ pick datum **D** from the model ⇒ diameter **.250** ⇒ **Enter** ⇒ **Shape** tab ⇒ **Through All** (both sides) [Fig. 4.15(a)] ⇒ ⇒ ⇒ [Fig. 4.15(b)] ⇒ ⇒ **MMB** ⇒ **LMB** to deselect

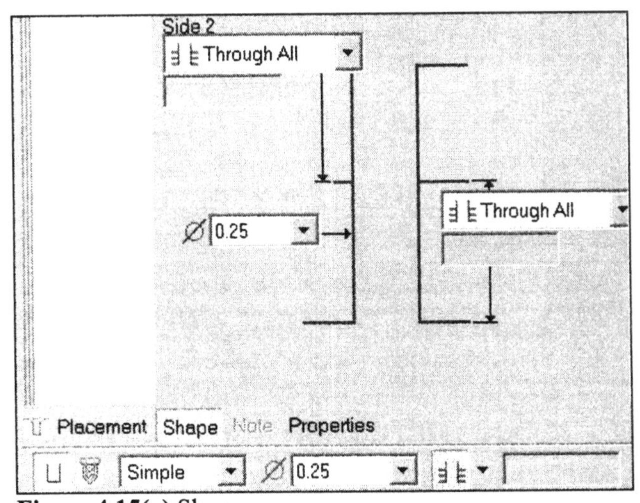

Figure 4.15(a) Shape

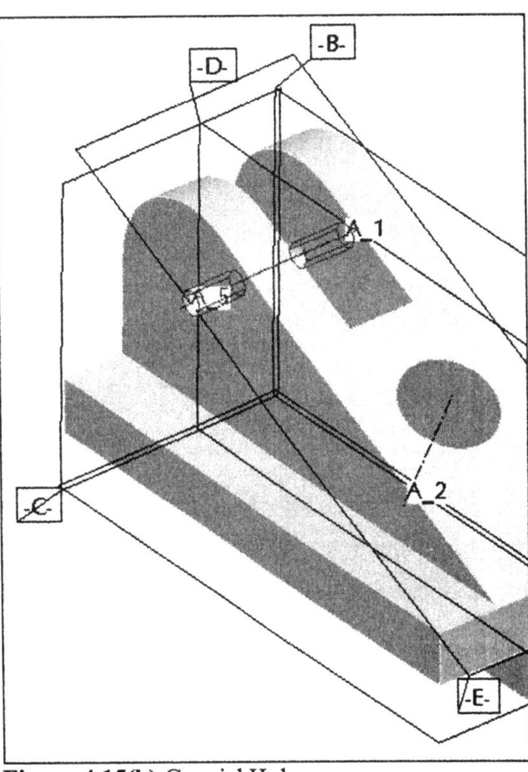

Figure 4.15(b) Coaxial Hole

Suppressing and Resuming Features

Next, you will create a new layer and add the two holes to it. The holes will then be selected in the Layer Tree and suppressed. Suppressed features are temporarily removed from the model along with their children (if any). In the example, you will notice that the holes and one axis (A_4) will be suppressed. Since axis A_1 is the parent of the small hole, it will not be suppressed. Suppressing features is like removing them from regeneration temporarily. However, you can "unsuppress" (resume) suppressed features at any time.

You can suppress features on a part to simplify the part model and decrease regeneration time. For example, while you work on one end of a shaft, it may be desirable to suppress features on the other end of the shaft. Similarly, while working on a complex assembly, you can suppress some of the features and components for which the detail is not essential to the current assembly process. Suppress features to do the following:

- Concentrate on the current working area by suppressing other areas
- Speed up a modification process because there is less to update
- Speed up the display process because there is less to display
- Temporarily remove features to try different design iterations

Unlike other features, the base feature cannot be suppressed. If you are not satisfied with your base feature, you can redefine the section of the feature, or start another part.

Click: **Show** in the **Navigator** ⇒ **Layer Tree** Layer Tree displays in place of the Model Tree ⇒ click on and highlight **Layers** in the Layer Tree ⇒ **RMB** ⇒ **New Layer** ⇒ Name: type **HOLES** as the name for the new layer *(do not press Enter)* ⇒ Selection Filter (bottom right-hand side below graphics window) click: **Feature** ⇒ select the two holes from the model [Fig. 4.16(a)] ⇒ **OK** ⇒ expand the **HOLES** layer and the **PRT_ALL_AXES** layer ⇒ press and hold down the **Ctrl** key and pick on the two holes in the **HOLES** layer [Fig. 4.16(b)] ⇒ **Edit** from the menu bar ⇒ **Suppress** ⇒ **OK** from the Suppress dialog box ⇒ **LMB** to deselect *(look at the Layer Tree)* [Fig. 4.16(c)] ⇒ **Edit** from the menu bar ⇒ **Resume** ⇒ **All** ⇒ **Show** ⇒ **Model Tree** ⇒ [icon] ⇒ **MMB** ⇒ **LMB** to deselect

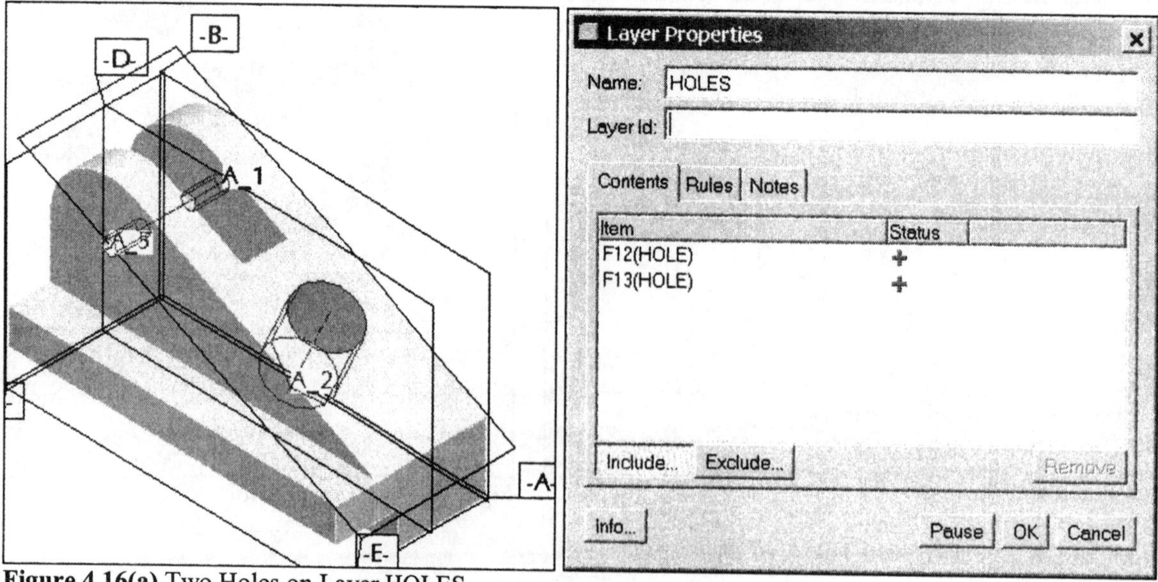

Figure 4.16(a) Two Holes on Layer HOLES

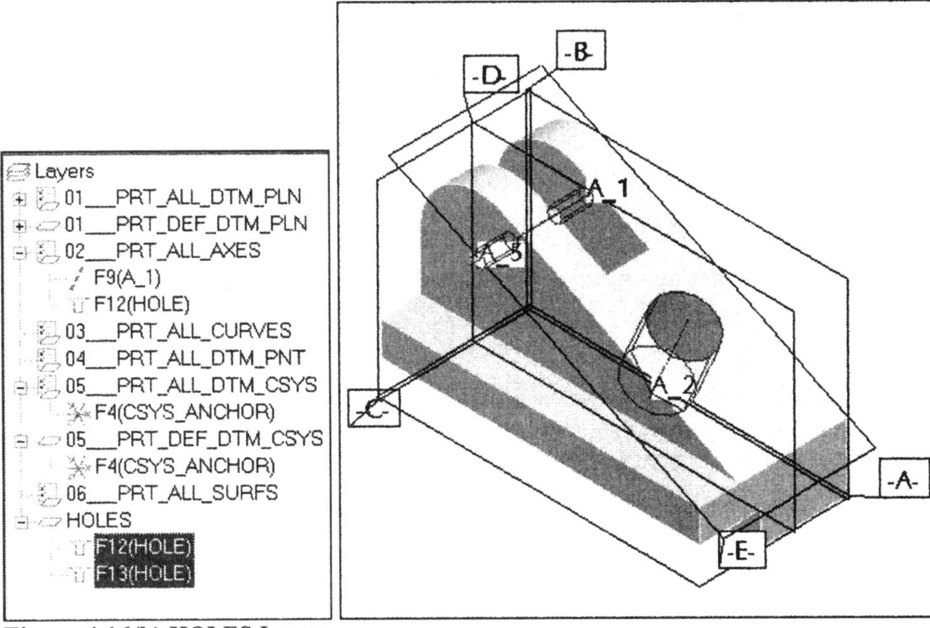

Figure 4.16(b) HOLES Layer

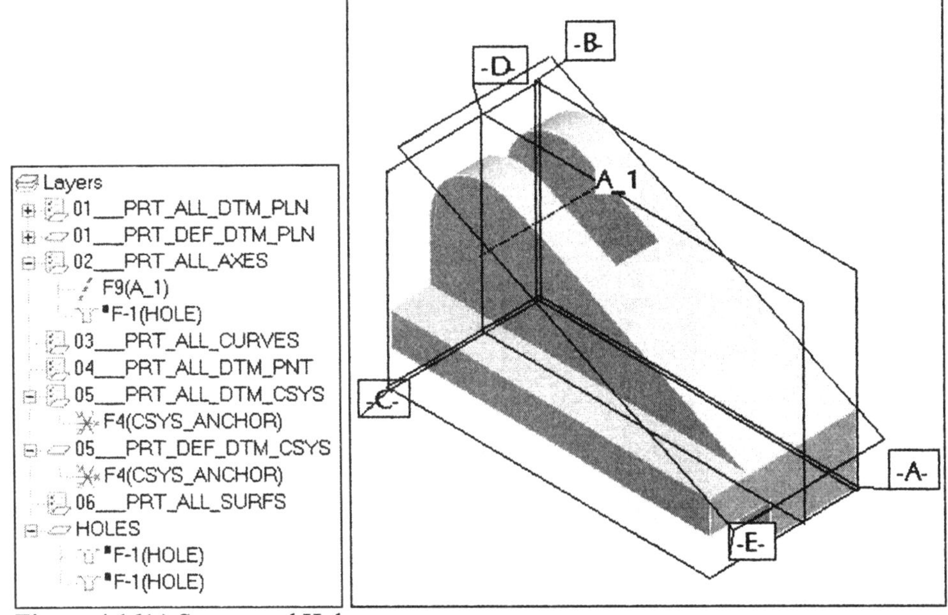

Figure 4.16(c) Suppressed Holes

Cross Sections

There are two types of cross sections: **planar** and **offset**. Planar cross sections can be crosshatched or filled, while offset cross sections can be crosshatched but not filled. You will be creating a planar cross section. Pro/E can create standard planar cross sections of models (parts or assemblies), offset cross sections of models (parts or assemblies), planar cross sections of datum surfaces or quilts (Part mode only) and planar cross sections that automatically intersect all quilts and all geometry in the current model.

Cross section cut planes do not intersect cosmetic features in a model.

Click: **Start the view manager** from the System Toolbar ⇒ View Manager dialog box displays [Fig. 4.17(a)] ⇒ **Xsec** tab [Fig. 4.17(b)] ⇒ **New** [Fig. 4.17(c)] ⇒ type name **A** [Fig. 4.17(d)] ⇒ **Enter** ⇒ **Planar** ⇒ **Single** ⇒ **MMB** ⇒ **Plane** ⇒ pick datum **D** ⇒ **Display** ⇒ **Show X-Hatching** ⇒ **Edit** ⇒ **Redefine** [Fig. 4.17(e)] ⇒ **Hatching** ⇒ **Fill** [Fig. 4.17(f)] ⇒ **Hatch** ⇒ **Spacing** ⇒ **Half** [Fig. 4.17(g)] ⇒ **MMB** ⇒ **MMB** ⇒ **Display** ⇒ **Normal** ⇒ **Close** ⇒ ⇒ **MMB**

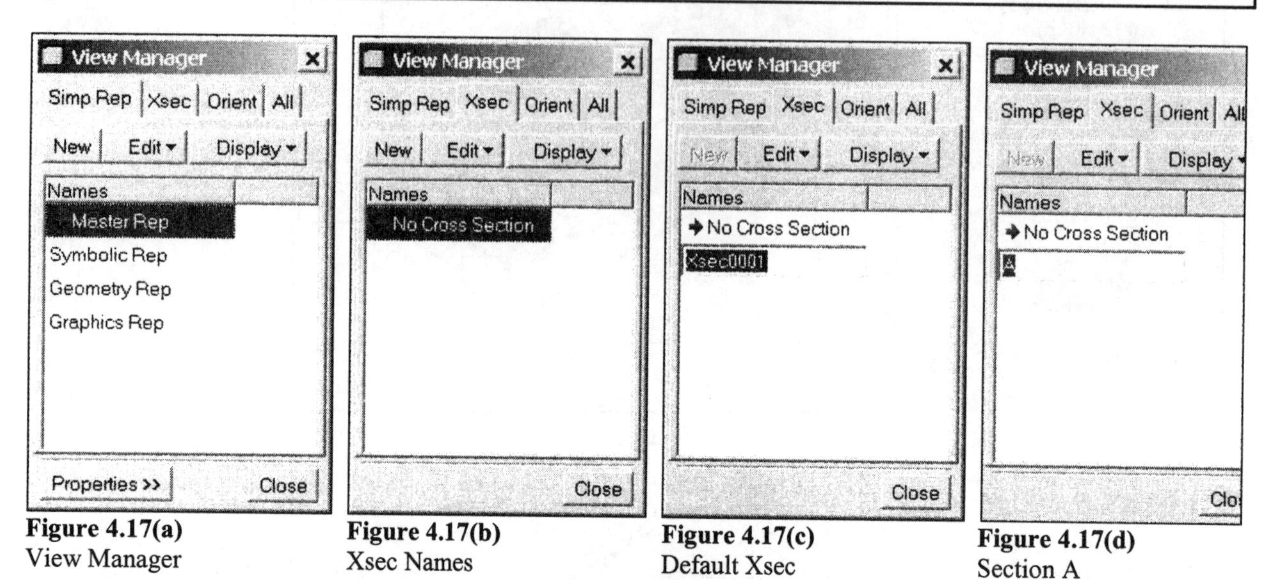

Figure 4.17(a) View Manager

Figure 4.17(b) Xsec Names

Figure 4.17(c) Default Xsec

Figure 4.17(d) Section A

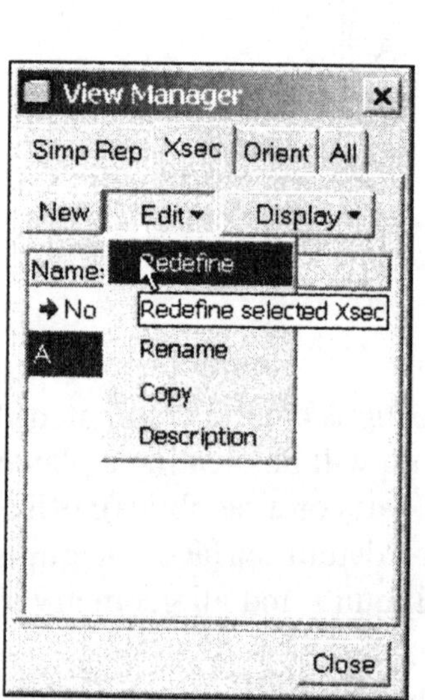

Figure 4.17(e) Redefine Xsec

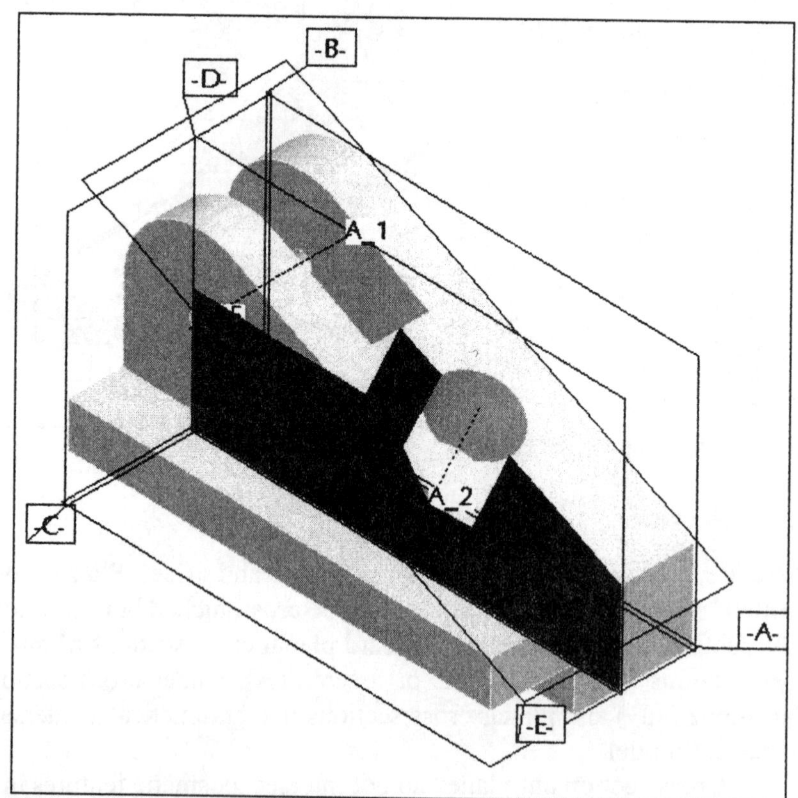

Figure 4.17(f) Model X-Section Fill

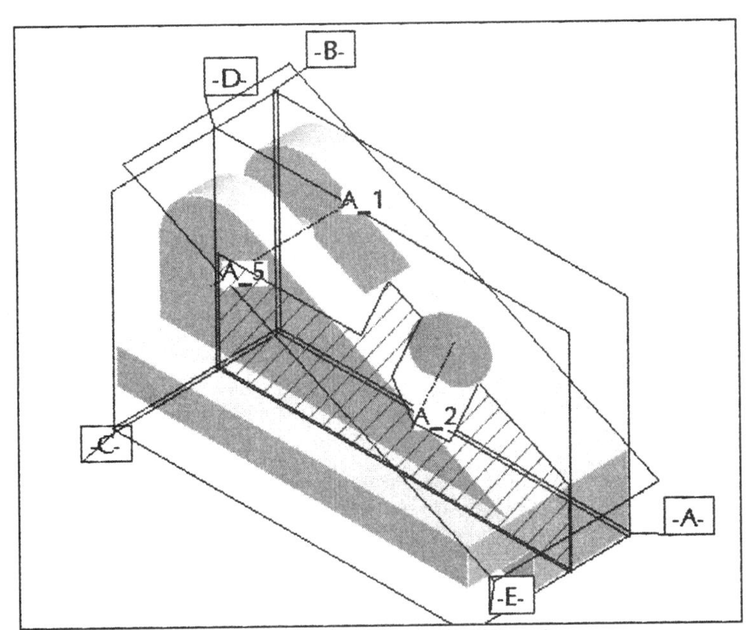

Figure 4.17(g) Model X-Section Hatch Half Spacing

The section passes through datum D. The slot is the *child* of datum D. If datum D moves, so will the slot and the small hole (and X-Section A). In order to ensure that the datum D stays centered on the upper portion of the part, you will need to create a relation to control the location of datum D. Relations will be covered in-depth in a later lesson. The first cut used a dimension from datum B for location (**1.875**). The relation should state that the distance from datum B to datum D will be one-half the value of the distance from datum B to the first cut. If the thickness of the upper portion of the part (**1.875**) changes, datum D will remain centered, as will the slot, and the X-Section. To start, you must first find out the feature dimension symbols (**d#**) required for the relation.

Select the first cut from the Model Tree (Extrude 2) ⇒ **RMB** ⇒ **Edit** [Fig. 4.18(a-b)] ⇒ **Info** from the menu bar ⇒ Switch Dimensions note the **d** symbol **d10** [Fig. 4.18(c)] ⇒ click on datum **D** in the **Model Tree** ⇒ **RMB** ⇒ **Edit** [Fig. 4.18(c)] ⇒ note the **d** symbol **d13** [Figs. 4.18(d-e)] *(if you have the incorrect symbol, your relation will not work) (Note: your "**d**" values may be different)*

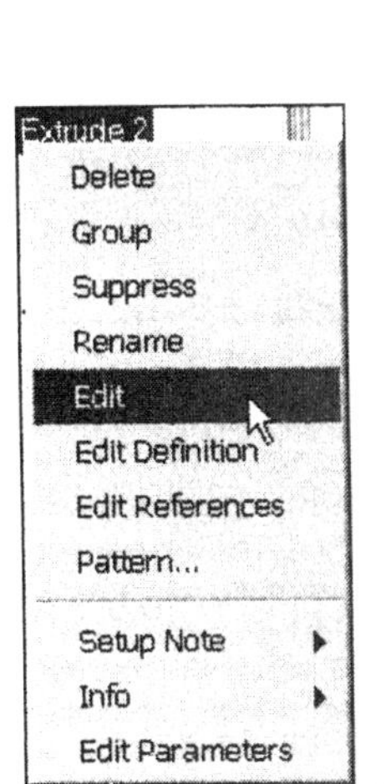

Figure 4.18(a) Edit

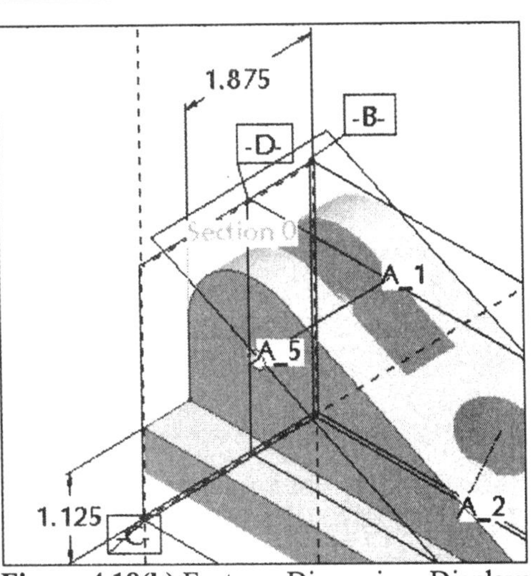

Figure 4.18(b) Features Dimensions Displayed

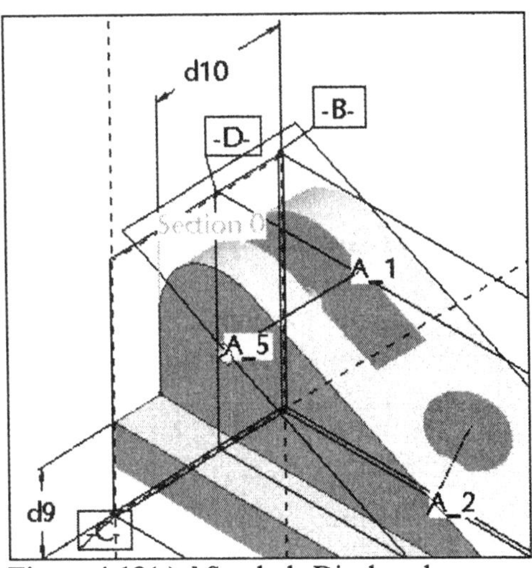

Figure 4.18(c) d Symbols Displayed

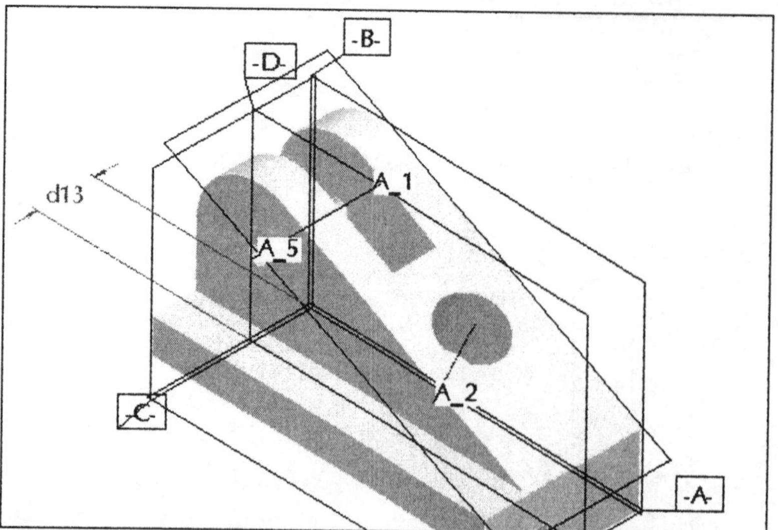

Figure 4.18(d) Using Edit to Display Datum D Dimension Symbol

Click: **Tools** from the menu bar ⇒ **Relations** ⇒ Relations dialog box displays ⇒ type **d13=d10/2** [Fig. 4.18(e)] *your "d" values may be different* ⇒ **Ok** ⇒ **Info** ⇒ **Switch Dimensions** ⇒ [icon] ⇒ **MMB**

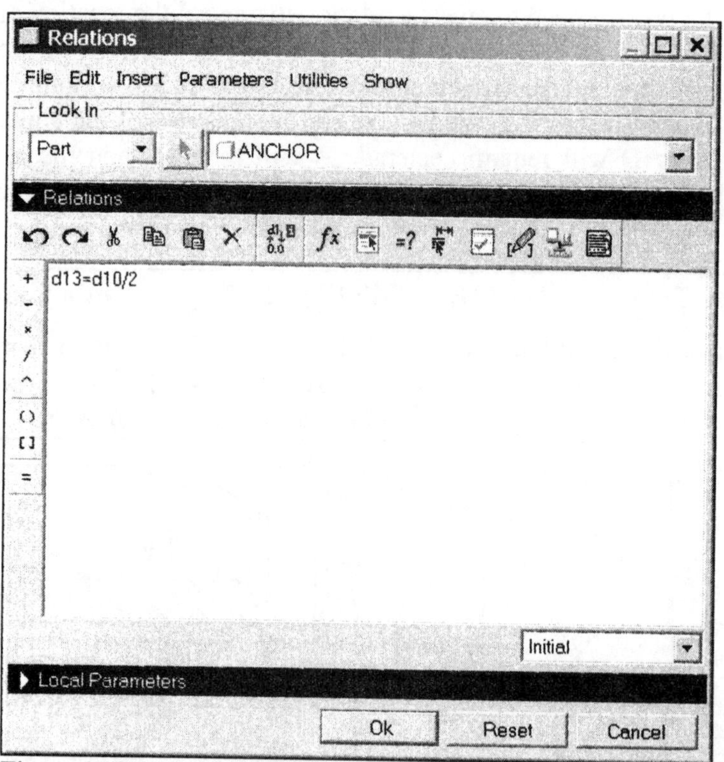

Figure 4.18(e) Relations **d13=d10/2** (Your d symbols may be different)

In the real world, you will seldom encounter a situation where the project is designed and modeled without a "design change" or **ECO** (Engineering Change Order). Therefore, let us assume that an ECO has been "issued" that states: *the location of the hole on the angled surface must be aligned with the center of the slot.*

Double-click on the large hole [Fig. 4.18(f)] ⇒ **Info** ⇒ **Switch Dimensions**- note the **d** symbol **d24** *(your "d" values may be different)* [Fig. 4.18(g)] (hole dimension from datum B) ⇒ **Tools** ⇒ **Relations** ⇒ below the first relation type **d24=d10/2** *(your "d" values may be different [Fig. 4.18(j)])* ⇒ **Ok** ⇒ **Info** ⇒ **Switch Dimensions** ⇒ double-click on the first cut on the model ⇒ double-click on **1.875** [Fig. 4.18(h)] and modify to **2.00** [Fig. 4.18(i)] ⇒ **Enter** ⇒ ⇒ ⇒ **TOP** [Fig. 4.18(k)]

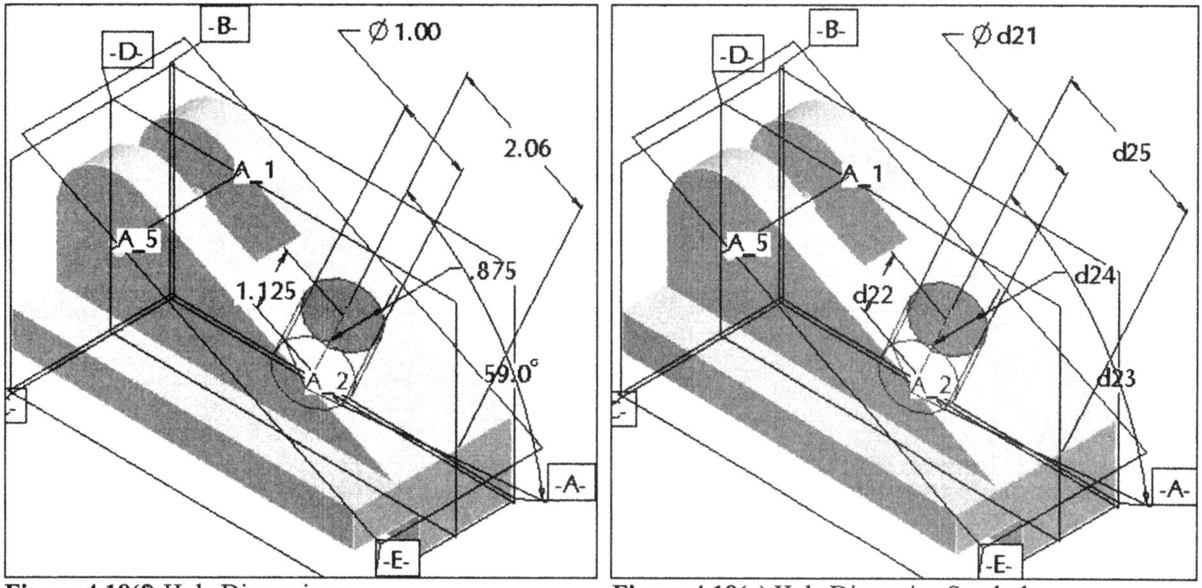

Figure 4.18(f) Hole Dimensions

Figure 4.18(g) Hole Dimension Symbols

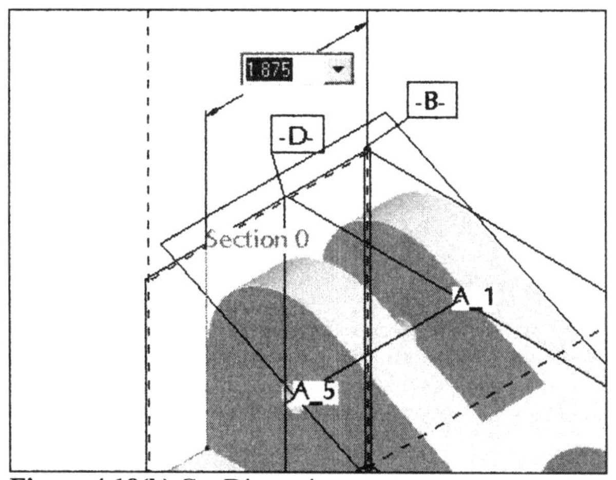

Figure 4.18(h) Cut Dimensions

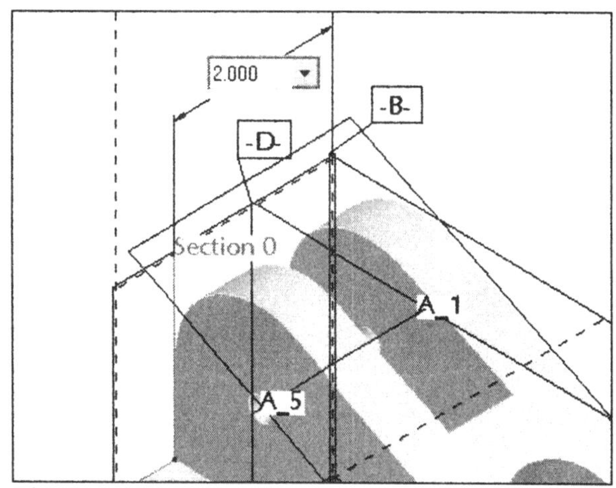

Figure 4.18(i) Edited Cut

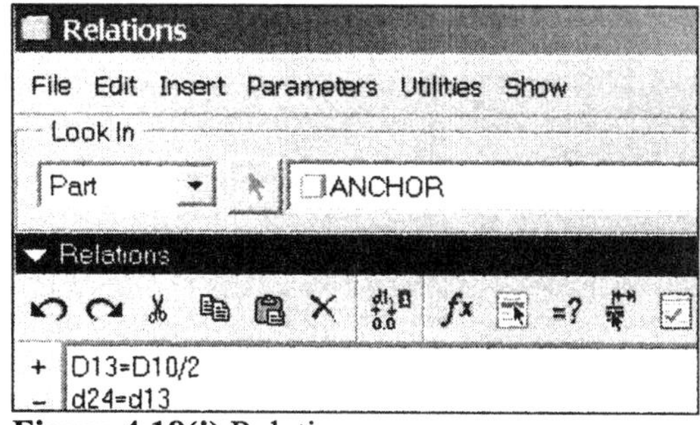

Figure 4.18(j) Relations

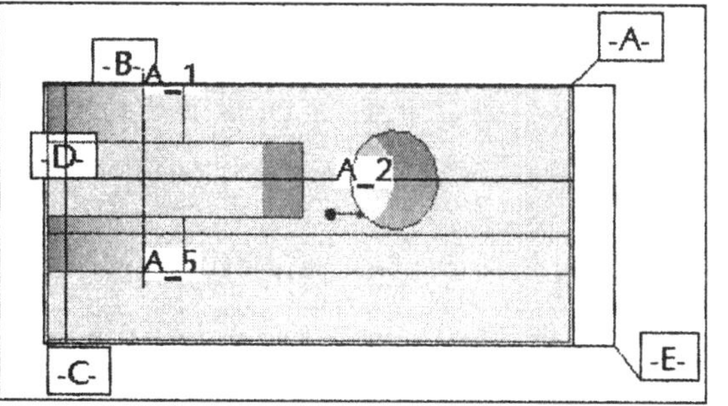

Figure 4.18(k) Top View, Slot, Hole, and Datum are Aligned

Click: ⇒ **Standard Orientation** ⇒ **Info** ⇒ **Switch Dimensions** ⇒ **Undo Edit Value** ⇒ **Info** ⇒ **Relations and Parameters** ⇒ click on **d10**, **d13**, and **d24** in the Browser in order to display them in the graphics window [Fig. 4.18(l)] ⇒ close the Browser with the quick sash ⇒ **Info** ⇒ **Switch Dimensions** ⇒ ⇒ **MMB** ⇒ **File** ⇒ **Delete** ⇒ **Old Versions** ⇒ **MMB** ⇒ **File** ⇒ **Close Window**

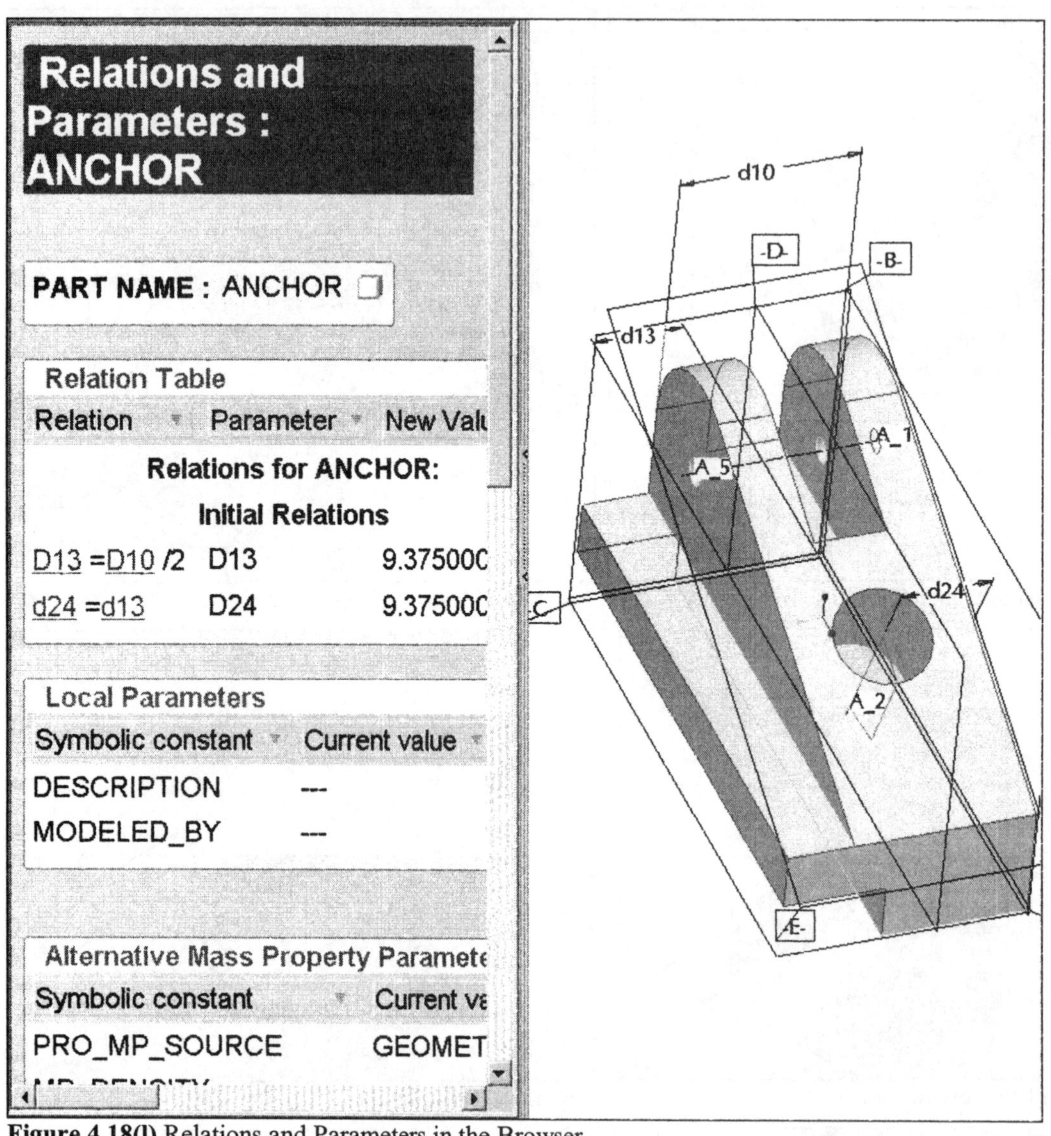

Figure 4.18(l) Relations and Parameters in the Browser

Lesson 5 Revolved Features

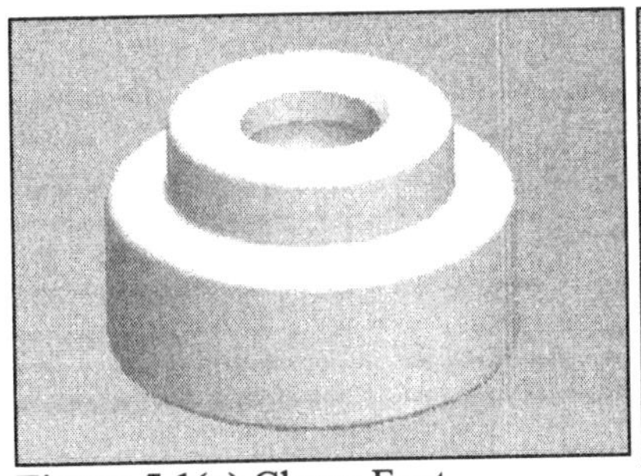
Figure 5.1(a) Clamp Foot

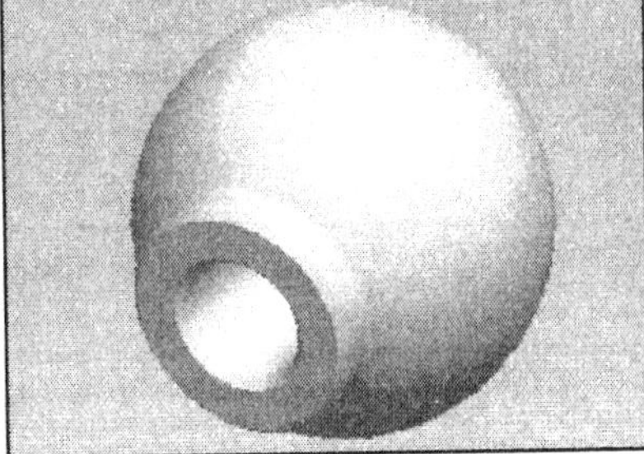
Figure 5.1(b) Clamp Swivel

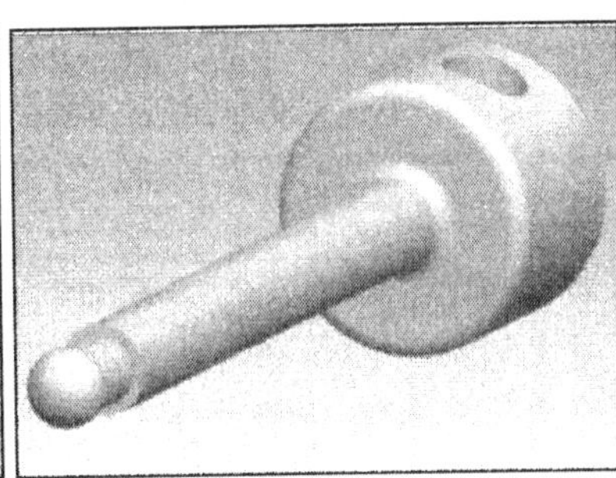
Figure 5.1(c) Clamp Ball

OBJECTIVES

- Master the **Revolve Tool** (Figs. 5.1(a-c)]
- Create **Chamfers** along part edges
- Learn how to **Sketch in 3D**
- Understand and use the **Navigation browser**
- Alter and set the **Items** and **Columns** displayed in the **Model Tree**
- Create standard **Tapped Holes**
- Create **Cosmetic Threads** and complete **tabular information** for threads
- Edit **Dimension Properties**
- Use the **Model Player** to extract information and dimensions
- Get a hard copy using the **Print** command

Revolved Features

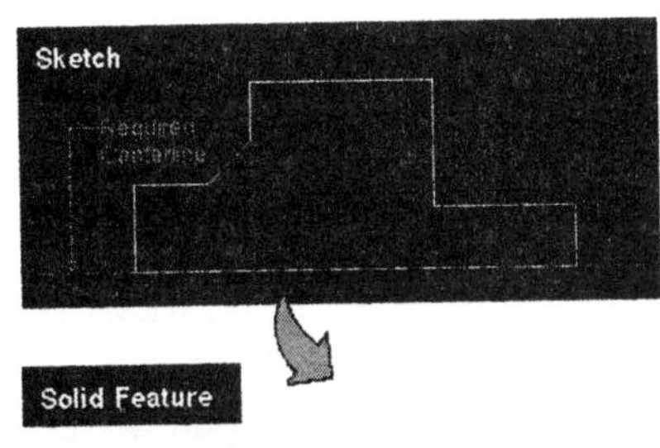

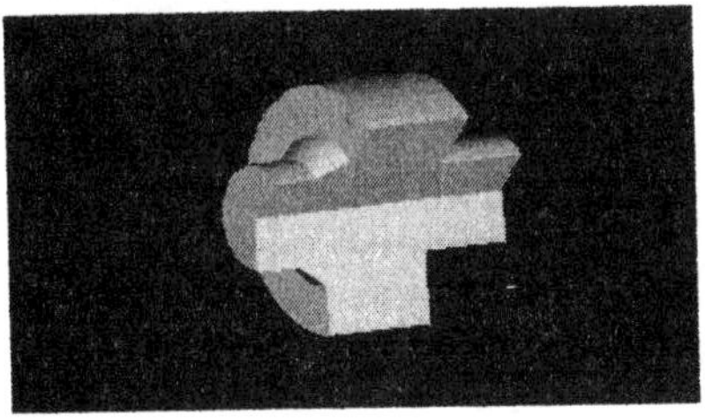
Figure 5.2 Revolved Protrusion
(CADTRAIN, COAch for Pro/ENGINEER)

The **Revolve Tool** creates a *revolved solid* or a *revolved cut* by revolving a sketched section around a centerline from the sketching plane (Fig. 5.2). You can have any number of centerlines in your sketch/section, but the first centerline will be the one used to rotate your section geometry. To create a revolved section, create a centerline and the geometry that will be revolved about that centerline. Rules for sketching a revolved feature include:

- The revolved section must have a centerline.
- By default Pro/E uses the first centerline sketched as the *axis of revolution* (you may select a different axis of revolution).
- The geometry must be sketched on only one side of the *axis of revolution.*
- The section must be closed for a solid (Fig. 5.2) but can be open for a cut.

A variety of geometric shapes and constructions are used on revolved features. For instance, **chamfers** are created at selected edges of the part. Chamfers are *pick-and-place* features.

Threads can be a *cosmetic feature* representing the *nominal diameter* or the *root diameter* of the thread. Information can be embedded in the feature. Threads show as a unique color. By putting cosmetic threads on a separate layer, you can hide, unhide, blank, or suppress them.

Chamfers

Chamfers are created between abutting edges of two surfaces on the solid model. An edge chamfer removes a flat section of material from a selected edge to create a beveled surface between the two original surfaces common to that edge. Multiple edges can be selected.

There are four basic dimensioning schemes for edge chamfers: (Fig. 5.3).

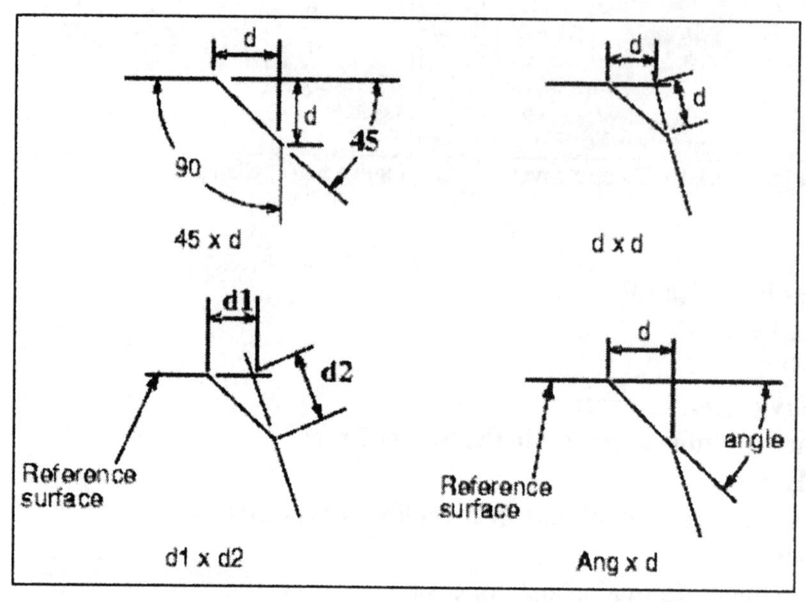

Figure 5.3 Chamfer Options

- **5 x d** Creates a chamfer that is at an angle of **45°** to both surfaces and a distance **d** from the edge along each surface. The distance is the only dimension to appear when edited. **45 x d** chamfers can be created only on an edge formed by the intersection of two *perpendicular* surfaces.
- **d x d** Creates a chamfer that is a distance **d** from the edge along each surface. The distance is the only dimension to appear when edited.
- **d1 x d2** Creates a chamfer at a distance **d1** from the selected edge along one surface and a distance **d2** from the selected edge along the other surface. Both distances appear along their respective surfaces.
- **Ang x d** Creates a chamfer at a distance **d** from the selected edge along one adjacent surface at an **Angle** to that surface.

Threads

Cosmetic threads are displayed with *magenta/purple* lines and circles. Cosmetic threads can be external or internal, blind or through. In the cylinder rod part, one end has external blind threads and the opposite end has internal blind threads. A cosmetic thread has a set of supported parameters that can be defined at its creation or later, when the thread is added.

Standard Holes

Standard holes are a combination of sketched and extruded geometry. It is based on industry-standard fastener tables. You can calculate either the tapped or clearance diameter appropriate to the selected fastener. You can use Pro/E supplied standard lookup tables for these diameters or create your own. Besides threads, standard holes can be created with chamfers.

Navigation Window

Besides using the File command and corresponding options (File ⇒ Set Working Directory), the *Navigation window* of the Pro/E main window can be used directly to access many of the same functions. Before you start the part and begin modeling, you will set the working directory using a different method, and explore some of the possibilities using the Navigator browser.

As you know, the working directory is a directory that you set up to contain Pro/E files. You must have read/write access to this directory. You usually start Pro/E from your working directory. A new working directory setting is not saved when you exit the current Pro/E session. By default, if you retrieve a file from a non-working directory, rename the file and then save it, the renamed file is saved to the directory from which it was originally retrieved, if you have read/write access to that directory. It is not saved in the current working directory, unless the config.pro option *save_object_in_current* is set to *yes*.

The navigation area is located on the left side of the Pro/E main window. It includes tabs for the Model Tree and Layer Tree, Folder Browser, Favorites, and Connections:

- **Model Tree** (default)
- **Layer Tree (Show ⇒ Layer Tree)**
- **Folder Browser**
- **Favorites**
- **Connections**

Folder Browser

The Folder browser (**Folder Browser**) is an expandable tree that lets you browse the file systems and other locations that are accessible from your computer. As you navigate the folder, the contents of the selected folder appear in the Pro/E browser as the Contents page. The Folder browser contains top-level nodes for accessing file systems and other locations that are known to Pro/E:

- **In Session** Pro/E objects that have been retrieved into local memory.
- **Shared Spaces** This is a shared file location accessed through the PTC Conference Center.
- **All registered servers** The browser lists all servers that you have registered with the Server Registry dialog box. These registered servers may include a Windchill server, a Pro/INTRALINK server, and an FTP server.
- **Node for the local file system** When you open the Folder browser, the local file system appears in the browser with the startup directory node expanded and highlighted.
- **Network Neighborhood** *(only for Windows)* The navigator shows computers on the networks to which you have access. The operations you can perform depend on your permissions to the remote computers.

Manipulating Folders

To work with folders, you can use the Folder browser toolbar or the shortcut menu. You can perform the following tasks with the toolbar:

- **Create a new folder**
- **Delete selected folders**
- **Open the Windchill Cabinets** This icon is available only when you are working with an active Windchill server
- **Open the Windchill Workspace** This icon is available only when you are working with an active Windchill server
- **Working directory**

Using the Shortcut Menu in the Folder browser

To open a shortcut menu, **RMB** click on an item in the Folder browser. The commands on the shortcut menu vary depending on the task and your permissions. The shortcut menu lists the following commands:

- **New Folder** Add a subfolder to the selected folder
- **Open** Open a folder in the Pro/ENGINEER browser
- **Expand** Expand a node (if not open), **Collapse** a node (if open)
- **Server Registry** Access the Server Registry dialog box
- **Set Working Directory** Designate the selected directory as the working directory
- **Rename** Rename a selected folder
- **Delete** Delete a selected folder and all subfolders

Click: **Folder Browser** in the Navigator [Fig. 5.4(a)] ⇒ click on the directory you wish to set as the working directory ⇒ **RMB** ⇒ **Set Working Directory** [Fig. 5.4(b)] *(Yours will be different)*

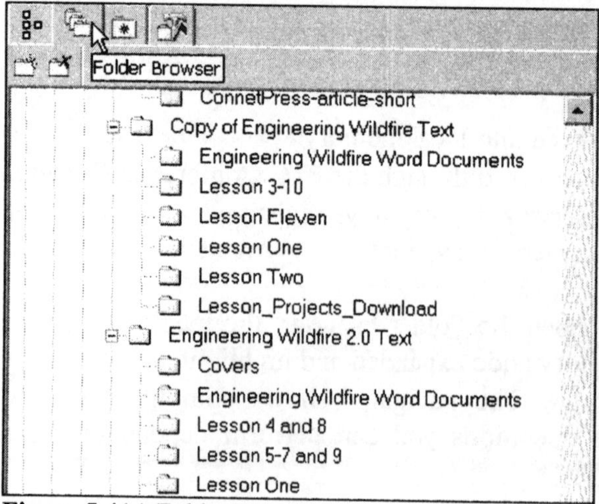

Figure 5.4(a) Folder Browser

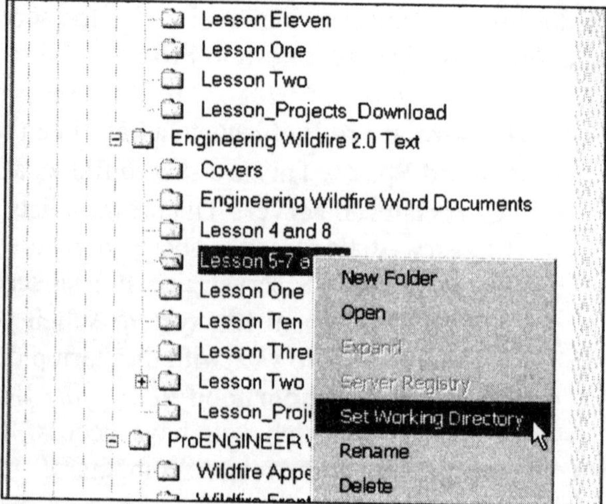

Figure 5.4(b) Folder Browser

Lesson 5 STEPS

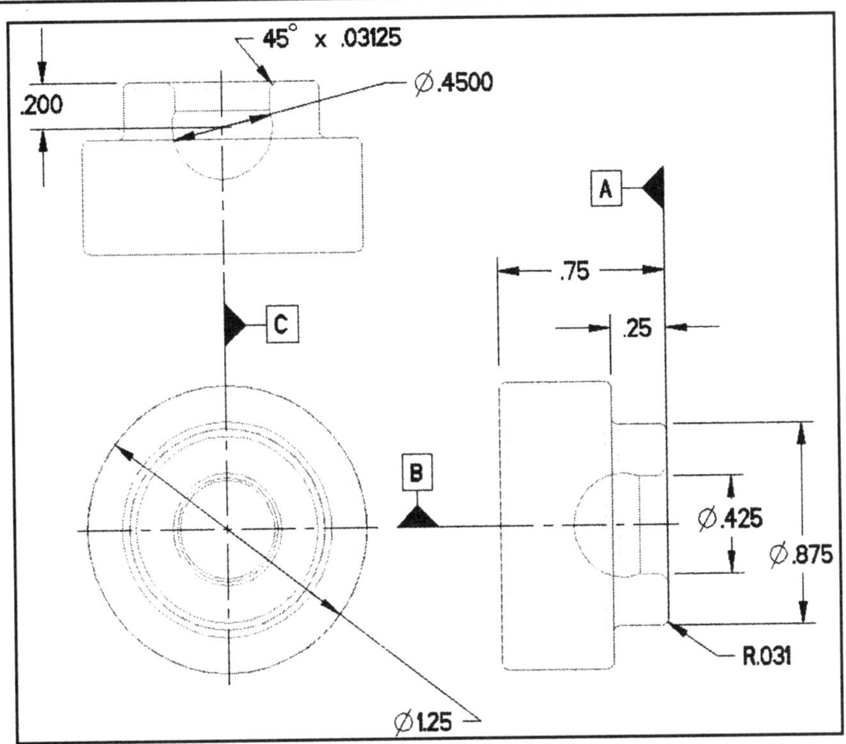

Figure 5.5 Clamp Foot Detail

Clamp Foot

The Clamp Foot (Fig. 5.5) is the first of three revolved parts created for Lesson 5. The Clamp Foot, Clamp Ball, and Clamp Swivel are three revolved parts needed for the Clamp assembly and drawings later in the text. The Clamp Foot requires a revolved extrusion (protrusion), a revolved cut, a chamfer and rounds.

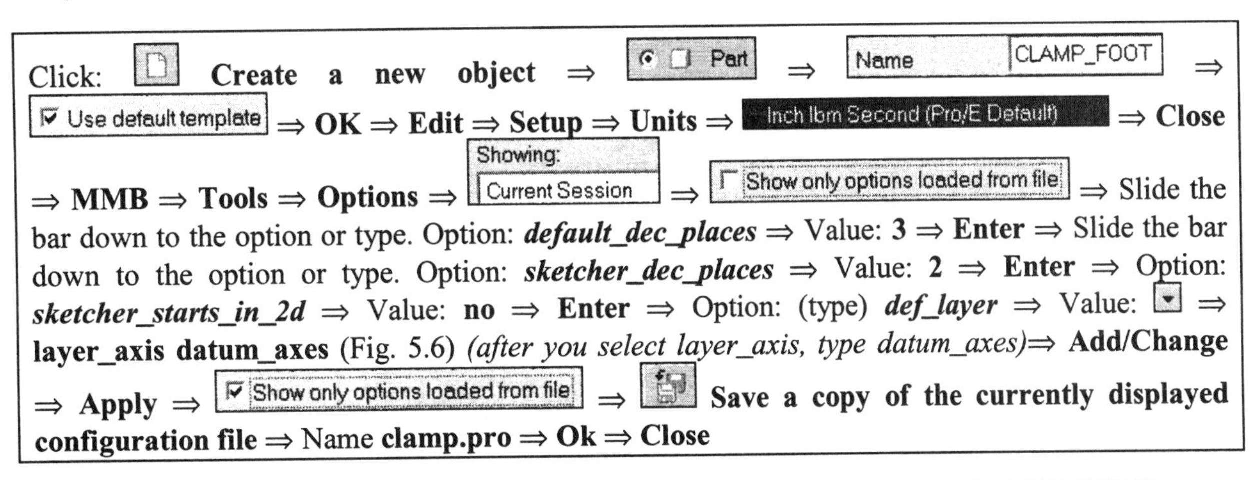

Click: **Create a new object** ⇒ Part ⇒ Name CLAMP_FOOT ⇒ Use default template ⇒ **OK** ⇒ **Edit** ⇒ **Setup** ⇒ **Units** ⇒ Inch lbm Second (Pro/E Default) ⇒ **Close** ⇒ **MMB** ⇒ **Tools** ⇒ **Options** ⇒ Showing: Current Session ⇒ Show only options loaded from file ⇒ Slide the bar down to the option or type. Option: ***default_dec_places*** ⇒ Value: **3** ⇒ **Enter** ⇒ Slide the bar down to the option or type. Option: ***sketcher_dec_places*** ⇒ Value: **2** ⇒ **Enter** ⇒ Option: ***sketcher_starts_in_2d*** ⇒ Value: **no** ⇒ **Enter** ⇒ Option: (type) ***def_layer*** ⇒ Value: ⇒ **layer_axis datum_axes** (Fig. 5.6) *(after you select layer_axis, type datum_axes)*⇒ **Add/Change** ⇒ **Apply** ⇒ Show only options loaded from file ⇒ **Save a copy of the currently displayed configuration file** ⇒ Name **clamp.pro** ⇒ **Ok** ⇒ **Close**

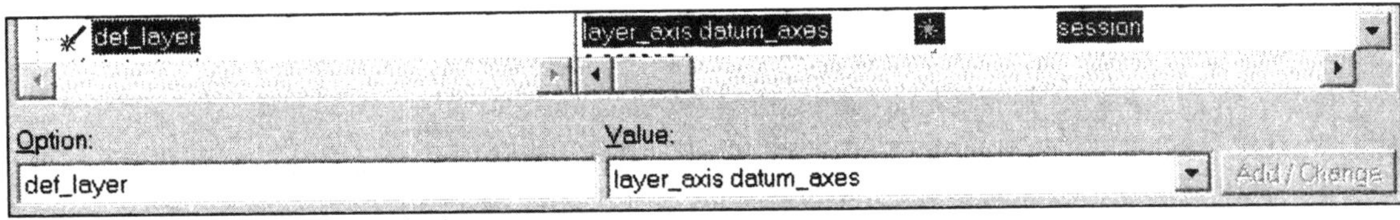

Figure 5.6 def_layer layer_axis datum_axes

The first protrusion is a revolved protrusion created with the Revolve Tool, you will be sketching the section in 3D since the configuration option; *sketcher_starts_in_2d,* was previously set to *no.* In the Environment dialog box, you can see that Use 2D Sketcher is deactivated.

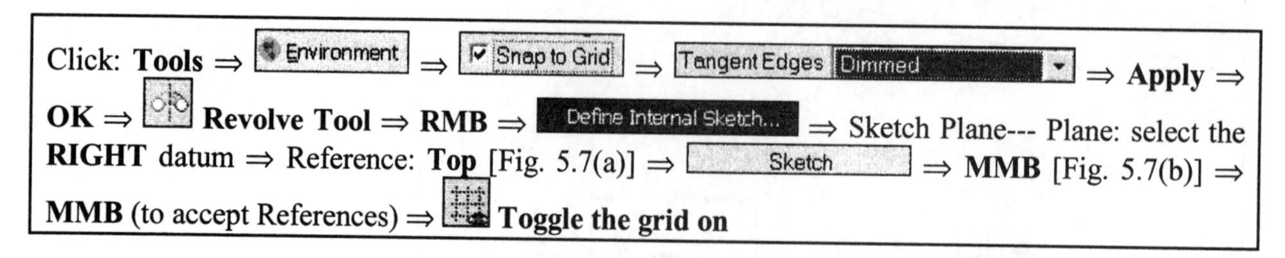

Click: **Tools** ⇒ Environment ⇒ Snap to Grid ⇒ Tangent Edges Dimmed ⇒ **Apply** ⇒ **OK** ⇒ **Revolve Tool** ⇒ **RMB** ⇒ Define Internal Sketch... ⇒ Sketch Plane--- Plane: select the **RIGHT** datum ⇒ Reference: **Top** [Fig. 5.7(a)] ⇒ Sketch ⇒ **MMB** [Fig. 5.7(b)] ⇒ **MMB** (to accept References) ⇒ **Toggle the grid on**

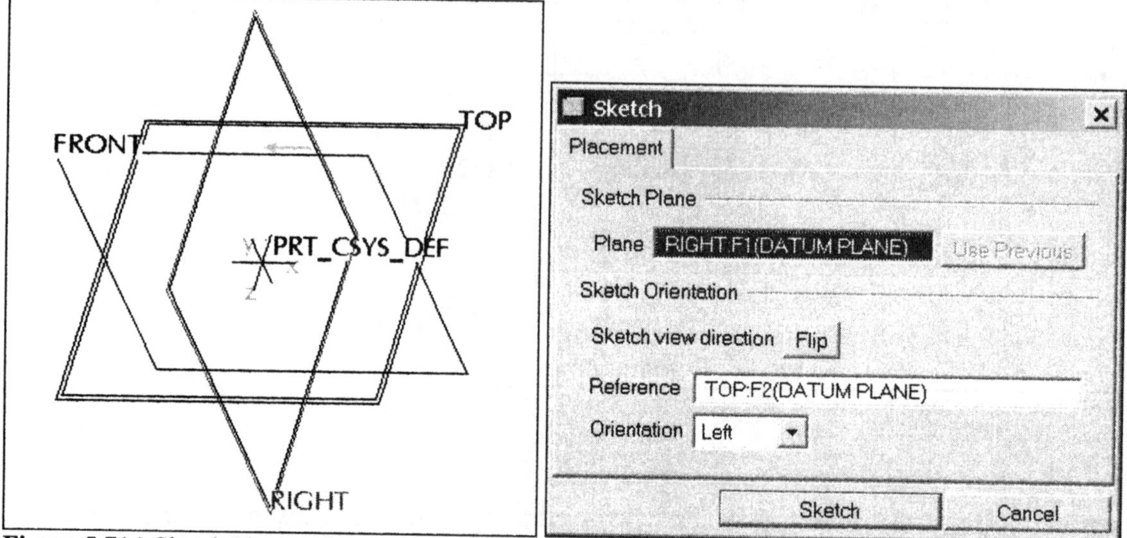

Figure 5.7(a) Sketch Plane and Reference

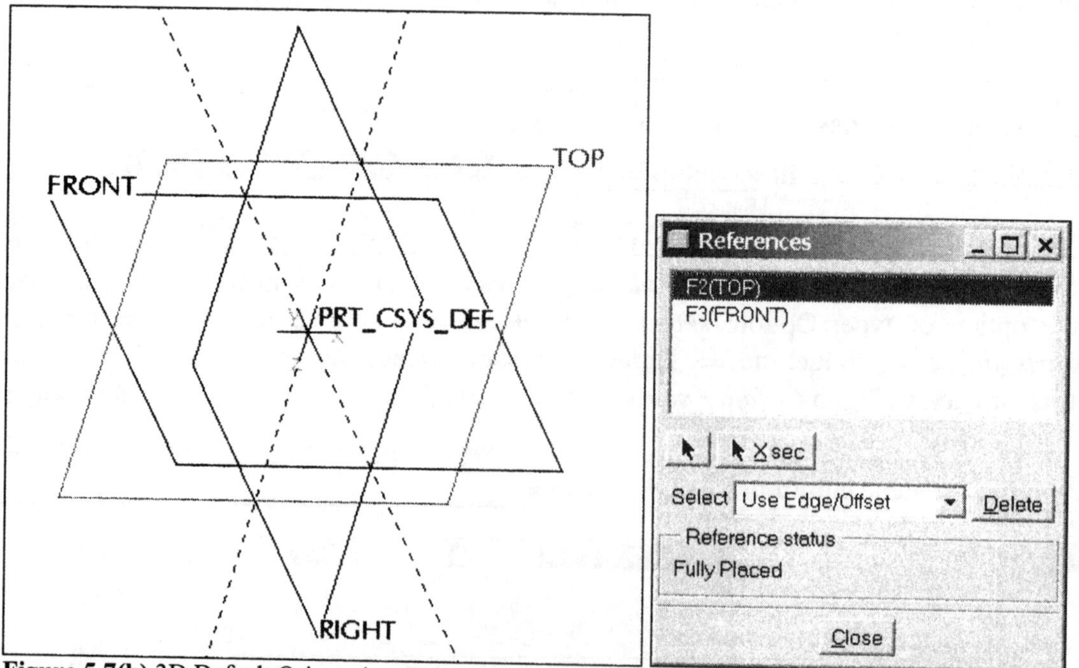

Figure 5.7(b) 3D Default Orientation (Trimetric) and References

Click: **RMB** ⇒ Centerline sketch a *vertical* centerline through the default coordinate system to be used as the axis of revolution [Fig. 5.7(c)] ⇒ **MMB** rotate the part to see the vertical plane clearly [Fig. 5.7(d)] ⇒ **RMB** ⇒ Line sketch the six lines on the RIGHT datum [Fig. 5.7(e)] ⇒ **MMB** ⇒ **MMB**

(Note: if you have difficulty, you can delete the lines and start again after clicking ***Orient the sketching plane parallel to the screen*** *and sketch the lines in 2D as in previous lessons)*

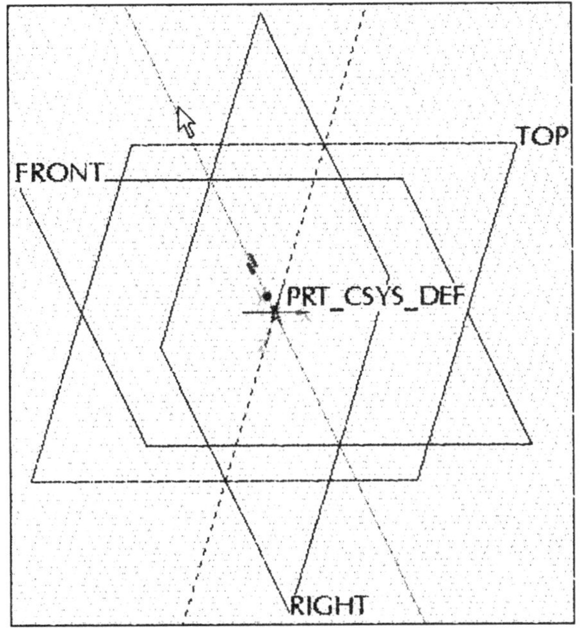

Figure 5.7(c) Sketch a Vertical Centerline.

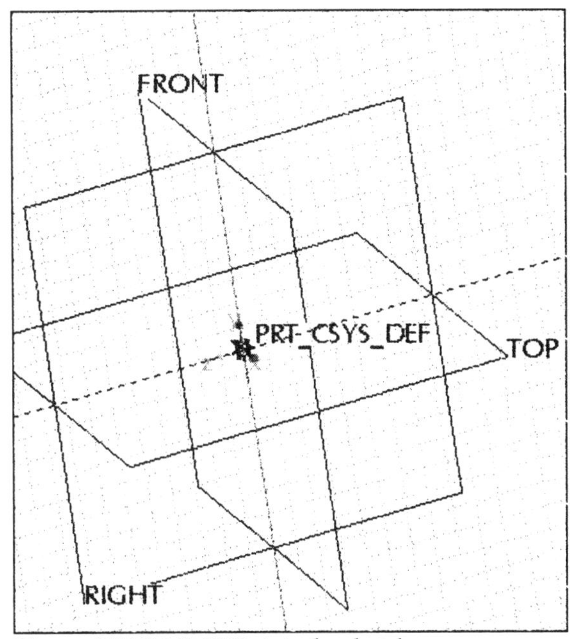

Figure 5.7(d) MMB Rotate the sketch

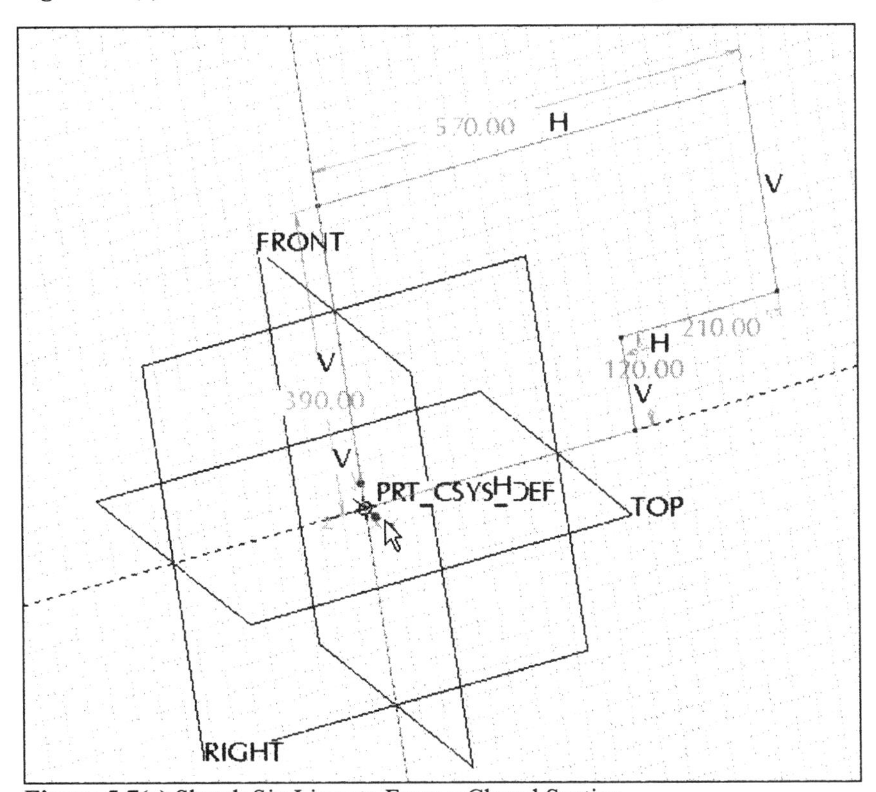

Figure 5.7(e) Sketch Six Lines to Form a Closed Section

Click: **Sketch** ⇒ **Options** ⇒ Snap To Grid [Fig. 5.7(f)] ⇒ ⇒ **off** ⇒ **Ctrl+R** ⇒ **RMB** ⇒ Dimension add the two vertical dimensions [Fig. 5.7(g)] ⇒ add a diameter dimension by picking the centerline, then pick the outer vertical edge line, and then pick the centerline again ⇒ **MMB** to place the dimension ⇒ repeat to dimension the other diameter [Fig. 5.7(h)] ⇒ **MMB** ⇒ **Ctrl+R**

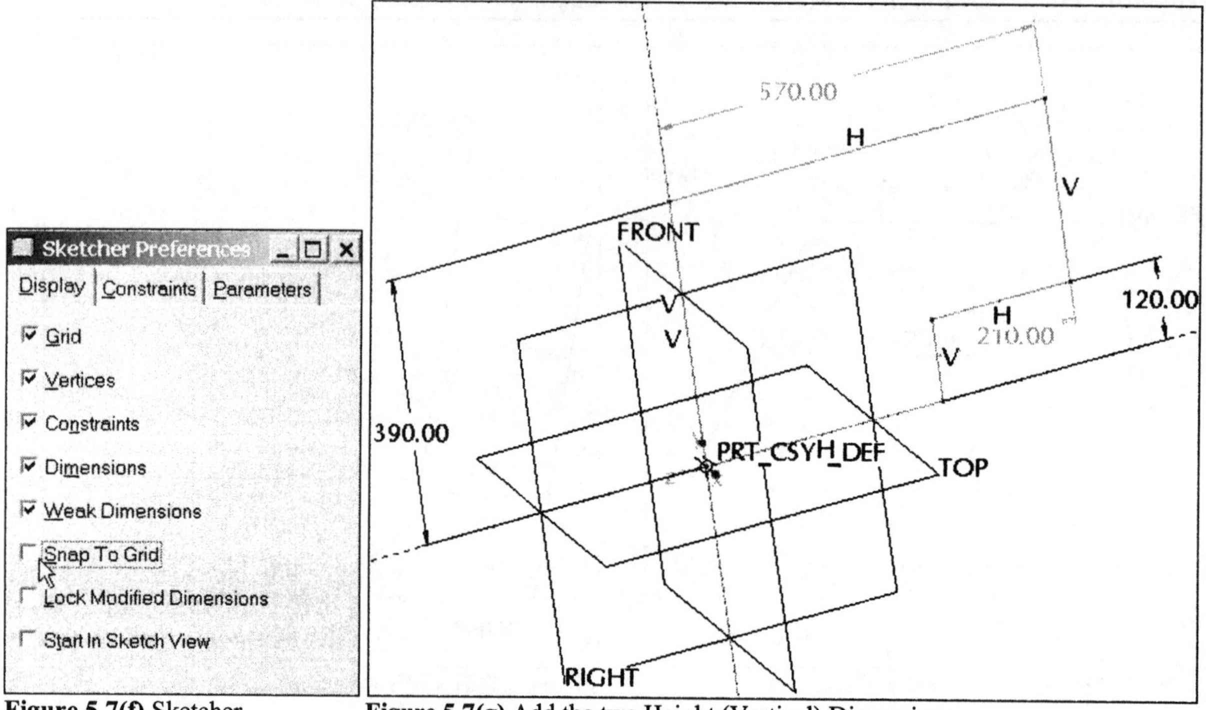

Figure 5.7(f) Sketcher Preferences

Figure 5.7(g) Add the two Height (Vertical) Dimensions

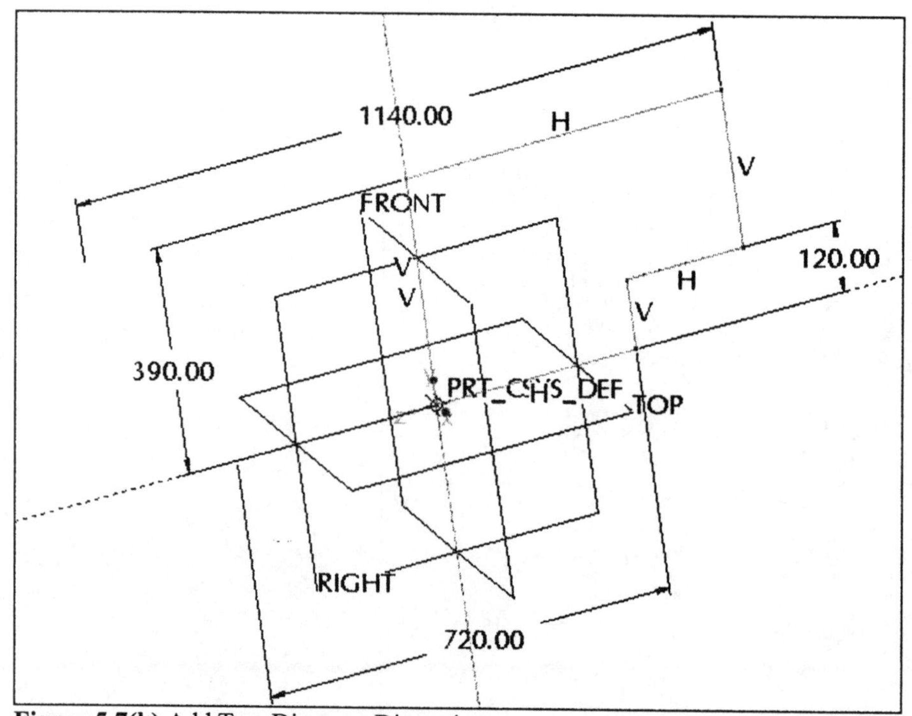

Figure 5.7(h) Add Two Diameter Dimensions

Modify the dimensions, click: [select icon] ⇒ window-in the sketch to capture all four dimensions ⇒ **RMB** ⇒ Modify... ⇒ Lock Scale ⇒ Regenerate [Fig. 5.7(i)] ⇒ modify the height to the design value (see Fig. 5.5) of **.750** ⇒ **Enter** ⇒ [check icon] [Fig. 5.7(j)] ⇒ **Ctrl+D** ⇒ **Ctrl+R**

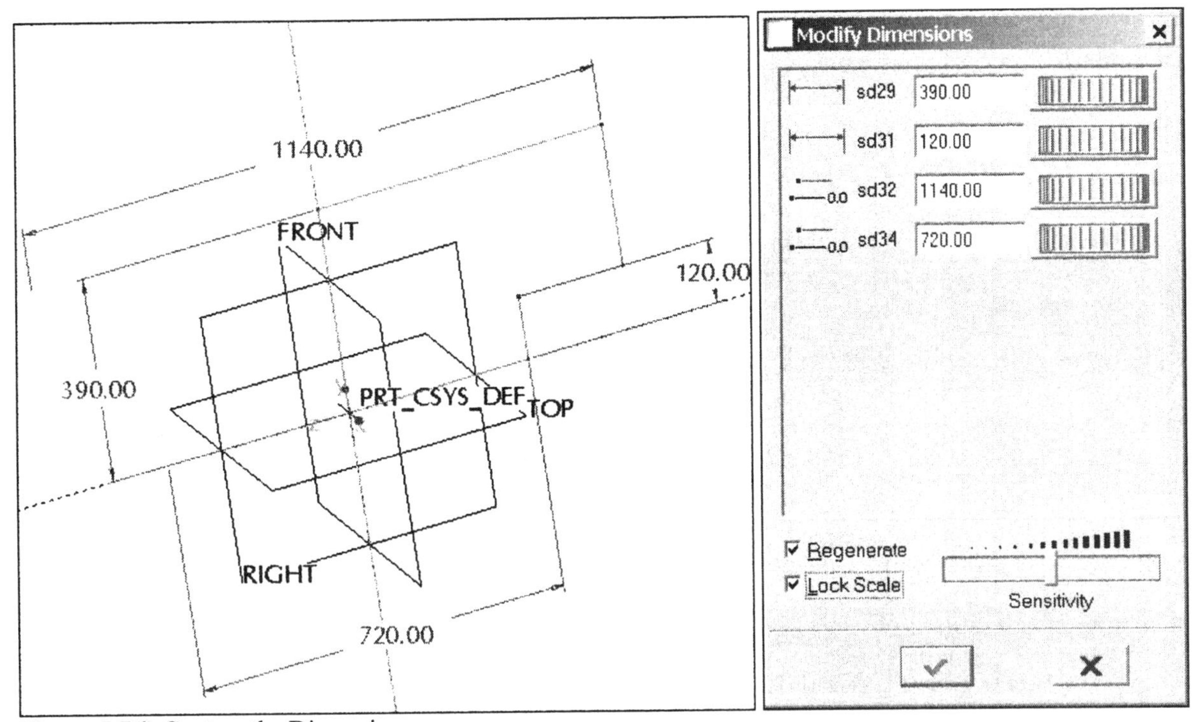

Figure 5.7(i) Capture the Dimensions

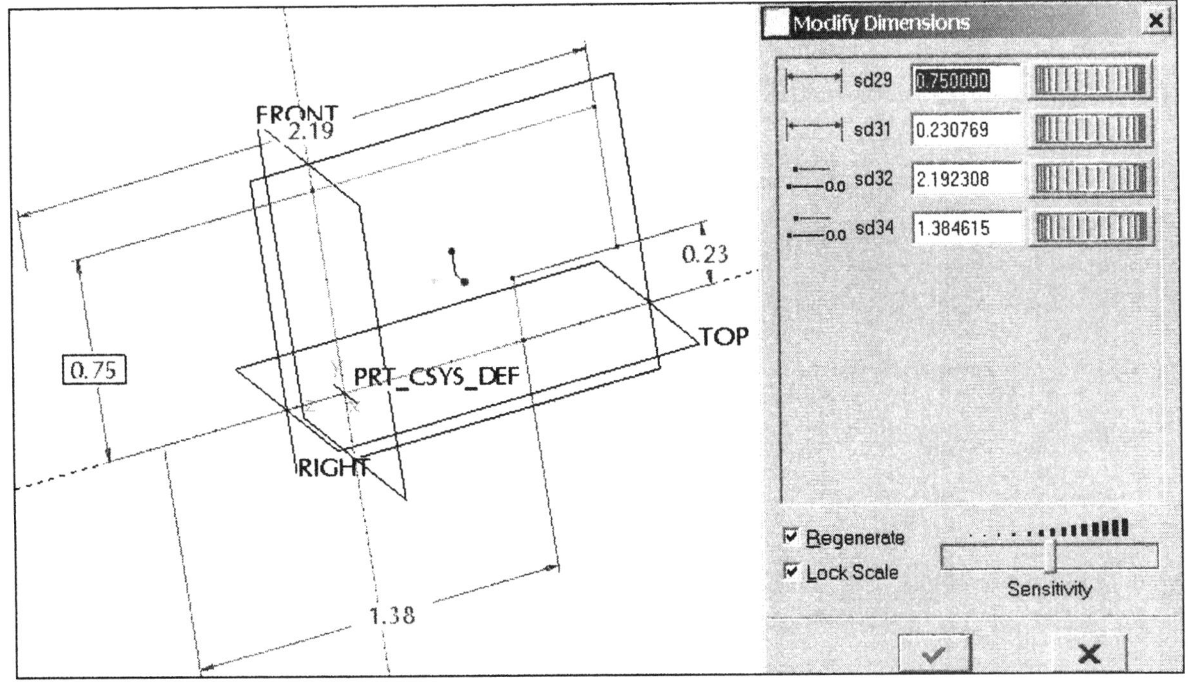

Figure 5.7(j) Modify the Height to **.750**

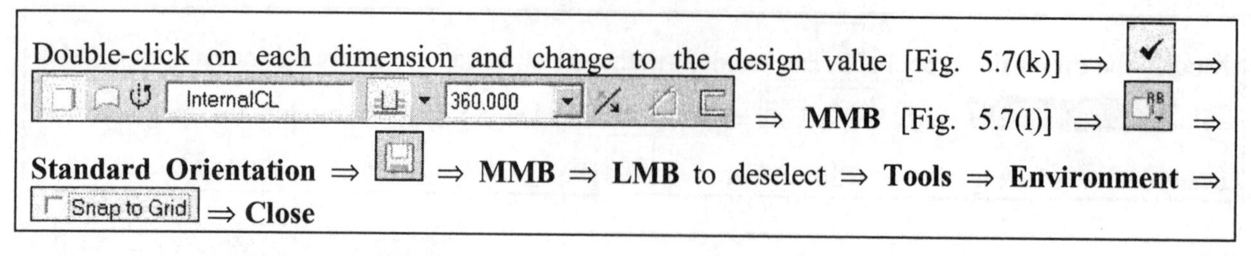

Double-click on each dimension and change to the design value [Fig. 5.7(k)] ⇒ ✔ ⇒ [InternalCL | 360.000] ⇒ **MMB** [Fig. 5.7(l)] ⇒ ⇒ **Standard Orientation** ⇒ ⇒ **MMB** ⇒ **LMB** to deselect ⇒ **Tools** ⇒ **Environment** ⇒ Snap to Grid ⇒ **Close**

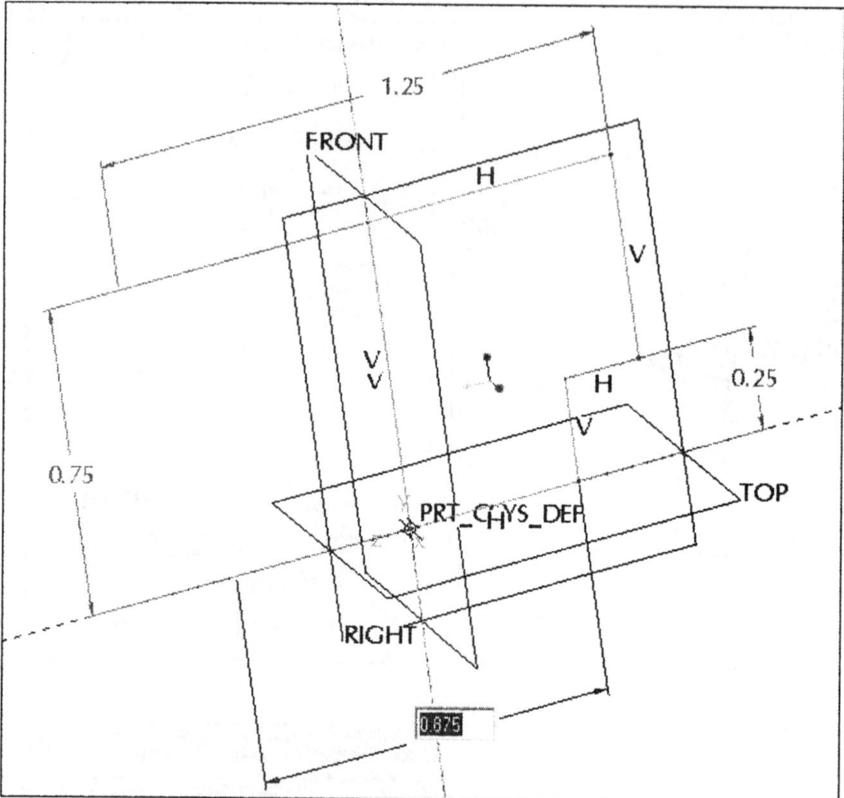

Figure 5.7(k) Modify the Remaining Dimensions

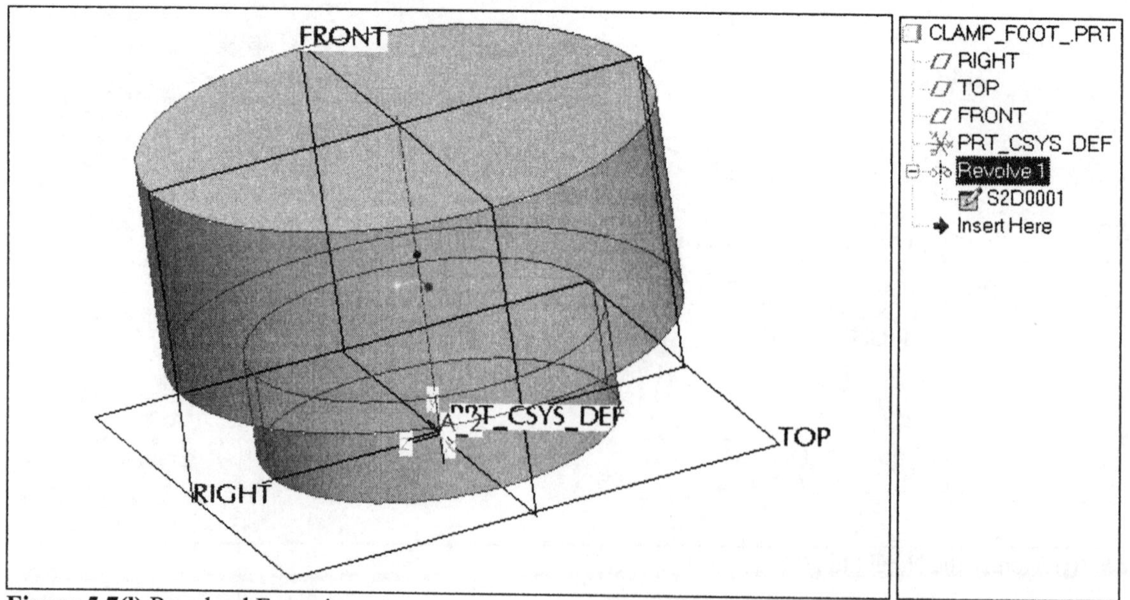

Figure 5.7(l) Revolved Extrusion

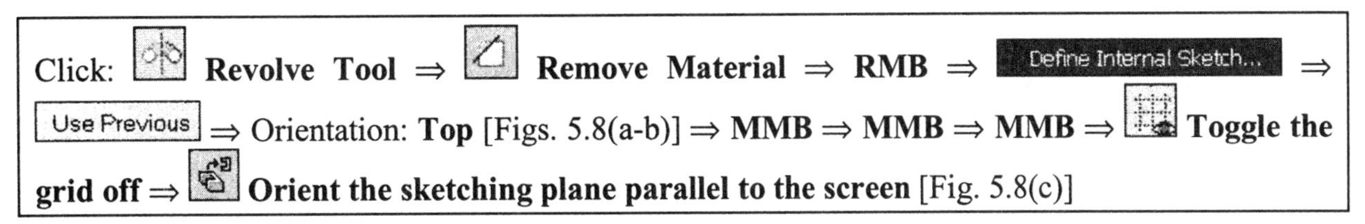

Click: **Revolve Tool** ⇒ **Remove Material** ⇒ **RMB** ⇒ Define Internal Sketch... ⇒ Use Previous ⇒ Orientation: **Top** [Figs. 5.8(a-b)] ⇒ **MMB** ⇒ **MMB** ⇒ **MMB** ⇒ **Toggle the grid off** ⇒ **Orient the sketching plane parallel to the screen** [Fig. 5.8(c)]

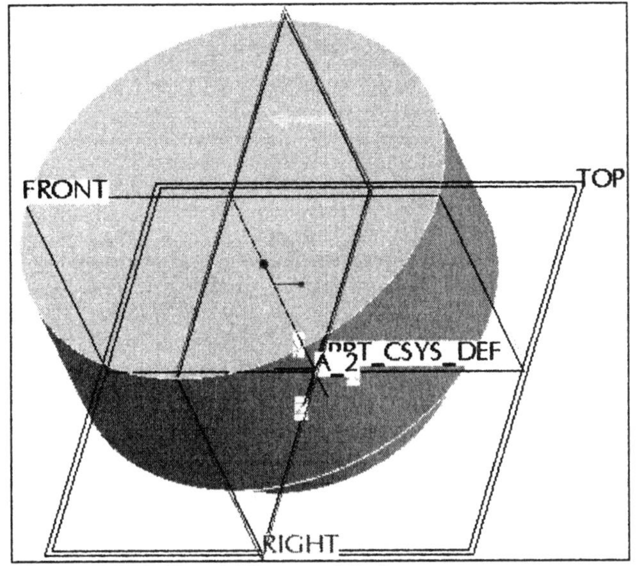

Figure 5.8(a) Sketch Plane and Reference

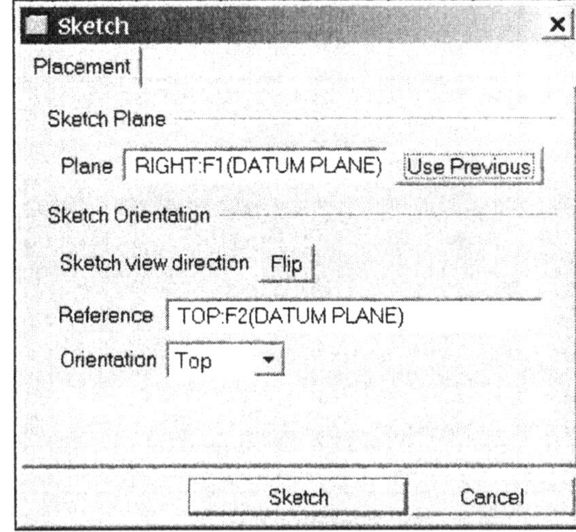

Figure 5.8(b) Sketch Dialog Box

Click: **RMB** ⇒ Centerline sketch a *vertical* centerline through the default coordinate system to be used as the axis of revolution [Fig. 5.8(d)] ⇒ **MMB** ⇒ [Fig. 5.8(e)]

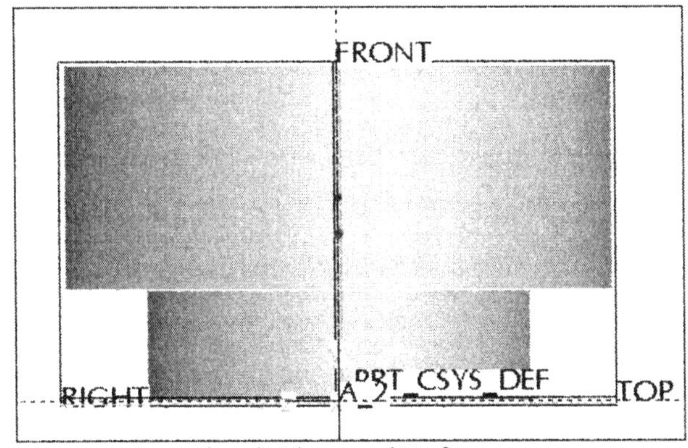

Figure 5.8(c) Sketch Plane and Reference

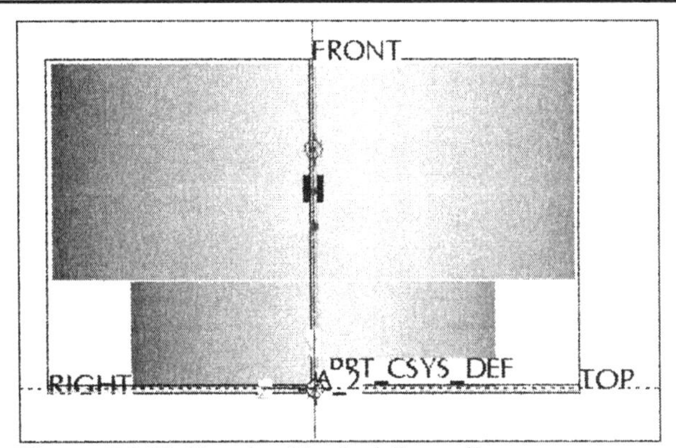

Figure 5.8(d) Sketch a Vertical Centerline

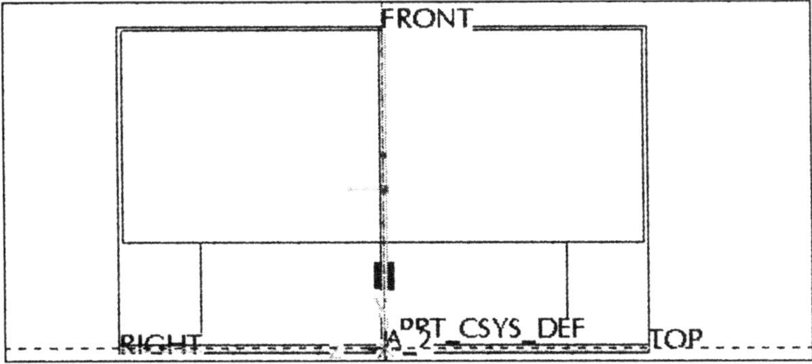

Figure 5.8(e) Hidden Line Display

From the Menu Bar click: **Sketch** ⇒ **Arc** ⇒ **Center and Ends** ⇒ sketch the arc [Fig. 5.8(f)] ⇒ **RMB** ⇒ Line sketch the three lines [Fig. 5.8(g)] *(it must be a closed section, so three lines are required)* ⇒ **MMB** ⇒ **MMB** ⇒ **off** ⇒ pick on each dimension and move to a better location ⇒ **Dynamically trim section entities** ⇒ press and hold **LMB** and draw a curve through the two elements you wish to remove [Fig. 5.8(h)] ⇒ release the **LMB** to complete the trim [Fig. 5.8(i)] *(if you delete the wrong items, click:* ***Undo sketcher operations*** *and try again)*

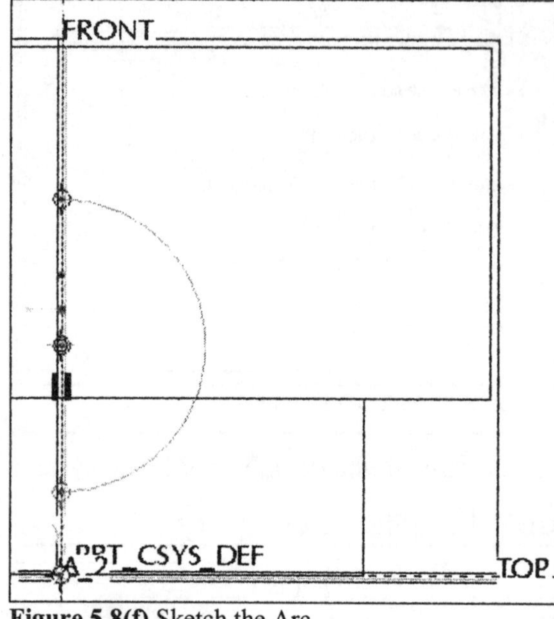

Figure 5.8(f) Sketch the Arc

Figure 5.8(g) Sketch the Lines

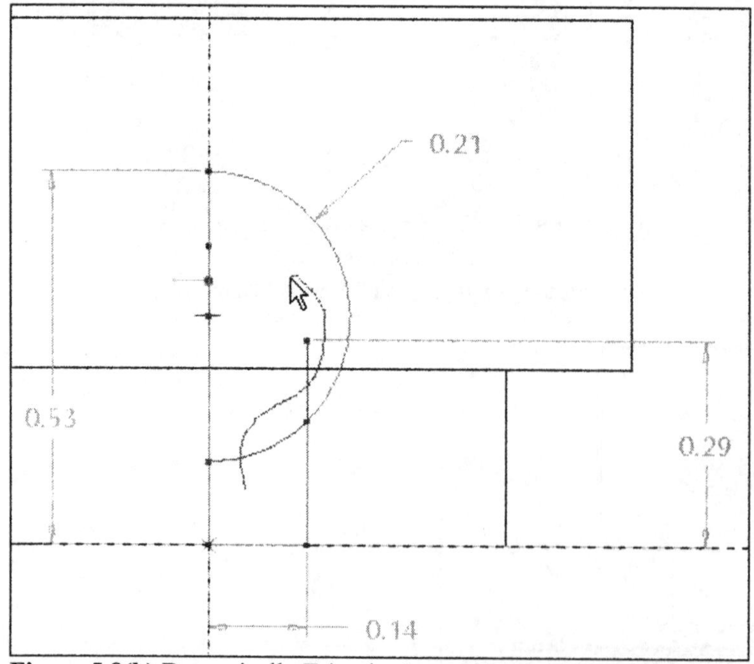

Figure 5.8(h) Dynamically Trim the Arc and Line

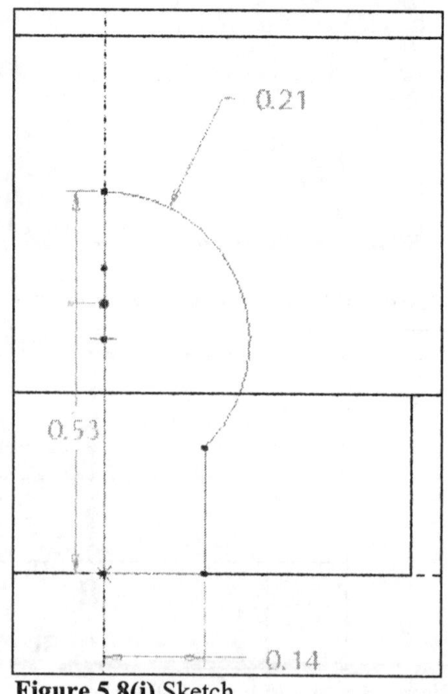

Figure 5.8(i) Sketch

Click: **RMB** ⇒ Dimension ⇒ double click on the arc ⇒ **MMB** to place the diameter dimension ⇒ add a dimension to the center of the arc from the bottom of the part ⇒ pick the centerline, then pick the vertical line, and then pick the centerline again ⇒ **MMB** to place the diameter dimension [Fig. 5.8(j)] ⇒ **MMB** to deselect ⇒ ⇒ window–in all three dimensions ⇒ **RMB** ⇒ **Modify** modify the dimensions to the design values (Fig. 5.5) [Fig. 5.8(k)] ⇒ ⇒ **on** ⇒ ⇒ ⇒ **Standard Orientation** ⇒ ⇒ InternalCL 360.000 ⇒ **MMB** ⇒ ⇒ **MMB** rotate the part to view the hole/cut [Fig. 5.8(l)] ⇒ ⇒ **MMB** ⇒ **LMB**

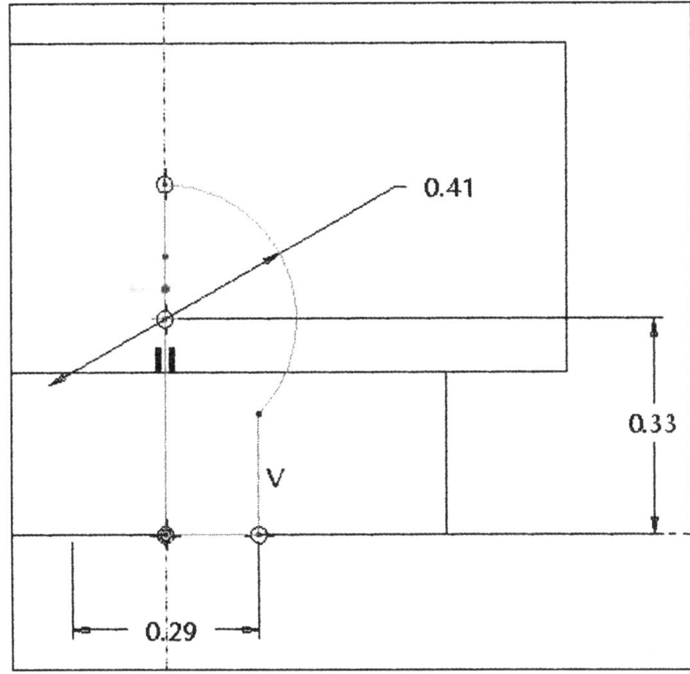

Figure 5.8(j) Correct Dimensioning Scheme

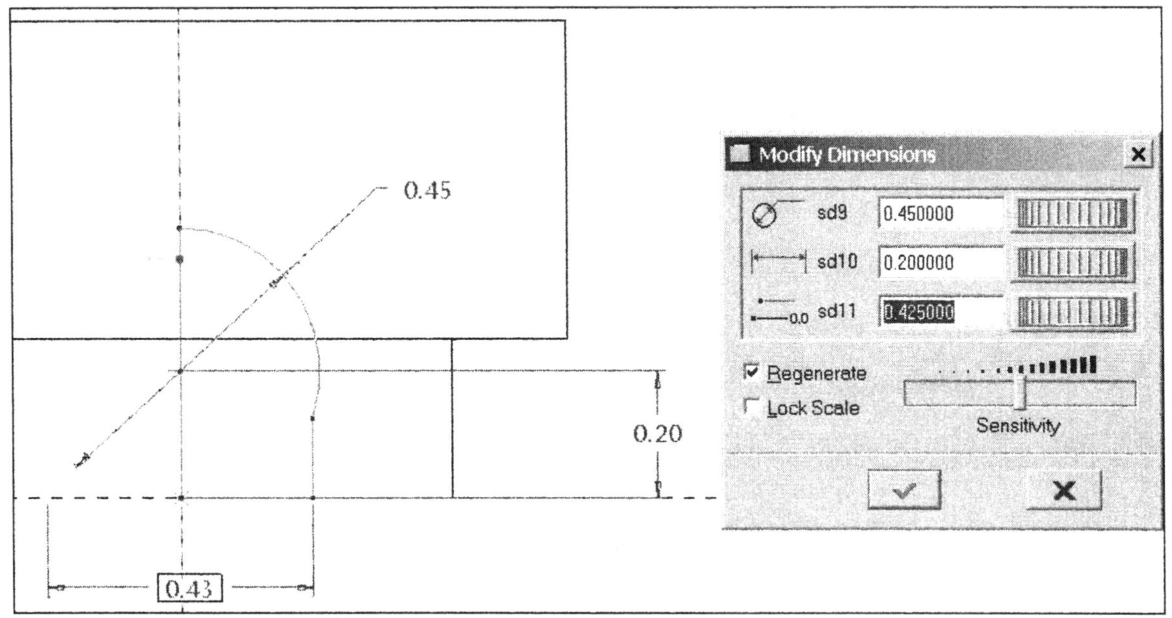

Figure 5.8(k) Modify Dimensions

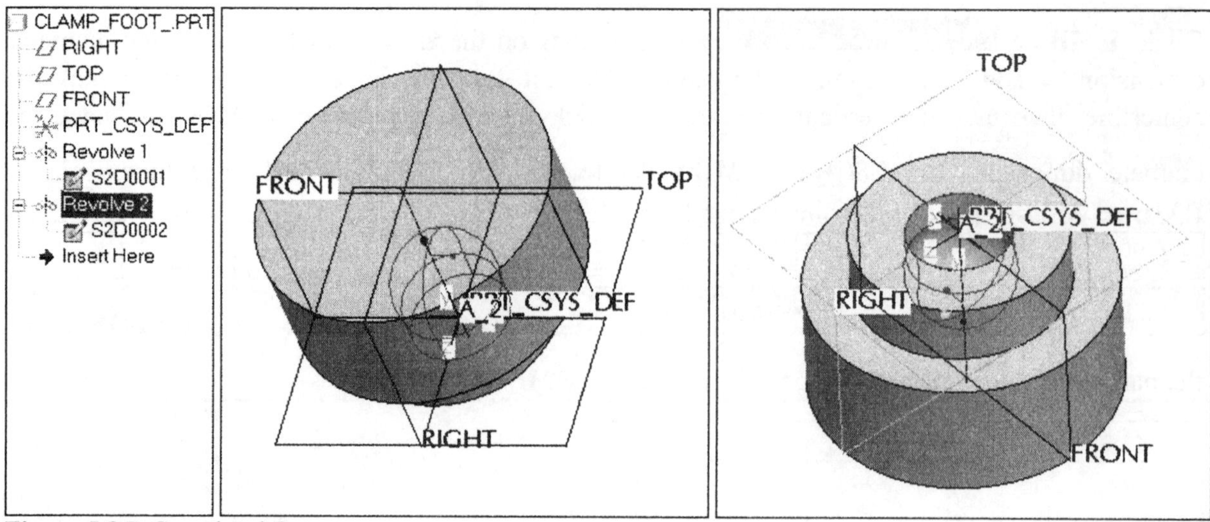

Figure 5.8(l) Completed Cut

Slowly click twice on the edge of the cut [Fig. 5.8(m)] ⇒ **Chamfer Tool** [Fig. 5.8(n)] ⇒ double-click on the dimension and modify to the design value of **.03125** [Fig. 5.8(o)] ⇒ **Enter** ⇒ **MMB** [Fig. 5.8(p)] ⇒ **LMB** to deselect ⇒ ⇒ **MMB** ⇒ **File** ⇒ **Delete** ⇒ **Old Versions** ⇒ **MMB**

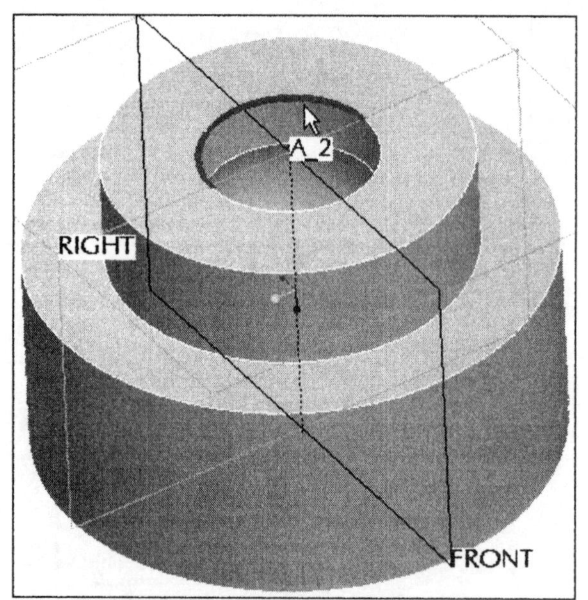

Figure 5.8(m) Pick on Edge

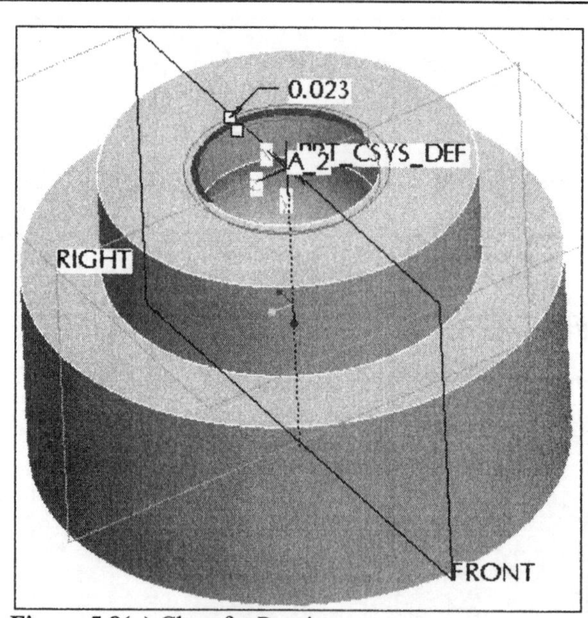

Figure 5.8(n) Chamfer Preview

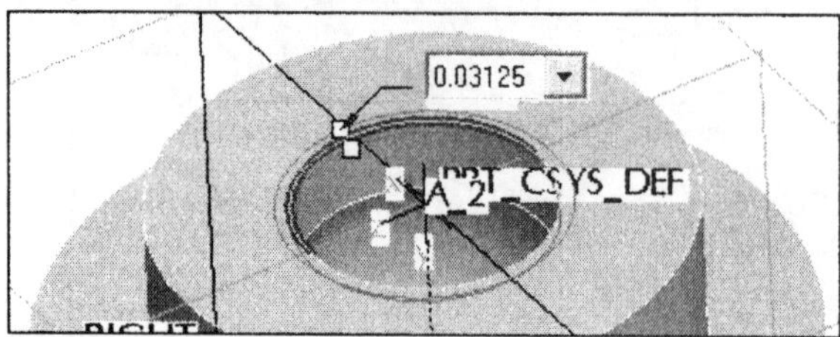

Figure 5.8(o) Chamfer Dimension

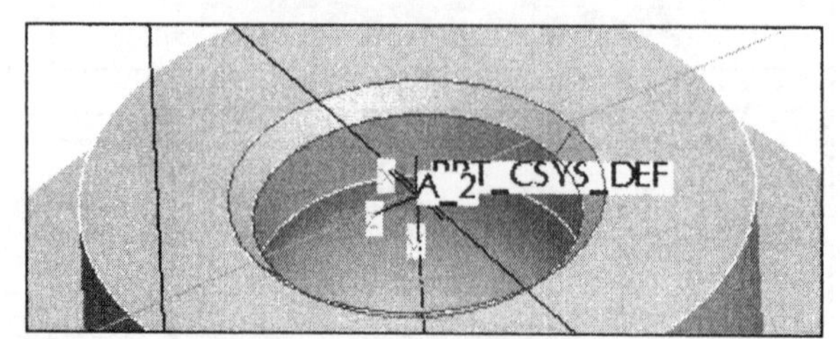
Figure 5.8(p) Completed Chamfer

Select one edge, then press and hold the **Ctrl** key and pick on the remaining three edges ⇒ **RMB** ⇒ **Round Edges** [Fig. 5.8(q)] ⇒ double-click on the dimension and modify to the design value of **.03125** [Fig. 5.8(r)] ⇒ **Enter** ⇒ **MMB** ⇒ [icon] ⇒ **MMB** ⇒ **LMB**

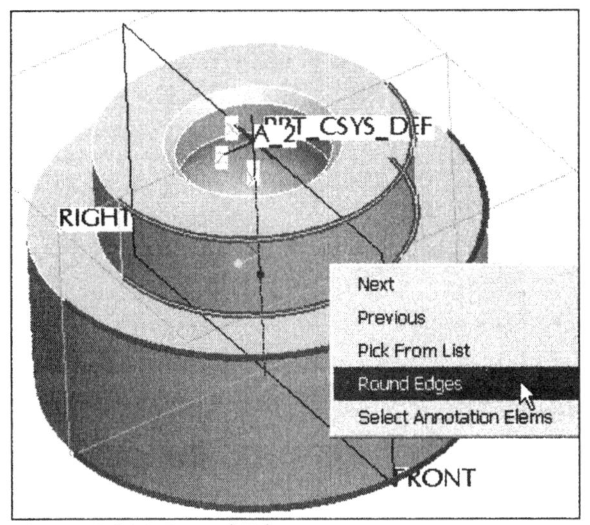

Figure 5.8(q) Round Edges

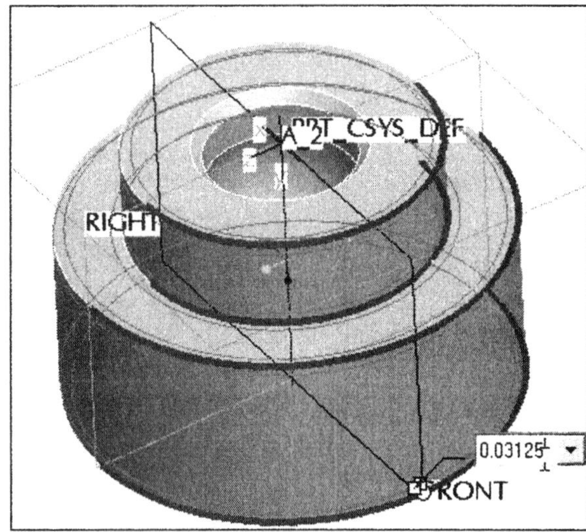

Figure 5.8(r) .03125 Round

Click: **Edit** ⇒ **Setup** ⇒ **Geom Tol** ⇒ **Set Datum** ⇒ pick **TOP** from the model ⇒ Name- type **A** ⇒ **OK** ⇒ pick **RIGHT** ⇒ Name- type **B** ⇒ **OK** ⇒ pick **FRONT** ⇒ Name- type **C** ⇒ **OK** [Fig. 5.8(s)] ⇒ **MMB** ⇒ **MMB** ⇒ **MMB** ⇒ [icon] ⇒ **Standard Orientation** ⇒ [icon] ⇒ **Edit** ⇒ **Setup** ⇒ **Material** ⇒ **Define** ⇒ type **Steel** ⇒ **Enter** ⇒ **File** ⇒ **Save** ⇒ **File** ⇒ **Exit** ⇒ **Assign** ⇒ pick **STEEL** ⇒ **Accept** ⇒ **MMB** ⇒ [icon] ⇒ **MMB** ⇒ **File** ⇒ **Delete** ⇒ **Old Versions** ⇒ **MMB** ⇒ **Window** ⇒ **Close**

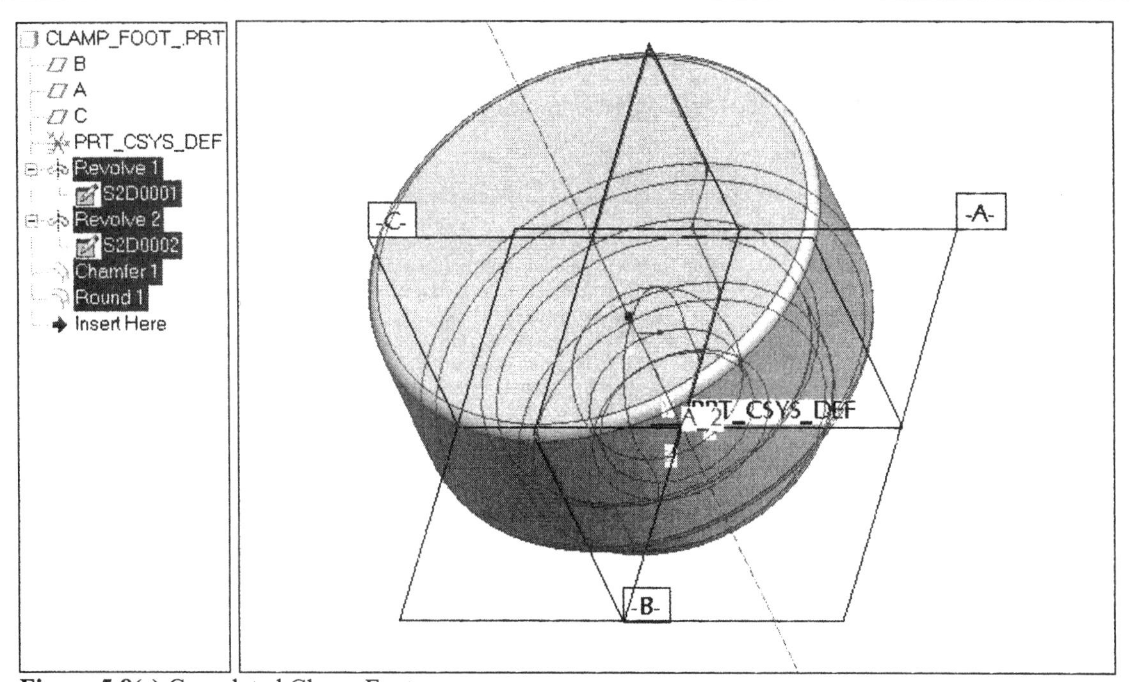

Figure 5.8(s) Completed Clamp Foot

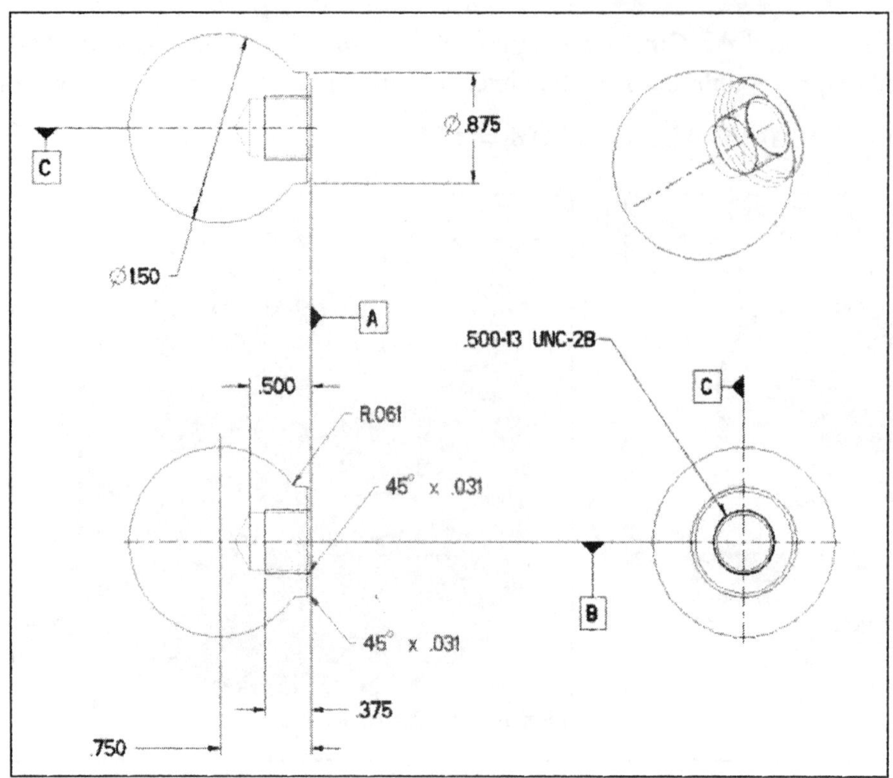

Figure 5.9 Clamp Ball Detail

Clamp Ball

The second part for this lesson is the Clamp Ball (Fig. 5.9). Much of the sketching is the same for this part as was done for the first revolved feature of the Clamp Foot. Instead of an internal revolved cut, you will create a standard hole. The Clamp Ball is made of *black plastic*.

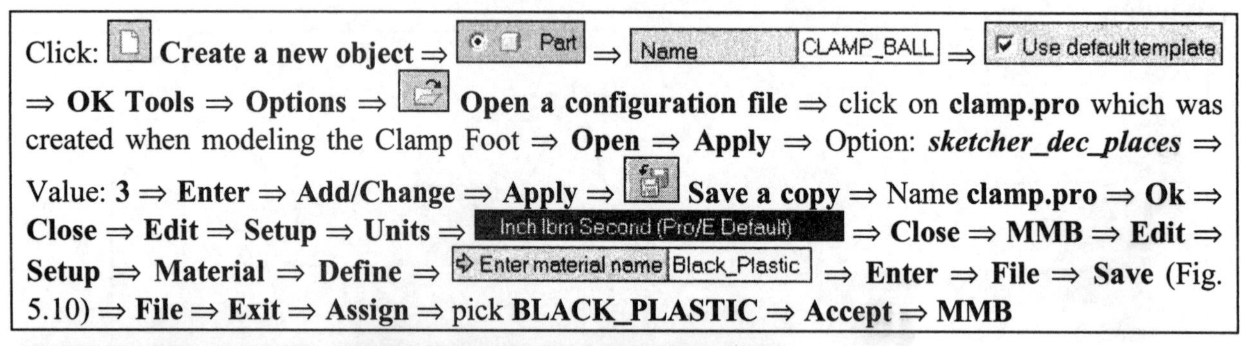

Click: **Create a new object** ⇒ Part ⇒ Name CLAMP_BALL ⇒ Use default template ⇒ **OK Tools** ⇒ **Options** ⇒ **Open a configuration file** ⇒ click on **clamp.pro** which was created when modeling the Clamp Foot ⇒ **Open** ⇒ **Apply** ⇒ Option: ***sketcher_dec_places*** ⇒ Value: **3** ⇒ **Enter** ⇒ **Add/Change** ⇒ **Apply** ⇒ **Save a copy** ⇒ Name **clamp.pro** ⇒ **Ok** ⇒ **Close** ⇒ **Edit** ⇒ **Setup** ⇒ **Units** ⇒ Inch lbm Second (Pro/E Default) ⇒ **Close** ⇒ **MMB** ⇒ **Edit** ⇒ **Setup** ⇒ **Material** ⇒ **Define** ⇒ Enter material name Black_Plastic ⇒ **Enter** ⇒ **File** ⇒ **Save** (Fig. 5.10) ⇒ **File** ⇒ **Exit** ⇒ **Assign** ⇒ pick **BLACK_PLASTIC** ⇒ **Accept** ⇒ **MMB**

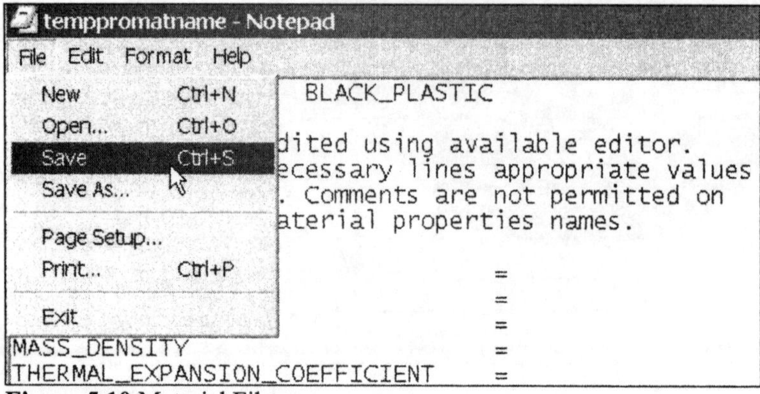

Figure 5.10 Material File

Model Tree

The Model Tree is a tabbed feature on the Pro/E navigator that contains a list of every feature or part in the current part, assembly, or drawing. The model structure is displayed in hierarchical (tree) format with the root object (the current part or assembly) at the top of its tree and the subordinate objects (parts or features) below. If you have multiple Pro/E windows open, the Model Tree contents reflect the file in the current "active" window. The Model Tree lists only the related feature- and part-level objects in a current file and does not list the entities (such as edges, surfaces, curves, and so forth) that comprise the features.

Each Model Tree item displays an icon that reflects its object type, for example, assembly, part, feature, or datum plane (also a feature). The icon can also show the display status for a feature, part, or assembly, for example, suppressed or hidden. The information in the Model Tree can be saved as a text (.txt) file.

Selection in the Model Tree is object-action oriented; you select objects in the Model Tree without first specifying what you intend to do with them. You can select components, parts, or features using the Model Tree. You cannot select the individual geometry that makes up a feature (entities). To select an entity, you must select it in the main graphics window.

You can add or remove items from the Model Tree column display using Settings in the Navigator:

- Select features, parts, or assemblies, and perform object-specific operations on them using the shortcut menu.
- Filter the display by item type, for example, showing or hiding datum features.
- Open a part within an assembly file by right-clicking the part in the Model Tree.
- Create or modify features and perform other operations such as deleting or redefining parts or features, and rerouting parts or features using the shortcut menu.
- Search the Model Tree for model properties or other feature information.
- Show the display status for an object, for example, suppressed or hidden.

Click: **Settings** in the Navigator [Fig. 5.11(a)] ⇒ **Tree Filters** ⇒ Annotations ⇒ Suppressed Objects [Fig. 5.11(b)] ⇒ **Apply** ⇒ **OK**

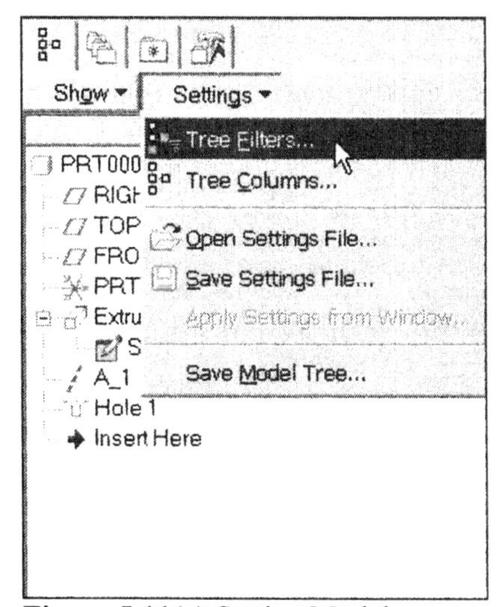

Figure 5.11(a) Setting Model Tree Filters

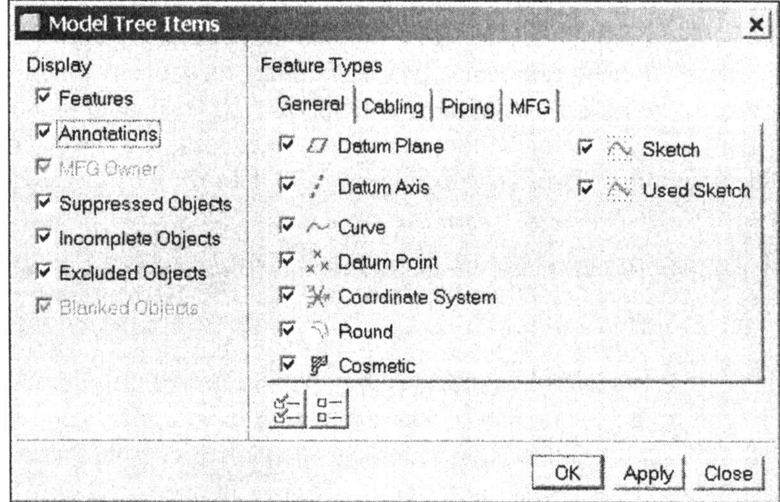

Figure 5.11(b) Model Tree Items Dialog Box

Click: **Settings** in the Navigator ⇒ Tree Columns... ⇒ Model Tree Columns dialog box opens: Type **Info** [Fig. 5.12(a)] ⇒ add types to the Displayed list, click: **Feat #** ⇒ >> ⇒ **Feat ID** ⇒ >> ⇒ Type ⇒ select **Layer** ⇒ **Layer Names** ⇒ >> ⇒ **Layer Status** ⇒ >> [Fig. 5.12(b)] ⇒ **Apply** ⇒ **OK**

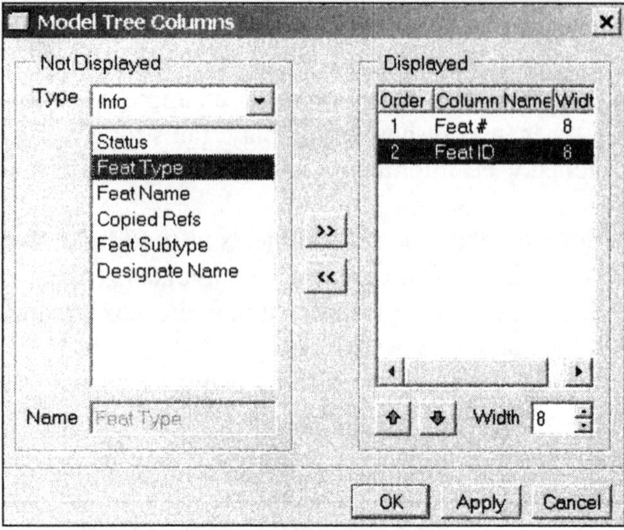

Figure 5.12(a) Model Tree Columns, Info

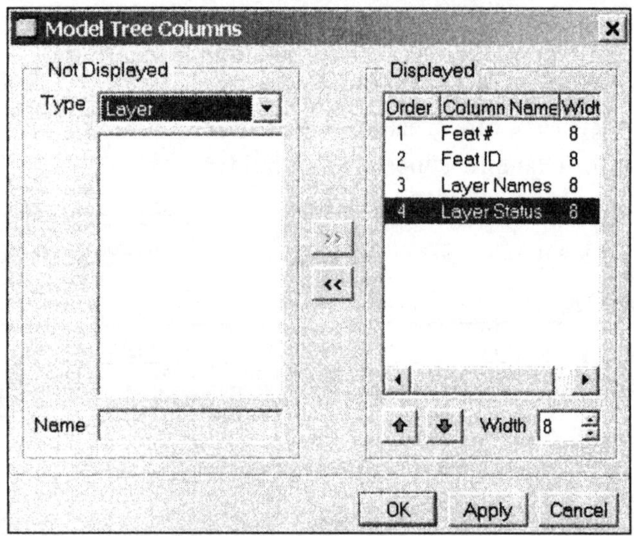

Figure 5.12(b) Model Tree Columns, Layer

Click on the sash ⇒ drag sash to expand Model Tree [Fig. 5.12(c)] ⇒ adjust the width of each column by dragging the column divider ⇒ drag the sash to the left to decrease the Model Tree size

	Feat #	Feat ID	Layer Names	Layer Status
CLAMP_BALL_PRT				
RIGHT	1	1	01__PRT_ALL_DTM_PLN, 01__PRT_DEF_DTM_PLN	Displayed
TOP	2	3	01__PRT_ALL_DTM_PLN, 01__PRT_DEF_DTM_PLN	Displayed
FRONT	3	5	01__PRT_ALL_DTM_PLN, 01__PRT_DEF_DTM_PLN	Displayed
PRT_CSYS_DEF	4	7	05__PRT_ALL_DTM_CSYS, 05__PRT_DEF_DTM_CSYS	Displayed

Figure 5.12(c) Expand the Columns

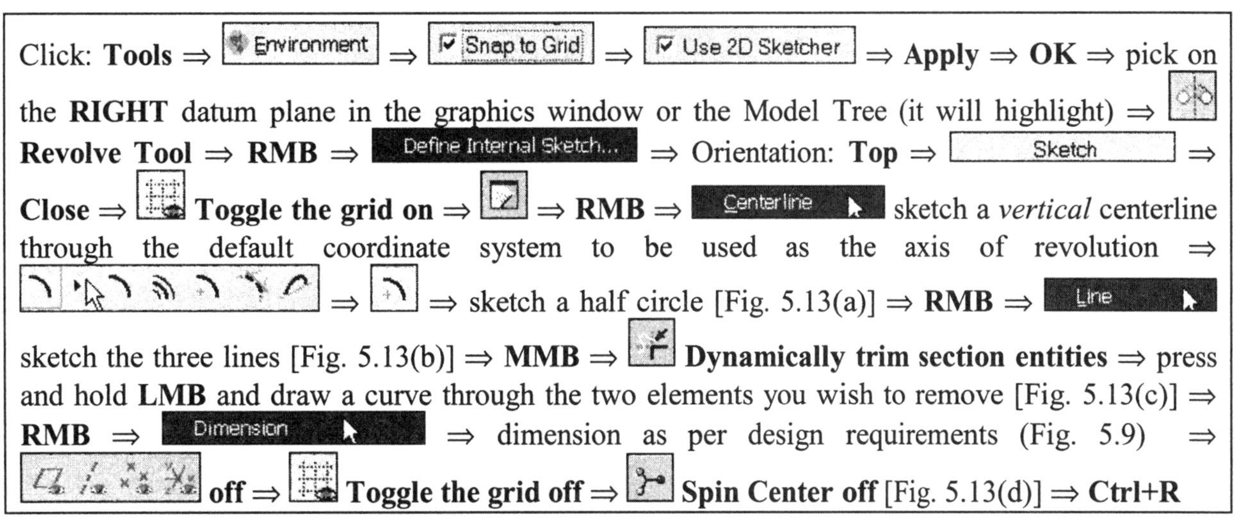
Click: **Tools** ⇒ Environment ⇒ Snap to Grid ⇒ Use 2D Sketcher ⇒ **Apply** ⇒ **OK** ⇒ pick on the **RIGHT** datum plane in the graphics window or the Model Tree (it will highlight) ⇒ **Revolve Tool** ⇒ **RMB** ⇒ Define Internal Sketch... ⇒ Orientation: **Top** ⇒ Sketch ⇒ **Close** ⇒ **Toggle the grid on** ⇒ ⇒ **RMB** ⇒ Centerline sketch a *vertical* centerline through the default coordinate system to be used as the axis of revolution ⇒ ⇒ ⇒ sketch a half circle [Fig. 5.13(a)] ⇒ **RMB** ⇒ Line sketch the three lines [Fig. 5.13(b)] ⇒ **MMB** ⇒ **Dynamically trim section entities** ⇒ press and hold **LMB** and draw a curve through the two elements you wish to remove [Fig. 5.13(c)] ⇒ **RMB** ⇒ Dimension ⇒ dimension as per design requirements (Fig. 5.9) ⇒ **off** ⇒ **Toggle the grid off** ⇒ **Spin Center off** [Fig. 5.13(d)] ⇒ **Ctrl+R**

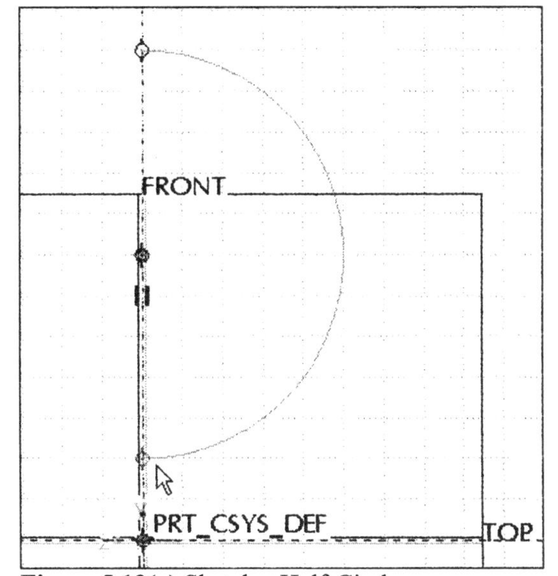

Figure 5.13(a) Sketch a Half-Circle

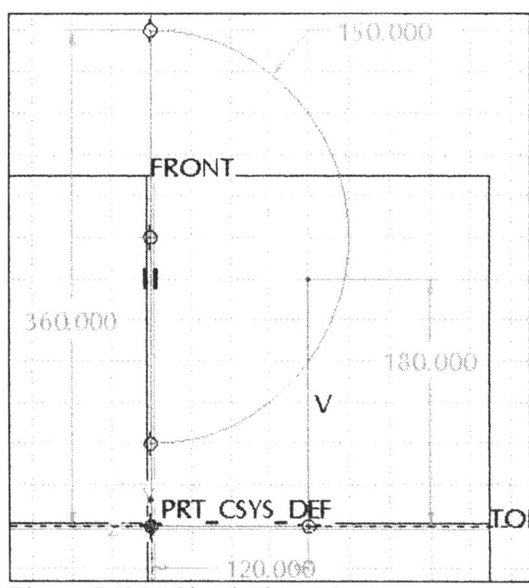

Figure 5.13(b) Sketch Three Lines

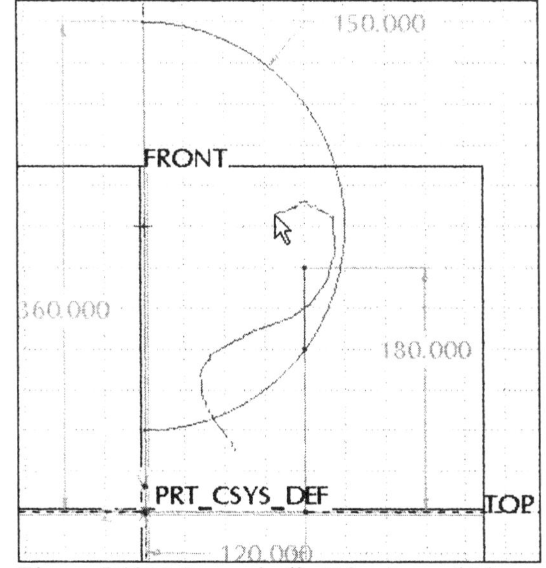

Figure 5.13(c) Dynamically Trim Arc and Line

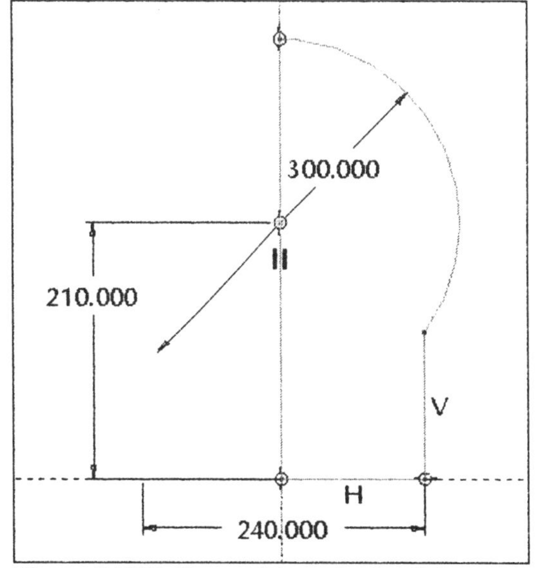

Figure 5.13(d) Correct Dimensioning Scheme

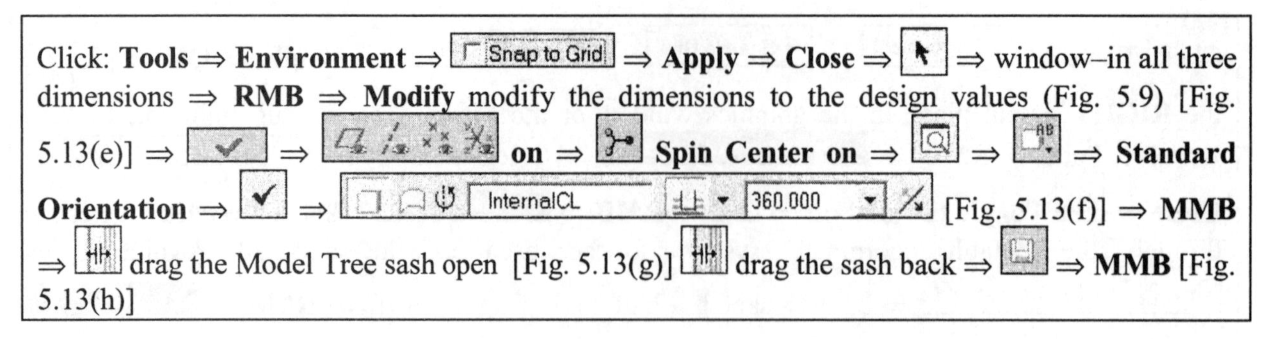

Click: **Tools** ⇒ **Environment** ⇒ Snap to Grid ⇒ **Apply** ⇒ **Close** ⇒ [icon] ⇒ window–in all three dimensions ⇒ **RMB** ⇒ **Modify** modify the dimensions to the design values (Fig. 5.9) [Fig. 5.13(e)] ⇒ [icon] ⇒ [icon] **on** ⇒ [icon] **Spin Center on** ⇒ [icon] ⇒ [icon] ⇒ **Standard Orientation** ⇒ [icon] ⇒ [InternalCL 360.000] [Fig. 5.13(f)] ⇒ **MMB** ⇒ [icon] drag the Model Tree sash open [Fig. 5.13(g)] [icon] drag the sash back ⇒ [icon] ⇒ **MMB** [Fig. 5.13(h)]

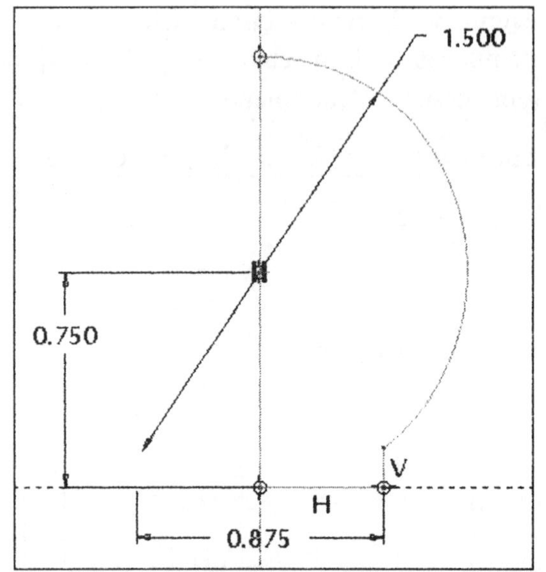

Figure 5.13(e) Modified Dimensions

Figure 5.13(f) Feature Preview

	Feat #	Feat ID	Layer Names	Layer Status
CLAMP_BALL_.PRT				
RIGHT	1	1	01___PRT_ALL_DTM_PLN, 01___PRT_DEF_DTM_PLN	Displayed
TOP	2	3	01___PRT_ALL_DTM_PLN, 01___PRT_DEF_DTM_PLN	Displayed
FRONT	3	5	01___PRT_ALL_DTM_PLN, 01___PRT_DEF_DTM_PLN	Displayed
PRT_CSYS_DEF	4	7	05___PRT_ALL_DTM_CSYS, 05___PRT_DEF_DTM_CSYS	Displayed
Revolve 1	5	39		Displayed

Figure 5.13(g) Revolve Feature Shown in Model Tree

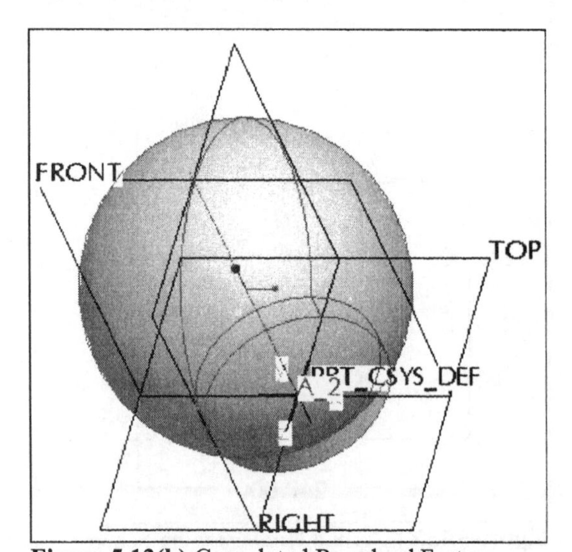

Figure 5.13(h) Completed Revolved Feature

Holes

Hole charts are used to lookup diameters for a given fastener size. You can create custom hole charts and specify their directory location with the configuration file option *hole_parameter_file_path.* UNC, UNF and ISO hole charts are supplied with Pro/E. Create a standard **.500-13 UNC-2B** hole, **.375** thread depth, **.50** tap drill. Include a standard chamfer.

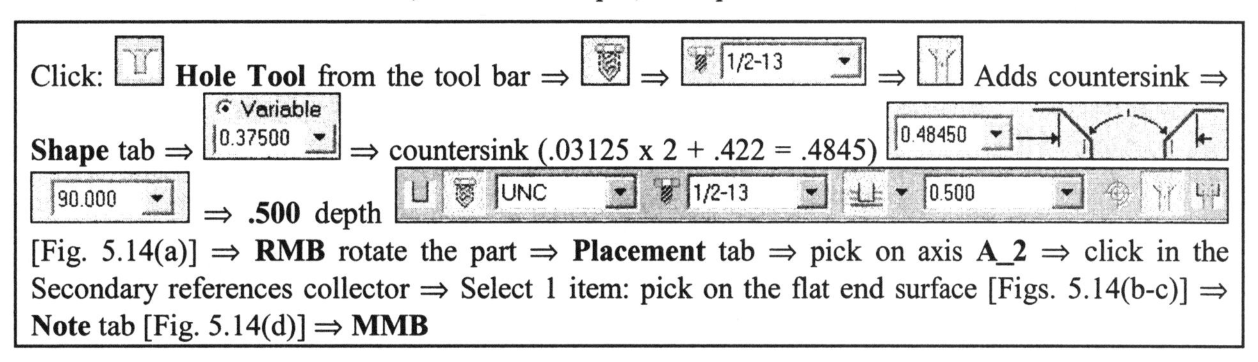

Click: **Hole Tool** from the tool bar ⇒ ⇒ 1/2-13 ⇒ Adds countersink ⇒ **Shape** tab ⇒ Variable 0.37500 ⇒ countersink (.03125 x 2 + .422 = .4845) 0.48450 90.000 ⇒ **.500** depth UNC 1/2-13 0.500 [Fig. 5.14(a)] ⇒ **RMB** rotate the part ⇒ **Placement** tab ⇒ pick on axis **A_2** ⇒ click in the Secondary references collector ⇒ Select 1 item: pick on the flat end surface [Figs. 5.14(b-c)] ⇒ **Note** tab [Fig. 5.14(d)] ⇒ **MMB**

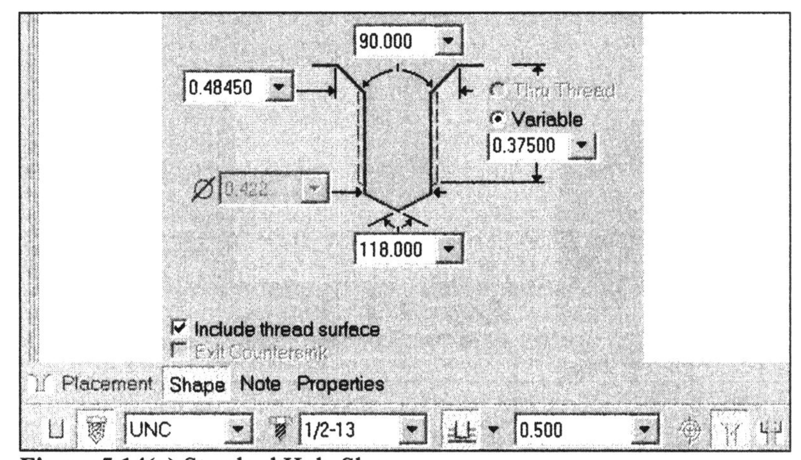

Figure 5.14(a) Standard Hole Shape

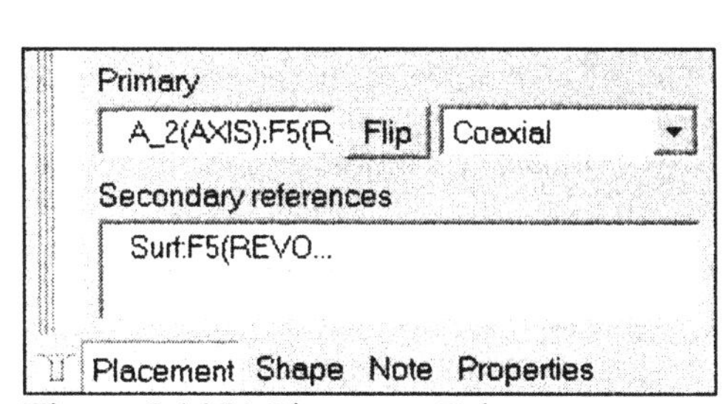

Figure 5.14(b) Placement Tab

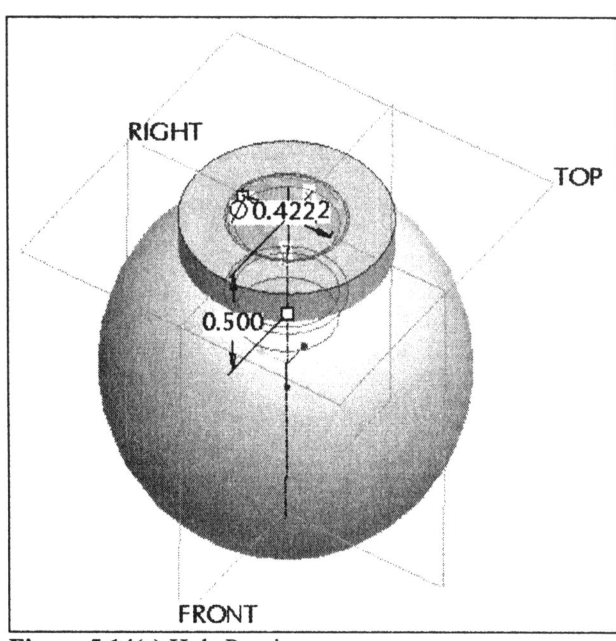

Figure 5.14(c) Hole Preview

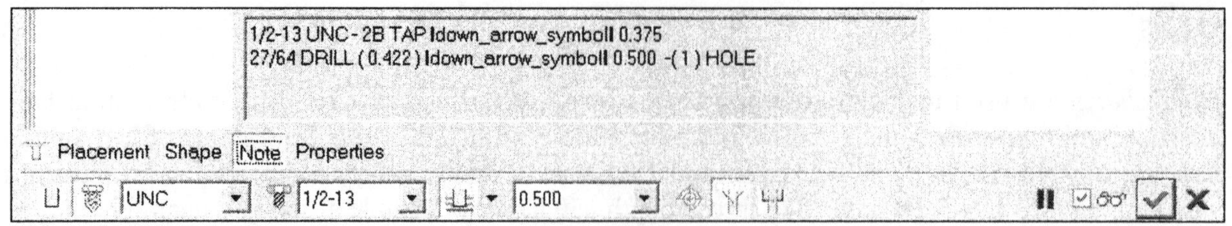

Figure 5.14(d) Hole Note

Click: ⇒ **Annotation** ⇒ pick on the note ⇒ **RMB** ⇒ **Properties** [Fig. 5.15(a)] ⇒ **Note** dialog box opens [Fig. 5.15(b)] ⇒ **OK** ⇒ ⇒ **Smart** ⇒ **LMB** to deselect ⇒ **Tools** ⇒ **Environment** ⇒ 3D Notes ⇒ **Apply** ⇒ **OK** ⇒ **Edit** ⇒ **Setup** ⇒ **Geom Tol** ⇒ **Set Datum** ⇒ pick **TOP** from the model ⇒ Name- type **A** ⇒ **OK** ⇒ pick **FRONT** ⇒ Name- type **B** ⇒ **OK** ⇒ pick **RIGHT** ⇒ Name- type **C** ⇒ **OK** ⇒ **MMB** ⇒ **MMB** ⇒ **MMB** ⇒ pick on the coordinate system in the Model Tree and rename CLAMP_BALL_CSYS *(do the same for the Clamp Foot if needed)* ⇒ **LMB** to deselect ⇒ ⇒ **MMB** ⇒ **File** ⇒ **Delete** ⇒ **Old Versions** ⇒ **MMB**

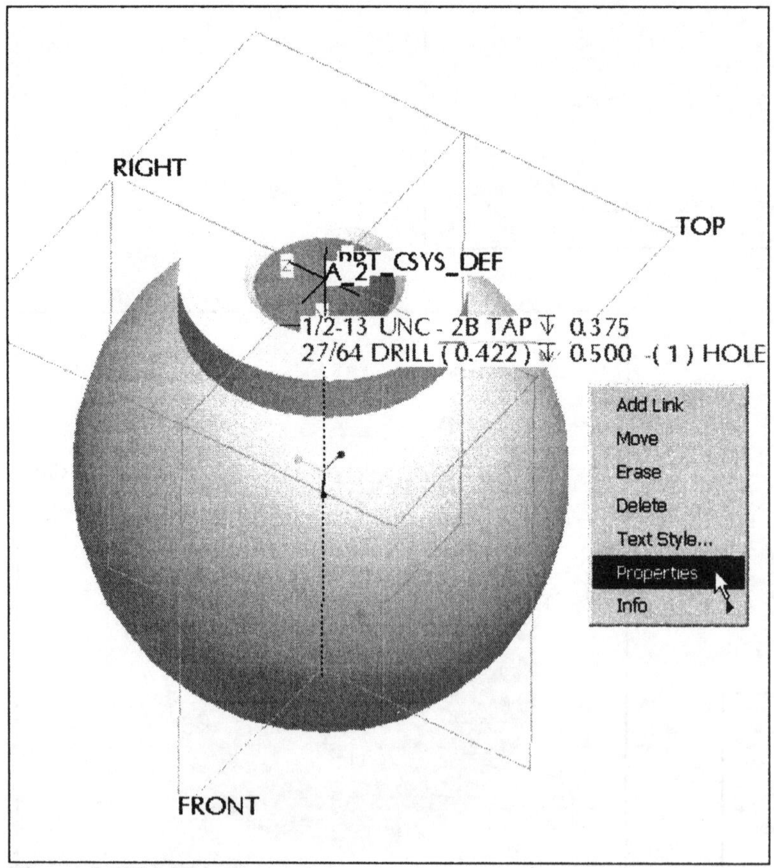

Figure 5.15(a) Note Properties

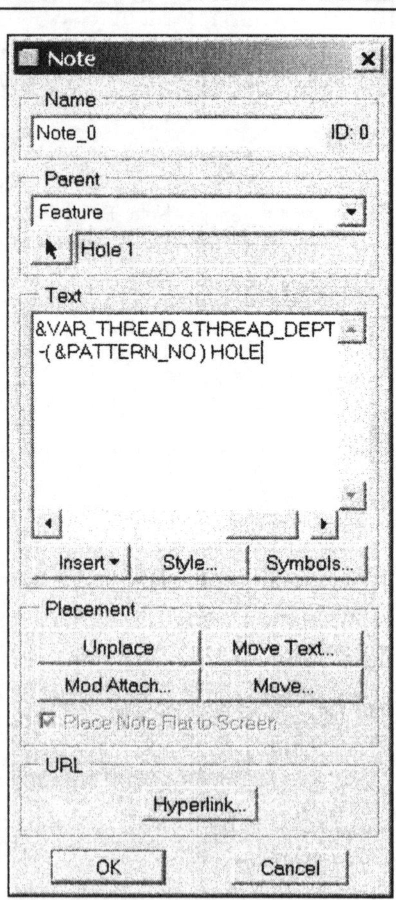

Figure 5.15(b) Note Dialog Box

Dimension Properties

Besides the value of a dimension and its position, there are several attributes that you can change using the Dimension Properties dialog box. The Properties tab provides options for *Value and tolerance* modification, *Format* options, and *Display* variables including *Flip Arrows*, which allows the toggling of dimension arrows inside or outside of its extension lines.

Move, which is available on all three tabs, moves the dimension itself, and the associated leader lines, to a new location. *Move Text* only moves the text associated with the dimension to a new location. *Text Symbol* provides a dialog box with symbology.

Double-click on the model to display the dimensions for the protrusion ⇒ pick on the **∅.875** dimension ⇒ **RMB** [Fig. 5.16(a)] ⇒ **Properties** Dimension Properties dialog box displays [Fig. 5.16(b)]

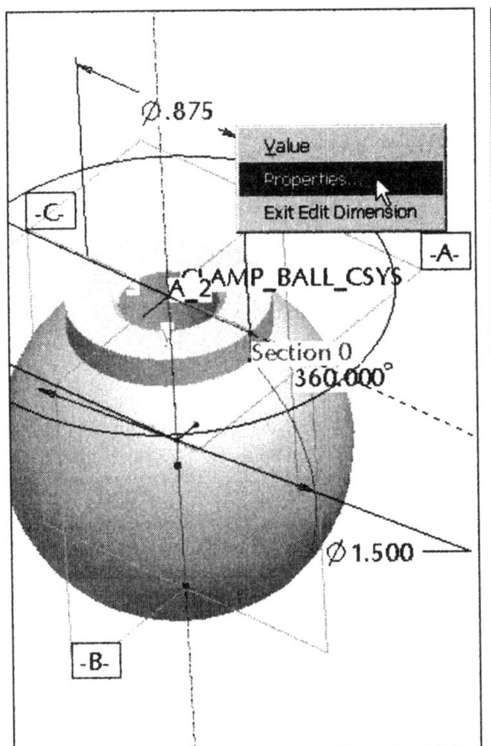

Figure 5.16(a) Dimension Displayed

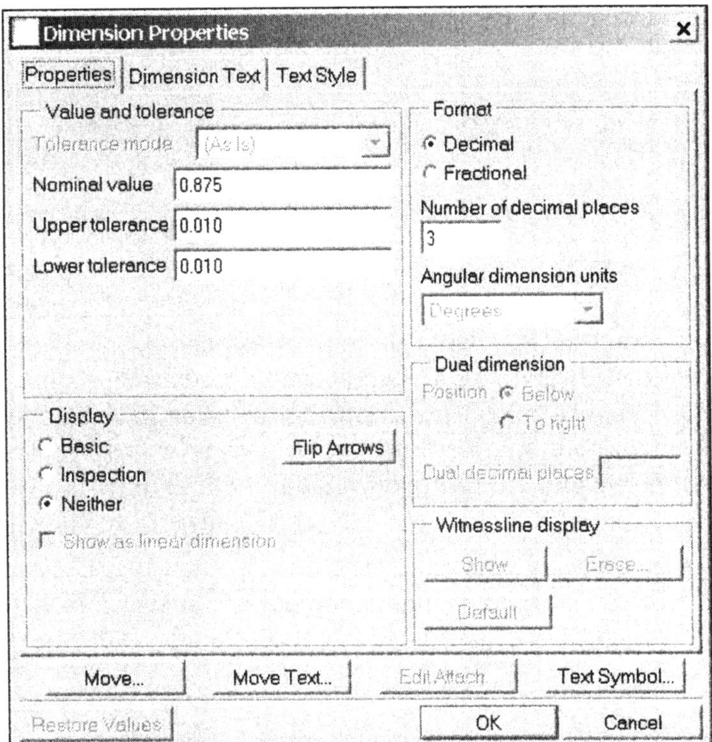

Figure 5.16(b) Dimension Properties Dialog Box

The Dimension Text tab [Fig. 5.16(c)] shows the parametric dimension symbol. The Text Style tab [Fig. 5.16(d)] provides options for Character, and Note/Dimension variations. Take some time and explore the options provided in the Dimension Properties dialog box.

Click: **Move** ⇒ select a new position for the **∅.875** dimension [Fig. 5.16(e)] ⇒ **OK** ⇒ repeat to move the other dimensions as required ⇒ **OK** ⇒ **LMB** to deselect ⇒ **Ctrl+S** ⇒ **MMB** ⇒ **File** ⇒ **Delete** ⇒ **Old Versions** ⇒ **MMB**

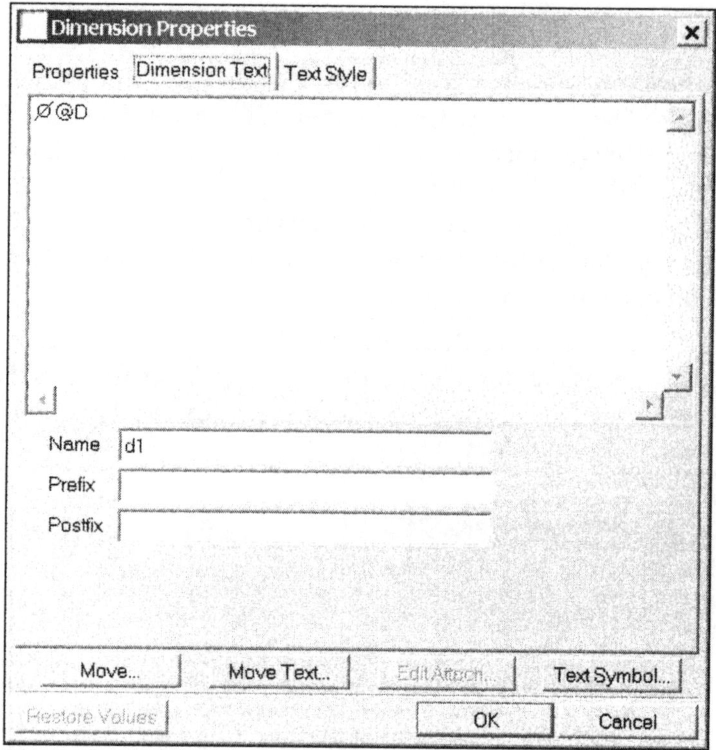

Figure 5.16(c) Dimension Properties, Dimension Text Tab

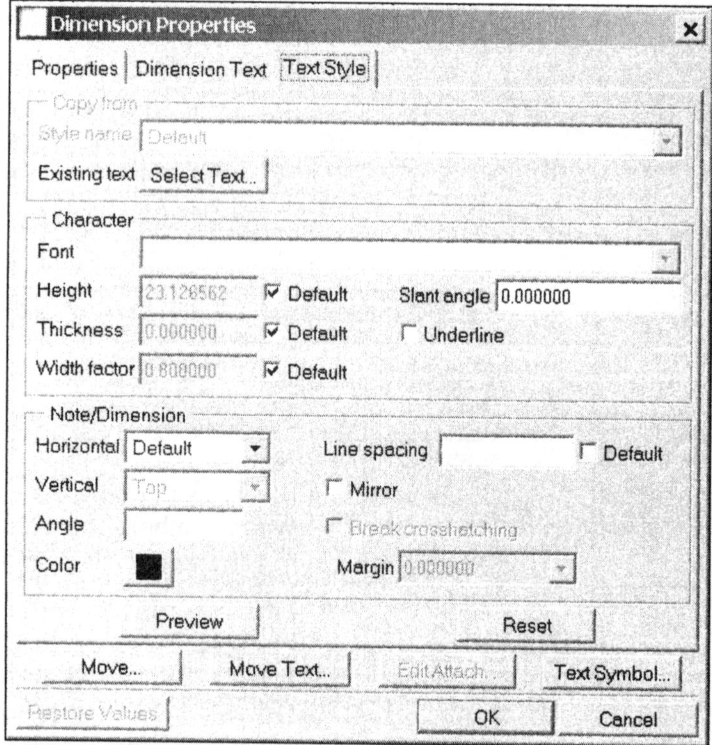

Figure 5.16(d) Dimension Properties, Text Style Tab

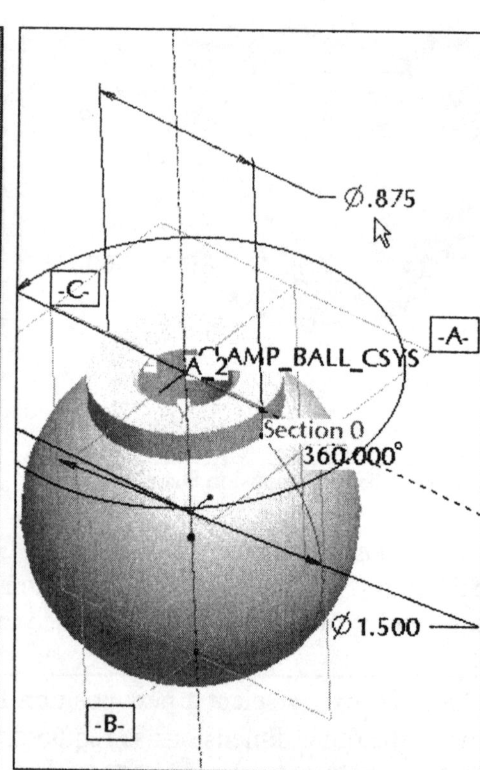

Figure 5.16(e) Moved Dimension

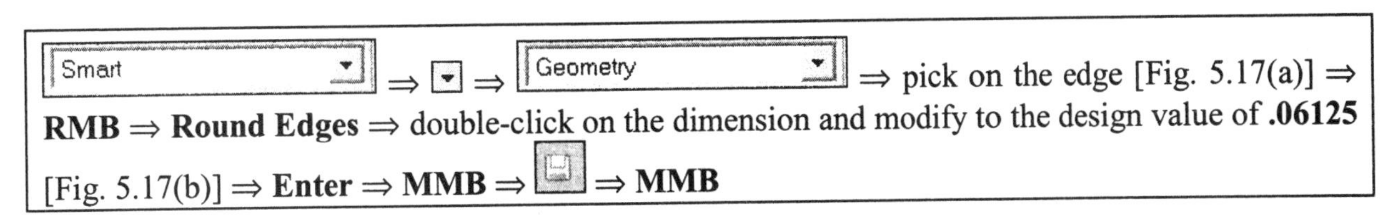

Smart ⇒ ▾ ⇒ Geometry ⇒ pick on the edge [Fig. 5.17(a)] ⇒ **RMB** ⇒ **Round Edges** ⇒ double-click on the dimension and modify to the design value of **.06125** [Fig. 5.17(b)] ⇒ **Enter** ⇒ **MMB** ⇒ [icon] ⇒ **MMB**

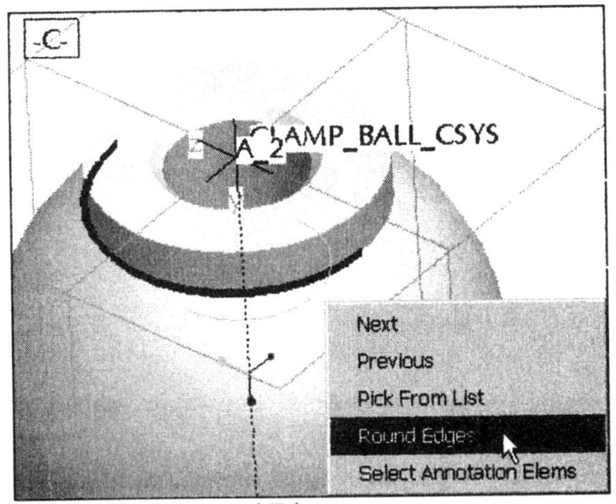

Figure 5.17(a) Round Edges

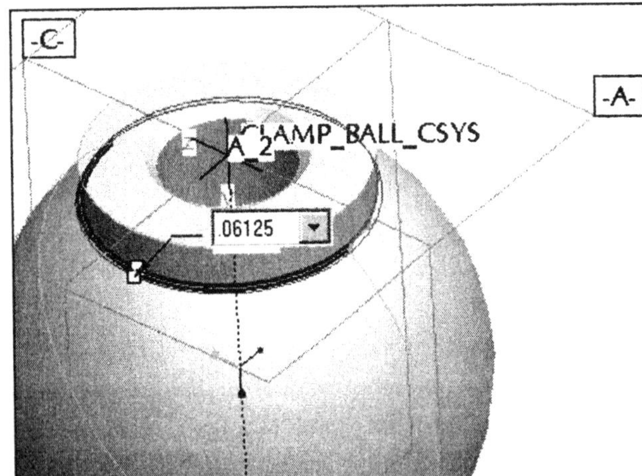

Figure 5.17(b) Radius **.06125**

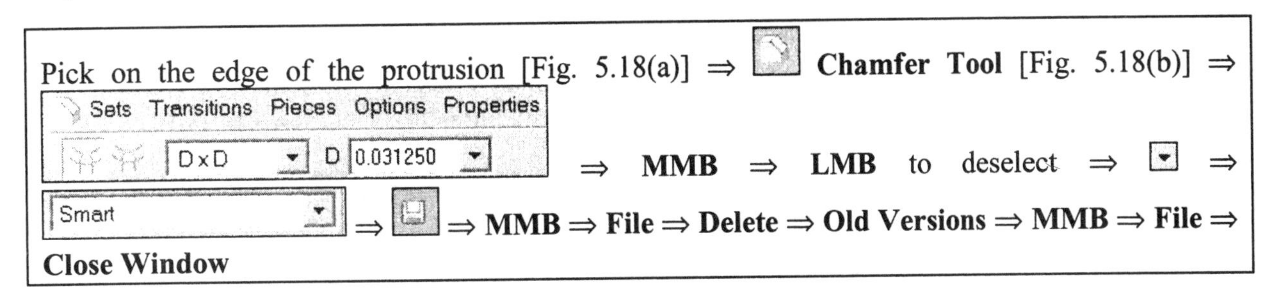

Pick on the edge of the protrusion [Fig. 5.18(a)] ⇒ [icon] **Chamfer Tool** [Fig. 5.18(b)] ⇒ Sets Transitions Pieces Options Properties DxD D 0.031250 ⇒ **MMB** ⇒ **LMB** to deselect ⇒ ▾ ⇒ Smart ⇒ [icon] ⇒ **MMB** ⇒ **File** ⇒ **Delete** ⇒ **Old Versions** ⇒ **MMB** ⇒ **File** ⇒ **Close Window**

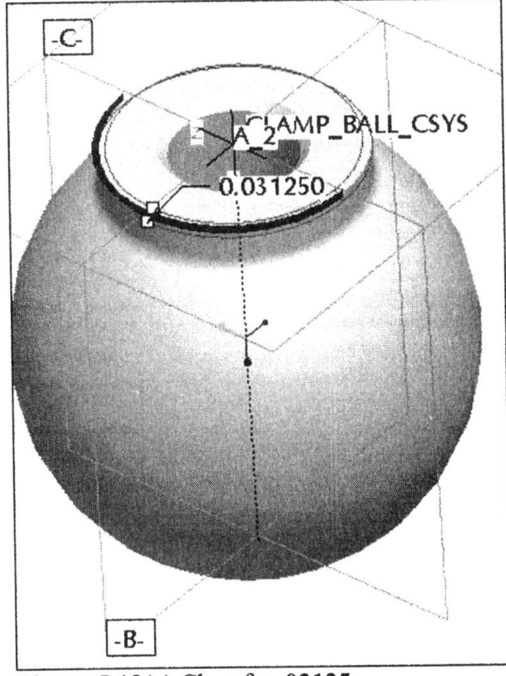

Figure 5.18(a) Chamfer **.03125**

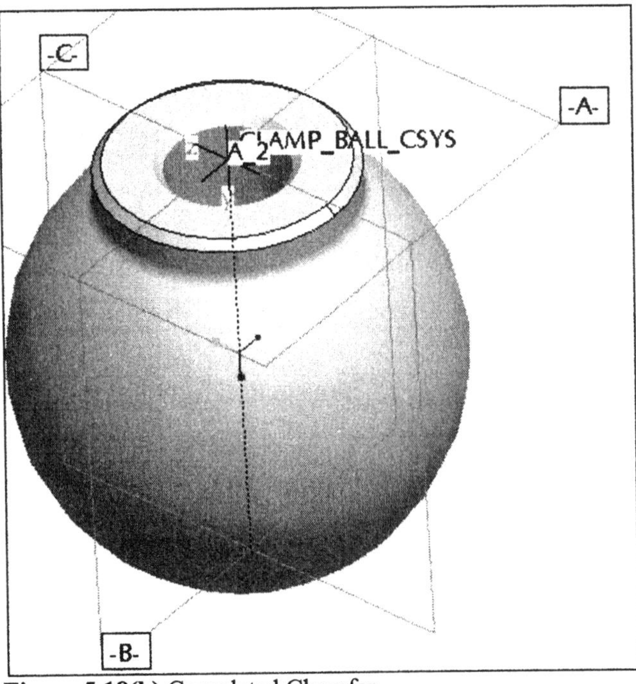

Figure 5.18(b) Completed Chamfer

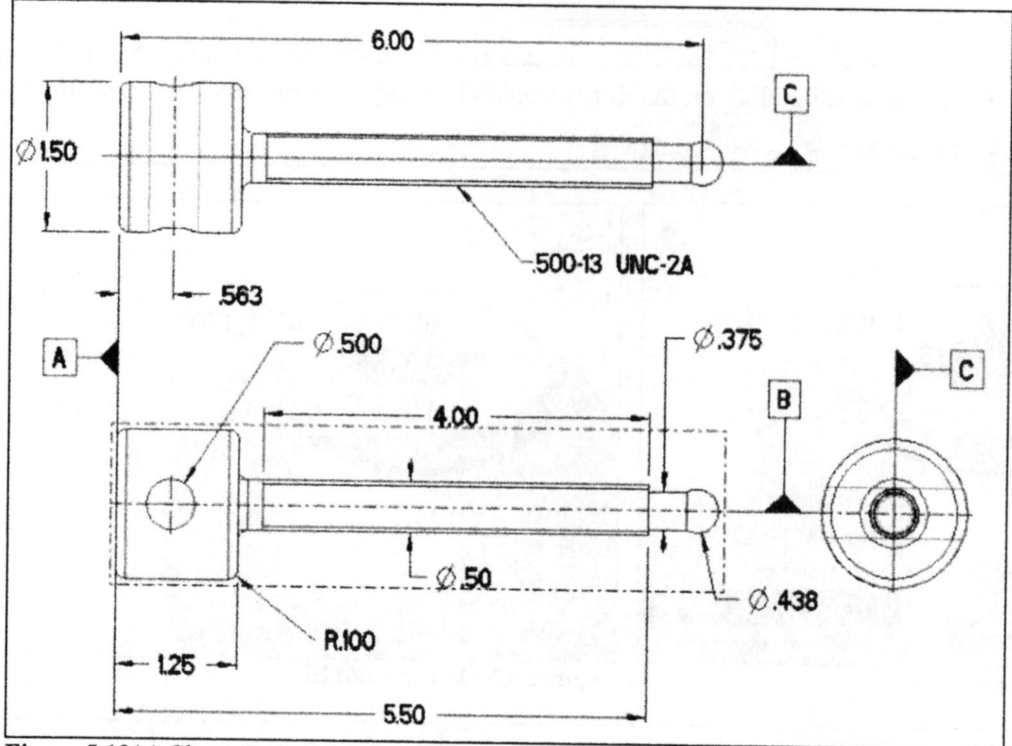

Figure 5.19(a) Clamp Swivel Detail

Clamp Swivel

The Clamp Swivel is the third part created by revolving one section about a centerline [Fig. 5.19(a)]. The Clamp Swivel is a component of the Clamp Assembly [Fig. 5.19(b)].

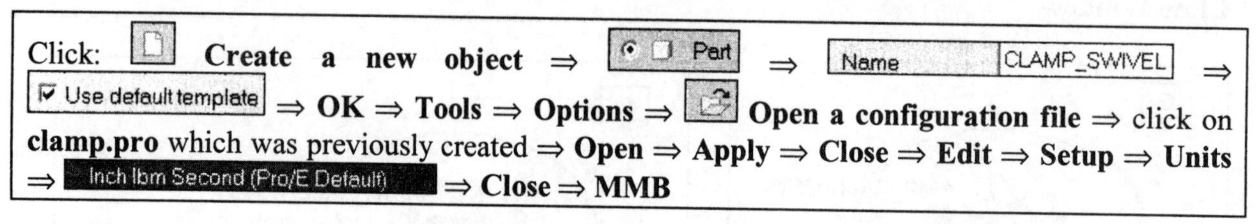

Click: **Create a new object** ⇒ Part ⇒ Name CLAMP_SWIVEL ⇒ Use default template ⇒ **OK** ⇒ **Tools** ⇒ **Options** ⇒ **Open a configuration file** ⇒ click on **clamp.pro** which was previously created ⇒ **Open** ⇒ **Apply** ⇒ **Close** ⇒ **Edit** ⇒ **Setup** ⇒ **Units** ⇒ Inch lbm Second (Pro/E Default) ⇒ **Close** ⇒ **MMB**

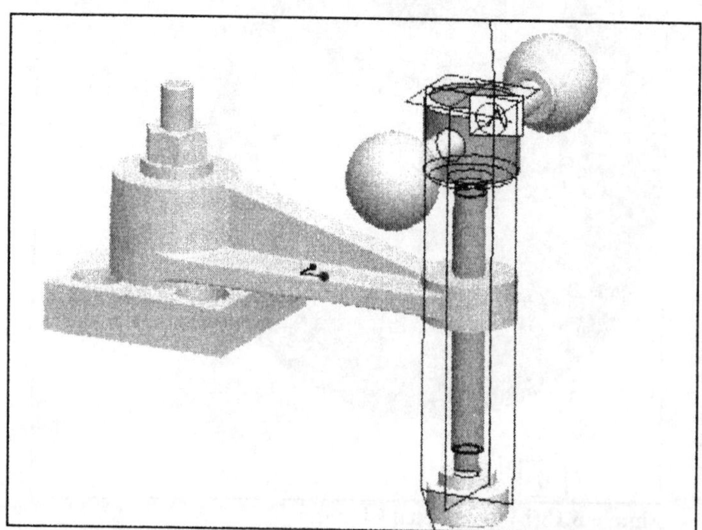

Figure 5.19(b) Clamp Swivel Highlighted in Clamp Assembly

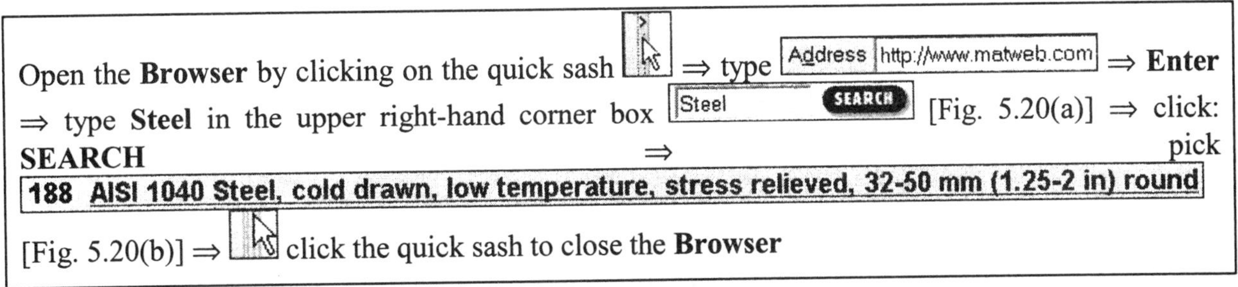

Open the **Browser** by clicking on the quick sash ⇒ type Address http://www.matweb.com ⇒ **Enter** ⇒ type **Steel** in the upper right-hand corner box Steel SEARCH [Fig. 5.20(a)] ⇒ click: **SEARCH** ⇒ pick **188 AISI 1040 Steel, cold drawn, low temperature, stress relieved, 32-50 mm (1.25-2 in) round** [Fig. 5.20(b)] ⇒ click the quick sash to close the **Browser**

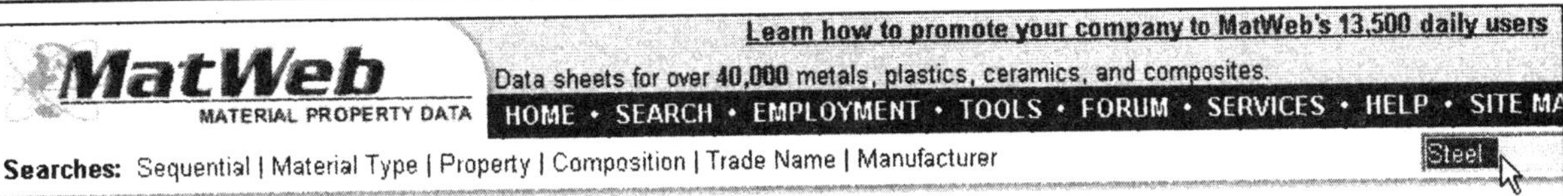

Searches: Sequential | Material Type | Property | Composition | Trade Name | Manufacturer Steel

MatWeb, Your Source for Materials Information

What is MatWeb?

The heart of MatWeb is a **searchable database of material data sheets**, including property information on thermoplastic and thermoset polymers such as ABS, nylon, polycarbonate, polyester, polyethylene and polypropylene; metals such as aluminum, cobalt, copper, lead, magnesium, nickel, steel, superalloys, titanium and zinc alloys; ceramics; plus semiconductors, fibers, and other engineering materials.

Feature
Pressu
Infiltrat
Casting

Have y
the
MatWe

MatWeb is freely available and does not require registration. You can still access all of the features that have always been available. However, advanced features are only available to our Registered and Premium users.

Figure 5.20(a) MatWeb.com

Mechanical Properties			
Hardness, Brinell	170	170	
Hardness, Knoop	191	191	Converted from Brinell har
Hardness, Rockwell B	86	86	Converted from Brinell har
Hardness, Rockwell C	6	6	Converted from Brinell har Value below normal HRC ran comparison purpose
Hardness, Vickers	178	178	Converted from Brinell har
Tensile Strength, Ultimate	585 MPa	84800 psi	
Tensile Strength, Yield	515 MPa	74700 psi	
Elongation at Break	10 %	10 %	in 5
Reduction of Area	30 %	30 %	
Modulus of Elasticity	200 GPa	29000 ksi	Typical fo
Bulk Modulus	140 GPa	20300 ksi	Typical for
Poisson's Ratio	0.29	0.29	Typical Fo
Izod Impact	49 J	36.1 ft-lb	as rolled, 45 J (33 ft-lb) anne 790°C (1450°F), 65 J (4 normalized at 900°C (1
Shear Modulus	80 GPa	11600 ksi	Typical for

Figure 5.20(b) Steel Properties

Use the material specifications from MatWeb [Fig. 5.20(b)] to complete as much of the material table as required (engineers, check your units) ⇒ **Edit** ⇒ **Setup** ⇒ **Material** ⇒ **Define** ⇒ type **Steel** ⇒ **Enter** ⇒ fill in the form as best you can [Fig. 5.21(a)] ⇒ **File** ⇒ **Save** ⇒ **File** ⇒ **Exit** ⇒ **Assign** ⇒ pick **STEEL** ⇒ **Accept** ⇒ **MMB**

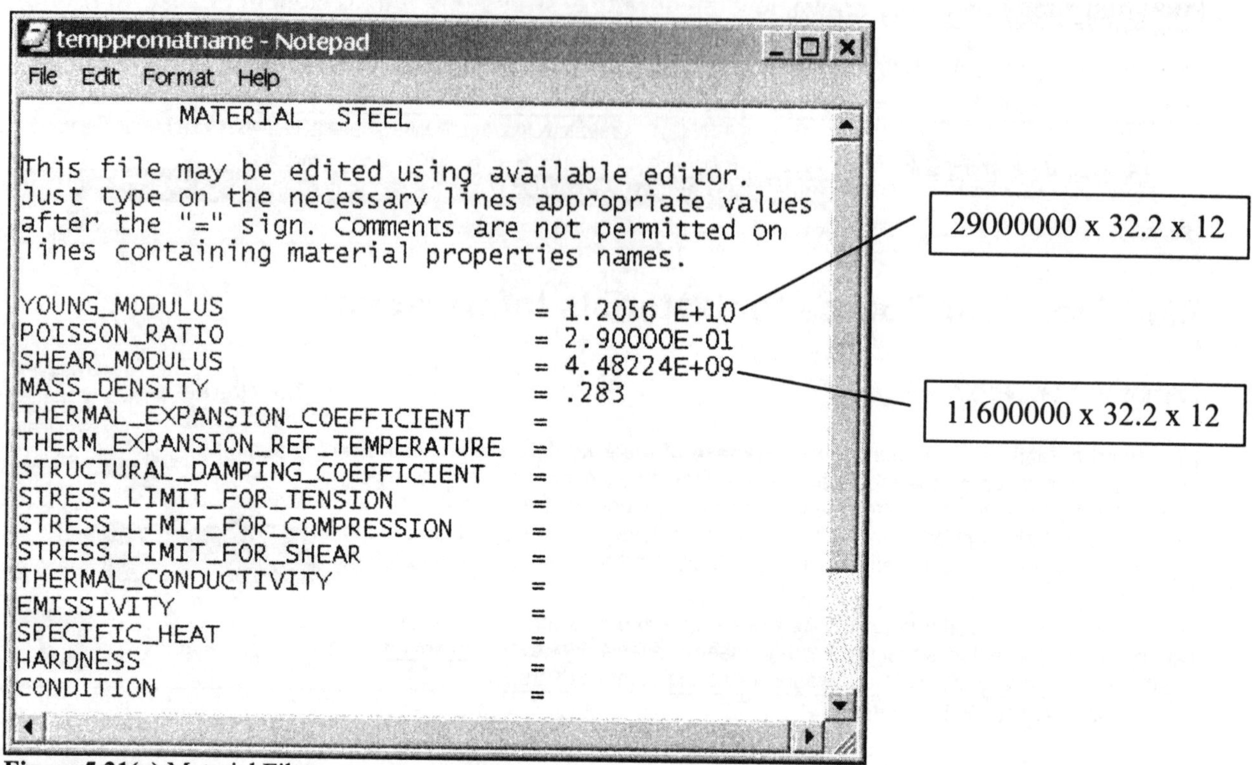

Figure 5.21(a) Material File

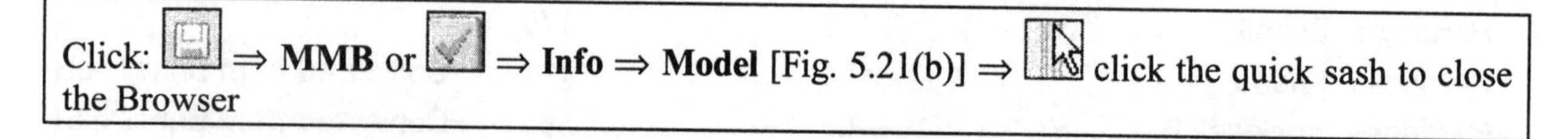

Click: ⇒ **MMB** or ⇒ **Info** ⇒ **Model** [Fig. 5.21(b)] ⇒ click the quick sash to close the Browser

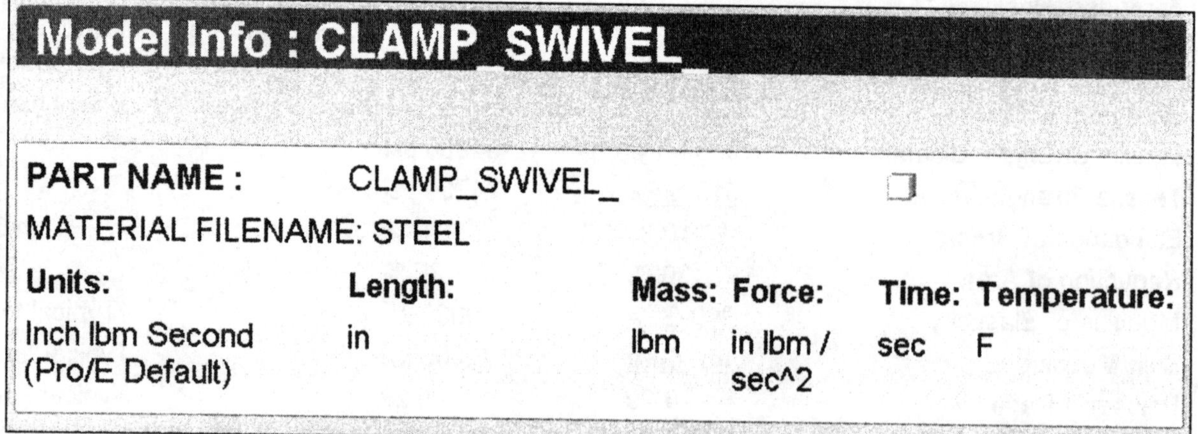

Figure 5.21(b) Model Info: Material Steel

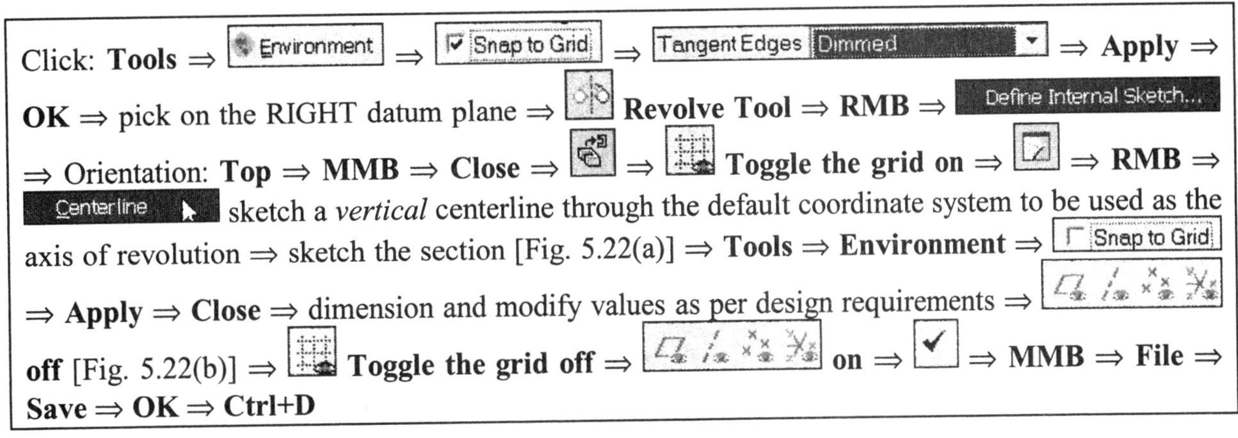

Click: **Tools** ⇒ Environment ⇒ Snap to Grid ⇒ Tangent Edges Dimmed ⇒ **Apply** ⇒ **OK** ⇒ pick on the RIGHT datum plane ⇒ **Revolve Tool** ⇒ **RMB** ⇒ Define Internal Sketch... ⇒ Orientation: **Top** ⇒ **MMB** ⇒ **Close** ⇒ ⇒ **Toggle the grid on** ⇒ ⇒ **RMB** ⇒ Centerline sketch a *vertical* centerline through the default coordinate system to be used as the axis of revolution ⇒ sketch the section [Fig. 5.22(a)] ⇒ **Tools** ⇒ **Environment** ⇒ Snap to Grid ⇒ **Apply** ⇒ **Close** ⇒ dimension and modify values as per design requirements ⇒ **off** [Fig. 5.22(b)] ⇒ **Toggle the grid off** ⇒ **on** ⇒ ⇒ **MMB** ⇒ **File** ⇒ **Save** ⇒ **OK** ⇒ **Ctrl+D**

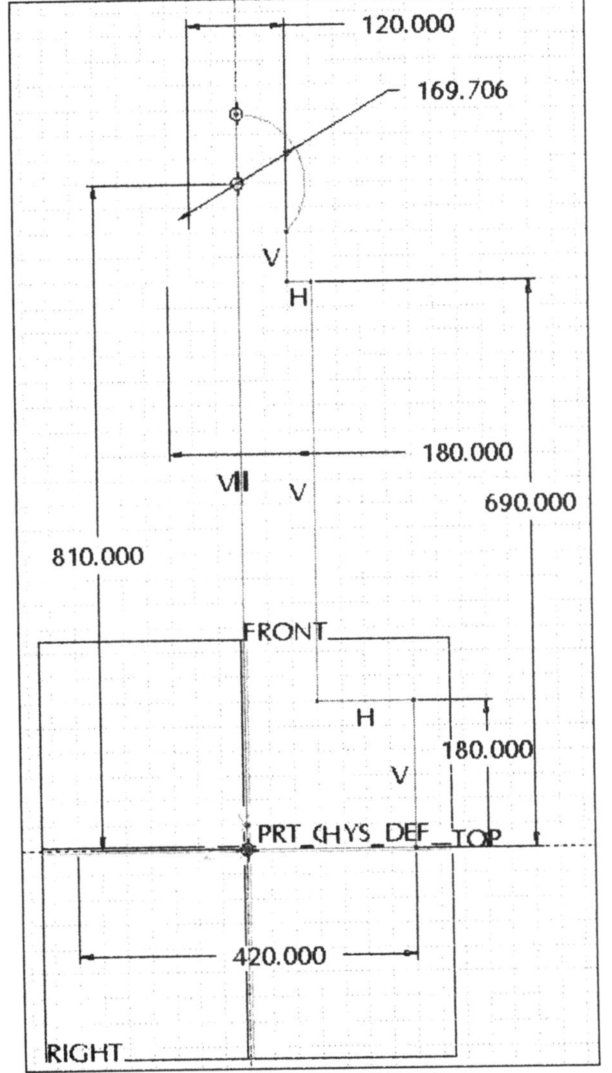

Figure 5.22(a) Sketch the Closed Section

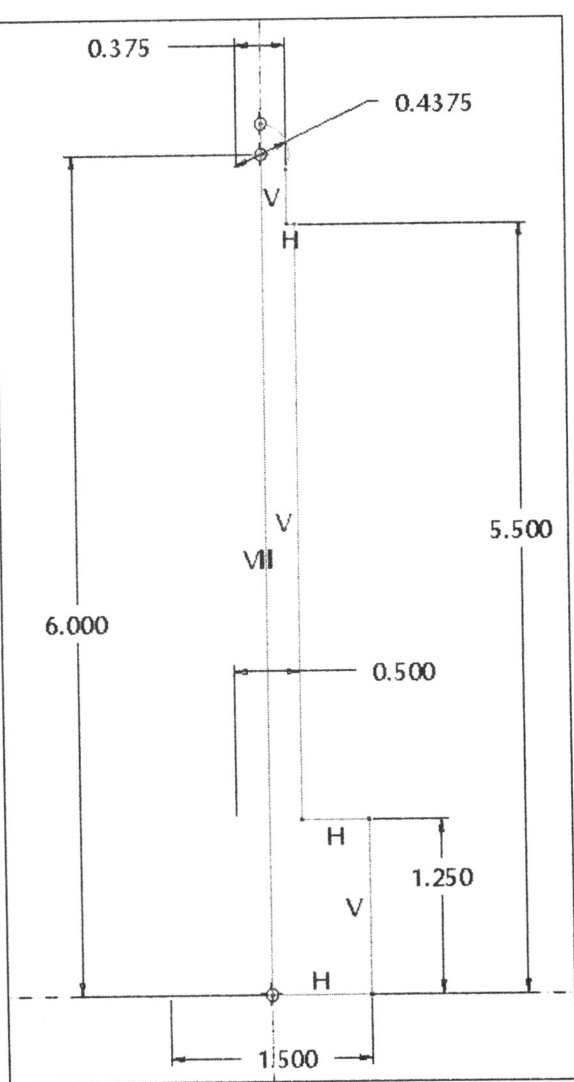

Figure 5.22(b) Dimension and Modify

Click: **Edit** ⇒ **Setup** ⇒ **Geom Tol** ⇒ **Set Datum** ⇒ pick **TOP** from the model ⇒ Name- type **A** ⇒ **MMB** ⇒ pick **RIGHT** ⇒ Name- type **B** ⇒ **MMB** ⇒ pick **FRONT** ⇒ Name- type **C** ⇒ **MMB** ⇒ **MMB** ⇒ **MMB** ⇒ **MMB** ⇒ rename the coordinate system CLAMP_SWIVEL_CSYS ⇒ ⇒ **MMB** (Fig. 5.23)

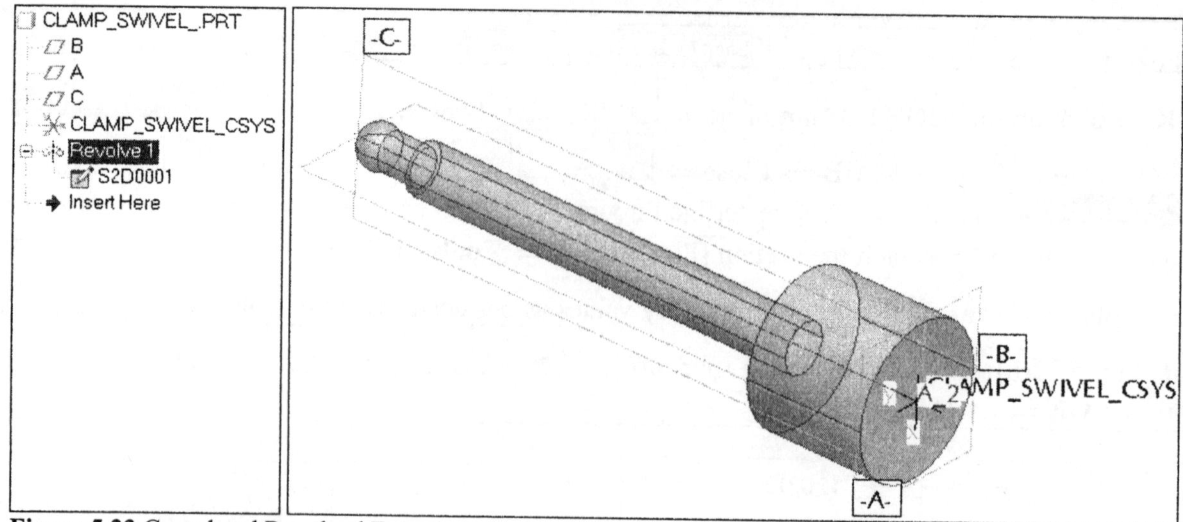

Figure 5.23 Completed Revolved Extrusion

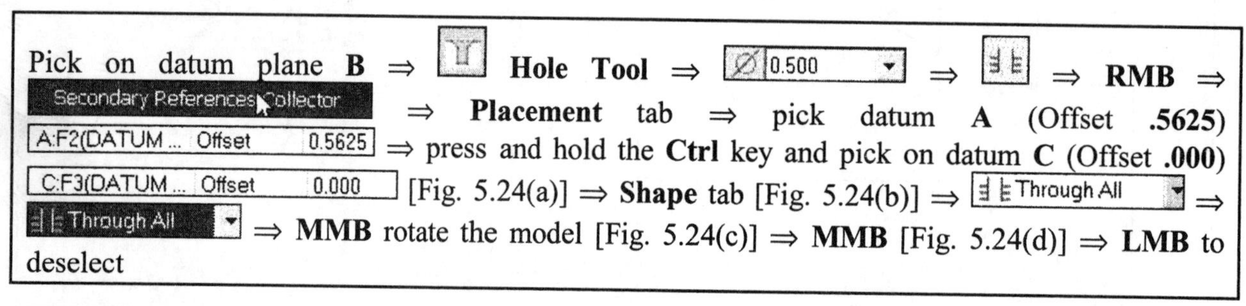

Pick on datum plane **B** ⇒ [Hole icon] **Hole Tool** ⇒ [Ø 0.500] ⇒ [depth icon] ⇒ **RMB** ⇒ [Secondary References Collector] ⇒ **Placement** tab ⇒ pick datum **A** (Offset **.5625**) [A:F2(DATUM ... Offset 0.5625] ⇒ press and hold the **Ctrl** key and pick on datum **C** (Offset **.000**) [C:F3(DATUM ... Offset 0.000] [Fig. 5.24(a)] ⇒ **Shape** tab [Fig. 5.24(b)] ⇒ [Through All] ⇒ [Through All] ⇒ **MMB** rotate the model [Fig. 5.24(c)] ⇒ **MMB** [Fig. 5.24(d)] ⇒ **LMB** to deselect

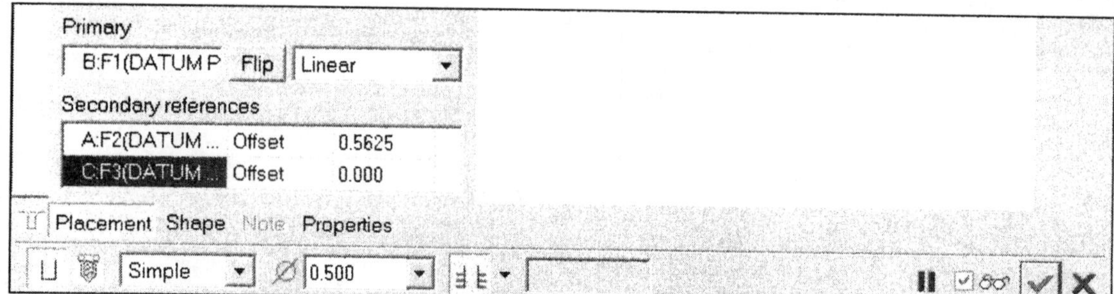

Figure 5.24(a) Placement Tab

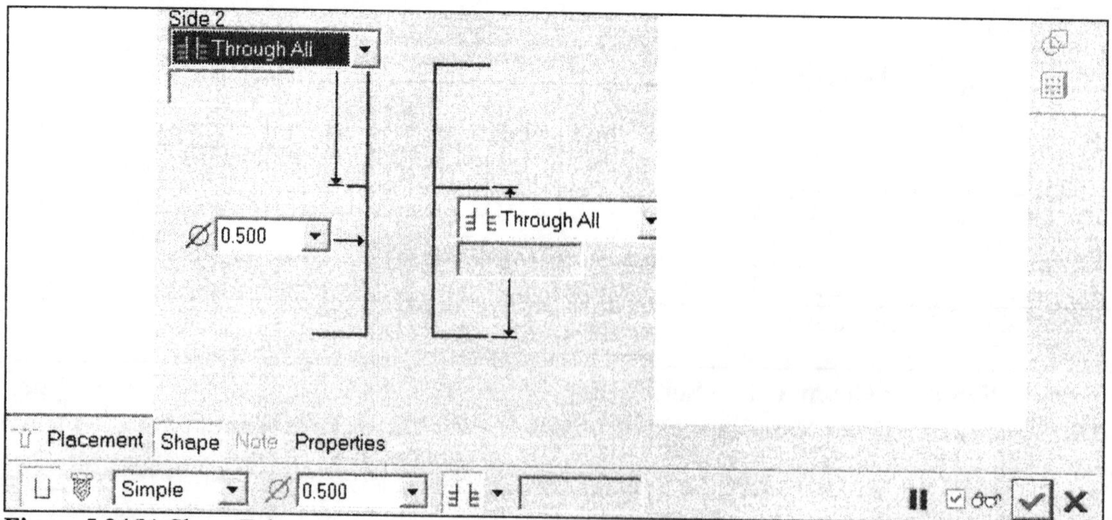

Figure 5.24(b) Shape Tab

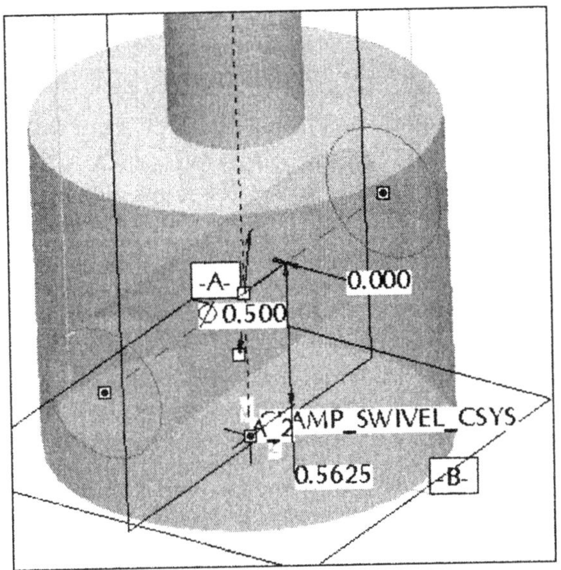

Figure 5.24(c) Hole Preview

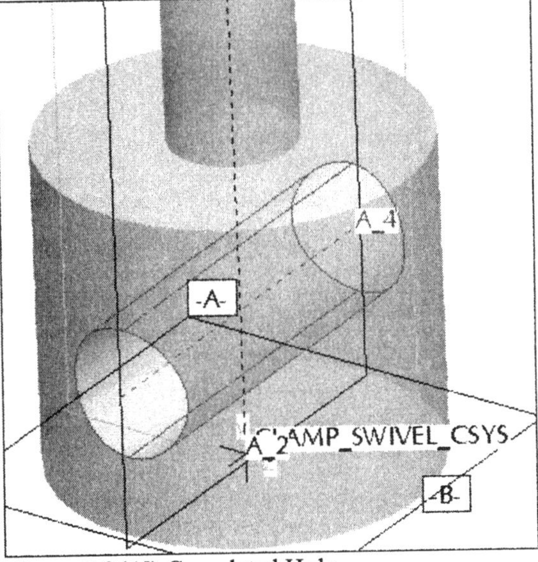

Figure 5.24(d) Completed Hole

Pick an edge, press and hold the **Ctrl** key and pick two more edges ⇒ **RMB** [Fig. 5.25(a)] ⇒ **Round Edges** ⇒ double-click on the dimension and modify to the design value of **.100** 0.100 ⇒ **Enter** ⇒ **MMB** [Fig. 5.25(b)] ⇒ ⇒ **MMB** ⇒ **LMB** to deselect

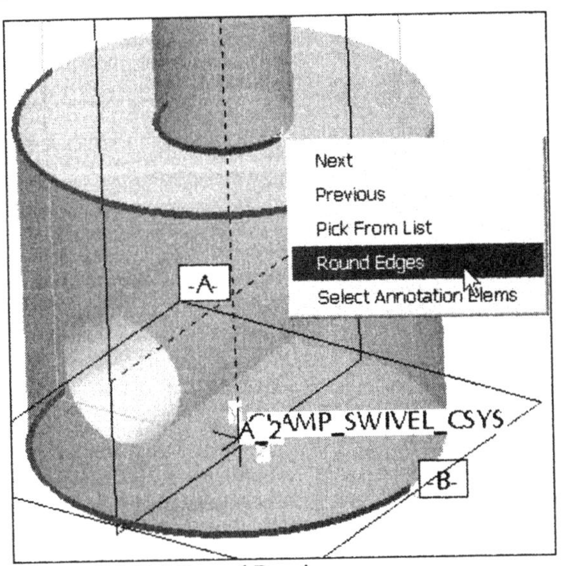

Figure 5.25(a) Round Preview

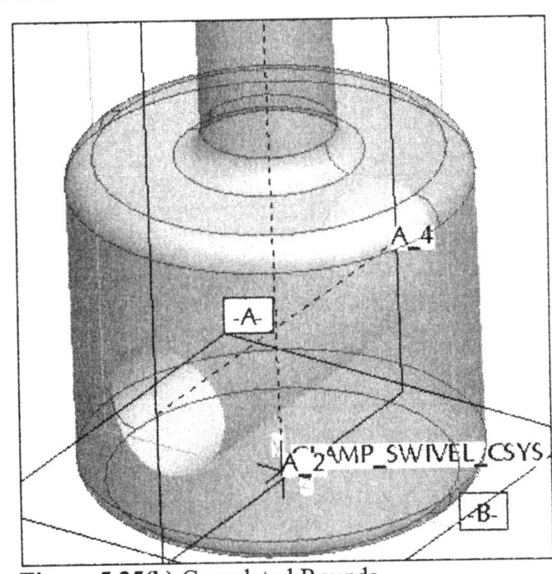

Figure 5.25(b) Completed Rounds

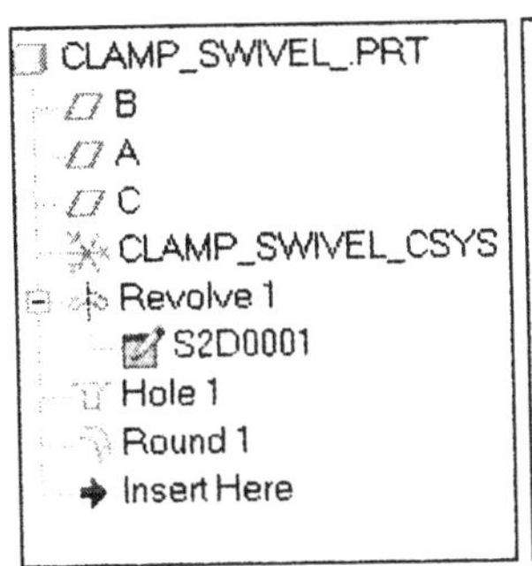

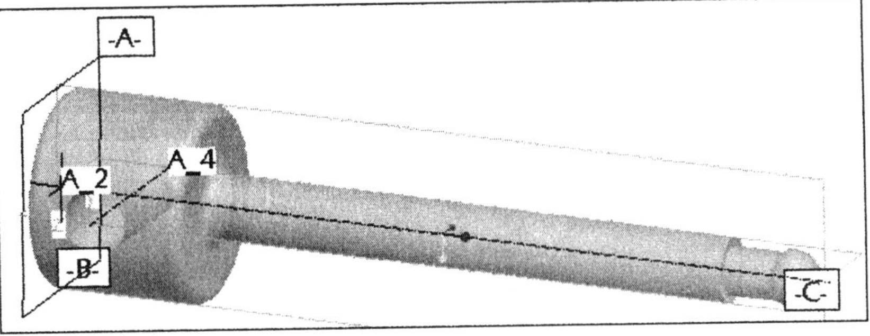

Figure 5.25(c) Completed Clamp_Swivel

Cosmetic Threads

A **cosmetic thread** is a feature that "represents" the diameter of a thread without having to show the actual threaded surfaces. Since a threaded feature is memory intensive, using cosmetic threads can save an enormous amount of visual memory on your computer. It is displayed in a unique default color. Unlike other cosmetic features, you cannot modify the line style of a cosmetic thread, nor are cosmetic threads affected by hidden line display settings in the ENVIRONMENT menu.

Threads are created with the default tolerance setting of limits. Cosmetic threads can be external or internal, and blind or through. You create cosmetic threads by specifying the minor or major diameter (for external and internal threads, respectively), starting surface, and thread length or ending edge.

For a starting surface, you can select a quilt surface, regular Pro/E surface, or split surface (such as a surface that belongs to a revolved feature, chamfer, round, or swept feature). For an "up to" surface, you can select any solid surface or a datum plane. A cosmetic thread that uses a depth parameter (a blind thread) cannot be defined from a non-planar surface.

The following table lists the parameters that can be defined for a cosmetic thread at its creation or later after the cosmetic thread has been added. In this table, "pitch" is the distance between two threads.

PARAMETER NAME	PARAMETER TYPE	PARAMETER DESCRIPTION
MAJOR_DIAMETER	Real Number	Thread major diameter
THREADS_PER_INCH	Real Number	Threads per inch (1/pitch)
THREAD FORM	String	Thread form
CLASS	String	Thread class
PLACEMENT	Character	Thread placement (A-external, B-internal)
METRIC	YesNo (True/False)	Thread is metric

Here, the external cosmetic thread is specified by the minor diameter (external threads), starting surface, and ending edge.

- **Internal cosmetic threads** are created automatically when holes are created using the Hole Tool (Standard- Tapped). In situations where the internal thread is unique and the Hole Tool cannot be used, internal cosmetic threads can be added to the hole.
- **External cosmetic threads** represent the *root diameter.* For threaded shafts you must create the external cosmetic threads.

After creating the cosmetic thread, edit the thread table. The thread size of an external thread must be changed to the nominal size from the root diameter defaulted on the Table. Create an external cosmetic thread (**.500-13 UNC-2A**) using the **∅.500** surface [Fig. 5.26(a)]. The thread starts at the "neck" and extends **4.00** along the swivel's shaft [Fig. 5.26(b)].

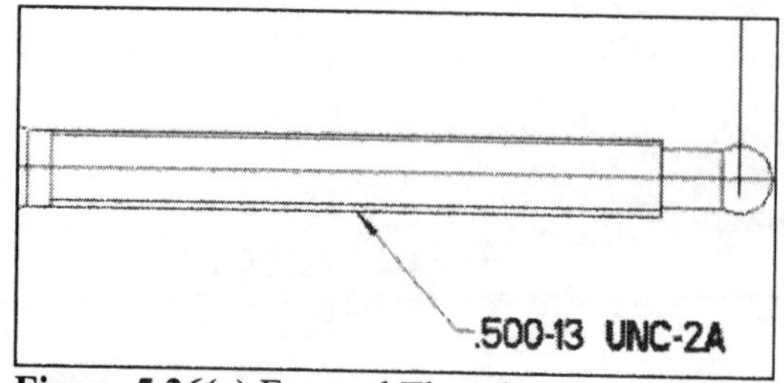

Figure 5.26(a) External Thread

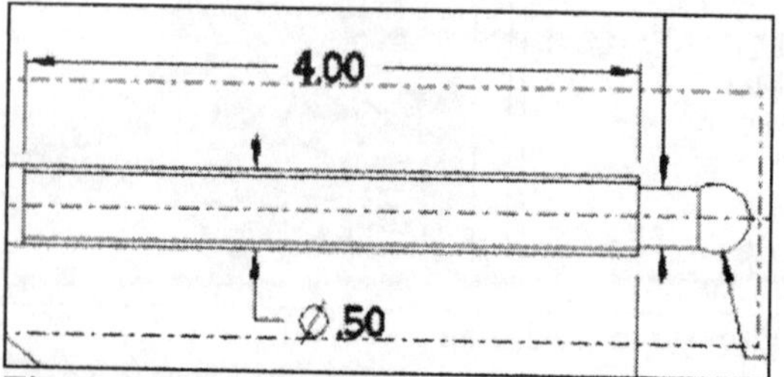

Figure 5.26(b) Thread Length

Click: **Insert** ⇒ **Cosmetic** ⇒ **Thread** ⇒ pick the cylindrical surface [Fig. 5.27(a)] ⇒ pick the thread start surface--the edge lip surface [Fig. 5.27(b)] ⇒ **Okay** [Fig. 5.27(c)] ⇒ **Blind** ⇒ **Done** ⇒ **4.00** [Enter depth 4.00] ⇒ **Enter** ⇒ type the value of the cosmetic thread root diameter: **.4485** [Enter DIAMETER .4485] ⇒ **Enter** [Fig. 5.27(d)] ⇒ **Mod Params** ⇒ edit the table

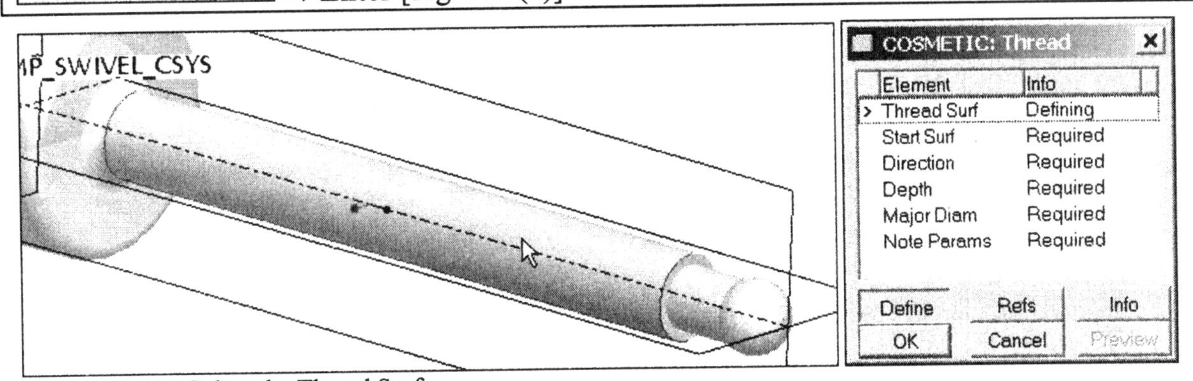

Figure 5.27(a) Select the Thread Surface

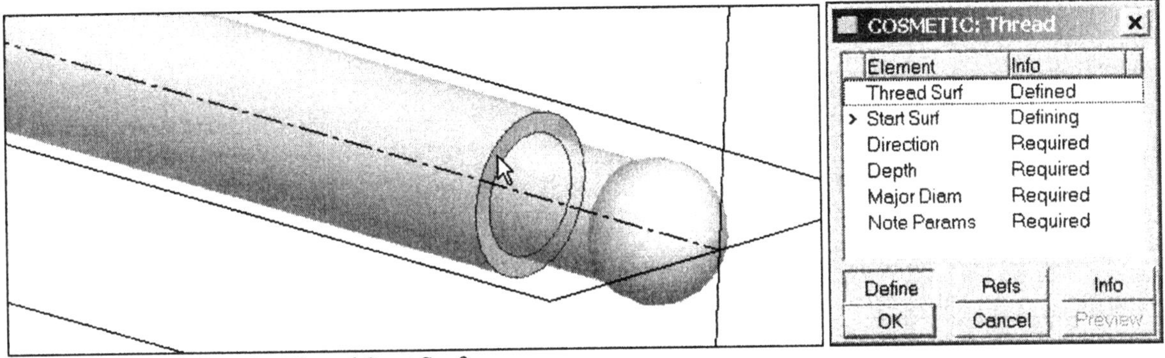

Figure 5.27(b) Select the Thread Start Surface

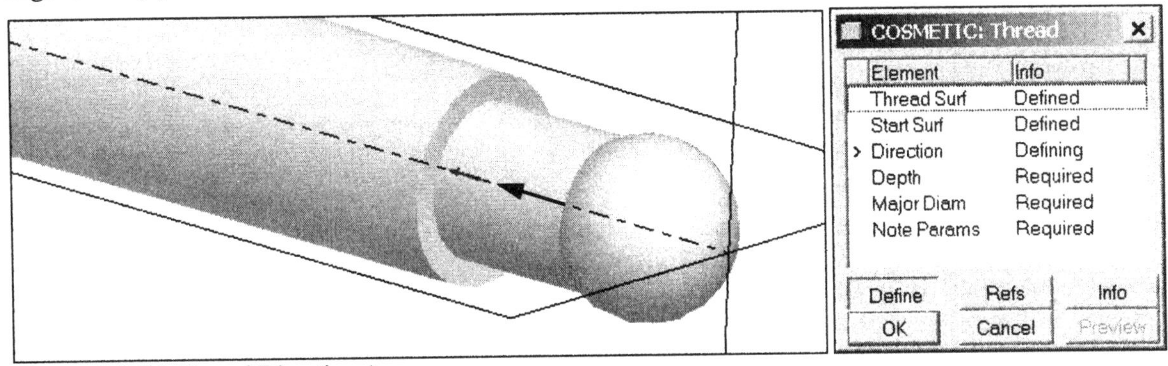

Figure 5.27(c) Thread Direction Arrow

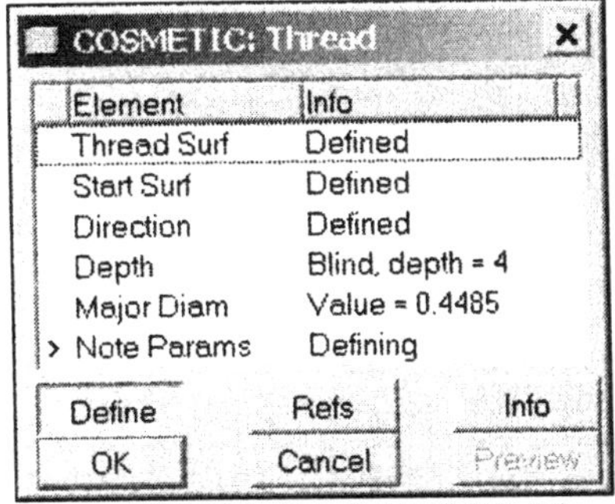

Figure 5.27(d) COSEMETIC: Thread Dialog Box

The Pro/TABLE thread parameters [Fig. 5.27(e)] shows the diameter as **0.4485**. Since you are cosmetically representing the *root diameter (.4485)* of the external thread on the model, the *thread diameter* is *smaller* than the *nominal (.500)* thread size.

Click in each table field and change the values [Fig. 5.27(f)] ⇒ THREADS_PER_INCH **13** ⇒ FORM **UNC** ⇒ CLASS **2** ⇒ PLACEMENT **A** ⇒ METRIC **False** ⇒ from Pro/TABLE, click: **File** ⇒ **Save** ⇒ **File** ⇒ **Exit** ⇒ **Show** [Fig. 5.27(g)] ⇒ **Close** ⇒ **MMB** [Fig. 5.27(h)] ⇒ **OK** [Fig. 5.27(i)] ⇒ **LMB**

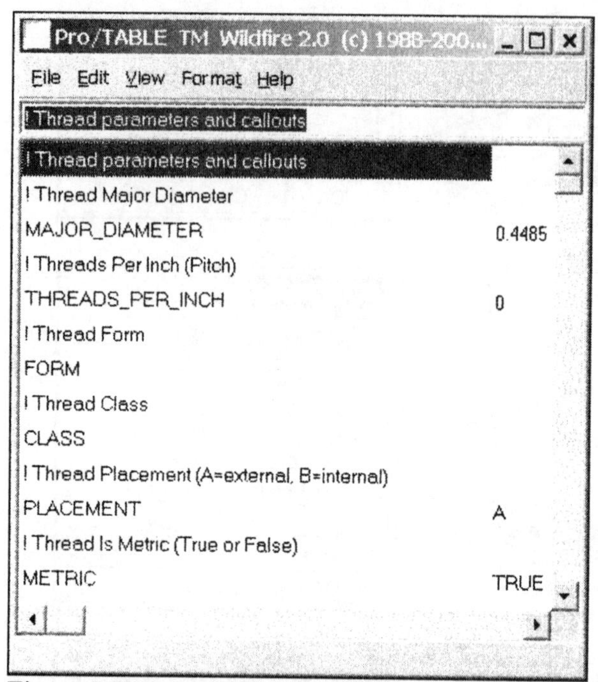

Figure 5.27(e) Pro/Table Thread Parameters

Figure 5.27(f) Edited Table Parameters

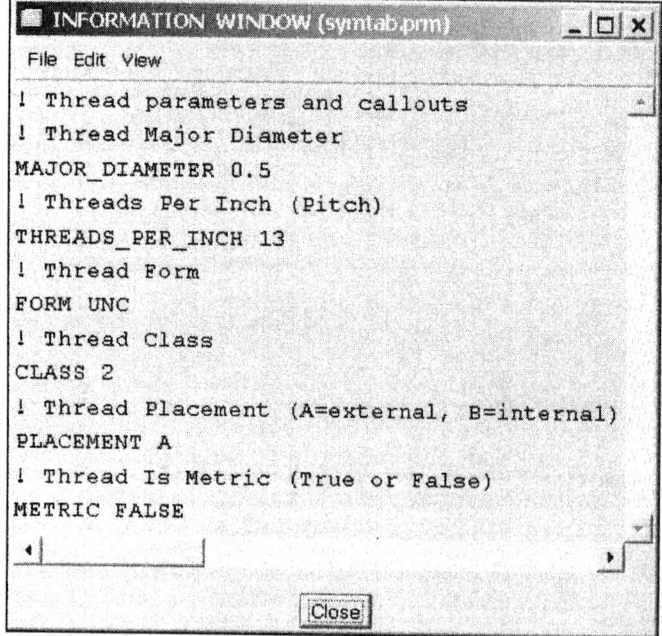

Figure 5.27(g) Thread Information Window

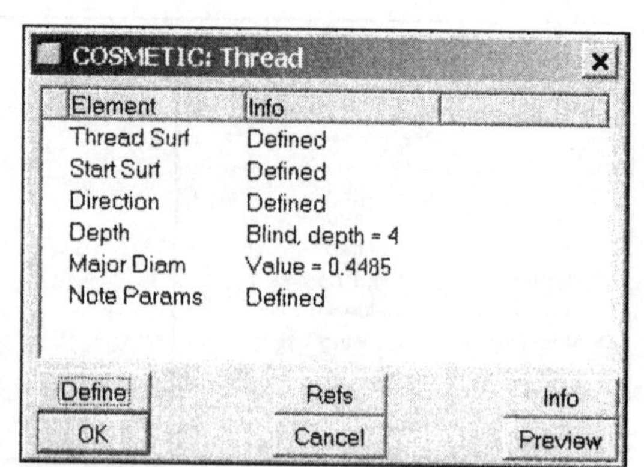

Figure 5.27(h) Completed Thread Elements

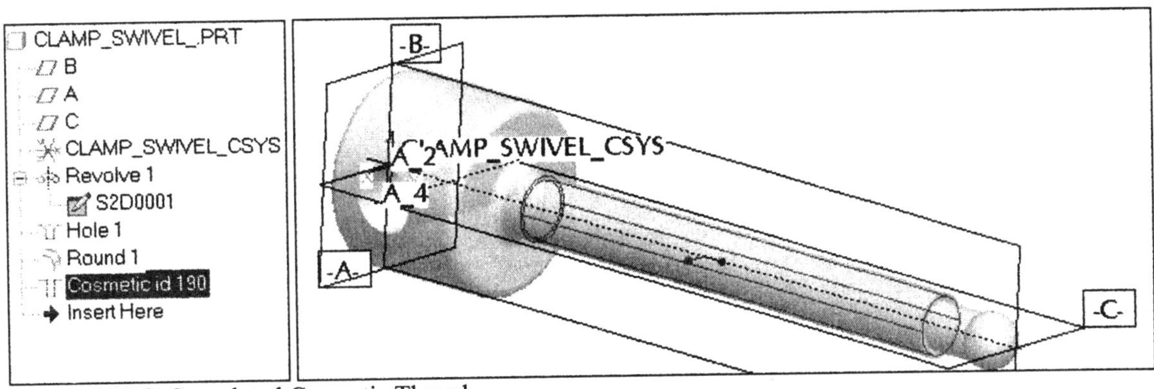

Figure 5.27(i) Completed Cosmetic Thread

Click on **Cosmetic id** in the Model Tree (Fig. 5.28) ⇒ **RMB** ⇒ **Info** ⇒ **Feature** ⇒ click on the quick sash to close the **Browser** ⇒ ⇒ **MMB**

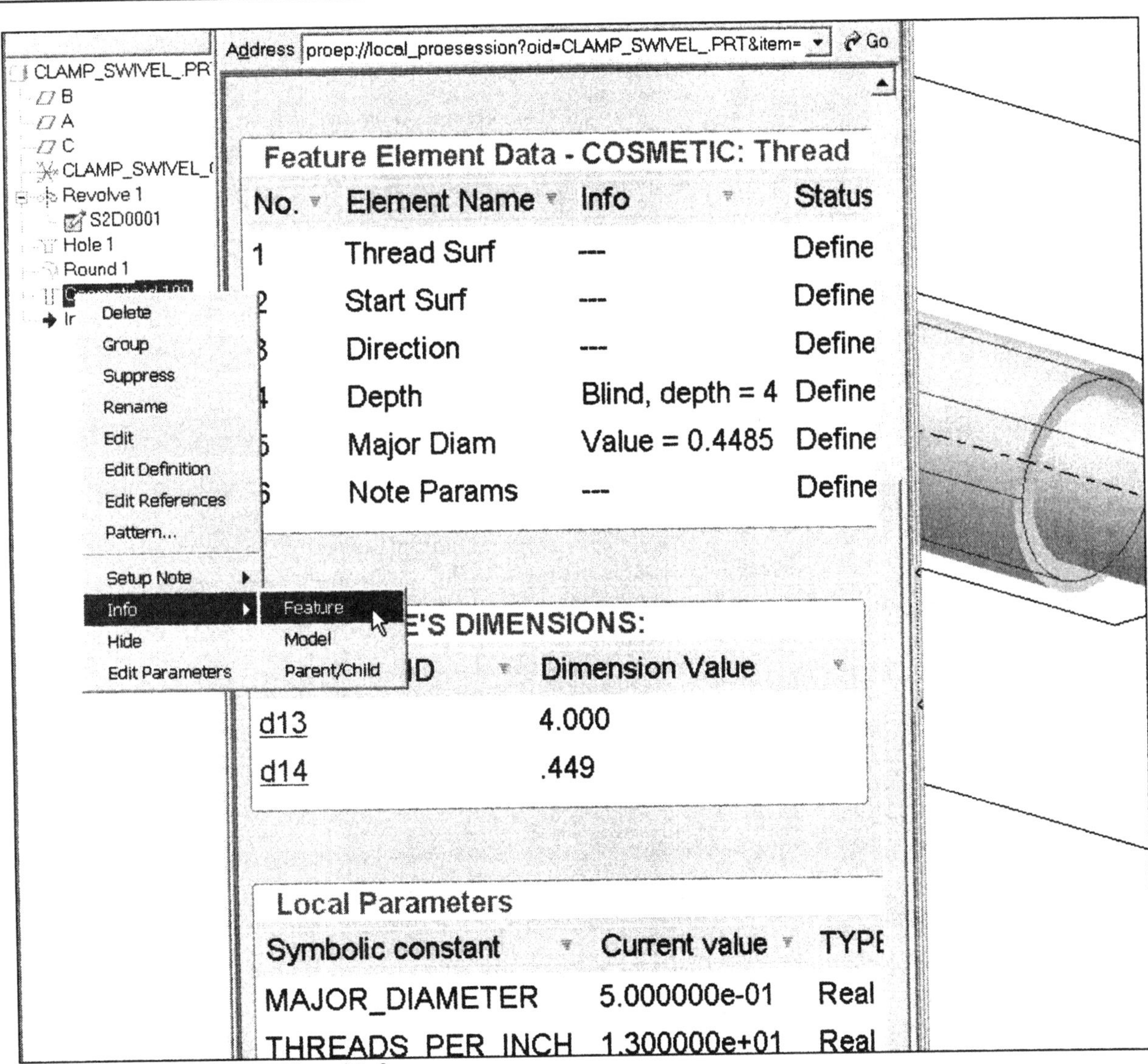

Figure 5.28 COSEMEIC Thread Info

Viewing Model Creation History

The **Model Player** (Fig. 5.29) option on the **Tools** menu lets you observe how a part is built. You can:

- Move backward or forward through the feature-creation history of the model in order to observe how the model was created. You can start the model playback at any point in its creation history
- Regenerate each feature in sequence, starting from the specified feature
- Display each feature as it is regenerated or rolled forward
- Update (regenerate all the features in) the entire display when you reach the desired feature.
- Obtain information about the current feature (you can show dimensions, obtain regular feature information, investigate geometry errors, and enter Fix Model mode)

Using the Model Player

To open the **Model Player** dialog box, click Tools ⇒ Model Player. You can select one of the following:

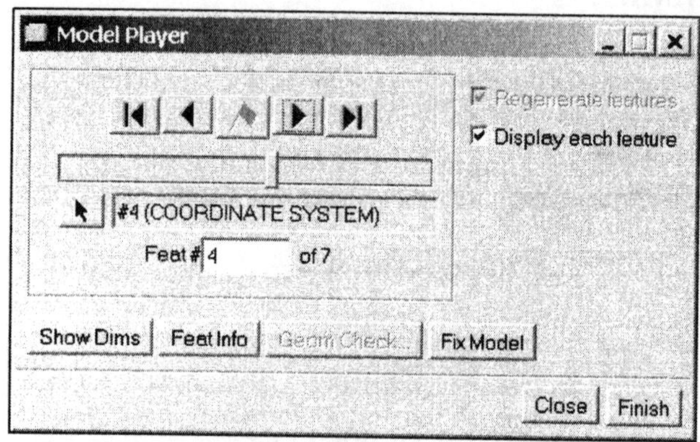

- **Regenerate features** Regenerates each feature in sequence, starting from the specified feature
- **Display each feature** Displays each feature in the graphics window as it is being regenerated
- **Compute CL** (Available in Manufacturing mode only) When selected, the CL data is recalculated for each NC sequence during regeneration.

Figure 5.29 Model Player at Feature #4 of Feature List

Select one of the following commands:

Go to the beginning of the model moves immediately to the beginning of the model

Step backward through the model one feature at-a-time and regenerates the preceding feature

Stop play

Step forward through the model one feature at-a-time and regenerates the next feature

Go to the last feature in the model moves immediately to the end of the model (resume all)

Slider Bar Drag the slider handle to the feature at which you want model playback to begin. The features are highlighted in the graphics window as you move through their position with the slider handle. The feature number and type are displayed in the selection panel [such as #4 (COORDINATE SYSTEM)] [Fig. 5.6(b)], and the feature number is displayed in the **Feat #** box.

Select feature from screen or model tree Lets you select a starting feature from the graphics window or the Model Tree. Opens the **SELECT FEAT** and **SELECT** menus. After you select a starting feature, its number and ID are displayed in the selection panel, and the feature number is displayed in the **Feat #** box.

Feat # 5 of 25 Lets you specify a starting feature by typing the feature number in the box. After you enter the feature number, the model immediately rolls or regenerates to that feature.

To stop playback, click the **Stop play** button. Use the following commands for information:

- **Show Dims** Displays the dimensions of the current feature
- **Feat Info** Provides regular feature information about the current feature in an Information window
- **Geom Check** Investigates the geometry error for the current feature.
- **Fix Model** Activates Resolve mode by forcing the current feature to abort regeneration.
- **Close** Closes the Model Player and enters Insert mode at the current feature
- **Finish** Closes the Model Player and returns to the last feature in the model

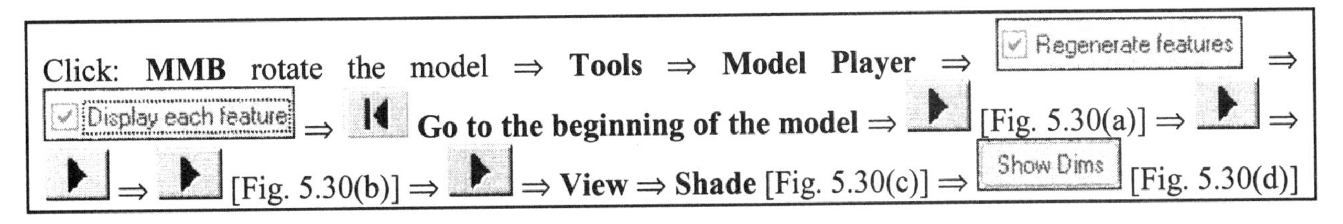

Click: **MMB** rotate the model ⇒ **Tools** ⇒ **Model Player** ⇒ Regenerate features ⇒ Display each feature ⇒ **Go to the beginning of the model** ⇒ [Fig. 5.30(a)] ⇒ ⇒ ⇒ [Fig. 5.30(b)] ⇒ ⇒ **View** ⇒ **Shade** [Fig. 5.30(c)] ⇒ Show Dims [Fig. 5.30(d)]

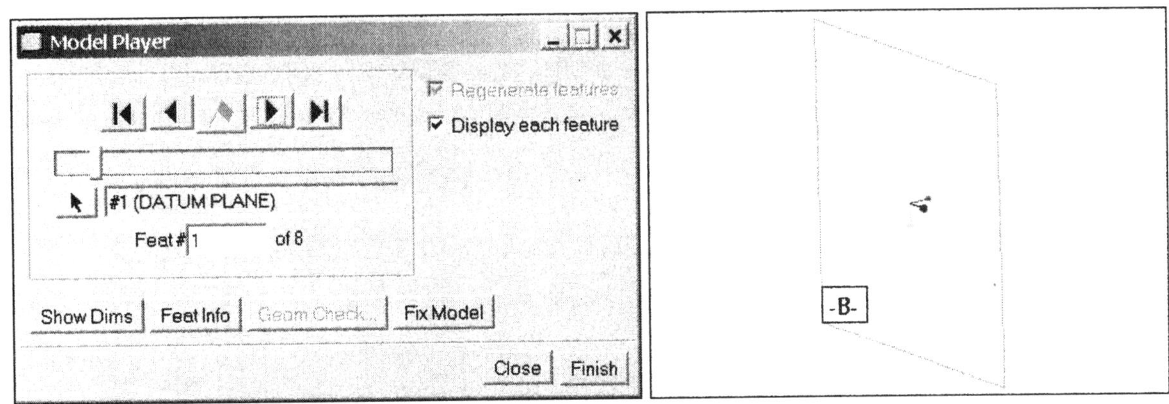

Figure 5.30(a) Regenerate First Feature

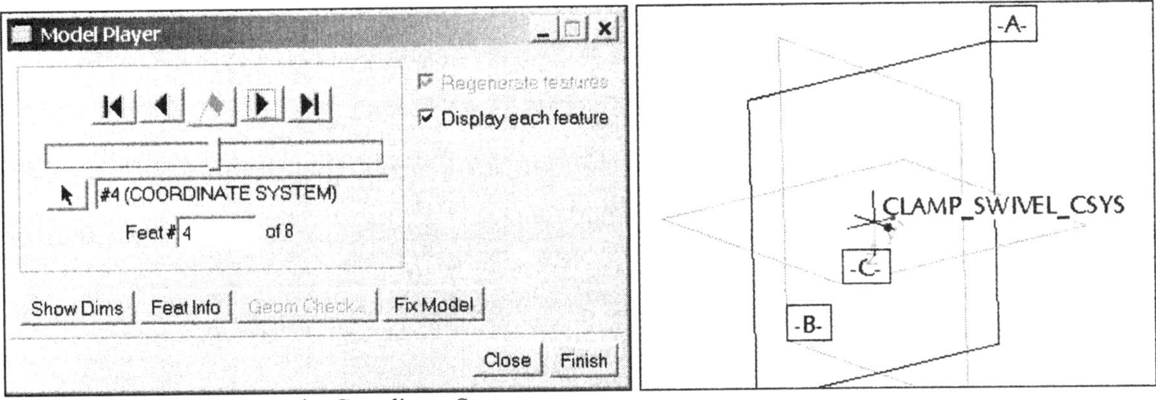

Figure 5.30(b) Regenerate the Coordinate System

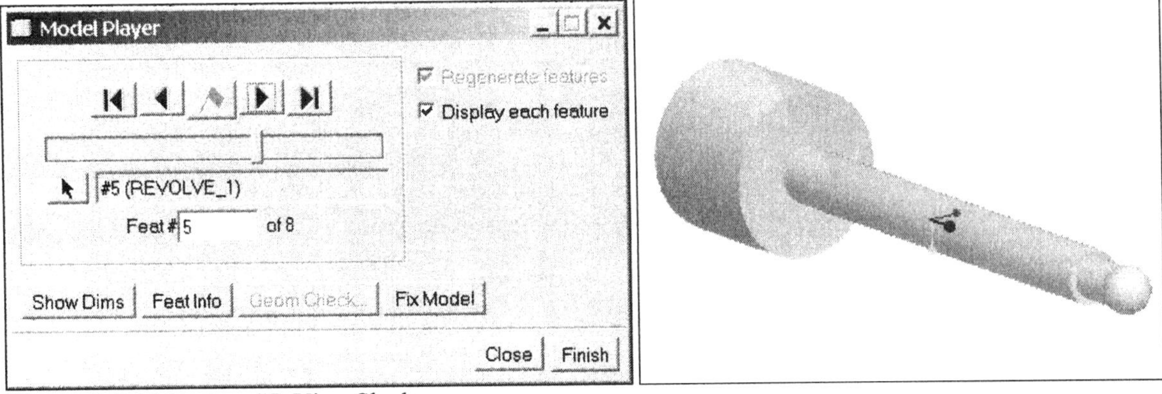

Figure 5.30(c) Feature #5, View Shade

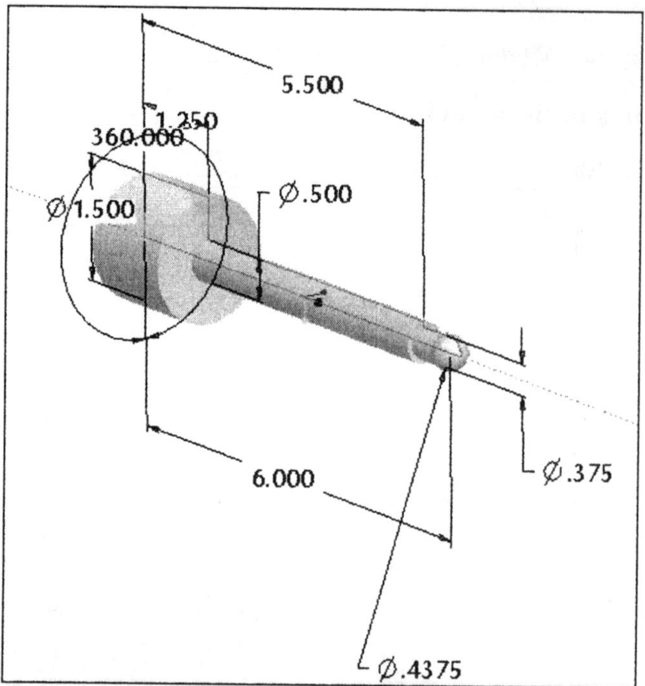

Figure 5.30(d) Feature #5, Revolve Dimensions

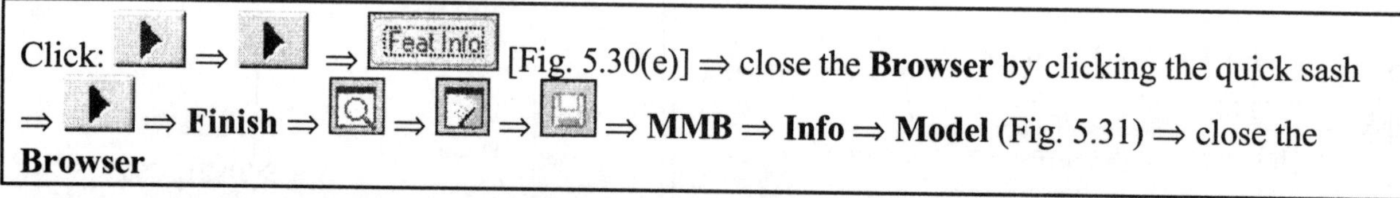
Click: ⇒ ⇒ Feat Info [Fig. 5.30(e)] ⇒ close the **Browser** by clicking the quick sash ⇒ ⇒ **Finish** ⇒ ⇒ ⇒ ⇒ **MMB** ⇒ **Info** ⇒ **Model** (Fig. 5.31) ⇒ close the **Browser**

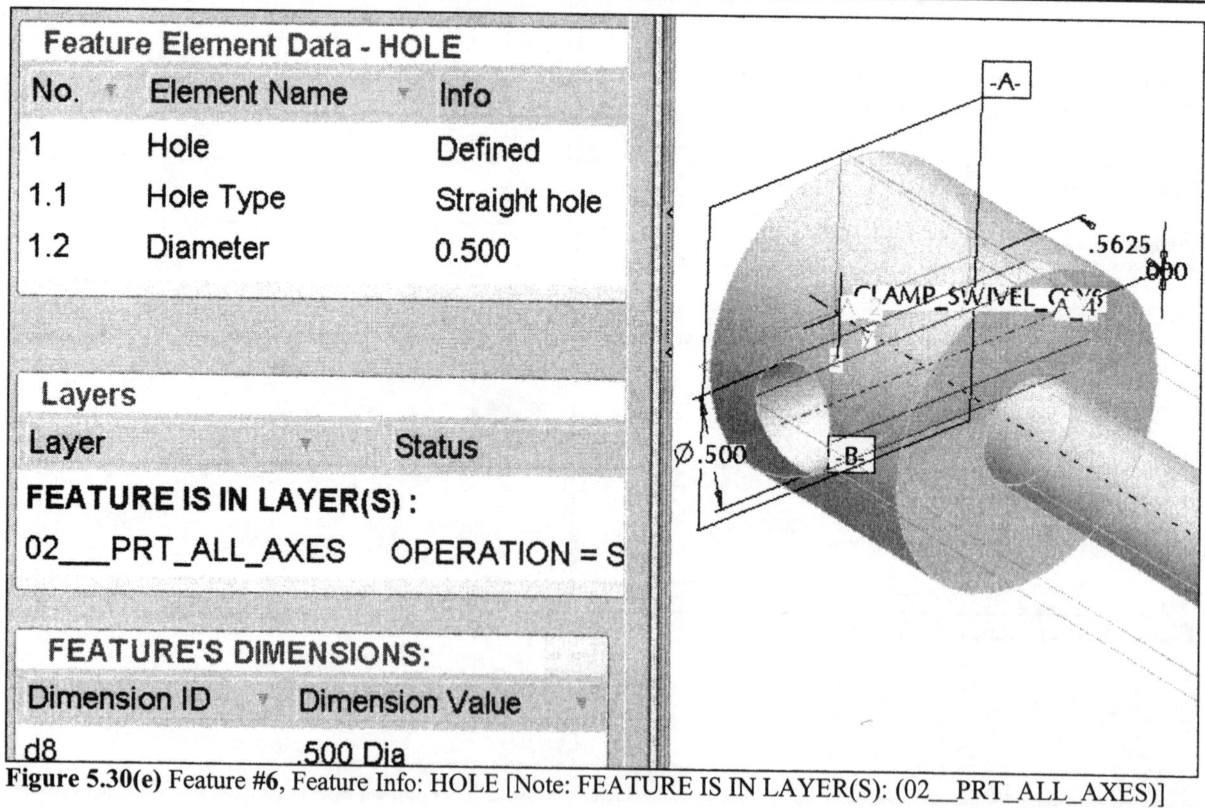

Figure 5.30(e) Feature **#6**, Feature Info: HOLE [Note: FEATURE IS IN LAYER(S): (02__PRT_ALL_AXES)]

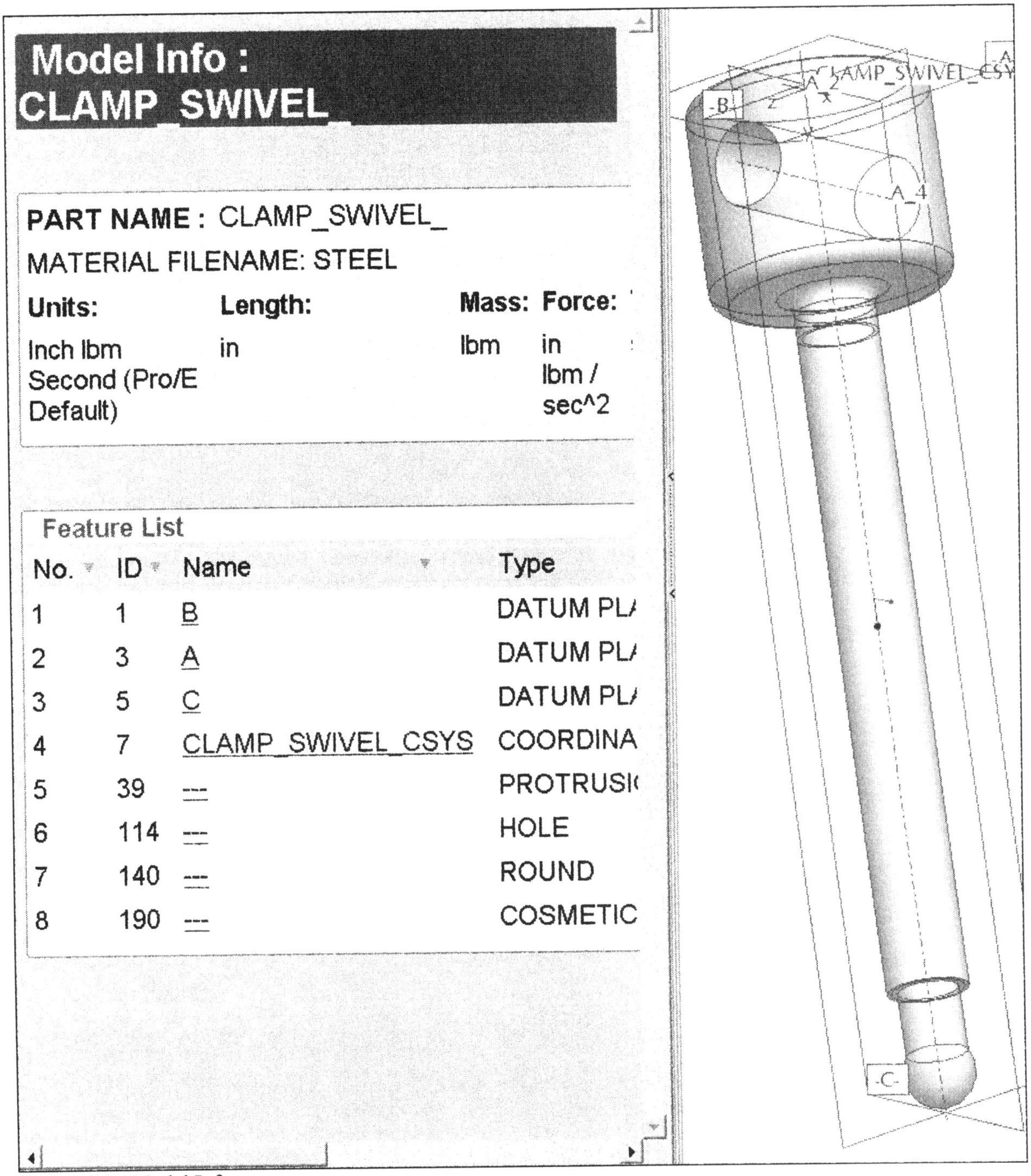

Model Info : CLAMP_SWIVEL_

PART NAME : CLAMP_SWIVEL_

MATERIAL FILENAME: STEEL

Units:	Length:	Mass:	Force:
Inch lbm Second (Pro/E Default)	in	lbm	in lbm / sec^2

Feature List

No.	ID	Name	Type
1	1	B	DATUM PL/
2	3	A	DATUM PL/
3	5	C	DATUM PL/
4	7	CLAMP_SWIVEL_CSYS	COORDINA
5	39	---	PROTRUSI(
6	114	---	HOLE
7	140	---	ROUND
8	190	---	COSMETIC

-B- CLAMP_SWIVEL_CSY A_2 A_4 -C-

Figure 5.31 Model Info

Printing and Plotting

From the File menu, you can print with the following options: scaling, clipping, displaying the plot on the screen, or sending the plot directly to the printer. Shaded images can also be printed from this menu. You can create plot files of the current object (sketch, part, assembly, drawing, or layout) and send them to the print queue of a plotter. The plotting interface to HPGL and PostScript formats is standard.

You can configure your printer using the Printer Configuration dialog box, available from the Print dialog box. If you are printing a shaded image, the Shaded Image Configuration dialog box opens instead of the Printer Configuration dialog box.

The following applies for plotting:

- Hidden lines appear as gray for a screen plot, but as dashed lines on paper.
- When Pro/E plots Pro/E line fonts, it scales them to the size of a sheet. It does not scale the user-defined line fonts, which do not plot as defined.
- You can use the configuration file option *use_software_linefonts* to make sure that the plotter plots a user-defined font exactly as it appears in Pro/E.
- You can plot a cross section from Part or Assembly mode.
- With a Pro/PLOT license, you can write plot files in a variety of formats.

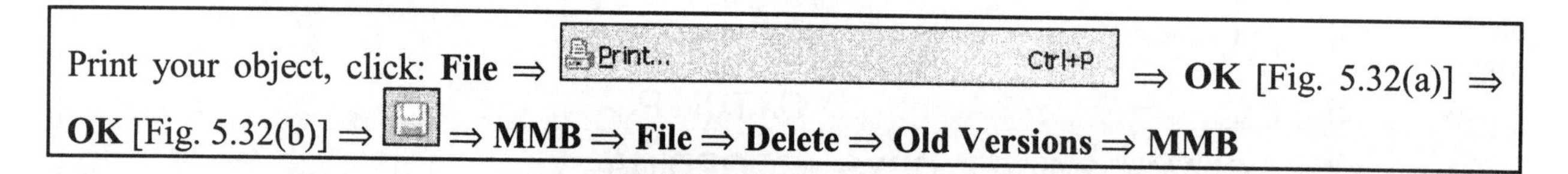

Print your object, click: **File** ⇒ Print... Ctrl+P ⇒ **OK** [Fig. 5.32(a)] ⇒ **OK** [Fig. 5.32(b)] ⇒ ⇒ **MMB** ⇒ **File** ⇒ **Delete** ⇒ **Old Versions** ⇒ **MMB**

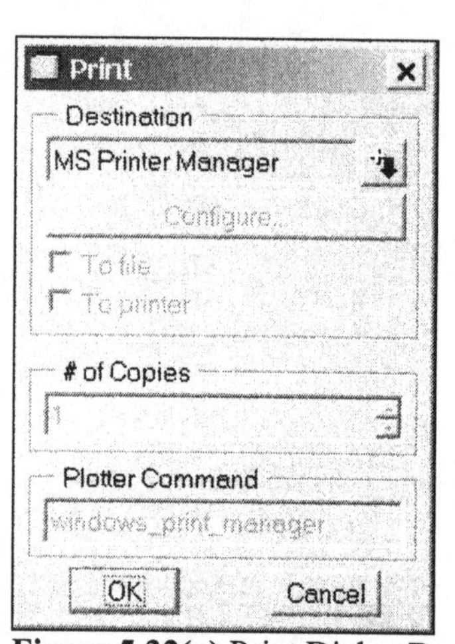

Figure 5.32(a) Print Dialog Box

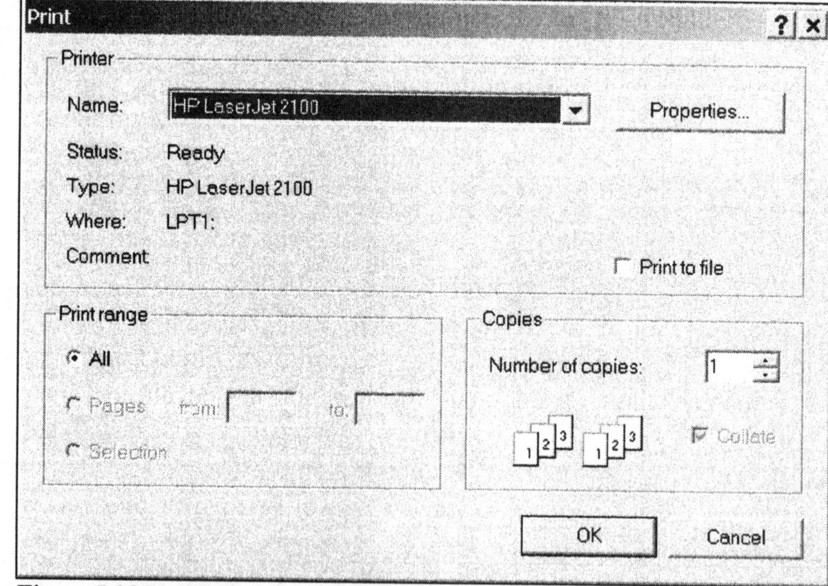

Figure 5.32(b) Windows Print Dialog Box

Lesson 5 is now complete. If you wish to model a project without instructions, a complete set of projects and illustrations are available at www.cad-resources.com ⇒ ***Downloads.***

Lesson 6 Feature Operations

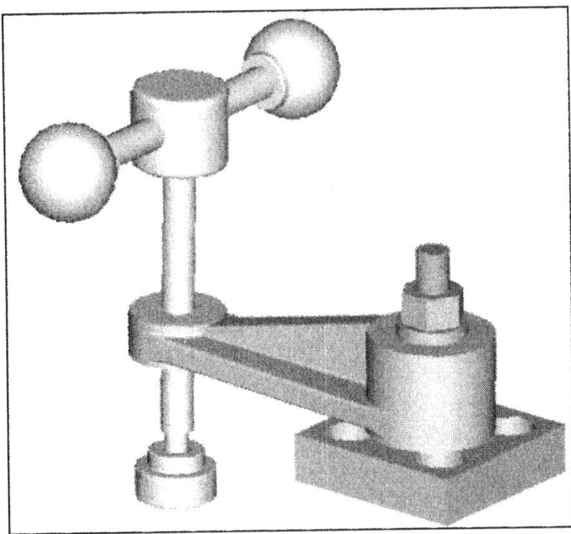

Figure 6.1 Swing Clamp Assembly

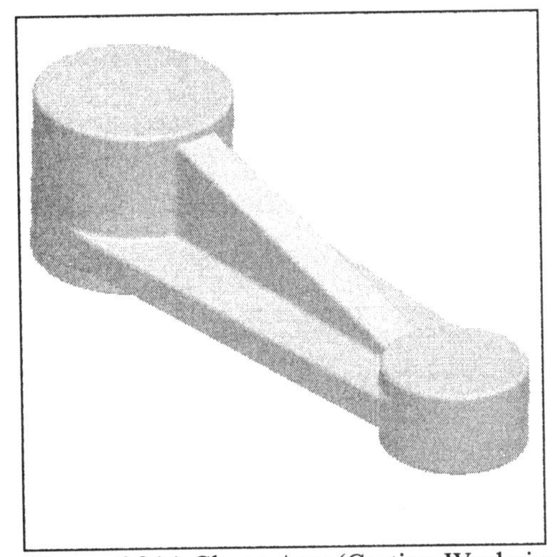

Figure 6.2(a) Clamp Arm (Casting-Workpiece)

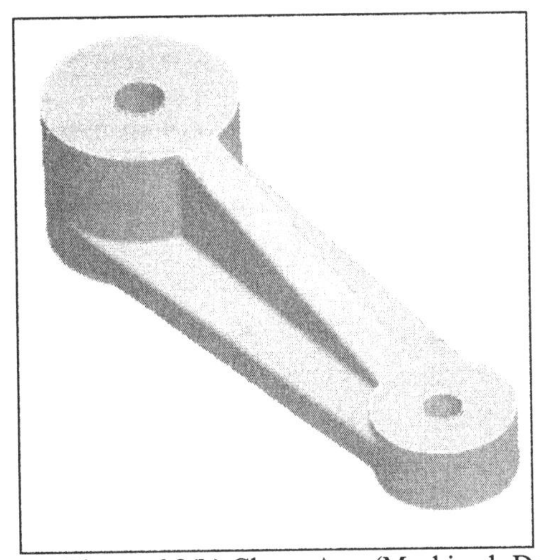

Figure 6.2(b) Clamp Arm (Machined- Design Part)

OBJECTIVES

- Use **Copy** and **Paste Special**
- Create **Ribs**
- Write **Relations** to control features
- Understand **Parameters**
- Understand and solve **Failures**
- Create a workpiece using a **Family Table**

Feature Operations

The Clamp Arm is used in the assembly (Fig. 6.1) of Lessons 7 and 9. Create two versions of the Clamp Arm; one with all cast surfaces [Fig. 6.2(a)] and the other with machined ends [Fig. 6.2(b)] (using a Family Table). This lesson will cover a wide range of Pro/E capabilities including: **Feature Operations**, **Copy** and **Paste Special**, **Relations**, **Parameters**, **Failures**, **Family Tables**, and the **Rib Tool**.

Ribs

A rib is a special type of protrusion designed to create a thin fin or web that is attached to a part. You always sketch a rib from a side view, and it grows about the sketching plane symmetrically or to either side. Because of the way ribs are attached to the parent geometry, they are always sketched as open sections.

When sketching an open section, Pro/E may be uncertain about the side to which to add the rib. Pro/E adds all material in the direction of the arrow. If the incorrect choice is made, toggle the arrow direction by clicking on the direction arrow on the screen.

A rib must "see" material everywhere it attaches to the part; otherwise, it becomes an unattached feature. There are two types of ribs: straight [Fig. 6.3(a)] and rotational [Fig. 6.3(b)]. The type is automatically set according to the attaching geometry (planar or curved).

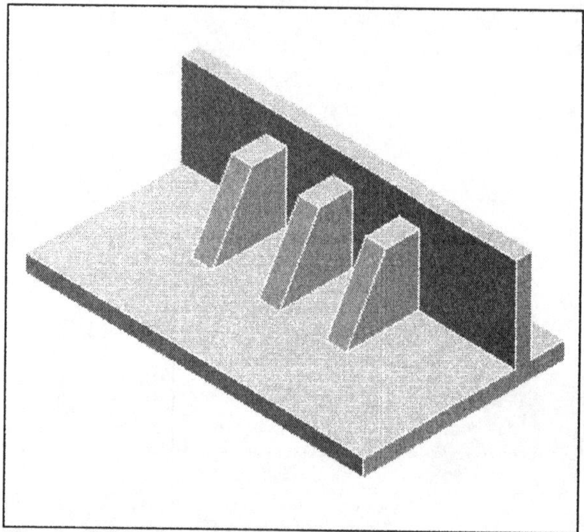

Figure 6.3(a) Straight Rib

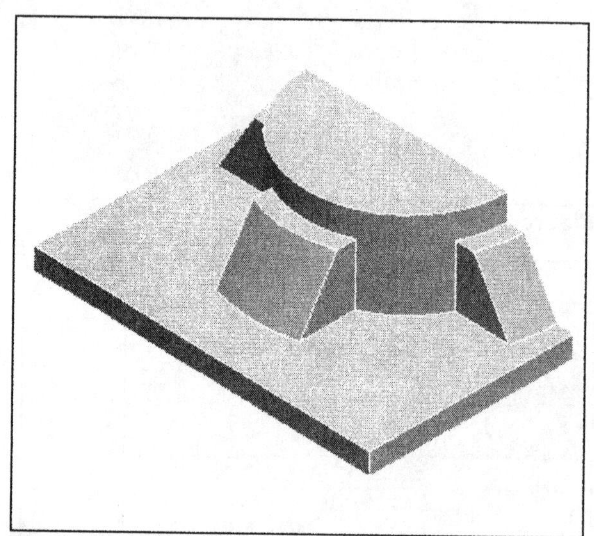

Figure 6.3(b) Rotational Rib

Relations

Relations (also known as parametric relations) are user-defined equations written between symbolic dimensions and parameters. Relations capture design relationships within features or parts, or among assembly components, thereby allowing users to control the effects of modifications on models.

Relations can be used to control the effects of modifications on models, to define values for dimensions in parts and assemblies, and to act as constraints for design conditions (for example, specifying the location of a hole in relation to the edge of a part). They are used in the design process to describe conditional relationships between different features of a part or an assembly.

Relations can be used to provide a value for a dimension. However, they can also be used to notify you when a condition has been violated, such as when a dimension exceeds a certain value. There are two basic types of relations, equality and comparison.

An equality relation equates a parameter on the left side of the equation to an expression on the right side. This type of relation is used for assigning values to dimensions and parameters. The following are a few examples of equality relations:

d2 = 25.500 **d8 = d4/2** **d7 = d1+d6/2** **d6 = d2*(sqrt(d7/3.0+d4))**

A comparison relation compares an expression on the left side of the equation to an expression on the right side. This type of relation is commonly used as a constraint or as a conditional statement for logical branching. The following are examples of comparison relations:

d1 + d2 > (d3 + 5.5)	Used as a constraint
IF (d1 + 5.5) > = d7	Used in a conditional statement

Parameter Symbols

Four types of parameter symbols are used in relations:

- **Dimensions** These are dimension symbols, such as **d8, d12**.
- **Tolerances** These are parameters associated with ± symmetrical and plus-minus tolerance formats. These symbols appear when dimensions are switched from numeric to symbolic.
- **Number of Instances** These are integer parameters for the number of instances in a direction of a pattern.
- **User Parameter** These can be parameters defined by adding a parameter or a relation (e.g., **Volume = d3 * d4 * d5**).

Operators and Functions

The following operators and functions can be used in equations and conditional statements:

Arithmetic Operators

+	**Addition**
–	**Subtraction**
/	**Division**
*	**Multiplication**
^	**Exponentiation**
( )	**Parentheses for grouping** [for example, **(d0 = (d1–d2)*d3)**]

Assignment Operator

=	Equal to

The = (equals) sign is an assignment operator that equates the two sides of an equation or relation. When it is used, the equation can have only a single parameter on the left side.

Comparison Operators

Comparison operators are used whenever a TRUE/FALSE value can be returned. For example, the relation **d1 >= 3.5** returns TRUE whenever d1 is greater than or equal to **3.5**. It returns FALSE whenever **d1** is less than **3.5**. The following comparison operators are supported:

==	Equal to
>	Greater than
>=	Greater than or equal to
!=, <>,~=	Not equal to
<	Less than

<table>
<tr><td><=</td><td>Less than or equal to</td></tr>
<tr><td>|</td><td>Or</td></tr>
<tr><td>&</td><td>And</td></tr>
<tr><td>~, !</td><td>Not</td></tr>
</table>

Mathematical Functions

The following operators can be used in relations, both in equations and in conditional statements. Relations may include the following mathematical functions:

cos ( )	cosine
tan ( )	tangent
sin ( )	sine
sqrt ( )	square root
asin ( )	arc sine
acos ( )	arc cosine
atan ( )	arc tangent
sinh ( )	hyperbolic sine
cosh ( )	hyperbolic cosine
tanh ( )	hyperbolic tangent

Failures

Sometimes model geometry cannot be constructed because features that have been modified or created conflict with or invalidate other features. This can happen when the following occurs:

- A protrusion is created that is unattached and has a one-sided edge.
- New features are created that are unattached and have one-sided edges.
- A feature is resumed that now conflicts with another feature (i.e. two chamfers on the same edge).
- The intersection of features is no longer valid because dimensional changes have moved the intersecting surfaces.
- A relation constraint has been violated.

Resolve Feature

After a feature fails, Pro/E enters Resolve Feature mode. Use the commands in the RESOLVE FEAT menu to fix the failed feature:

- **Undo Changes** Undo the changes that caused the failed regeneration attempt, and return to the last successfully regenerated model.
- **Investigate** Investigate the cause of the regeneration failure using the Investigate submenu.
- **Fix Model** Roll the model back to the state before failure and select commands to fix the problem.
- **Quick Fix** Choose an option from the **QUICK FIX** menu, the options are as follows:
    - **Redefine** Redefine the failed feature.
    - **Reroute** Reroute the failed feature.
    - **Suppress** Suppress the failed feature and its children.
    - **Clip Supp** Suppress the failed feature and all the features after it.
    - **Delete** Delete the failed feature.

Failed Features

If a feature fails during creation and it does not use the dialog box interface, Pro/E displays the FEAT FAILED menu with the following options:

- **Redefine** Redefine the feature
- **Show Ref** Display the SHOW REF menu so you can see the references of the failed feature. Pro/E displays the reference number in the Message Window.
- **Geom Check** Check for problems with overlapping geometry, misalignment, and so on
- **Feat Info** Get information about the feature

If a feature fails, you can redisplay the part with all failed geometry highlighted in different colors. Pro/E displays the corresponding error messages in an Information Window. Features can fail during creation for the following reasons:

- **Overlapping geometry** A surface intersects itself. If Pro/E finds a self-intersecting surface, it does not perform any further surface checks.
- **Surface has edges that coincide** The surface has no area. Pro/E highlights the surface in red.
- **Inverted geometry** Pro/E highlights the inverted geometry in purple and displays an error message.
- **Bad edges** Pro/E highlights bad edges in blue and displays an error message.

Family Tables

Family Tables are effective for two main reasons: they provide a beneficial tool, and they are easy to use. You need to understand the functionality of Family Tables, and you must understand when a Family Table is required and what circumstances should promote its use.

To determine whether a model is a candidate for a Family Table: establish whether the original and the variation would ever have to co-exist at the same time (both in the same assembly, both shown in the same drawing, both with an independent Bill of Materials) and whether they should be tied together (most of the same dimensions, features, and parameters). If so, the component is a candidate for the creation of a Family Table [Figs. 6.4(a-b)], otherwise, the model may be a candidate for copying to an independent model.

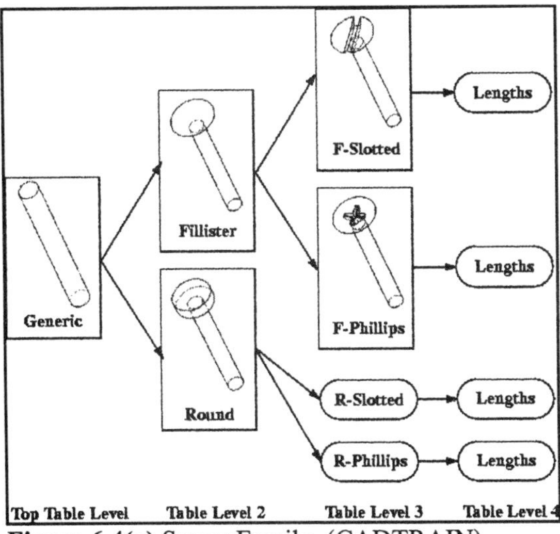

Figure 6.4(a) Screw Family (CADTRAIN)

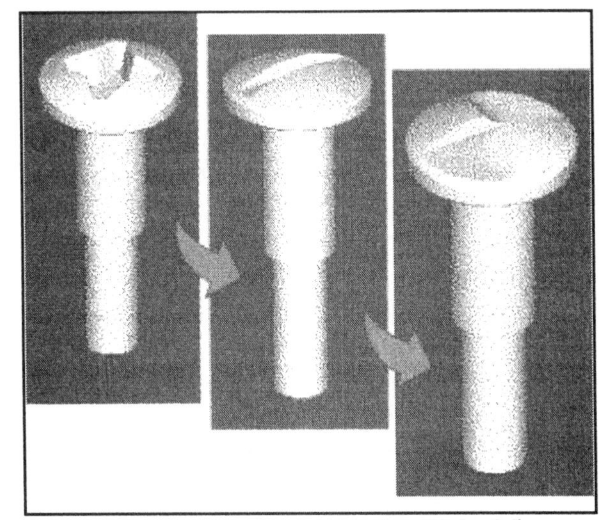

Figure 6.4(b) Screws (CADTRAIN)

Lesson 6 STEPS

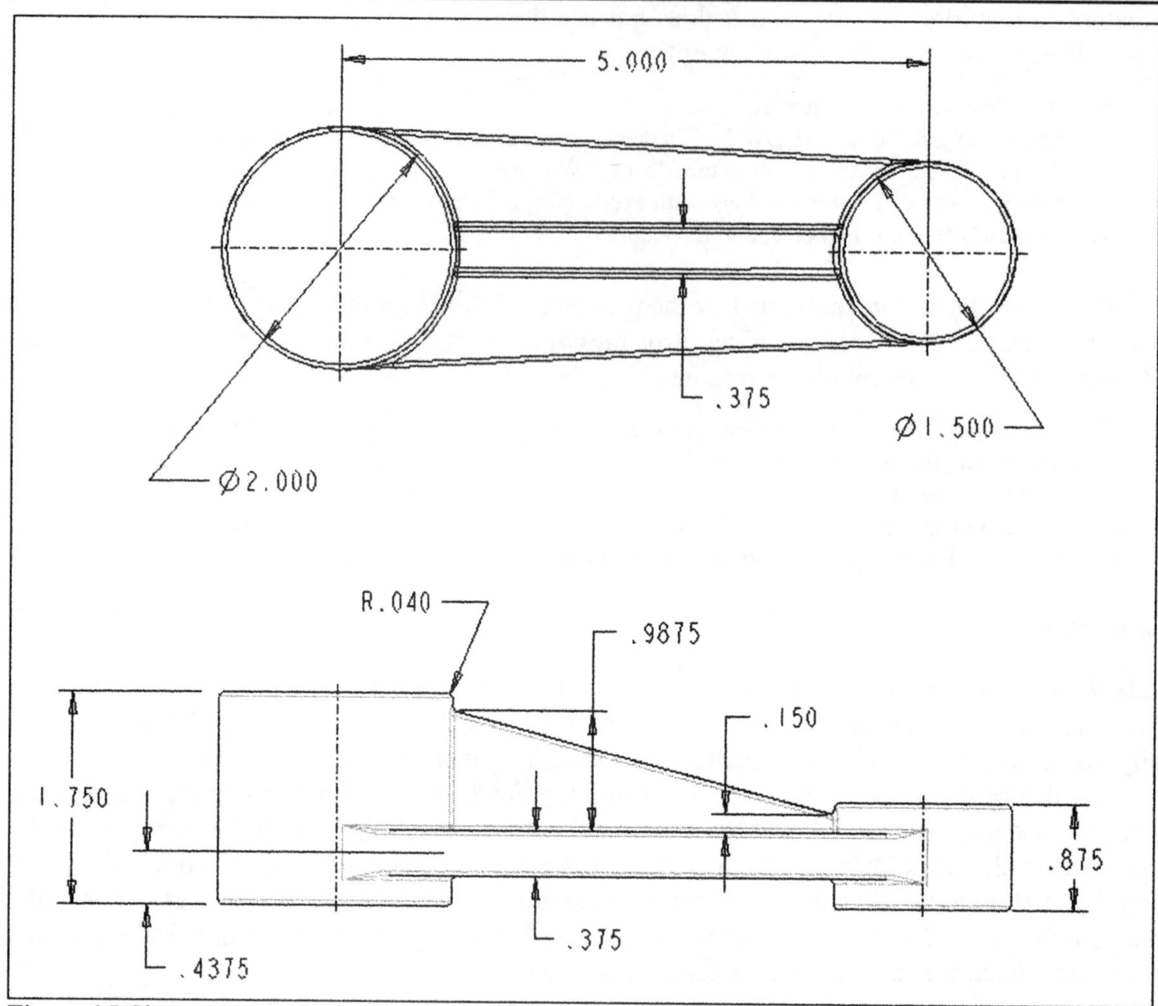

Figure 6.5 Clamp Arm Workpiece Drawing

Clamp Arm

The Clamp_Arm is modeled in two stages. Model the casting (Fig. 6.5) using the casting detail, and then use the machine detail to complete the last (machined) features. The last step will be to create a Family Table with an instance that suppresses the machined features. By having a *casting part* (which is called a ***workpiece*** in **Pro/NC**) and a separate but almost identical *machined part* (which is called a ***design part*** in **Pro/NC**), you can create an operation for machining and an NC sequence. During the manufacturing process, *you merge the workpiece into the design part* and create a ***manufacturing model***.

The difference between the two files is the difference between the volume of the workpiece/casting part and the volume of the design part/machined part. The removed volume can be seen as *material removal* when you are performing an **NC Check** operation on the manufacturing model. If the machining process gouges the part, the gouge will display the interference. The cutter location can also be displayed as an animated machining process.

For additional tutorials on Pro/MANUFACTURING: Pro/NC, and Expert Machinist, see: *www.cad-resources.com and click* ***Downloads***.

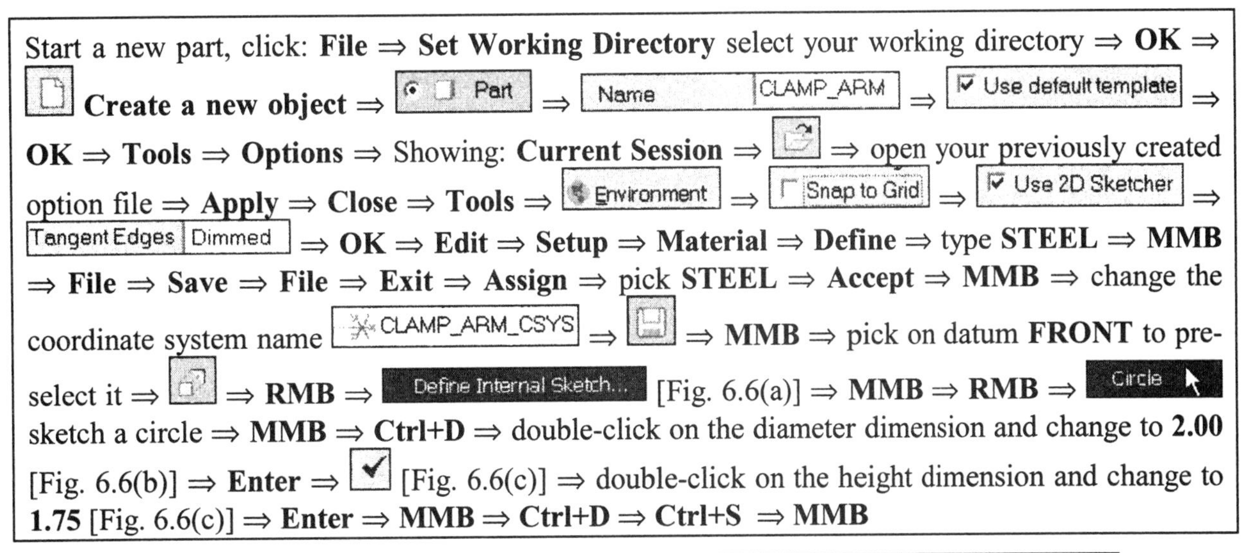

Start a new part, click: **File** ⇒ **Set Working Directory** select your working directory ⇒ **OK** ⇒ [icon] **Create a new object** ⇒ [Part] ⇒ [Name CLAMP_ARM] ⇒ [Use default template] ⇒ **OK** ⇒ **Tools** ⇒ **Options** ⇒ Showing: **Current Session** ⇒ [icon] ⇒ open your previously created option file ⇒ **Apply** ⇒ **Close** ⇒ **Tools** ⇒ [Environment] ⇒ [Snap to Grid] ⇒ [Use 2D Sketcher] ⇒ [Tangent Edges Dimmed] ⇒ **OK** ⇒ **Edit** ⇒ **Setup** ⇒ **Material** ⇒ **Define** ⇒ type **STEEL** ⇒ **MMB** ⇒ **File** ⇒ **Save** ⇒ **File** ⇒ **Exit** ⇒ **Assign** ⇒ pick **STEEL** ⇒ **Accept** ⇒ **MMB** ⇒ change the coordinate system name [CLAMP_ARM_CSYS] ⇒ [icon] ⇒ **MMB** ⇒ pick on datum **FRONT** to pre-select it ⇒ [icon] ⇒ **RMB** ⇒ [Define Internal Sketch...] [Fig. 6.6(a)] ⇒ **MMB** ⇒ **RMB** ⇒ [Circle] sketch a circle ⇒ **MMB** ⇒ **Ctrl+D** ⇒ double-click on the diameter dimension and change to **2.00** [Fig. 6.6(b)] ⇒ **Enter** ⇒ [icon] [Fig. 6.6(c)] ⇒ double-click on the height dimension and change to **1.75** [Fig. 6.6(c)] ⇒ **Enter** ⇒ **MMB** ⇒ **Ctrl+D** ⇒ **Ctrl+S** ⇒ **MMB**

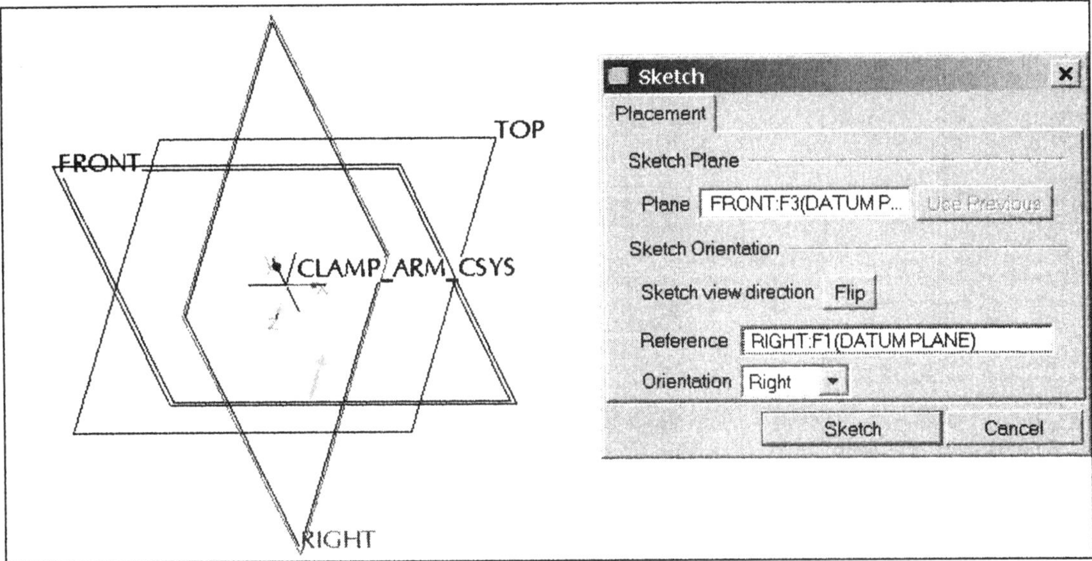

Figure 6.6(a) Sketch Plane and Orientation

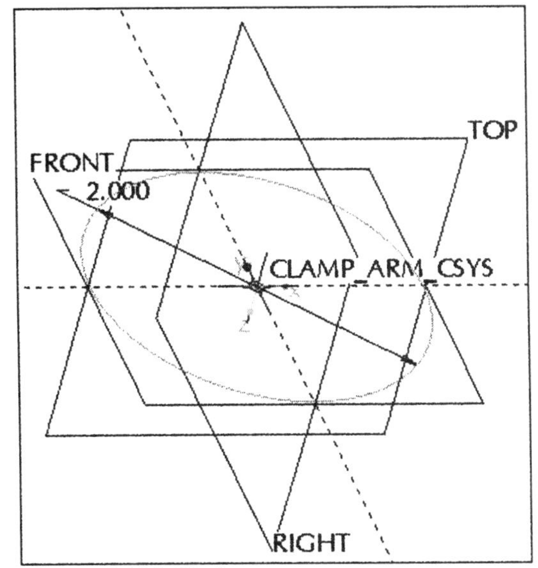

Figure 6.6(b) Circle

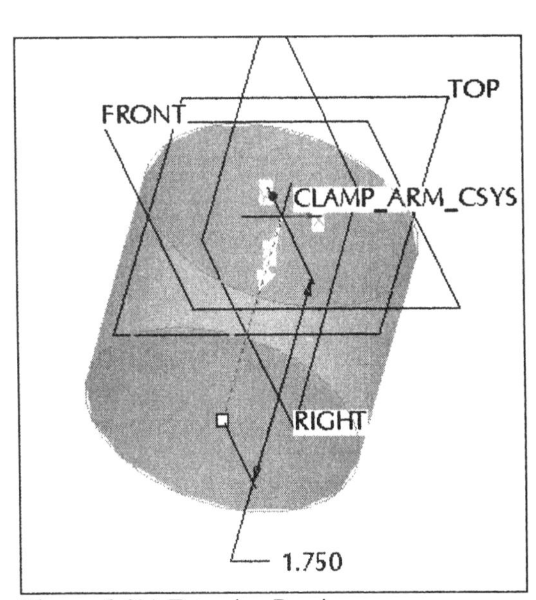

Figure 6.6(c) Extrusion Preview

With the extrusion still selected/highlighted ⇒ **Ctrl+C** ⇒ **Edit** ⇒ Paste Special ⇒ Apply Move/Rotate transformations to copies ⇒ **OK** ⇒ Transformations ⇒ pick datum plane **Right** as the **Direction Reference** ⇒ move the Drag Handle to the right until **5.00** [Fig. 6.7(a)] ⇒ **MMB** [Fig. 6.7(b)] ⇒ **Ctrl+S** ⇒ **Enter** ⇒ **LMB** to deselect

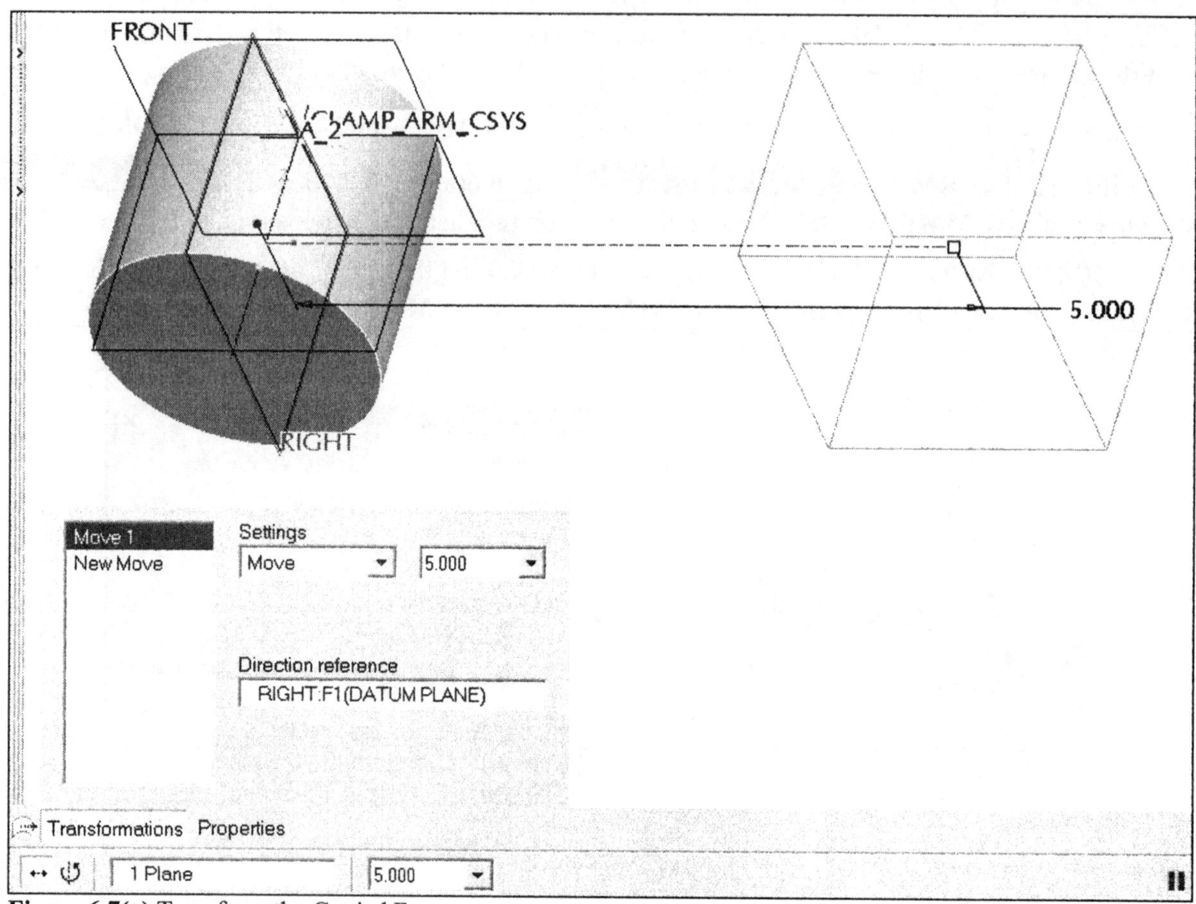

Figure 6.7(a) Transform the Copied Feature

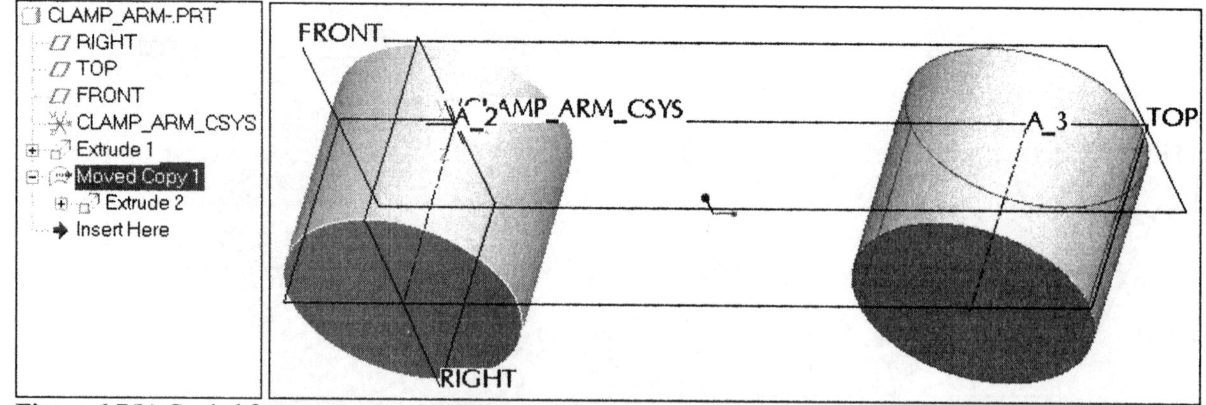

Figure 6.7(b) Copied feature

Double-click on the copied feature and change the diameter dimension to **1.50** and the height dimension to **.875** [Fig. 6.7(c)] ⇒ **Regenerates Model** [Fig. 6.7(d)] ⇒ **Ctrl+S** ⇒ **MMB**

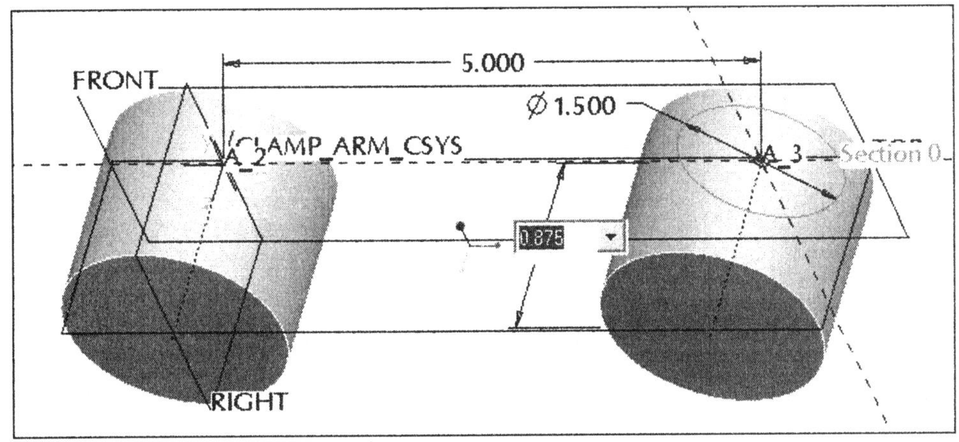

Figure 6.7(c) Modify the Dimensions

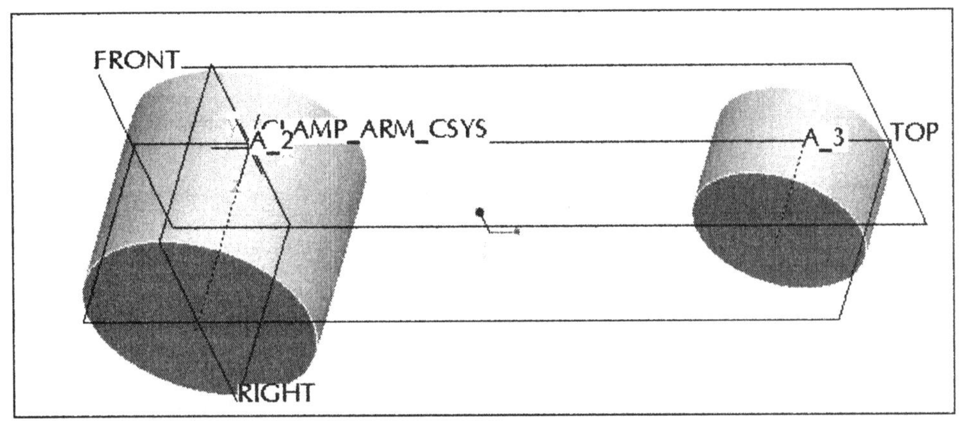

Figure 6.7(d) Regenerated Model

Click on the **FRONT** datum plane ⇒ **Datum Plane Tool** ⇒ Translation 0.4375 (Fig. 6.8) ⇒ **OK**

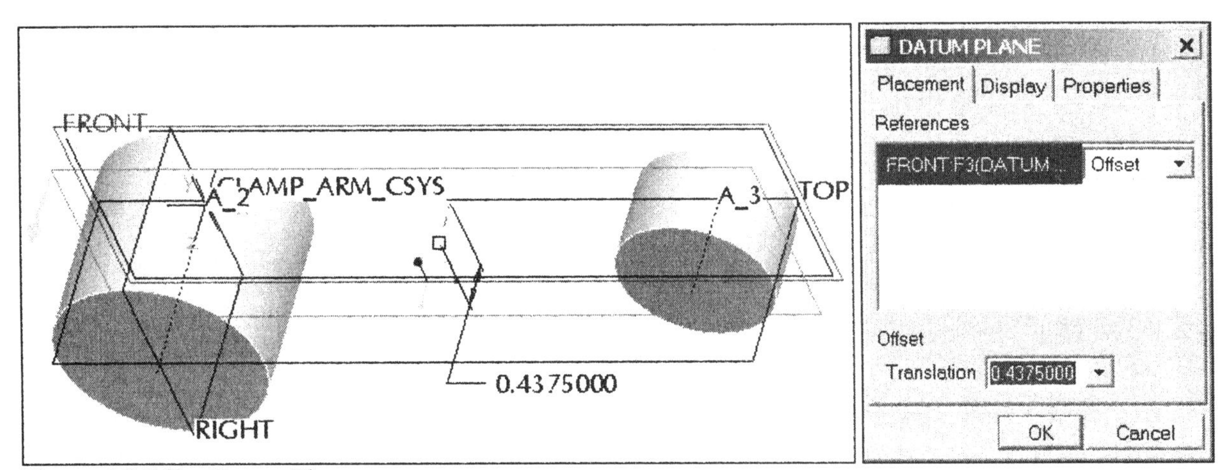

Figure 6.8 Offset Datum Plane

With datum plane **DTM1** still selected/highlighted, click: **Extrude Tool** ⇒ **RMB** ⇒ Define Internal Sketch... ⇒ **MMB** ⇒ delete the datum references and add one arc from both circles [Fig. 6.9(a)] ⇒ **MMB** ⇒ **MMB** ⇒ **RMB** ⇒ **Centerline** ⇒ create a horizontal centerline ⇒ **RMB** ⇒ **Line** ⇒ create a line that is tangent to both circular extrusions, pick tangent starting position [Fig. 6.9(b-c)] ⇒ complete the line by picking on the opposite circular reference [Fig. 6.9(d)] ⇒ **MMB** [Fig. 6.9(e)] ⇒ create the bottom tangent line ⇒ **MMB** ⇒ **Hidden Line** [Fig. 6.9(f)]

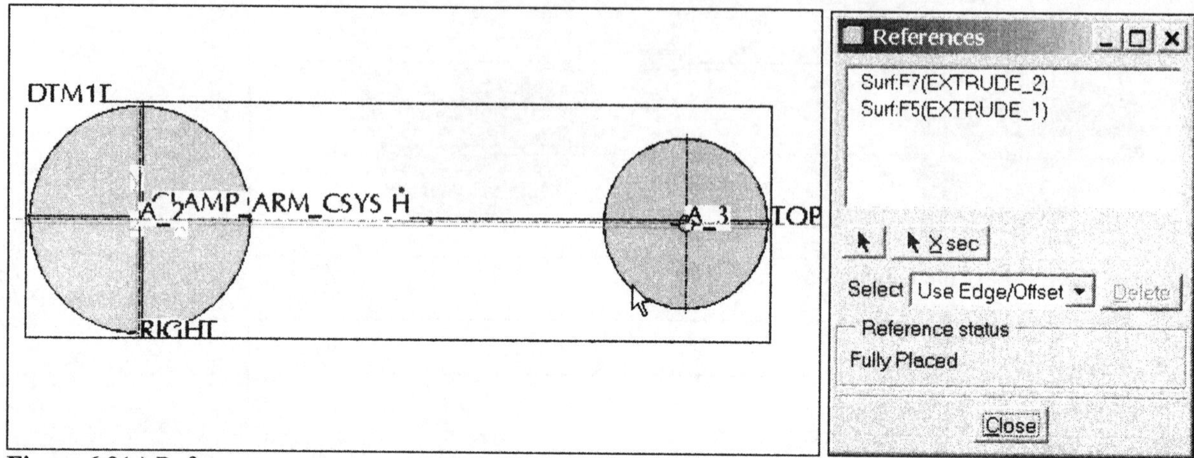

Figure 6.9(a) References

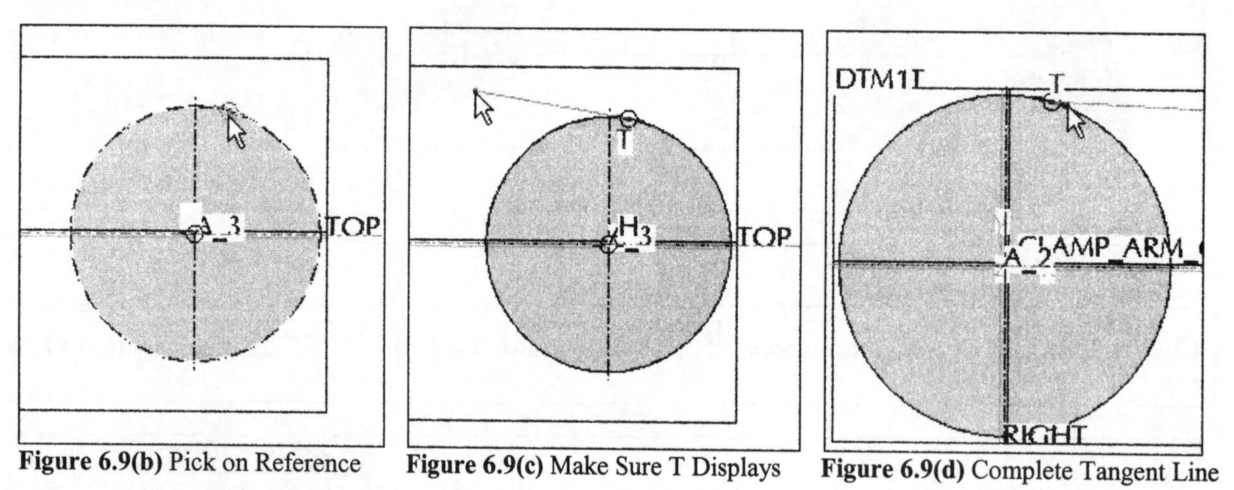

Figure 6.9(b) Pick on Reference **Figure 6.9(c)** Make Sure T Displays **Figure 6.9(d)** Complete Tangent Line

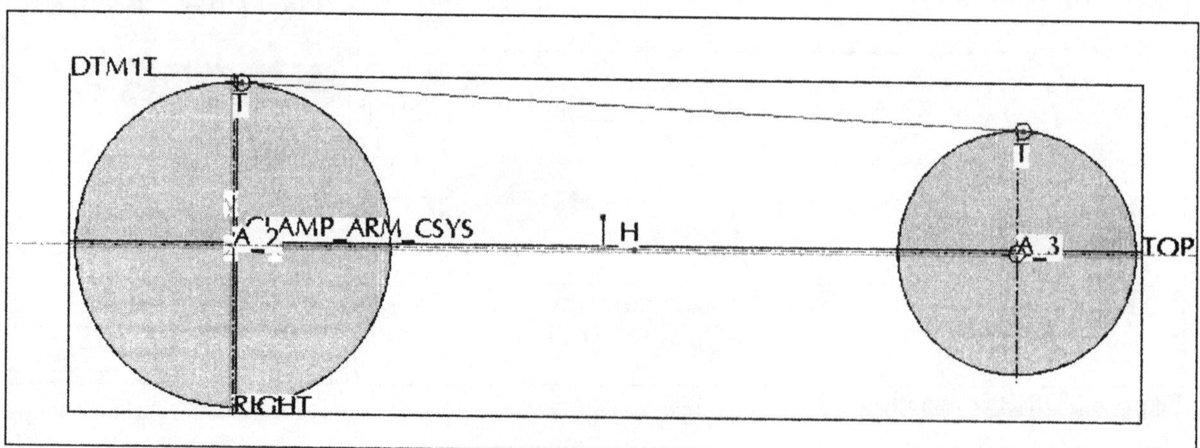

Figure 6.9(e) Tangent Line

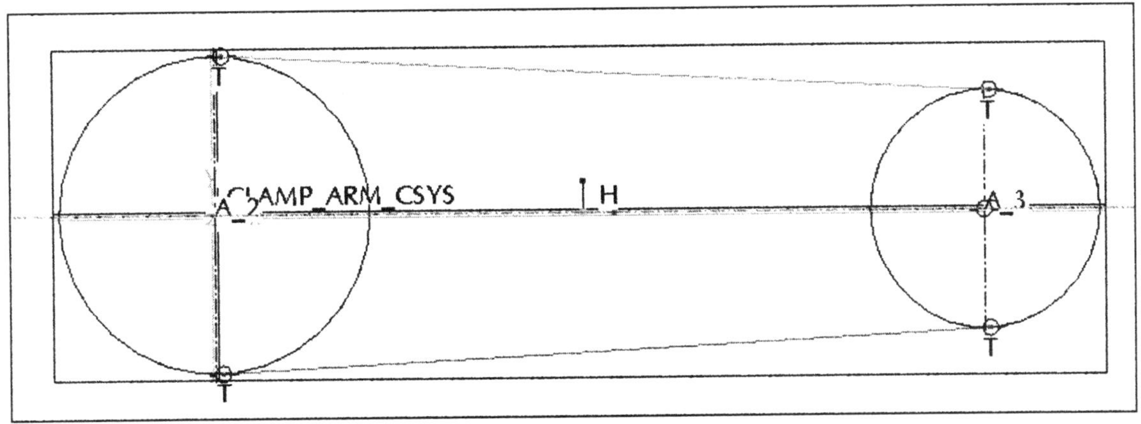

Figure 6.9(f) Tangent Lines

Click: **Create an arc** ⇒ pick the center and ends (from each tangent end of the line) [Fig. 6.9(g)] ⇒ ⇒ repeat on the opposite end [Fig. 6.9(h)] ⇒ **MMB** ⇒ **LMB** to deselect ⇒ [Fig. 6.9(i)]

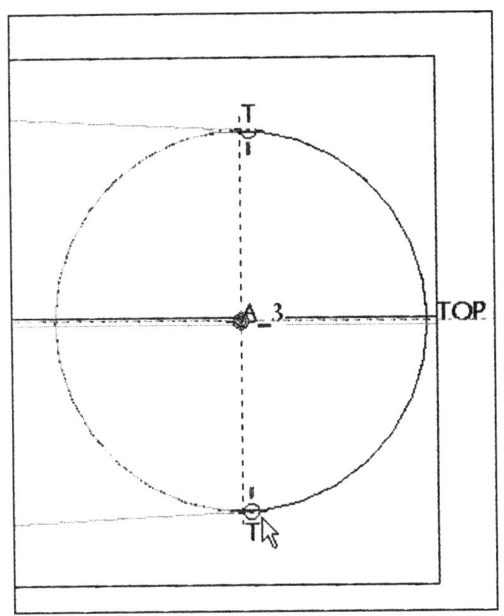

Figure 6.9(g) Arc

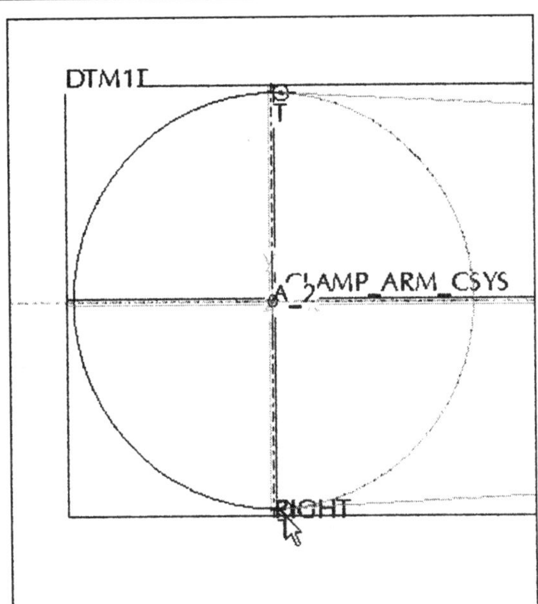

Figure 6.9(h) Arc Opposite End

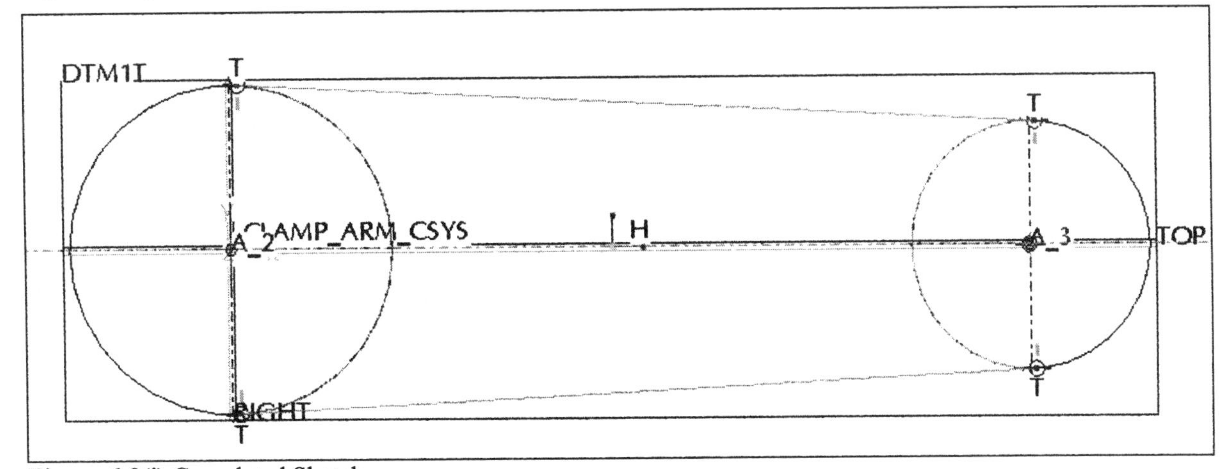

Figure 6.9(i) Completed Sketch

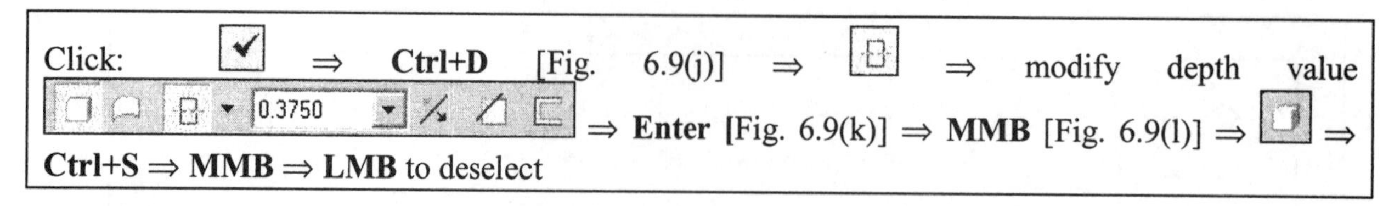

Click: ✓ ⇒ **Ctrl+D** [Fig. 6.9(j)] ⇒ [icon] ⇒ modify depth value [0.3750] ⇒ **Enter** [Fig. 6.9(k)] ⇒ **MMB** [Fig. 6.9(l)] ⇒ [icon] ⇒ **Ctrl+S** ⇒ **MMB** ⇒ **LMB** to deselect

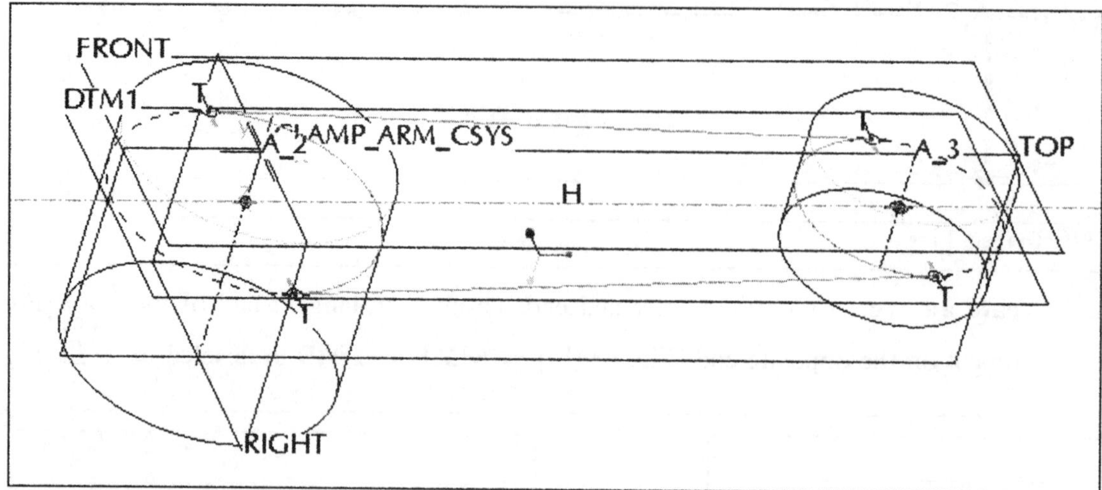

Figure 6.9(j) 3D Sketch

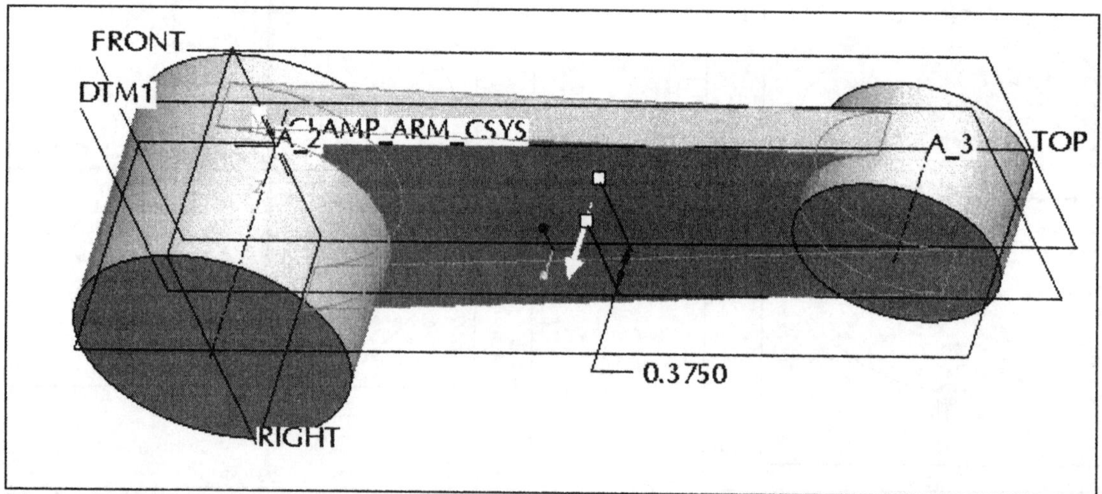

Figure 6.9(k) Preview

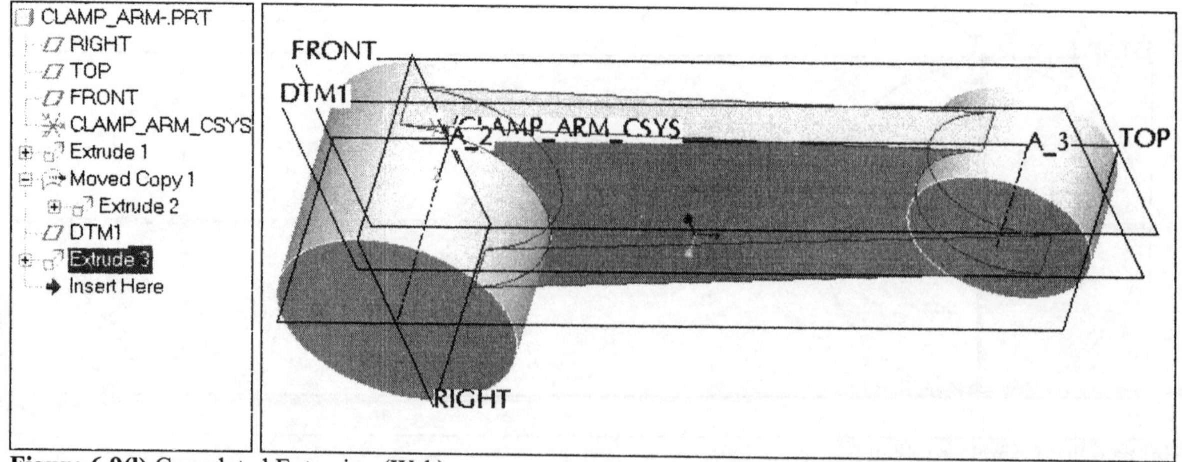

Figure 6.9(l) Completed Extrusion (Web)

Click: **Rib Tool** ⇒ **RMB** ⇒ Define Internal Sketch... ⇒ Sketch Plane- Plane: pick **TOP** [Fig. 6.10(a)] ⇒ flip the viewing direction arrow by picking on it [Fig. 6.10(b)] ⇒ **Sketch** ⇒ delete the two existing references ⇒ **off** ⇒ pick on the inside vertical edge of both cylindrical extrusions and the top edge of the web extrusion as the references [Fig. 6.10(c)] ⇒ **Close** ⇒ **on** ⇒ **RMB** ⇒ **Line** ⇒ **Ctrl+MMB** to zoom in ⇒ draw one angled line from one vertical reference to the other [Fig. 6.10(d)] ⇒ **MMB** ⇒ **RMB** ⇒ **Dimension** add the vertical dimensions [Fig. 6.10(e)] ⇒ **MMB** ⇒ pick a dimension ⇒ **RMB** ⇒ **Modify** change the dimension value [Fig. 6.10(f)] ⇒ repeat for other dimension

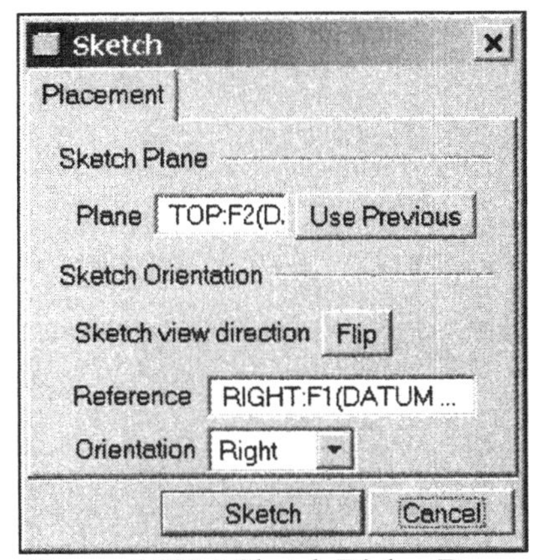

Figure 6.10(a) Sketch Dialog Box

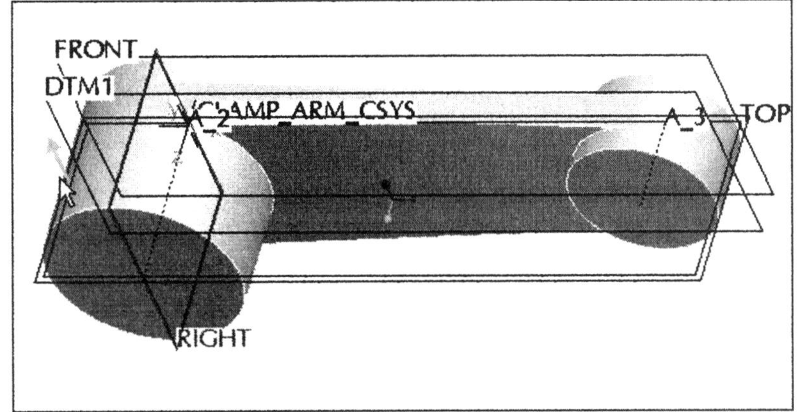

Figure 6.10(b) Flip the Viewing Direction Arrow

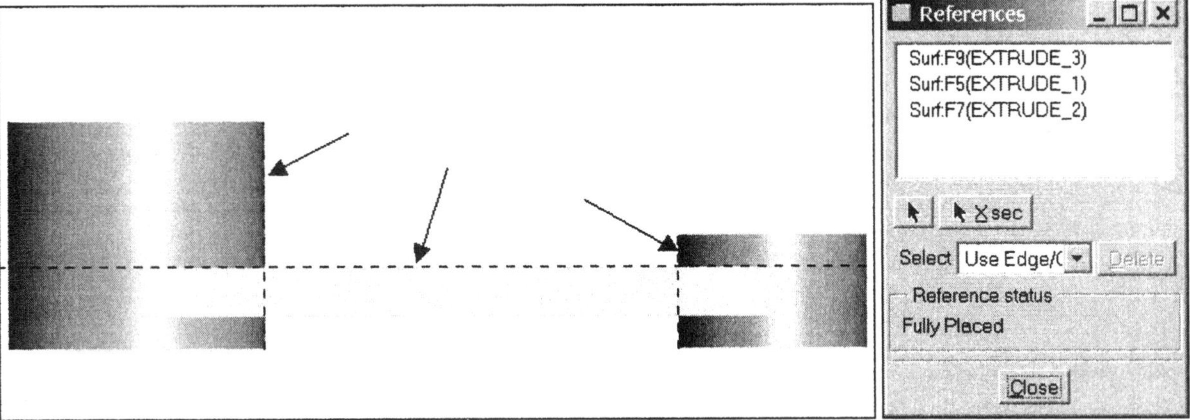

Figure 6.10(c) New References

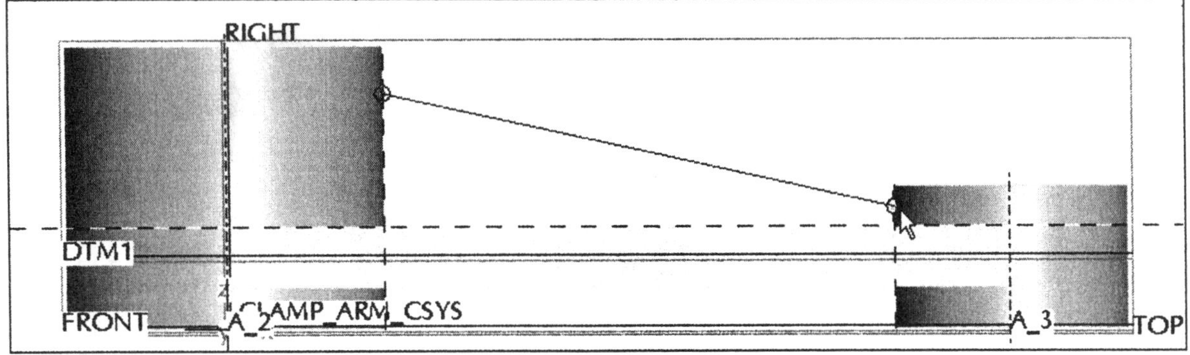

Figure 6.10(d) Sketch the Line

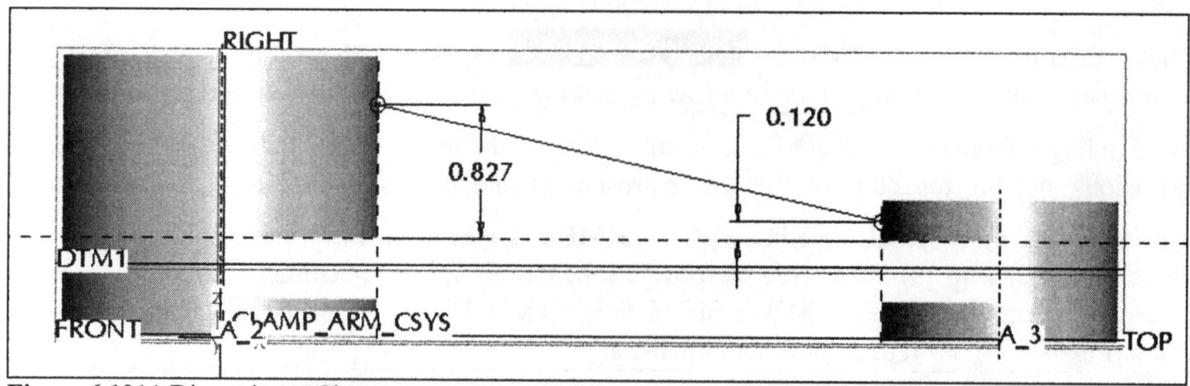

Figure 6.10(e) Dimension as Shown

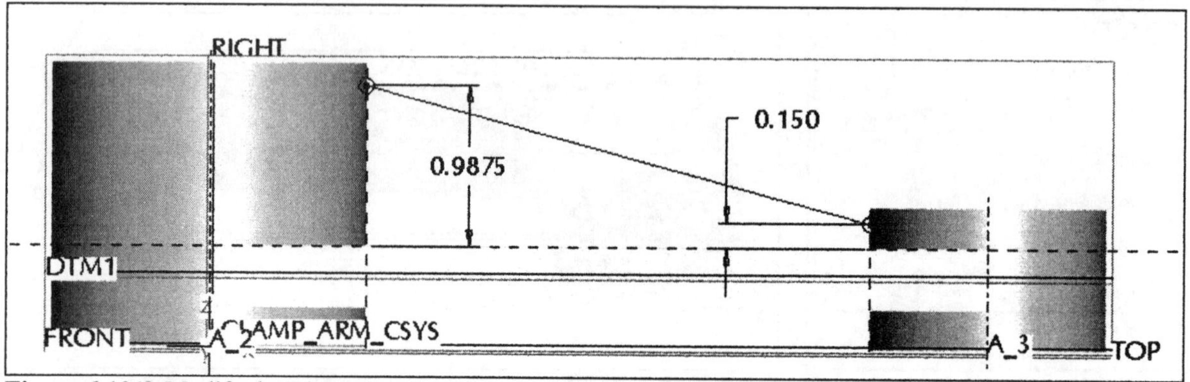

Figure 6.10(f) Modify the Dimensions

Click: **Ctrl+D** ⇒ ✔ ⇒ pick on the yellow arrow to flip the direction of the rib creation towards the part ⇒ move the drag handle until a rib thickness of **.375** is displayed [Fig. 6.10(g)] [0.375] ⇒ **MMB** ⇒ **Ctrl+S** ⇒ **MMB** ⇒ **LMB** to deselect

(Note that clicking [icon] *in the **Dashboard** of the **Rib Tool** toggles the rib from centered, to the right, or to the left of the datum plane).*

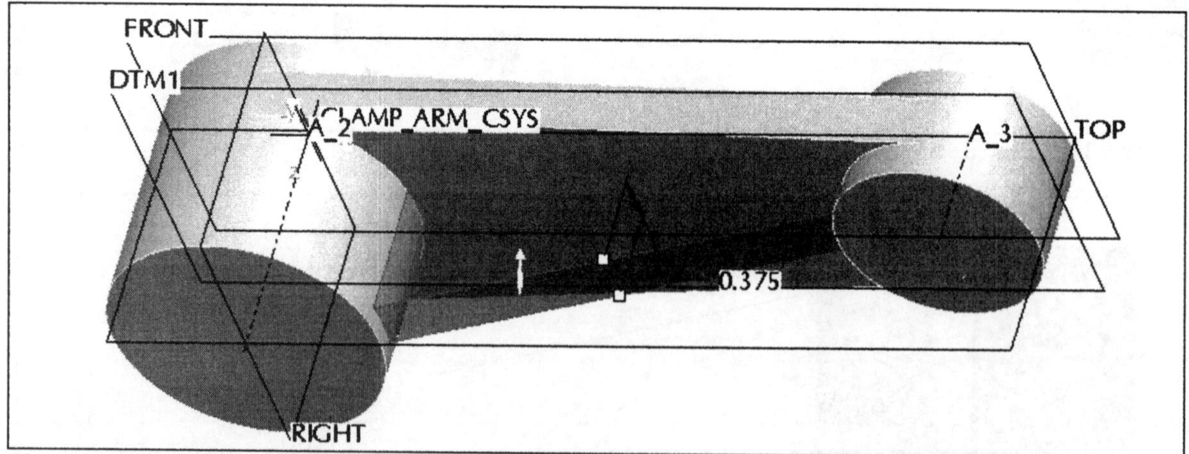

Figure 6.10(g) Previewed Rib

Create the rounds, click on the edge until it highlights in red ⇒ press and hold the **Ctrl** key and pick on the other edges until they highlight ⇒ **RMB** ⇒ **Round Edges** [Fig. 6.11(a)] ⇒ modify the radius to **.04** ⇒ **MMB** [Fig. 6.11(b)] ⇒ **off** ⇒ select the six edges [Fig. 6.11(c)] ⇒ **RMB** ⇒ **Round Edges** ⇒ **MMB** ⇒ **Ctrl+S** ⇒ **MMB** ⇒ **LMB** to deselect

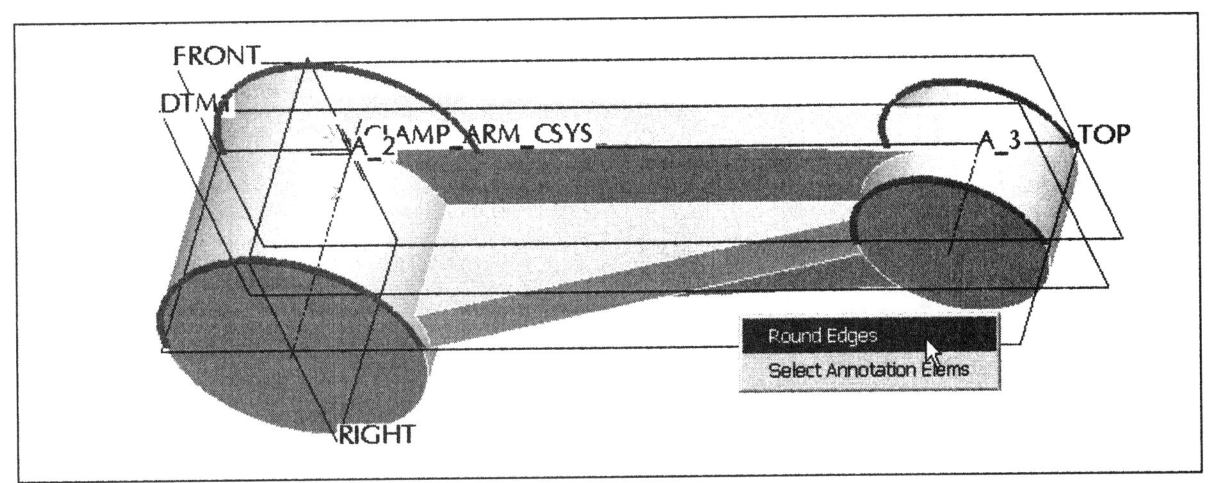

Figure 6.11(a) Previewed Rounds

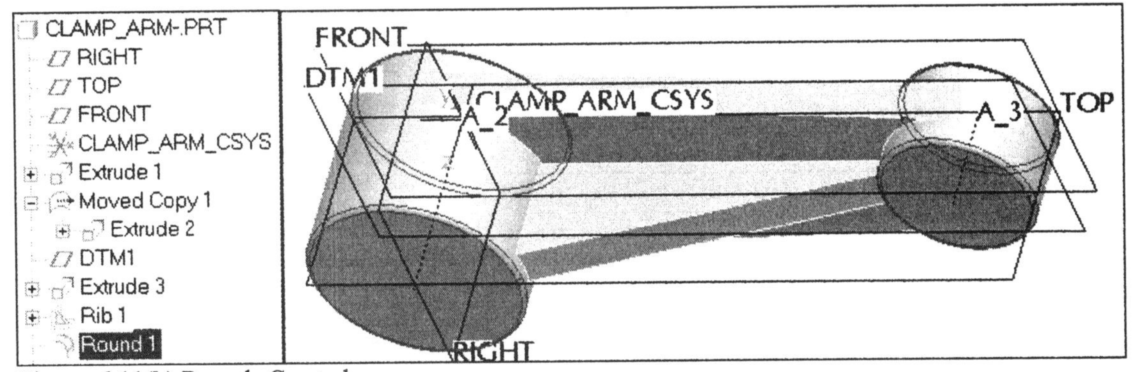

Figure 6.11(b) Rounds Created

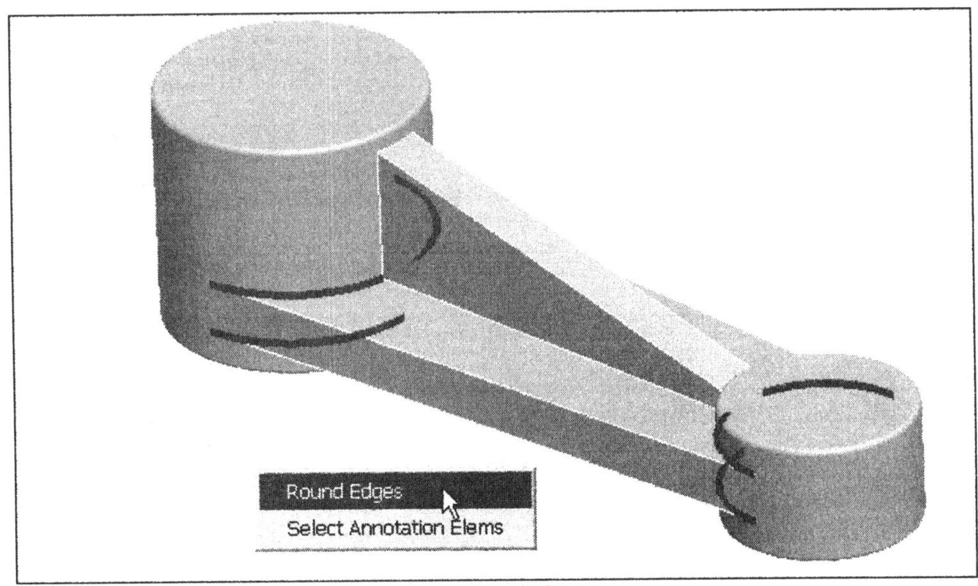

Figure 6.11(c) Select the Six Edges to be Rounded

Pick on the six edges [Fig. 6.11(d)] ⇒ **RMB** ⇒ **Round Edges** ⇒ **MMB** ⇒ select the edges ⇒ **RMB** ⇒ **Round Edges** [0.040] ⇒ **MMB** ⇒ click on the edge [Fig. 6.11(e)] ⇒ **RMB** ⇒ **Round Edges** ⇒ **MMB** [Fig. 6.11(f)] ⇒ [icons] **on** ⇒ **Ctrl+S** ⇒ **MMB** ⇒ **LMB** to deselect

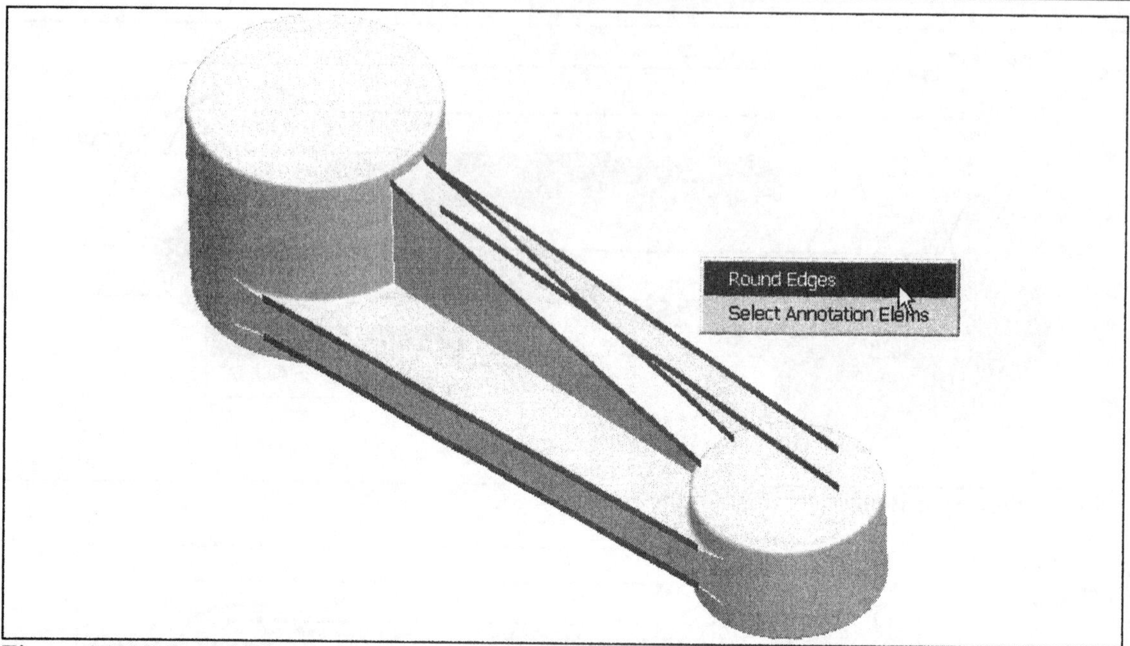

Figure 6.11(d) Round Edges

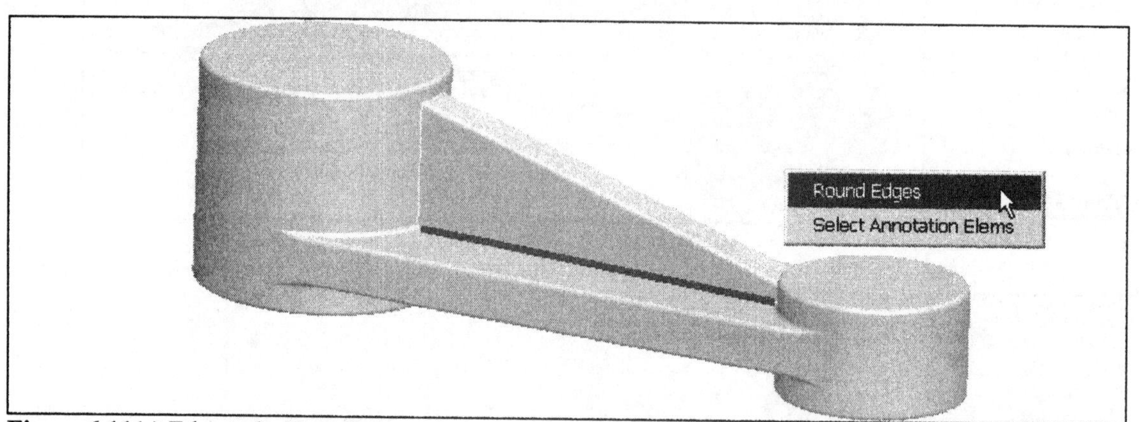

Figure 6.11(e) Edge to be Rounded

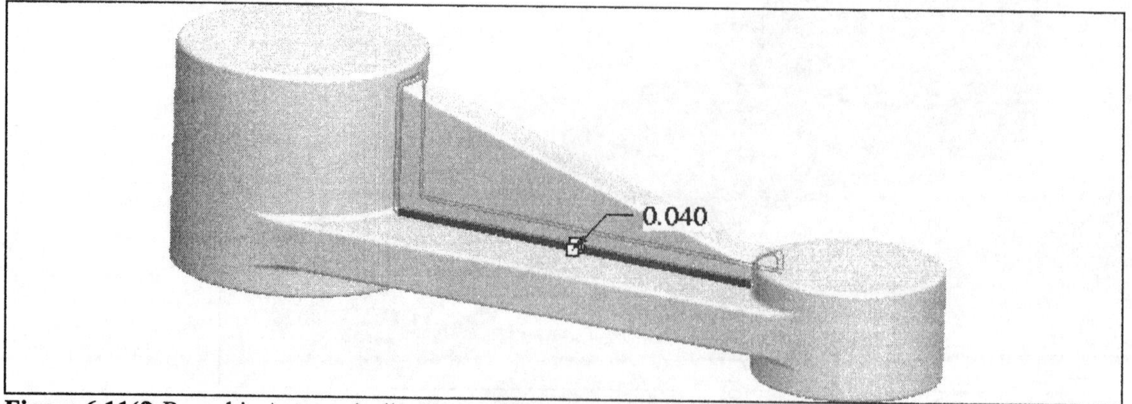

Figure 6.11(f) Round is Automatically Generated Around the Rib

The "machined" features of the part can be created with cuts to "face" the ends of the cylindrical protrusions. A standard tapped hole, and a thru hole with two countersinks are also added as "machined" features. You will be using the dimensions from the machine drawing (Fig. 6.12).

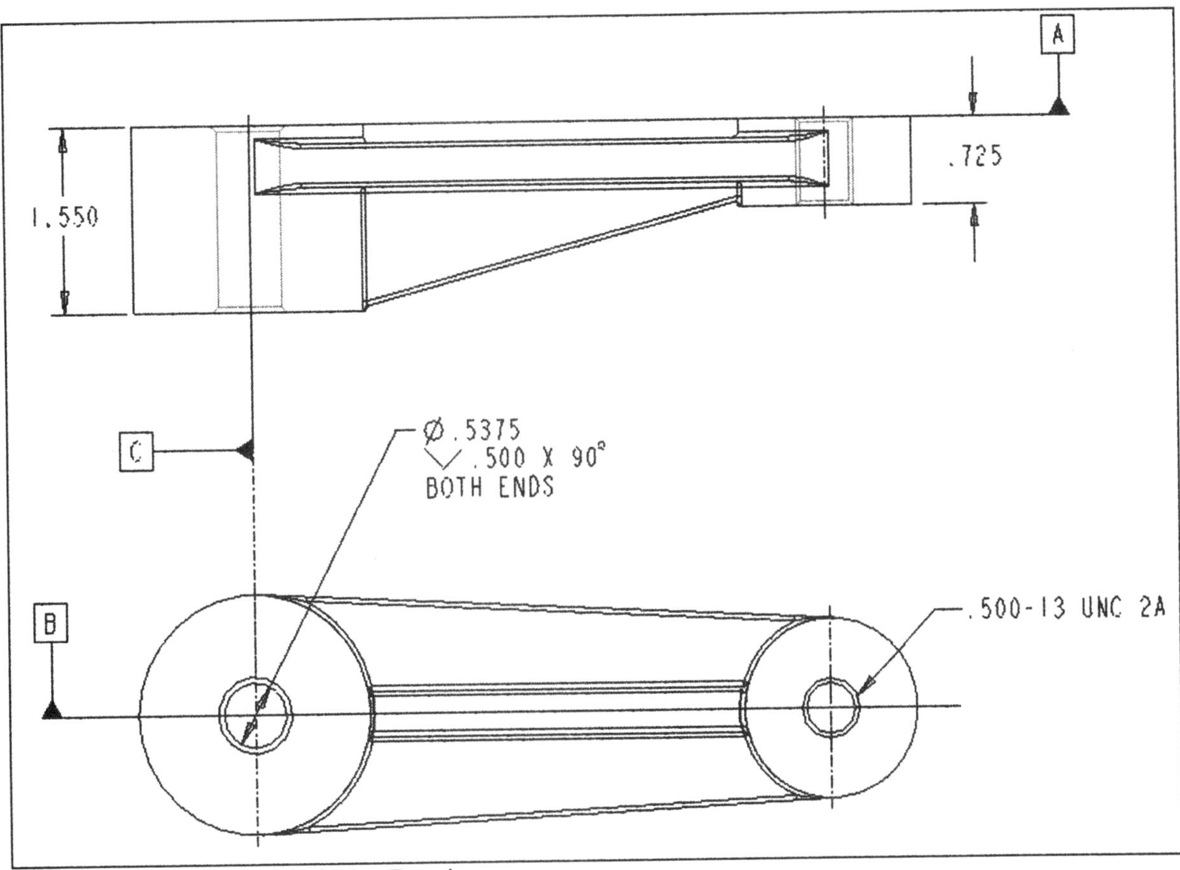

Figure 6.12 Clamp Arm Machining Drawing

Pick on datum **TOP** to pre-select it ⇒ [icon] ⇒ **RMB** ⇒ **Remove Material** ⇒ **RMB** ⇒ Define Internal Sketch... ⇒ pick on the yellow arrow to flip the direction of viewing [Fig. 6.13(a)] ⇒ **MMB**

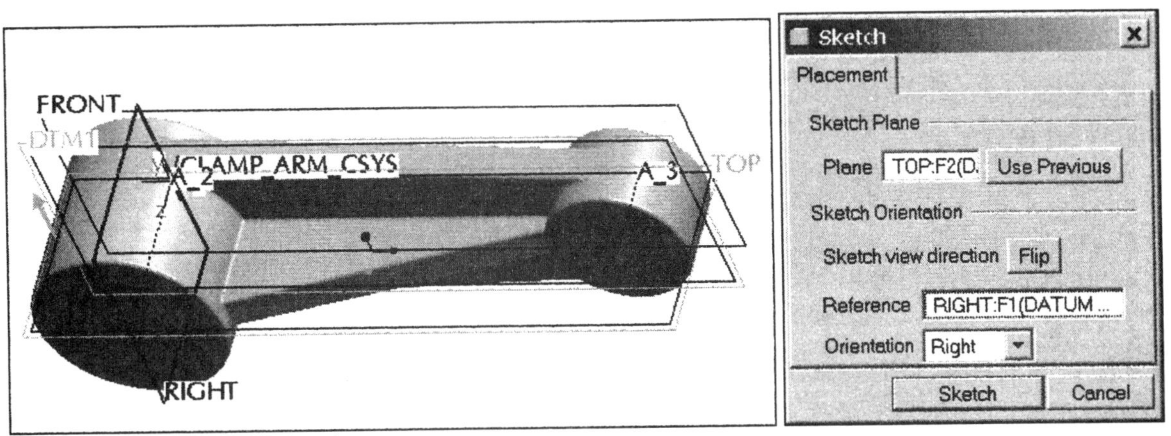

Figure 6.13(a) Viewing Direction

Click: [icons] **off** ⇒ pick on the outside vertical edge of both cylindrical extrusions as references [Fig. 6.13(b)] ⇒ **Close** ⇒ **RMB** ⇒ **Line** ⇒ **Ctrl+MMB** to zoom in ⇒ draw one horizontal line from one vertical reference to the other ⇒ pick the dimension ⇒ **RMB** ⇒ **Modify** ⇒ **.100** [Fig. 6.13(c)] ⇒ [icons] **on** ✓ ⇒ **Ctrl+D** ⇒ Options ⇒ Side 1 Through All ⇒ Side 2 Through All ⇒ flip material direction arrow if necessary [Fig. 6.13(d)]

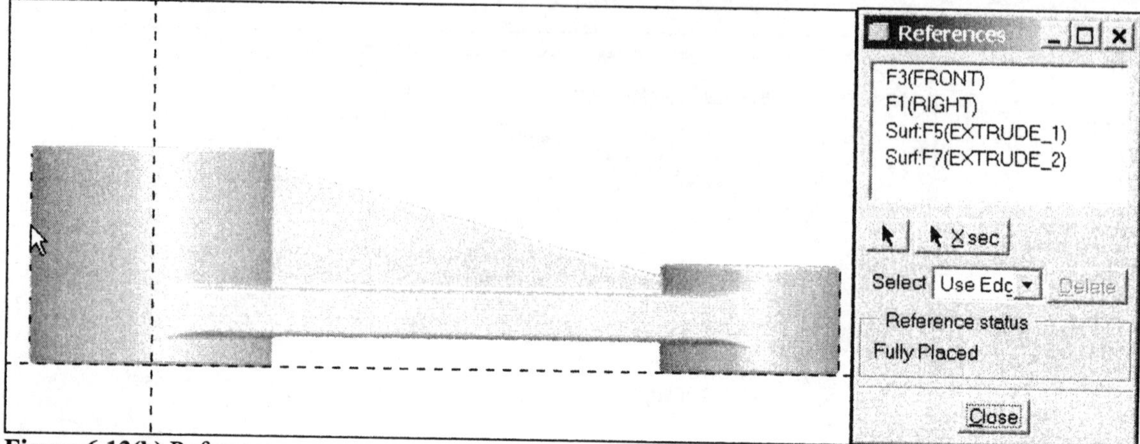

Figure 6.13(b) References

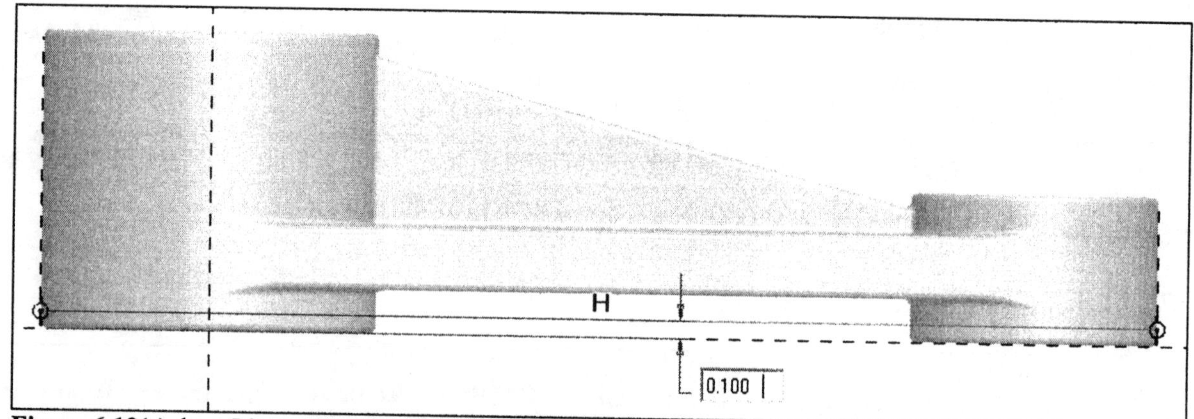

Figure 6.13(c) draw Line and Modify the Dimension

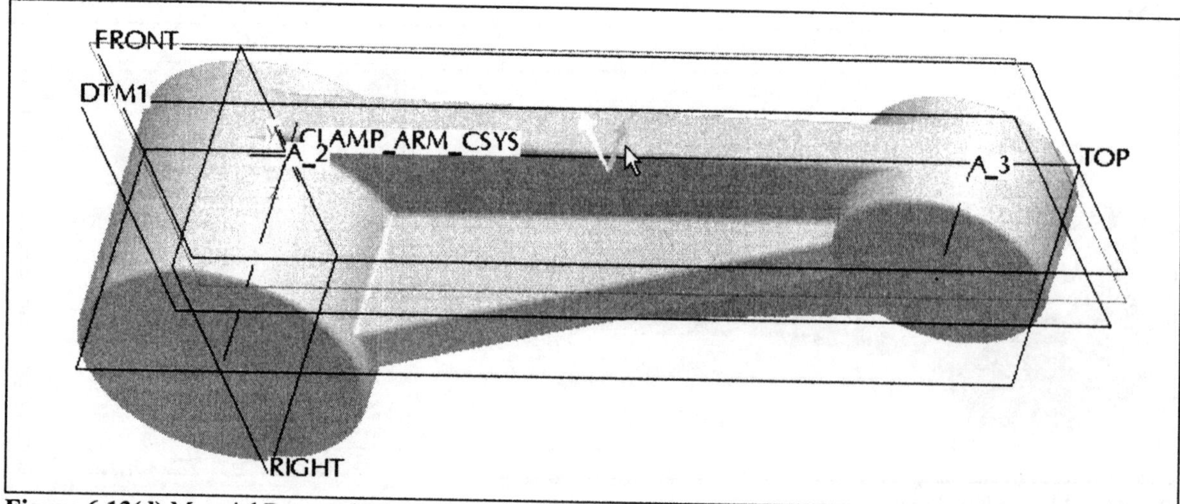

Figure 6.13(d) Material Removal Direction

Click: ⇒ ⇒ **MMB** ⇒ **RMB** ⇒ **Edit** [Figs. 6.13(e-f)] ⇒ **LMB** ⇒ **Ctrl+S** ⇒ **MMB**

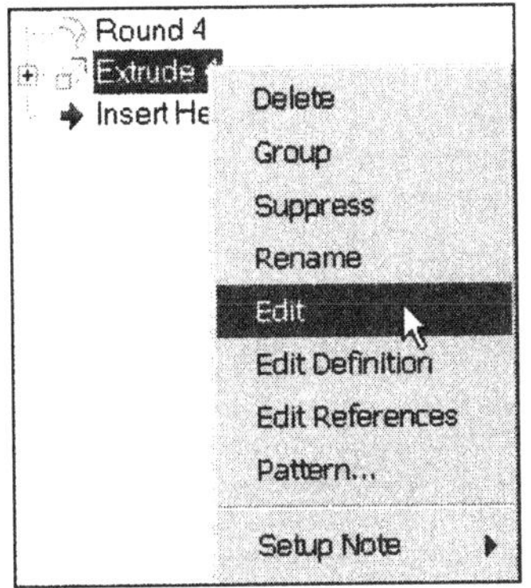

Figure 6.13(e) Edit

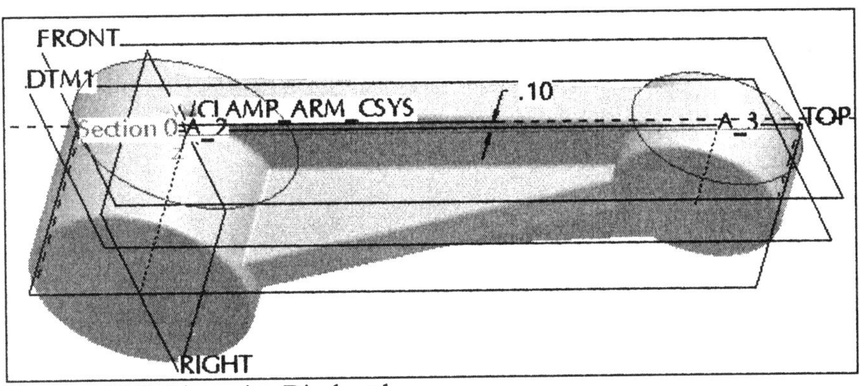

Figure 6.13(f) Dimension Displayed

Click: ⇒ **RMB** ⇒ Remove Material ⇒ **RMB** ⇒ Define Internal Sketch... ⇒ pick the top surface of the large circular protrusion ⇒ [Fig. 6.14(a)] ⇒ **Sketch** ⇒ **MMB** rotate the part slightly ⇒ pick on the circular edge of the large protrusion to add as a reference [Fig. 6.14(b)] ⇒ **Close** ⇒ ⇒ **RMB** ⇒ Circle sketch a circle [Fig. 6.14(c)] ⇒ **Ctrl+D** ⇒ ✔

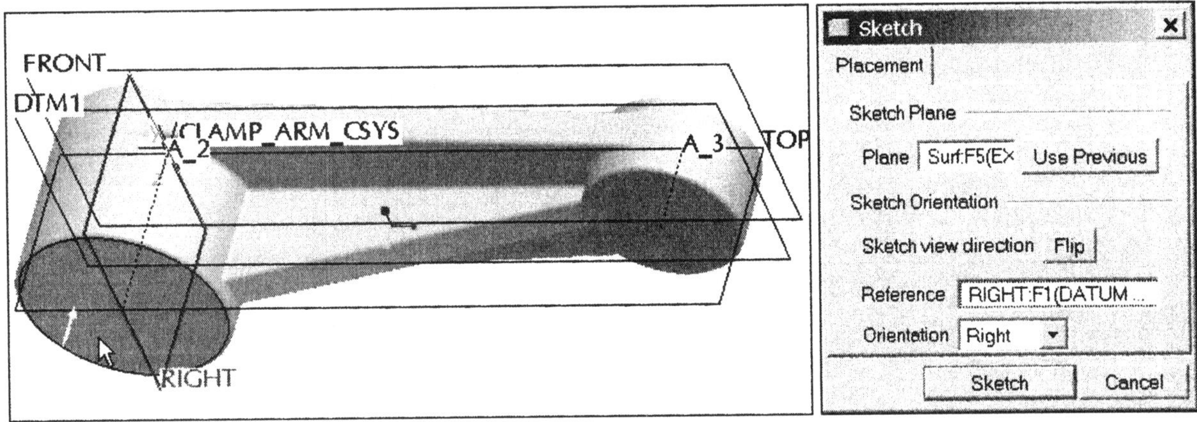

Figure 6.14(a) Sketch Plane and Orientation

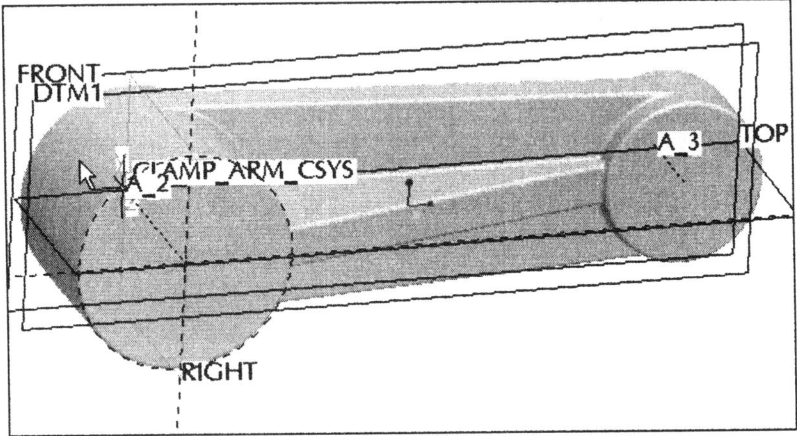

Figure 6.14(b) Add the Reference

Figure 6.14(c) Sketch a Circle

Type depth 0.10 [Fig. 6.14(d)] ⇒ **Enter** ⇒ ☑ ⇒ **MMB** ⇒ **MMB** [Fig. 6.14(e)] ⇒ repeat the process and cut **.050** from the top of the small end [Fig. 6.14(f)] ⇒ **MMB** ⇒ **Ctrl+S** ⇒ **MMB**

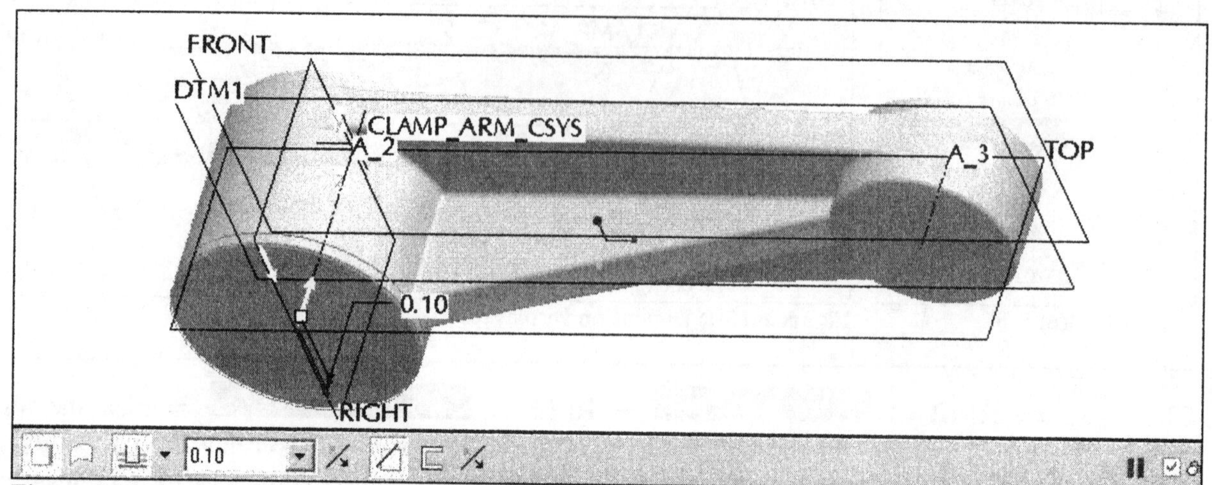

Figure 6.14(d) .10 Cut

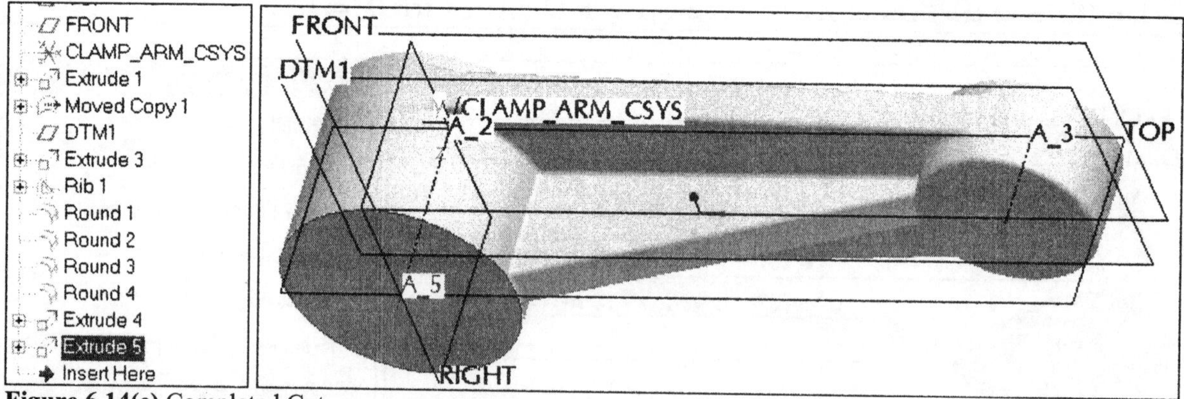

Figure 6.14(e) Completed Cut

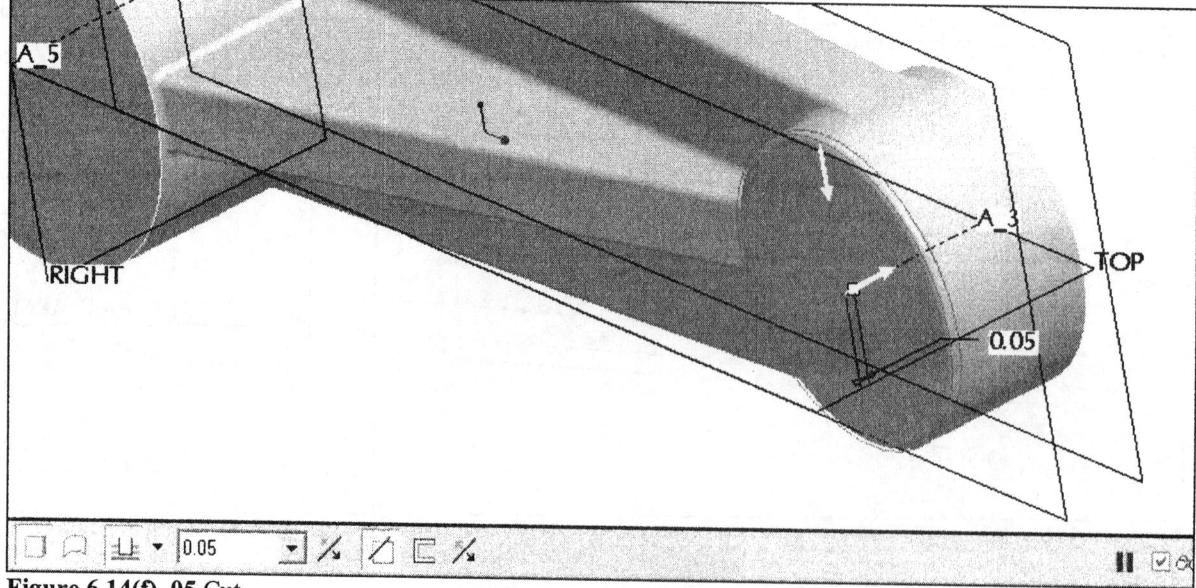

Figure 6.14(f) .05 Cut

Flexing the Model

During the design of a typical part there are many modifications made to the design. The ability to make changes without causing failures is important. "Flexing" the model, changing and editing dimension values to see if the model integrity withstands these modifications, establishes how robust your design is. Modify the part's geometry and observe the change, undo after each edit. If you get a failure, **Undo Changes**.

Pick on the rib ⇒ **RMB** ⇒ **Edit** ⇒ double-click on the **.375** dimension and change to **1.00** ⇒ **Enter** ⇒ [Fig. 6.15(a)] ⇒ pick on the first circular extrusion ⇒ **RMB** ⇒ **Edit** ⇒ double-click on the **1.75** dimension and change to **2.50** ⇒ **Enter** ⇒ [Fig. 6.15(b)] ⇒ pick on the second circular extrusion ⇒ **RMB** ⇒ **Edit** ⇒ double-click on the **1.50** dimension and change to **2.75** ⇒ **Enter** ⇒ [Fig. 6.15(b)] ⇒ ⇒ ⇒ ⇒ **Ctrl+S** ⇒ **MMB**

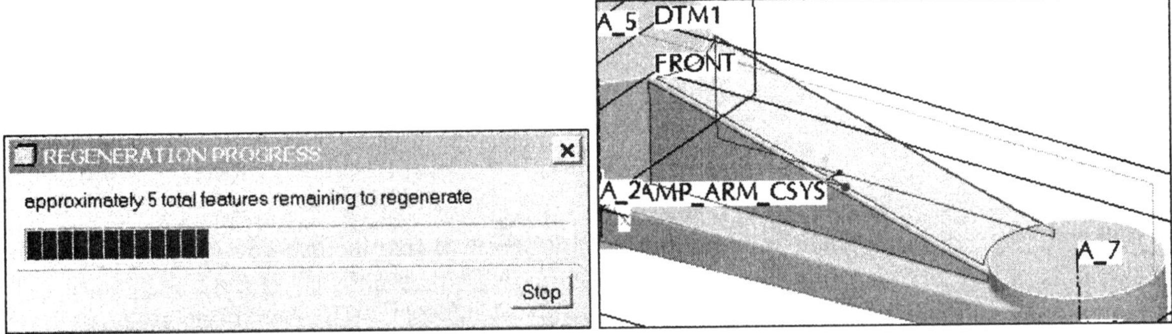

Figure 6.15(a) Regenerate the Rib

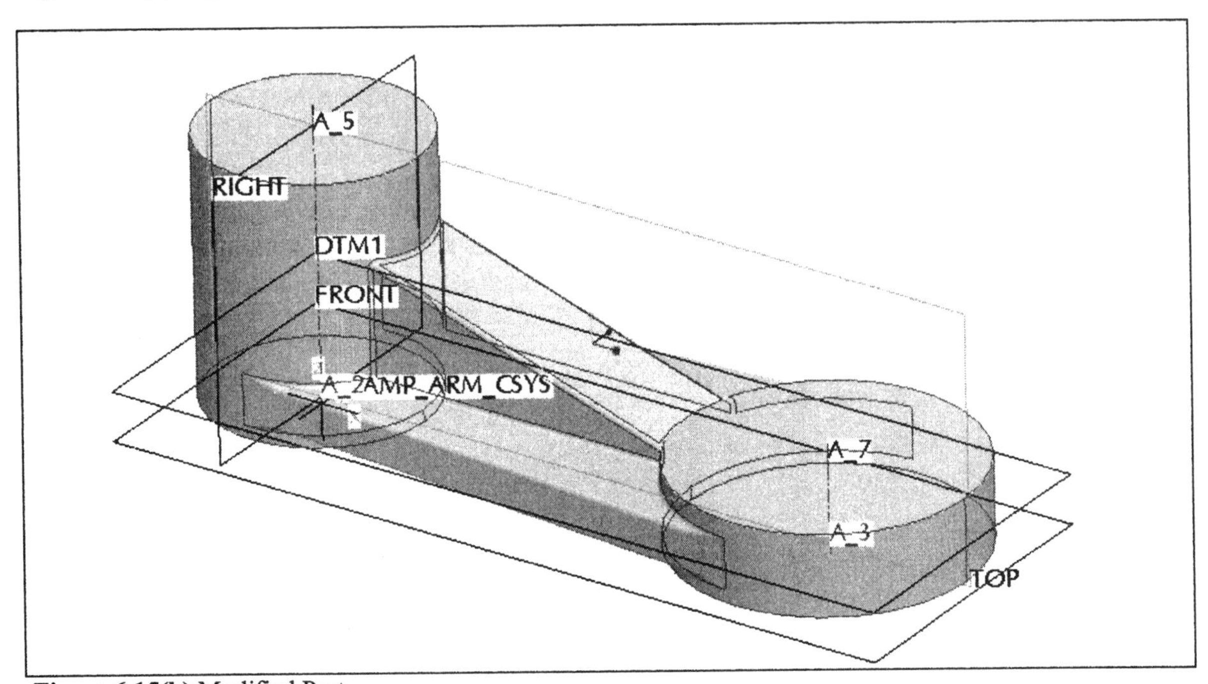

Figure 6.15(b) Modified Part

Measuring Geometry

Using analysis measure, you can measure model geometry with one of the following commands:

- **Distance** Displays the distance between two entities
- **Curve Length** Displays the length of the curve or edge
- **Angle** Displays the angle between two entities
- **Area** Displays the area of the selected surface, quilt, facets, or an entire model
- **Diameter** Displays the diameter of the surface
- **Transform** Displays a note showing the transformation matrix values between two coordinate systems

Click: **Analysis** ⇒ **Measure** ⇒ Type **Distance** ⇒ pick the top of the large circular protrusion ⇒ place your cursor over the bottom of the protrusion and click **RMB** until it highlights ⇒ **LMB** to select [Fig. 6.16(a)] ⇒ measure the distance between the two cylinders using axis to axis [Fig. 6.16(b)] ⇒ **Close**

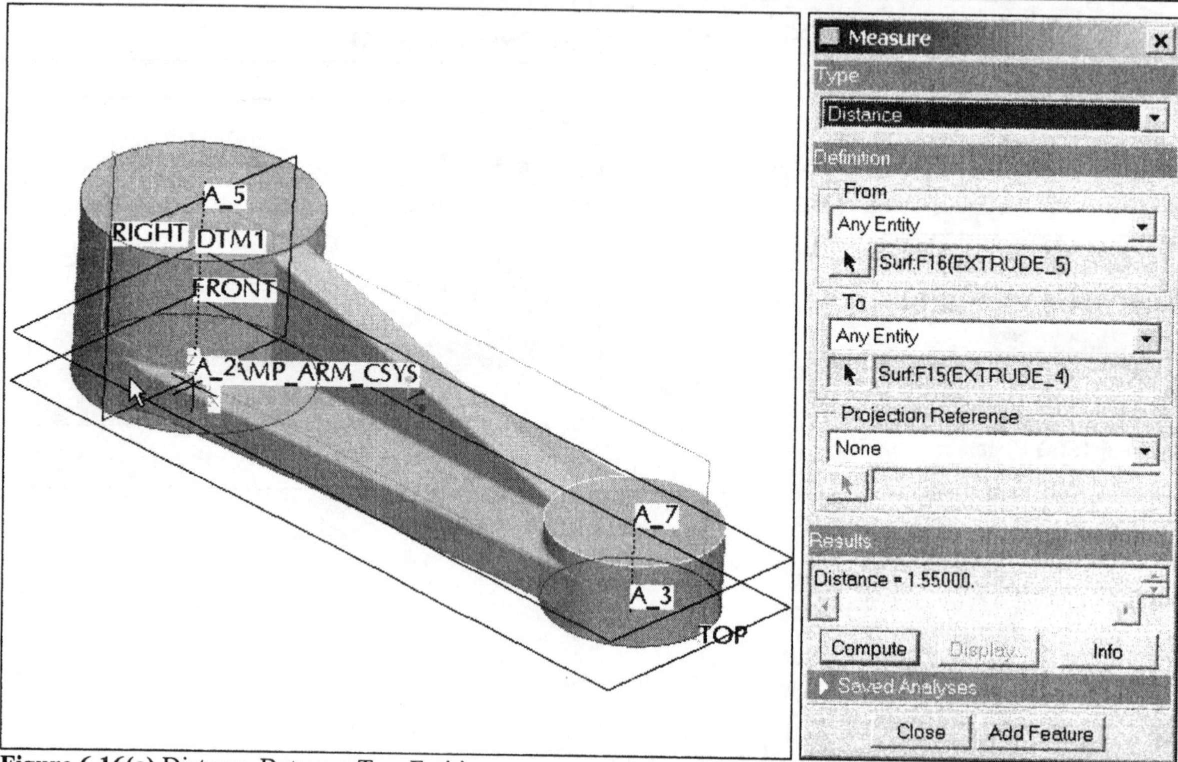

Figure 6.16(a) Distance Between Two Entities

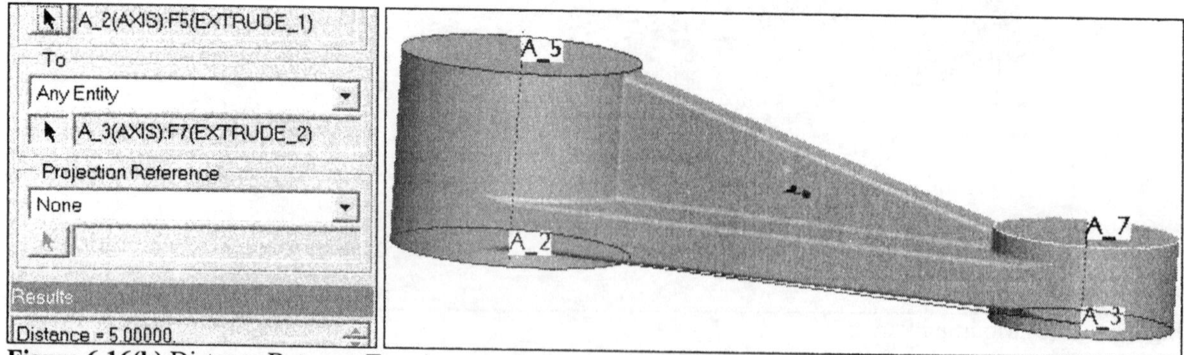

Figure 6.16(b) Distance Between Two Axes

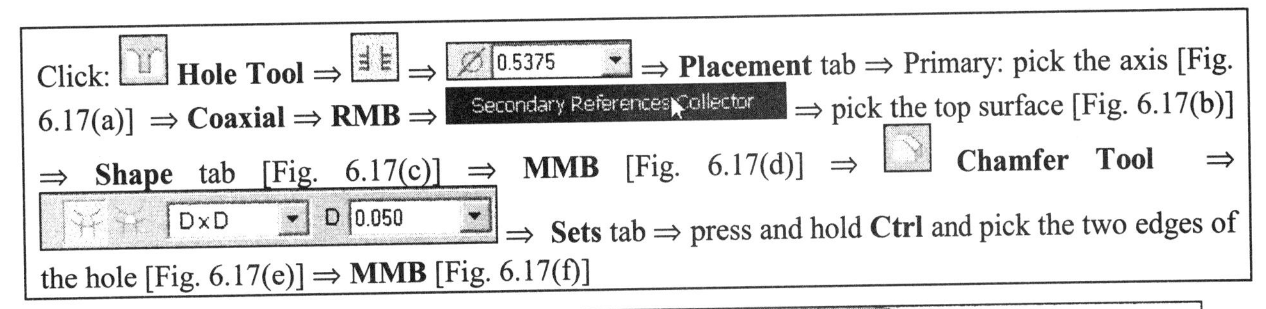

Click: **Hole Tool** ⇒ ⇒ 0.5375 ⇒ **Placement** tab ⇒ Primary: pick the axis [Fig. 6.17(a)] ⇒ **Coaxial** ⇒ **RMB** ⇒ Secondary References Collector ⇒ pick the top surface [Fig. 6.17(b)] ⇒ **Shape** tab [Fig. 6.17(c)] ⇒ **MMB** [Fig. 6.17(d)] ⇒ **Chamfer Tool** ⇒ DxD D 0.050 ⇒ **Sets** tab ⇒ press and hold **Ctrl** and pick the two edges of the hole [Fig. 6.17(e)] ⇒ **MMB** [Fig. 6.17(f)]

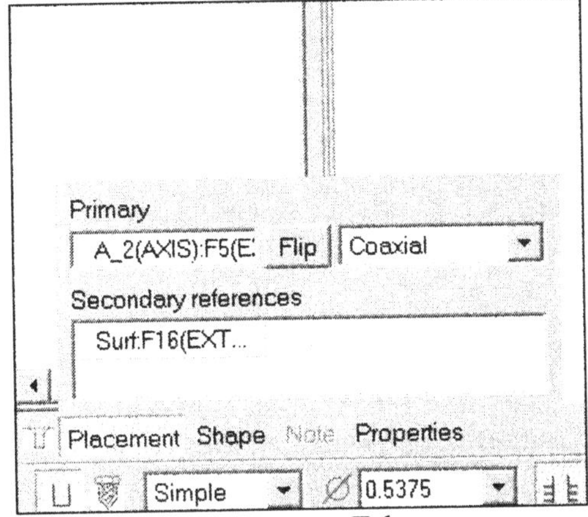

Figure 6.17(a) Placement Tab

Figure 6.17(b) Primary and Secondary References

Figure 6.17(c) Shape Tab

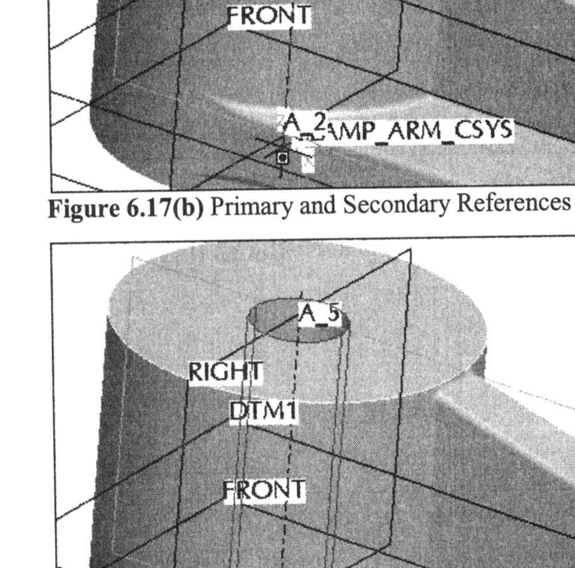

Figure 6.17(d) Hole

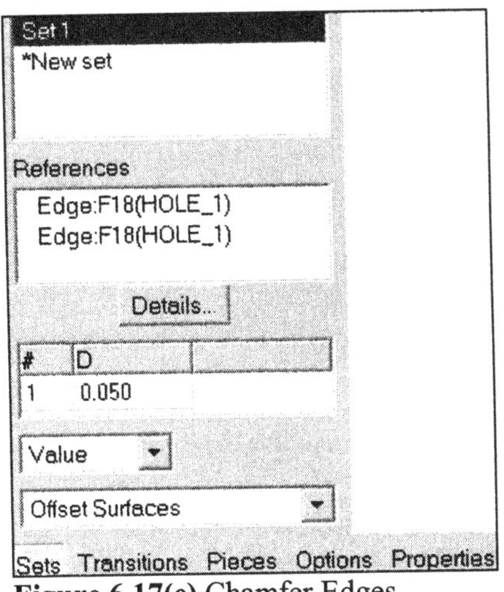

Figure 6.17(e) Chamfer Edges

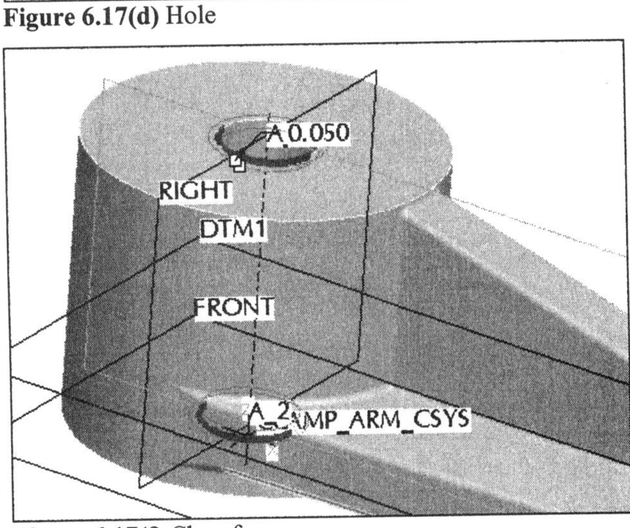

Figure 6.17(f) Chamfers

Click: **Hole Tool** ⇒

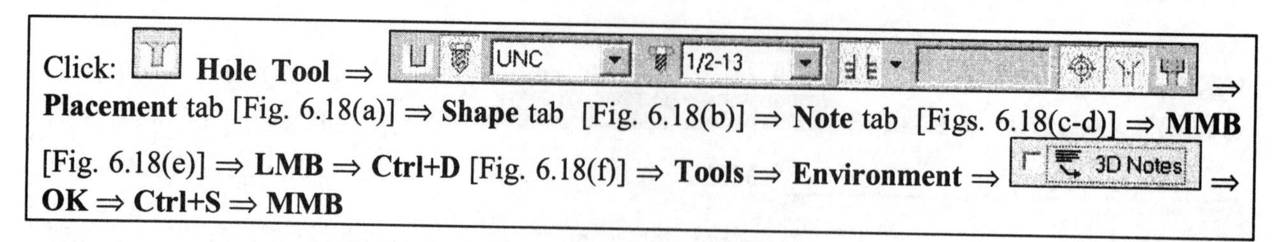

⇒ **Placement** tab [Fig. 6.18(a)] ⇒ **Shape** tab [Fig. 6.18(b)] ⇒ **Note** tab [Figs. 6.18(c-d)] ⇒ **MMB** [Fig. 6.18(e)] ⇒ **LMB** ⇒ **Ctrl+D** [Fig. 6.18(f)] ⇒ **Tools** ⇒ **Environment** ⇒ 3D Notes ⇒ **OK** ⇒ **Ctrl+S** ⇒ **MMB**

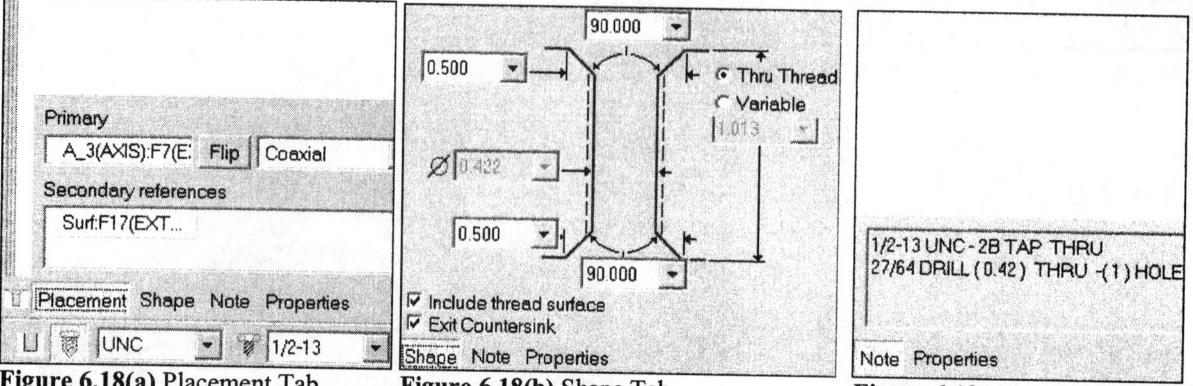

Figure 6.18(a) Placement Tab **Figure 6.18(b)** Shape Tab **Figure 6.18(c)** Note Tab

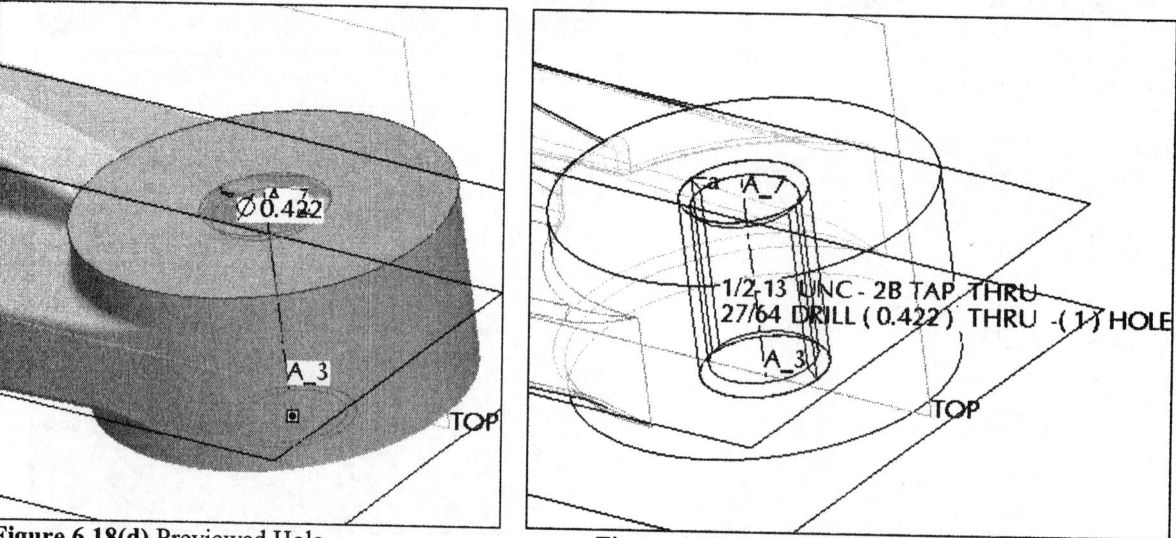

Figure 6.18(d) Previewed Hole **Figure 6.18(e)** Standard Hole

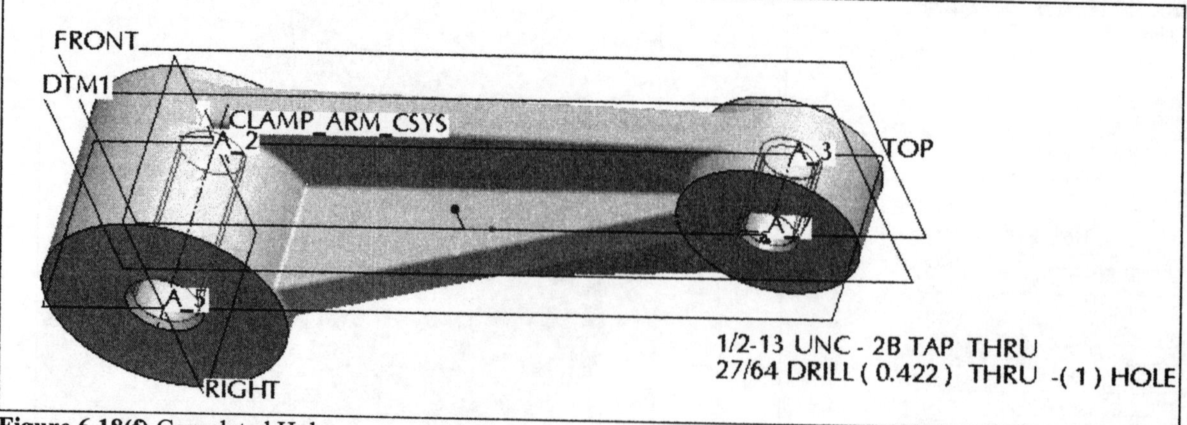

Figure 6.18(f) Completed Holes

Family Tables

Family Tables are used any time a part or assembly has several unique iterations developed from the original model. The iterations must be considered as separate models, not just iterations of the original model. In this lesson, you will add instances of a family table to create a machined part and a casting of the Clamp Arm (a version without machined cuts or holes).

You will be creating a Family Table from the generic model. The (base) model is the **Generic**. Each variation is referred to as an **Instance**. When you create a Family Table, Pro/E allows you to *select dimensions,* which can vary between instances. You can also *select features* to add to the Family Table. Features can vary by being suppressed or resumed in an instance. When you are finished selecting items (e.g., dimensions, features, and parameters), the Family Table is automatically generated.

When adding features to the table, enter an **N** to suppress the feature, or a **Y** to resume the feature. Each instance must have a unique name.

Family tables are spreadsheets, consisting of columns and rows. *Rows* contain instances and their corresponding values; *columns* are used for items. The column headings include the *instance name* and the names of all of the *dimensions, parameters, features, members,* and *groups* that were selected to be in the table. The Family Table dialog box is used to create and modify family tables.

Family tables include:

- The base object (generic object or *generic*) on which all members of the family are based.
- Dimensions, parameters, feature numbers, user-defined feature names, and assembly member names that are selected to be table-driven (*items*).
    - o **Dimensions** are listed by name (for example, **d125**) with the associated symbol name (if any) on the line below it (for example, depth).
    - o **Parameters** are listed by name (dim symbol).
    - o **Features** are listed by feature number with the associated feature type (for example, [cut]) or feature name on the line below it. The generic model is the first row in the table. Only modifying the actual part, suppressing, or resuming features can change the table entries belonging to the generic; *you cannot change the generic model by editing its row entries in the family table.*
- Names of all family members (*instances*) created in the table and the corresponding values for each of the table-driven items

Use the following commands to create a family table, click: **Tools** ⇒ **Family Table**--the Family Table: dialog box opens [Fig. 6.19(a)]

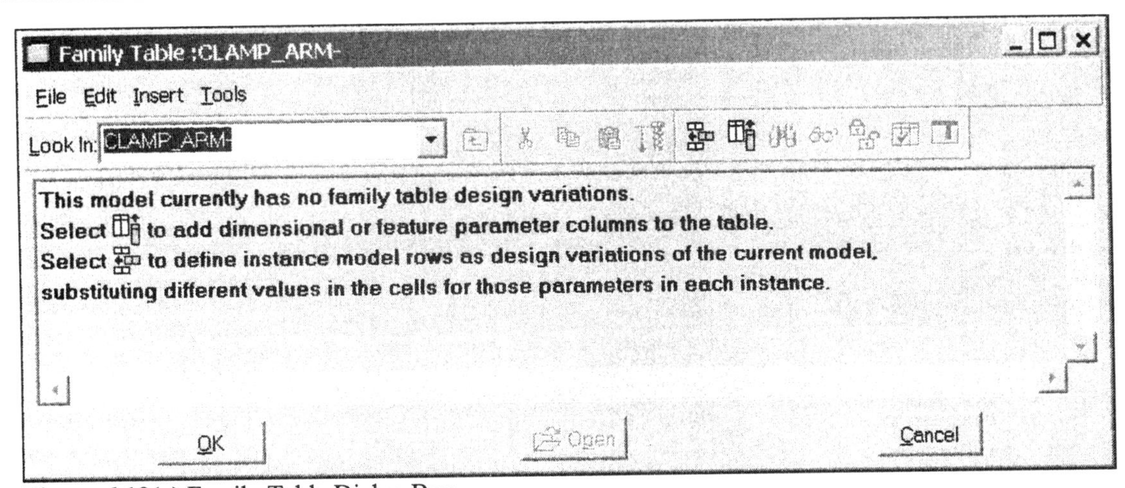

Figure 6.19(a) Family Table Dialog Box

Click: **Add/delete the table columns** ⇒ Feature (from the Add Item options) ⇒ select the cuts, holes, and chamfers from the model or the Model Tree [Fig. 6.19(b)] ⇒ **OK** ⇒ **MMB** ⇒ **MMB** [Fig. 6.19(c)] ⇒ **Insert a new instance at the selected row** ⇒ click on the name of the new instance **CLAMP_ARM_INST** ⇒ type **CLAMP_ARM_DESIGN** ⇒ **Enter** [Fig. 6.19(d)]

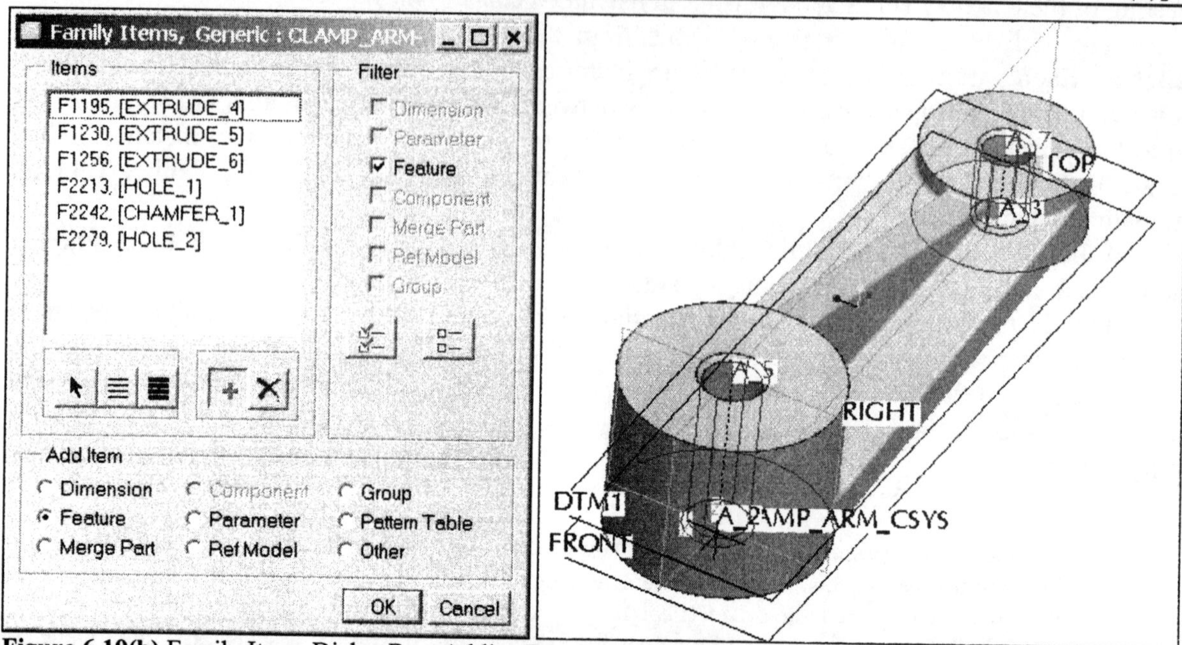

Figure 6.19(b) Family Items Dialog Box, Adding Features

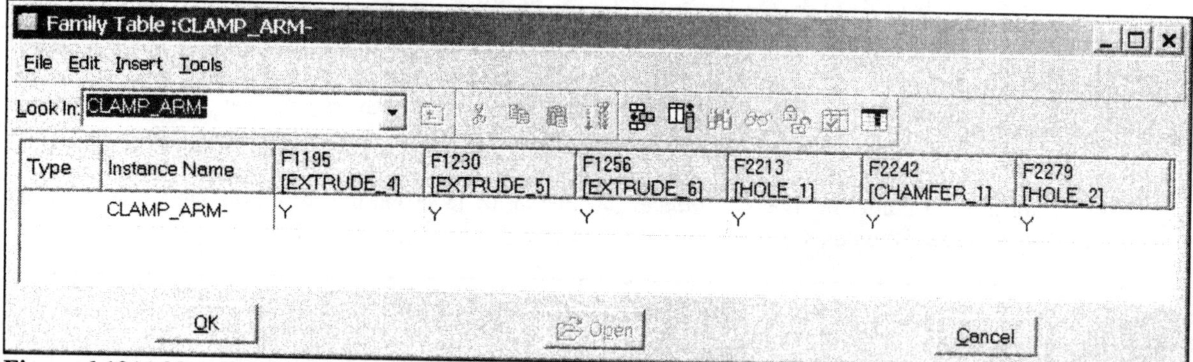

Figure 6.19(c) New Family Table

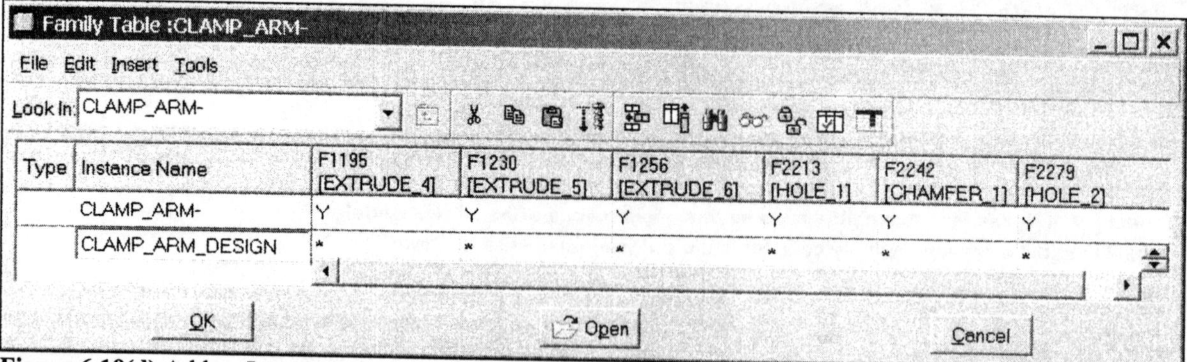

Figure 6.19(d) Add an Instance

The second instance **CLAMP_ARM_INST** should be highlighted ⇒ type **CLAMP_ARM_WORKPIECE** [Fig. 6.19(e)] ⇒ click in the cell of the first feature and change to **N** (not used) [Fig. 6.19(f)] ⇒ change all cells for the **CLAMP_ARM_WORKPIECE** to **N** [Fig. 6.19(g)] ⇒ **Verify Instances of the family** ⇒ **Verify** ⇒ **Close** ⇒ **OK** from Family Table dialog box

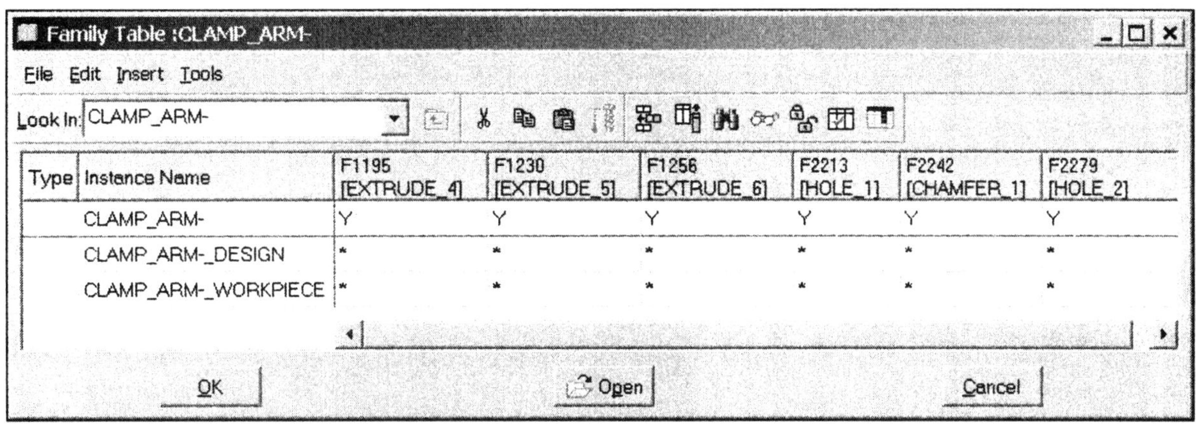

Figure 6.19(e) Add a Second Instance

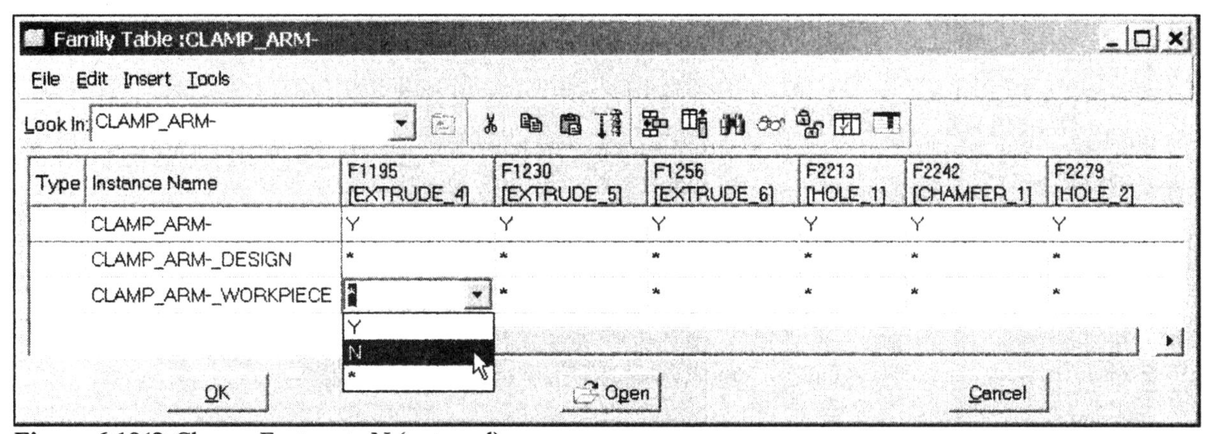

Figure 6.19(f) Change Feature to N (not used)

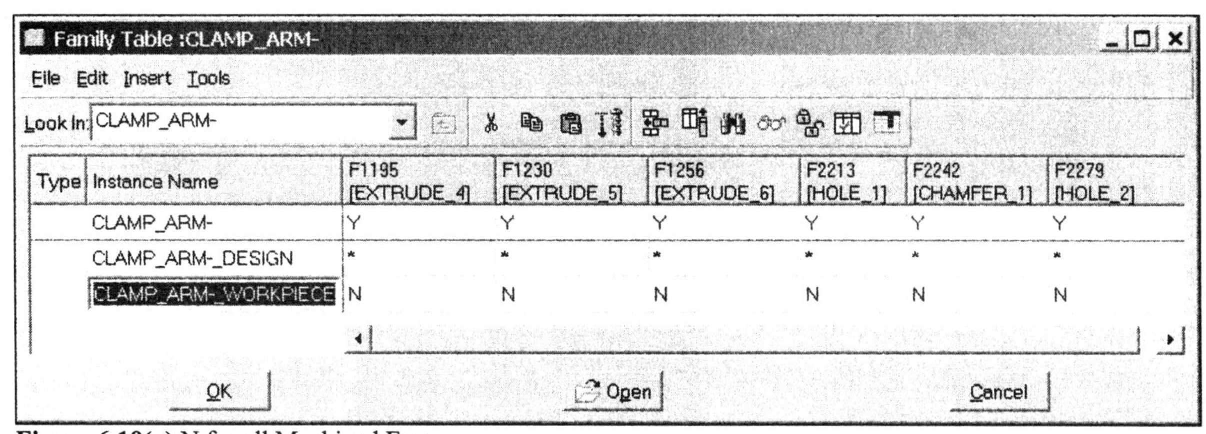

Figure 6.19(g) N for all Machined Features

A Family Table controls whether a feature is present or not for a given design instance, not whether a feature is displayed. The Generic is the base model [Fig. 6.19(h)].

Click: **Tools** ⇒ **Family Table** ⇒ click on **CLAMP_ARM_DESIGN** ⇒ Open [Fig. 6.19(i)] ⇒ adjust your windows to view both models ⇒ from Generic part window, click: **Tools** ⇒ **Family Table** ⇒ click on **CLAMP_ARM_WORKPIECE** ⇒ **RMB** ⇒ Open [Fig. 6.19(j)] ⇒ **Window** ⇒ **Close** ⇒ **Window** from the menu bar of **CLAMP_ARM_DESIGN** ⇒ **Activate** ⇒ **Window** ⇒ **Close** ⇒ **Window** ⇒ **Activate** ⇒ **Ctrl+S** ⇒ **MMB**

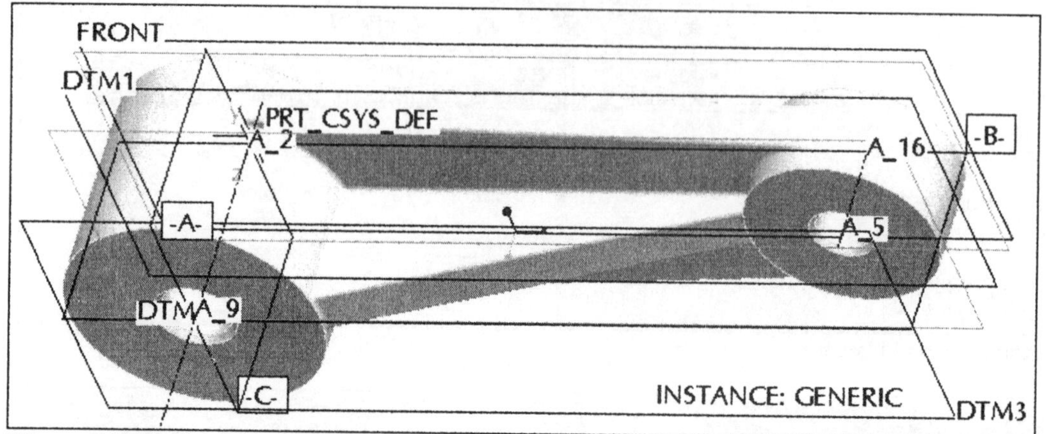

Figure 6.19(h) Instance: GENERIC

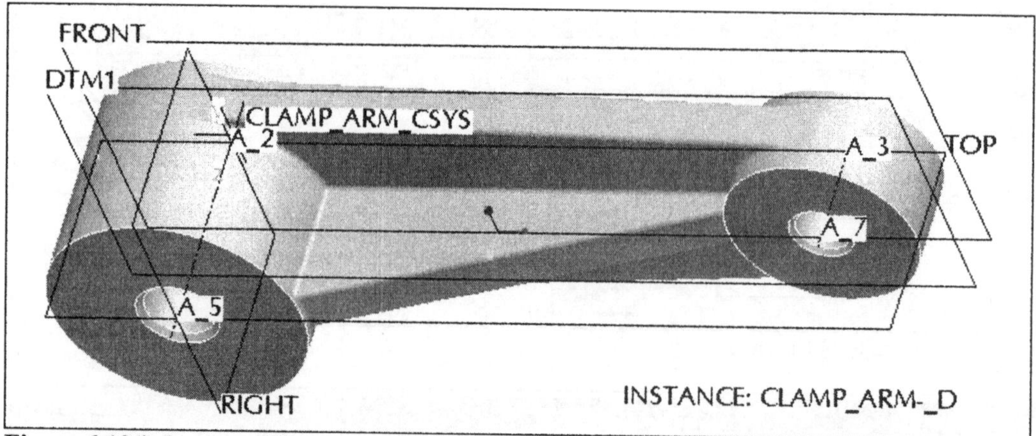

Figure 6.19(i) Instance: CLAMP_ARM_DESIGN

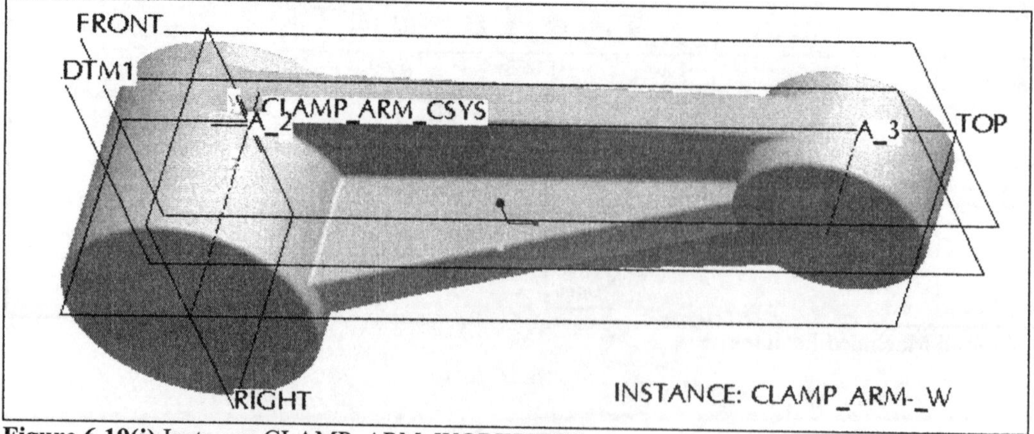

Figure 6.19(j) Instance: CLAMP_ARM_WORKPIECE

Write a relation to keep the thickness of the rib the same as that for the "web", click: **off** ⇒ with the **Ctrl** key pressed, pick on the "web" extrusion and the rib ⇒ **RMB** ⇒ **Edit** [Fig. 6.20(a)] ⇒ **Info** ⇒ Switch Dimensions ⇒ pick on a **"d"** symbol ⇒ **RMB** ⇒ **Properties** ⇒ **Move** ⇒ pick a new location for the dimension ⇒ **OK** ⇒ repeat for the other dimension [Fig. 6.20(b)] ⇒ **Tools** ⇒ **Relations** ⇒ type **d8=d7** [Fig. 6.20(c)] ⇒ **Ok** ⇒ **Info** ⇒ Switch Dimensions ⇒ **Ctrl+R** ⇒ pick on the "web" ⇒ **RMB** ⇒ **Edit** ⇒ double-click on the **.375** dimension and change to **.60** ⇒ **Enter** ⇒ **Edit** ⇒ **Regenerate** [Fig. 6.20(d)] ⇒ **Undo Changes** ⇒ **Confirm** ⇒ modify the value to **.20** ⇒ **Edit** ⇒ **Regenerate** [Fig. 6.20(e)] ⇒ modify to the design value of **.375** ⇒**Edit** ⇒ **Regenerate** ⇒ **on** ⇒ **Ctrl+S** ⇒ **MMB**

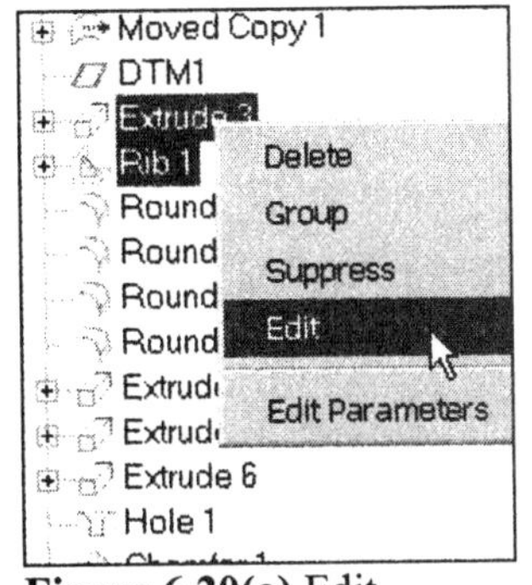

Figure 6.20(a) Edit

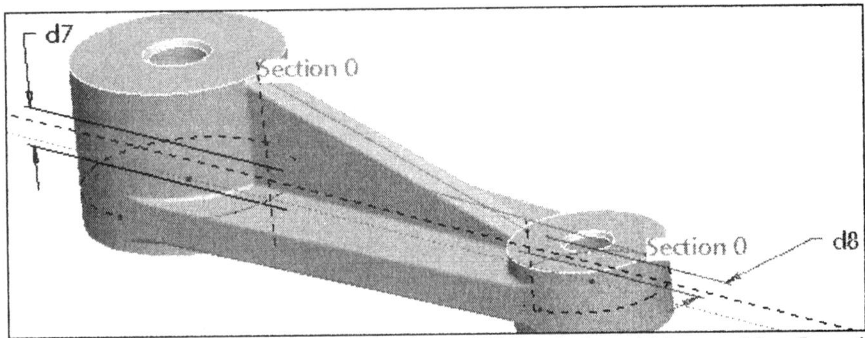

Figure 6.20(b) Switch Dimensions to show **"d"** Symbols and Move to New Location

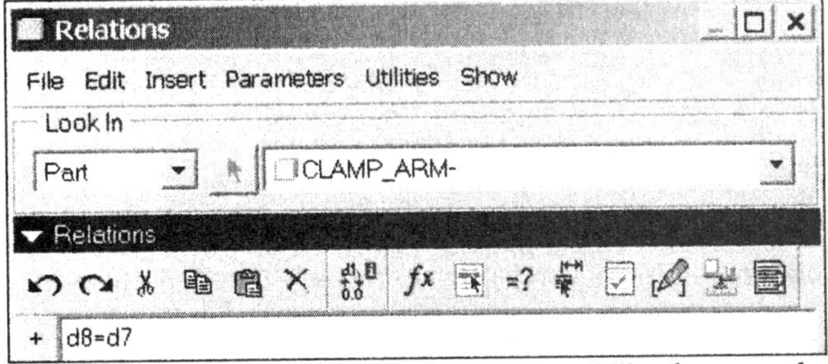

Figure 6.20(c) Relations Dialog Box (your **"d"** values may be different)

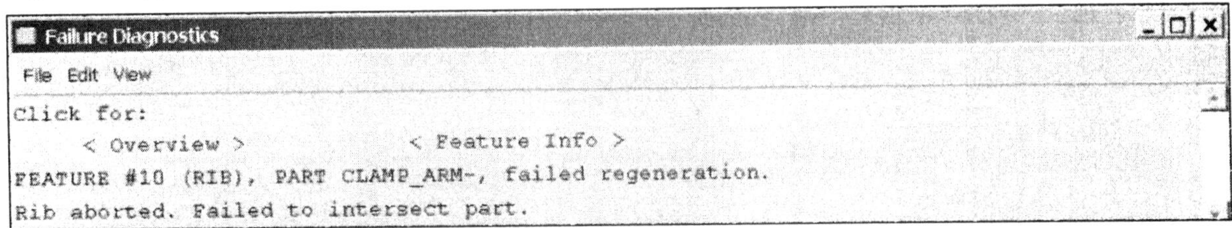

Figure 6.20(d) Failure Diagnostics Dialog Box

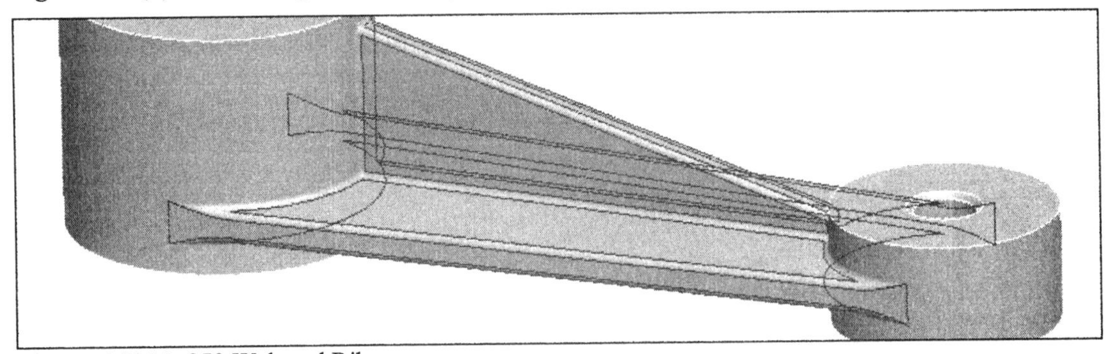

Figure 6.20(e) .250 Web and Rib

Pro/MANUFACTURING

You now have two separate models, a casting (workpiece) and a machined part (design part). During the manufacturing process, the workpiece is merged (assembled) into the design part thereby creating a manufacturing model [Fig. 6.21(a)]. The difference between the two files is the difference between the volume of the casting and the volume of the machined part. The manufacturing model is used to machine the part [Fig. 6.21(b)].

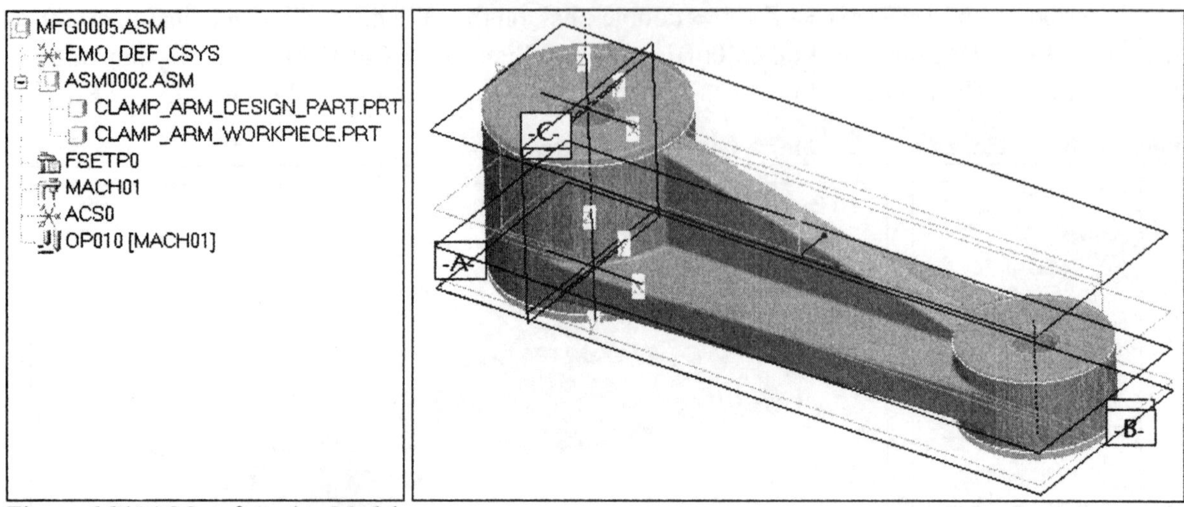

Figure 6.21(a) Manufacturing Model

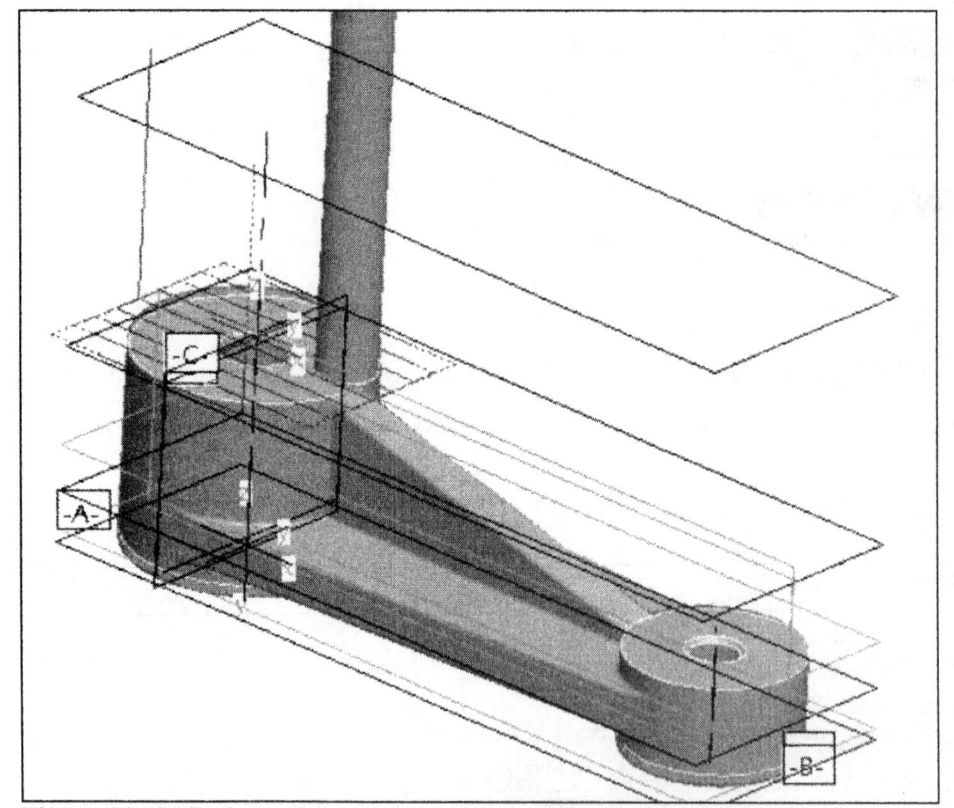

Figure 6.21(b) Facing

Lesson 6 is now complete. If you wish to model a project without instructions, a complete set of projects and illustrations are available at www.cad-resources.com ⇒ ***Downloads.***

Lesson 7 Assemblies

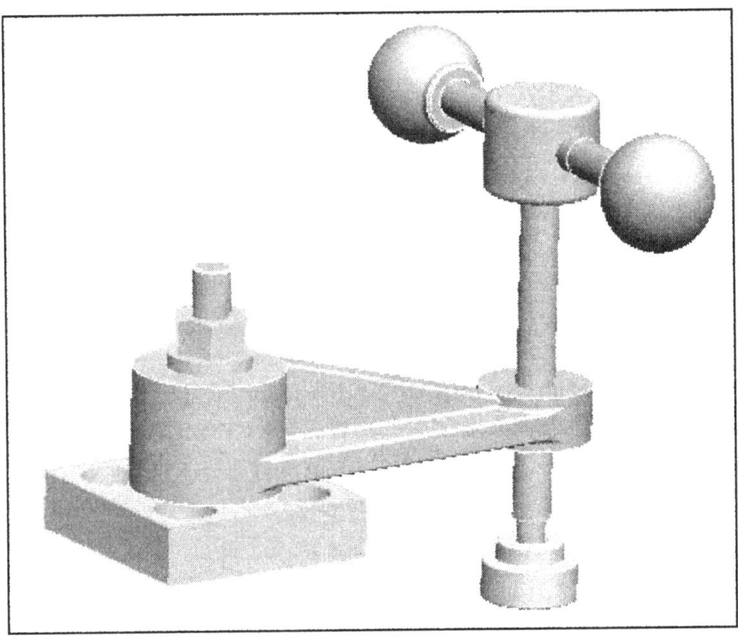

Figure 7.1(a) Swing Clamp Assembly

OBJECTIVES

- **Assemble** components to form an assembly
- Create a **subassembly**
- Understand and use a variety of **Assembly Constraints**
- **Modify** a component constraint
- **Edit** a constraint value
- Check for **clearance** and **interference**

Assembly Constraints

Assembly mode allows you to place together components and subassemblies to form an assembly [Fig. 7.1(a)]. Assemblies [Fig. 7.1(b)] can be modified, reoriented, documented, or analyzed. An assembly can be assembled into another assembly, thereby becoming a subassembly.

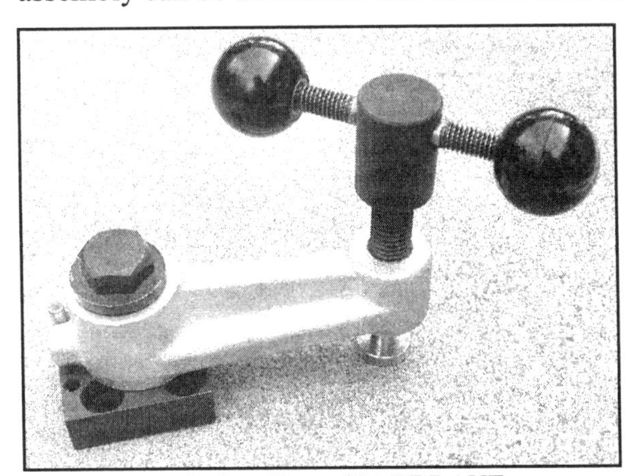

Figure 7.1(b) Swing Clamp (**CARRLANE** at www.carrlane.com)

Placing Components

To assemble components you simply use or Insert ⇒ Component ⇒ Assemble. After selecting a component from the Open dialog box, the Component Placement dialog box [Fig. 7.1(g)] opens and the component appears in the assembly window. If the component being assembled has previously defined interfaces, the Select Interface dialog box opens to allow you to select an interface. Alternatively, you can select a component from a browser window and drag it into the Pro/E window. If there is an assembly in the window, Pro/E will begin to assemble the component into the current assembly. Either the Component Placement or the Select Interface dialog boxes will come up as previously described.

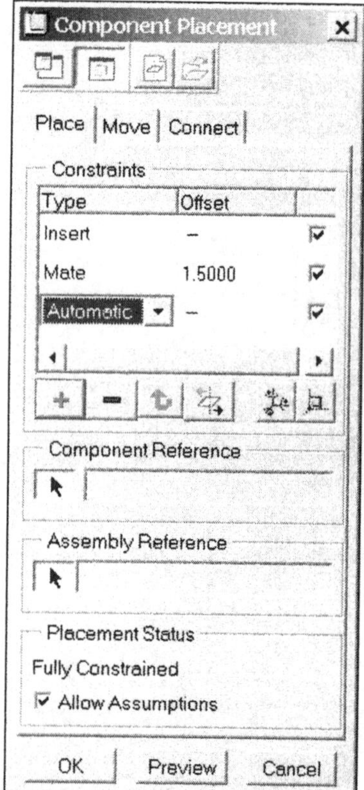

Figure 7.1(g) Component Placement Dialog Box

Using icons in the toolbar at the top of the dialog box, specify the screen window in which the component is displayed while you position it. You can change windows at any time using:

- **Show component in a separate window while specifying constraints**
- **Show component in the assembly window while specifying constraints**

The Automatic placement constraint is selected by default when a new component is introduced into an assembly for placement. Do one of the following: Select a reference on the component and a reference on the assembly, in either order, to define a placement constraint. After you select a pair of valid references from the component and the assembly, Pro/E automatically selects a constraint type appropriate to the specified references.

Before selecting references, you may change the type of constraint by selecting a type from the Constraints Type list. Clicking the current constraint in the Type box shows the list.

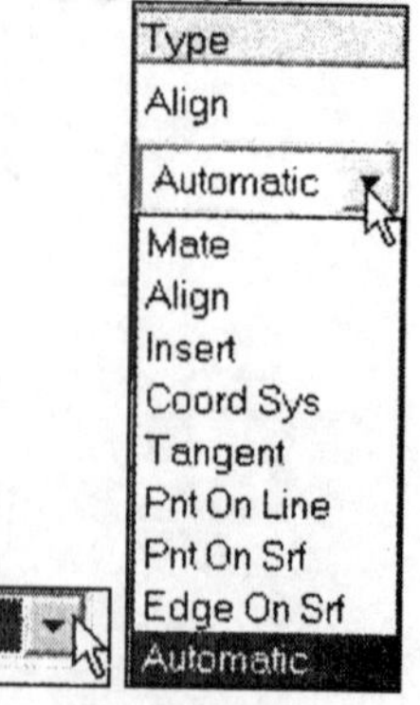

You may also select an offset type from the Offset list. **Coincident** is the default offset, but you may also choose 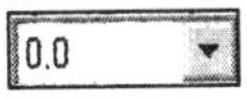 or **Oriented** from the list. If you choose **0.0**, type in the value of the offset in the cell  and then press Enter. You can add, remove, change, fix, and convert constraints using:

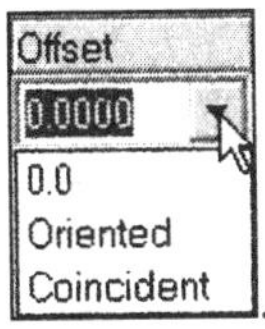

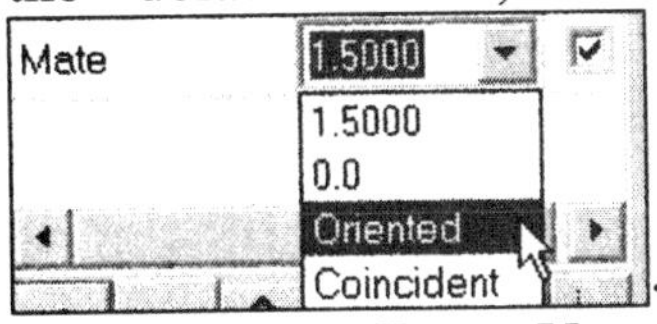

- After you define a constraint, **Specify a new constraint** is automatically selected, and you can define another constraint. You can define as many constraints as you want (up to Pro/E limit of 50 constraints). As you define constraints, each constraint is listed under the Constraints area and the status of the component is reported in the Placement Status area as you select constraint references.
- You can select one of the constraints listed in the Constraints area at any time and change the constraint type, flip between Mate and Align with **Change orientation of constraint**, modify the offset value, reset and Align constraint to forced or unforced, or switch between allowing and disallowing Pro/E assumptions.
- Use **Assemble component at default location** to align the default Pro/E -created coordinate system of the component to the default Pro/E -created coordinate system of the assembly. Pro/E places the component at the assembly origin. Use **Fix component to current position** to fix the current location of the component that was moved or packaged.
- Use **Remove the selected constraint** to delete a placement constraint for the component, select one of the constraints listed in the Constraints area, and then click.
- Use **Convert constraints to Mechanism connection or vice versa** to change the existing constraint to a Mechanism constraint. You can use different types of mechanism connections to assemble your model depending on component placement. In addition, you can convert placement constraints to mechanism connections to allow for kinematic movement of components.

Click Preview **Preview component placement with current constraints** to show the location of the component, as it would be with the current placement constraints. Click OK when the Placement Status of the component is shown as Fully Constrained, or Fully Constrained / Allow Assumptions.

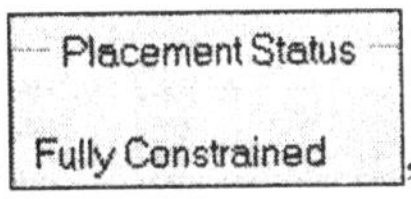

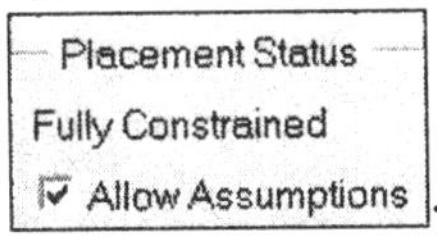

If the Placement Status is Constraints Invalid, you should correct or complete the constraint definition.

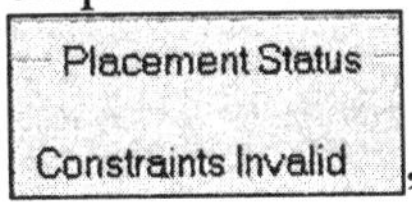

If constraints are incomplete (Partially Constrained), you can leave the component as *packaged*. A packaged component is one that is included in the assembly but not fully placed Partially Constrained. Packaged components follow the behavior dictated by the configuration file option *package constraints*. The warning "You are leaving this component as packaged." will display if you leave the component partially constrained, and the model tree 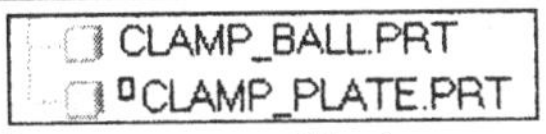 will display the component with a small box in front of its name. If constraints are conflicting, you can restart or continue placing the component. Restarting erases all previously defined constraints for the component. You can also uncheck a constraint in the Constraints table to make it inactive.

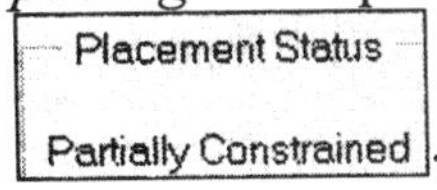

Lesson 7 STEPS

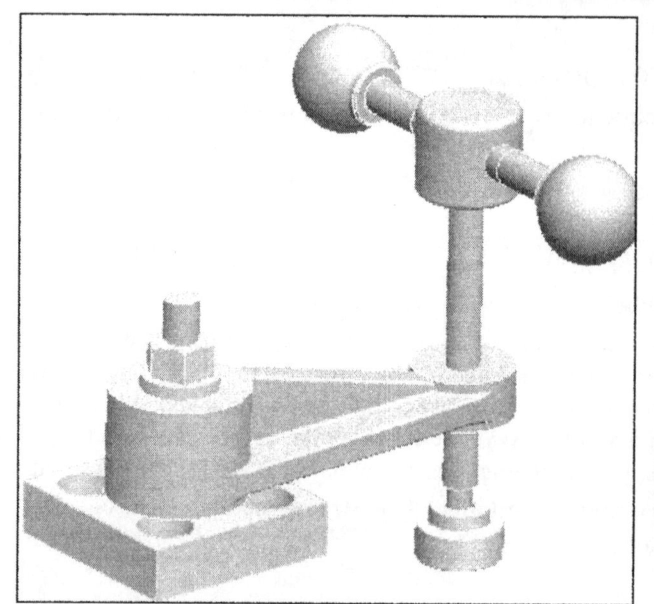

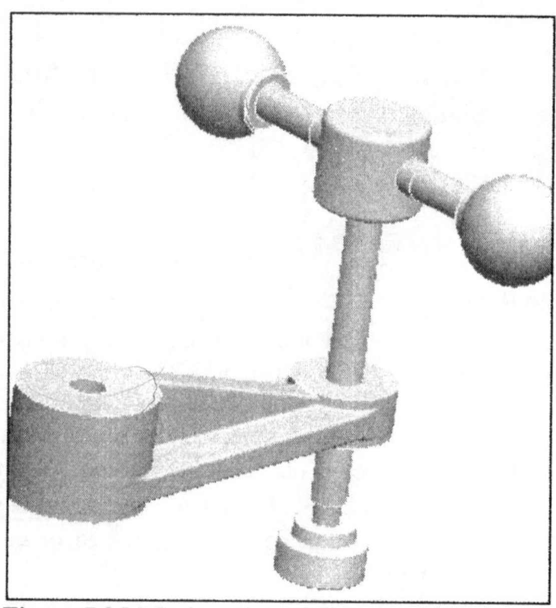

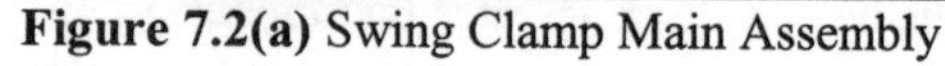
Figure 7.2(a) Swing Clamp Main Assembly

Figure 7.2(b) Swing Clamp Sub-Assembly

Swing Clamp Assembly

The parts required in this lesson are from this text. *If you have not modeled these parts previously, please do so before you start the following systematic instructions.* The other components required for the assembly are standard *off-the-shelf* hardware items that you can get from Pro/E by accessing the library. If your system does not have a Pro/LIBRARY license for the Basic and Manufacturing libraries, model the parts using the detail drawings provided here or download them from the PTC online Catalog *(the Student Edition and the Tryout Edition do not allow access to the catalog parts).* The **Flange Nut**, the **3.50 Double-ended Stud**, and the **5.00 Double-ended Stud** are standard. The **Clamp Plate** component is the first component of the main assembly and will be modeled later when completing the main assembly [Fig. 7.2(a)] and using the *top-down design* approach.

Because you will be creating the sub-assembly [Fig. 7.2(b)] using the *bottom-up design* approach, all the components must be available before any assembling starts. *Bottom-up design* means that existing parts are assembled, one by one, until the assembly is complete. The assembly starts with a set of datum planes and a coordinate system. The parts are constrained to the datum features of the assembly. The sequence of assembly will determine the parent-child relationships between components.

Top-down design is the design of an assembly where one or more component parts are created in Assembly mode as the design unfolds. Some existing parts are available, such as standard components and a few modeled parts. The remaining design evolves during the assembly process. The main assembly will involve creating one part using the *top-down design* approach.

Regardless of the design method, the assembly datum planes and coordinate system should be on their own separate *assembly layer*. Each part should also be placed on separate assembly layers; the part's datum features should already be on *part layers*.

Before starting the assembly, you will be modeling each part or retrieving *standard parts* from the library and saving them under unique names in *your* directory. *Unless instructed to do so, do not use the library parts directly in the assembly.* Start this process by retrieving the standard parts from Pro/Library [Figs. 7.3(a-c)] or the online PTC Catalog.

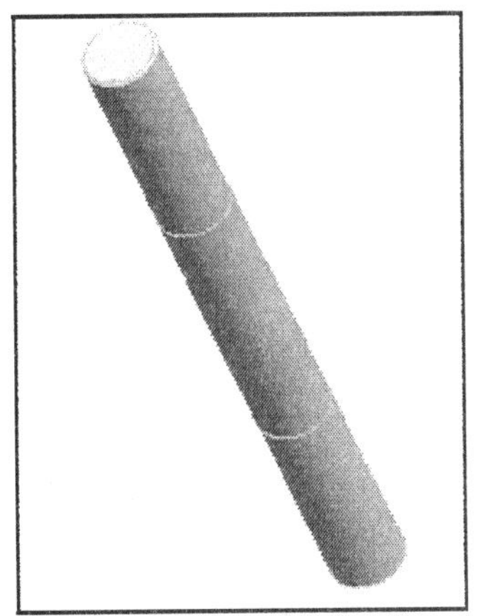

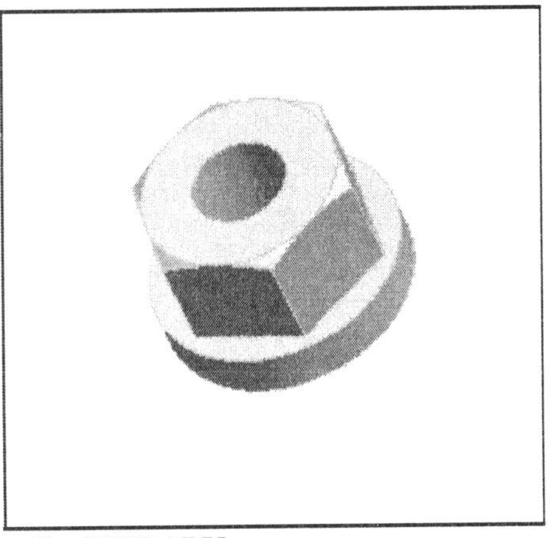

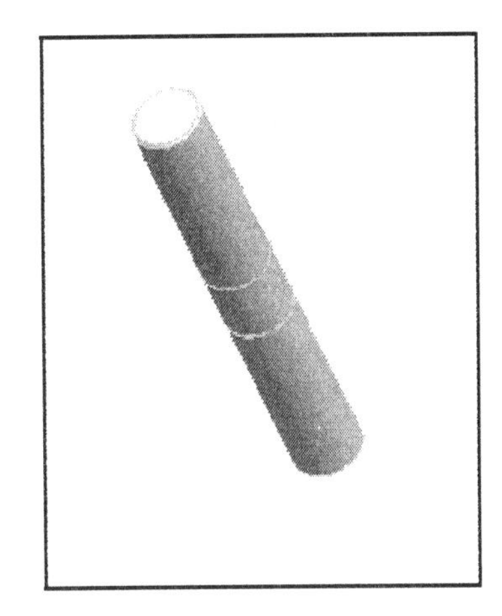

Figure 7.3(a-c) Standard Parts from Pro/LIBRARY

File ⇒ **Set Working Directory** ⇒ select your working directory ⇒ **OK** ⇒ **File** ⇒ **Open** ⇒ navigate to **Pro Library** ⇒ **mfglib** ⇒ **Open** ⇒ **fixture_lib** ⇒ **Open** ⇒ **nuts_bolts_screws** ⇒ **Open** ⇒ **st.prt** ⇒ **Open** ⇒ **By Parameter** tab (from Select Instance dialog box) ⇒ **d0,thread_dia** ⇒ **.500** ⇒ **d8,stud_length** ⇒ **3.500** [Figs. 7.14(a-b)] ⇒ **Open** ⇒ **File** ⇒ **Save a Copy** ⇒ **CLAMP_STUD35** ⇒ **OK** ⇒ **File** ⇒ **Erase** ⇒ **Current** ⇒ **Yes**

Alternative method: *(if you have an Academic Version or Commercial Version of Wildfire 2.0, but no access to Pro/LIBRARY, you may use this method): Click* ***Connections*** Connections ⇒ Catalogs ⇒ *under the logo for CAD Register.com* CADRegister *click:* ***Search the catalog*** *(click OK if prompted)* ⇒ CADRegister ⇒ Cair Lane Mfg. Co. ⇒ Tooling Components & Clamps ⇒ Clamps and Accessories ⇒ Studs ⇒ 3D ⇒ *if needed, click Start Evaluation* ⇒ 1/2-13 CL-1/2-13X3.50-STUD ⇒ ***Yes*** *to install* ⇒ *Click:* ***Drag to Pro/ENGINEER Wildfire*** ⇒ ***Open*** *from the WinZip dialog box* ⇒ ***I Agree*** ⇒ *drag and drop the neutral file into the graphics window* ⇒ ***OK*** ⇒ ***File*** ⇒ ***Save a Copy*** ⇒ *type new name* ***CLAMP_STUD35*** ⇒ ***OK*** ⇒ ***File*** ⇒ ***Erase*** ⇒ ***Current*** ⇒ ***Yes*** *(Note: This model will not have dimensions and will be all one diameter.)*

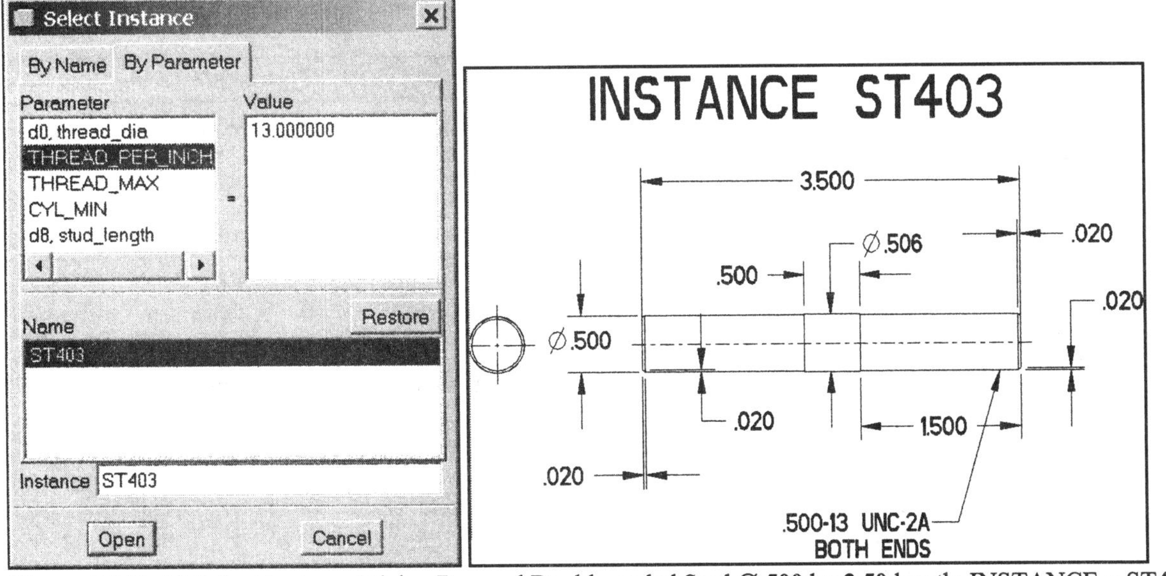

Figure 7.14(a-b) Select Instance Dialog Box and Double-ended Stud ∅**.500** by **3.50** length: INSTANCE = ST403

File ⇒ **Open** ⇒ navigate to **Pro Library** ⇒ **mfglib** ⇒ **Open** ⇒ **fixture_lib** ⇒ **Open** ⇒ **nuts_bolts_screws** ⇒ **Open** ⇒ **st.prt** ⇒ **Open** ⇒ **By Parameter** (from Select Instance dialog box) ⇒ **d0,thread_dia** ⇒ **.500** ⇒ **d8,stud_length** ⇒ **5.00** [Figs. 7.14(c-d)] ⇒ **Open** ⇒ **File** ⇒ **Save a Copy** ⇒ **CLAMP_STUD5** ⇒ **OK** ⇒ **File** ⇒ **Erase** ⇒ **Current** ⇒ **Yes** *(or download the part from the CARR Lane Catalog)*

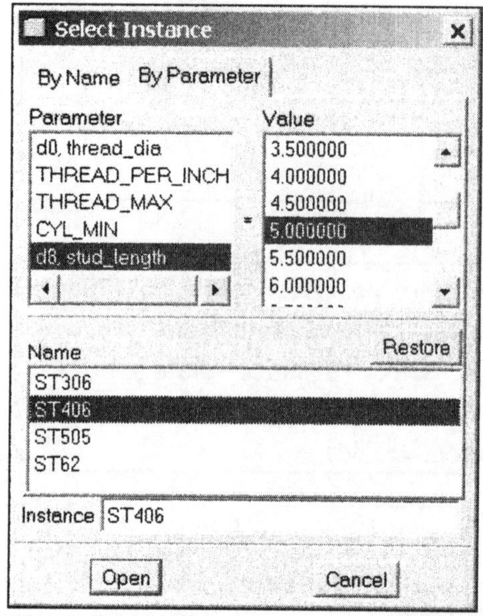

Figure 7.14(c) Select Instance Dialog Box

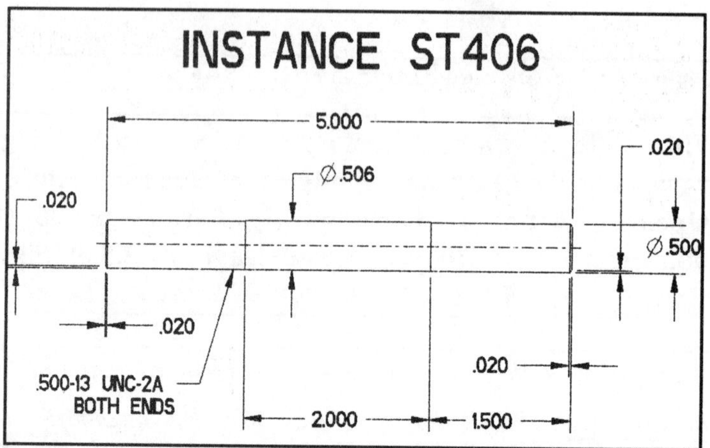

Figure 7.14(d) Double-ended Stud Ø**.500** by **5.00** length: INSTANCE = ST406

File ⇒ **Open** ⇒ navigate to **Pro Library** ⇒ **mfglib** ⇒ **Open** ⇒ **fixture_lib** ⇒ **Open** ⇒ **nuts_bolts_screws** ⇒ **Open** ⇒ **fn.prt** ⇒ **Open** ⇒ **By Parameter** (from Select Instance dialog box) ⇒ **d4,thread_dia** ⇒ **.500** ⇒ **Open** ⇒ **File** ⇒ **Save a Copy** ⇒ **CLAMP_FLANGE_NUT** [Figs. 7.14(e-f)] ⇒ **OK** ⇒ **File** ⇒ **Erase** ⇒ **Current** ⇒ **Yes** *(or use the Alternative method and download the part from the CARR Lane Catalog or build it from the dimensions shown)*

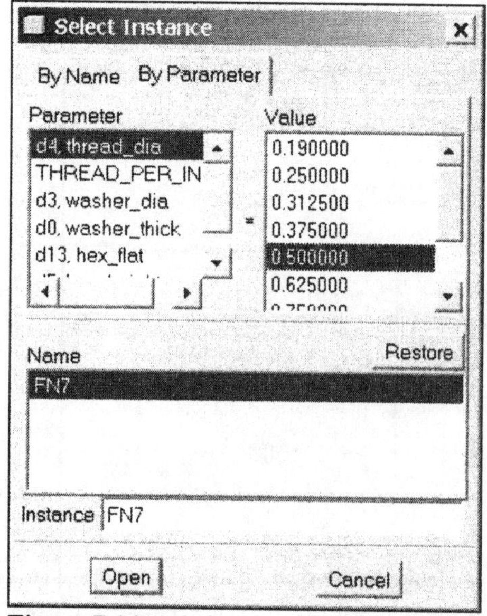

Figure 7.14(e) Select Instance Dialog Box

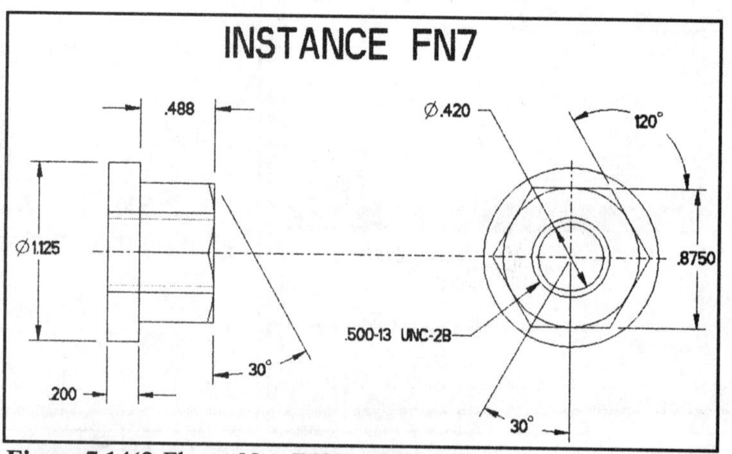

Figure 7.14(f) Flange Nut: INSTANCE=FN7

Open the Clamp_Arm [Fig. 7.15(a)], the Clamp_Swivel [Fig. 7.15(b)], the Clamp_Ball [Fig. 7.15(c)] and the Clamp_Foot [Fig. 7.15(d)] ⇒ review the components and standard parts for correct color, layering, coordinate system naming, and datum planes ⇒ **Save a Copy** of each to your working directory.

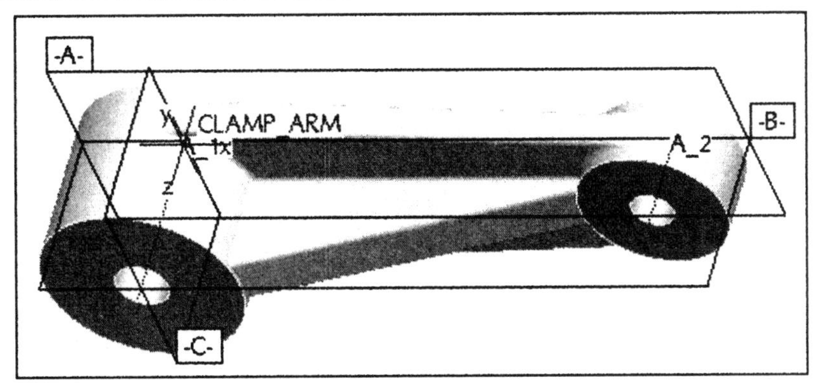

Figure 7.15(a) Clamp_Arm

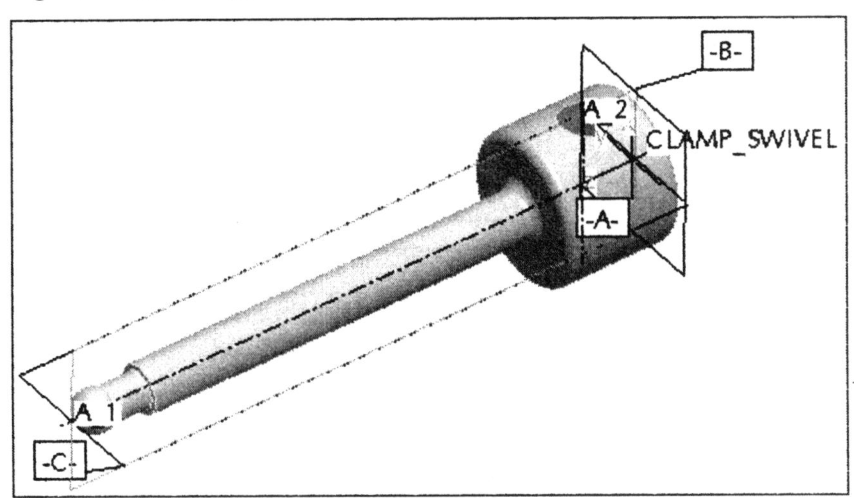

Figure 7.15(b) Clamp_Swivel

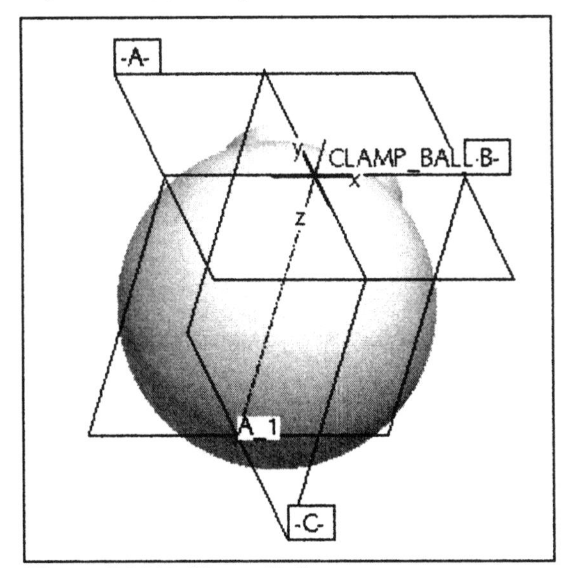

Figure 7.15(c) Clamp_Ball

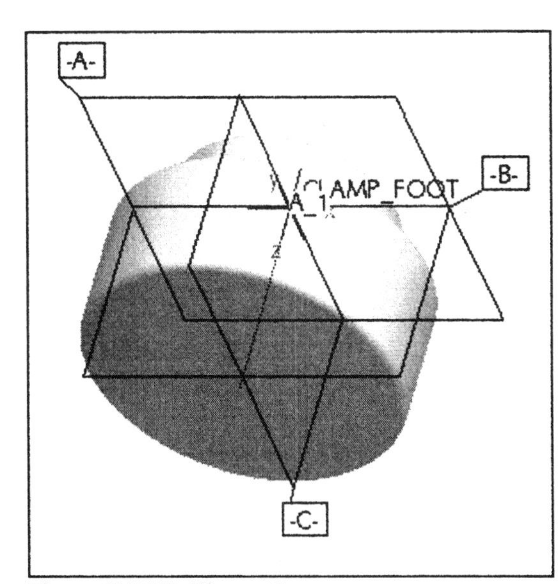

Figure 7.15(d) Clamp_Foot

You now have eight components (two identical Clamp_Ball components are used) required for the assembly. The ninth component- the Clamp_Plate (Fig. 7.16) will be created using *top-down design* procedures when you start the main assembly. *All parts must be in the same working directory used for the assembly.*

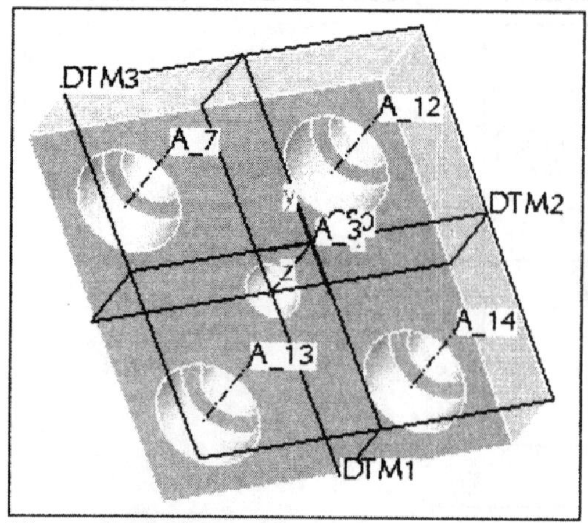

Figure 7.16 Clamp_Plate

A subassembly will be created first. The main assembly is created second. The subassembly will be added to the main assembly to complete the project. Note: the assembly and the components can have different units. Therefore, you must check and correctly set the assembly units before creating or assembling components or sub-assemblies.

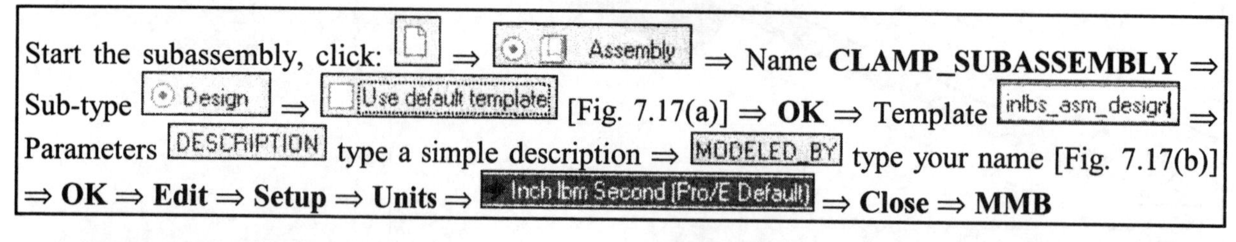
Start the subassembly, click: [new] ⇒ Assembly ⇒ Name **CLAMP_SUBASSEMBLY** ⇒ Sub-type Design ⇒ Use default template [Fig. 7.17(a)] ⇒ **OK** ⇒ Template inlbs_asm_design ⇒ Parameters DESCRIPTION type a simple description ⇒ MODELED_BY type your name [Fig. 7.17(b)] ⇒ **OK** ⇒ **Edit** ⇒ **Setup** ⇒ **Units** ⇒ Inch lbm Second (Pro/E Default) ⇒ **Close** ⇒ **MMB**

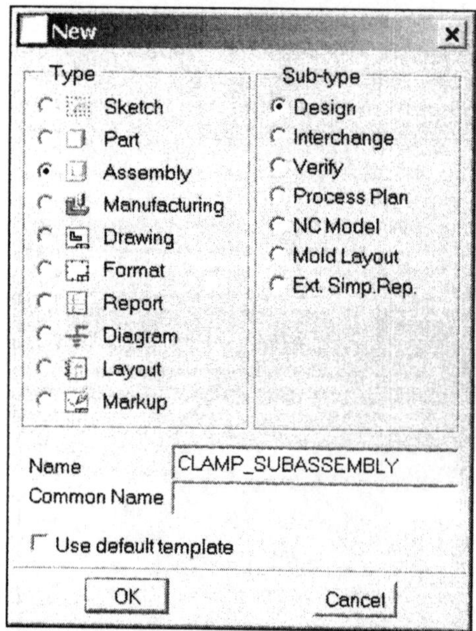

Figure 7.17(a) New Dialog Box

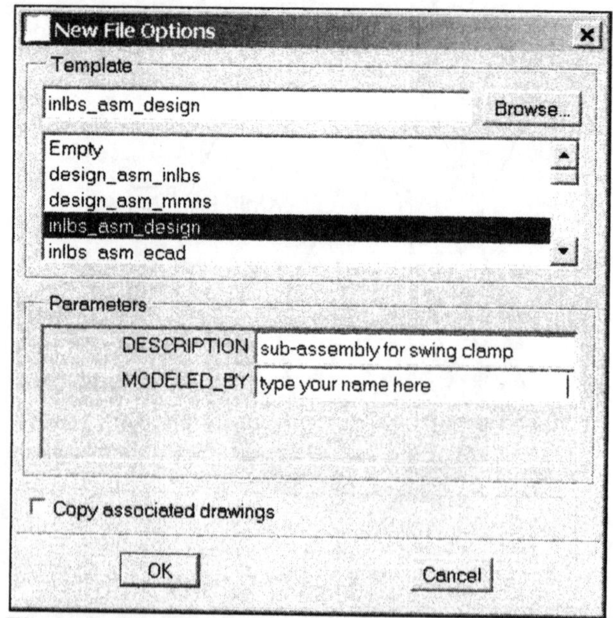

Figure 7.17(b) New File Options Dialog Box

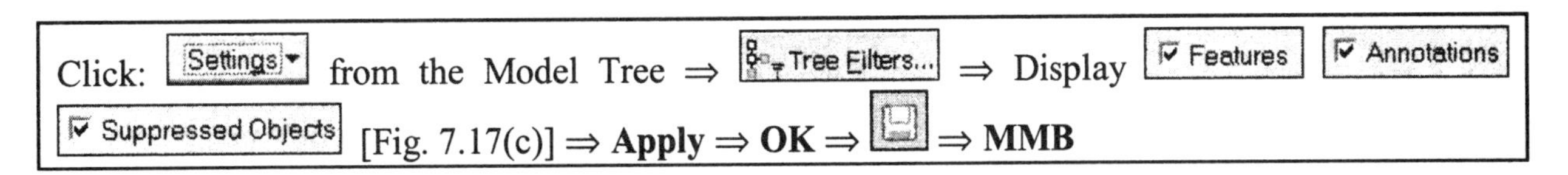

Click: Settings from the Model Tree ⇒ Tree Filters... ⇒ Display Features Annotations Suppressed Objects [Fig. 7.17(c)] ⇒ **Apply** ⇒ **OK** ⇒ ⇒ **MMB**

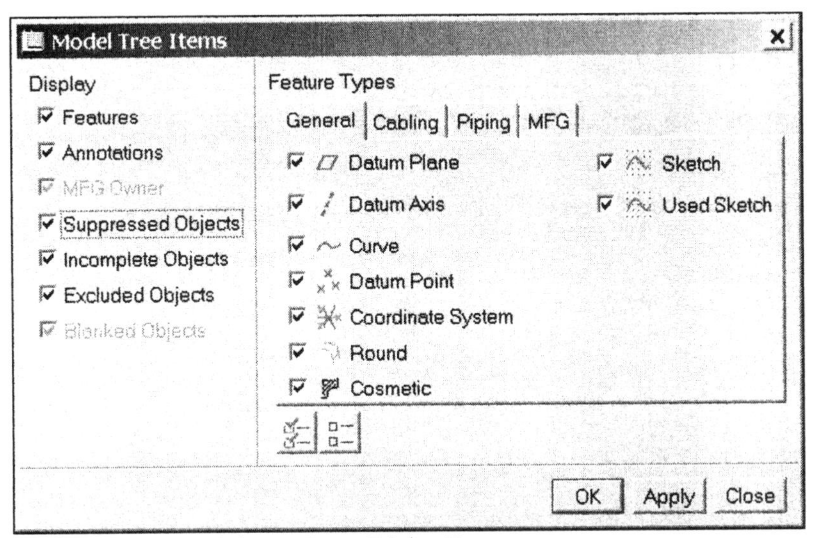

Figure 7.17(c) Model Tree Items Dialog Box

Datum planes and the coordinate system are created per the template provided by Pro/E. The datum planes will have the names, **ASM_RIGHT**, **ASM_TOP**, and **ASM_FRONT**.

Change the coordinate system name: slowly double-click on ASM_DEF_CSYS in the Model Tree ⇒ type new name **SUB_ASM_CSYS** SUB_ASM_CSYS ⇒ click in the graphics window [Fig. 7.17(d)]

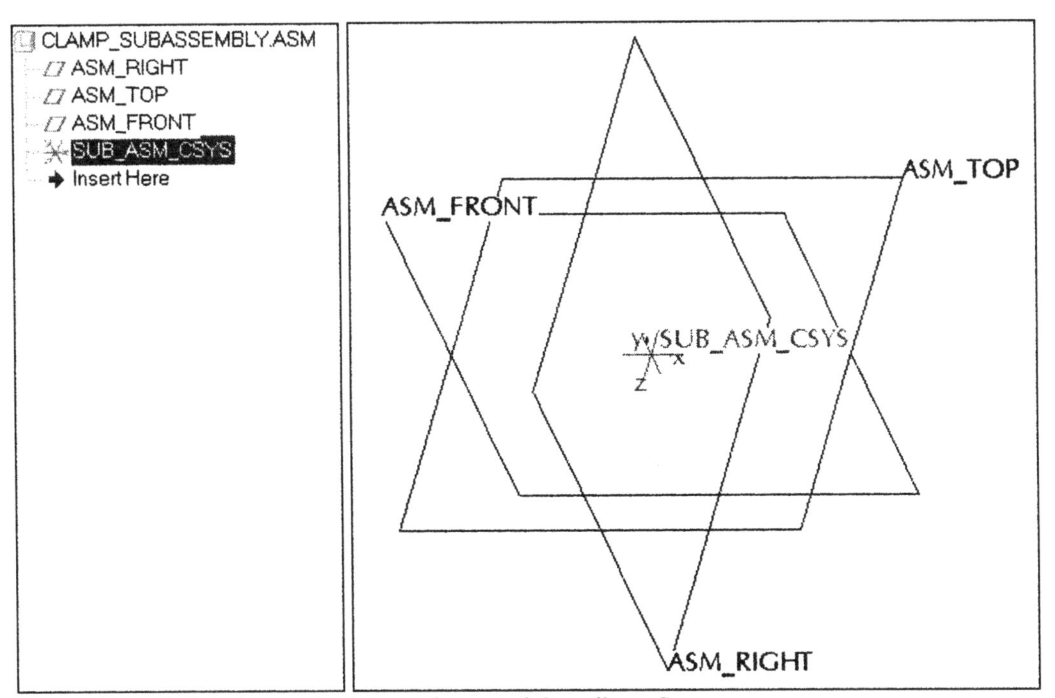

Figure 7.17(d) Sub-Assembly Datum Planes and Coordinate System

Regardless of the design methodology, the assembly datum planes and coordinate system should be on their own separate *assembly layer*. Each part should also be placed on separate assembly layers; the part's datum features should already be on *part layers*. Look over the default template for assembly layering.

Click: Show ⇒ Layer Tree ⇒ expand the branches [Fig. 7.17(e)] ⇒ Show ⇒ Model Tree

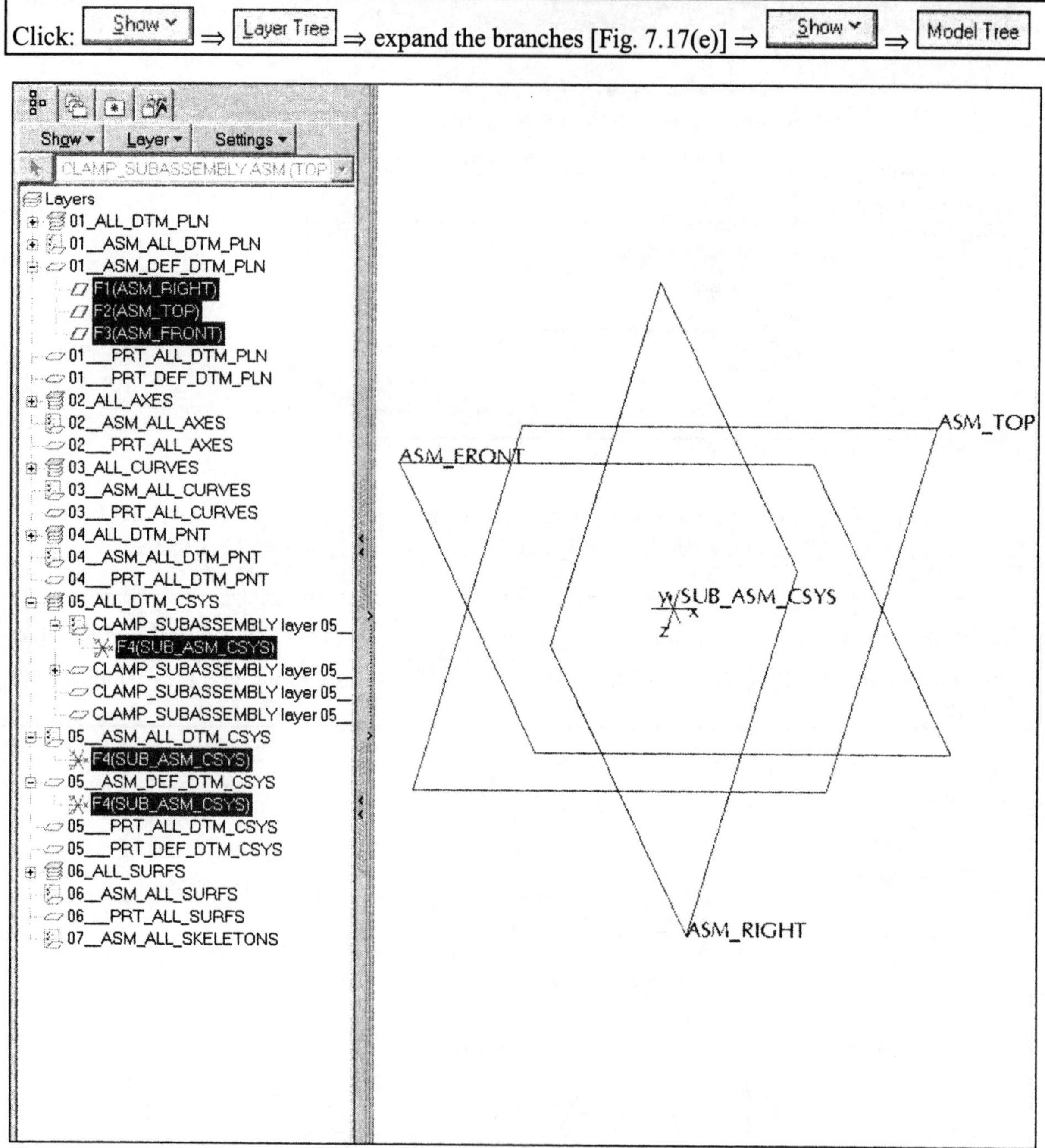

Figure 7.17(e) Default Template for Assembly Layering

The first component to be assembled to the subassembly is the Clamp_Arm. The simplest and quickest method of adding a component to an assembly is to match the coordinate systems. The first component assembled is usually where this *constraint* is used, because after the first component is established, few if any of the remaining components are assembled to the assembly coordinate system (with the exception of *top-down design*) or, for that matter, other parts' coordinate systems. Make sure all your models are in the same working directory before you start the assembly process.

If you have named your models to something that is not listed here, pick the appropriate part model as requested.

Click: **Add component to the assembly** ⇒ pick the **clamp_arm.prt** from the Open dialog box [the machined part (generic), not the casting (instance)] ⇒ **Preview** [Fig. 7.18(a)] ⇒ **Open** ⇒ **Show component in a separate window while specifying constraints** [Fig. 7.18(b)] *(you may need to resize and reposition your windows)*

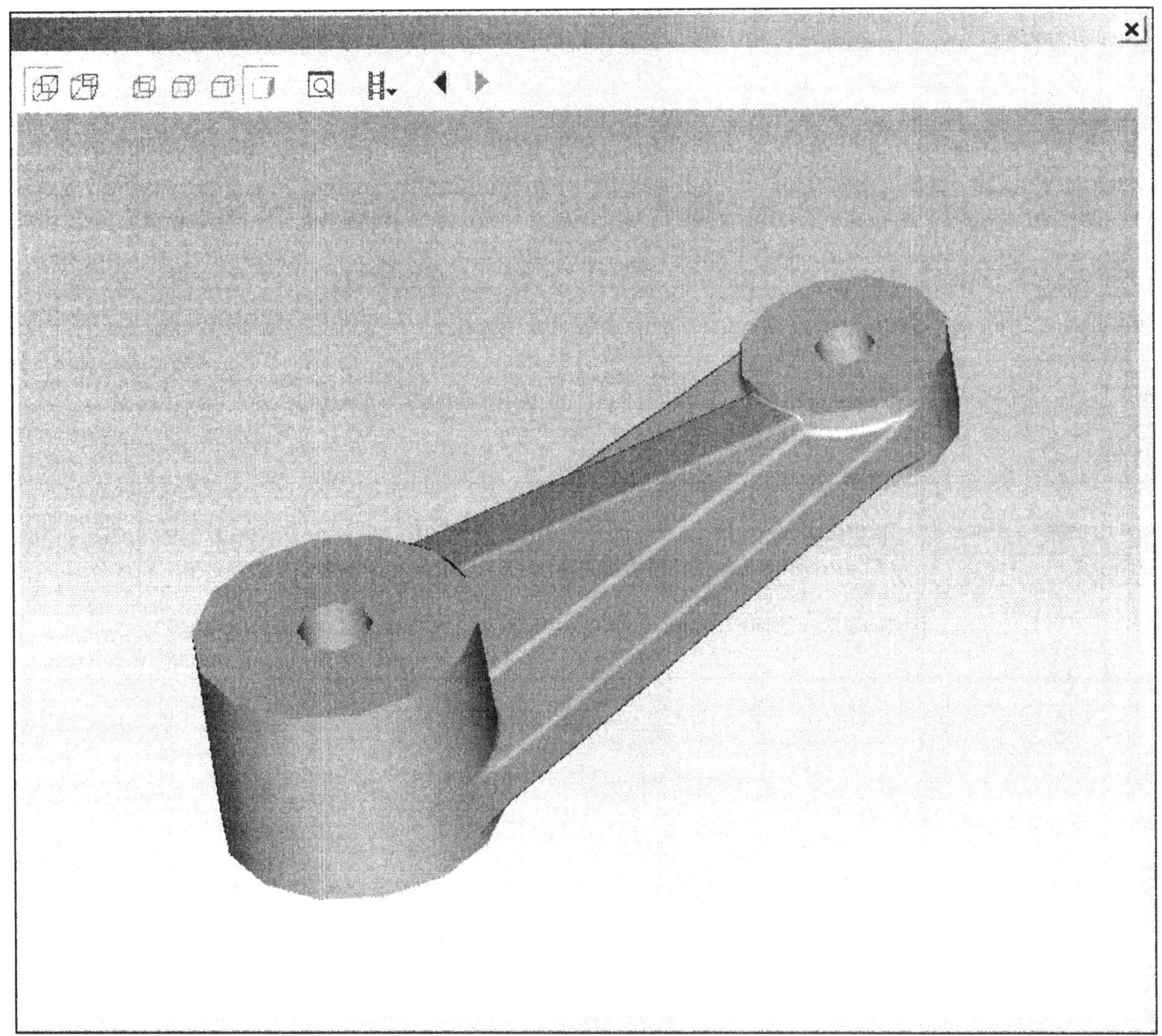

Figure 7.18(a) Previewed Clamp_Arm Component

Click: Constraints **Assemble component at default location** [Figs. 7.18(c-e)] ⇒ **OK** [Fig. 7.18(f)]

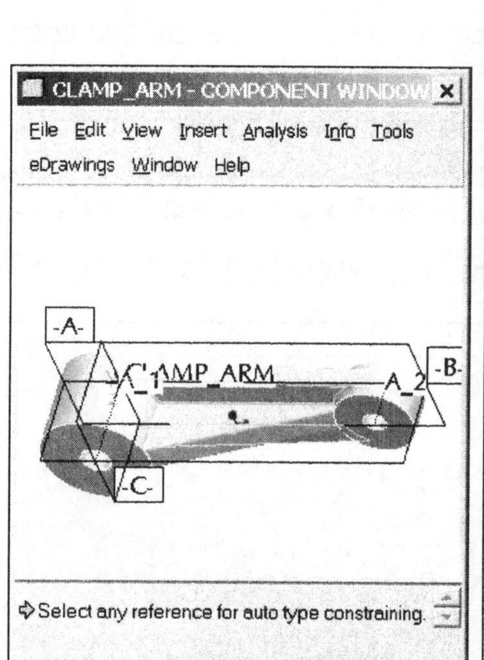

Figure 7.18(b) Component Window

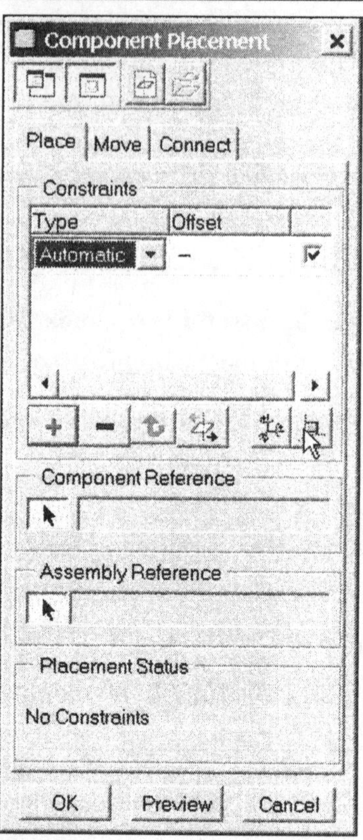

Figure 7.18(c) Automatic

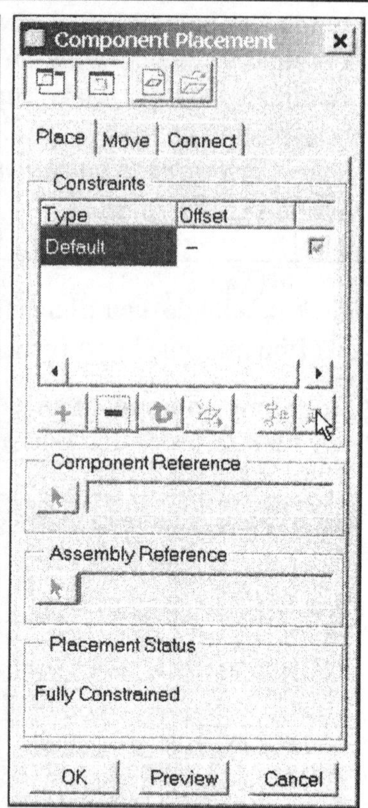

Figure 7.18(d) Default

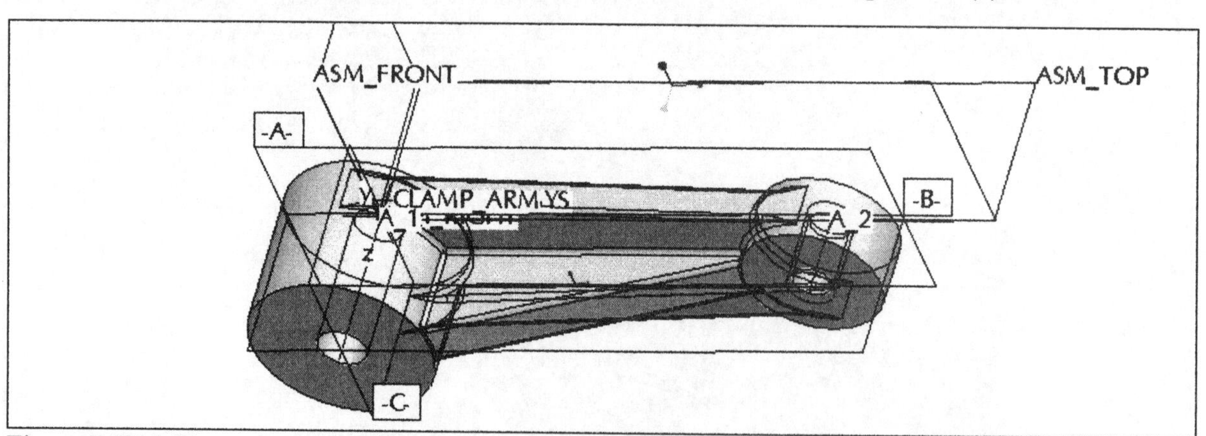

Figure 7.18(e) Clamp_Arm Assembled at Default Location

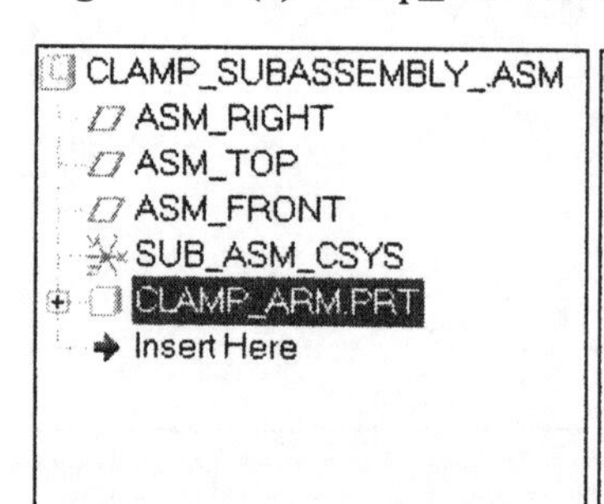

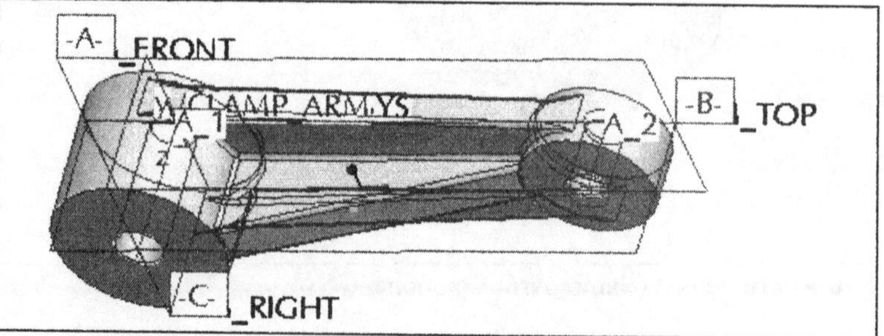

Figure 7.18(f) First Component Assembled

The Clamp_Arm has been assembled to the default location. This means to align the default Pro/E-created coordinate system of the component to the default Pro/E-created coordinate system of the assembly. Pro/E places the component at the assembly origin. By using the constraint, Coord Sys and selecting the assembly and then the component's coordinate systems would have accomplished the same thing, but with more picks. There will be situations where this constraint is useful, as when one or more of the coordinate systems used in the assembly is not the default coordinate system created when the default template is selected.

The next component to be assembled is the Clamp_Swivel. Two constraints will be used with this component: Insert and Mate (Offset). *Placement constraints* are used to specify the relative position of a *pair of surfaces/references* between two components. The Mate, Align, Insert, commands are placement constraints. The two surfaces/references must be of the same type. When using a datum plane as a placement constraint, specify Mate or Align. When using Mate (Offset) or Align (Offset), enter the *offset distance*. The *offset direction* is displayed with an arrow in the graphics window. If you need an offset in the opposite direction, enter a negative value.

Click: **Add component to the assembly** ⇒ pick the **clamp_swivel.prt** from the Open dialog box ⇒ **Preview** [Fig. 7.19(a)] ⇒ **Open** ⇒ *(clicking on the active radio button, toggles off the component window)* ⇒ **Tools** ⇒ **Environment** ⇒ Standard Orient Isometric ⇒ **Apply** ⇒ **OK**

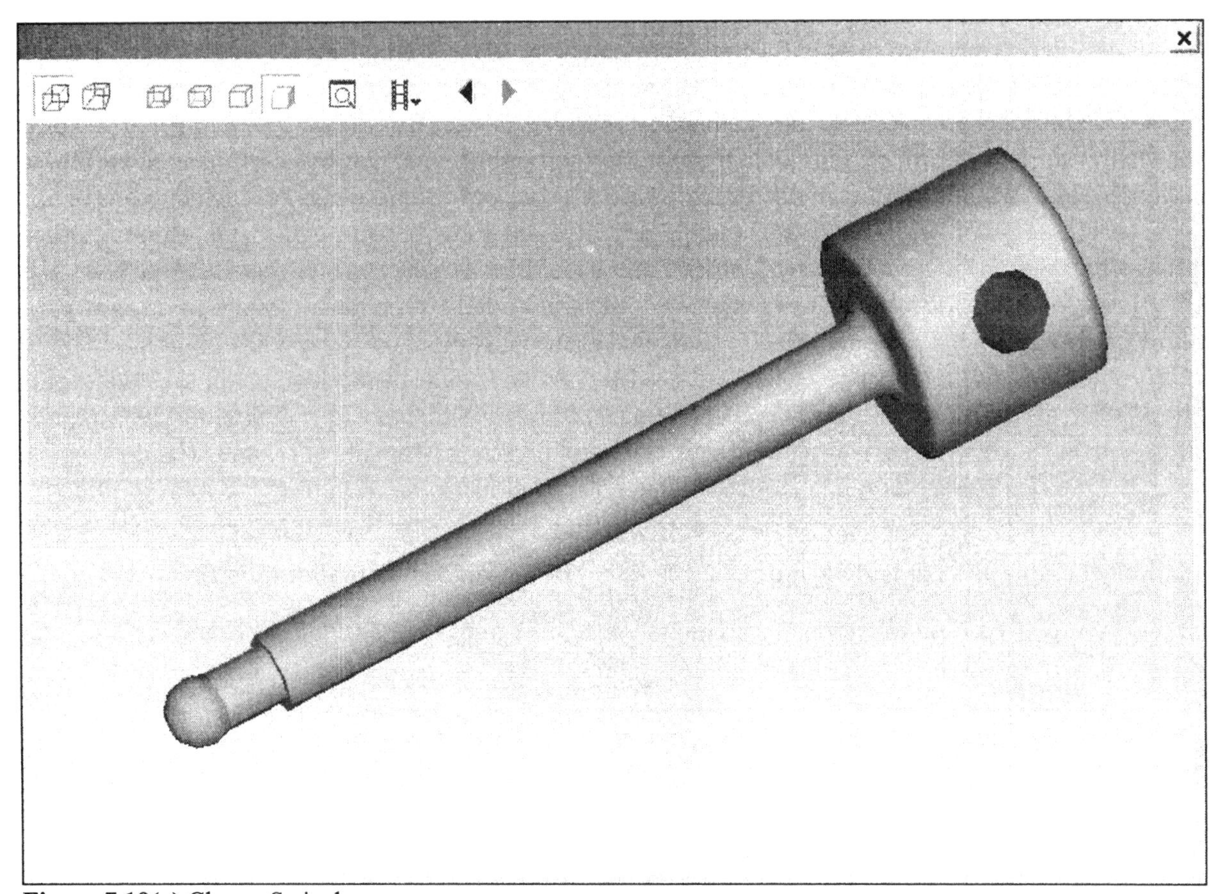

Figure 7.19(a) Clamp_Swivel

Click: (**Automatic**) ⇒ pick the cylindrical surface of the Clamp_Swivel [Fig. 7.19(b)] ⇒ pick the hole surface of the Clamp_Arm [Fig. 7.19(c)] *constraint type becomes **Insert*** [Fig. 7.19(d)]

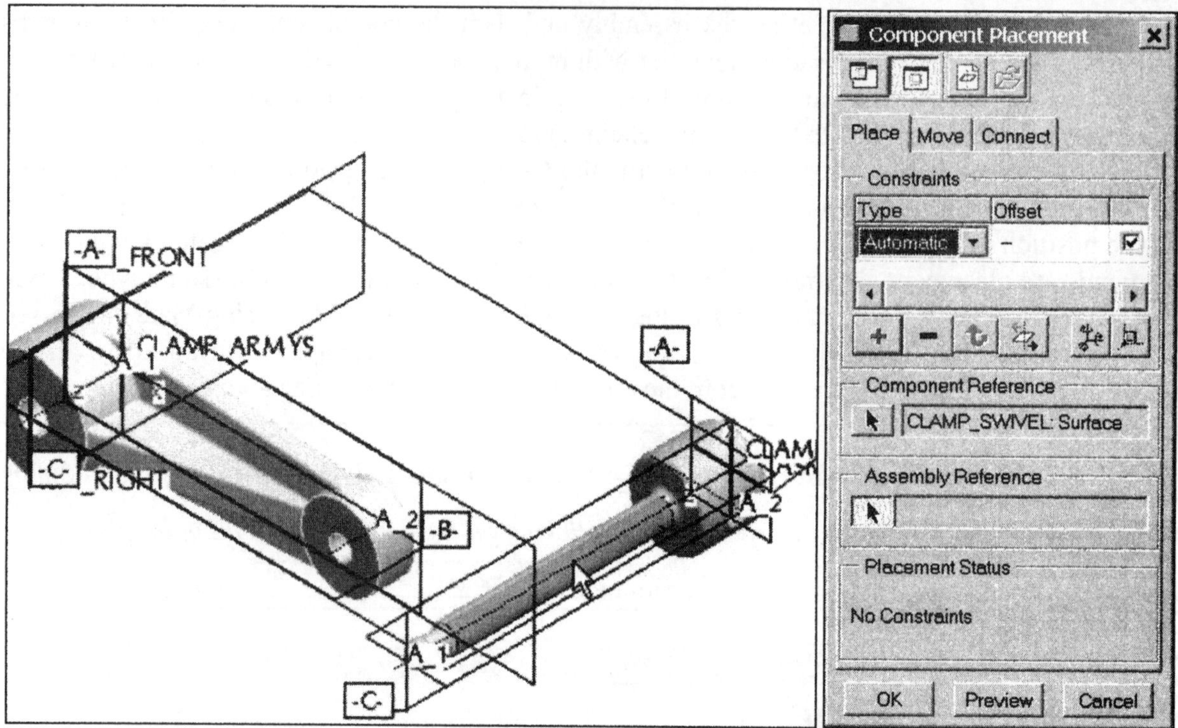

Figure 7.19(b) Pick on the Clamp_Swivel Surface

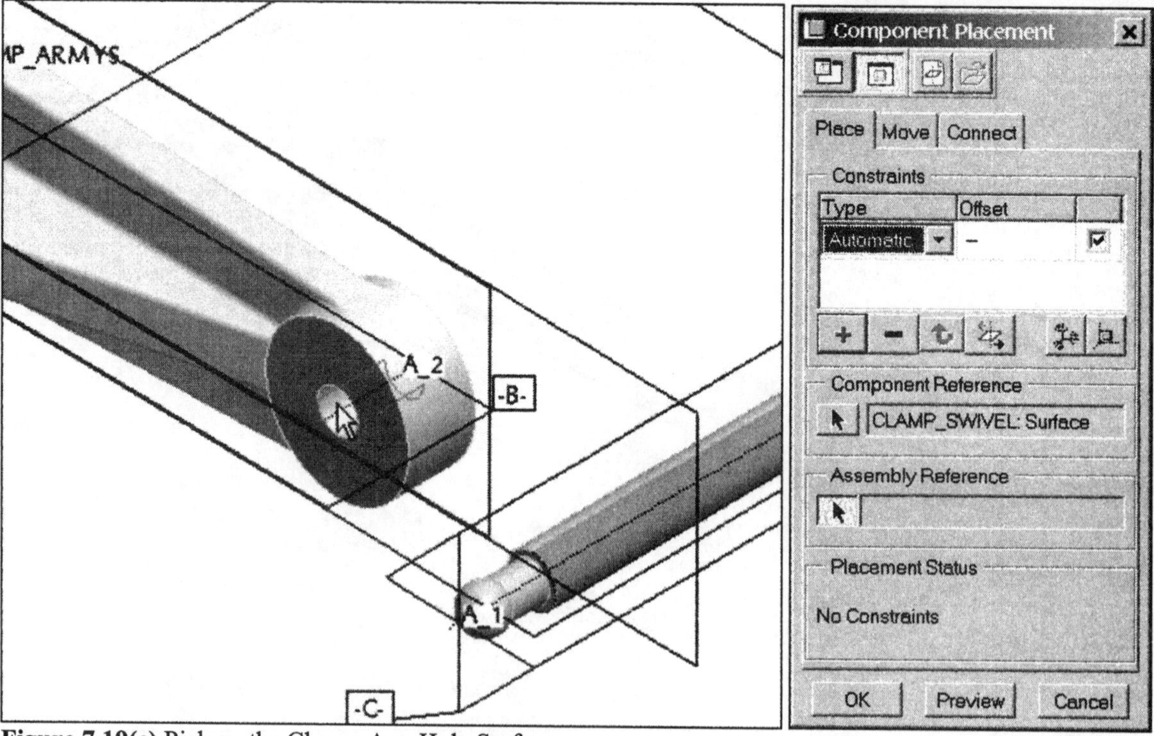

Figure 7.19(c) Pick on the Clamp_Arm Hole Surface

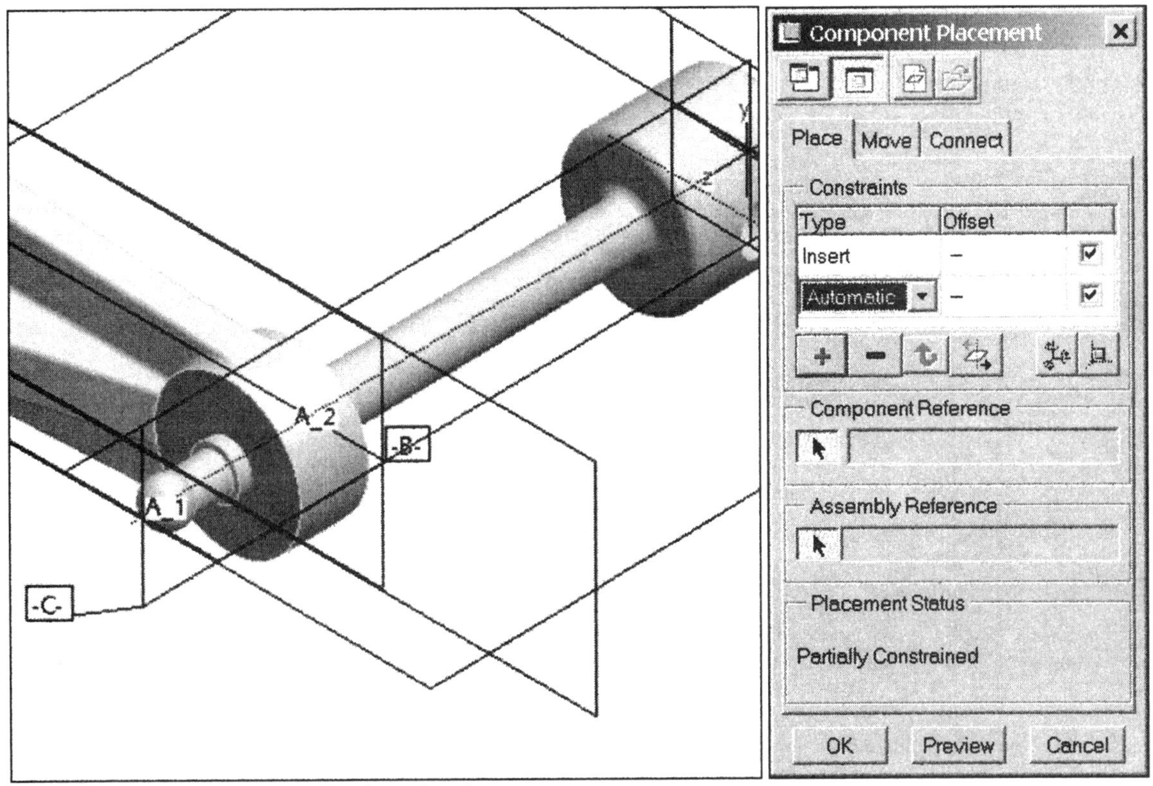

Figure 7.19(d) Insert Constraint Completed

Constraints [Automatic] pick the underside surface of the Clamp_Swivel [Fig. 7.19(e)] ⇒ pick the top surface of the Clamp_Arm [Fig. 7.19(f)] ⇒ Offset (ins) in indicated direction: type **1.50** ⇒ [✓] ⇒ [Align] ⇒ [▾] ⇒ [Mate] [Fig. 7.19(g)] *the Clamp_Swivel reverses* [Fig. 7.19(h)]

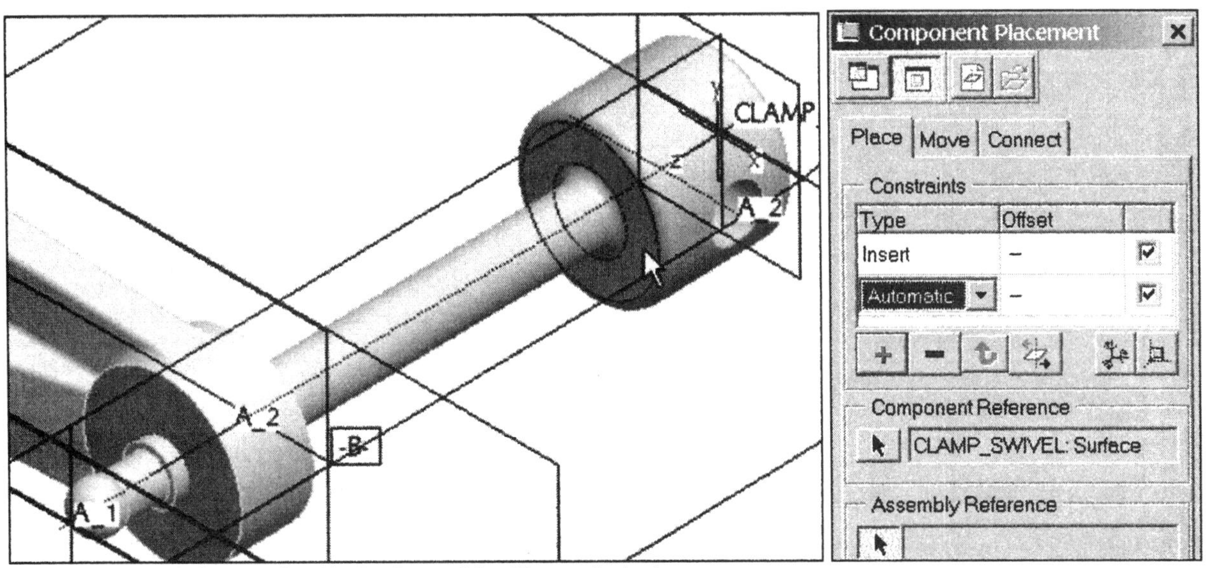

Figure 7.19(e) Select Underside Surface of the Clamp_Swivel

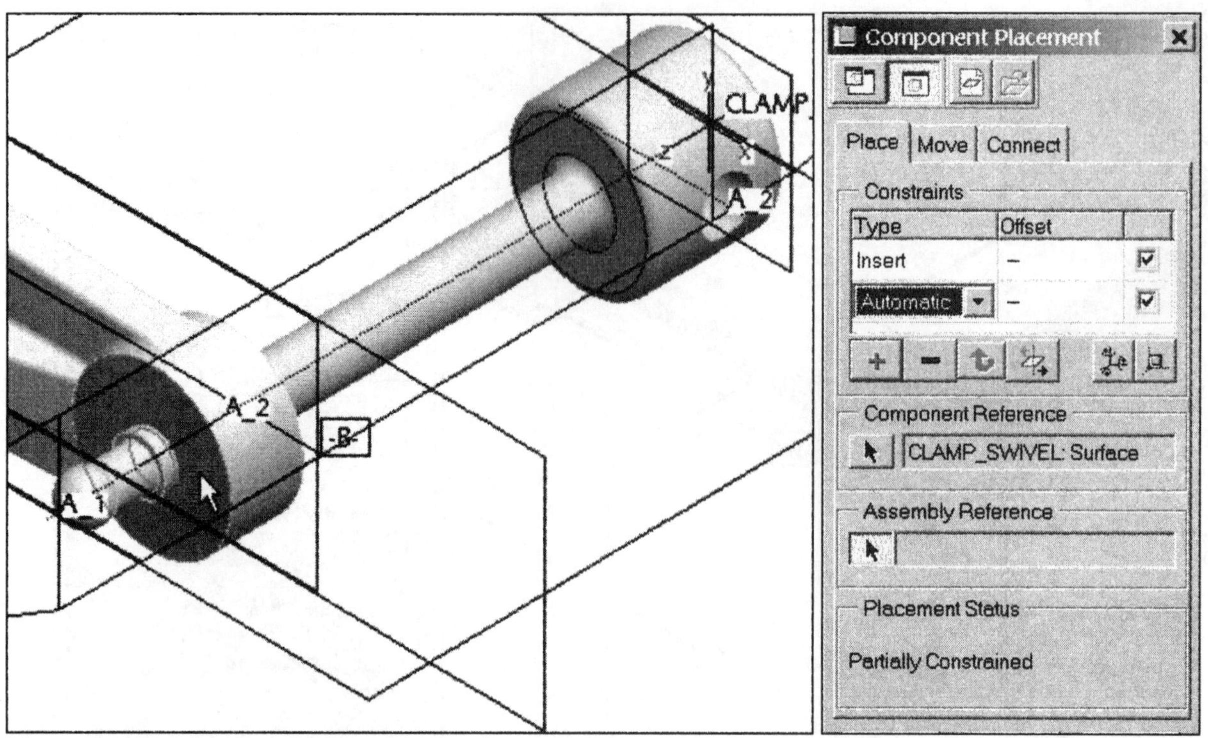

Figure 7.19(f) Select Top Surface of the Small Circular Protrusion

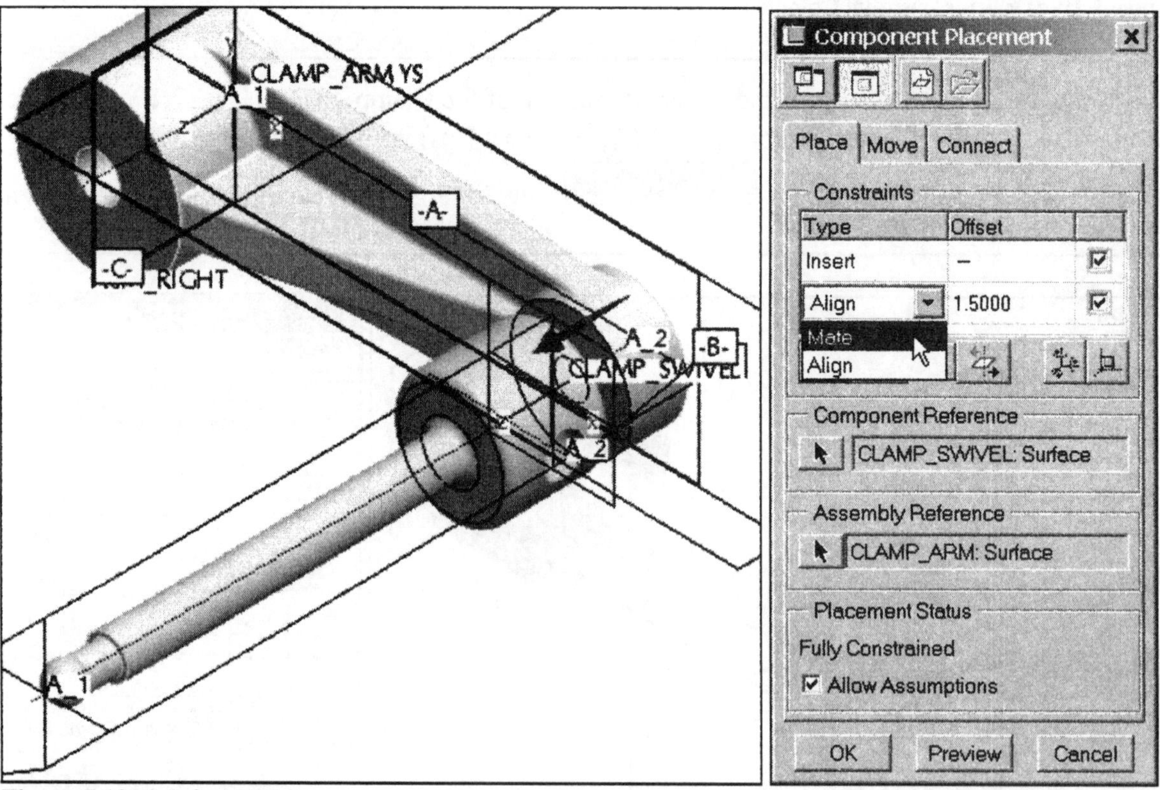

Figure 7.19(g) Select Mate

The Placement Status shows the component is Fully Constrained since Allow Assumptions is checked by default [Fig. 7.19(h)] ⇒ **OK** *the second component is now assembled* (Fig. 7.20)

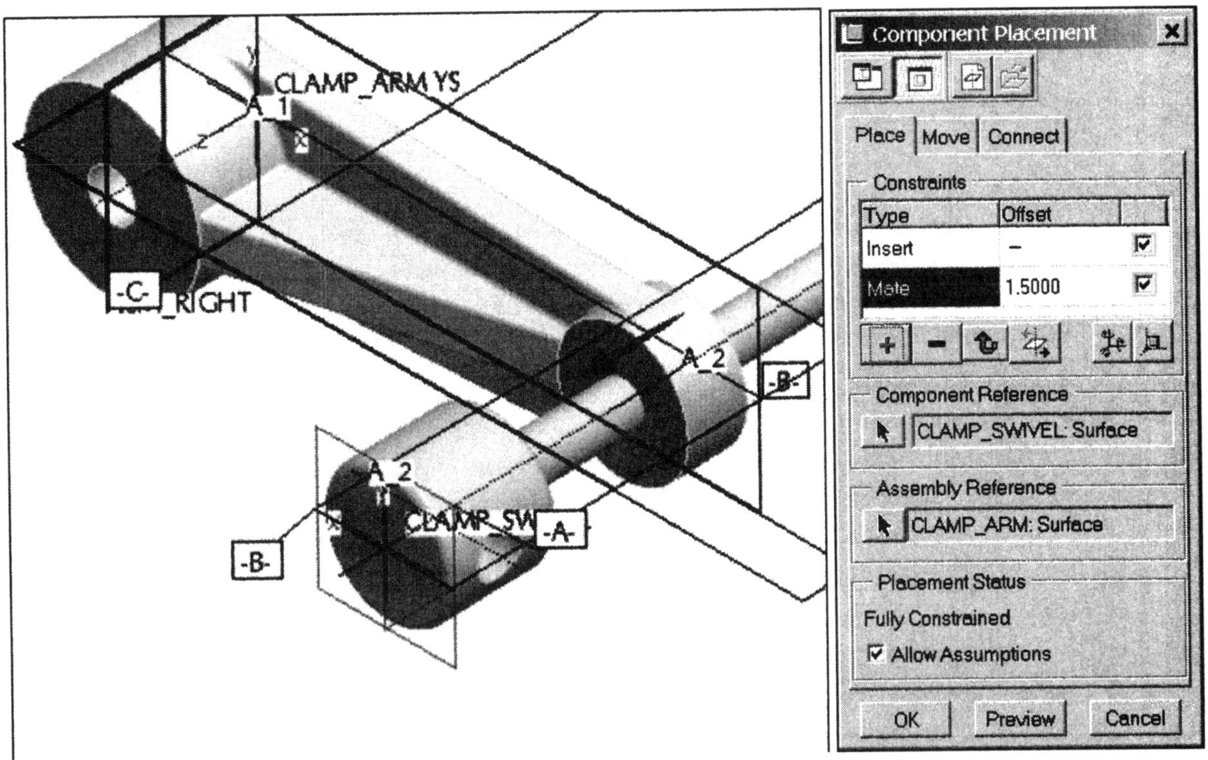

Figure 7.19(h) Showing **1.50** for the Mate Offset Distance

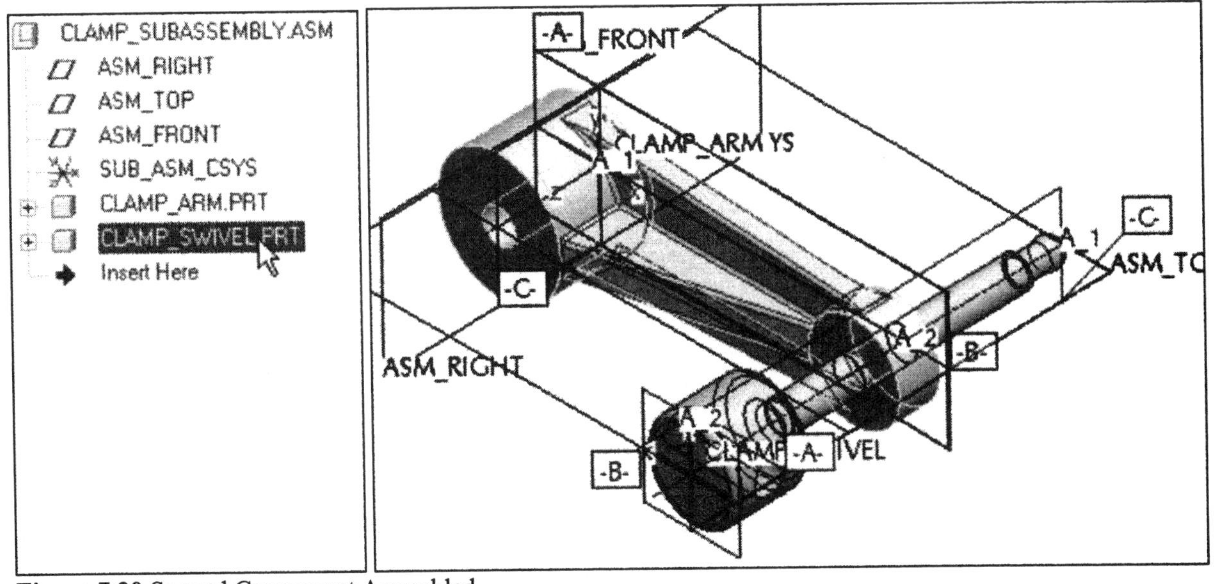

Figure 7.20 Second Component Assembled

Regenerating Models

You can use Regenerate to find bad geometry, broken parent-child relationships, or any other problem with a part feature or assembly component. In general, it is a good idea to regenerate the model every time you make a change, so that you can see the effects of each change in the graphics window as you build the model. By regenerating often, it helps you stay on course with your original design intent by helping you to resolve failures as they happen.

When Pro/E regenerates a model, it recreates the model feature by feature, in the order in which each feature was created, and according to the hierarchy of the parent-child relationship between features.

In an assembly, component features are regenerated in the order in which they were created, and then in the order in which each component was added to the assembly. Pro/E regenerates a model automatically in many cases, including when you open, save, or close a part or assembly or one of its instances, or when you open an instance from within a Family Table. You can also use the Regenerate command to manually regenerate the model.

The Regenerate command, located on the Edit menu or using the icon **Regenerates Model**, lets you recalculate the model geometry, incorporating any changes made since the last time the model was saved. If no changes have been made, Pro/E informs you that the model has not changed since the last regeneration.

The Custom Regenerate command (Assembly Mode), located on the Edit menu or using the icon **Specify the list of modified features or components to regenerate**, opens the Regeneration Manager dialog box *if there are features or components that have been changed that require regeneration* (Fig. 7.21).

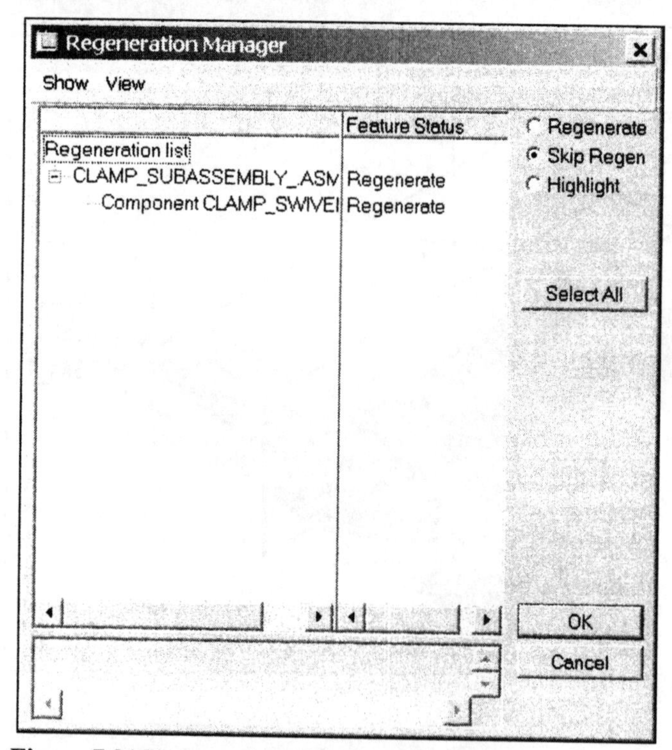

Figure 7.21 Regeneration Manager Dialog Box

Initially, Pro/E expands the tree to the first level in the Regeneration list column. To display features, choose Show from the menu bar. To expand or collapse the tree to any level, choose View from the menu bar. Information appears below the Regeneration list.

In the dialog box do one of the following:

- To select all features/components for regeneration, select Regenerate and then Select All.
- To omit all features/components from regeneration, select Skip Regen and then Select All.
- To determine the reason an object requires regeneration, select Highlight and select an entry in the Regeneration list.

A column next to the Regeneration list indicates which entries you have selected to skip regeneration or to be regenerated.

Click: Specify the list of modified features or components to regenerate ⇒ Regen Manager Info dialog box opens (Fig. 7.22) ⇒ click **OK** *(since there are no objects to be regenerated at this time)*

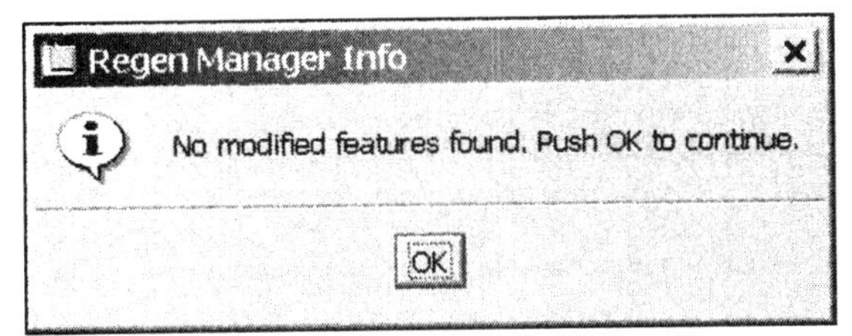

Figure 7.22 Regen Manager Info Dialog Box

The next component to be assembled is the Clamp_Foot.

Click: **Add component to the assembly** ⇒ pick the **clamp_foot.prt** from the Open dialog box ⇒ **Preview** [Fig. 7.23(a)] ⇒ **Open** ⇒ **Show component in a separate window while specifying constraints** ⇒ spin and zoom in on your model and resize your windows as desired ⇒ Constraints Automatic pick the internal cylindrical surface of the Clamp_Foot [Fig. 7.23(b)] ⇒ pick the external cylindrical surface of the Clamp_Swivel [Fig. 7.23(c)] *constraint becomes **Insert*** ⇒ Constraints Automatic pick the spherical hole of the Clamp_Foot [Fig. 7.23(d)] ⇒ pick the spherical end of the Clamp_Swivel [Fig. 7.23(e)] *constraint becomes **Mate*** ⇒ **Preview** ⇒ **View** ⇒ **Shade** ⇒ **View** ⇒ **Repaint** [Fig. 7.23(f)] ⇒ **OK** ⇒ ⇒ **MMB**

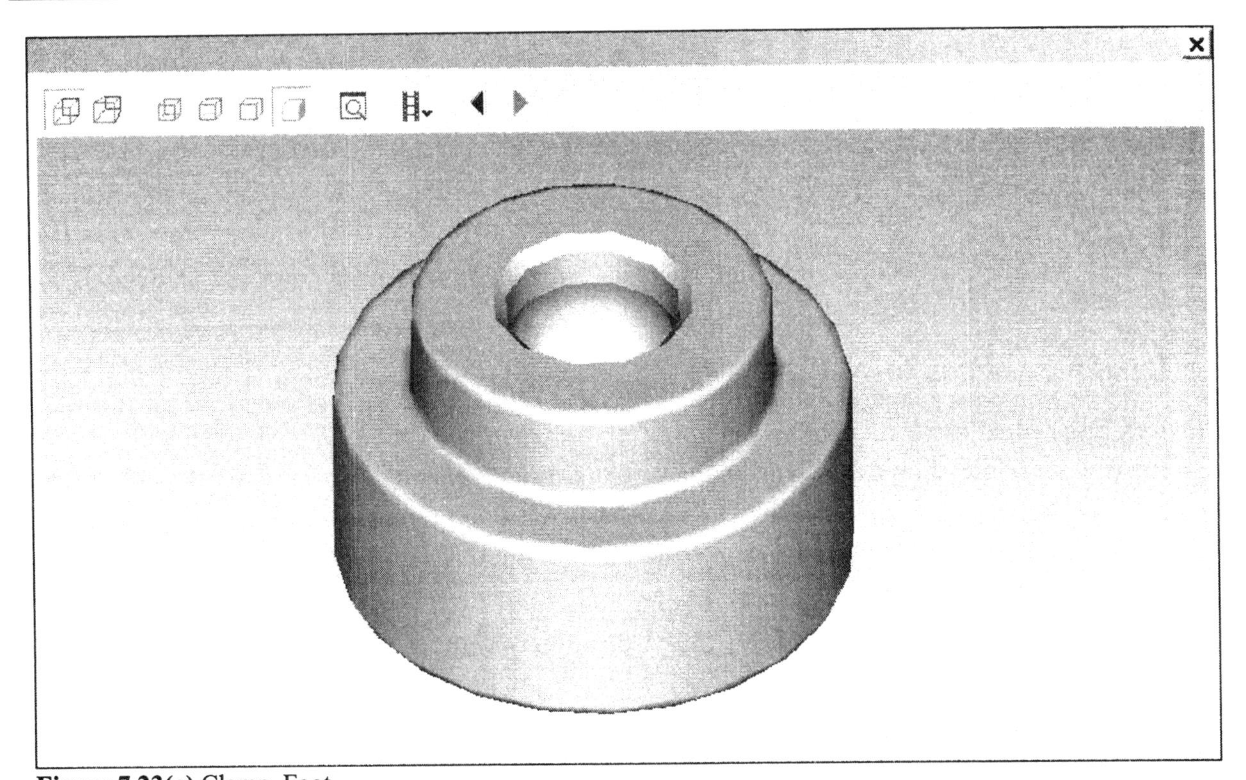

Figure 7.23(a) Clamp_Foot

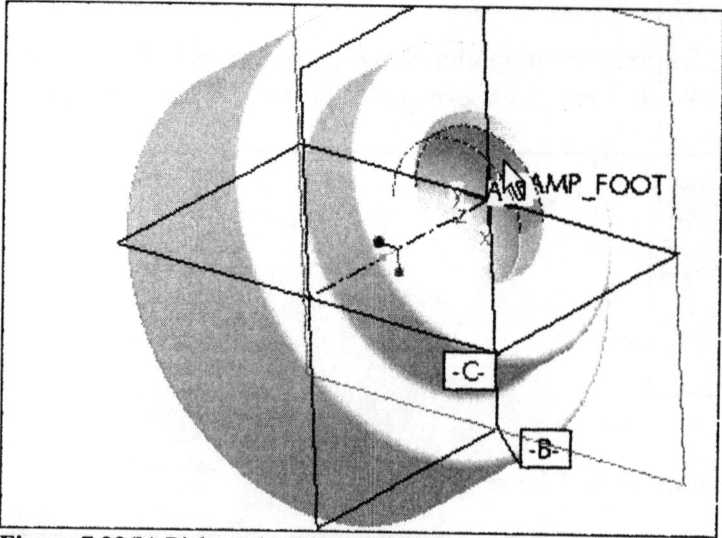

Figure 7.23(b) Pick on the Internal Cylindrical Surface

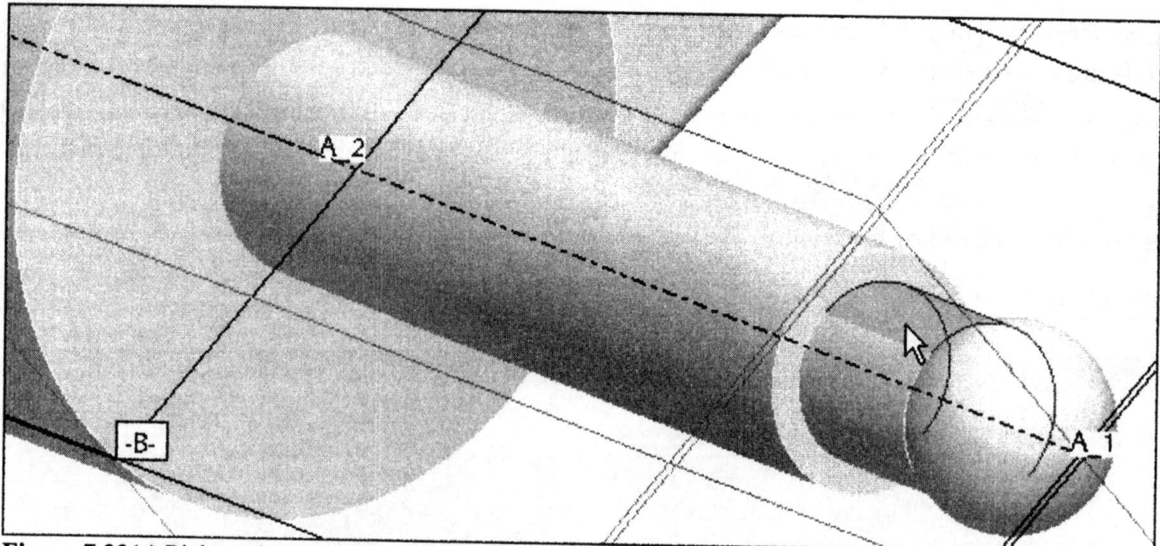

Figure 7.23(c) Pick on the External Cylindrical Surface

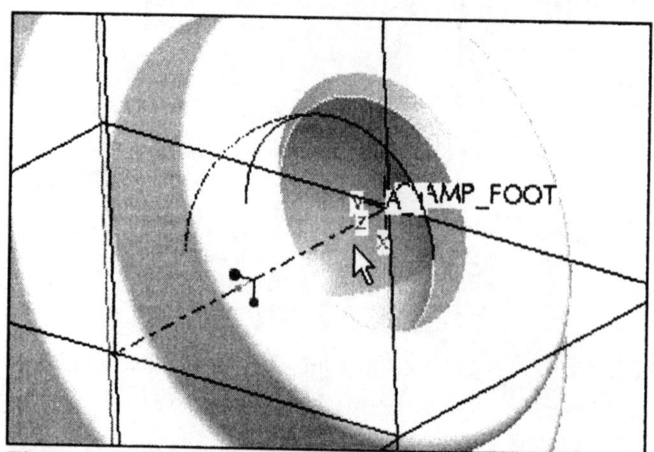

Figure 7.23(d) Pick on the Internal Spherical Hole Surface

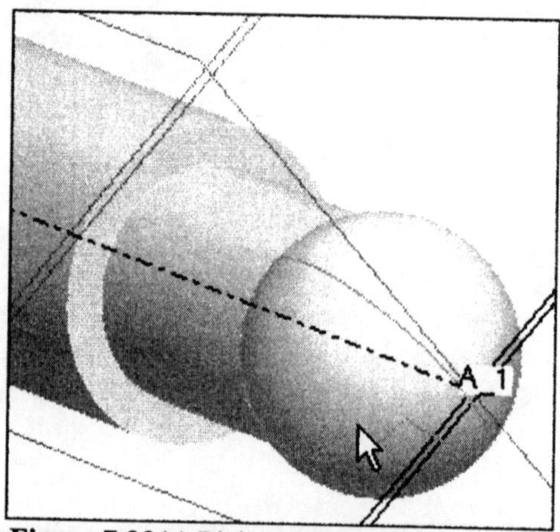

Figure 7.23(e) Pick on the External Spherical Surface

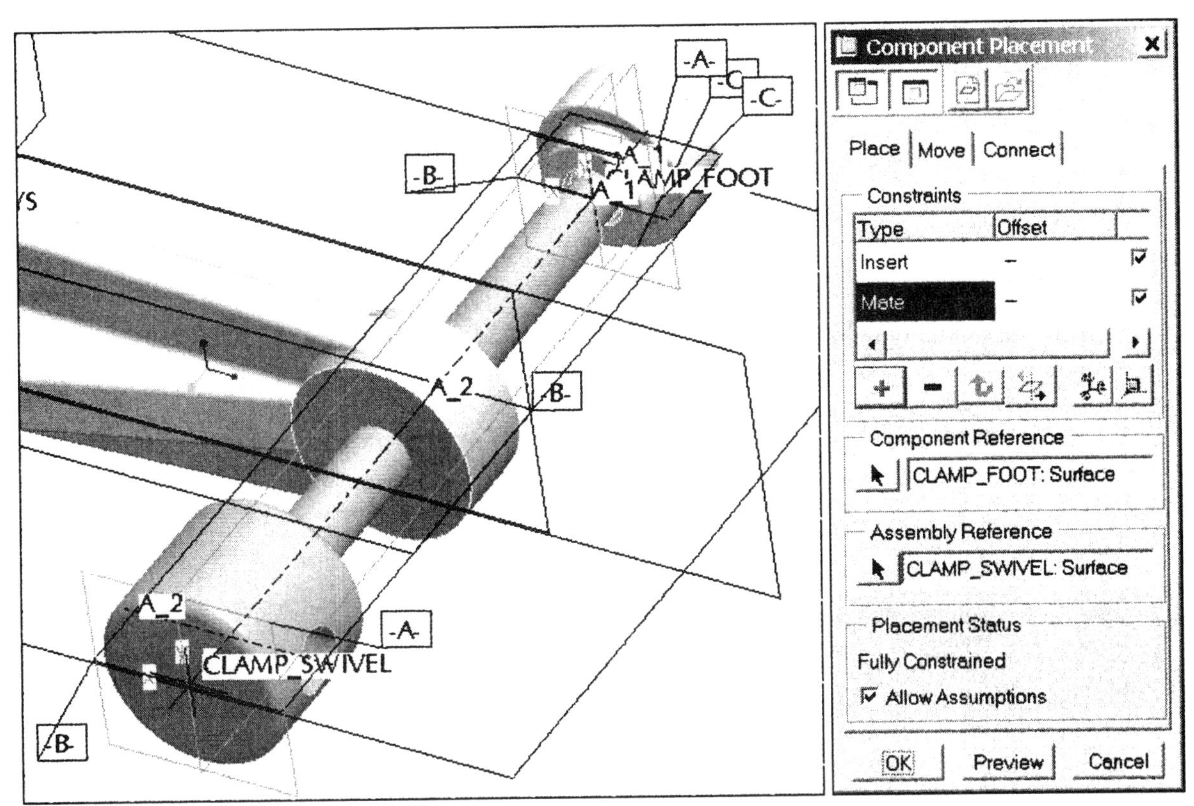

Figure 7.23(f) Fully Constrained Clamp_Foot

Assemble the **5.00** double-ended stud. Click: **Add component to the assembly** ⇒ pick **clamp_stud5.prt** ⇒ **Preview** [Fig. 7.24(a)]

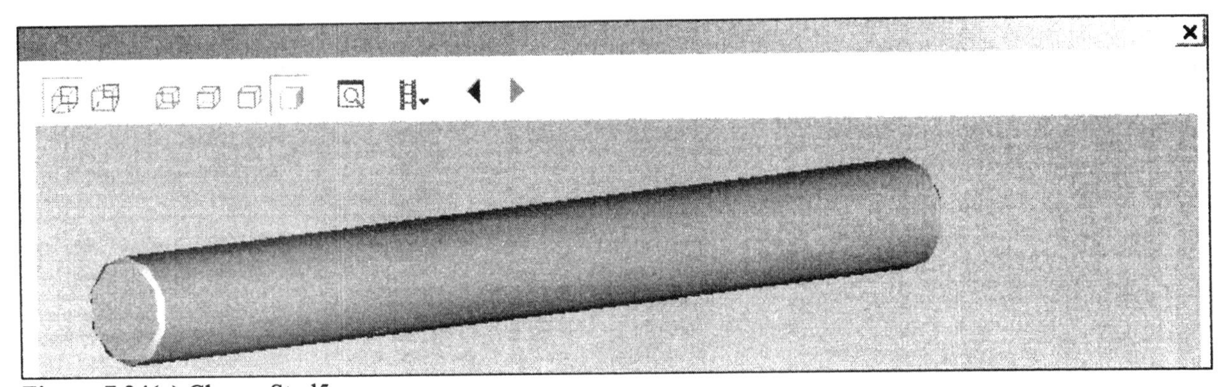

Figure 7.24(a) Clamp_Stud5

Click: **Open** ⇒ clicking on the active button, toggles off the component window ⇒ Constraints Automatic pick the external cylindrical surface of the Clamp_Stud5 [Fig. 7.24(b)] ⇒ pick the internal cylindrical surface of the hole of the Clamp_Swivel [Fig. 7.24(b)] *constraint becomes **Insert*** ⇒ Constraints Automatic pick the end surface of the Clamp_Stud5 [Fig. 7.24(c)] ⇒ to pick the appropriate datum of the Clamp_Swivel, place the cursor on/near the appropriate datum reference [Fig. 7.24(d)] ⇒ **RMB** ⇒ **Pick From List** ⇒ pick: C.F2(DATUM PLANE):CLAMP_SWIVEL [Fig. 7.24(e)] ⇒ **OK**

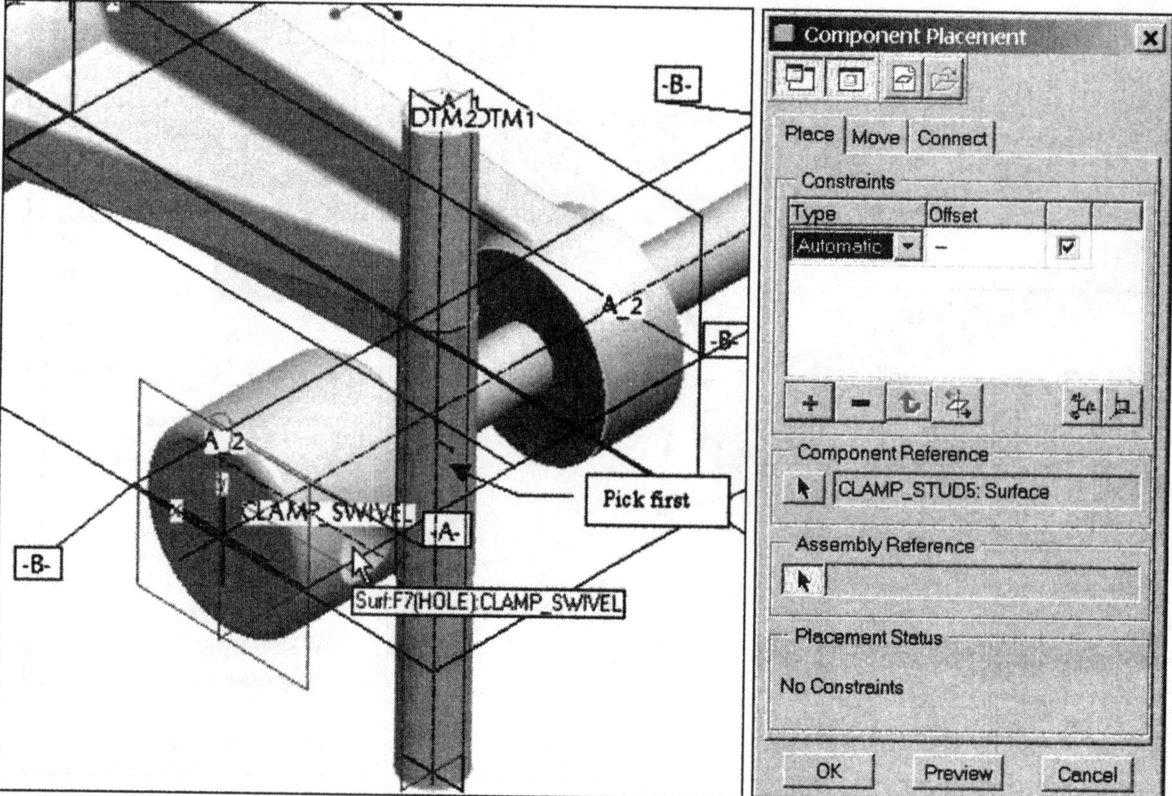

Figure 7.24(b) Select the Two References for the First Constraint

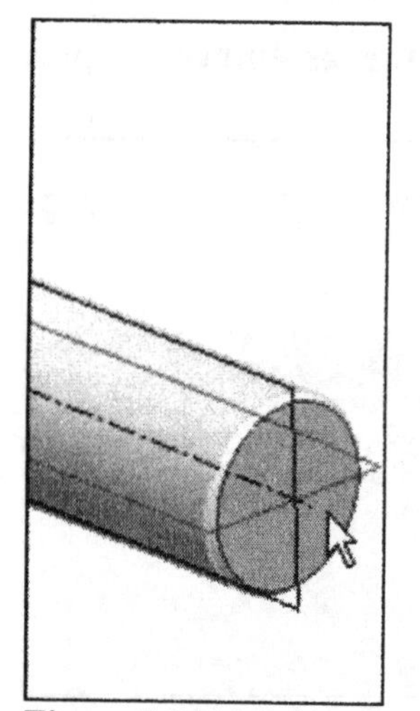

Figure 7.24(c) Select the End Surface

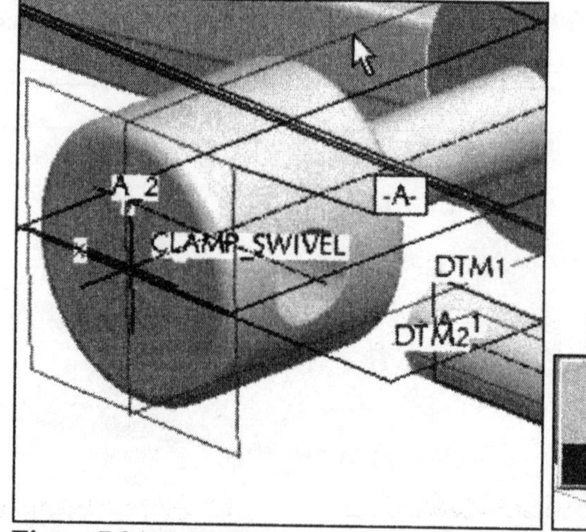

Figure 7.24(d) RMB on Datum C of the Clamp_Swivel

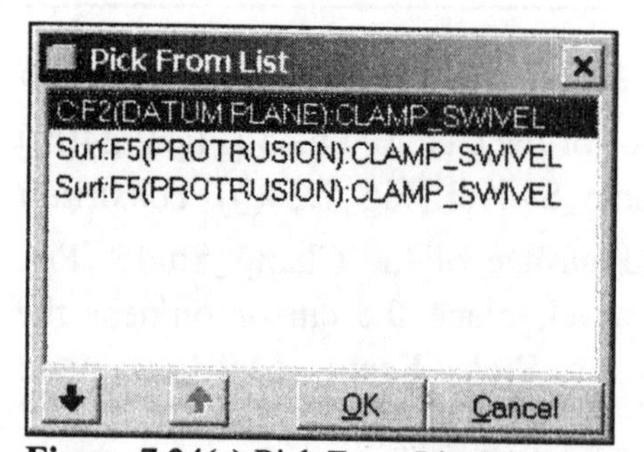

Figure 7.24(e) Pick From List Dialog Box

Offset (ins) in indicated direction: type **–2.50** ⇒ **Enter** [Mate -2.5000] *constraint becomes **Mate*** [Fig. 7.24(f)] ⇒ **OK** ⇒ [save icon] ⇒ **MMB** ⇒ click: **Settings** from the Model Tree ⇒ **Tree Columns** ⇒ use [>>] to add the columns names to be Displayed as [Fig. 7.25(a)] ⇒ **Apply** ⇒ **OK** ⇒ resize the Model Tree column widths [Fig. 7.25(b)]

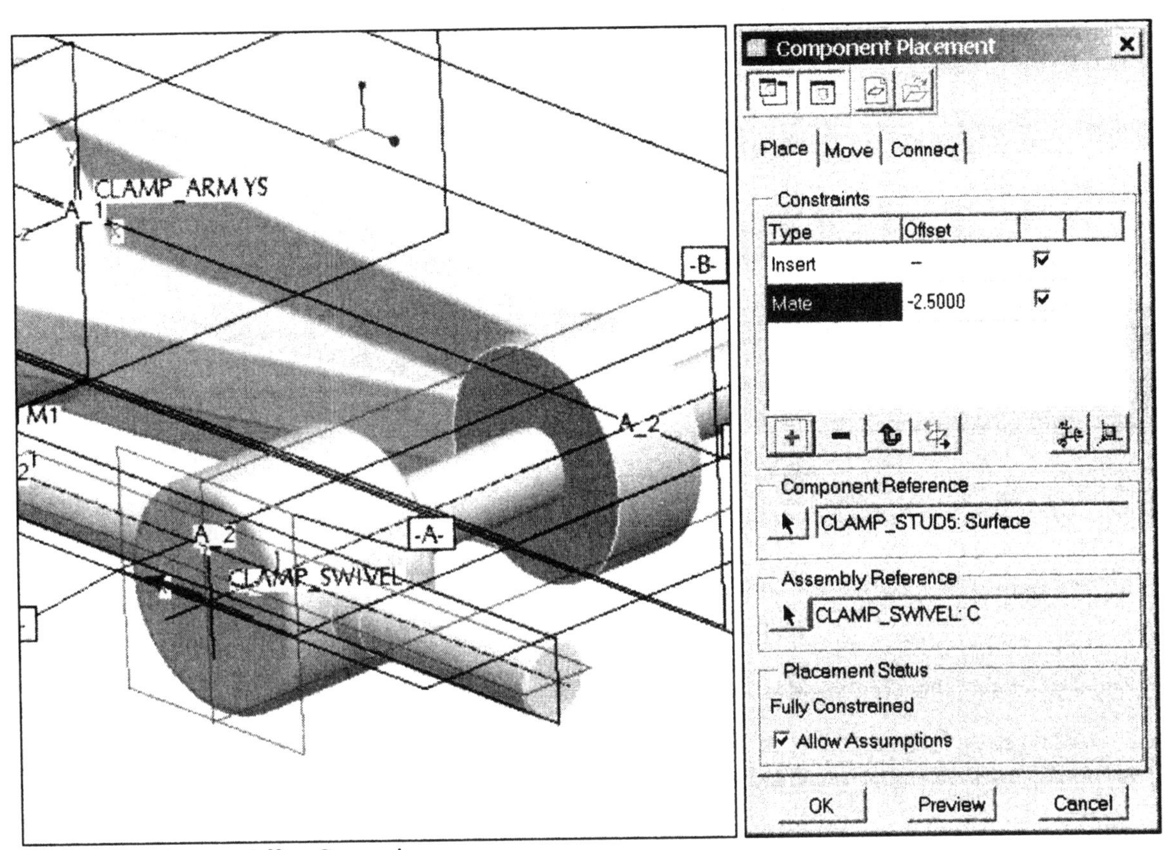

Figure 7.24(f) Mate Offset Constraint

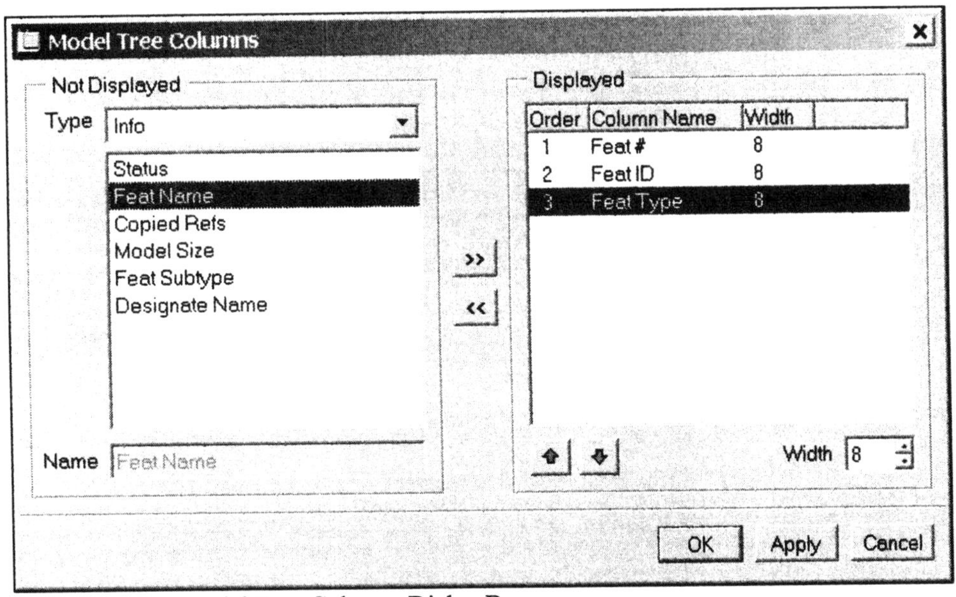

Figure 7.25(a) Model Tree Columns Dialog Box

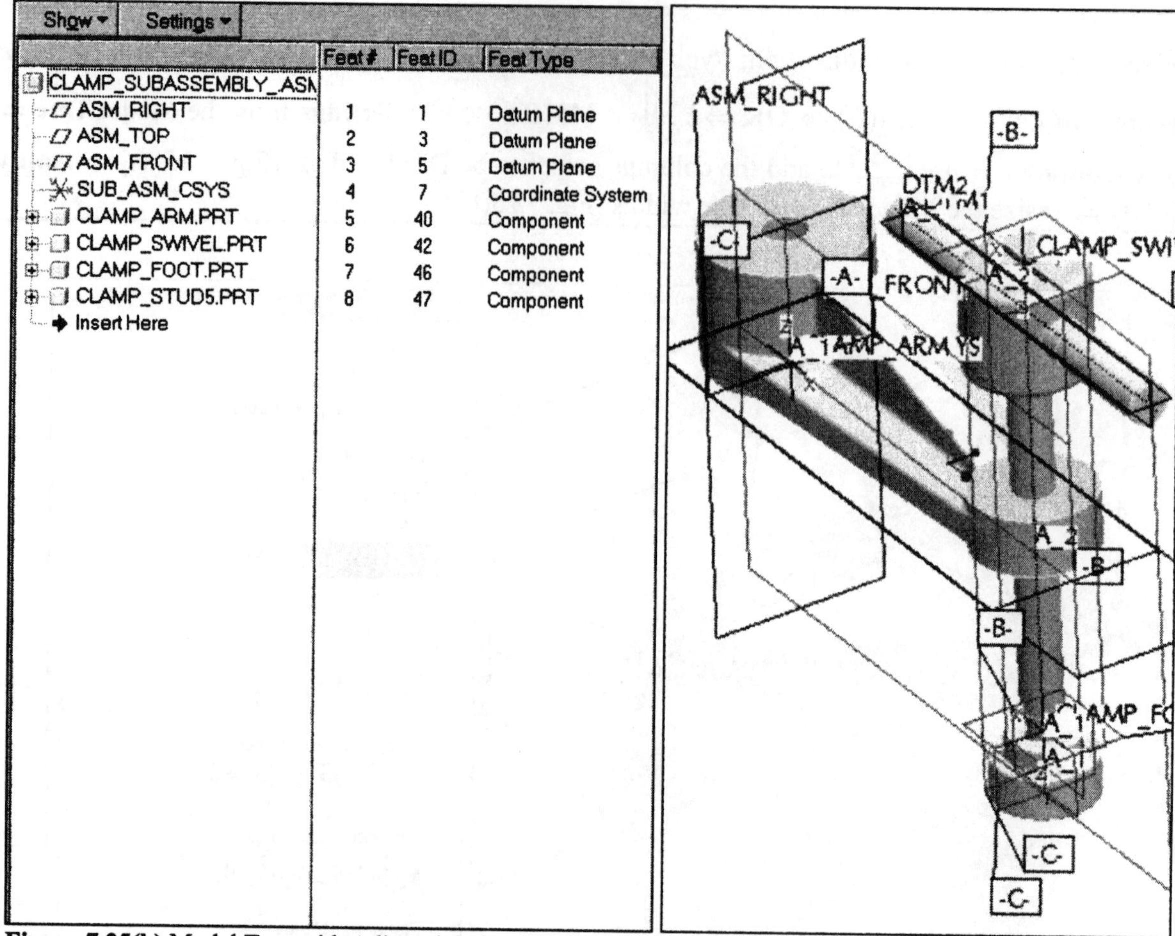

Figure 7.25(b) Model Tree with Adjusted Columns

The Clamp_Ball handles are the last components of the Clamp_Subassembly. Instructions are provided for assembling one Clamp_Ball; you must assemble the other on your own.

Click: **Add component to the assembly** ⇒ pick the **clamp_ball.prt** ⇒ **Preview** ⇒ ⇒ **MMB** spin the model to see the hole ⇒ **Open** [Fig. 7.26(a)]

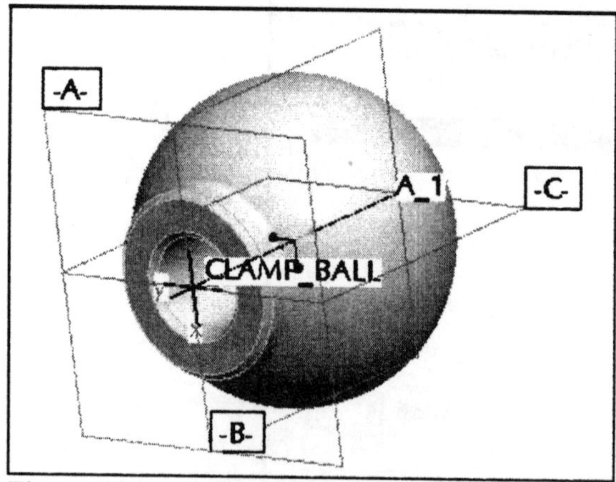

Figure 7.26(a) Clamp_Ball

Spin and resize the model as needed [Fig. 7.26(b)] ⇒ Constraints Automatic pick the internal cylindrical surface of the hole of the Clamp_Ball [Fig. 7.26(c)] ⇒ pick the external cylindrical surface of the Clamp_Stud5 [Fig. 7.26(c)] *constraint becomes* ***Insert***

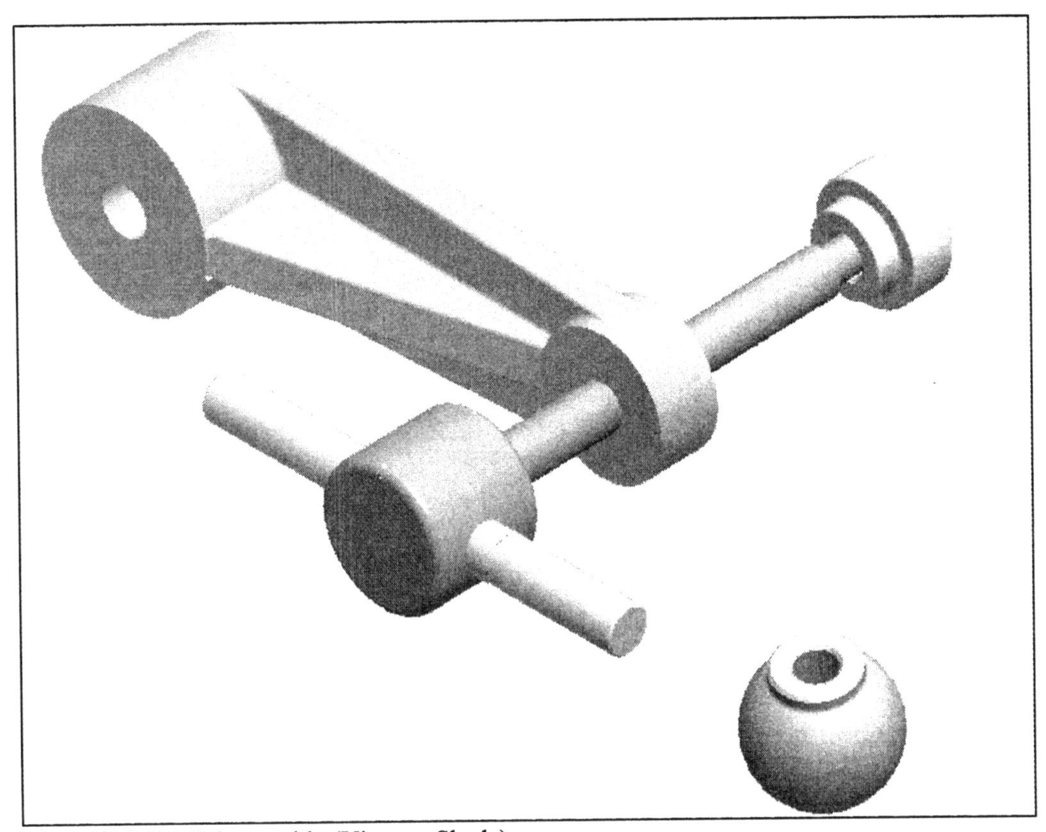

Figure 7.26(b) Subassembly (View ⇒ Shade)

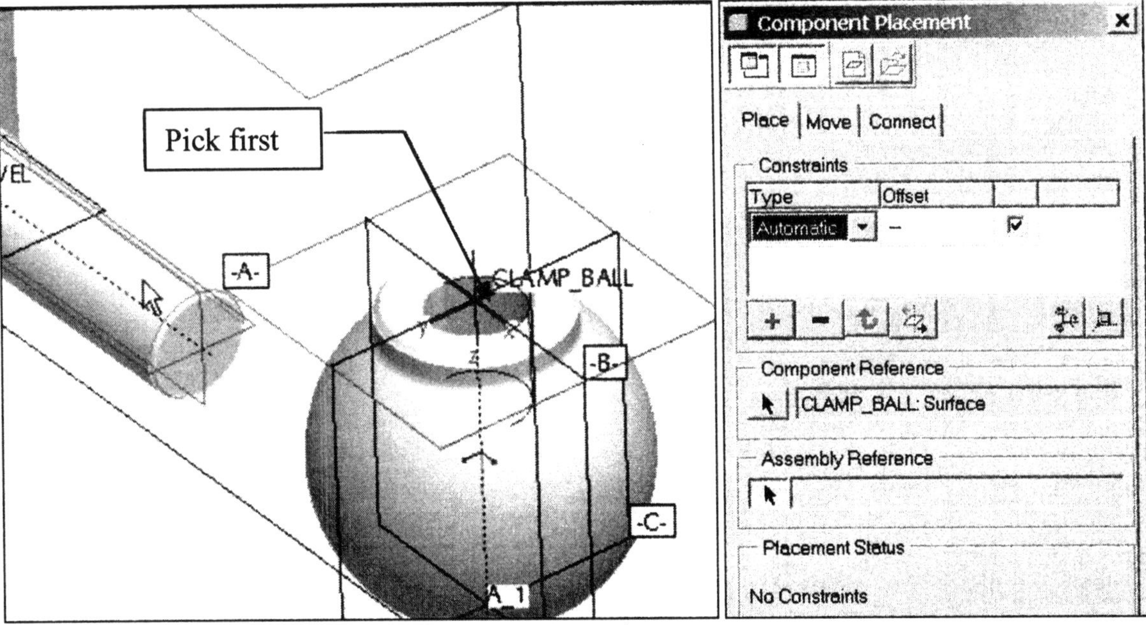

Figure 7.26(c) Select Two References for the First Constraint

Constraints [Automatic] pick the flat end of the Clamp_Ball [Fig. 7.26(d)] ⇒ pick the end surface of the Clamp_Stud5 [Figs. 7.26(d-e)] ⇒ type **-.500** for the Offset distance ⇒ **Enter** ⇒ in the Constraints box, click: [Align] ⇒ click: [▼] for drop down choices ⇒ click: **Mate** [Mate] [Figs. 7.26(f-g)] ⇒ **OK** ⇒ [Save] ⇒ **MMB**

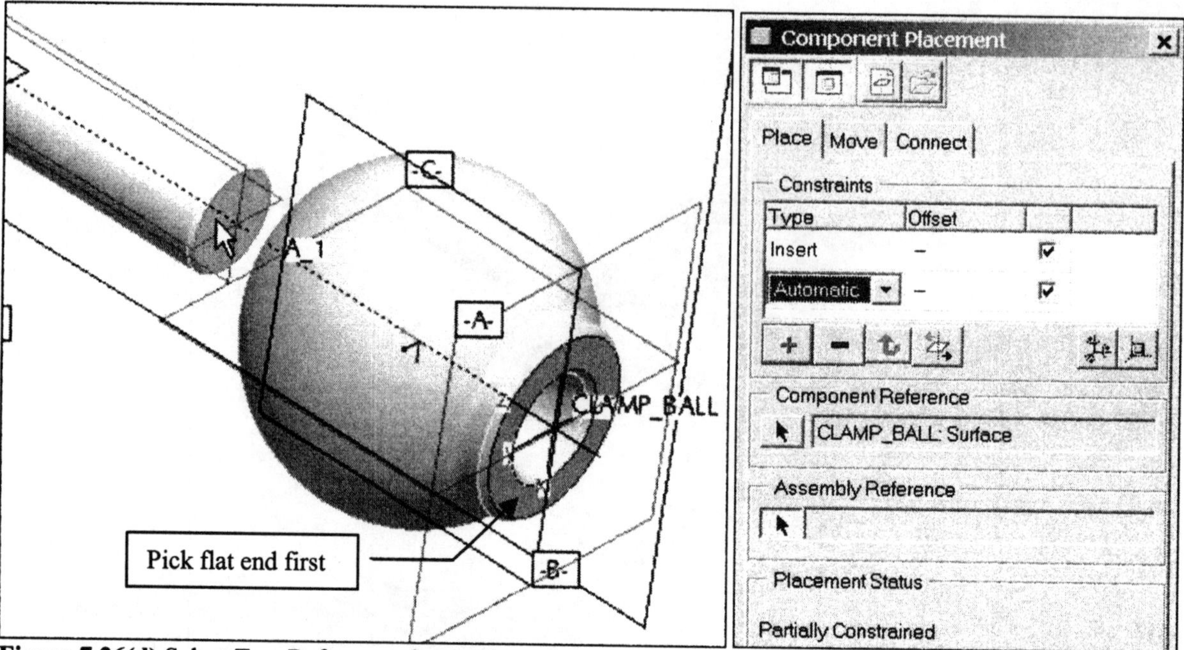

Figure 7.26(d) Select Two References for the Second Constraint

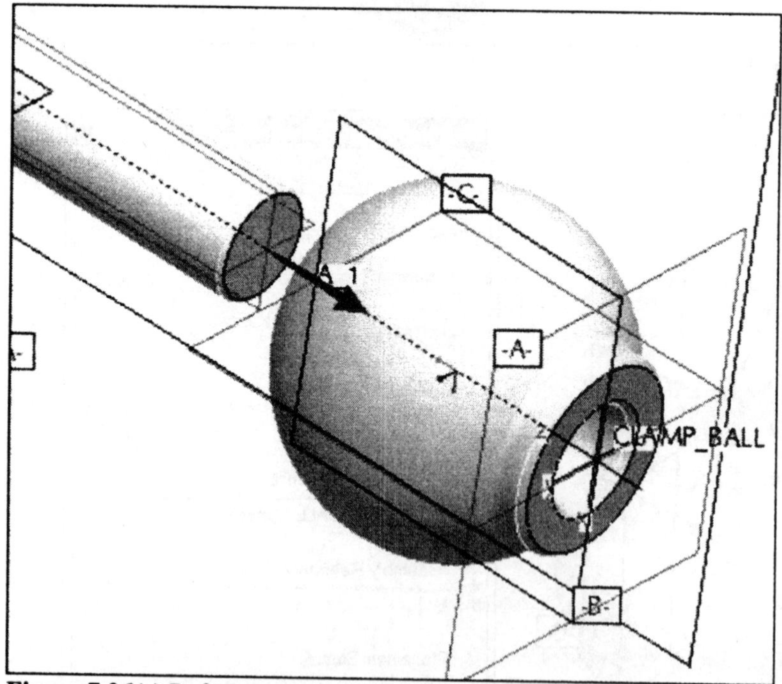

Figure 7.26(e) Reference Defaults to Align

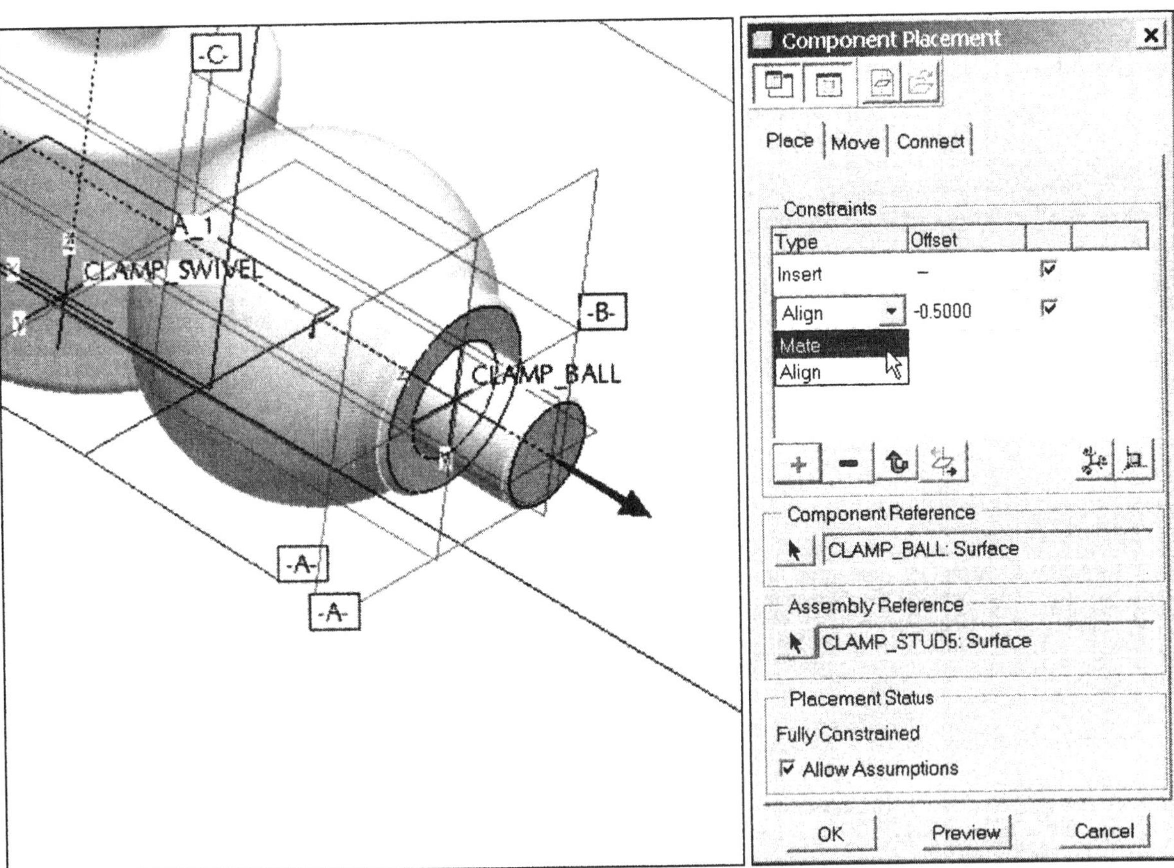

Figure 7.26(f) Change Offset to **-.500** and Select Mate as the Constraint Type

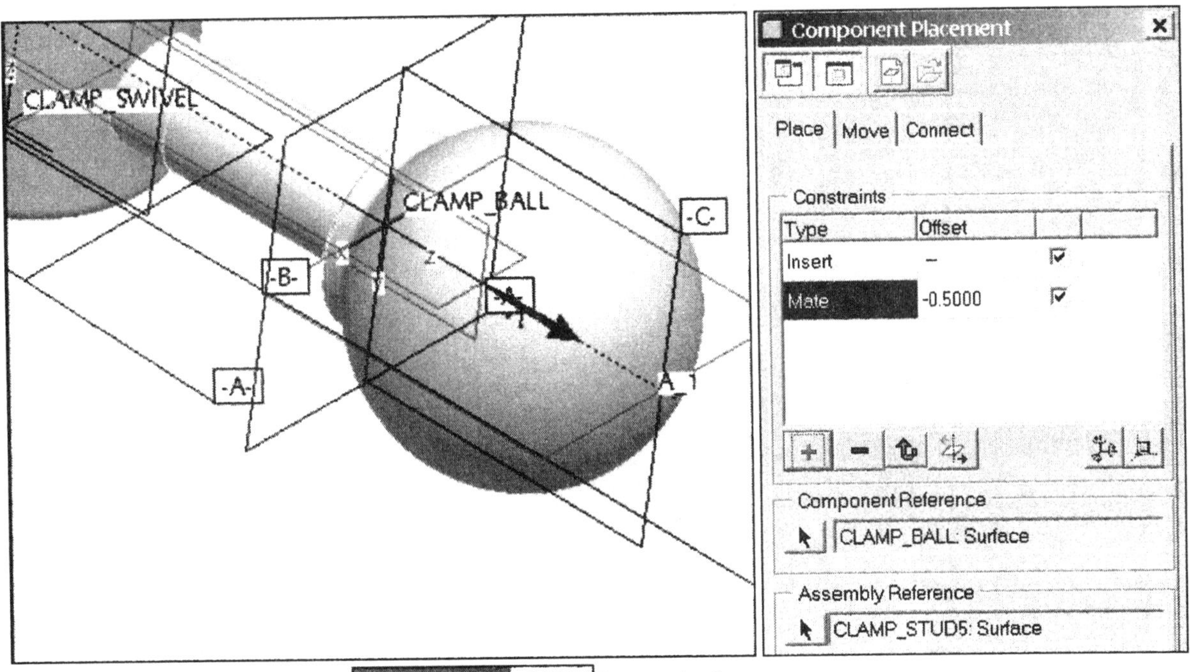

Figure 7.26(g) Mate Constraint Mate -0.5000 is now Active

Assemble the second Clamp_Ball using Copy and Paste ⇒ pick on the **Clamp_Ball** component to select it *(highlights in red)* [Fig. 7.26(h)] ⇒ **Ctrl+C** ⇒ **Ctrl+V** ⇒ Dim 1 [Fig. 7.26(i)] ⇒ **Done** ⇒ Enter Dim 1 .4375 to change the variable dimension ⇒ **Enter**

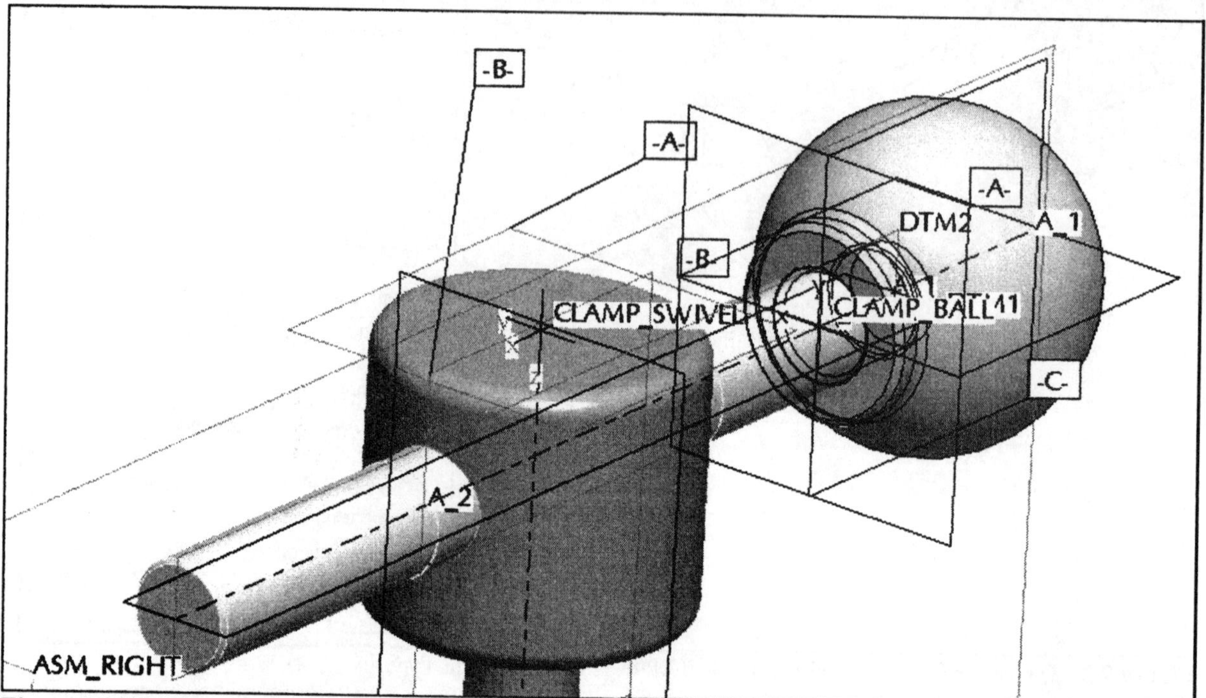

Figure 7.26(h) First Clamp_Ball Selected

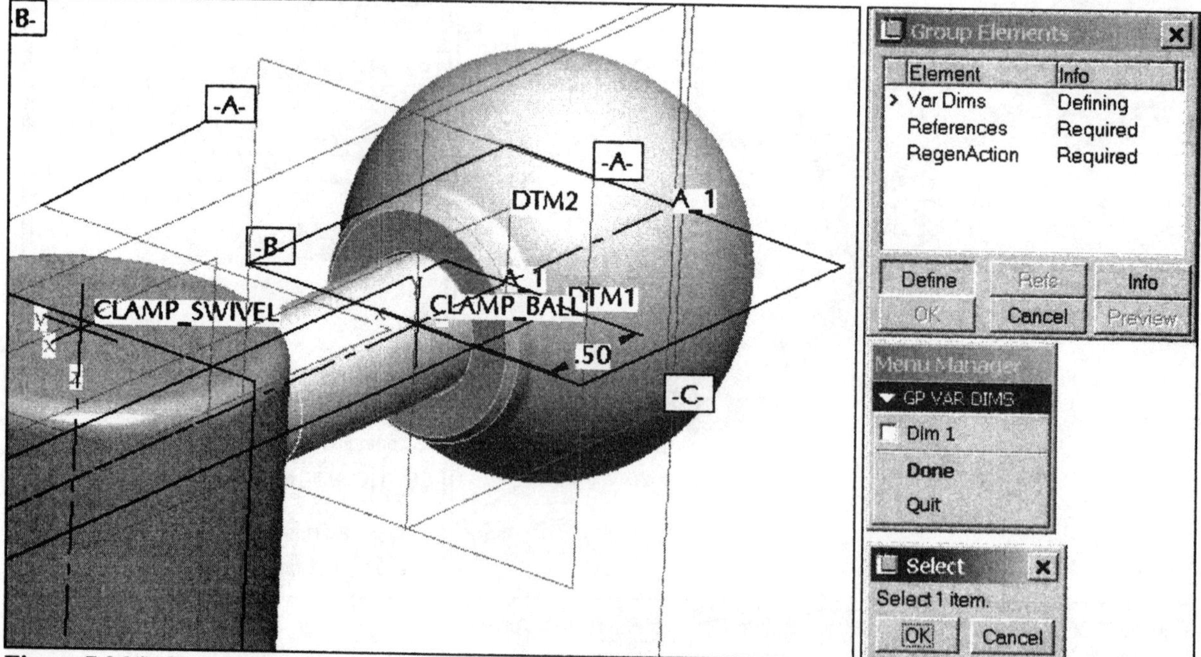

Figure 7.26(i) Group Elements Dialog Box

Alternate [Fig. 7.26(j)] ⇒ pick the opposite end surface of the Clamp_Stud5 [Fig. 7.26(k)] ⇒ **Same** [Fig. 7.26(l)] ⇒ **Done** ⇒ **View** ⇒ **Repaint** ⇒ double-click on the first Clamp_Ball component ⇒ double-click on **.500** dimension [0.50] ⇒ type **.4375** [.4375] to modify the distance from the end of the shaft so it does not bottom-out in the hole ⇒ **Enter** [Fig. 7.26(m)] ⇒ **Edit** ⇒ **Regenerate** (Fig. 7.27)

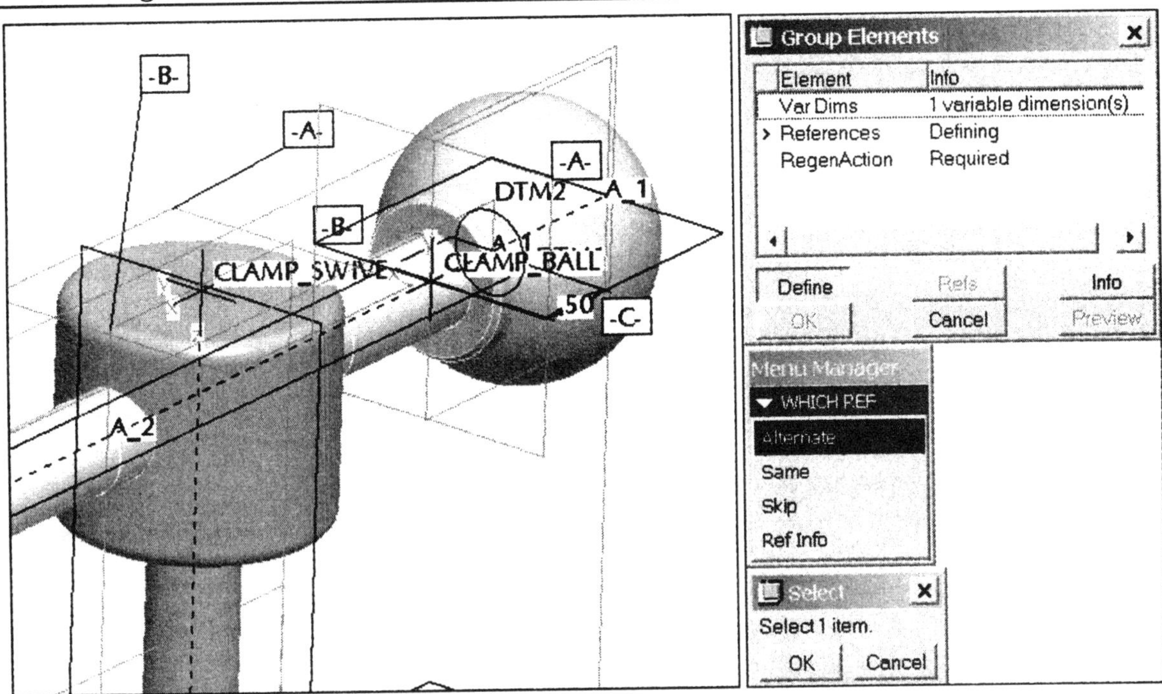

Figure 7.26(j) Alternate References

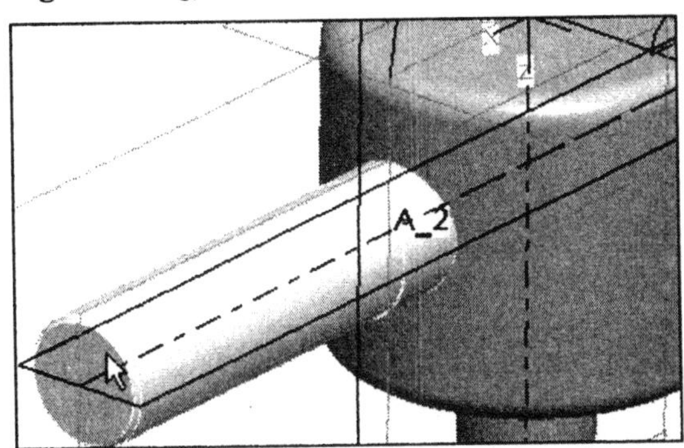

Figure 7.26(k) Pick the Opposite End Surface

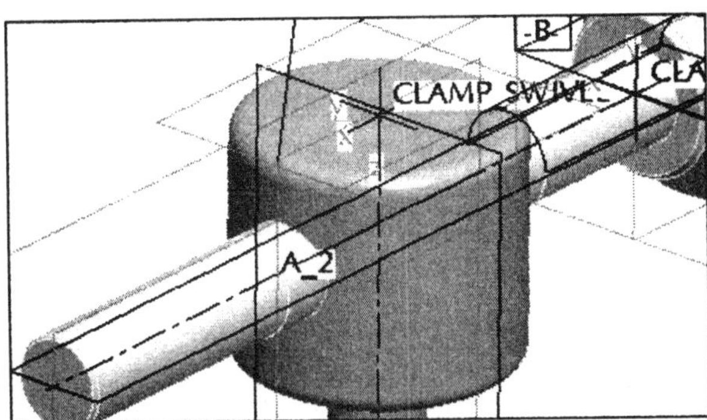

Figure 7.26(l) Second Reference Remains the Same

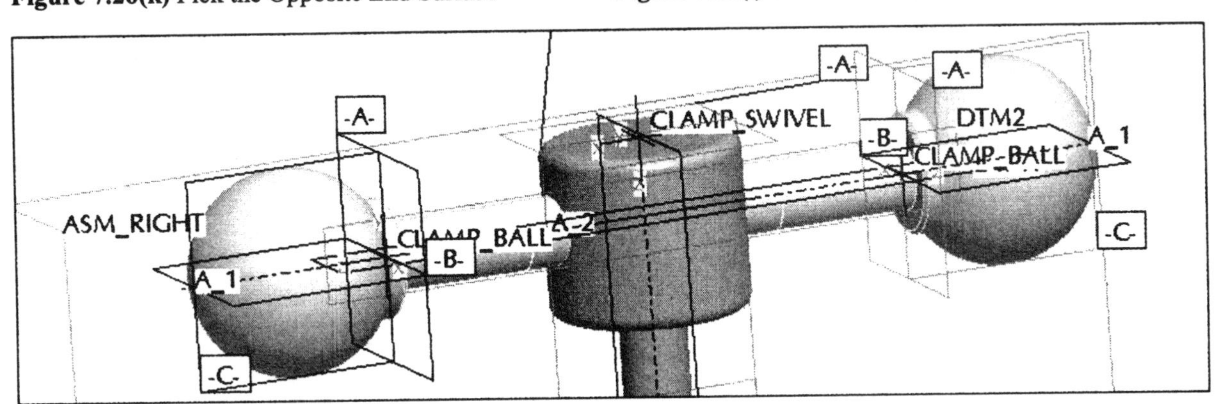

Figure 7.26(m) Clamp_Balls Assembled

Click: **Info** from the menu bar ⇒ **Bill of Materials** ⇒ Top Level (Fig. 7.28) ⇒ **OK** (Fig. 7.29)

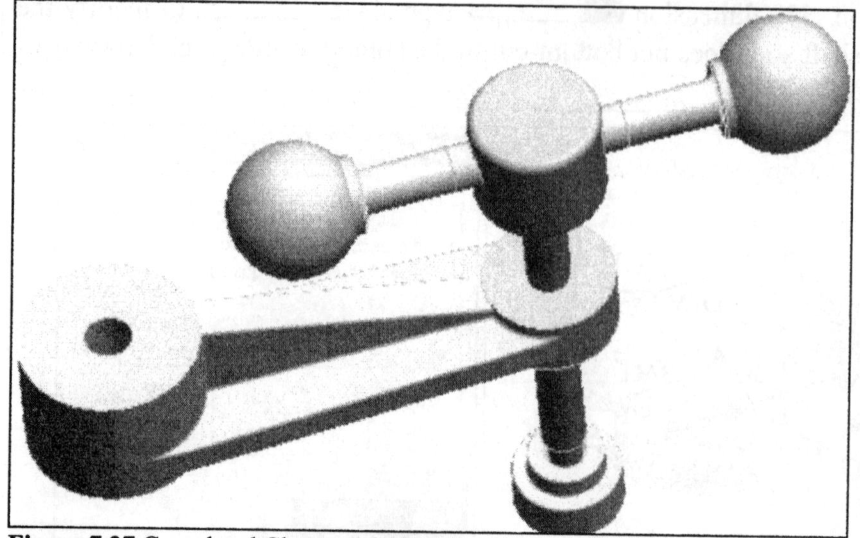

Figure 7.27 Completed Clamp_Subassembly

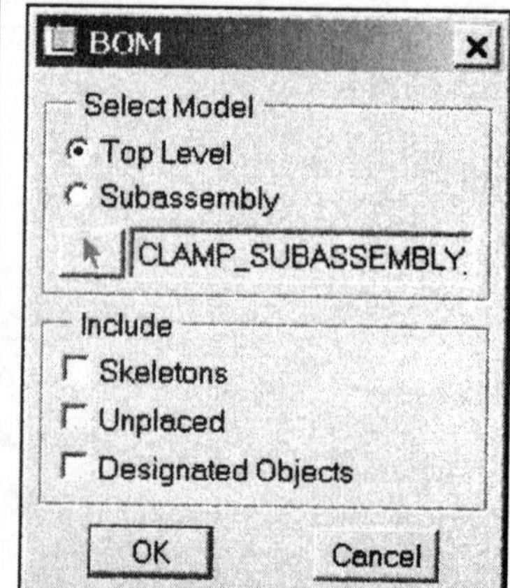

Figure 7.28 BOM

Bom Report : CLAMP_SUBASSEMBLY_

Assembly CLAMP_SUBASSEMBLY_ contains:

Quantity	Type	Name	Actions
1	Part	CLAMP_ARM	
1	Part	CLAMP_SWIVEL	
1	Part	CLAMP_FOOT	
1	Part	CLAMP_STUD5	
2	Part	CLAMP_BALL	

Summary of parts for assembly CLAMP_SUBASSEMBLY_:

Quantity	Type	Name	Actions
1	Part	CLAMP_ARM	
1	Part	CLAMP_SWIVEL	
1	Part	CLAMP_FOOT	
1	Part	CLAMP_STUD5	
2	Part	CLAMP_BALL	

Figure 7.29 Bill of Materials (BOM)

Swing Clamp Assembly

The first features for the main assembly will be the default datum planes and coordinate system. The first part assembled on the main assembly will be the Clamp_Plate, which will be created using *top-down design*; where the assembly is active, and the component is created within the assembly mode. The subassembly is still *"in session-in memory"* even though it will not show on the screen after its window is closed. For the two standard parts of the main assembly, you will use specific types of selected constraints instead of using Automatic, which allows Pro/E to default to an appropriate constraint.

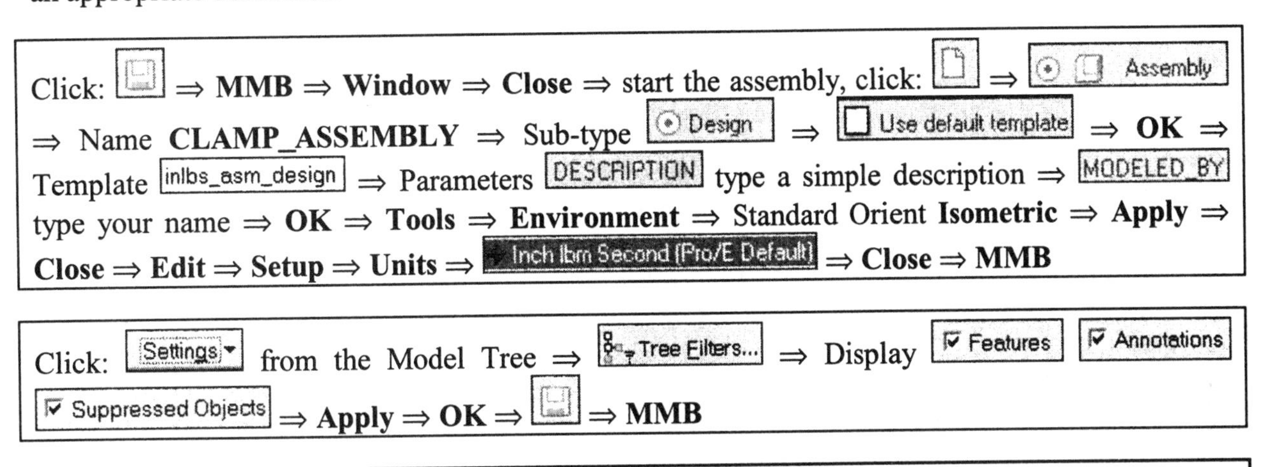

Click: ⇒ **MMB** ⇒ **Window** ⇒ **Close** ⇒ start the assembly, click: ⇒ Assembly ⇒ Name **CLAMP_ASSEMBLY** ⇒ Sub-type Design ⇒ Use default template ⇒ **OK** ⇒ Template inlbs_asm_design ⇒ Parameters DESCRIPTION type a simple description ⇒ MODELED_BY type your name ⇒ **OK** ⇒ **Tools** ⇒ **Environment** ⇒ Standard Orient **Isometric** ⇒ **Apply** ⇒ **Close** ⇒ **Edit** ⇒ **Setup** ⇒ **Units** ⇒ Inch lbm Second (Pro/E Default) ⇒ **Close** ⇒ **MMB**

Click: Settings from the Model Tree ⇒ Tree Filters... ⇒ Display Features Annotations Suppressed Objects ⇒ **Apply** ⇒ **OK** ⇒ ⇒ **MMB**

Change the coordinate system name: slowly double-click on ASM_DEF_CSYS in the Model Tree ⇒ type new name **ASM_CSYS** ⇒ click anywhere in the graphics window ⇒ slowly double-click on each of the datum identifiers in the Model Tree and add **CL_** as a prefix for each (i.e. **CL_ASM_TOP**) (Fig. 7.30) ⇒ **Environment** ⇒ Standard Orient **Trimetric** ⇒ **Apply** ⇒ **OK**

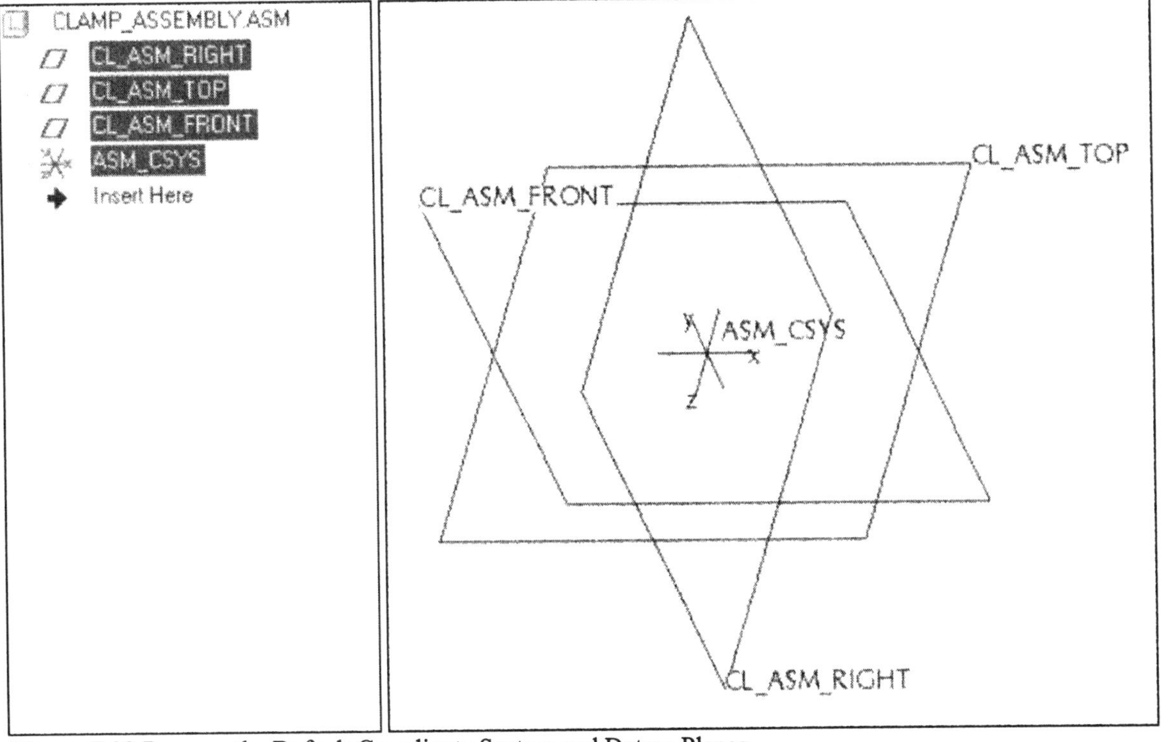

Figure 7.30 Rename the Default Coordinate System and Datum Planes

Creating Components in the Assembly Mode

The Clamp_Subassembly is now complete. The subassembly will be added to the assembly after the Clamp_Plate is created and assembled. Using the Component Create dialog box (Fig. 7.31), you can create different types of components: parts, subassemblies, skeleton models, and bulk items. You cannot reroute components created in the Assembly mode.

The following methods allow component creation in the context of an assembly without requiring external dependencies on the assembly geometry:

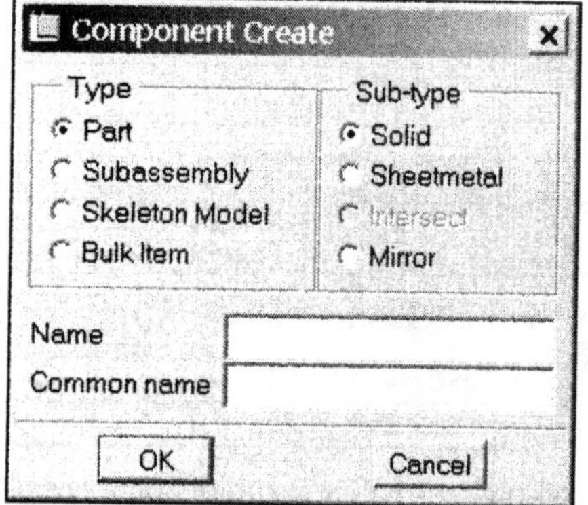

Figure 7.31 Component Create Dialog Box

- Create a component by copying another component or existing start part or start assembly
- Create a component with default datums
- Create an empty component
- Create the first feature of a new part; this initial feature is dependent on the assembly
- Create a part from an intersection of existing components
- Create a mirror copy of an existing part or subassembly
- Create Solid or Sheetmetal components
- Mirror Components

Main Assembly, Top-Down Design

The Clamp_Plate component is the first component of the main assembly. The Clamp_Plate is a new part. You will be modeling the plate *"inside"* the assembly using *top-down design*. Figure 7.32 provides the dimensions necessary to model the Clamp_Plate.

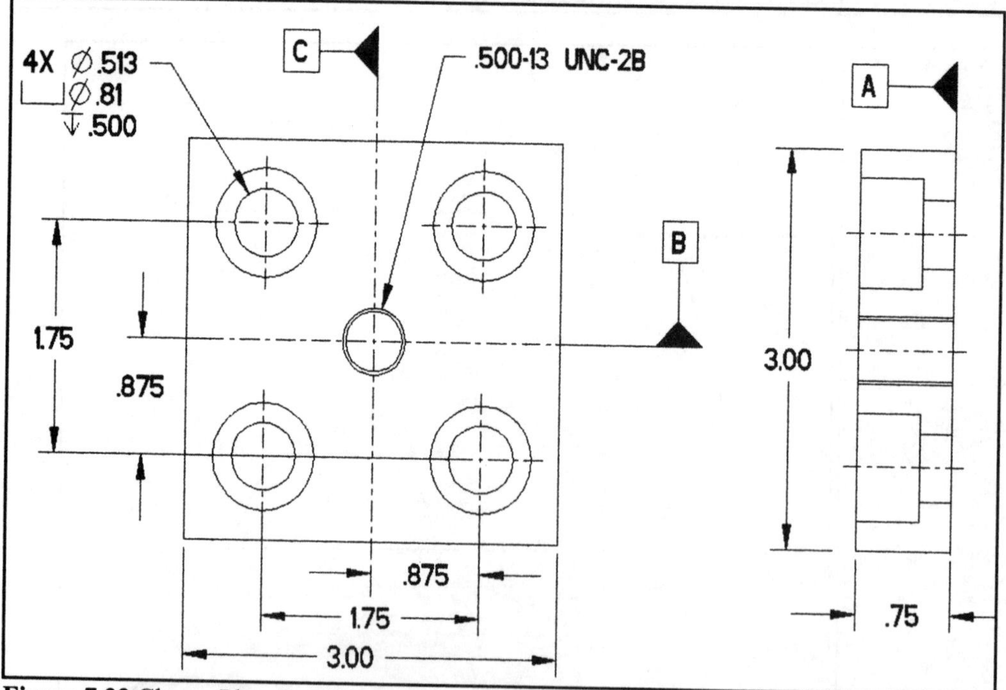

Figure 7.32 Clamp_Plate Detail Drawing

Click: **Create a component in assembly mode** Component Create dialog box displays [Fig. 7.33(a)] ⇒ Type Part ⇒ Sub-type Solid ⇒ Name **CLAMP_PLATE** ⇒ **OK** Creation Options dialog box displays [Fig. 7.33(b)] ⇒ Locate Default Datums ⇒ Align Csys To Csys ⇒ **OK** ⇒ pick **ASM_CSYS** from the graphics window and CLAMP_PLATE.PRT displays in the Model Tree [Fig. 7.33(c)], *the small green symbol/icon indicates that this component is now active* ⇒ ⇒ **MMB**
(If you cannot save, RMB on the CLAMP_ASSEMBLY.ASM in the Model Tree ⇒ Activate ⇒ RMB on the CLAMP_PLATE.PRT ⇒ Activate ⇒ File ⇒ Save⇒ MMB)

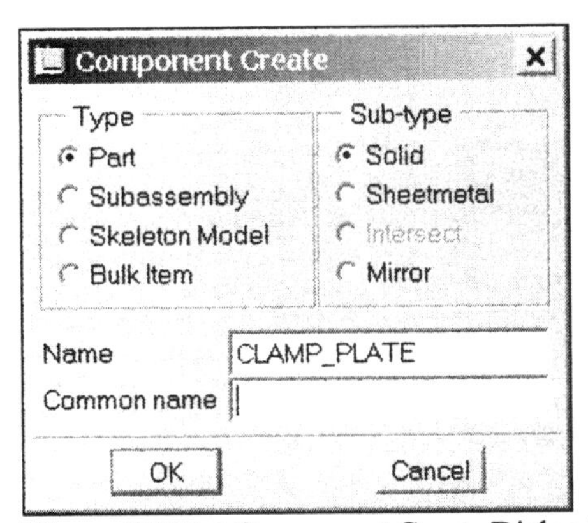

Figure 7.33(a) Component Create Dialog Box

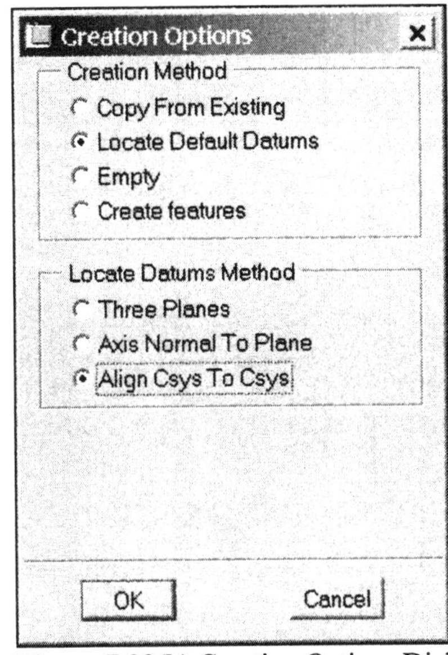

Figure 7.33(b) Creation Options Dialog Box

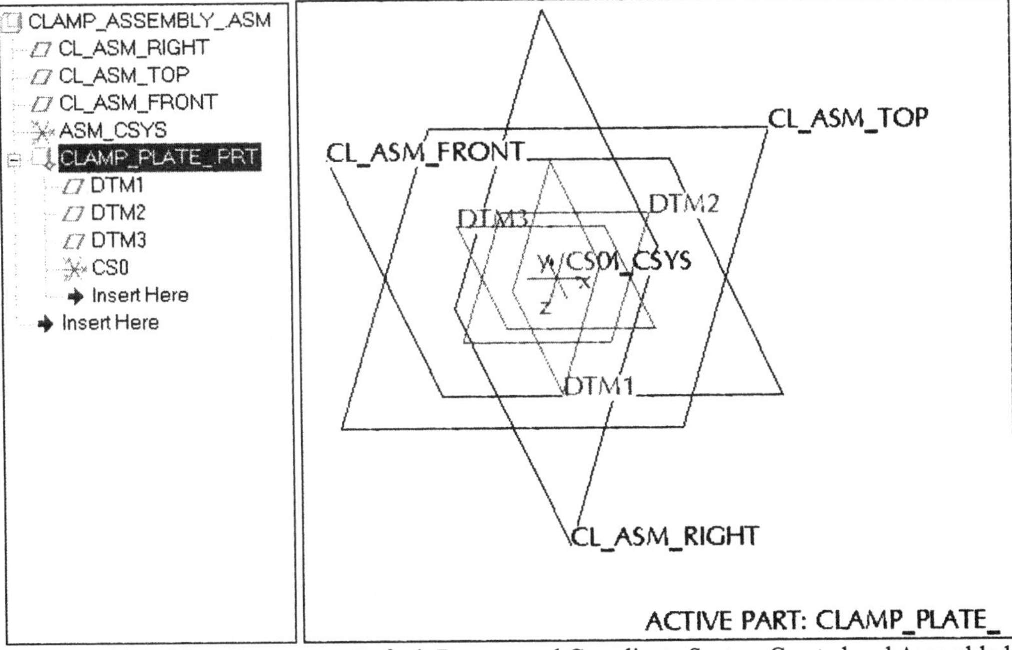

Figure 7.33(c) New Component Default Datums and Coordinate System Created and Assembled

Assembly Tools are now *unavailable* in the right column Toolbar. The current component is the Clamp_Plate. *You are now effectively in the Part mode*, except that you can see the assembly features and components. Be sure to reference only part features as you model (part datum planes and part coordinate system), otherwise you will create unwanted external references. There are many situations where external references are needed and desired, but in this case, you are simply modeling a new part.

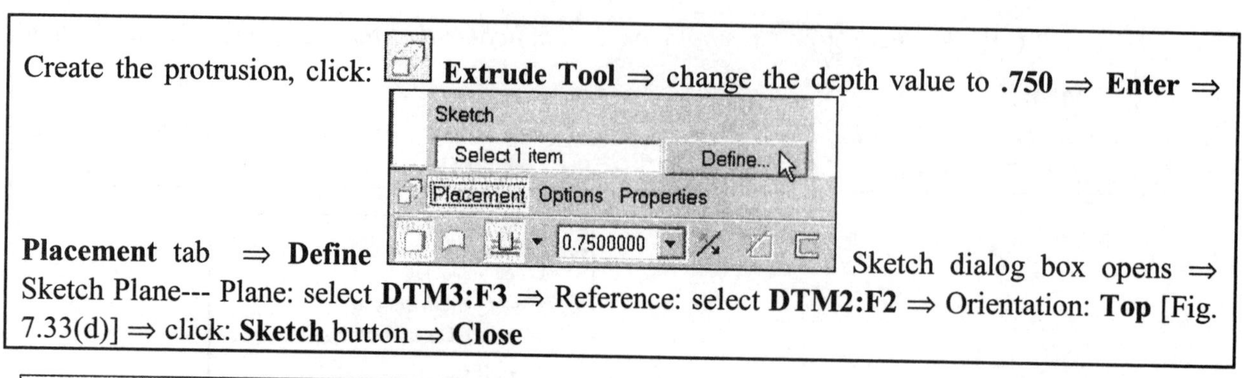

Create the protrusion, click: **Extrude Tool** ⇒ change the depth value to **.750** ⇒ **Enter** ⇒ **Placement** tab ⇒ **Define** Sketch dialog box opens ⇒ Sketch Plane--- Plane: select **DTM3:F3** ⇒ Reference: select **DTM2:F2** ⇒ Orientation: **Top** [Fig. 7.33(d)] ⇒ click: **Sketch** button ⇒ **Close**

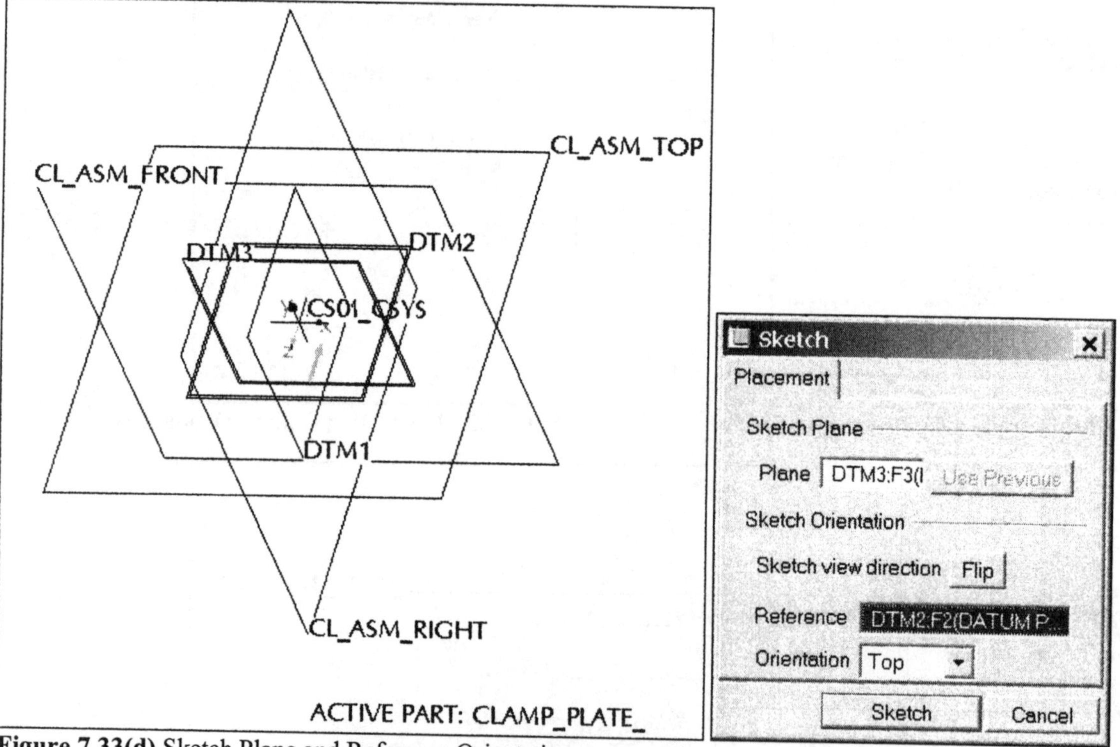

Figure 7.33(d) Sketch Plane and Reference Orientation

Click: **Create 2 point centerlines** ⇒ sketch a vertical centerline ⇒ sketch a horizontal centerline ⇒ **Create rectangle** ⇒ pick the two corners of the rectangle *(if you select the corners carefully you will get a square with only one dimension, if not, then use constraints to achieve the same result)* [Fig. 7.33(e)] ⇒ **Modify the values of dimensions** ⇒ change the value to be **3.00** ⇒ **Enter** ⇒ ⇒ ⇒ ⇒ **Standard Orientation** [Fig. 7.33(f)] ⇒ **MMB** ⇒ ⇒ **MMB** [Fig. 7.33(g)] *the active part is the Clamp_Plate*

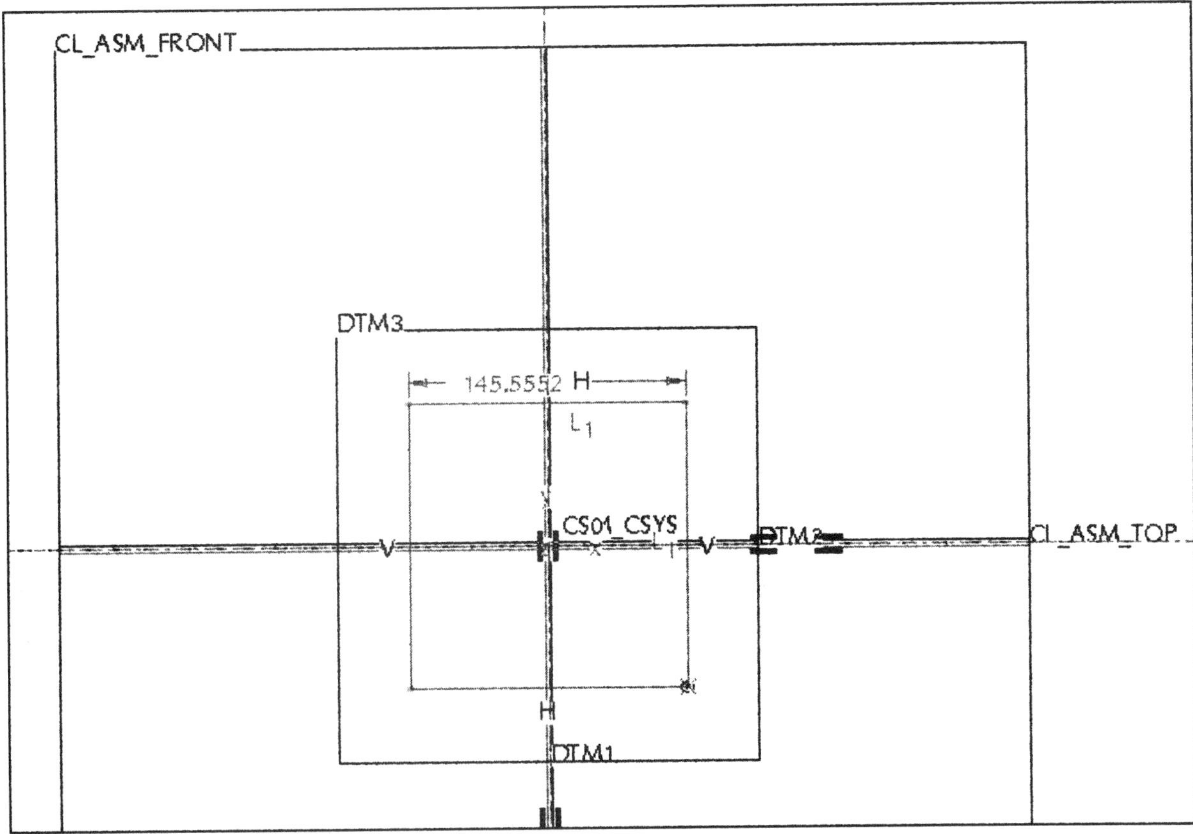

Figure 7.33(e) Sketch a Square Section

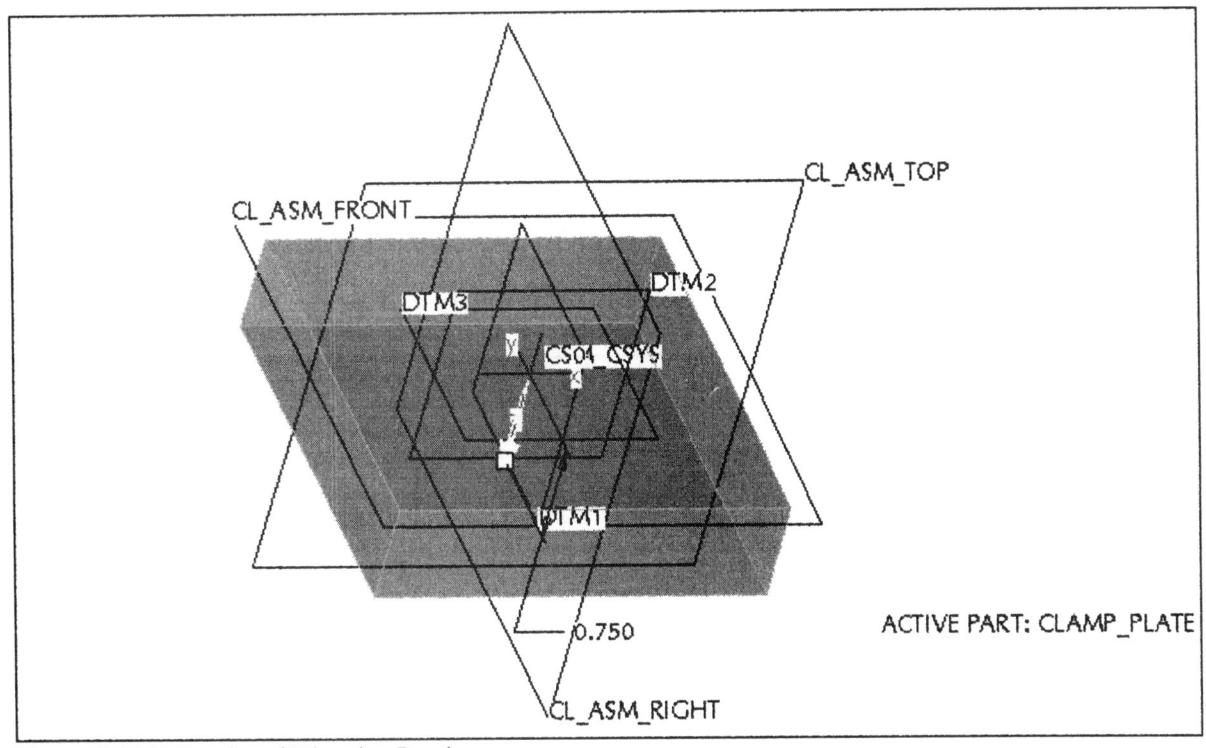

Figure 7.33(f) Depth and Direction Preview

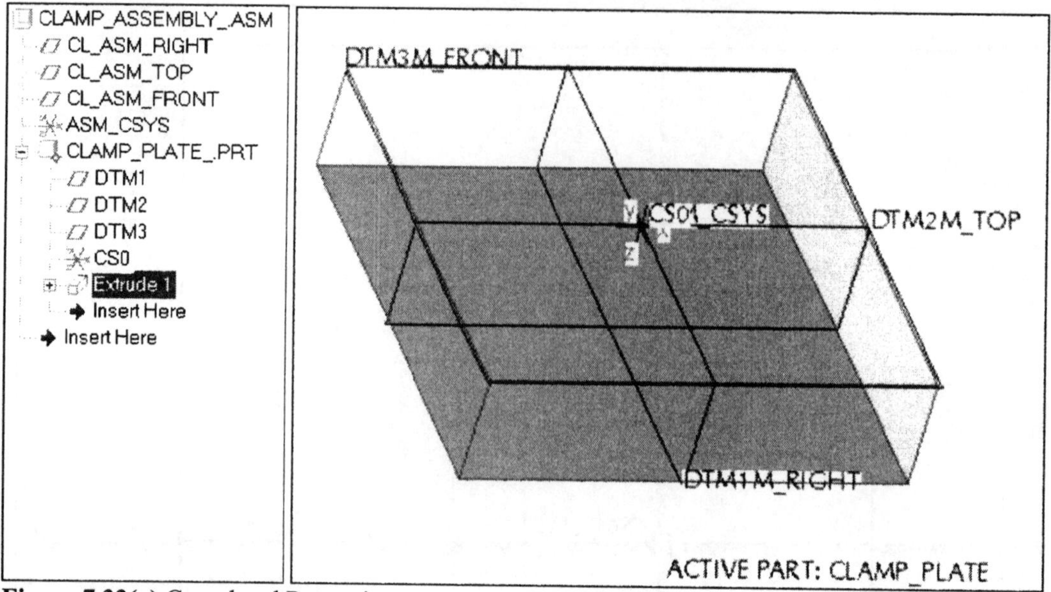

Figure 7.33(g) Completed Protrusion

Throughout the lesson, follow the steps as given, even if it seems that your picks create something different than what is described. Drag the handles to the appropriate references and let Pro/E select the references automatically even if you think the selections are incorrect. Later we will address these inconsistencies, but if you correct them on your own, steps described later will not work (or be needed).

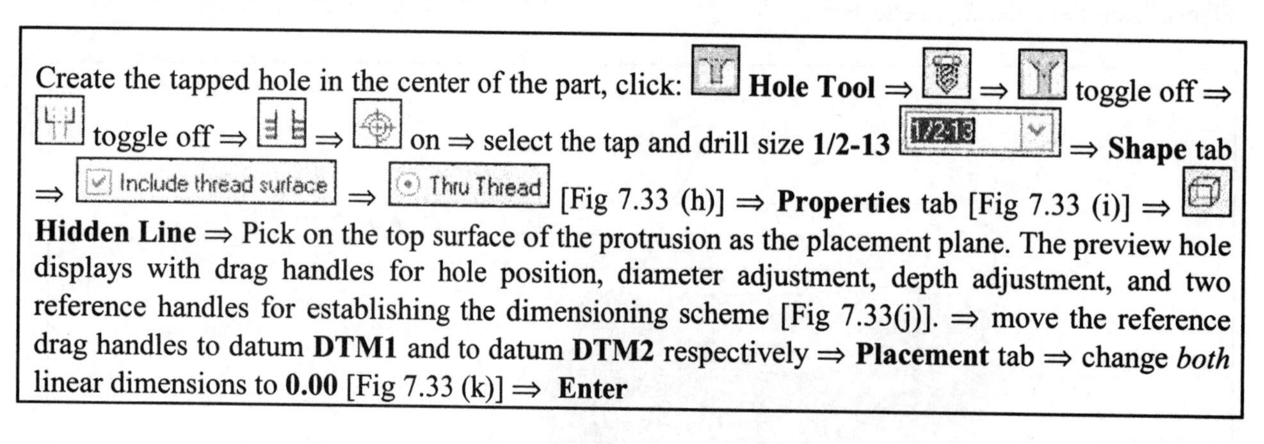

Create the tapped hole in the center of the part, click: **Hole Tool** ⇒ ⇒ toggle off ⇒ toggle off ⇒ ⇒ on ⇒ select the tap and drill size **1/2-13** ⇒ **Shape** tab ⇒ Include thread surface ⇒ Thru Thread [Fig 7.33 (h)] ⇒ **Properties** tab [Fig 7.33 (i)] ⇒ **Hidden Line** ⇒ Pick on the top surface of the protrusion as the placement plane. The preview hole displays with drag handles for hole position, diameter adjustment, depth adjustment, and two reference handles for establishing the dimensioning scheme [Fig 7.33(j)]. ⇒ move the reference drag handles to datum **DTM1** and to datum **DTM2** respectively ⇒ **Placement** tab ⇒ change *both* linear dimensions to **0.00** [Fig 7.33 (k)] ⇒ **Enter**

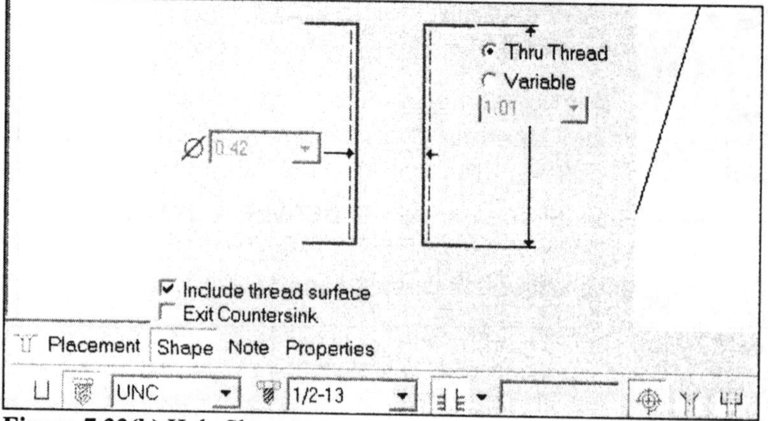

Figure 7.33(h) Hole Shape

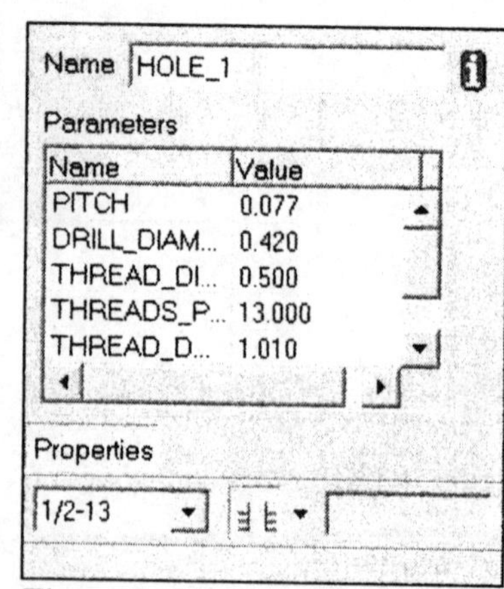

Figure 7.33(i) Hole Properties

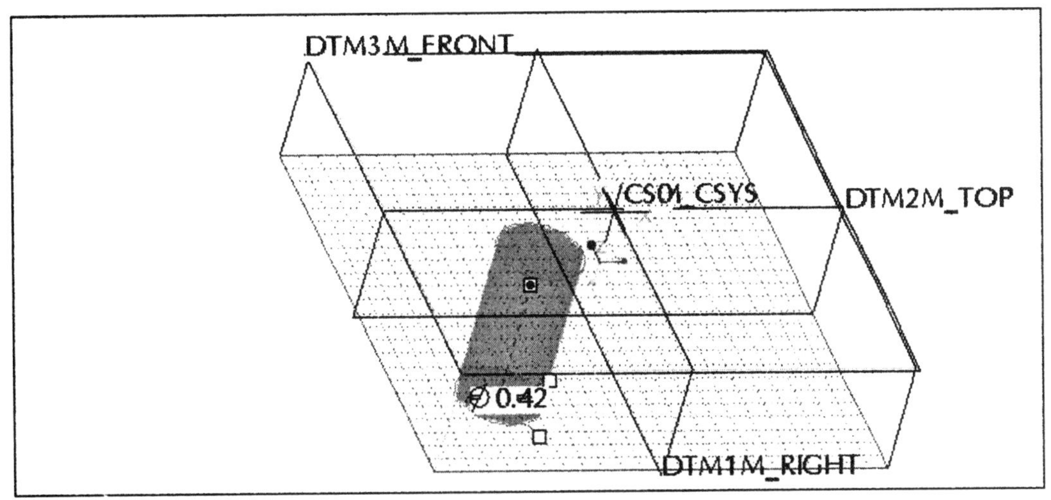

Figure 7.33(j) Initial Hole Placement

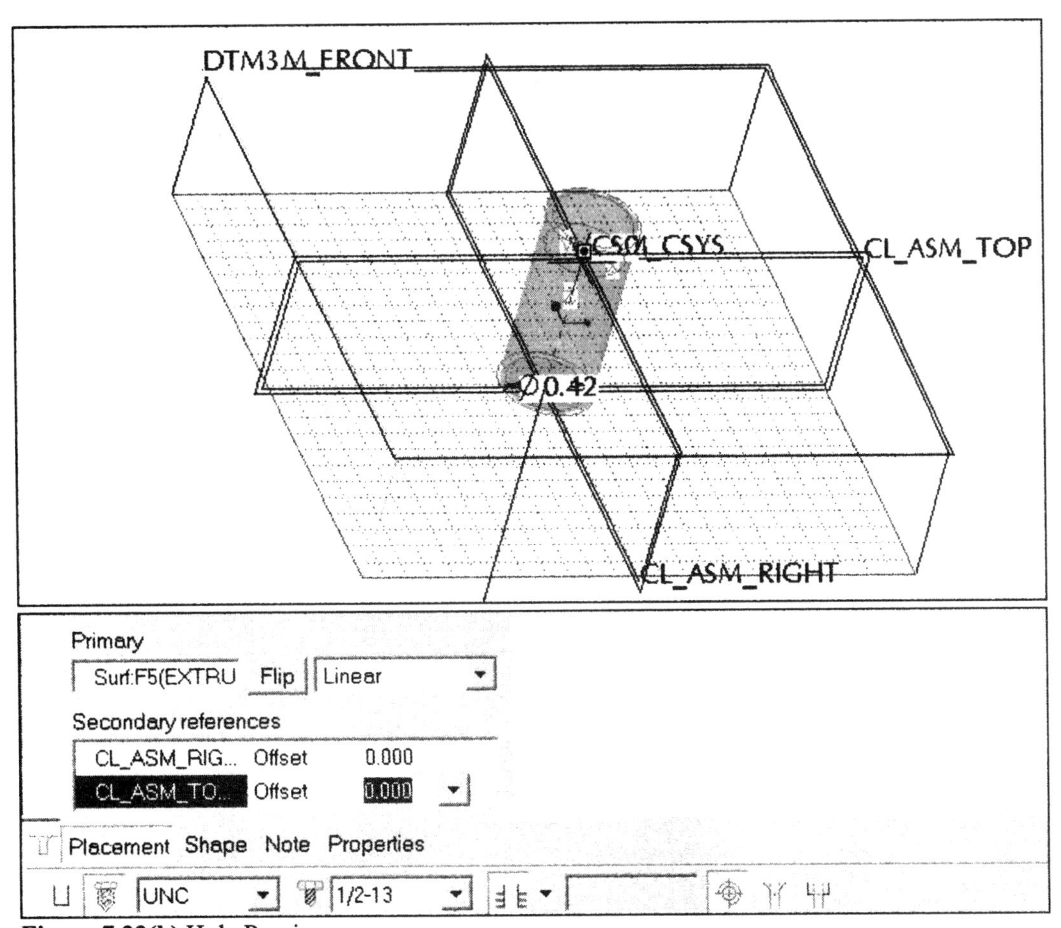

Figure 7.33(k) Hole Preview

Click: ⇒ ⇒ **LMB** in the graphics window to deselect ⇒ ⇒ **MMB** [Fig. 7.33(l)]

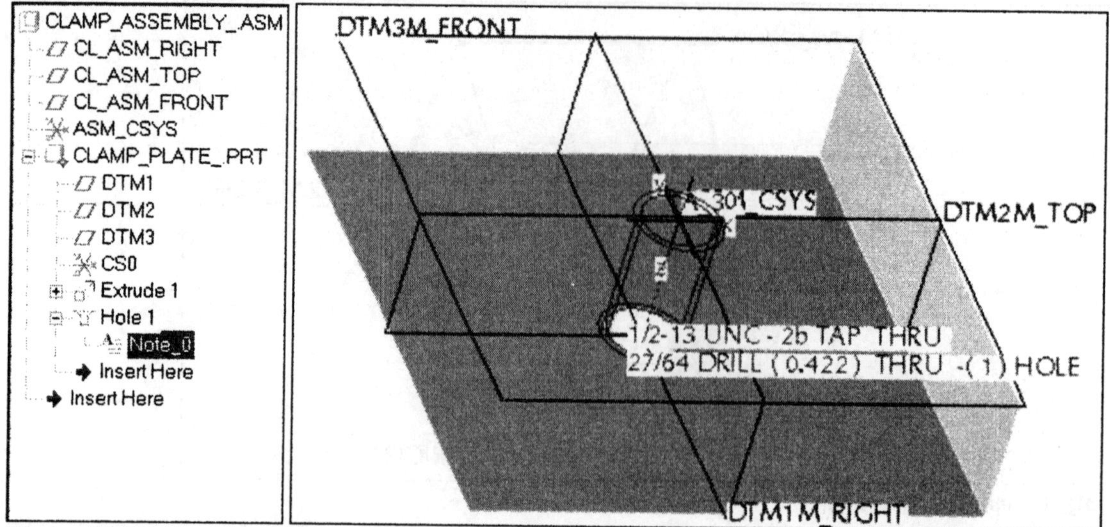

Figure 7.33(l) Completed Tapped Hole

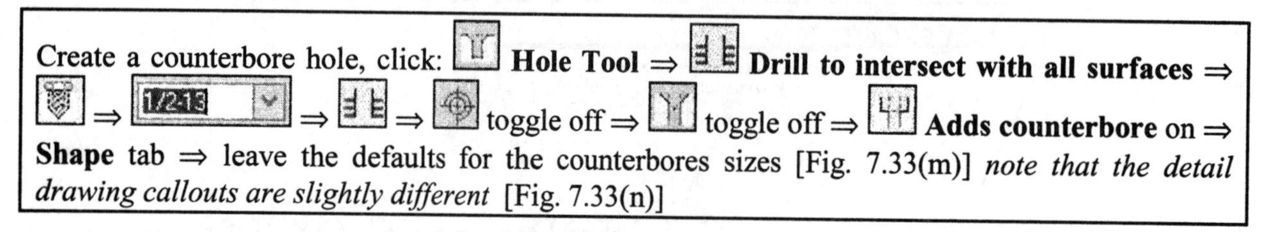

Create a counterbore hole, click: **Hole Tool** ⇒ **Drill to intersect with all surfaces** ⇒ ⇒ 1/2-13 ⇒ ⇒ toggle off ⇒ toggle off ⇒ **Adds counterbore** on ⇒ **Shape** tab ⇒ leave the defaults for the counterbores sizes [Fig. 7.33(m)] *note that the detail drawing callouts are slightly different* [Fig. 7.33(n)]

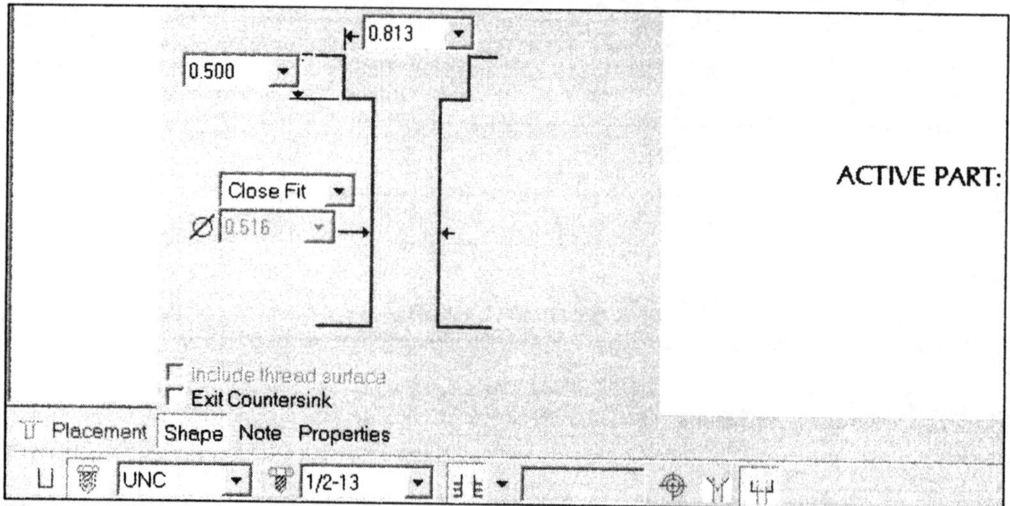

Figure 7.33(m) Counterbore Specifications

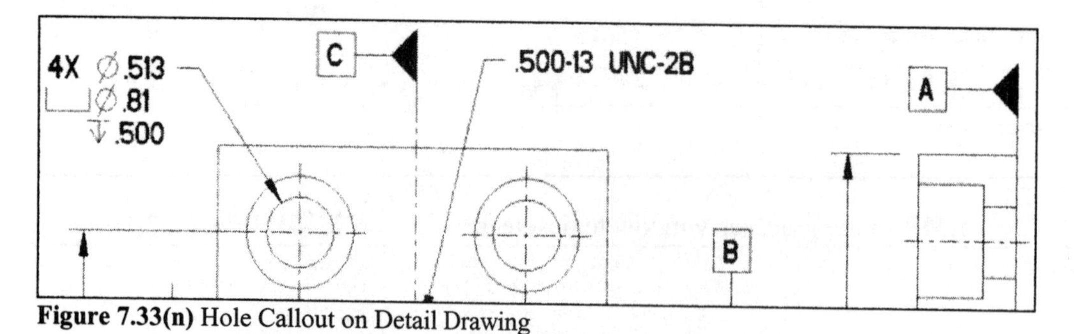

Figure 7.33(n) Hole Callout on Detail Drawing

Click: Hidden Line ⇒ pick on the top surface for the placement plane ⇒ move the reference drag handles to datum **DTM1** and to datum **DTM2** respectively ⇒ **Placement** tab [Fig 7.33(o)] ⇒ change linear dimensions to **-.875** *(negative .875* -0.875*)* and **.875** (0.875) ⇒ **Enter** ⇒ [Fig. 7.33(p)]

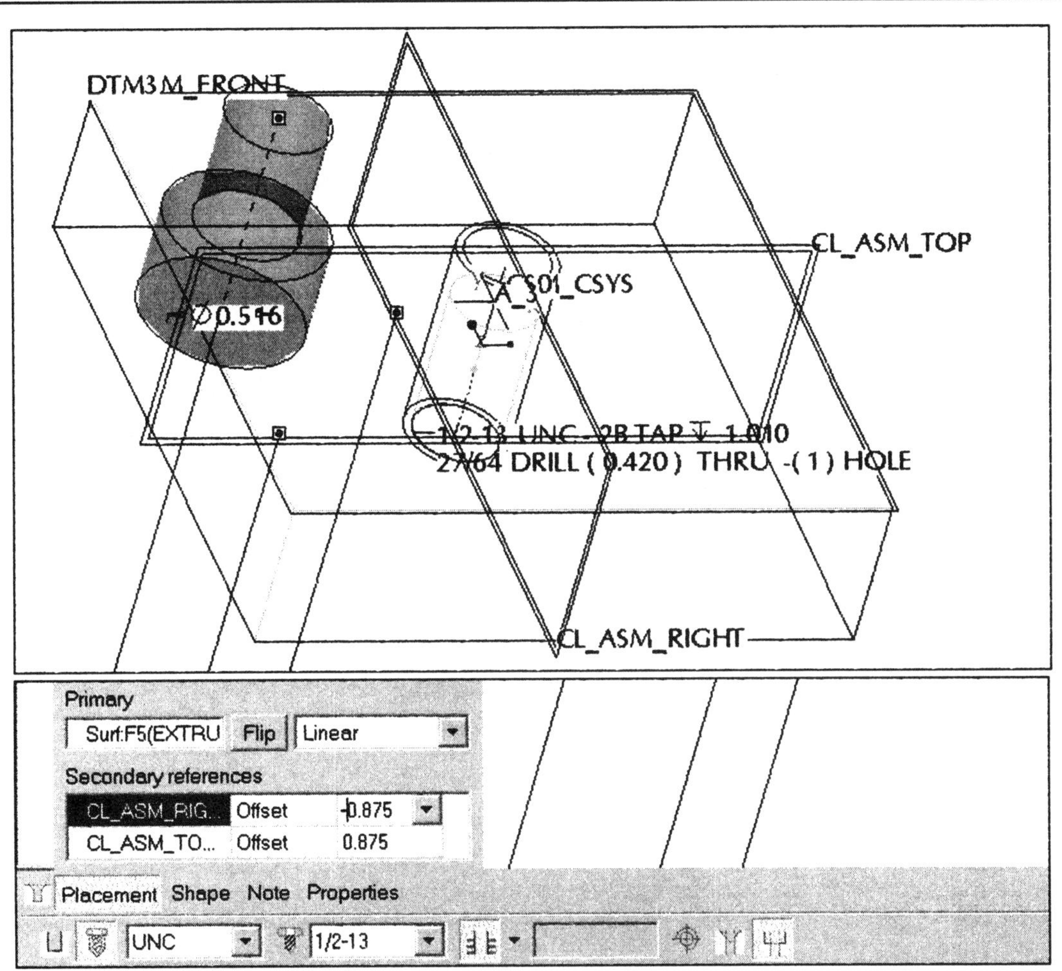

Figure 7.33(o) Hole Specifications

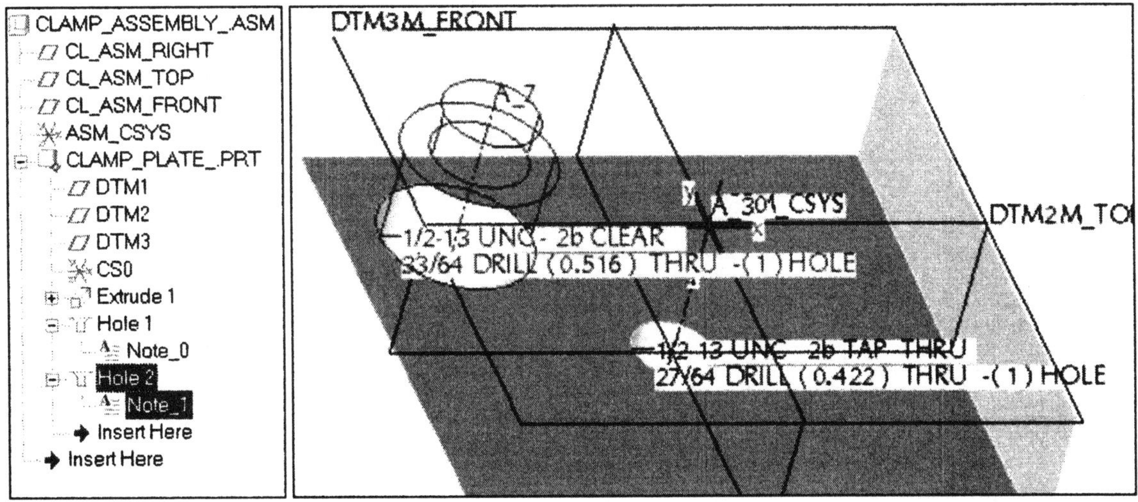

Figure 7.33(p) Completed Counterbore

The counterbore "seems" correct, but still check for external references.

Click: **Info** ⇒ **Global Reference Viewer** [Fig 7.33(q)] ⇒ **Show Filters** tab ⇒ activate all Display Objects [Fig. 7.33(r)] ⇒ **Apply** ⇒ **OK** ⇒ expand as necessary [Fig. 7.33(s)] ⇒ **File** ⇒ **Close**

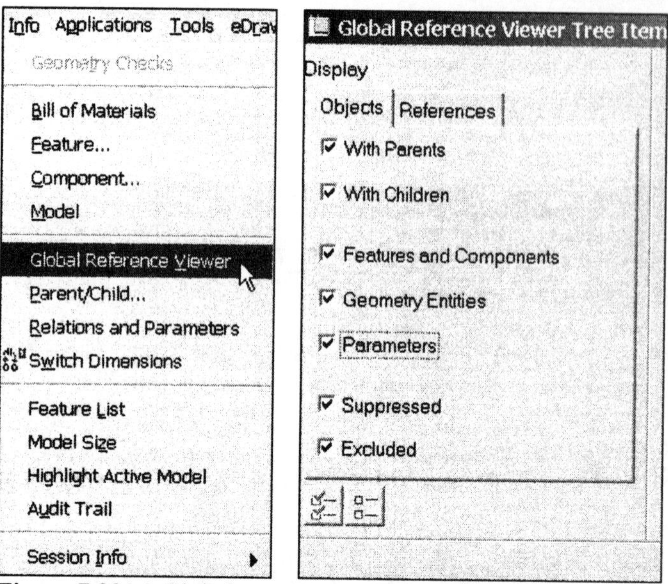

Figure 7.33(q) Global

Figure 7.33(r) Show Filters

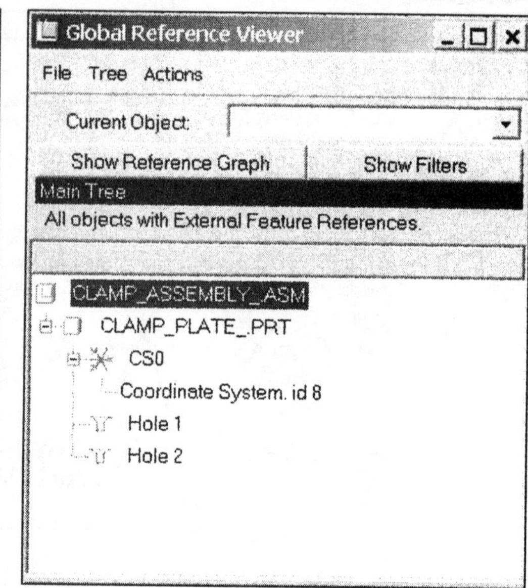

Figure 7.33(s) Global Reference Viewer

If you look back at Figure 7.33(k) and Figure 7.33(o) you will see that the assembly datums were used to place the holes instead of DTM1 and DTM2. The following commands will show you how to edit the references of one of the holes. Redo the references of the other hole using similar steps.

Click: **RMB** on the hole in the Model Tree ⇒ **Edit Definition** [Fig. 7.33(t)]

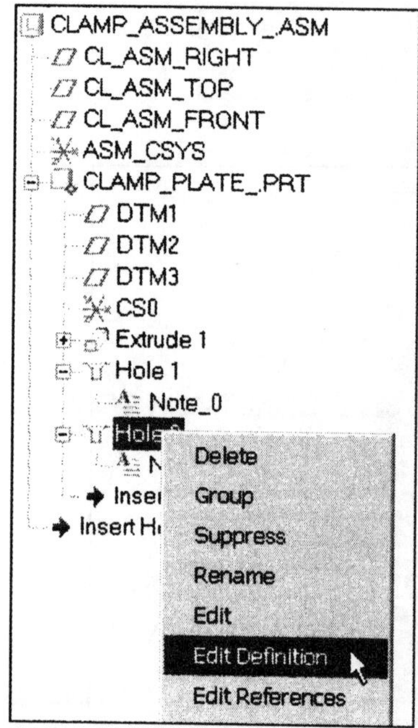

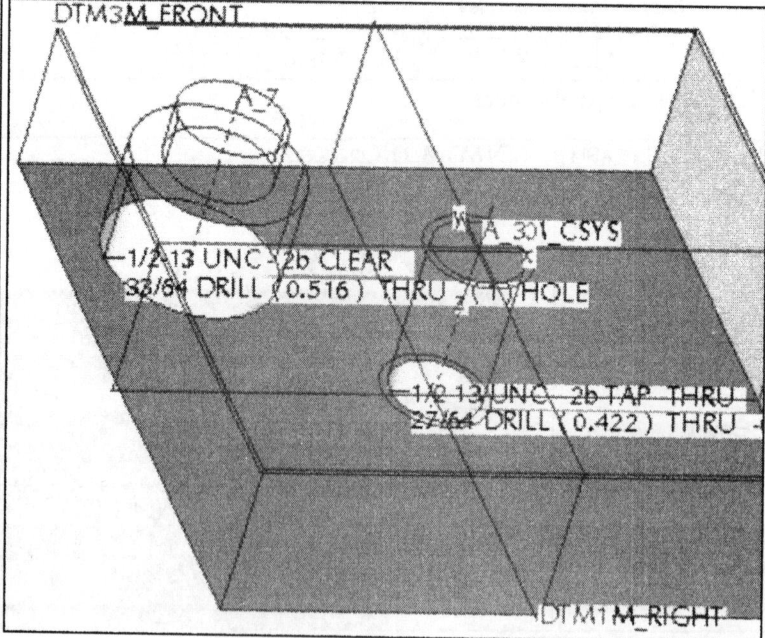

Figure 7.33(t) Edit Definition

Click: **Placement** tab ⇒ click on the first Secondary reference ⇒ **RMB** ⇒ **Remove** [Fig. 7.33(u)] ⇒ **RMB** ⇒ **Remove** removes the second reference ⇒ move one of the drag handles near **DTM1** and drop it ⇒ with the datum highlighted, click **RMB** ⇒ **Pick From List** [Fig. 7.33(v)] ⇒ click on **DTM1** from the list [Fig. 7.33(w)] ⇒ **OK** ⇒ press and hold **Ctrl** key

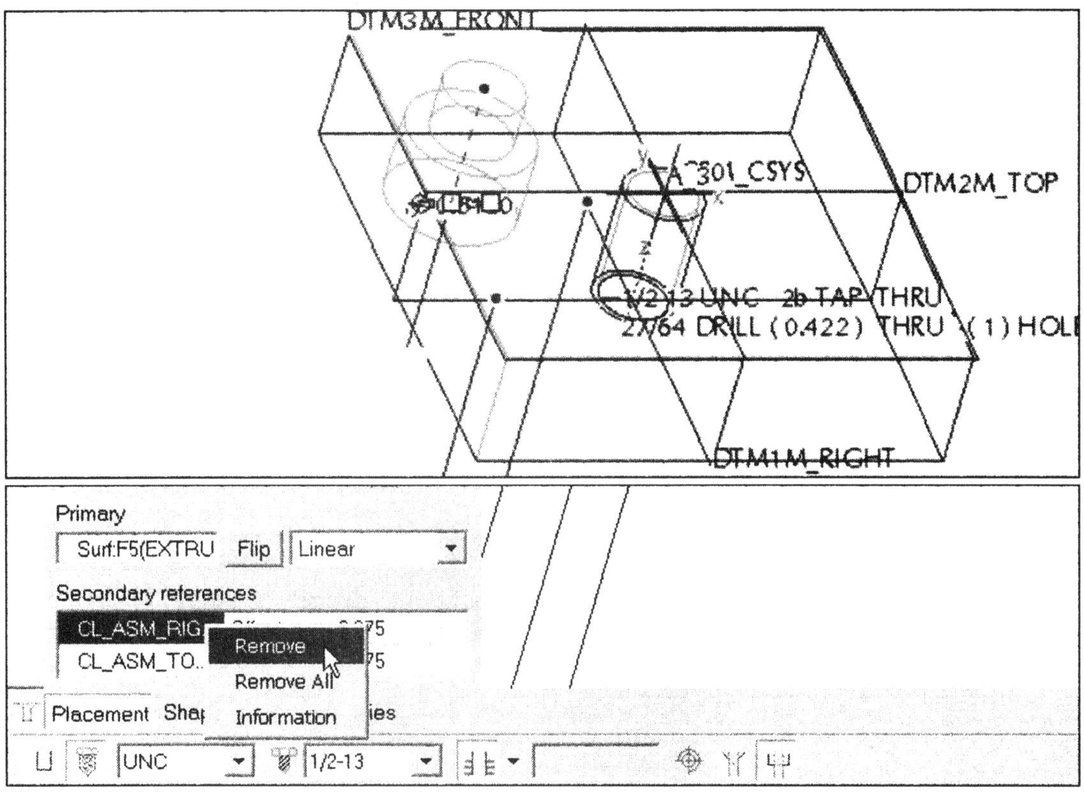

Figure 7.33(u) Remove Secondary References

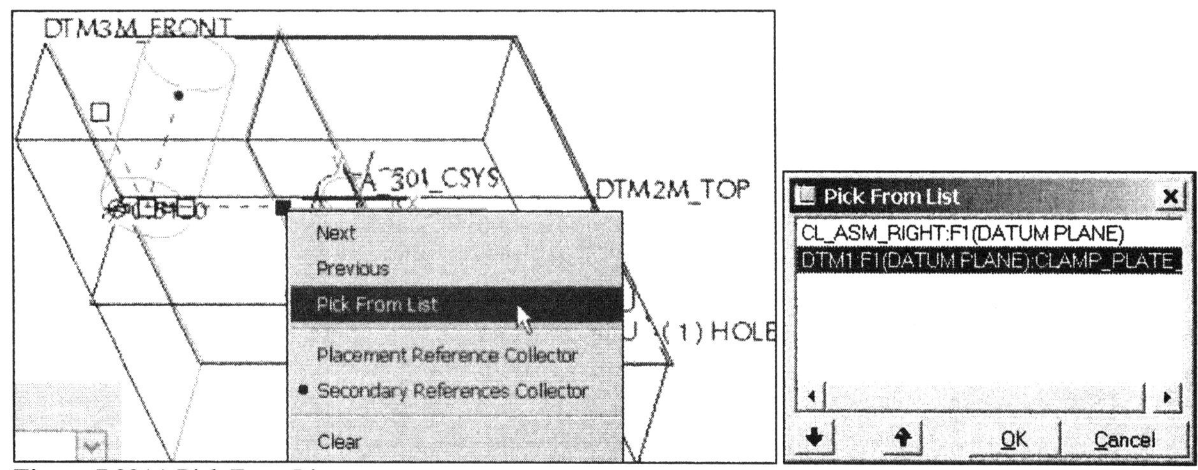

Figure 7.33(v) Pick From List

If you have difficulty with this method, simply click on the three assembly datum planes in the Model Tree ⇒ RMB ⇒ Hide. Then go and redo the references. Unhide the assembly datums after you are finished changing the references.

Move the other drag handle near **DTM2** and drop it ⇒ with the datum highlighted, **RMB** ⇒ **Pick From List** [Fig. 7.33(w)] ⇒ click on **DTM2** from the list [Fig. 7.33(w)] ⇒ release the **Ctrl** key ⇒ **OK** [Fig. 7.33(x)] ⇒

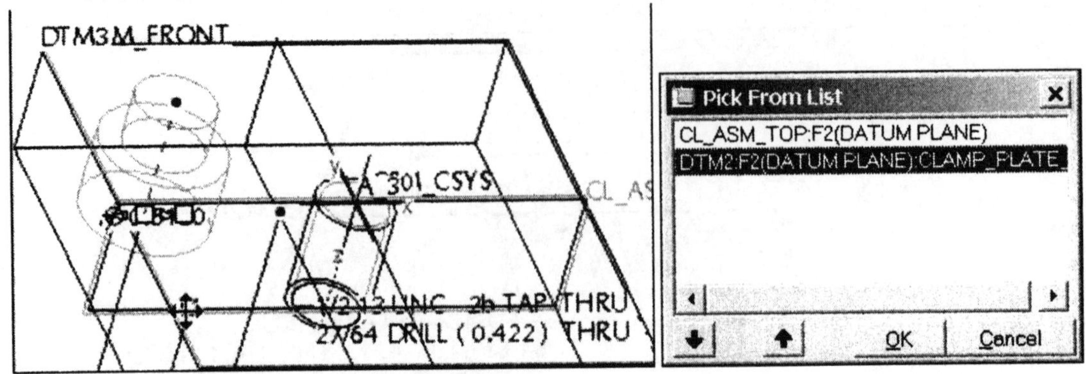

Figure 7.33(w) Redo Next Secondary Reference

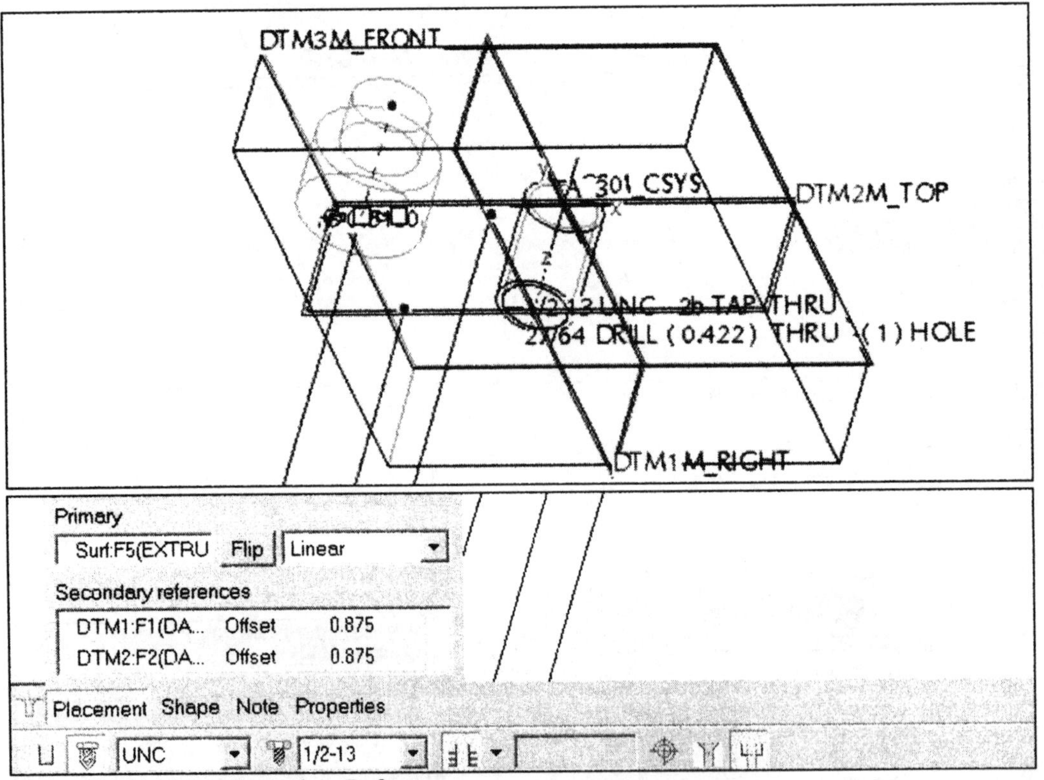

Figure 7.33(x) New Secondary References

Click: **Info** ⇒ **Global Reference Viewer** [Fig. 7.33(y)] *only one hole shows as having an external reference* ⇒ ⇒ repeat this process for the tapped hole in the center of the part [Fig. 7.33 (z)]

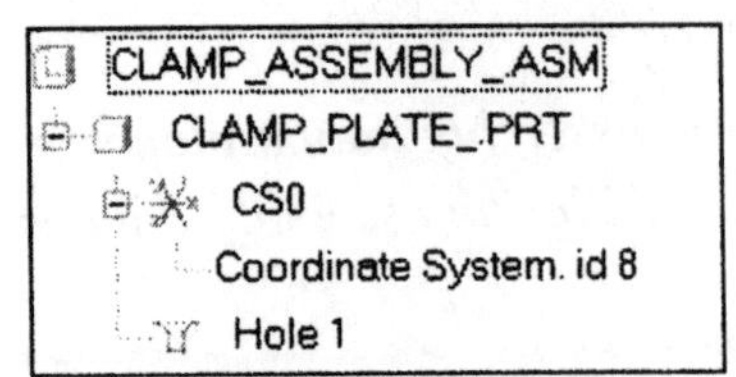

Figure 7.33(y) One Hole Left with External References

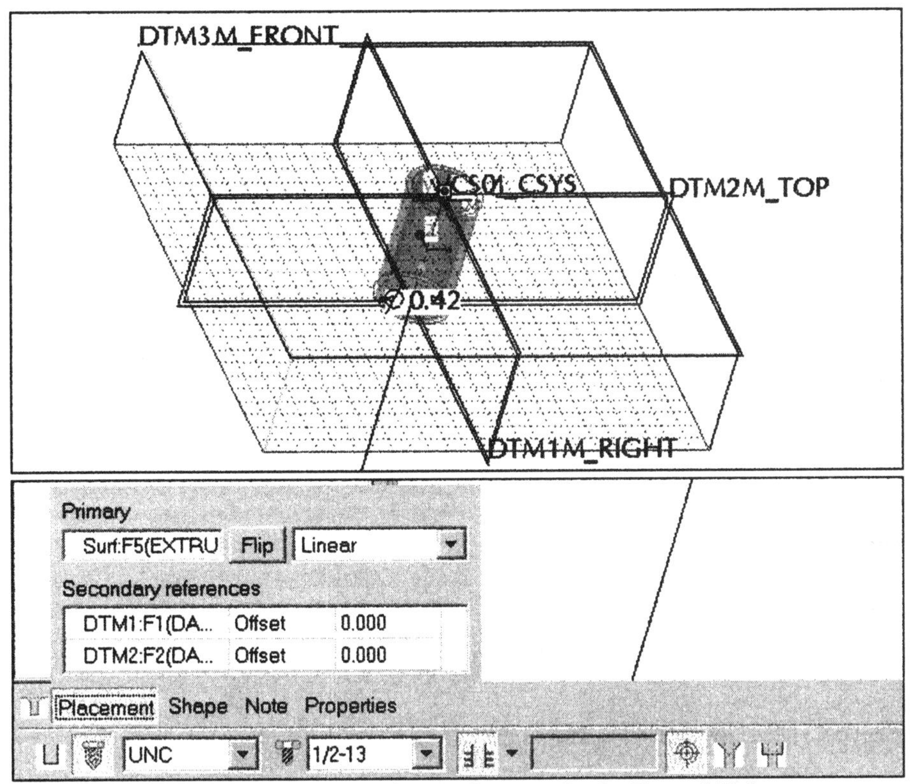

Figure 7.33(z) Changing the Secondary References for the Tapped Hole

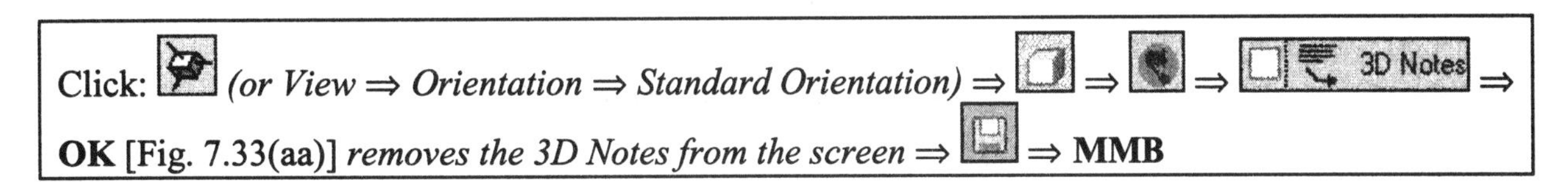

Click: *(or View ⇒ Orientation ⇒ Standard Orientation)* ⇒ ⇒ ⇒ 3D Notes ⇒ **OK** [Fig. 7.33(aa)] *removes the 3D Notes from the screen* ⇒ ⇒ **MMB**

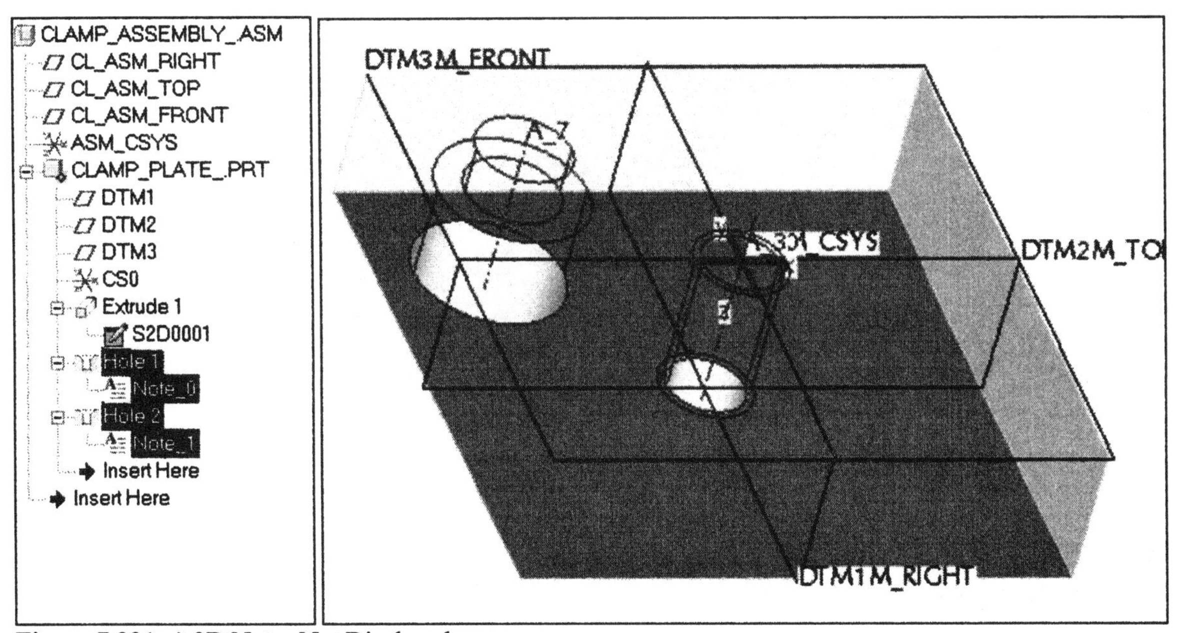

Figure 7.33(aa) 3D Notes Not Displayed

Pattern the counterbore holes, click on the last hole in the Model Tree Hole id ⇒ **RMB** ⇒ **Pattern** [Fig. 7.34(a)] ⇒ **Dimensions** tab ⇒ pick on the horizontal **.875** dimension [Fig. 7.34(b)] ⇒ highlight value and type **–1.75** ⇒ **Enter** ⇒ in Direction 2 collector box, pick Click here to a... ⇒ pick on the vertical **.875** dimension [Fig. 7.34(c)] ⇒ highlight value and type **–1.75** ⇒ **Enter** ⇒ **MMB** ⇒ **Tools** ⇒ **Environment** ⇒ Standard Orient Isometric ⇒ **Apply** ⇒ **OK** [Fig. 7.34(d)] ⇒ ⇒ ⇒ **MMB**

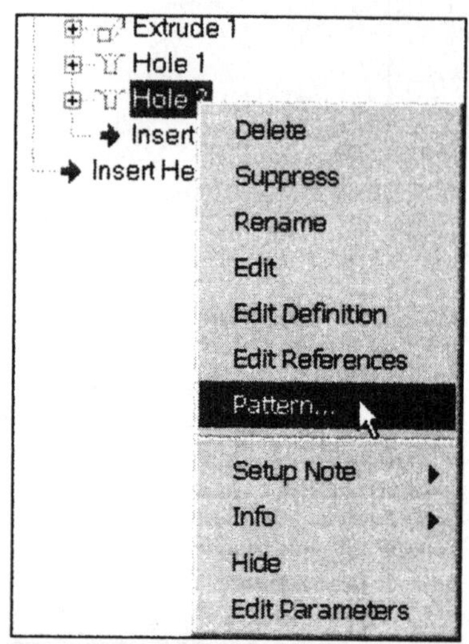

Figure 7.34(a) Pattern the Counterbore Hole

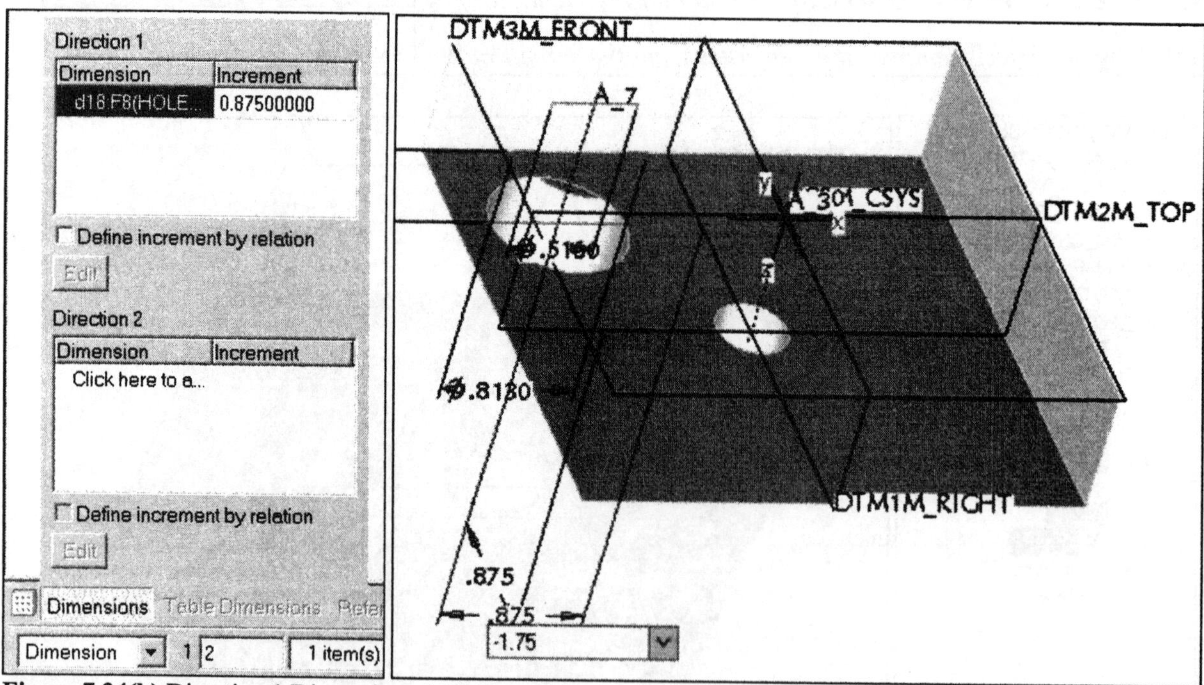

Figure 7.34(b) Direction 1 Dimension **–1.75**, Two Items

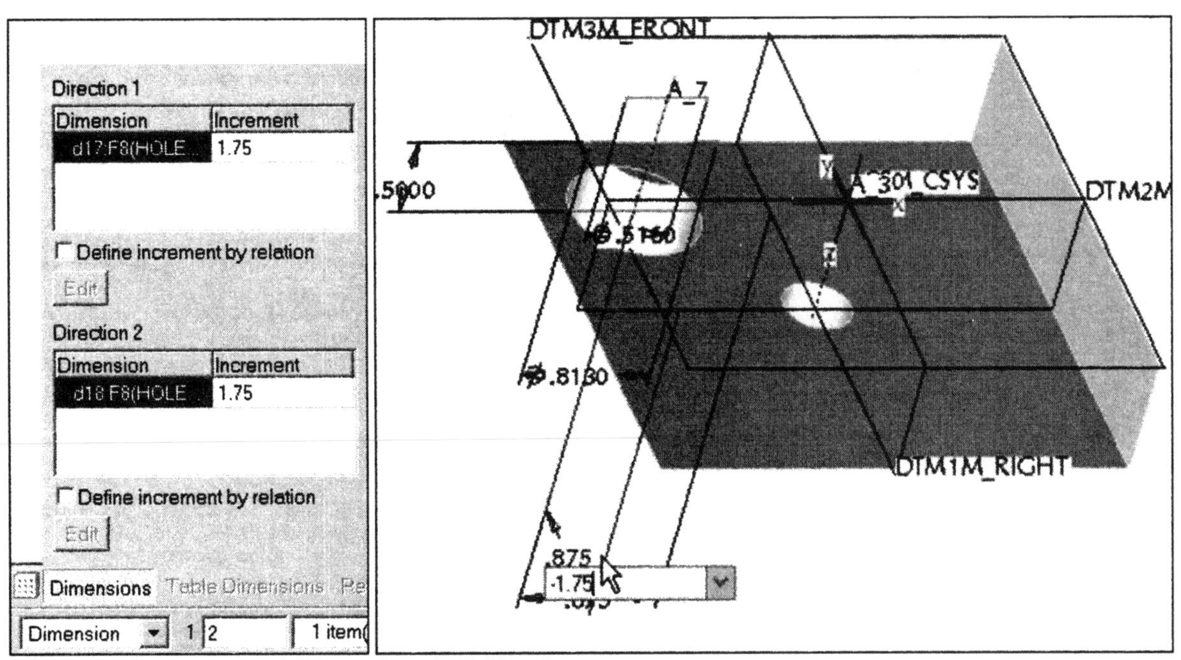

Figure 7.34(c) Direction 2 Dimension **–1.75**, Two Items

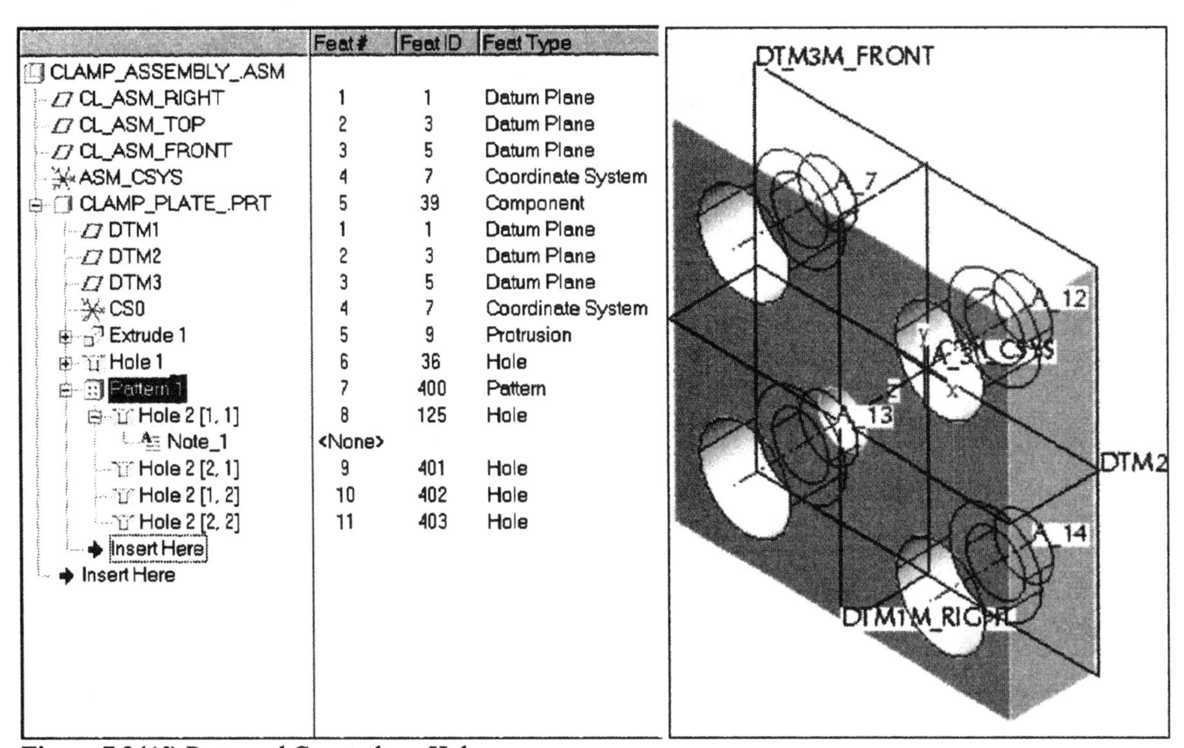

Figure 7.34(d) Patterned Counterbore Holes

Click CLAMP_ASSEMBLY.ASM in the Model Tree ⇒ **RMB** ⇒ **Activate** ⇒ pick datum tag CL_ASM_FRONT in the graphics window [Fig. 7.35(a)] ⇒ **RMB** ⇒ Move Datum Tag ⇒ pick a new position [Fig. 7.35(b)] ⇒ repeat this process and move the other assembly datum tags ⇒ **Add component to the assembly** ⇒ pick the **clamp_subamssembly.asm** ⇒ **Preview** [Fig. 7.36(a)] ⇒ **Open**

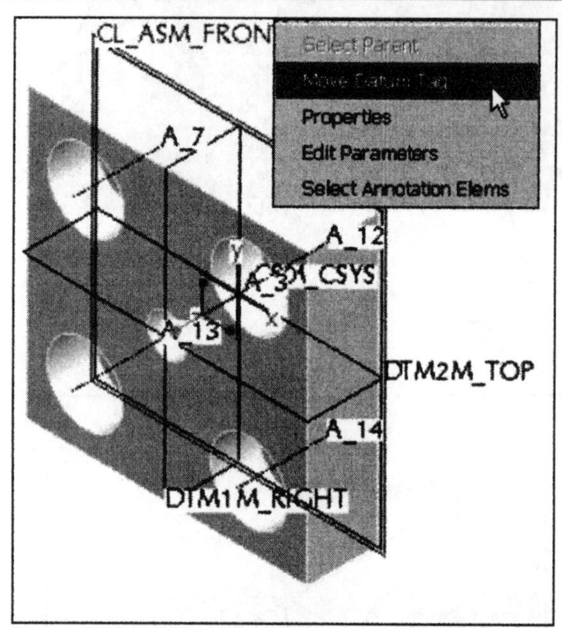

Figure 7.35(a) Move Datum Tag

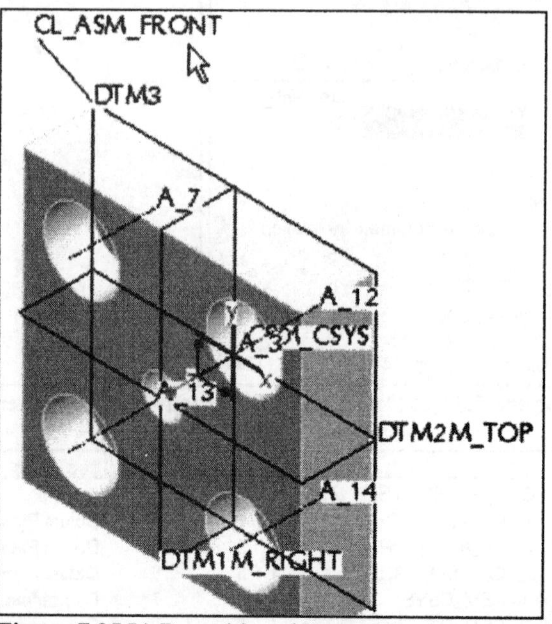

Figure 7.35(b) Repositioned Datum Tag

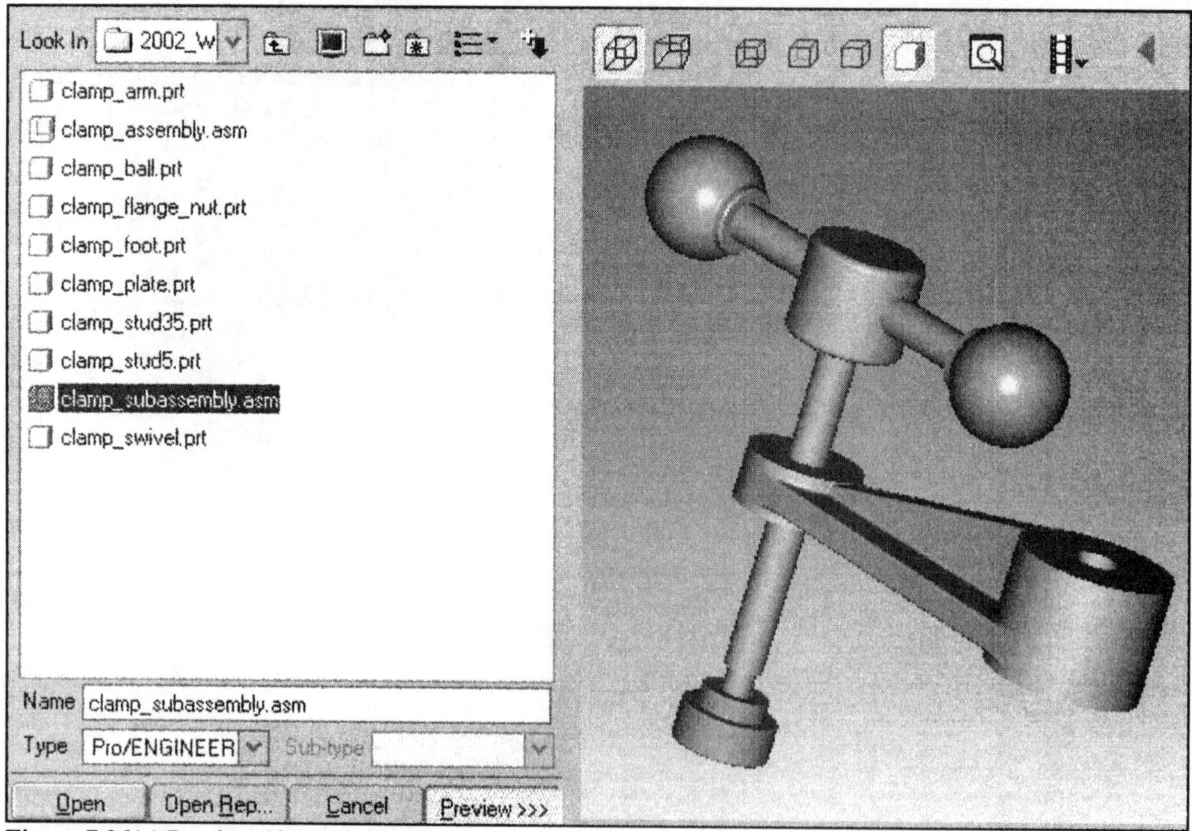

Figure 7.36(a) Preview Clamp_Subassembly

Click: toggle off component window ⇒ Constraints **Mate** [Fig. 7.36(b)] ⇒ spin the model ⇒ **View** ⇒ **Shade** ⇒ pick the bottom surface of the Clamp_Arm (of the Clamp_Subassembly) [Fig. 7.36(c)]

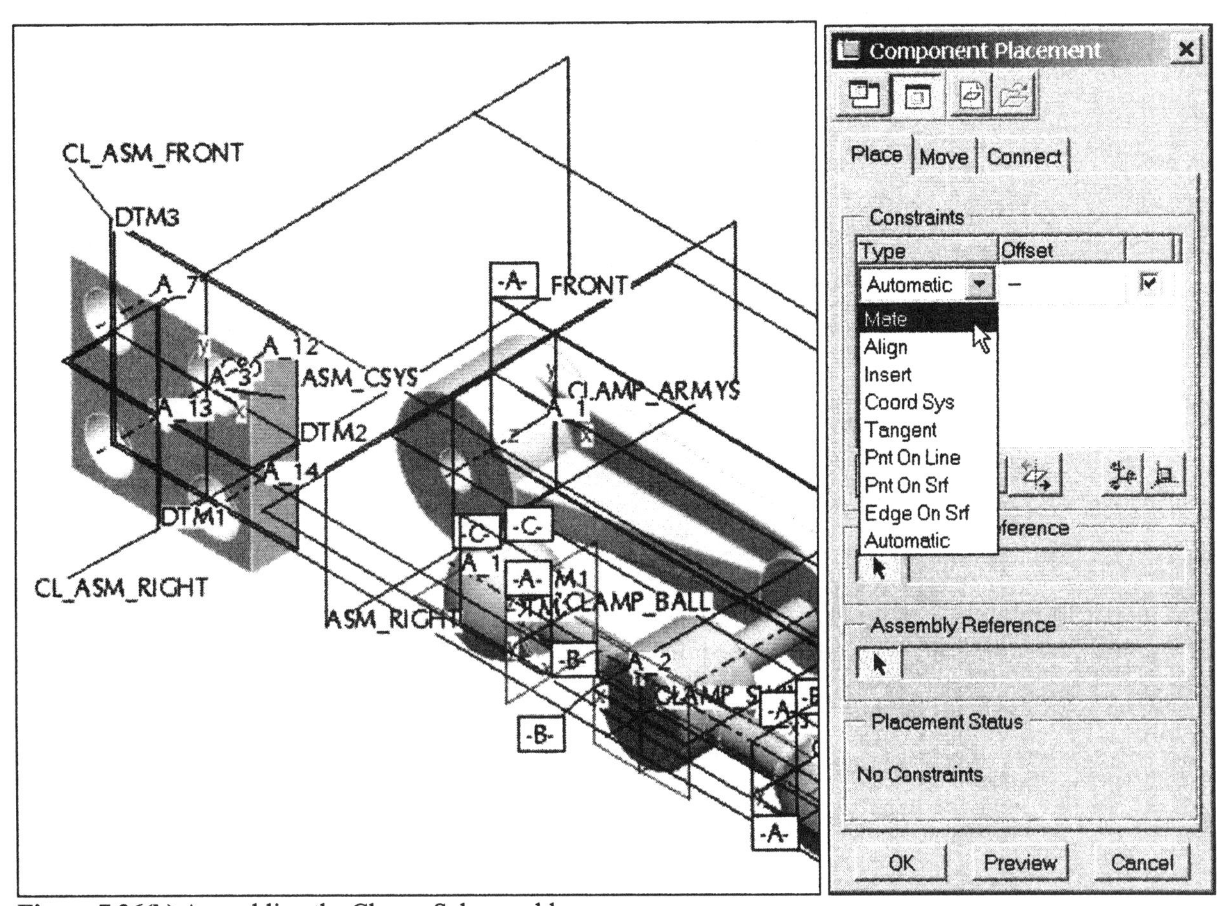

Figure 7.36(b) Assembling the Clamp_Subassembly

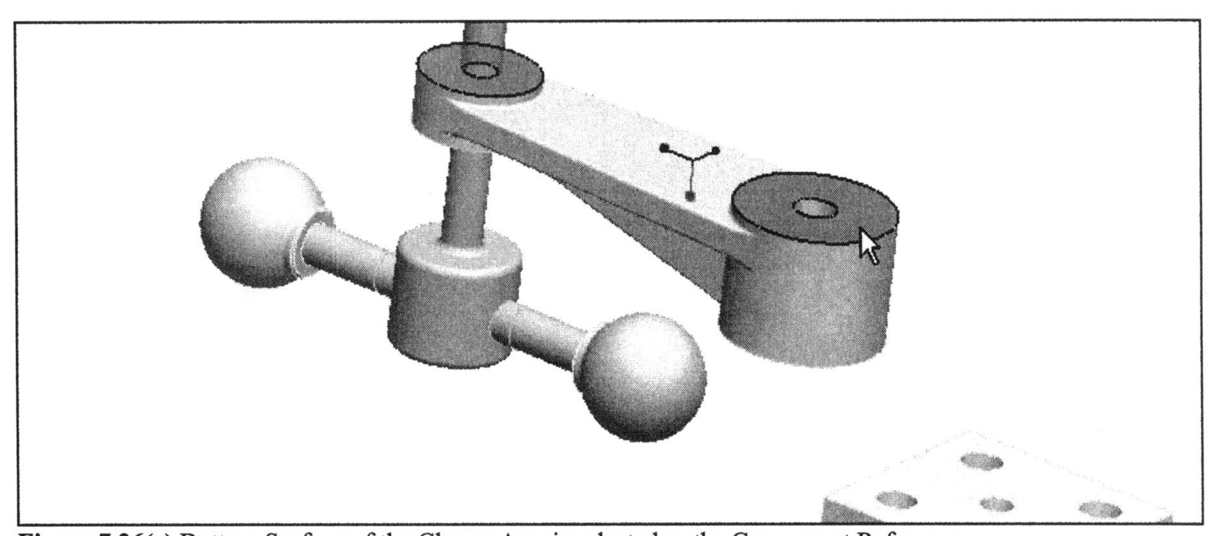

Figure 7.36(c) Bottom Surface of the Clamp_Arm is selected as the Component Reference

Spin the model ⇒ **View** ⇒ **Shade** ⇒ pick the top surface of the Clamp_Plate [Fig. 7.36(d)] ⇒ **Specify a new constraint** ⇒ Automatic ⇒ ⇒ **Align** ⇒ pick the hole axis of the Clamp_Arm [Fig. 7.36(e)] ⇒ pick the center tapped hole axis of the Clamp_Plate [Fig. 7.36(f)] *(note that you could have picked the two respective hole surfaces and achieve the same result with an Insert Constraint)*

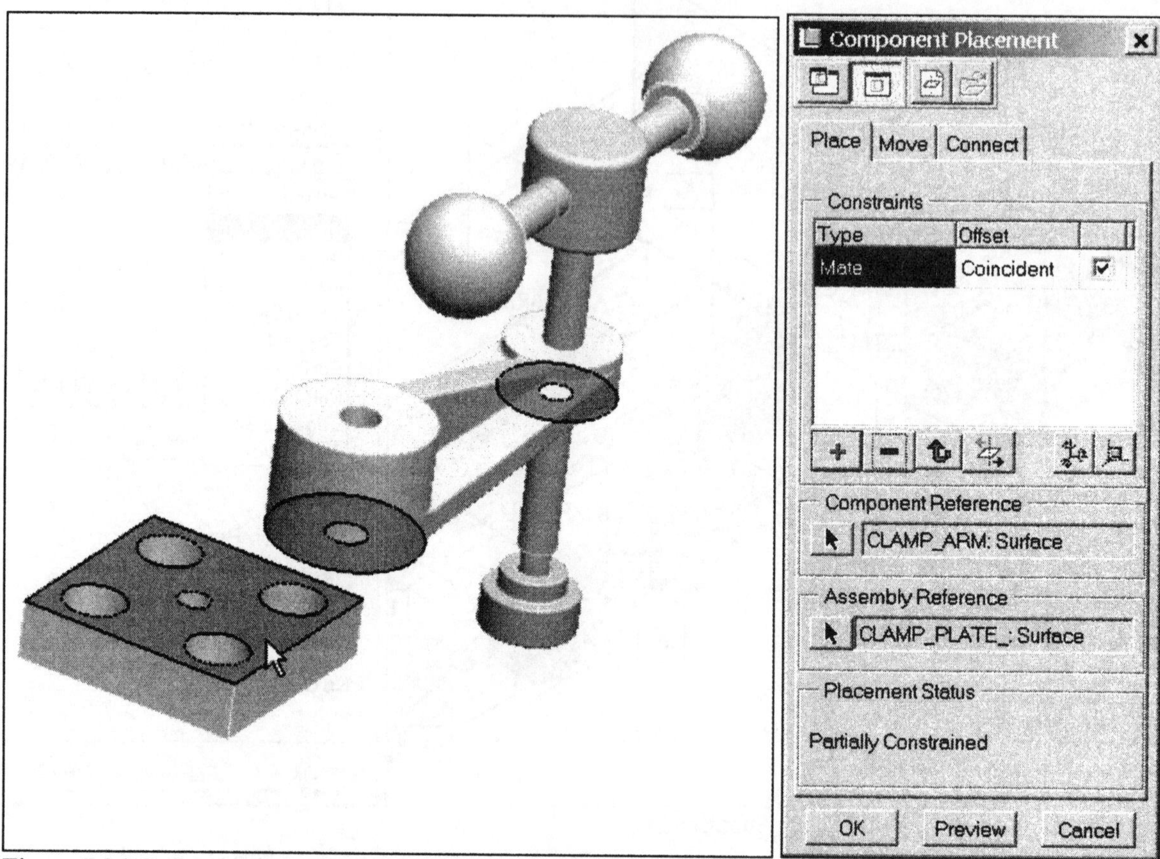

Figure 7.36(d) Top Surface of the Clamp_Plate is selected as the Assembly Reference

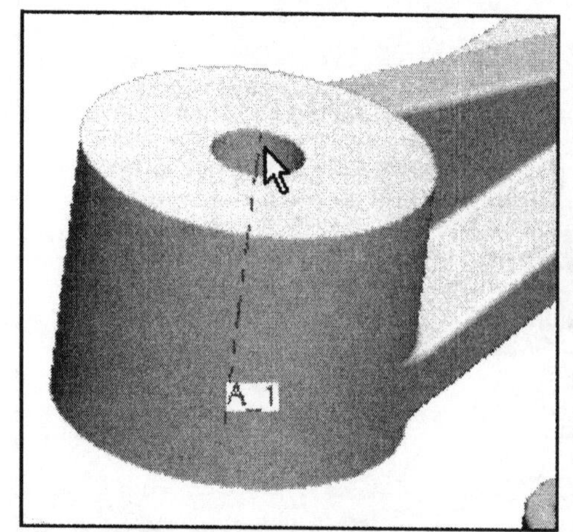

Figure 7.36(e) Pick the Axis on the Clamp_Arm

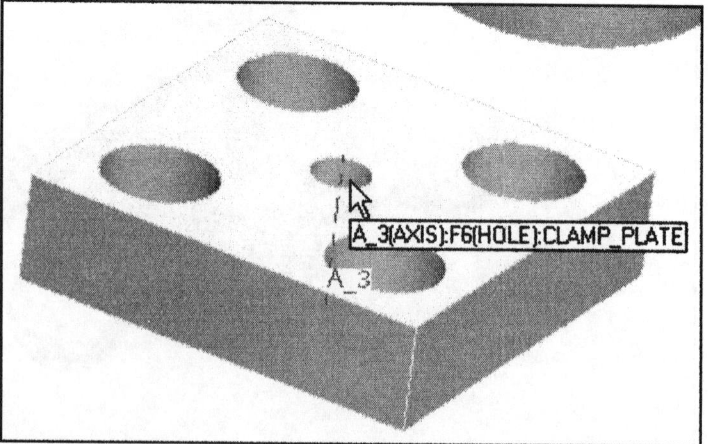

Figure 7.36(f) Pick the Tapped Hole Axis on the Clamp_Plate

Click: **OK** [Fig. 7.36(g)] ⇒ [icon] ⇒ **FRONT** (Fig. 7.37) ⇒ **View** ⇒ **Shade** ⇒ [icon] *(or View ⇒ Orientation ⇒ Standard Orientation)* ⇒ [icon] ⇒ **MMB**

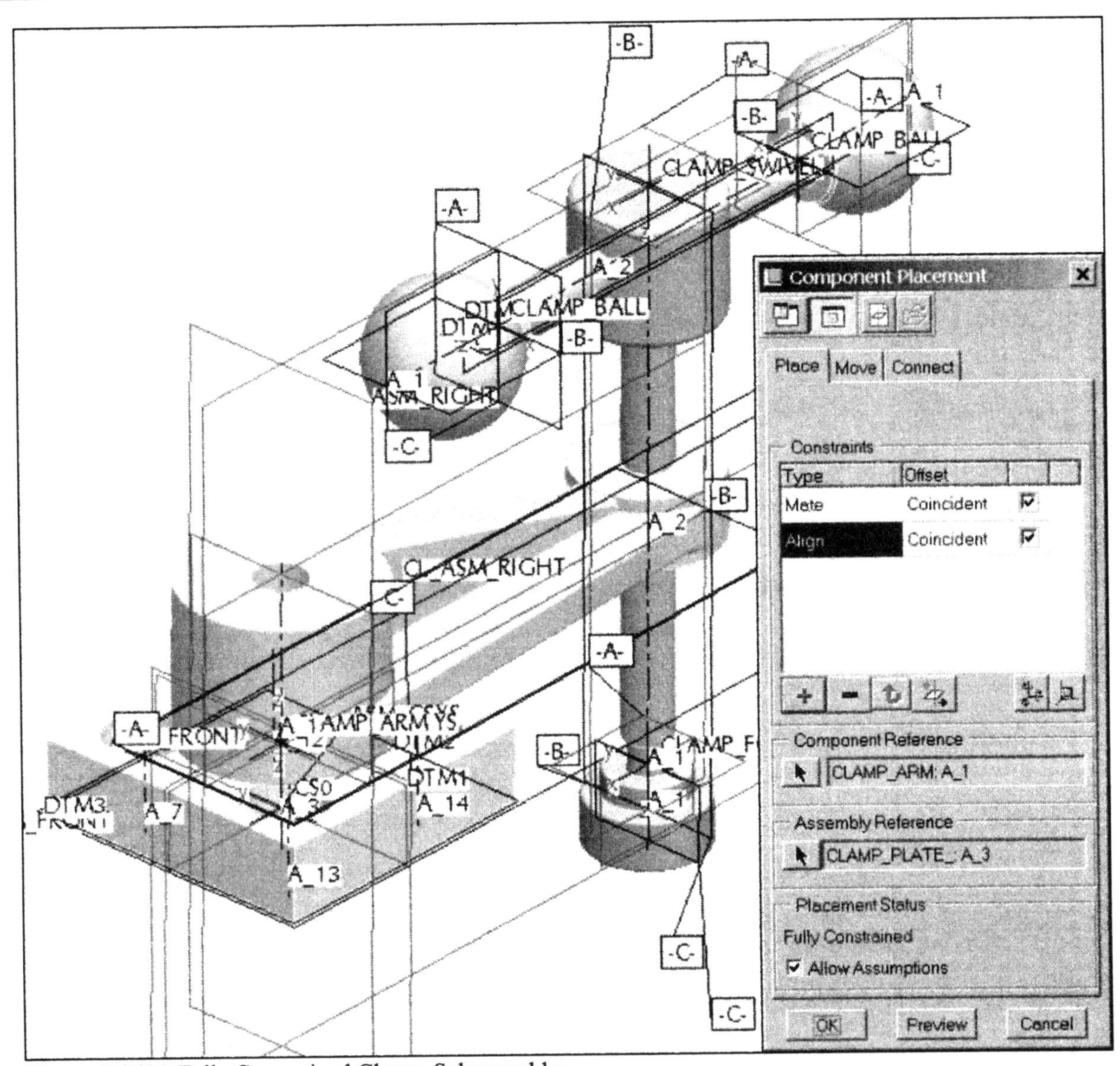

Figure 7.36(g) Fully Constrained Clamp_Subassembly

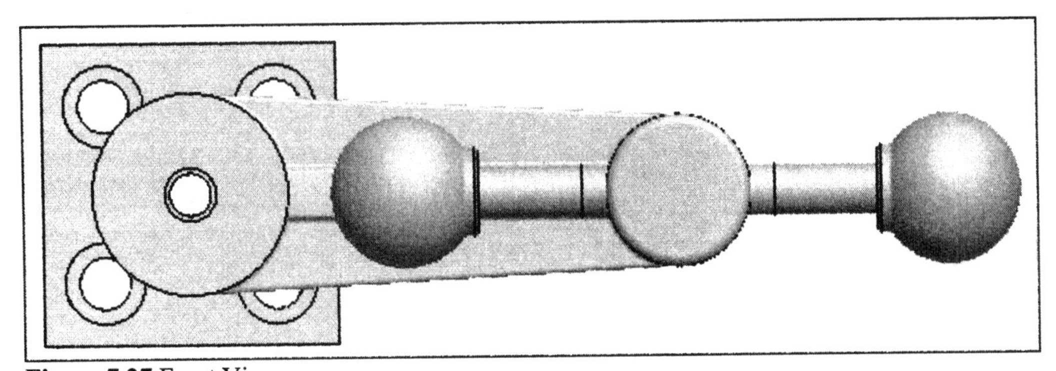

Figure 7.37 Front View

Freeform Mouse-Driven Component Manipulation

The last two components of the Clamp_Assembly will now be added. In both cases, you will use Freeform Mouse-Driven Component Manipulation. Whenever the Component Placement dialog box (Fig. 7.38) is available for placing a component or redefining placement constraints, a spin center (when active) for the active component is always visible. You can manipulate the active component using a combination of mouse and keyboard commands *(Ctrl+Alt pressed at the same time)*. Manipulating a component is easier than moving to a separate tab in the Component Placement dialog box to package-move a component around on the screen. You can switch between full view navigation and component manipulation easily with the Ctrl+Alt keys. You can perform translation and rotation adjustments while you establish constraints. The component motion respects any constraints as they are established, as is the case with *Move tab* functionality. The spin icon appears during the entire component placement operation and defaults to the bounding box center. You can modify this location, using the Preferences dialog box, accessed from the Orientation dialog box.

Regardless of the selected tab- Place or Move: you can translate or rotate the component dynamically using:

- **Translate** Press and hold: **Ctrl+Alt** with **RMB**
- **Rotate** Press and hold: **Ctrl+Alt** with **MMB**
- **Z-axis** Press and hold: **Ctrl+Alt** with **LMB**

Because of the orthographic projection used by Pro/E, motion in the Z-axis, or screen normal, is noticeable only if objects intersect. Thus, while a component is moving in the Z-axis, the camera angle is adjusted to provide a noticeable effect.

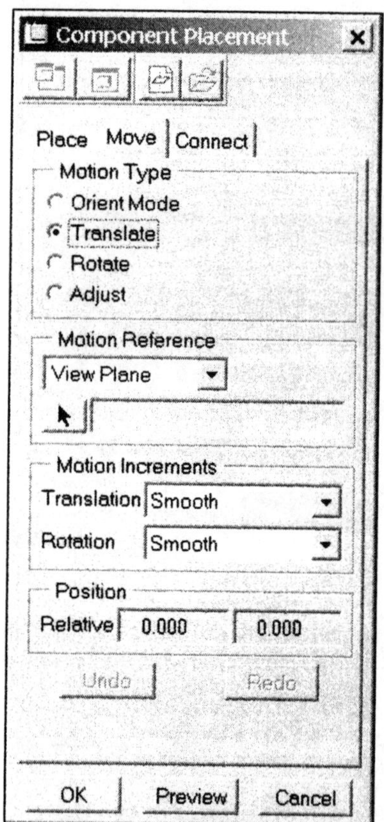

Figure 7.38 Component Placement Dialog Box (Move Tab Active)

Click: **Add component to the assembly** ⇒ pick the **clamp_stud35.prt** ⇒ **Preview** [Fig. 7.39(a)] ⇒ **Open**

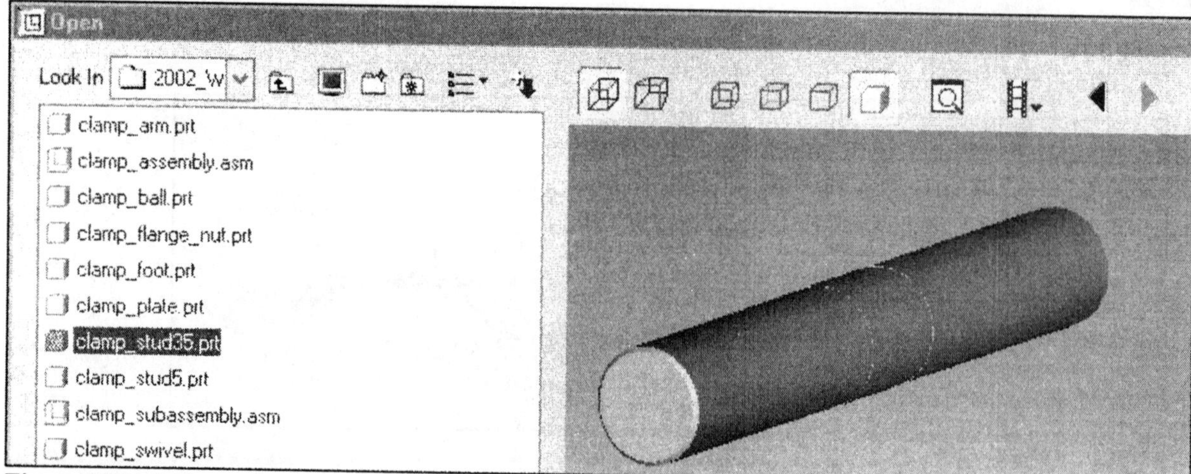

Figure 7.39(a) Clamp_Stud35

Click: **Datum planes off** ⇒ Constraints Automatic, click ⇒ Insert ⇒ pick the cylindrical surface of the Clamp_Stud35 ⇒ pick the hole surface of the Clamp_Arm [Figs. 7.39(b-c)]

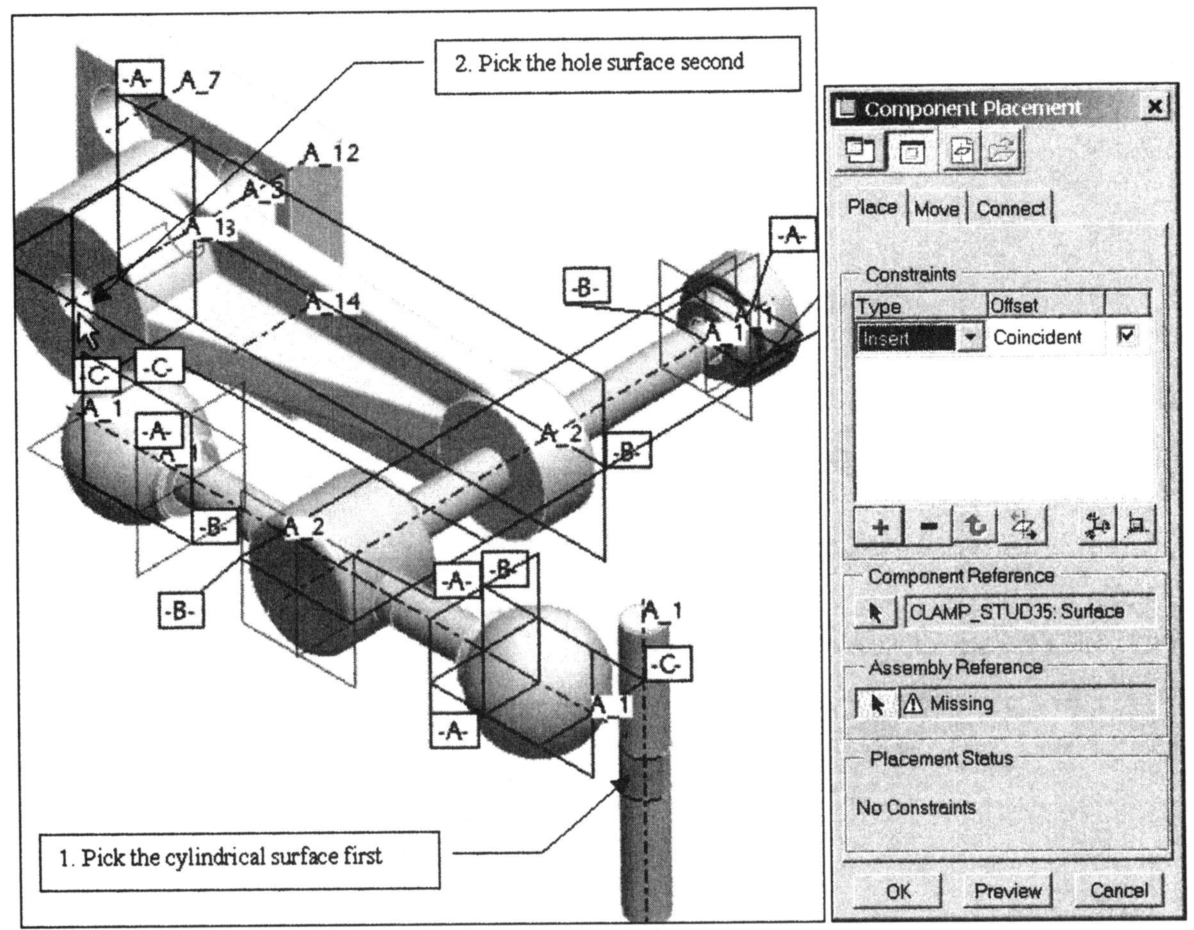

Figure 7.39(b) Component Placement Dialog Box, Place Tab

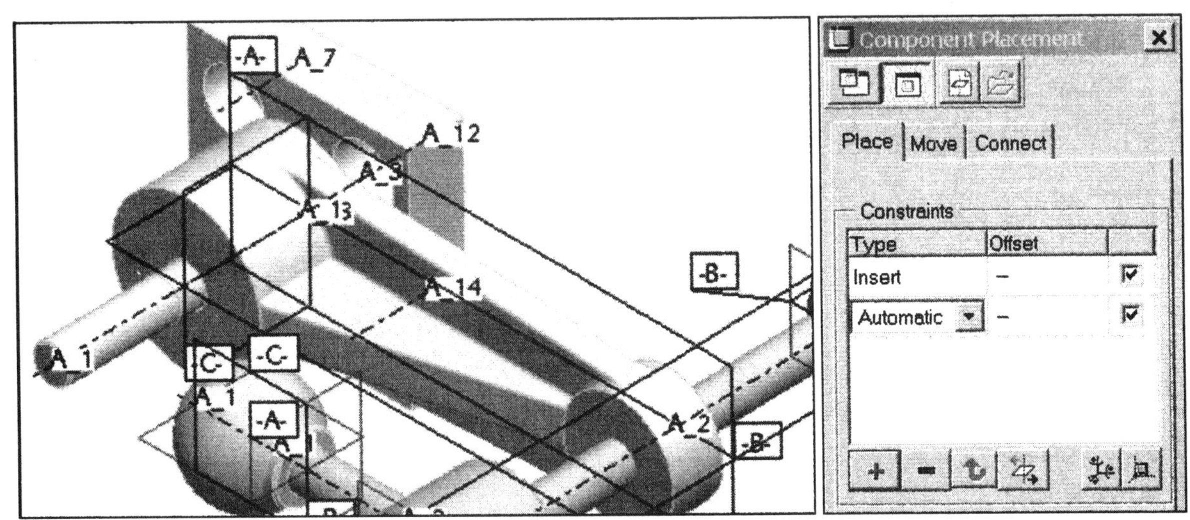

Figure 7.39(c) Insert Constraint

Click: ⇒ **LEFT** ⇒ **Move** tab ⇒ Motion Type ⇒ pick the Clamp_Stud35 and slide it deeper into the hole ⇒ **LMB** [Fig. 7.39(d)] ⇒ ⇒ **TOP** [Fig. 7.39(e)] ⇒ **Hidden Line** ⇒ pick Clamp_Stud35 and translate it until it is slightly inside the Clamp_Plate [Figs. 7.39(f-g)] ⇒ **LMB**

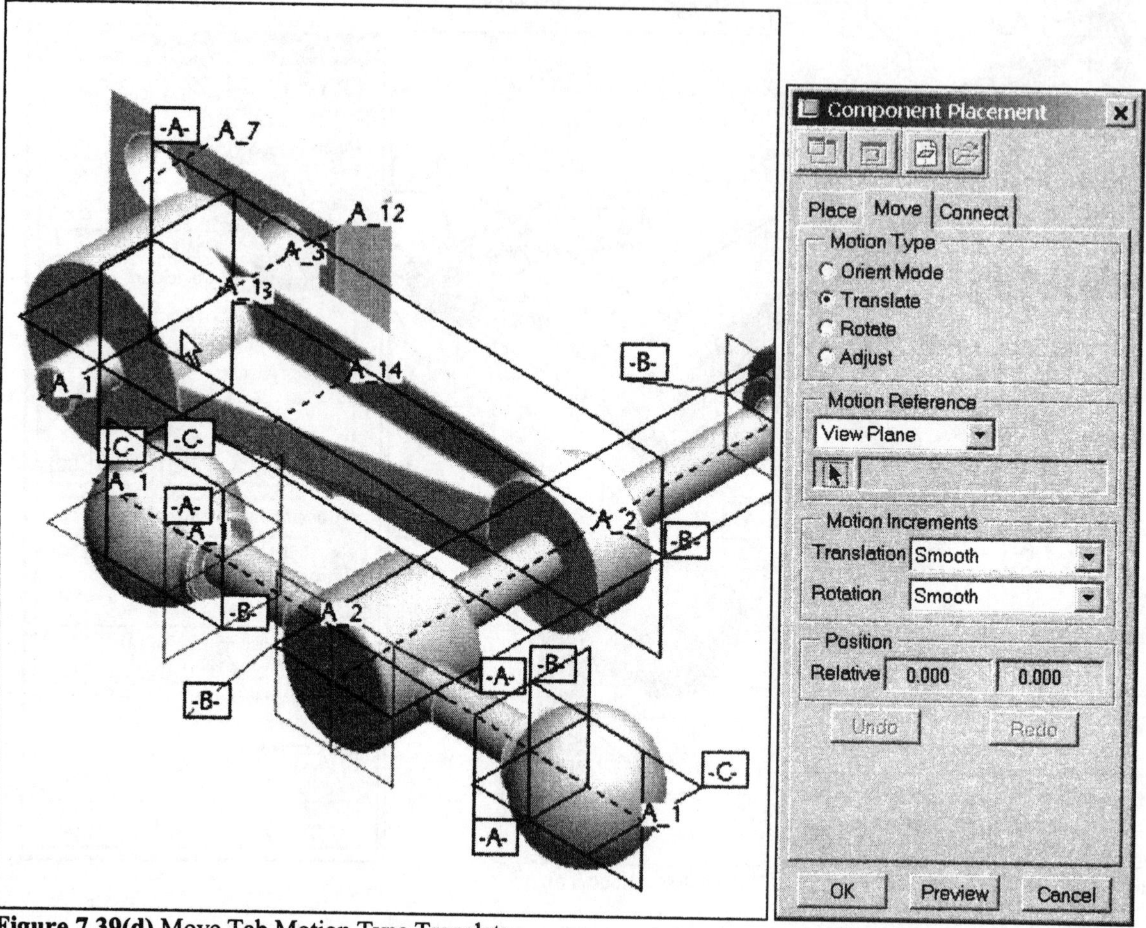

Figure 7.39(d) Move Tab Motion Type Translate

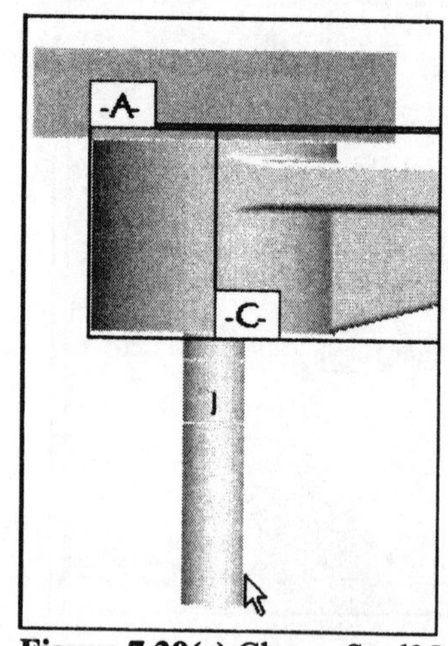

Figure 7.39(e) Clamp_Stud35

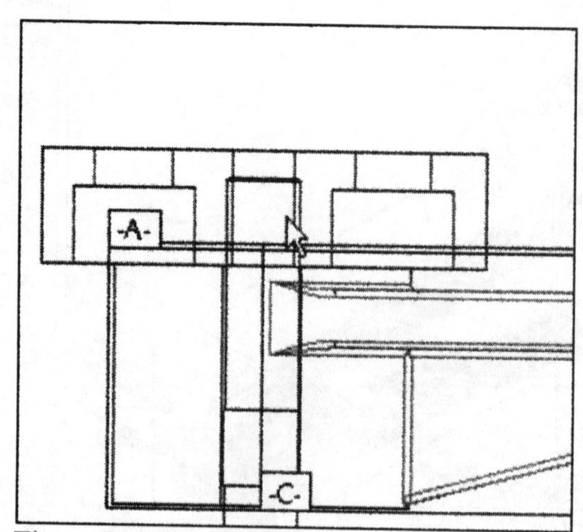

Figure 7.39(f) Translated Position

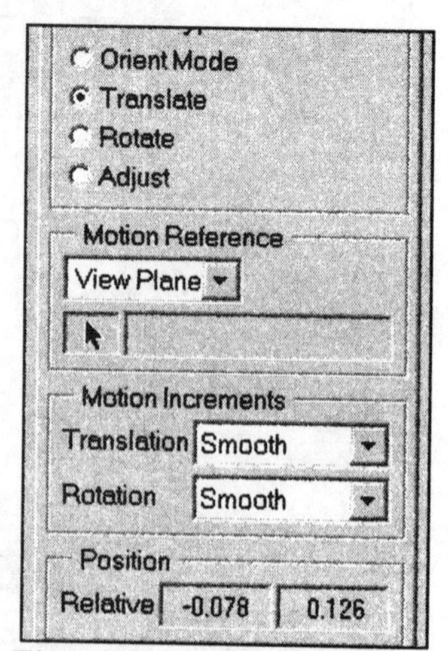

Figure 7.39(g) Placement

Click: **Place** tab ⇒ ⇒ **Standard Orientation** ⇒ **Shading** ⇒ **Fix component to current position** [Fig. 7.39(h)] ⇒ **OK** ⇒ *(or View ⇒ Orientation ⇒ Standard Orientation)* ⇒ ⇒ **MMB**

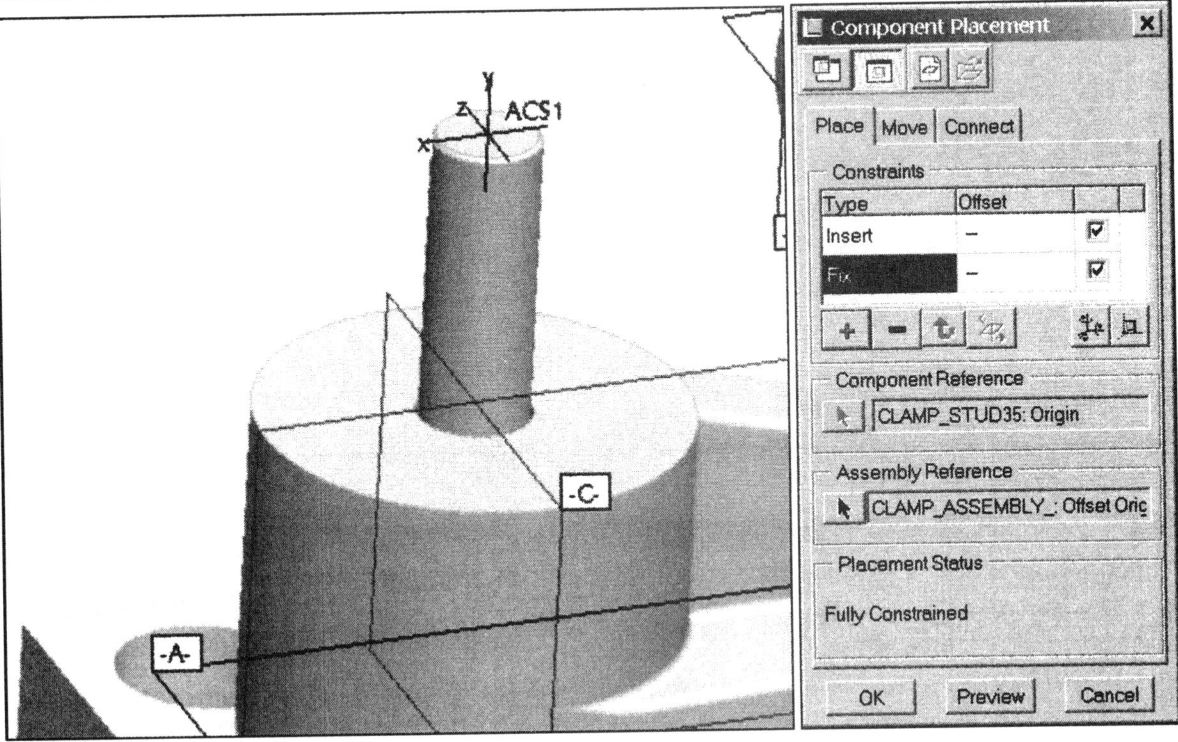

Figure 7.39(h) Add Fix Constraint

Click: **Add component to the assembly** ⇒ pick the **clamp_flange_nut.prt** ⇒ **Preview** [Fig. 7.40(a)] ⇒ **Open** ⇒ **Move** tab ⇒ Motion Type **Adjust** ⇒ Motion Reference **Sel Plane** ⇒ pick the top surface of the Clamp_Arm [Fig. 7.40(b)]

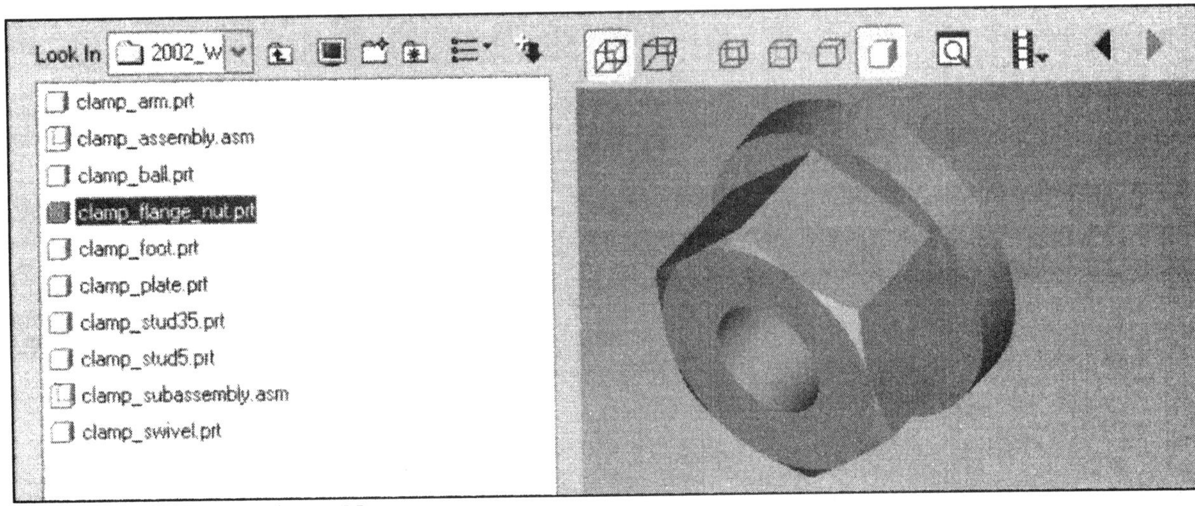

Figure 7.40(a) Clamp_Flange_Nut

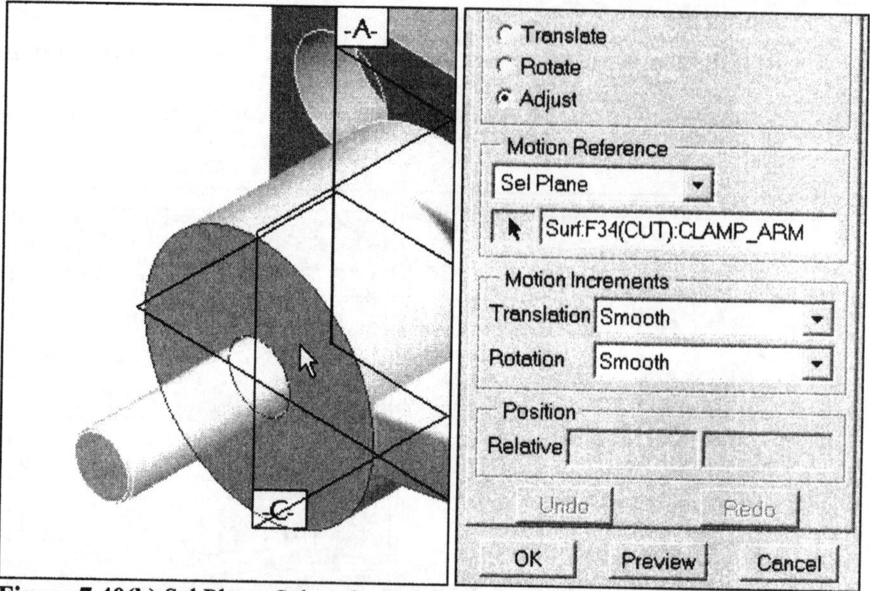

Figure 7.40(b) Sel Plane, Select the Top Surface of the Clamp_Arm

Pick the top surface of the Clamp_Flange_Nut [Fig. 7.40(c)] ⇒ **Align Offset** ⇒ type **2.00** ⇒ **Enter** ⇒ Motion Type **Translate** ⇒ Motion Reference **View Plane** ⇒ [icon] ⇒ **FRONT** ⇒ pick the Clamp_Flange_Nut and slide it near the Clamp_Stud35 [Figs. 7.40(d-e)] ⇒ **LMB** again to place [Fig. 7.40(f)] ⇒ [icon] *(or View ⇒ Orientation ⇒ Standard Orientation)*

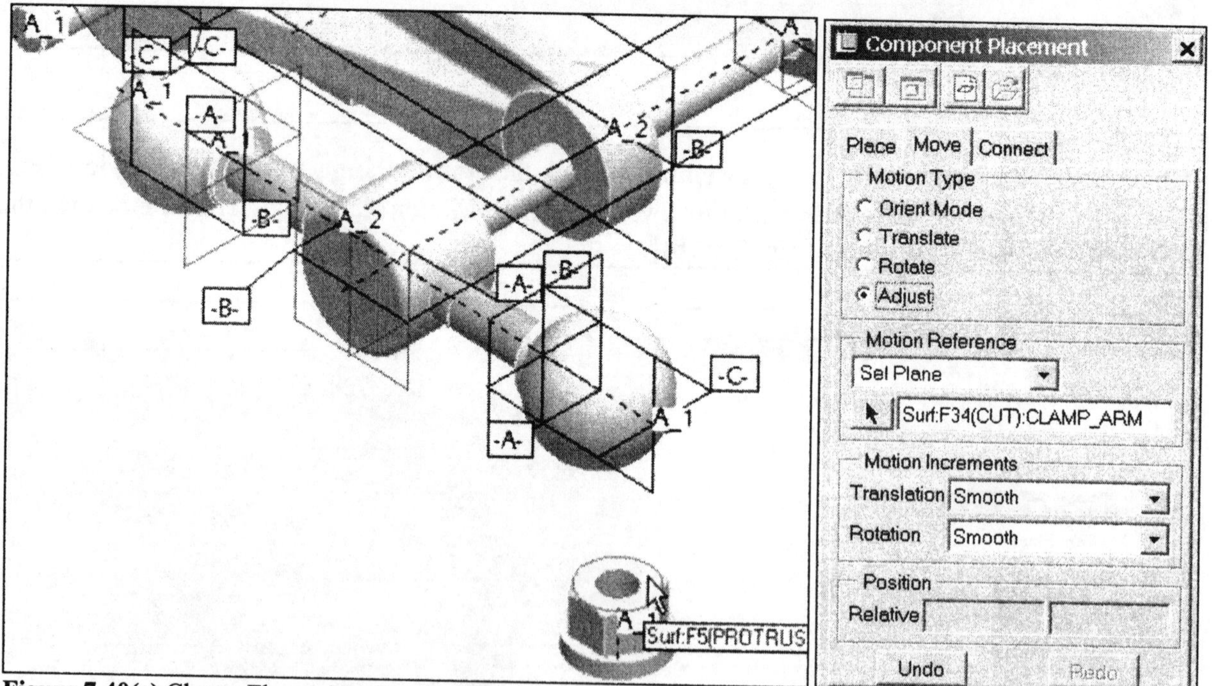

Figure 7.40(c) Clamp_Flange_Nut Top Surface

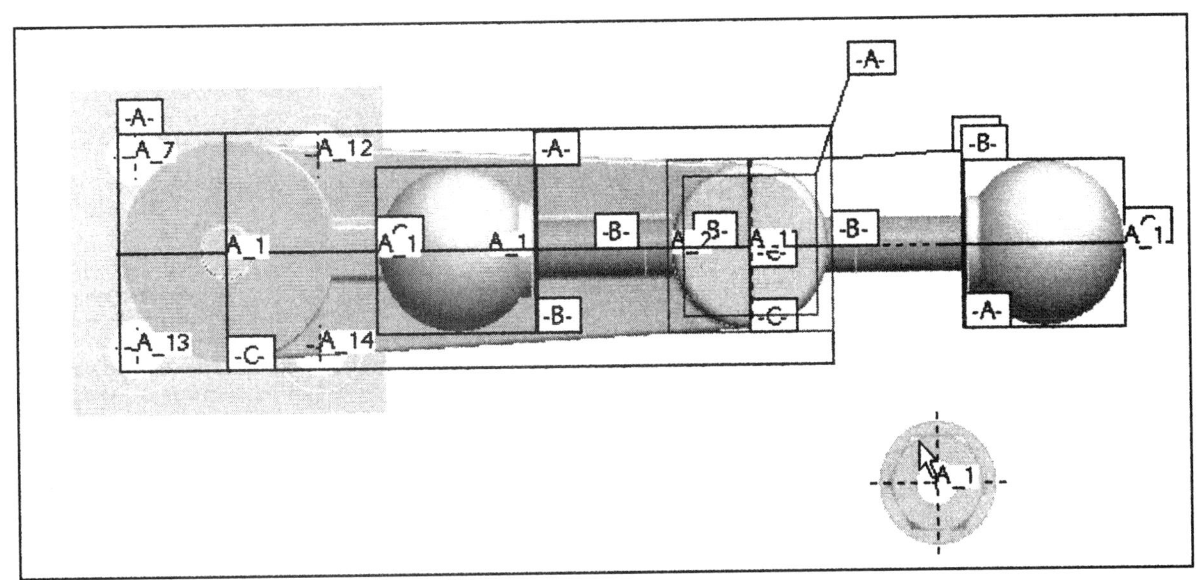

Figure 7.40(d) Select the Clamp_Flange_Nut

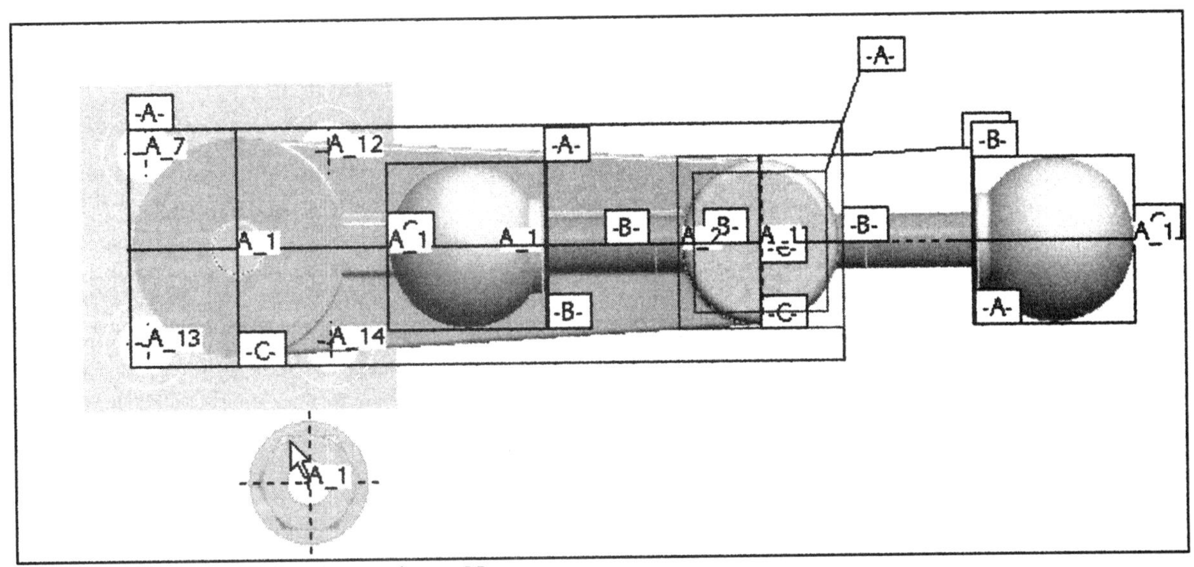

Figure 7.40(e) Moving the Clamp_Flange_Nut

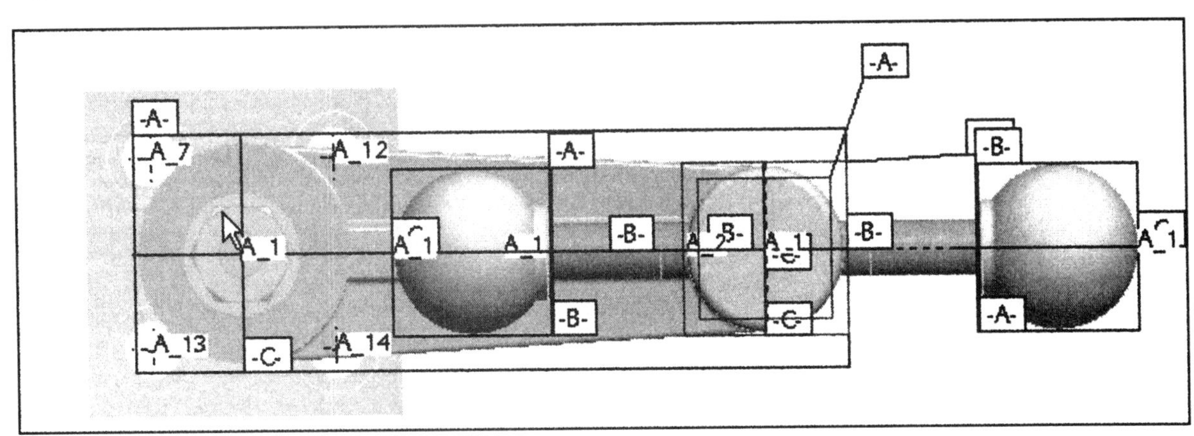

Figure 7.40(f) Position the Clamp_Flange_Nut

Click: **Place** tab Automatic ⇒ pick the hole's surface of the Clamp_Flange_Nut ⇒ pick on the shaft of the Clamp_Stud35 [Fig. 7.40(g)] *constraint becomes* ***Insert*** ⇒ Automatic pick on the upper surface of the Clamp_Arm ⇒ place the cursor on the Clamp_Flange_Nut [Fig. 7.40(h)] ⇒ **RMB** ⇒ **Next** ⇒ **LMB** to select [Fig. 7.40(i)] *constraint becomes* ***Mate*** ⇒ type **0** ⇒ **Enter** [Fig. 7.40(j)] ⇒ **OK** ⇒ ⇒ **Standard Orientation** ⇒ ⇒ **MMB**

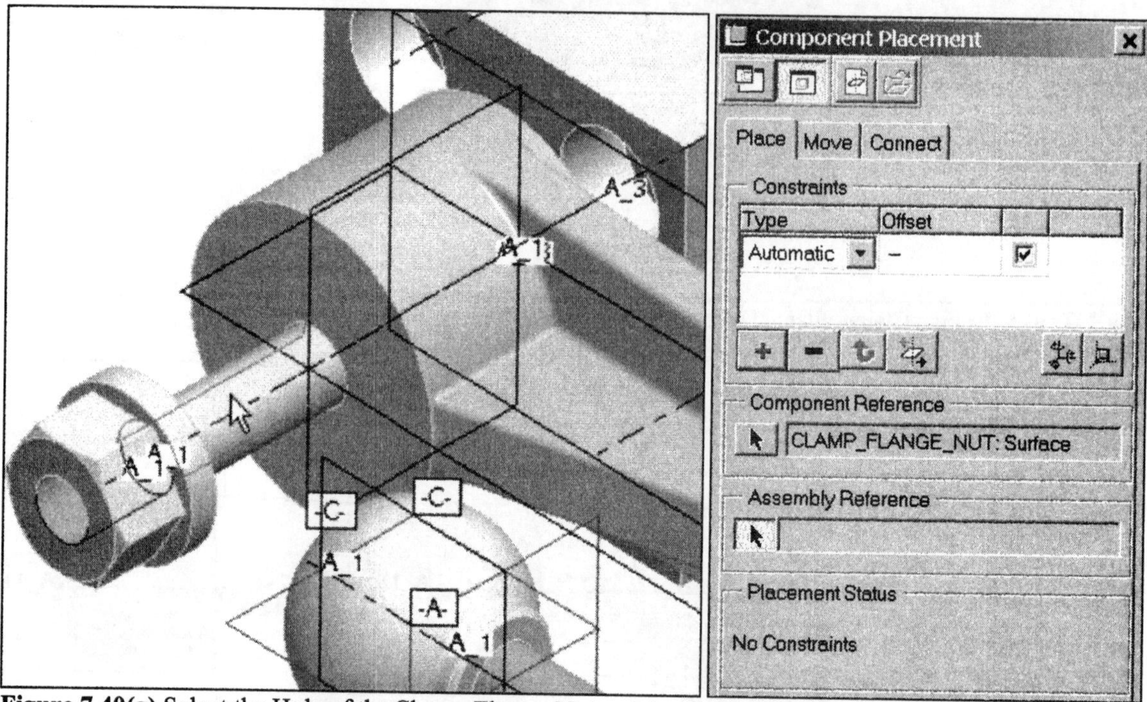

Figure 7.40(g) Select the Hole of the Clamp_Flange_Nut and the Shaft of the Clamp_Stud35

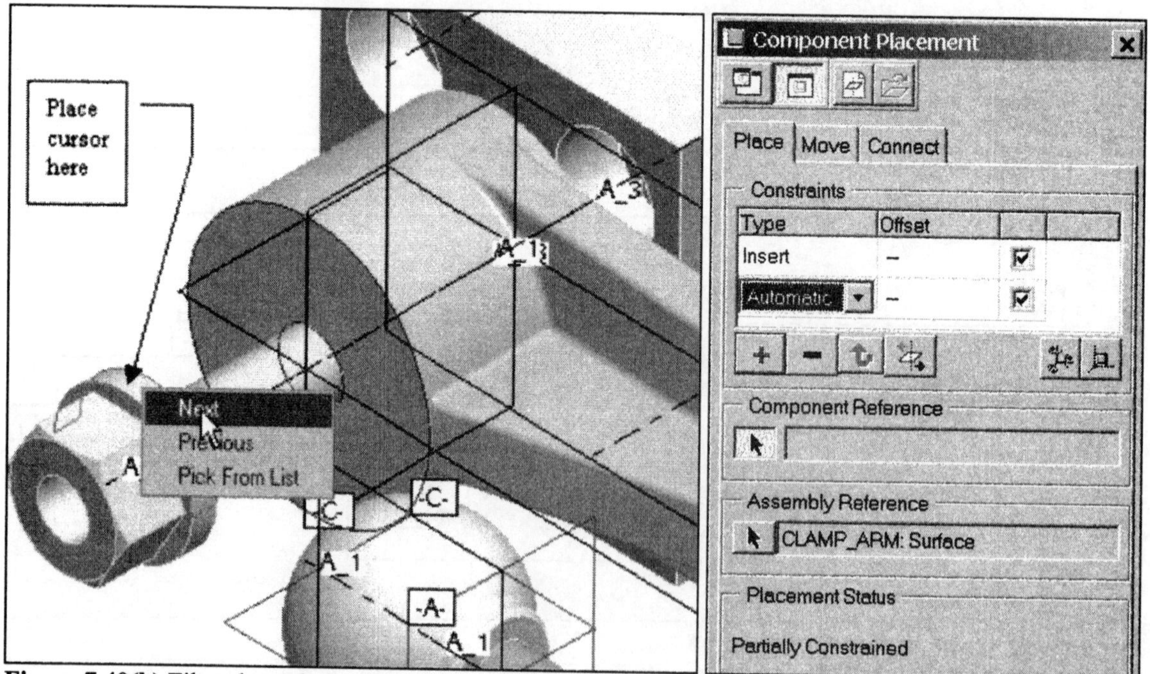

Figure 7.40(h) Filter through the Selections to Highlight the Bottom Surface of the Clamp_Flange_Nut

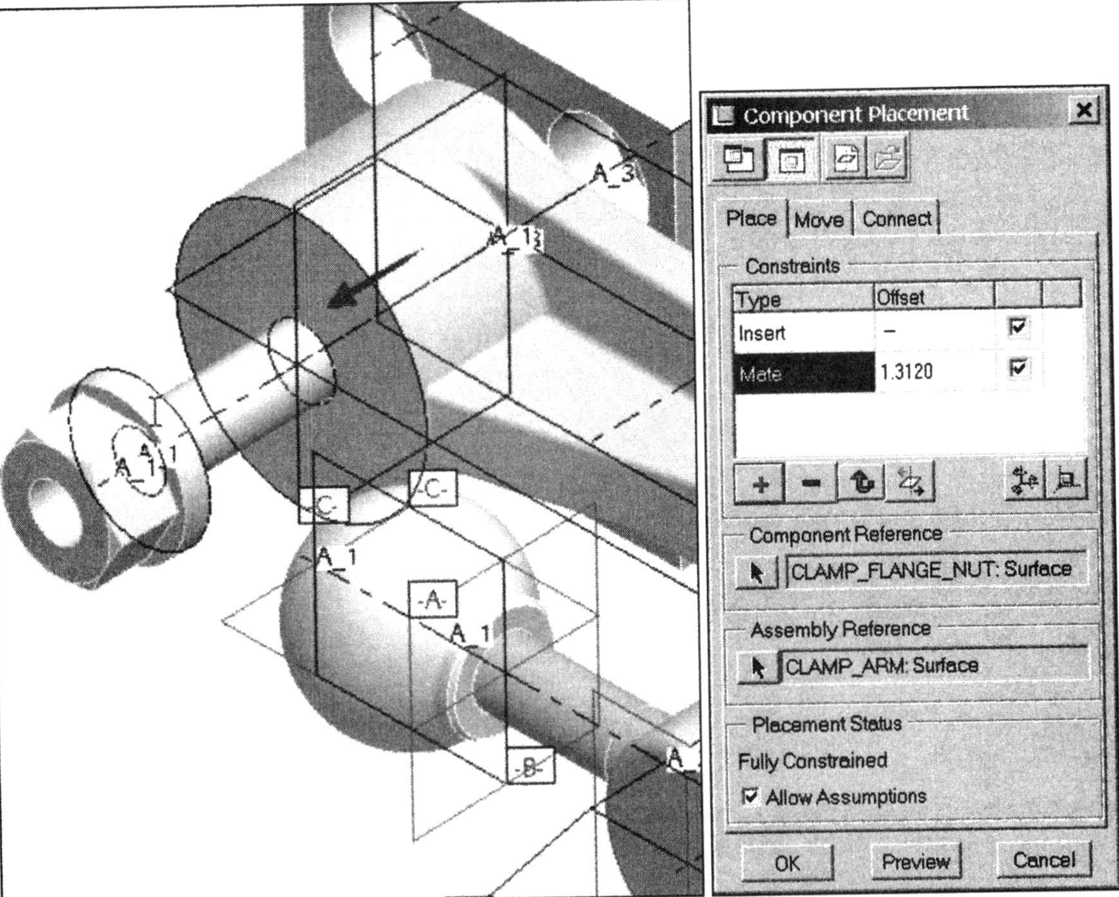

Figure 7.40(i) Select Bottom Surface of the Clamp_Flange_Nut. Constraint becomes Mate

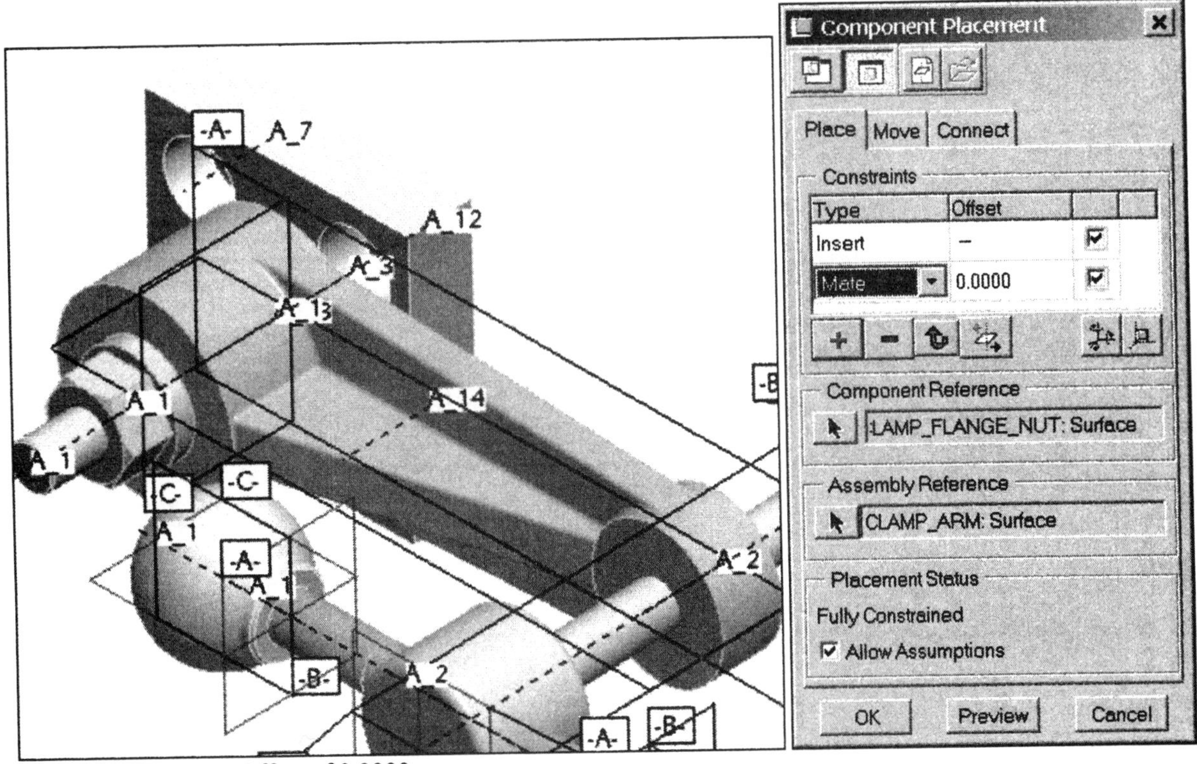

Figure 7.40(j) Mate Offset of **0.0000**

Perform a check on the assembly using the **Analysis** command. If you look at the long **(5.00)** stud's detail drawing [see Figure 7.14(d)], you will see that the shaft diameter is greater than **.500** at the center (**.506**) of the stud and that each end has a **.500-13 UNC** thread. The hole in the swivel is **.500**. This means that there should be a slight interference (Press Fit) between the two components. *(Note: if you used the CARR Lane part from the PTC catalog, the neutral file part has a constant .500 diameter)*. Check the clearance and interference between these components.

Click: **Analysis** ⇒ **Model Analysis** ⇒ Type **Pairs Clearance** ⇒ Definition- From **Whole Part** ⇒ pick the Clamp_Swivel ⇒ Definition- To **Whole Part** ⇒ pick the Clamp_Stud5 (Fig. 7.41) ⇒ Results Volume of interference is 0.00690137. *(Note: the interference will be different or zero if you used a PTC Catalog CARR Lane Stud)* ⇒ **Close**

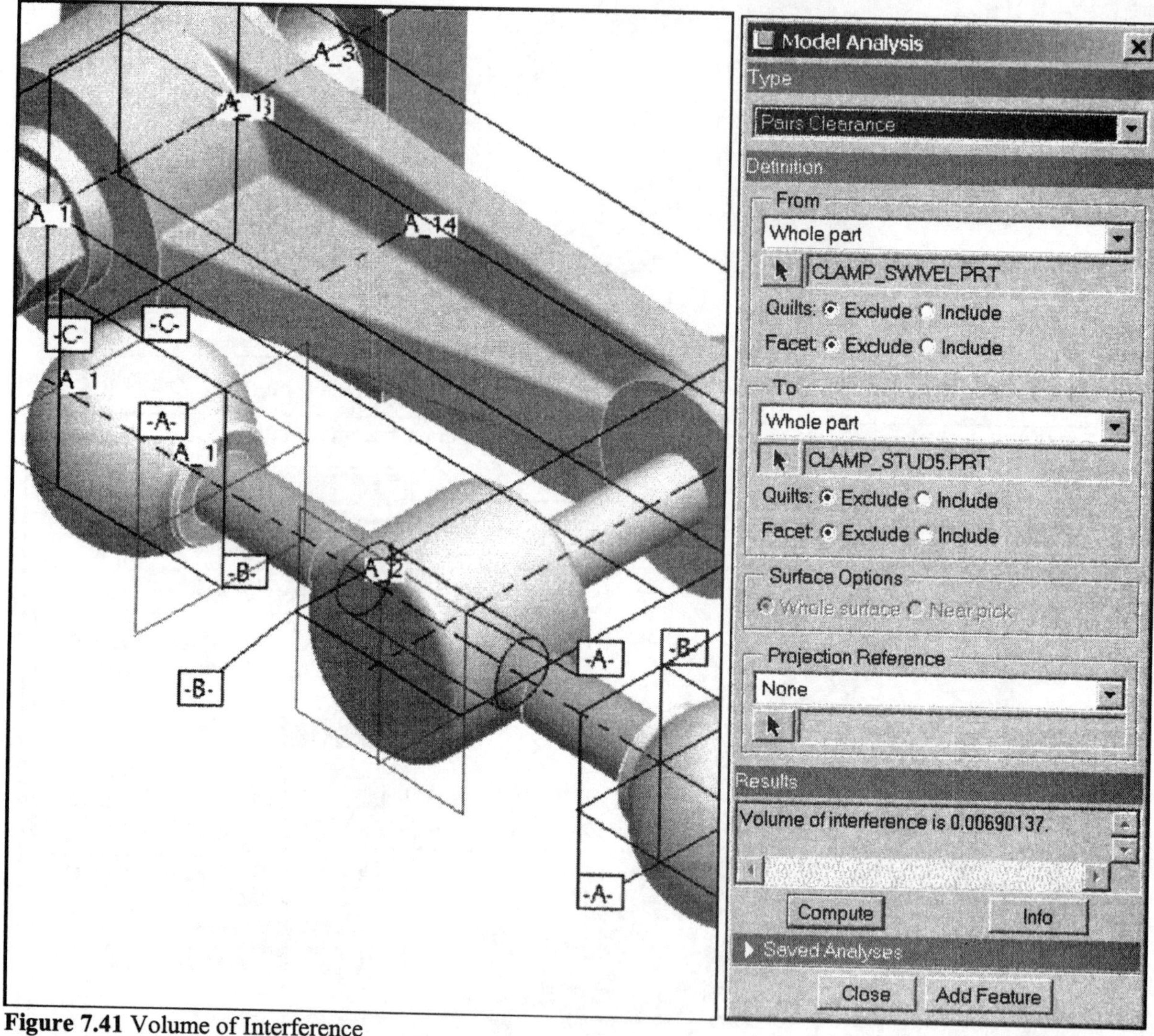

Figure 7.41 Volume of Interference

Rotating Components of an Assembly

To rotate an existing component or set of components of an assembly (or subassembly), you select a coordinate system to use as a reference, pick one or more components, and give the rotation angle about a chosen axis of the reference coordinate system. The Clamp_Subassembly has a Clamp_Swivel, Clamp_Foot, Clamp_Stud5, and two Clamp_Ball components that can be rotated about the Clamp_Arm during normal operation of the assembly. You will rotate these components so that the Clamp_Stud5 and two Clamp_Ball components are perpendicular to the Clamp_Arm. This position looks better when you are displaying the assembly as exploded (and in its unexploded state). The components will be rotated in the subassembly, and the change will be propagated to the assembly.

Click: CLAMP_SUBASSEMBLY.ASM in the Model Tree [Fig. 7.42(a)] ⇒ **RMB** ⇒ **Open** [Fig. 7.42(b)] ⇒ **Window** in the Clamp_Subassembly menu bar ⇒ **Activate** to make sure you are working on the correct object ⇒ **Datum planes off** ⇒ **Redraw the current view**

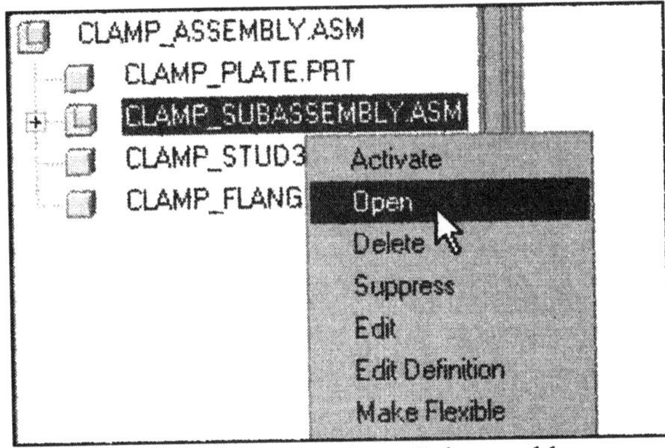

Figure 7.42(a) Open the Clamp Subassembly

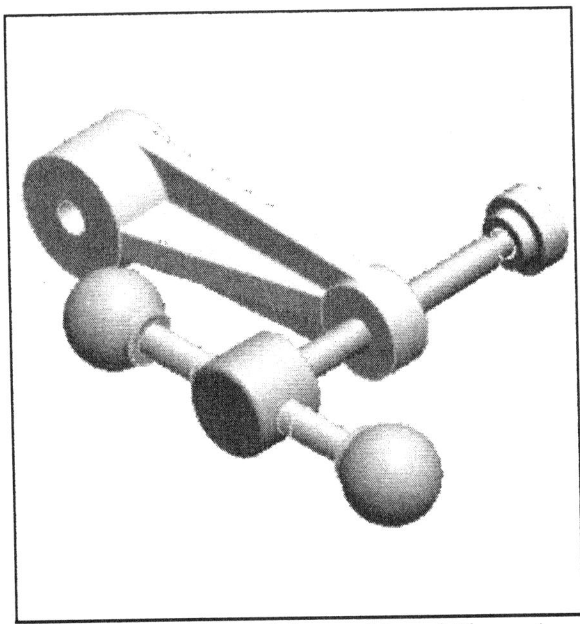

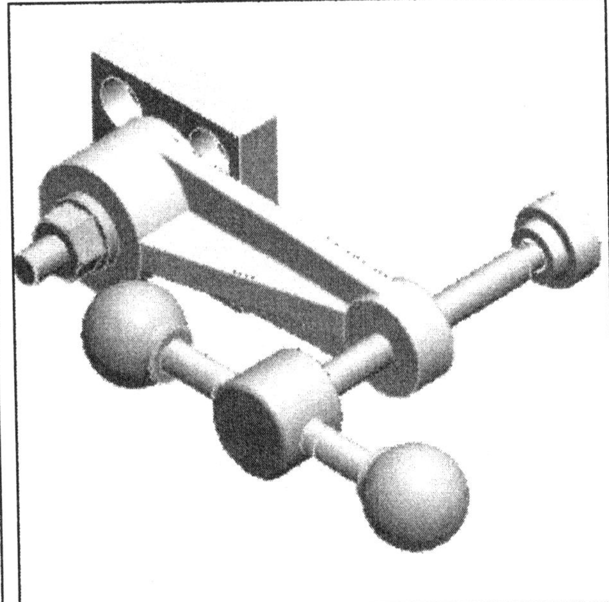

Figure 7.42(b) Clamp_Subassembly and Clamp_Assembly are Open in Different Windows

Click: **Edit** ⇒ **Component Operations** ⇒ ▼ ⇒ **Transform** ⇒ Select coordinate system: pick coordinate system **CLAMP_SWIVEL** [Fig. 7.42(c)] *or choose the coordinate system from the Model Tree* ⇒ Select components to move: pick the **CLAMP_SWIVEL.PRT** component [Fig. 7.42(d)] *you may also choose the component from the Model Tree* ⇒ **MMB** ⇒ **Rotate** ⇒ **Z Axis** *check your model, if you modeled it according to the instructions in Lesson Project 5, the Z axis of the coordinate system will run along the protrusion axis of the Clamp_Swivel.prt, if not, then select the appropriate axis on your component* ⇒ type **90** as the angle of rotation ⇒ **Enter** ⇒ **MMB** ⇒ **MMB** ⇒ **Regenerates Model** [Fig. 7.42(e)]

Only the Clamp_Swivel was selected, but the Clamp_Stud5, both Clamp_Balls, and the Clamp_Foot were rotated **90** degrees. These other components were children of the Clamp_Swivel and therefore rotated through the same **90** degree angle.

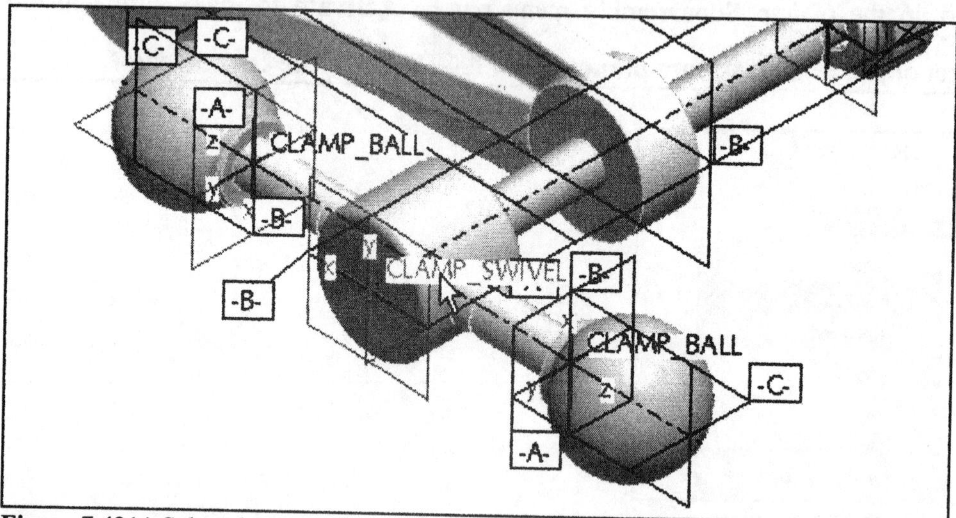

Figure 7.42(c) Select the Coordinate System CLAMP_SWIVEL

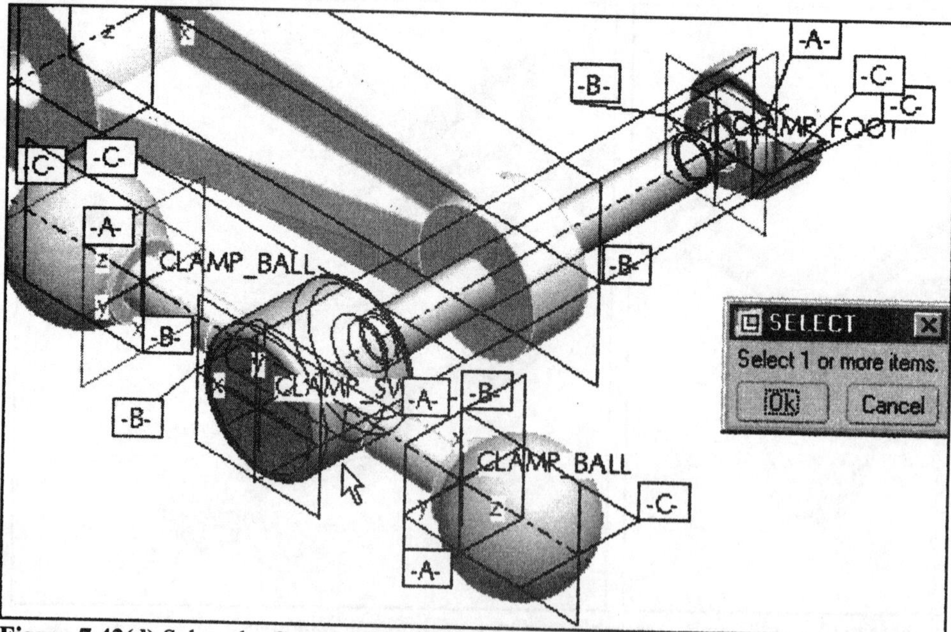

Figure 7.42(d) Select the Component CLAMP_SWIVEL.PRT

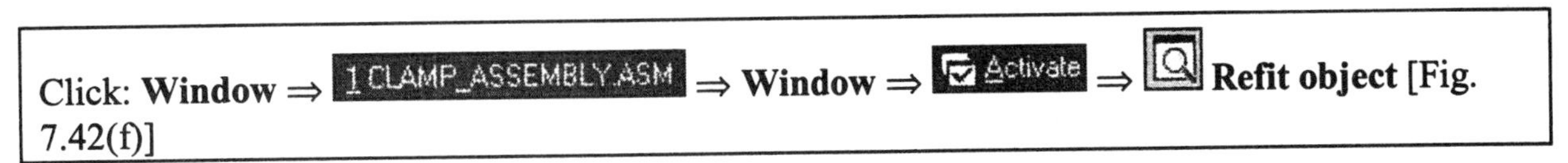
Click: **Window** ⇒ 1 CLAMP_ASSEMBLY.ASM ⇒ **Window** ⇒ Activate ⇒ **Refit object** [Fig. 7.42(f)]

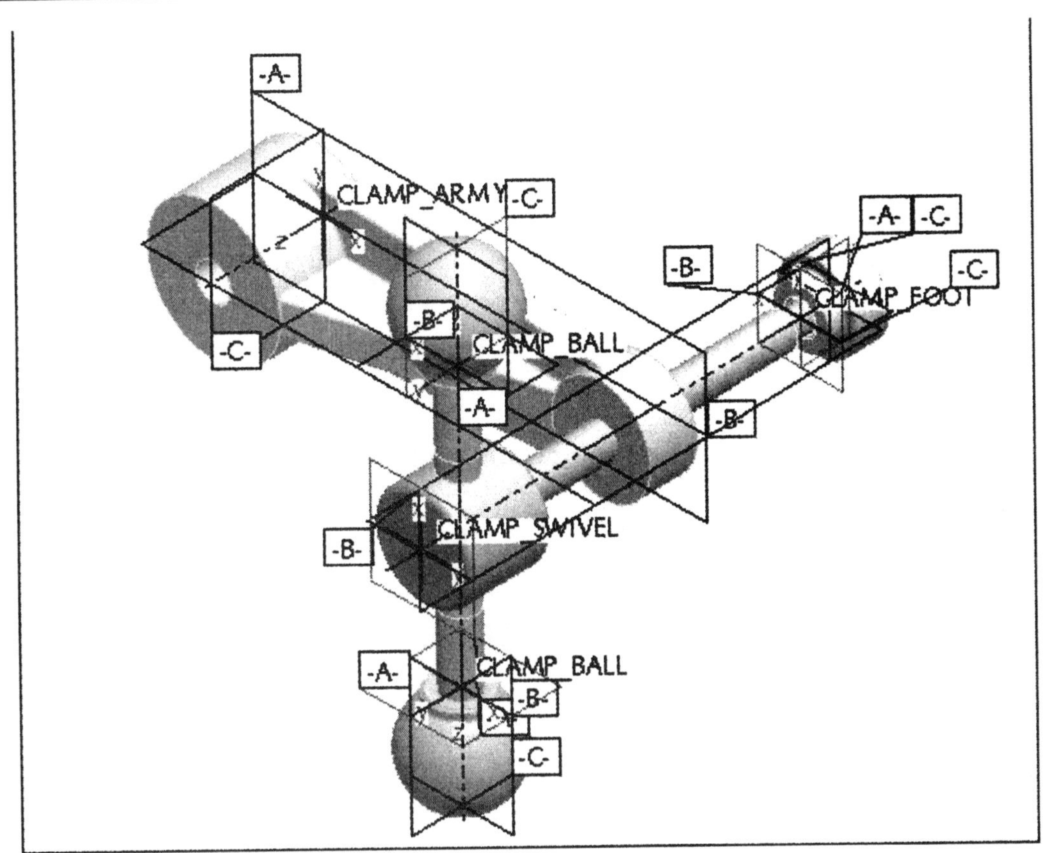

Figure 7.42(e) Rotated Swivel and Component Children

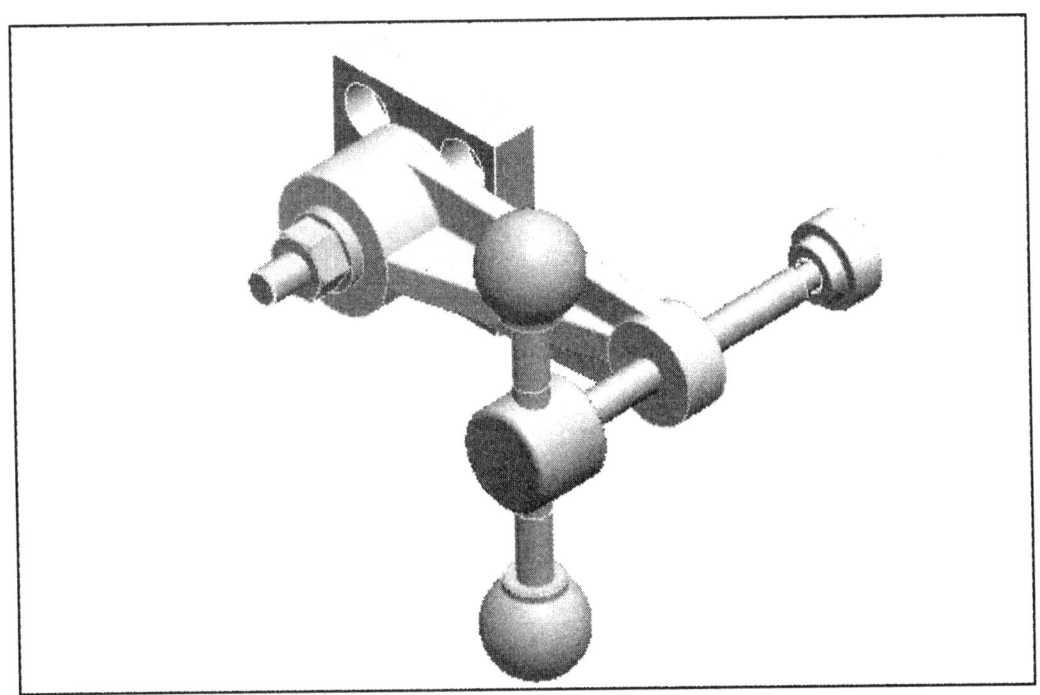

Figure 7.42(f) Clamp Assembly is now Active

Bill of Materials

The Bill of Materials (BOM) lists all parts and part parameters in the current assembly or assembly drawing. It can be displayed in HTML or text format and is separated into two parts: breakdown and summary. The Breakdown section lists what is contained in the current assembly or part. The BOM HTML breakdown section lists quantity, type, name (hyperlink), and three actions (highlight, information and open) about each member or sub-member of your assembly:

- **Quantity** Lists the number of components or drawings
- **Type** Lists the type of the assembly component (part or sub-assembly)
- **Name** Lists the assembly component and is hyper-linked to that item. Selecting this hyperlink highlights the component in the graphics window
- **Action** is divided into three areas:
    - **Highlight** Highlights the selected component in the assembly graphics window
    - **Information** Provides model information on the relevant component
    - **Open** Opens the component in another Pro/ENGINEER window

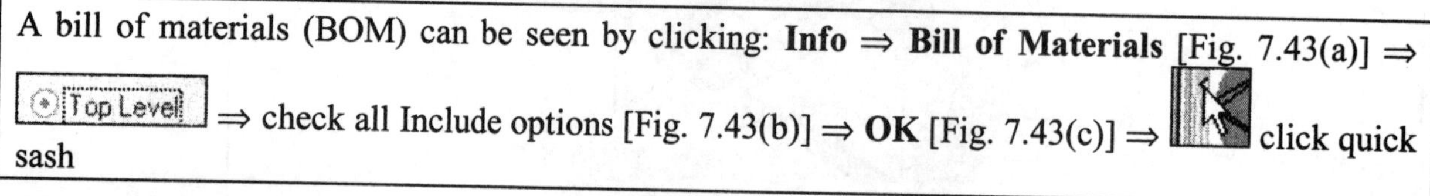
A bill of materials (BOM) can be seen by clicking: **Info** ⇒ **Bill of Materials** [Fig. 7.43(a)] ⇒ Top Level ⇒ check all Include options [Fig. 7.43(b)] ⇒ **OK** [Fig. 7.43(c)] ⇒ click quick sash

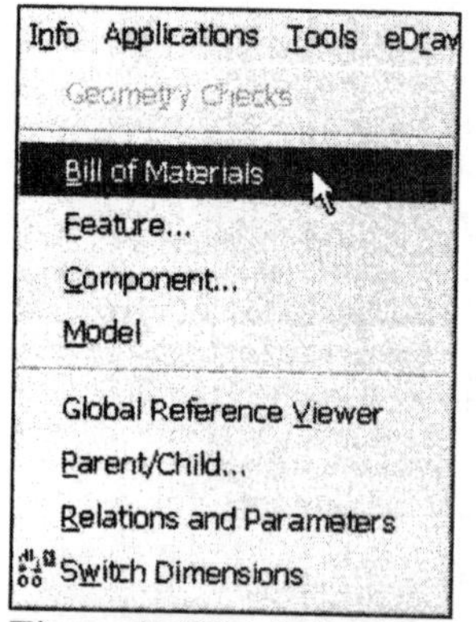

Figure 7.43(a) Info Bill of Materials

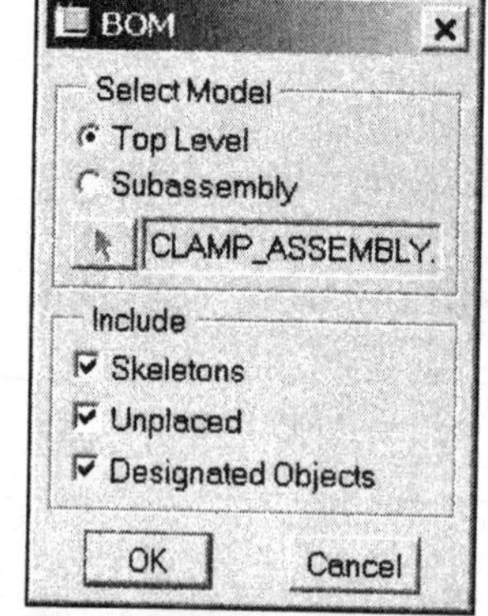

Figure 7.43(b) BOM Dialog Box

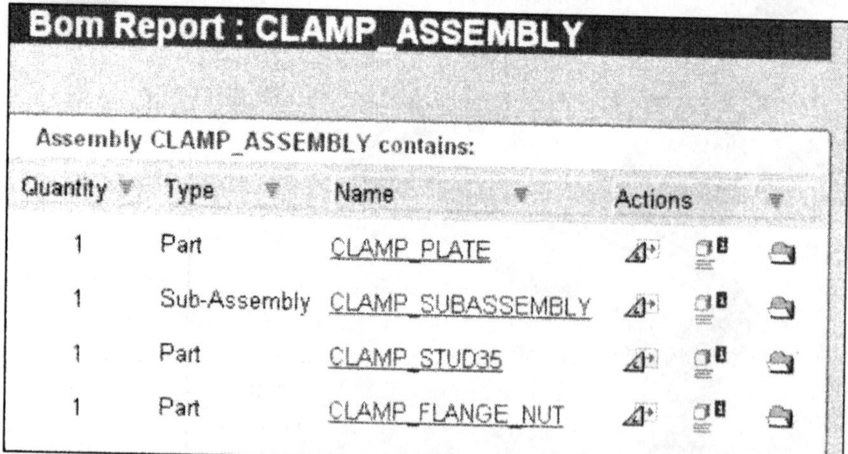
Bom Report : CLAMP_ASSEMBLY

Assembly CLAMP_ASSEMBLY contains:

Quantity	Type	Name	Actions
1	Part	CLAMP_PLATE	
1	Sub-Assembly	CLAMP_SUBASSEMBLY	
1	Part	CLAMP_STUD35	
1	Part	CLAMP_FLANGE_NUT	

Figure 7.43(c) BOM

Lesson 7 is now complete. If you wish to model another assembly project without step-by-step instructions, a complete set of projects and illustrations are available at ***www.cad-resources.com*** ⇒ ***Downloads***.

Lesson 8 Part Drawings

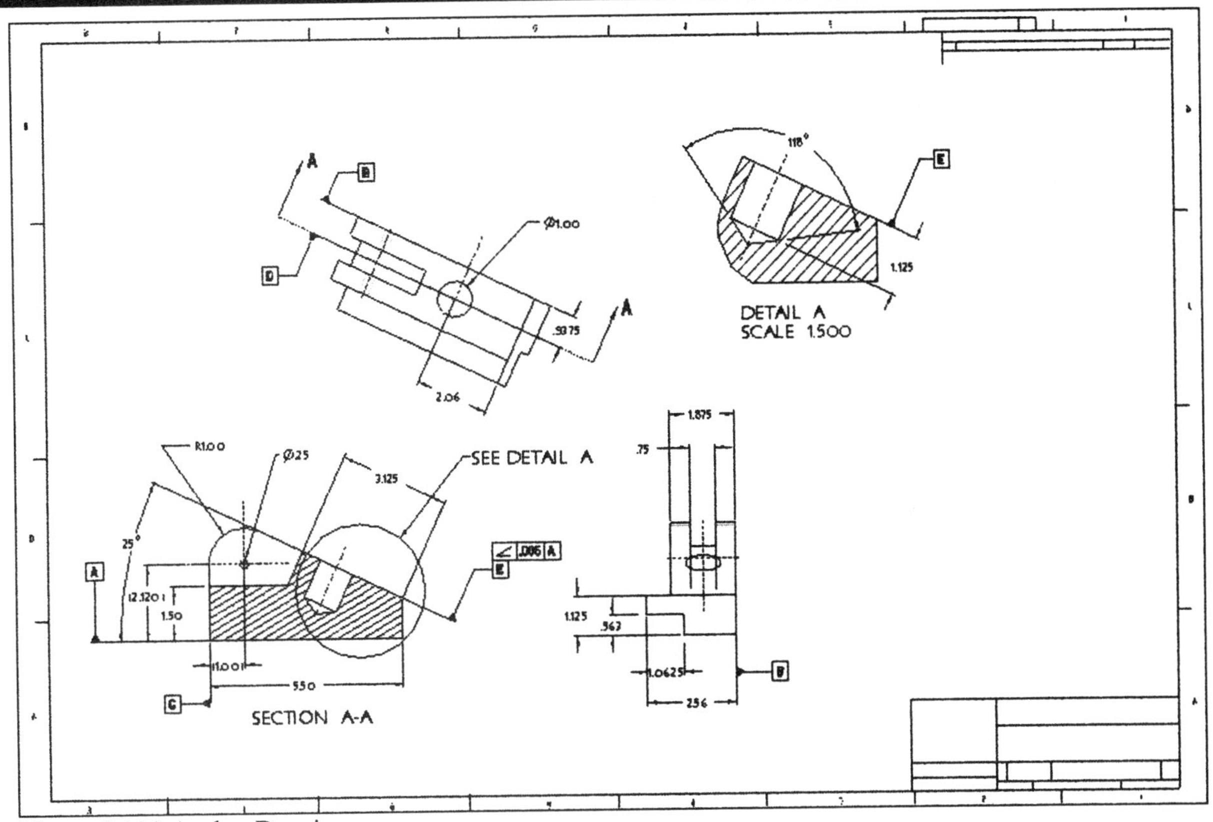

Figure 8.1(a) Anchor Drawing

OBJECTIVES

- Establish a **Drawing Options** file to use when detailing
- Identify the need for **views** to clarify interior features of a part
- Create **Cross Sections** using datum planes
- Produce **Auxiliary Views**
- Create **Detail Views**
- Use **multiple drawing sheets**
- Apply **standard drafting conventions** and linetypes to illustrate interior features

Part Drawings

Designers and drafters use drawings to convey design and manufacturing information. Drawings consist of a **Format** and views of a part (or assembly). Standard views, sectional views, detail views, and auxiliary views are utilized to describe the objects' features and sizes. **Sectional views,** also called **sections**, are employed to clarify and dimension the internal construction of an object. Sections are needed for interior features that cannot be clearly described by hidden lines in conventional views. **Auxiliary views** are used to show the *true shape/size* of a feature or the relationship of features that are not parallel to any of the principal planes of projection. Many objects have inclined surfaces and features that cannot be adequately displayed and described by using principal views alone. To provide a clearer description of these features, it is necessary to draw a view that will show the *true shape/size*.

Lesson 8 STEPS

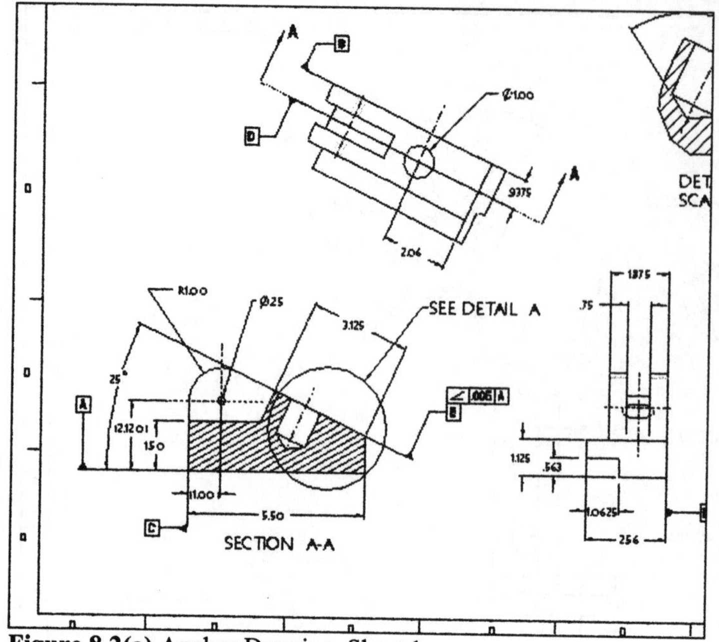

Figure 8.2(a) Anchor Drawing, Sheet 1

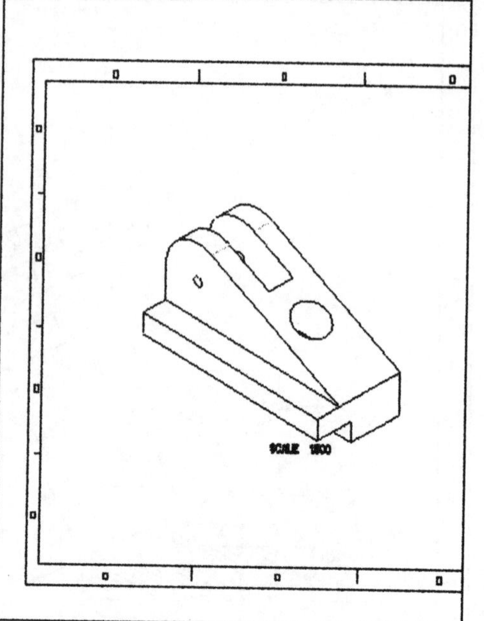

Figure 8.2(b) Anchor Drawing, Sheet 2

Anchor Drawing

You will be creating a multiple sheet detail drawing of the Anchor [Figs. 8.2(a-b)]. The front view will be a full section. A right side view and an auxiliary view are required to detail the part. Views will be displayed according to visibility requirements per ANSI standards, such as no hidden lines in sections. The part is to be dimensioned according to ASME Y14.5M. You will add a standard format. Detailed views of other parts will be introduced to show the wide variety of view capabilities.

Click: **File** ⇒ **Set Working Directory** ⇒ select the directory where the **anchor.prt** was saved ⇒ **OK** ⇒ [icon] ⇒ [Drawing] ⇒ Name **ANCHOR** ⇒ [Use default template] *(note: if you keep the "Use default template" checked, the Front, Top, and Right views will be automatically created for you)* ⇒ **OK** ⇒ Default Model **Browse** ⇒ pick **anchor.prt** ⇒ **Open** ⇒ Standard Size **D** [Fig. 8.2(c)] ⇒ **OK** ⇒ **File** ⇒ **Page Setup** ⇒ [D Size] [Fig. 8.2(d)] ⇒ [icon] ⇒ **Browse** ⇒ System Formats Open dialog box opens [Fig. 8.2(e)] ⇒ pick **d.frm** ⇒ **Open** ⇒ **OK** ⇒ [icon] ⇒ **OK** [Fig. 8.2(f)] ⇒ **RMB** ⇒ **Properties** ⇒ **Drawing Options** the Options dialog box opens ⇒ create a new **.dtl** file, *(or* [icon] *⇒ pick a previously saved **.dtl** from your directory list ⇒ Open)*

Change the following options to the values listed below:

Option: ***drawing_text_height*** Value: ***.25*** ⇒ **Add/Change**
Option: ***default_font*** Value: ***filled*** ⇒ **Add/Change**
Option: ***draw_arrow_style*** Value: ***filled*** ⇒ **Add/Change**
Option: ***allow_3d_dimensions*** Value: ***yes*** ⇒ **Add/Change**

Click: **Apply** ⇒ [icon] **Save a copy of the currently displayed configuration file** [Fig. 8.2(g)] ⇒ type a unique name for your file ⇒ **Ok** ⇒ **Close** ⇒ **MMB**

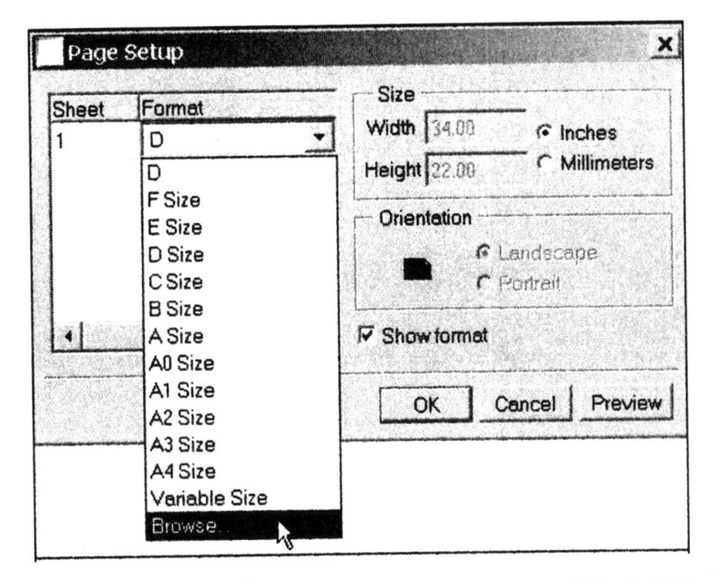

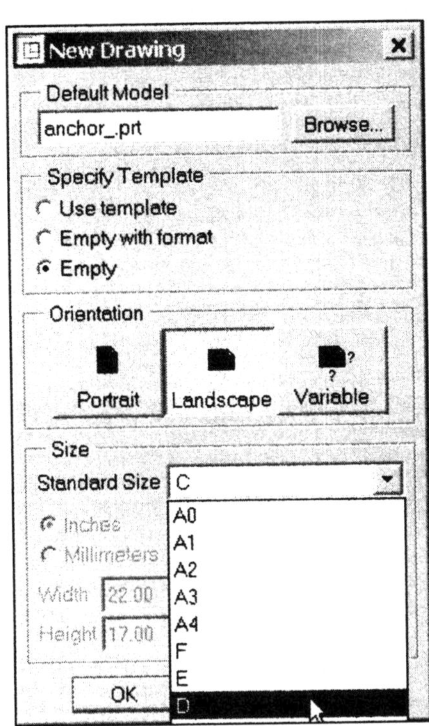

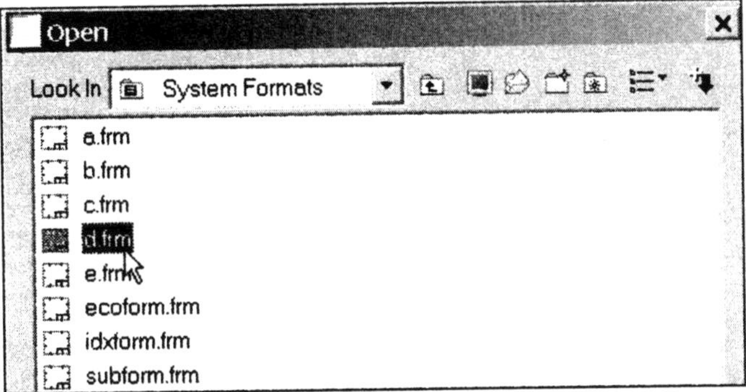

Figure 8.2(c) Standard Size D **Figure 8.2[d (top)]** Page Setup **Figure 8.2[e (bottom)]** Formats

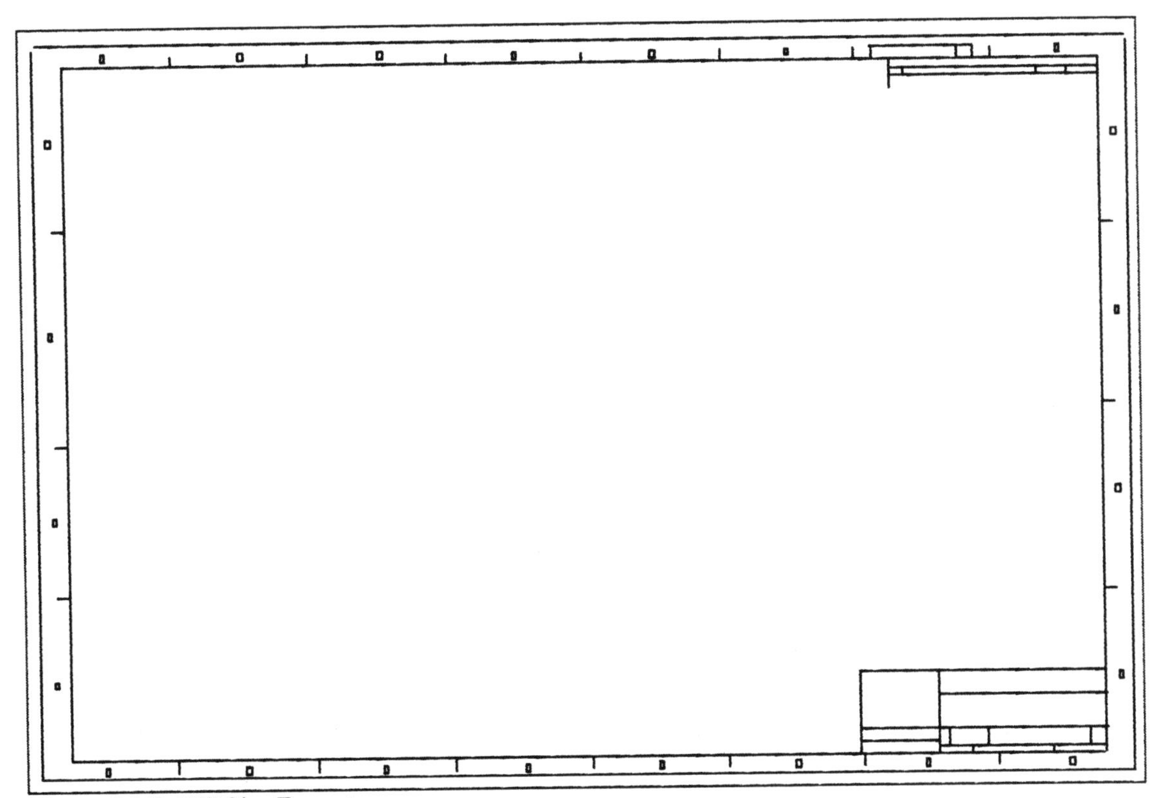

Figure 8.2(f) "D" Size Format

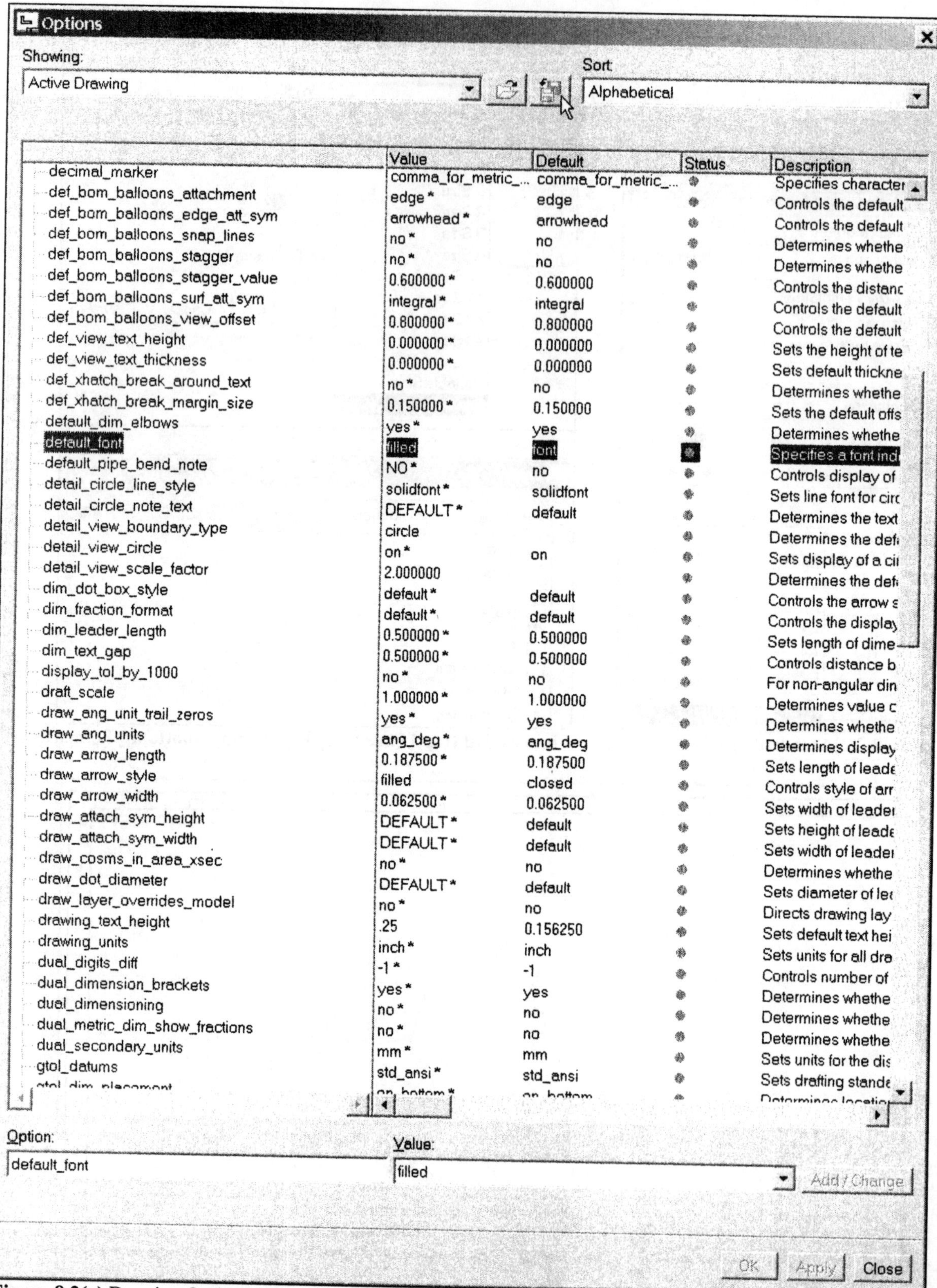

Figure 8.2(g) Drawing Options File

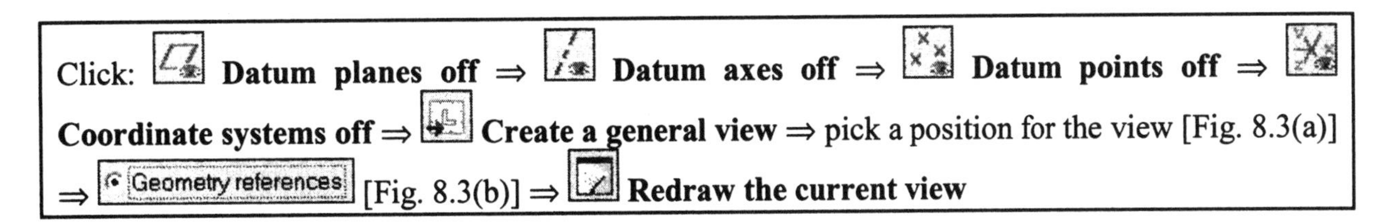

Click: **Datum planes off** ⇒ **Datum axes off** ⇒ **Datum points off** ⇒ **Coordinate systems off** ⇒ **Create a general view** ⇒ pick a position for the view [Fig. 8.3(a)] ⇒ Geometry references [Fig. 8.3(b)] ⇒ **Redraw the current view**

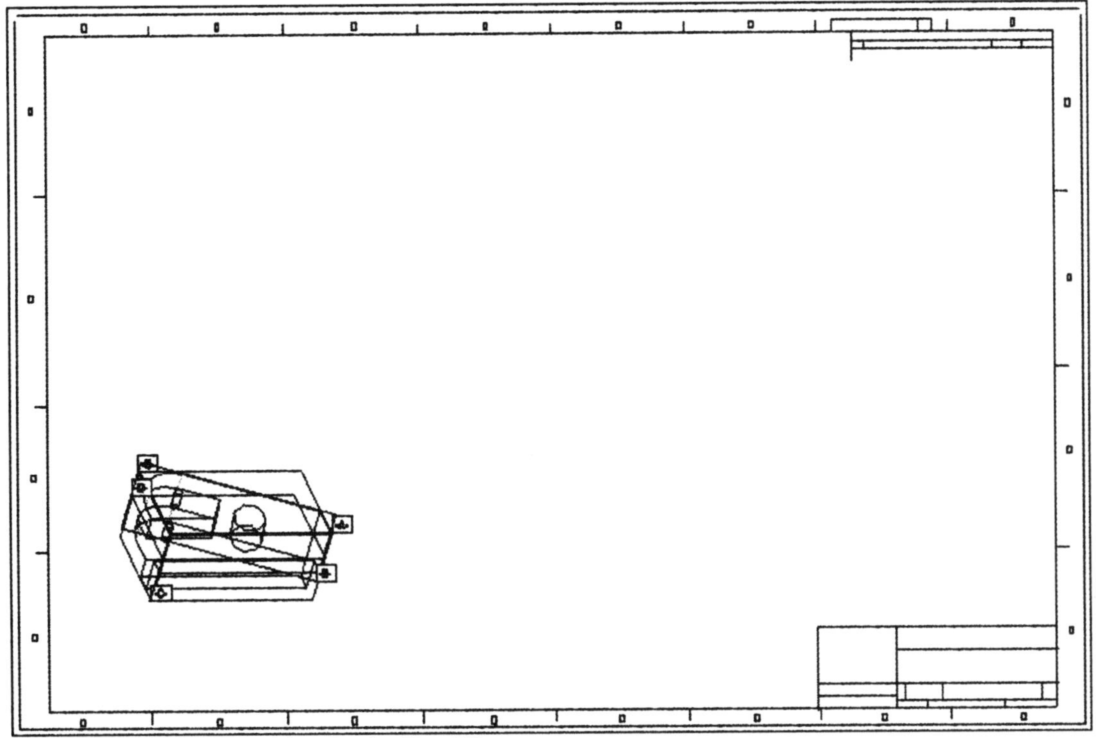

Figure 8.3(a) Pick a Position for the First View

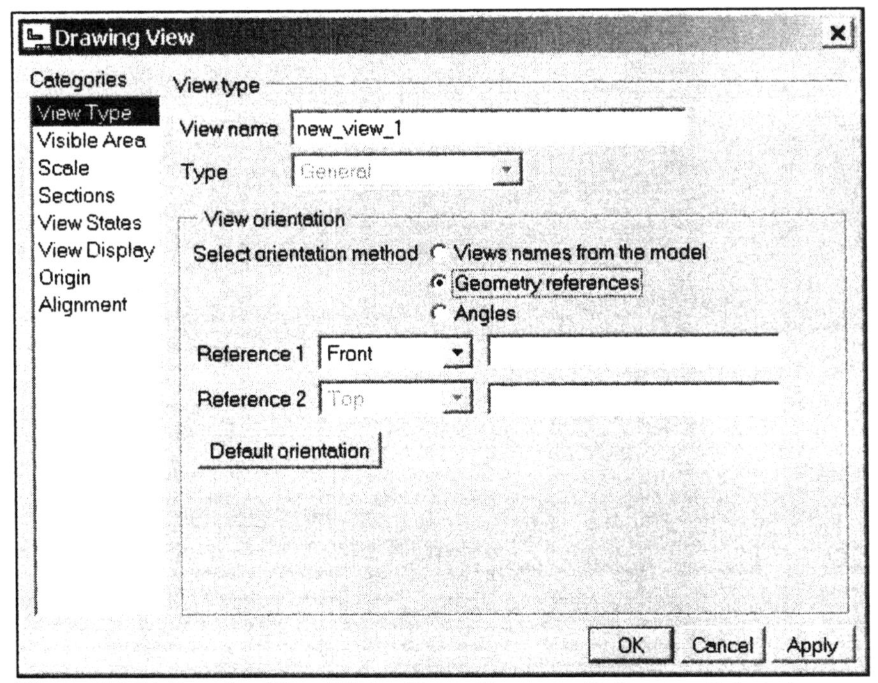

Figure 8.3(b) Drawing View Dialog Box, •Geometry References Selected

Reference 1: **Front** ⇒ pick datum **B** [Fig. 8.3(c)] ⇒ Reference 2: **Top** [Fig. 8.3(d)] ⇒ pick datum **A** ⇒ **OK** [Fig. 8.3(e)] ⇒ ⇒ ⇒ **MMB** ⇒ **LMB** to deselect

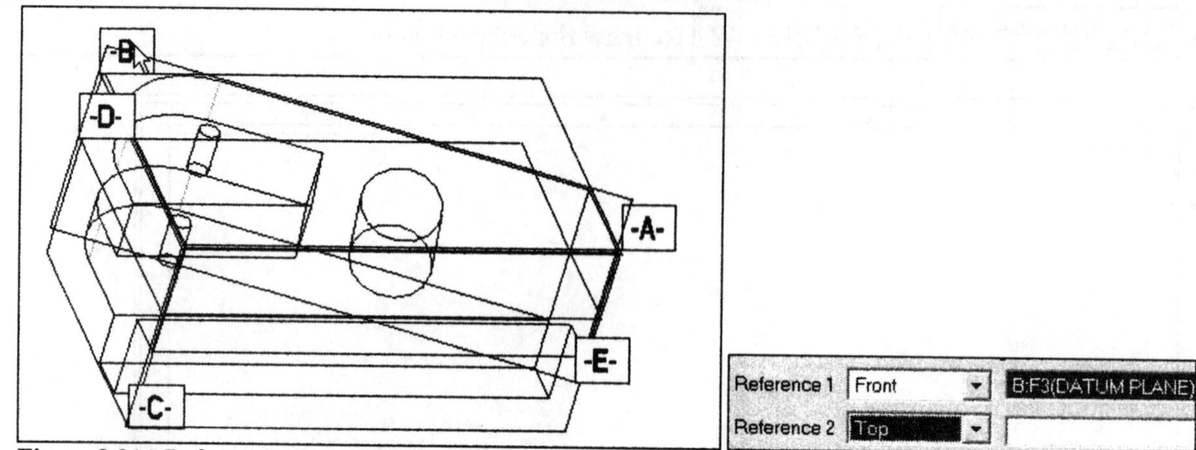

Figure 8.3(c) Reference 1 Front, Datum B

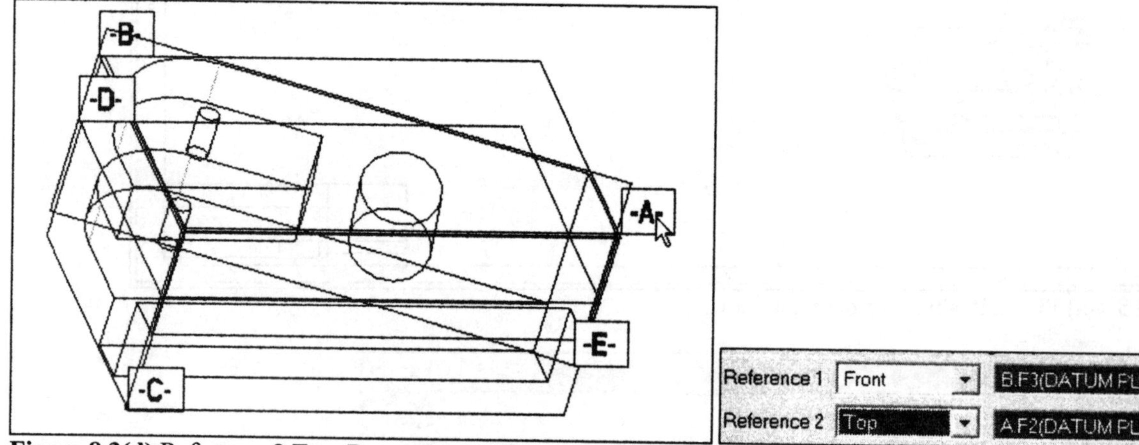

Figure 8.3(d) Reference 2 Top, Datum A

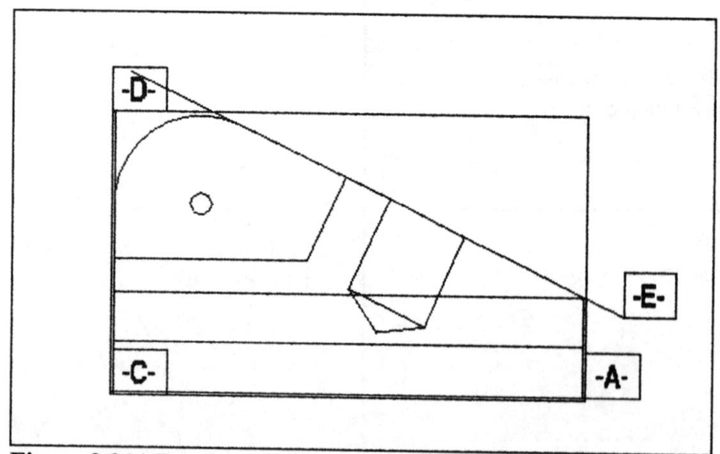

Figure 8.3(e) Reoriented View

Add two more views, click: **Insert** ⇒ **Drawing View** ⇒ **Projection** ⇒ Select CENTER POINT for drawing view. pick a position for the right side view [Fig. 8.3(f)] ⇒ **LMB** to deselect ⇒ pick on the Front view ⇒ **RMB** ⇒ **Insert Projection View** ⇒ Select CENTER POINT for drawing view. pick a position for the top view [Fig. 8.3(g)] ⇒ **LMB** ⇒ unlock ⇒ pick on a view, hold down the **LMB**, and reposition as needed ⇒ **LMB** to deselect ⇒ **Update** ⇒ ⇒ ⇒ **MMB** [Fig. 8.3(h)]

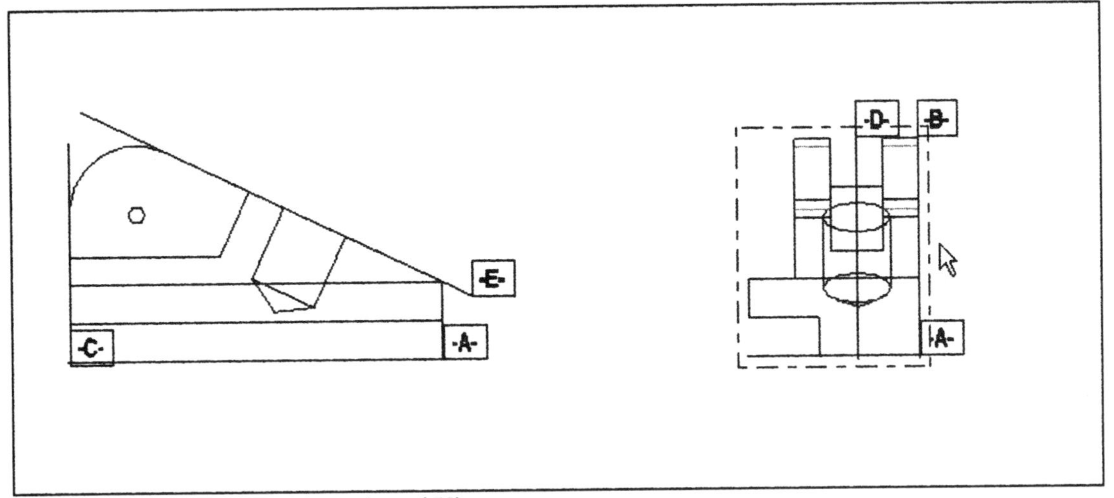

Figure 8.3(f) Add Right Side Projected View

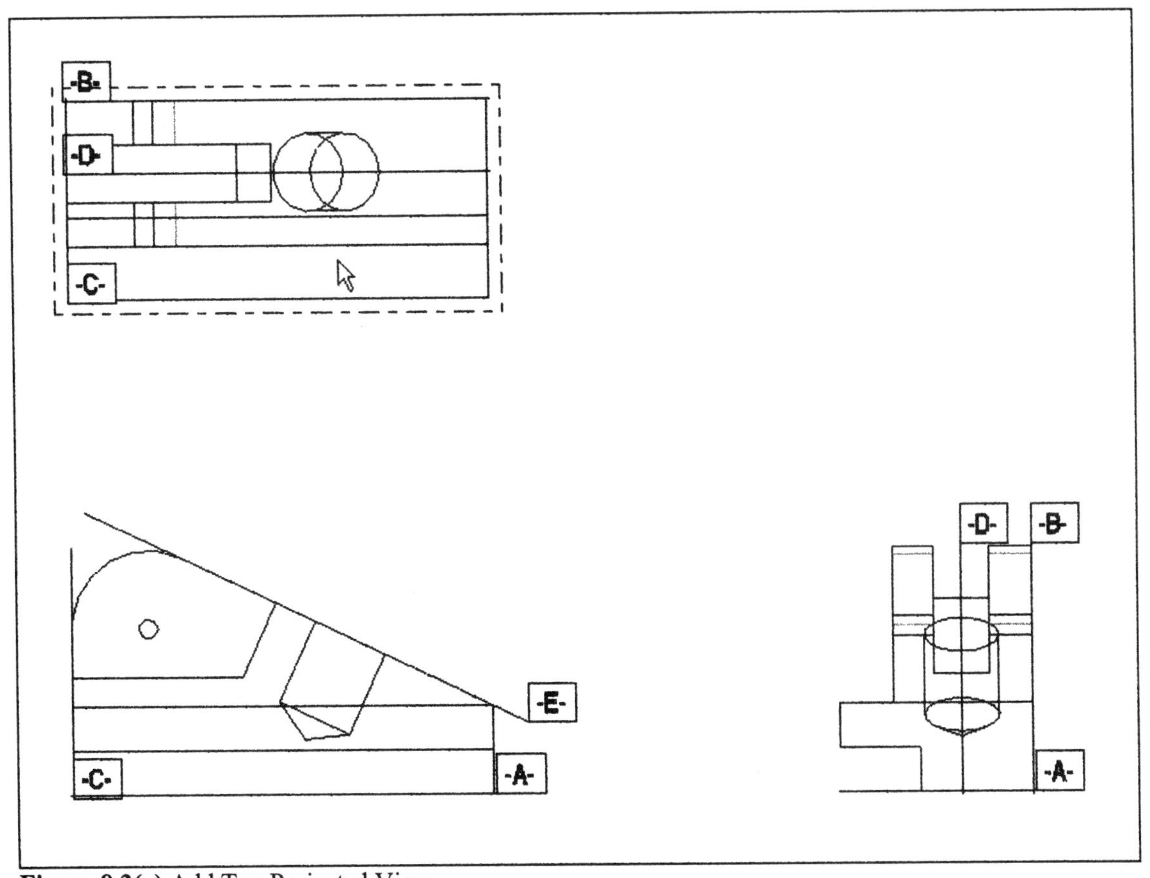

Figure 8.3(g) Add Top Projected View

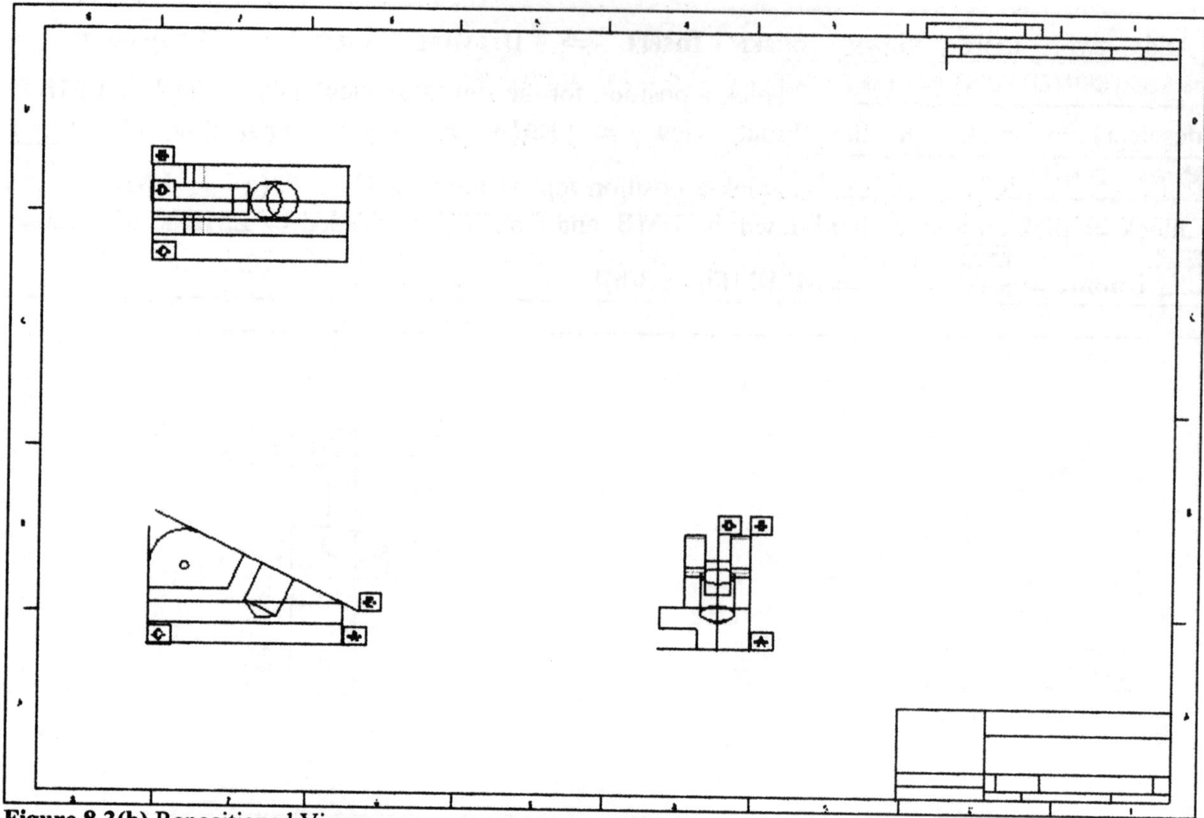

Figure 8.3(h) Repositioned Views

The top view does not help in the description of the part's geometry. Delete the top view before adding an auxiliary view that will display the true shape of the angled surface.

Pick on the top view ⇒ **RMB** [Fig. 8.3(i)] ⇒ **Delete** [Fig. 8.3(j)] ⇒ ⇒ ⇒ **MMB**

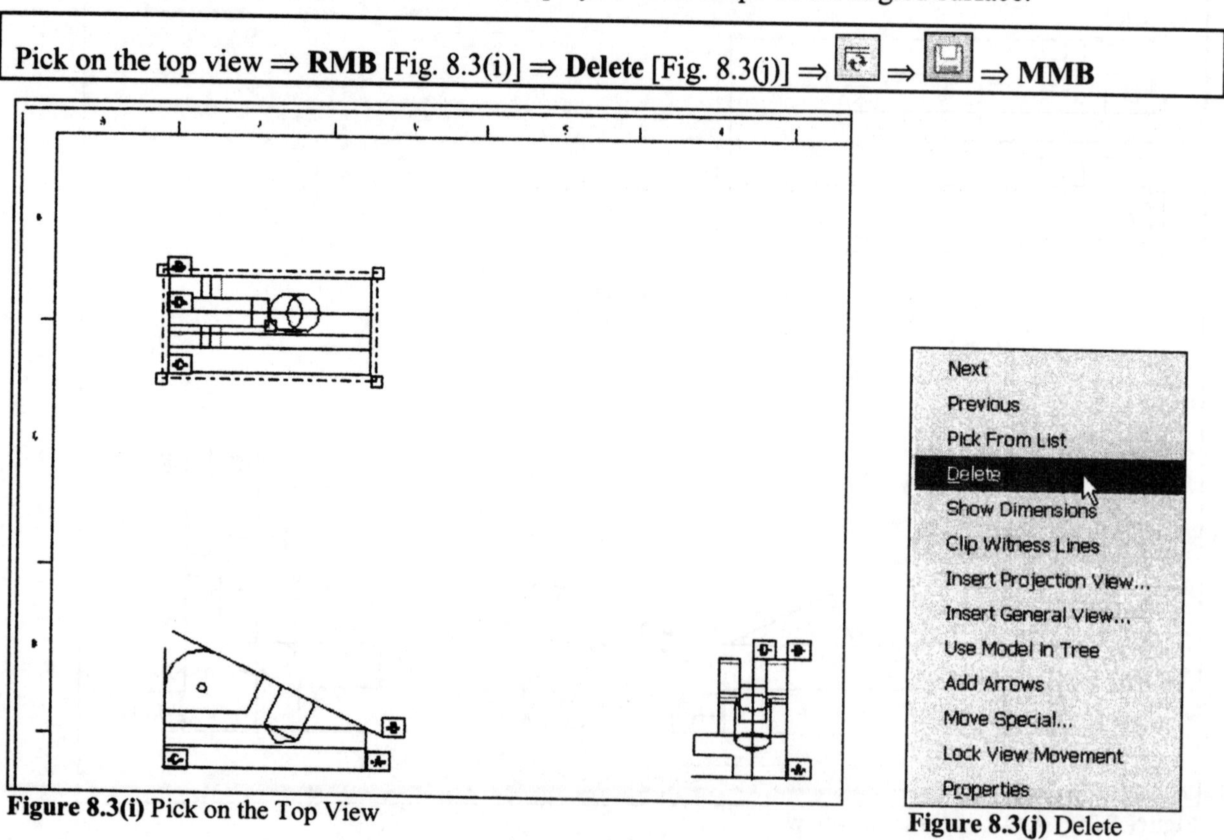

Figure 8.3(i) Pick on the Top View

Figure 8.3(j) Delete

Click: **Insert ⇒ Drawing View ⇒ Auxiliary ⇒** Select edge of or axis through, or datum plane as, front surface on main view. pick on datum **E** [Fig. 8.3(k)] ⇒ Select CENTER POINT for drawing view. [Fig. 8.3(l)] ⇒ **LMB** to place the view [Fig. 8.3(m)] ⇒ **LMB** to deselect ⇒ **RMB ⇒ Update Sheet ⇒** [icon] ⇒ **MMB**

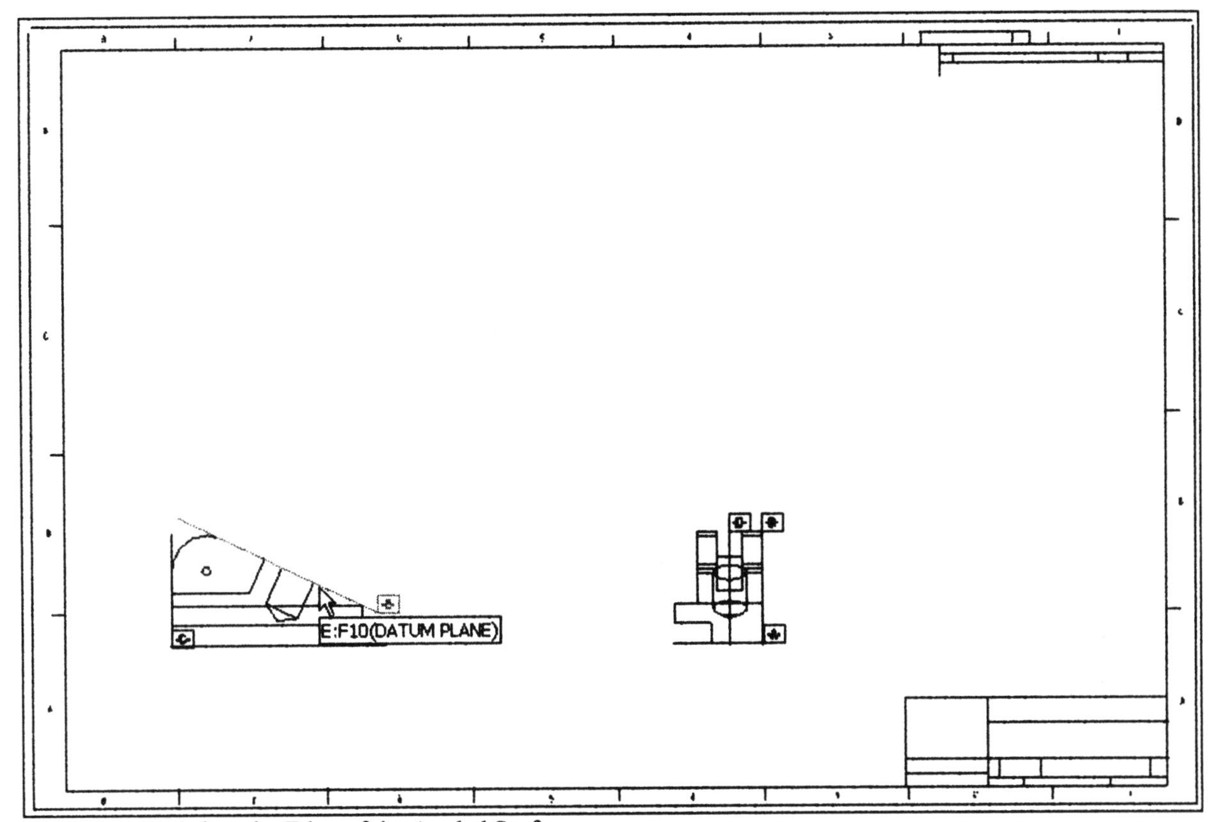

Figure 8.3(k) Select the Edge of the Angled Surface

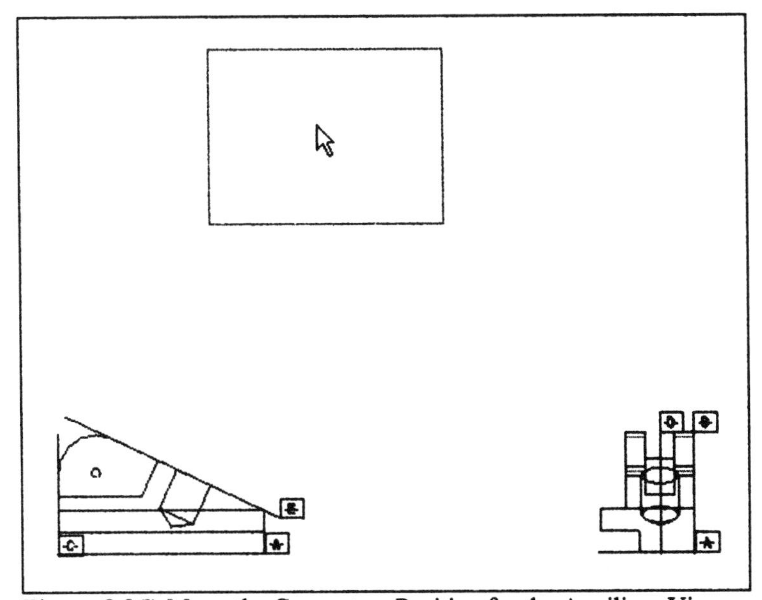

Figure 8.3(l) Move the Cursor to a Position for the Auxiliary View

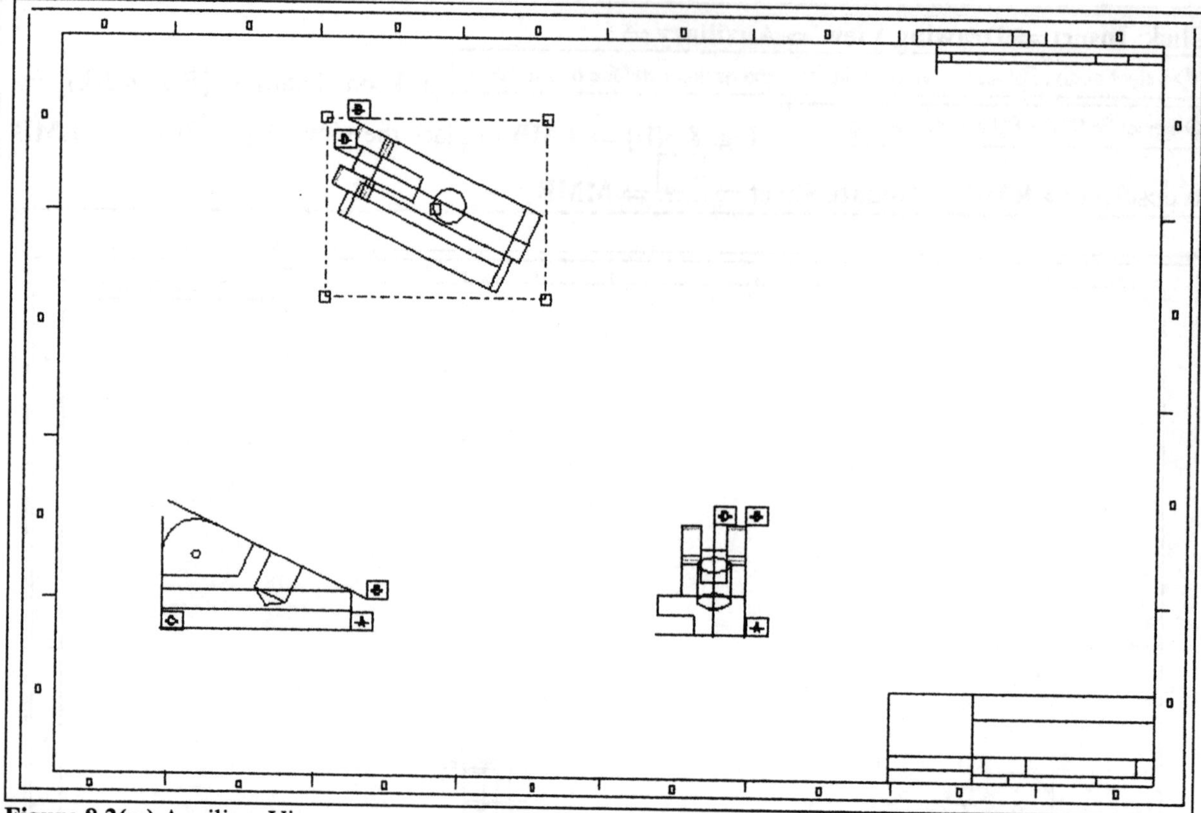

Figure 8.3(m) Auxiliary View

The correct standard was not selected when you made changes to the Drawing Options. ASME symbols for datum planes (gtol_datums) are the correct style standard used on drawings. The ANSI style [Fig. 8.3(n)] was discontinued in 1994, though retained by some companies as an "in house" standard. For outside vendors and for manufacturing internationally, the ISO-ASME standards should be applied to all manufacturing drawings.

Click: **RMB** ⇒ **Properties** ⇒ **Drawing Options** ⇒ gtol_datums std_asme ⇒ **Add/Change** ⇒ **Apply** ⇒ **Close** ⇒ **MMB** ⇒ ⇒ ⇒ [Fig. 8.3(o)] ⇒ ⇒ **MMB** ⇒ **Sketch** ⇒ **Sketcher Preferences** ⇒ Grid intersection off ⇒ **Close**

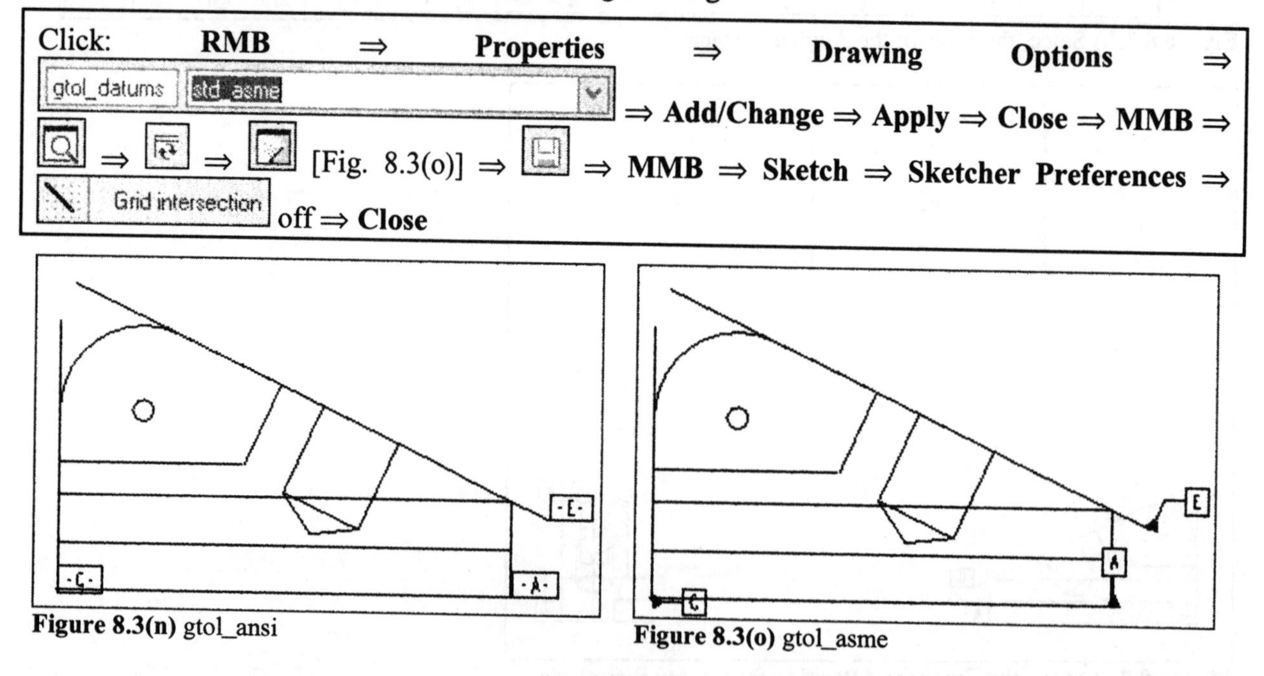

Figure 8.3(n) gtol_ansi

Figure 8.3(o) gtol_asme

Next, change the front view into a sectional view. The section **A** was created in Part mode.

Pick on the front view as the view to be modified [Fig. 8.4(a)] ⇒ **RMB** ⇒ **Properties** [Fig. 8.4(b)] ⇒ Categories **Sections** ⇒ **•2D cross-section** ⇒ [+] **Add cross-section to view** ⇒ pick section **A** from the Name list [Fig. 8.4(c)] ⇒ **Apply** [Fig. 8.4(d)] ⇒ **Close** ⇒ **RMB** ⇒ **Add Arrows** [Fig. 8.4(e)] ⇒ pick the auxiliary view [Fig. 8.4(f)] ⇒ **LMB** ⇒ [icon] ⇒ [icon] ⇒ [icon] ⇒ [icon] ⇒ **MMB**

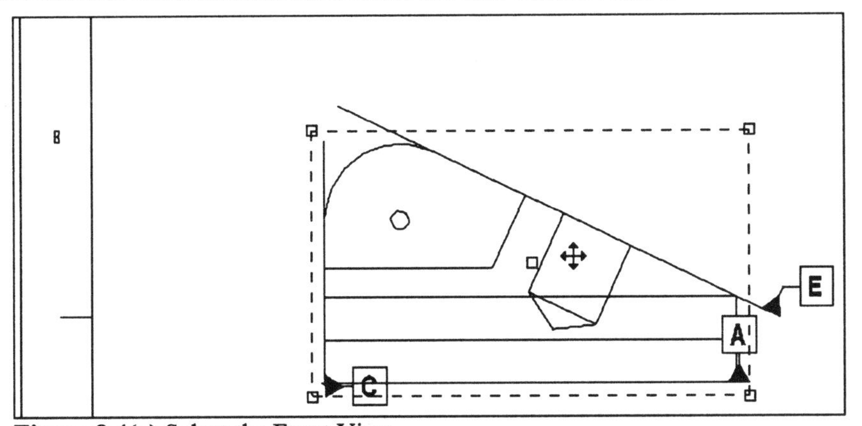

Figure 8.4(a) Select the Front View

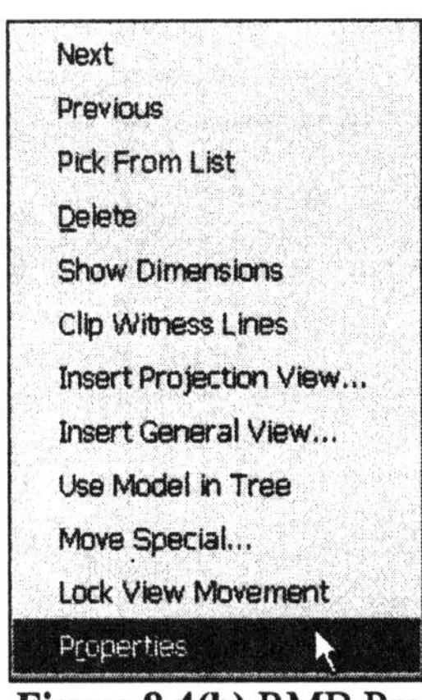

Figure 8.4(b) RMB Properties

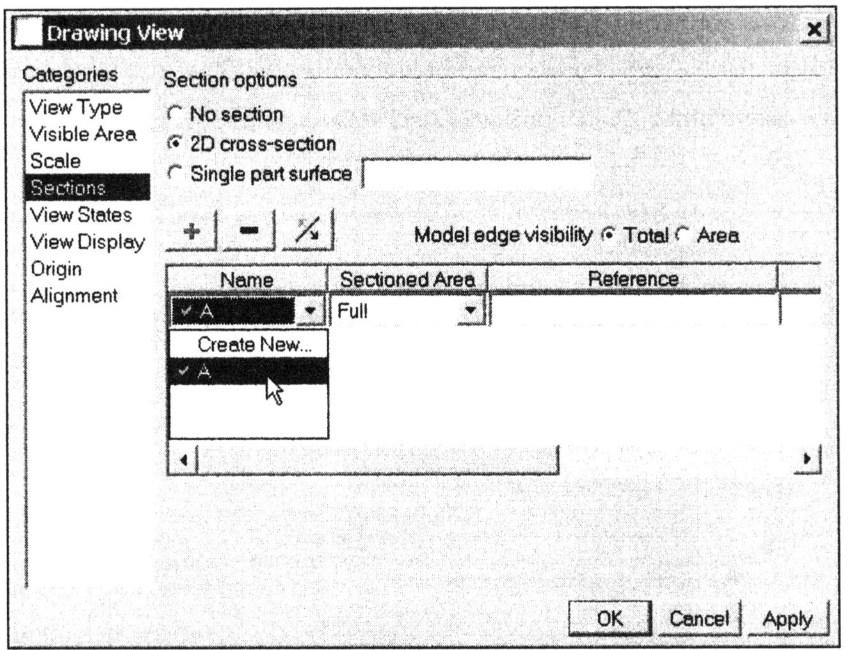

Figure 8.4(c) Drawing View Dialog Box

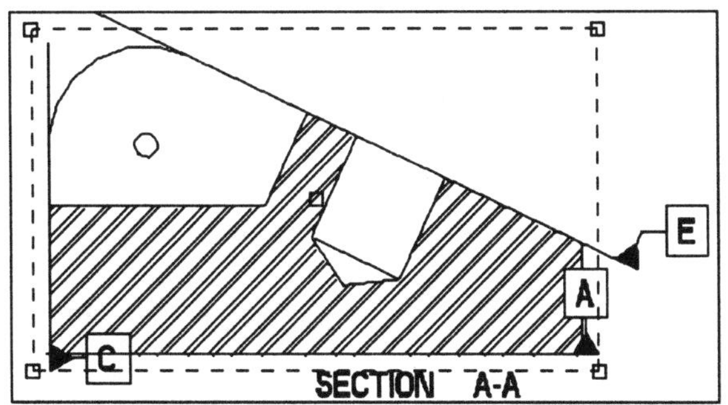

Figure 8.4(d) Section A-A

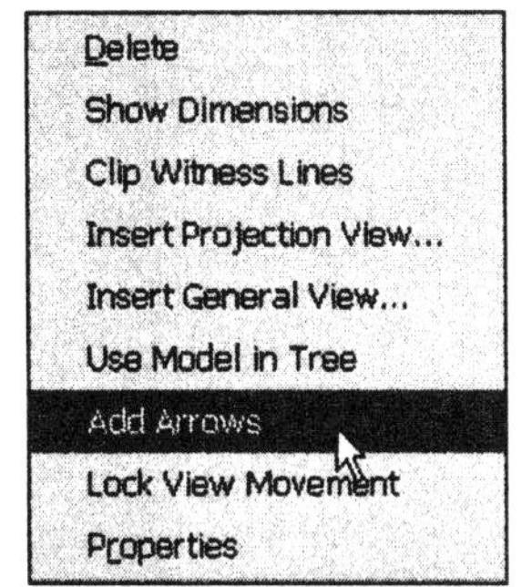

Figure 8.4(e) RMB Add Arrows

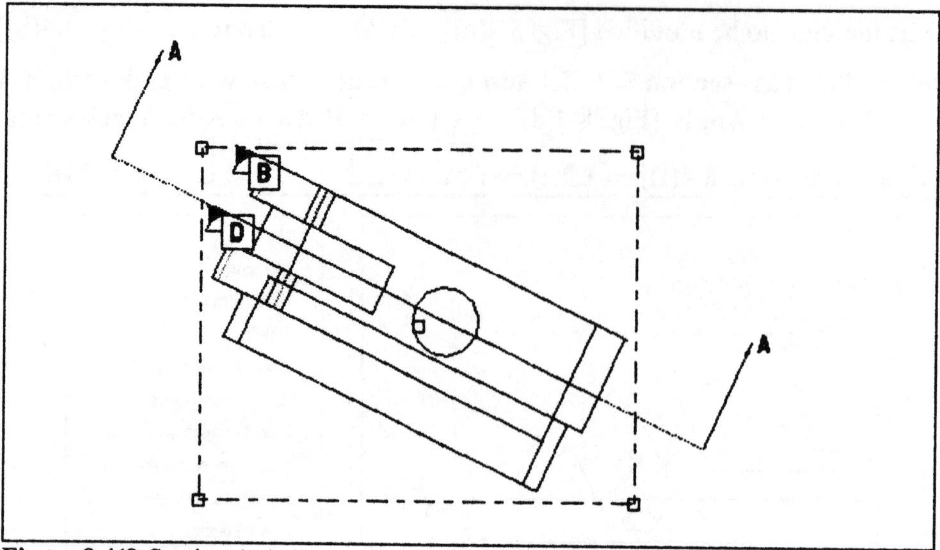

Figure 8.4(f) Section A-A Arrows

Modify the visibility of the views to remove all hidden lines. While pressing and holding down the **Ctrl** key, pick on all three views [Fig. 8.5(a)] ⇒ click **RMB** with the cursor outside of any view outlines ⇒ **Properties** [Fig. 8.5(b)] ⇒ Display style **No Hidden** [Fig. 8.5(c)] ⇒ Tangent edges display style **Dimmed** [Fig. 8.5(d)] ⇒ **OK** ⇒ **LMB** [Fig. 8.5(e)]

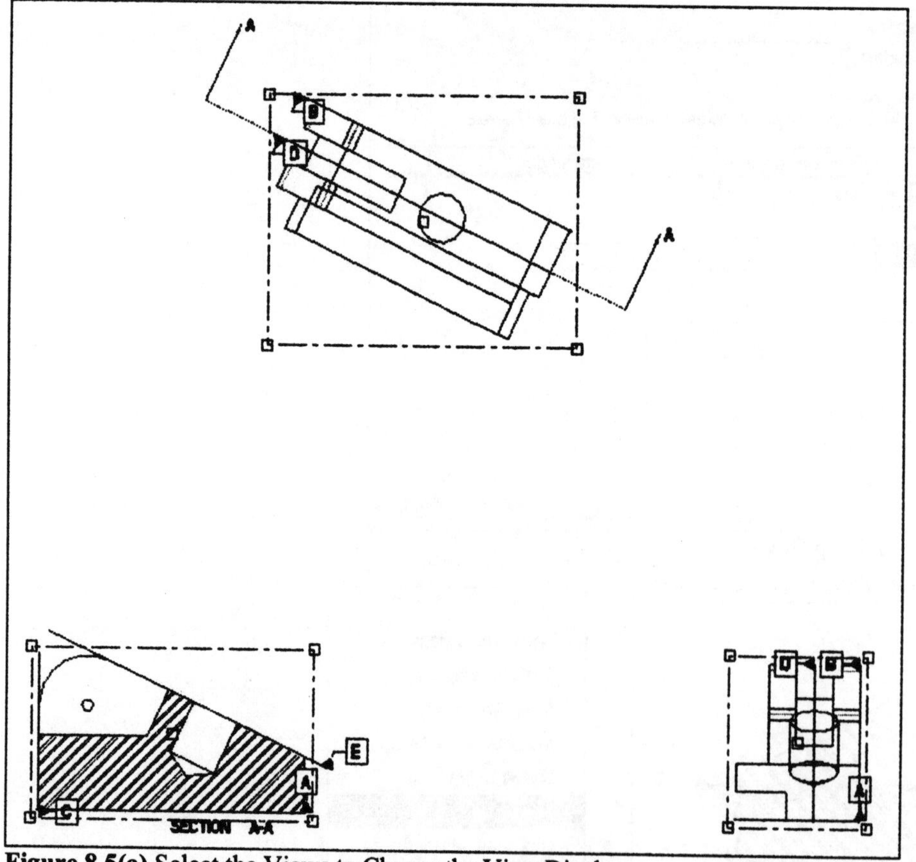

Figure 8.5(a) Select the Views to Change the View Display

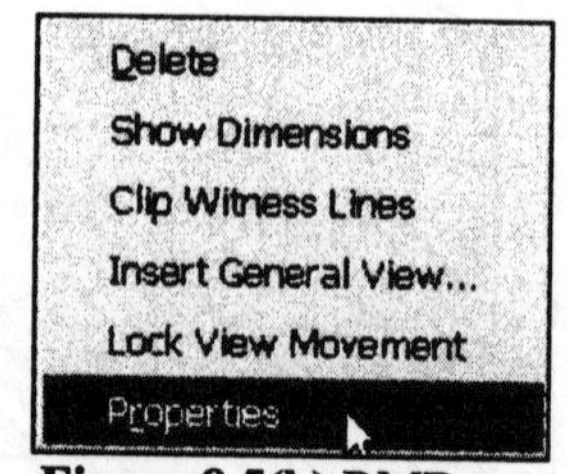

Figure 8.5(b) RMB Properties

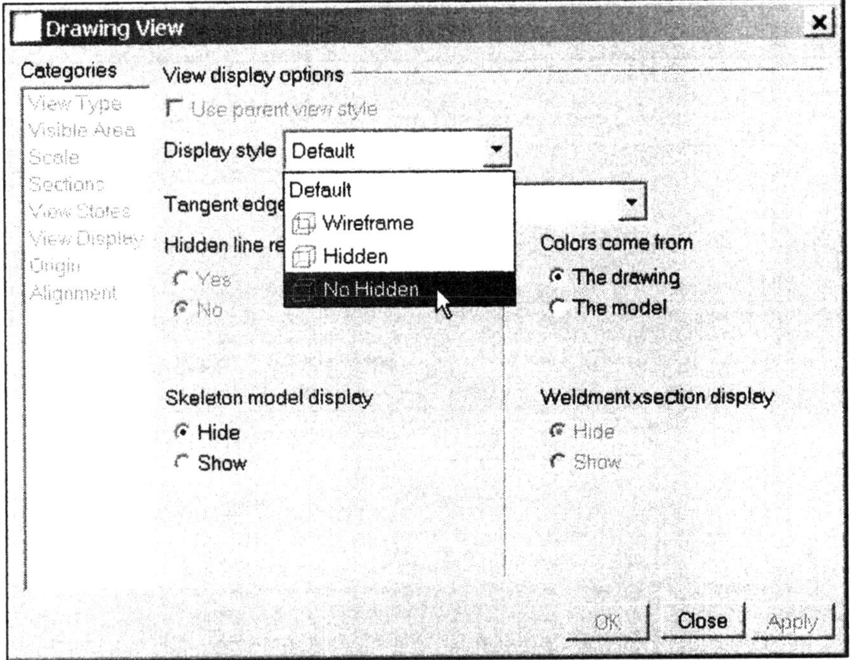

Figure 8.5(c) Drawing View Dialog Box, No Hidden

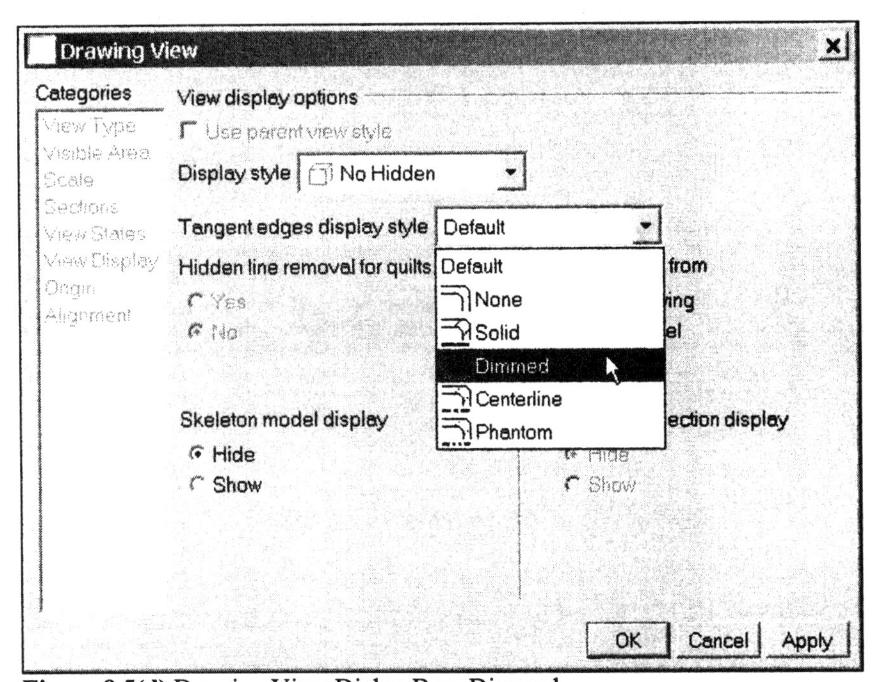

Figure 8.5(d) Drawing View Dialog Box, Dimmed

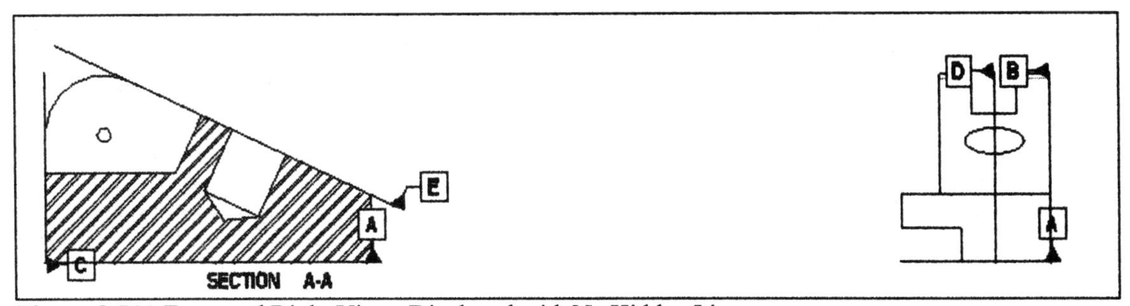

Figure 8.5(e) Front and Right Views Displayed with No Hidden Lines

Show all dimensions and axes (centerlines), click: **Open the Show/Erase dialog box** ⇒ Show ⇒ **Dimension** **Axis** **Note** **Datum Plane** [Fig. 8.6(a)] ⇒ **Show All** ⇒ **Yes** ⇒ **Accept All** ⇒ **Close** ⇒ **LMB** [Fig. 8.6(b)]

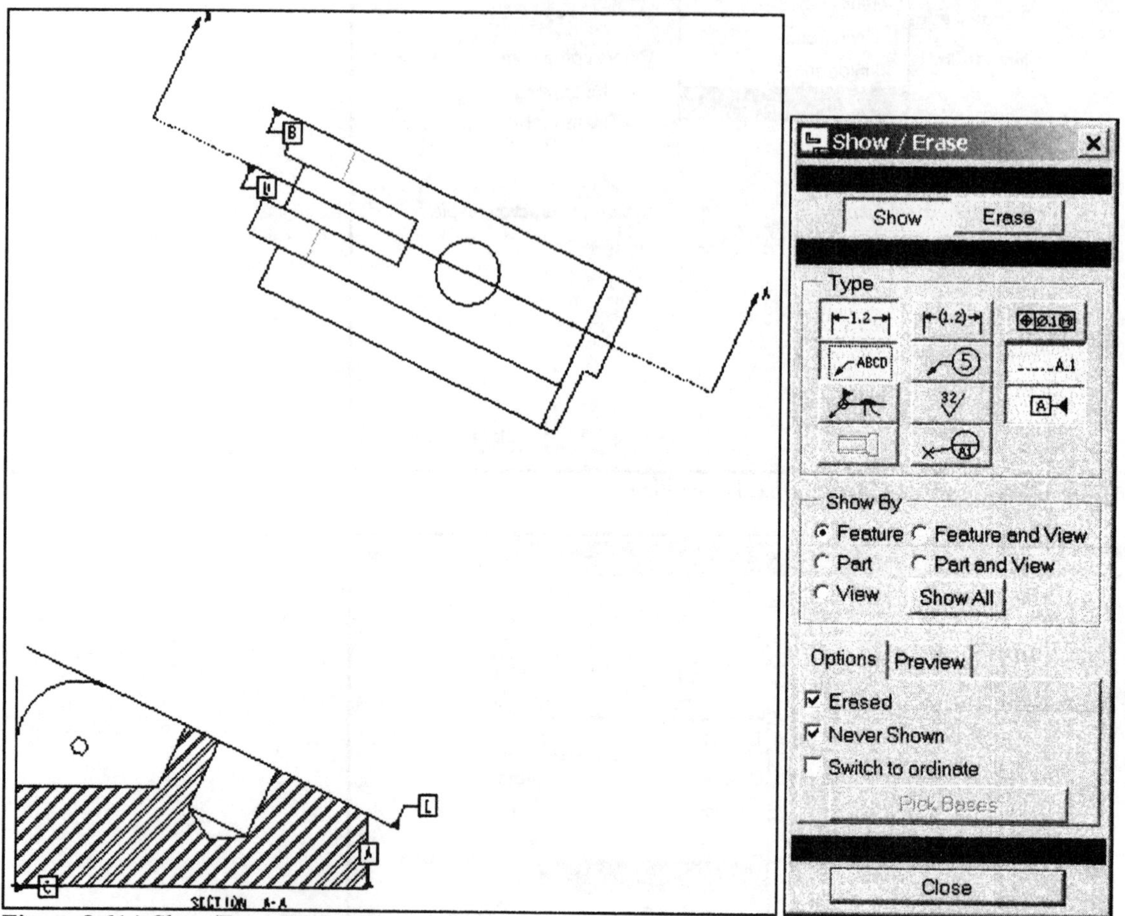

Figure 8.6(a) Show/Erase Dialog Box

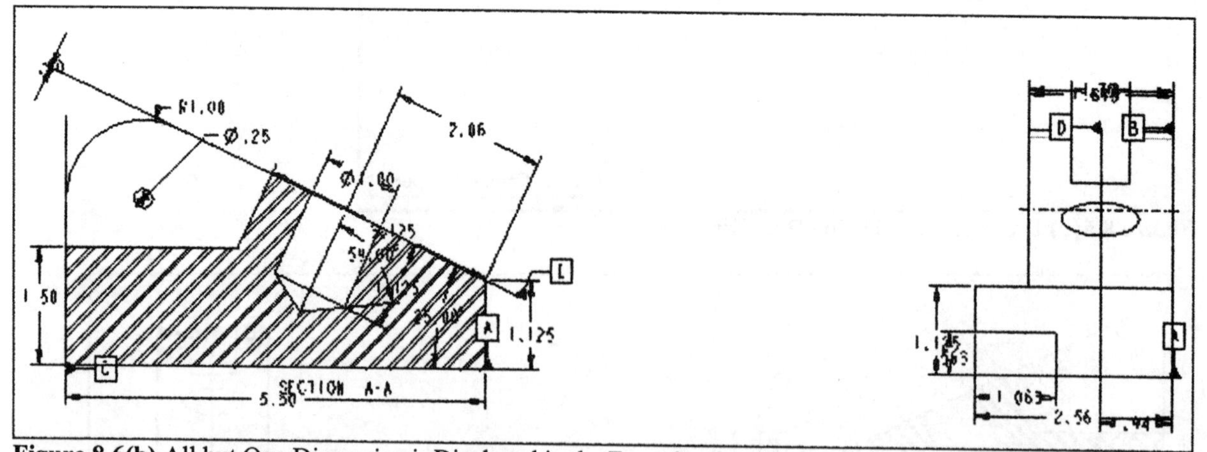

Figure 8.6(b) All but One Dimension is Displayed in the Front Section View and the Right Side View

Click: **Clean up position of dimensions around a view** ⇒ select all dimensions by enclosing them in a selection box ⇒ **MMB** ⇒ Create Snap Lines ⇒ Increment .500 [Fig. 8.6(c)] ⇒ **Apply** [Fig. 8.6(d)] ⇒ **Close** ⇒ use **Ctrl +MMB** to zoom as needed ⇒ **Shift+MMB** to pan as needed

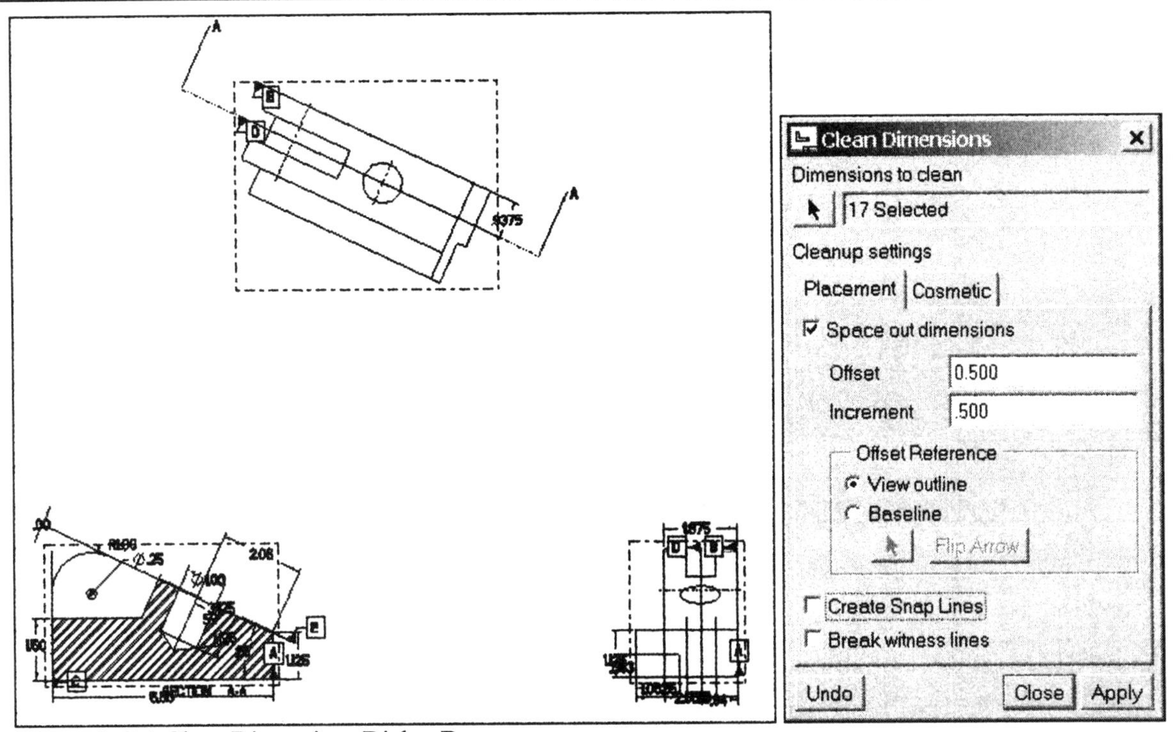

Figure 8.6(c) Clean Dimensions Dialog Box

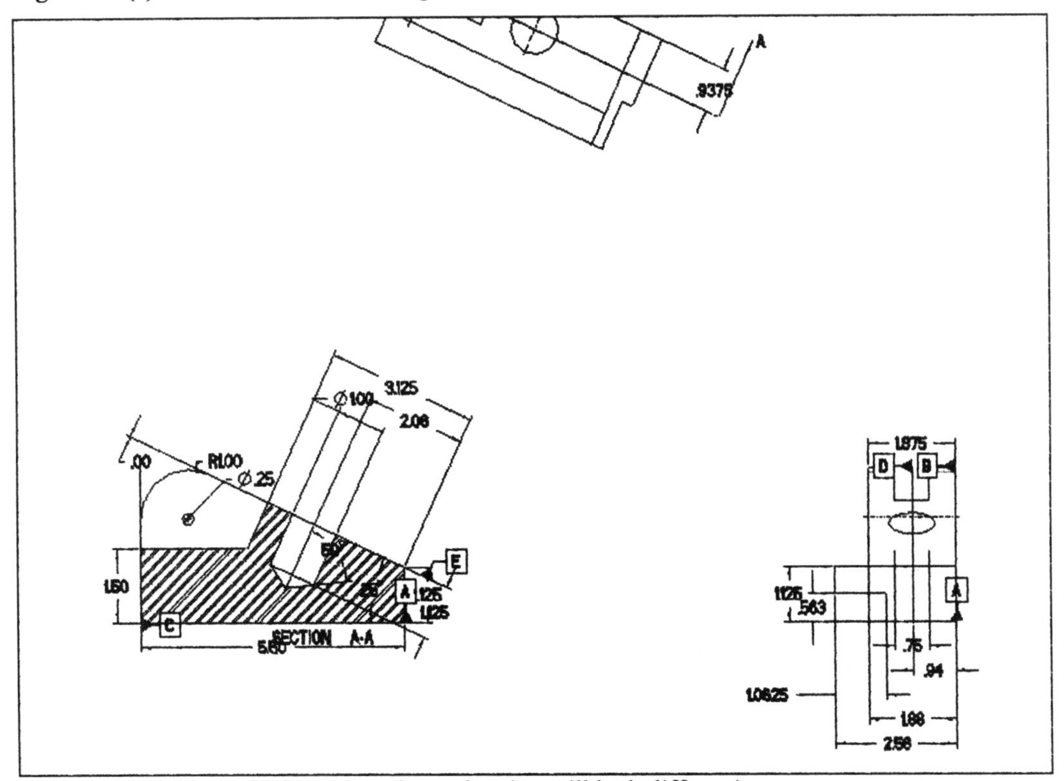

Figure 8.6(d) Cleaned Dimensions (your drawing will look different)

Pick on the **1.00** diameter hole dimension [Fig. 8.7(a)] ⇒ **RMB** ⇒ **Move Item to View** [Fig. 8.7(b)] ⇒ pick on the auxiliary view [Fig. 8.7(c)] ⇒ pick on and reposition the **1.00** dimension [Fig. 8.7(d)] ⇒ using the dimension handles reposition, move dimension text, clip extension lines, or flip arrows where appropriate ⇒ move, erase, or clip axes and datums as necessary ⇒ add a reference dimension to the small hole ⇒ **Insert** ⇒ **Reference Dimension** ⇒ **New References** ⇒ pick datum **A** ⇒ pick the horizontal axis of the small hole ⇒ **MMB** to place the reference dimension ⇒ **MMB** ⇒ **LMB**

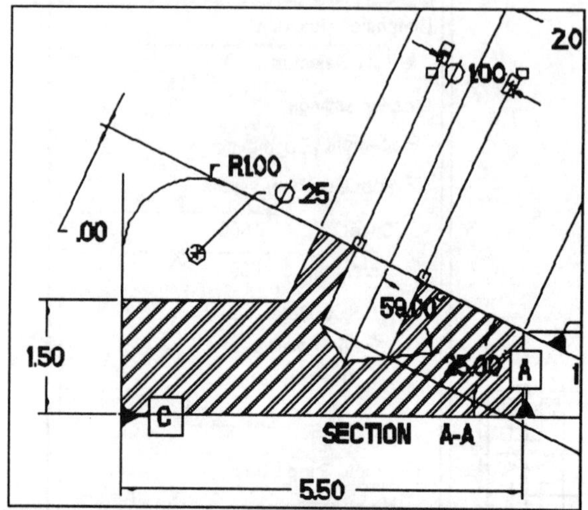

Figure 8.7(a) Pick on the **1.00** Diameter Dimension

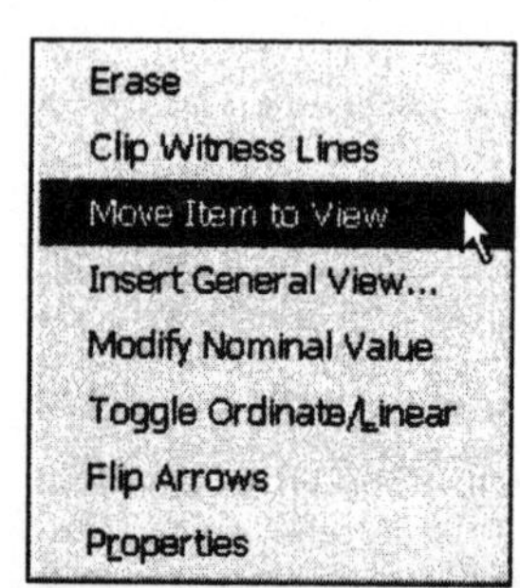

Figure 8.7(b) Move Item to View

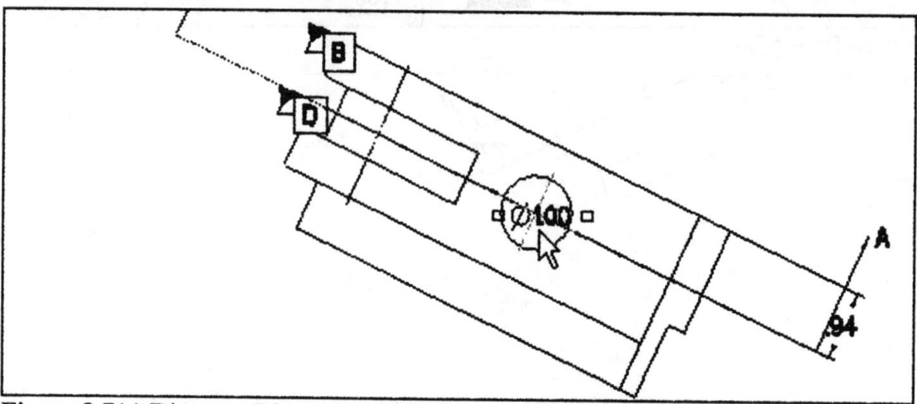

Figure 8.7(c) Diameter Dimension **1.00** Moved to Auxiliary View

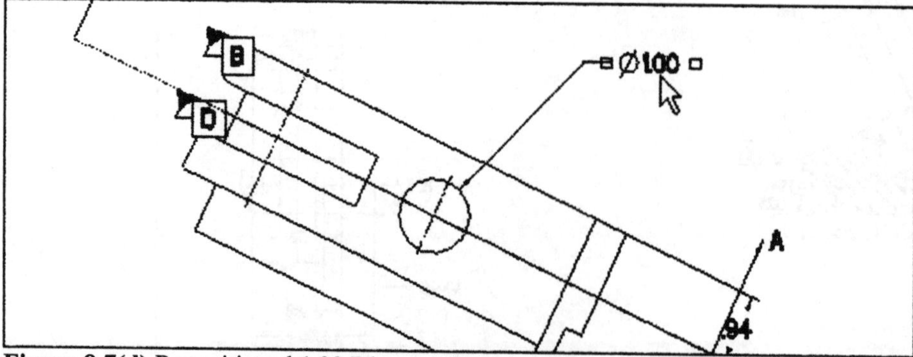

Figure 8.7(d) Repositioned **1.00** Diameter Dimension

Change the arrow length and width to be in proportion to the text height. Click: **RMB** ⇒ **Properties** ⇒ **Drawing Options** opens an Options dialog box ⇒ ***draw_arrow_length .25***, ***draw_arrow_width .10*** ⇒ **Add/Change** ⇒ **Apply** ⇒ **Close** ⇒ **MMB** [Figs. 8.8(a-c)] ⇒ **Update the display of all views in the active sheet**

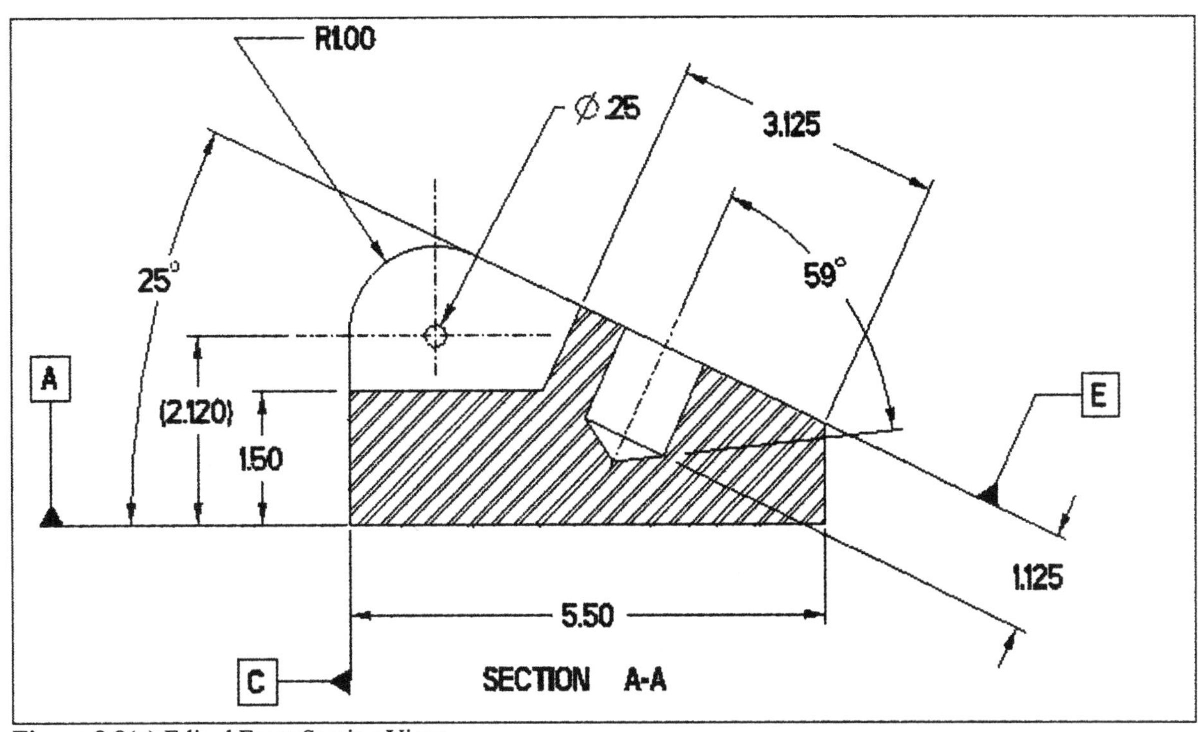

Figure 8.8(a) Edited Front Section View

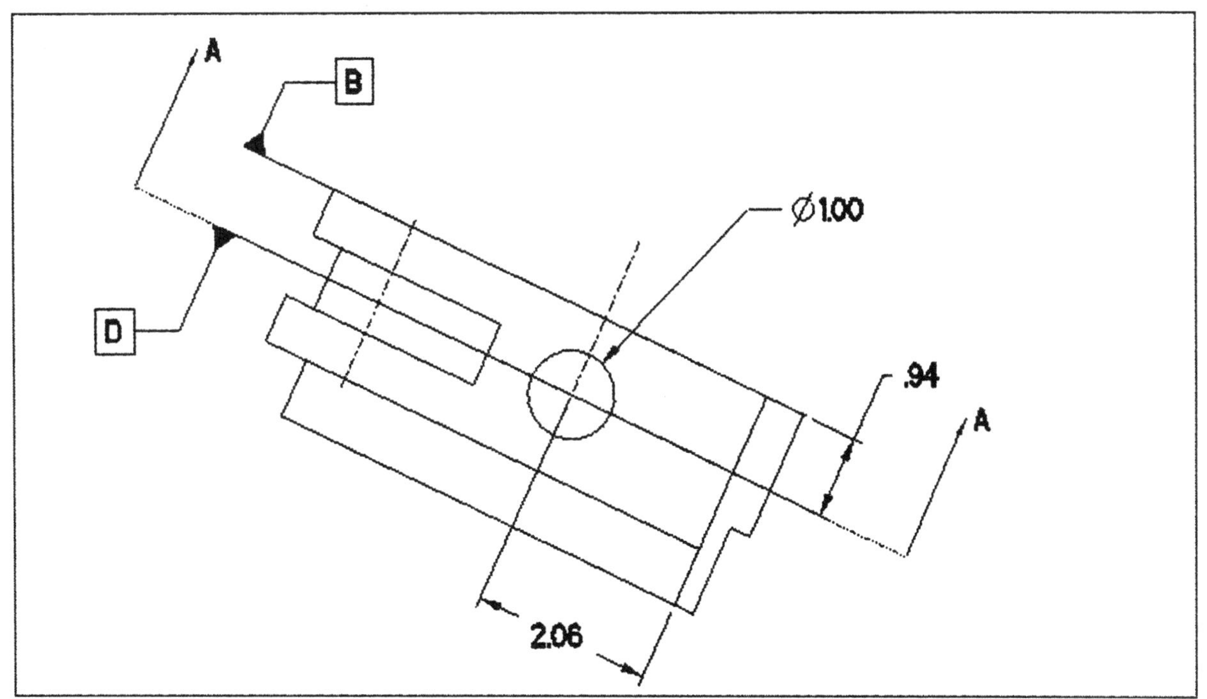

Figure 8.8(b) Edited Auxiliary View

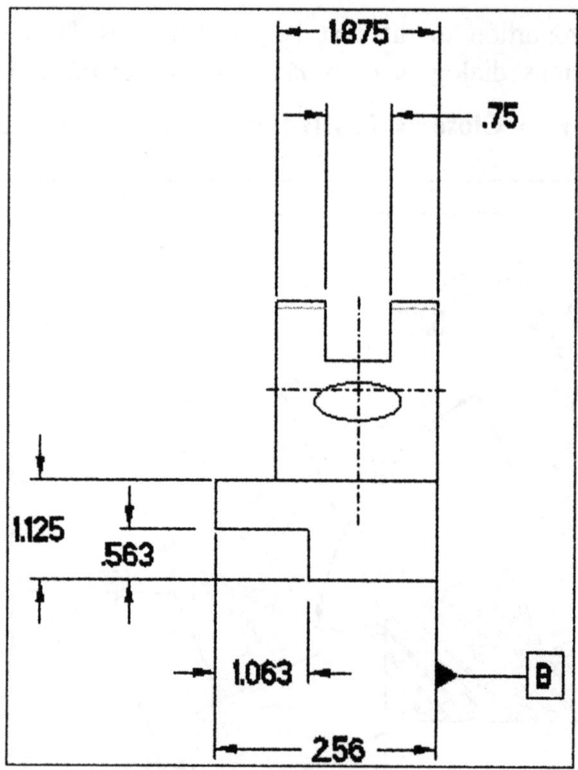

Figure 8.8(c) Edited Right Side View

Add the edges of the small thru hole back into the auxiliary view. The edges will display in the graphics window as light gray, but print as dashed on the drawing plot.

Click: **View** from Menu bar ⇒ **Drawing Display** ⇒ **Edge Display** ⇒ **Hidden Line** ⇒ press and hold the **Ctrl** key and pick near where the small thru hole would show as hidden in the auxiliary view [Fig. 8.9(a)] ⇒ pick the opposite edge [Fig. 8.9(b)] ⇒ **MMB** ⇒ **Done** ⇒ ⇒ ⇒ ⇒ ⇒ **MMB** [Figs. 8.9(c-d)]

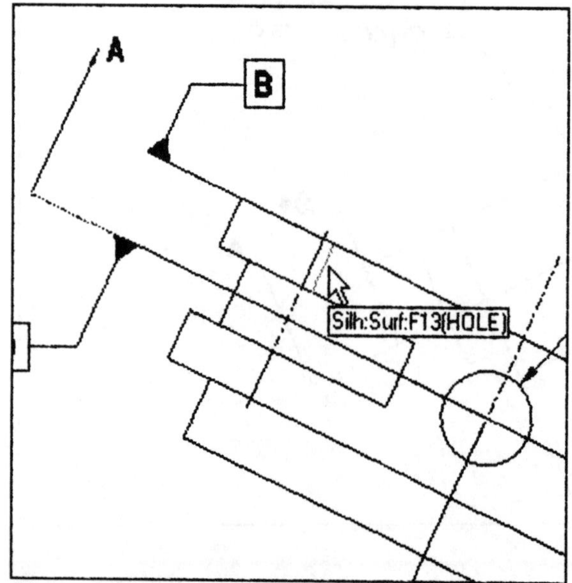

Figure 8.9(a) Pick near the Hole Edge

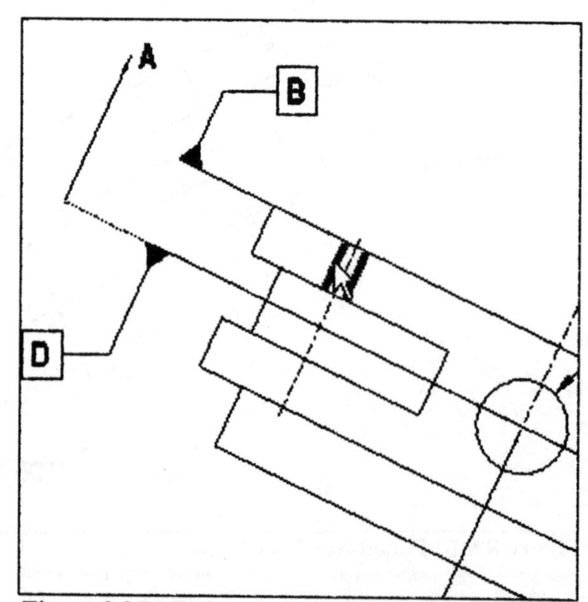

Figure 8.9(b) Pick near the Opposite Edge of Hole

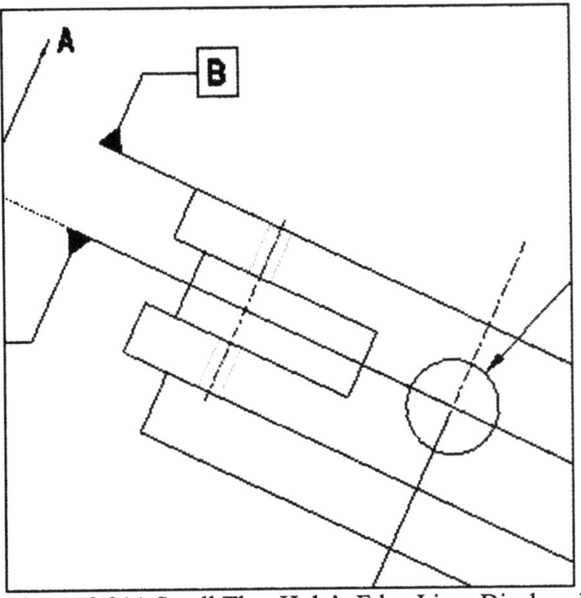

Figure 8.9(c) Small Thru Hole's Edge Lines Displayed

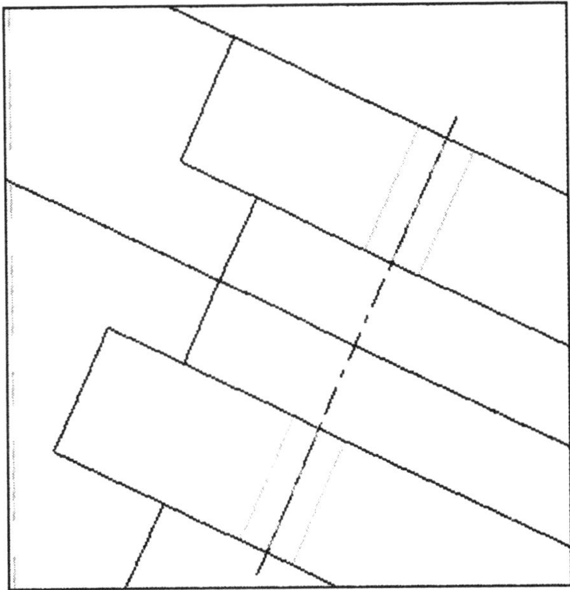
Figure 8.9(d) Close-up of the Thru Hole's Edge Lines

To increase the clarity of this drawing, you will need to master a number of capabilities. Partial views, detail views, using multiple sheets, and modifying section lining (crosshatch lines) are just a few of the many options available in Drawing mode. Create a Detail View.

Click: **Insert** ⇒ **Drawing View** ⇒ **Detailed** ⇒ [Select center point for detail on an existing view.] pick the top edge of the hole in the front section view [Fig. 8.10(a)] ⇒ [Sketch a spline, without intersecting other splines, to define an outline.] each **LMB** pick adds a point to the spline ⇒ **MMB** to end the spline [Fig. 8.10(b)] ⇒ [Select CENTER POINT for drawing view.] pick in the upper right of the drawing sheet to place the view ⇒ **RMB** on view ⇒ **Properties** [Fig. 8.10(c)] ⇒ **Scale** ⇒ •Custom scale **1.500** [Fig. 8.10(d)] ⇒ **Apply** ⇒ **Close** ⇒ **LMB** to deselect

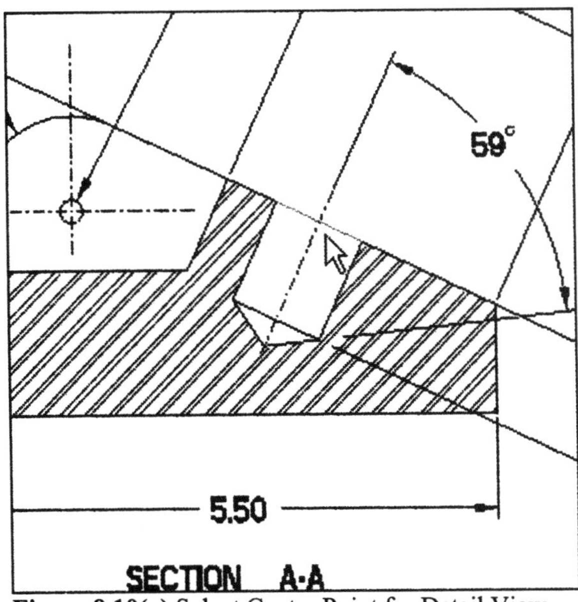

Figure 8.10(a) Select Center Point for Detail View

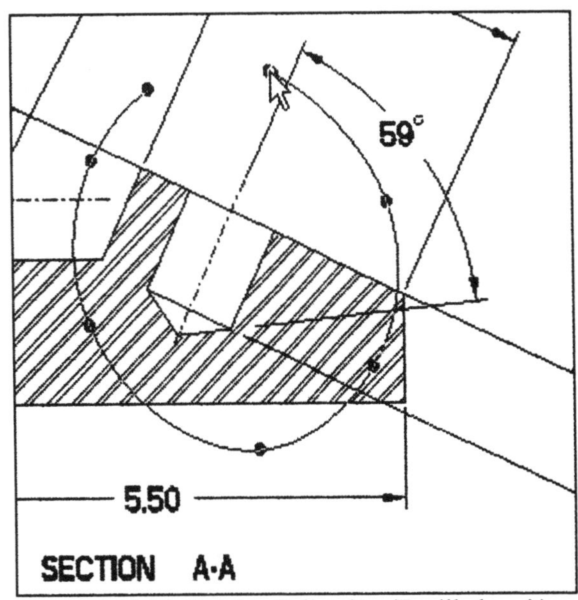

Figure 8.10(b) Sketch a Spline (Pro/E will close it)

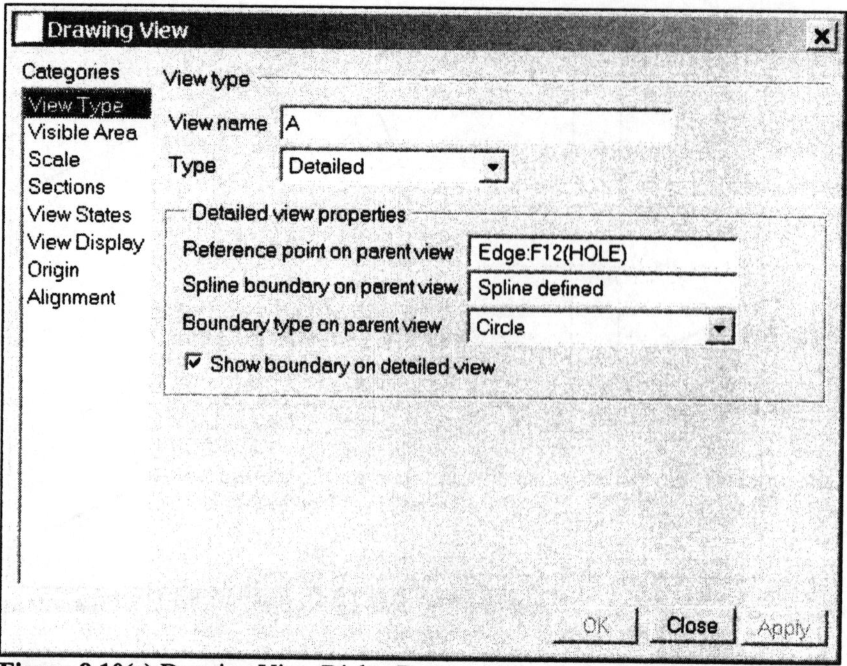

Figure 8.10(c) Drawing View Dialog Box

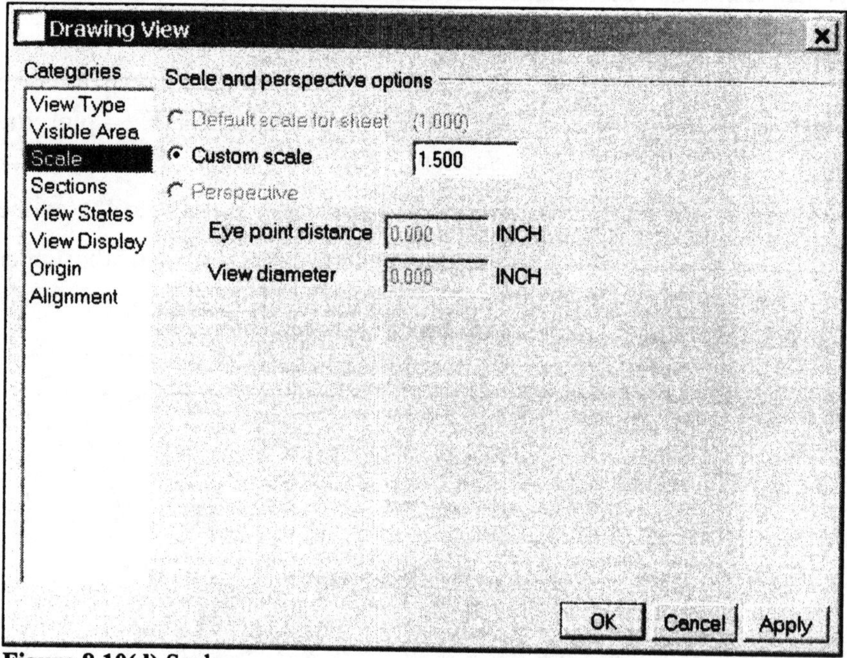

Figure 8.10(d) Scale

Add an axis to the detail of the hole; click **View** from Menu bar ⇒ **Show and Erase** ⇒ Show ⇒ only -.-.-A.1 **Axis** active ⇒ •**Feature and View** [Fig. 8.10(e)] ⇒ pick on the hole in the detail view ⇒ **Show All** ⇒ **Yes** ⇒ **Accept All** [Fig. 8.10(f)] ⇒ **Close** [Fig. 8.10(g)] ⇒ **LMB**

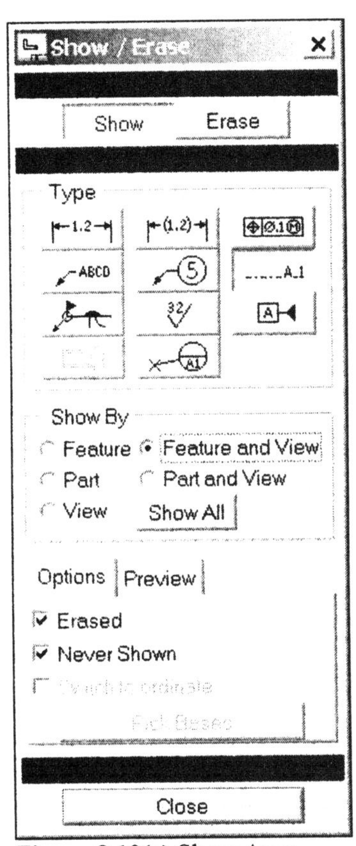

Figure 8.10(e) Show Axes

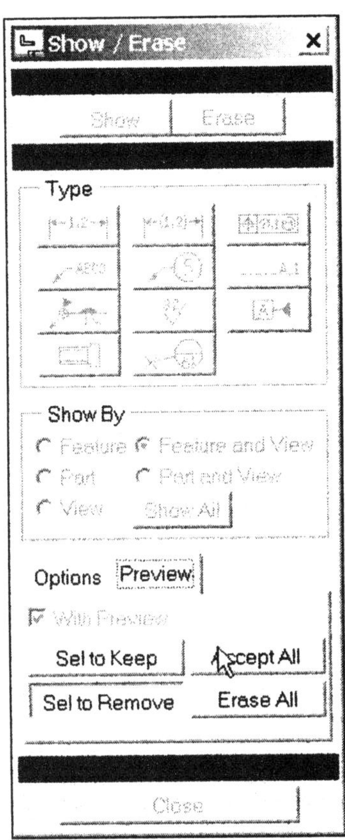

Figure 8.10(f) Accept All

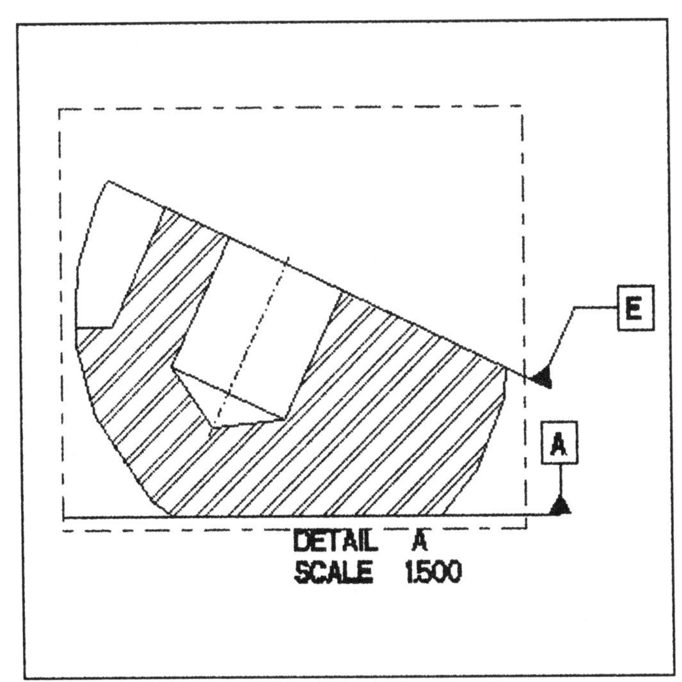

Figure 8.10(g) Axis Displayed

In DETAIL A, erase datum **A** and clip datum **E**. Also, clip the axis. ⇒ with the **Ctrl** key pressed, pick on the text items: **SECTION A-A, SEE DETAIL A** and **DETAIL A SCALE 1.500** [Fig. 8.11(a)] ⇒ **RMB** ⇒ **Text Style** ⇒ Default ⇒ Height **.375** [Fig. 8.11(b)] ⇒ **Apply** ⇒ **OK** [Fig. 8.11(c)]

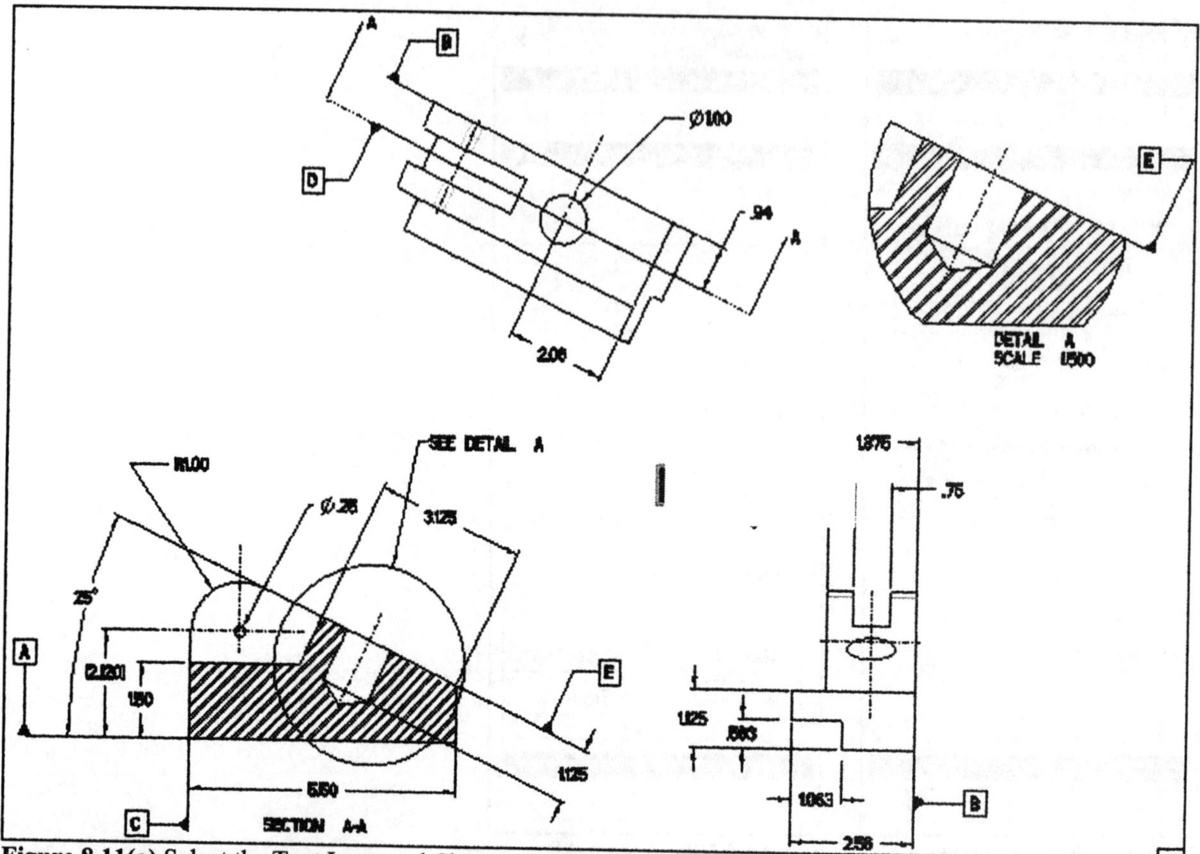

Figure 8.11(a) Select the Text Items and Change Their Height

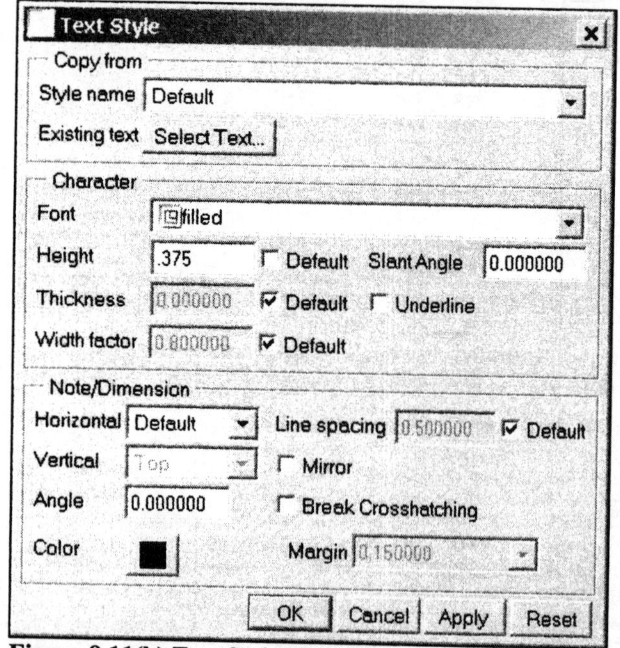

Figure 8.11(b) Text Style Dialog Box

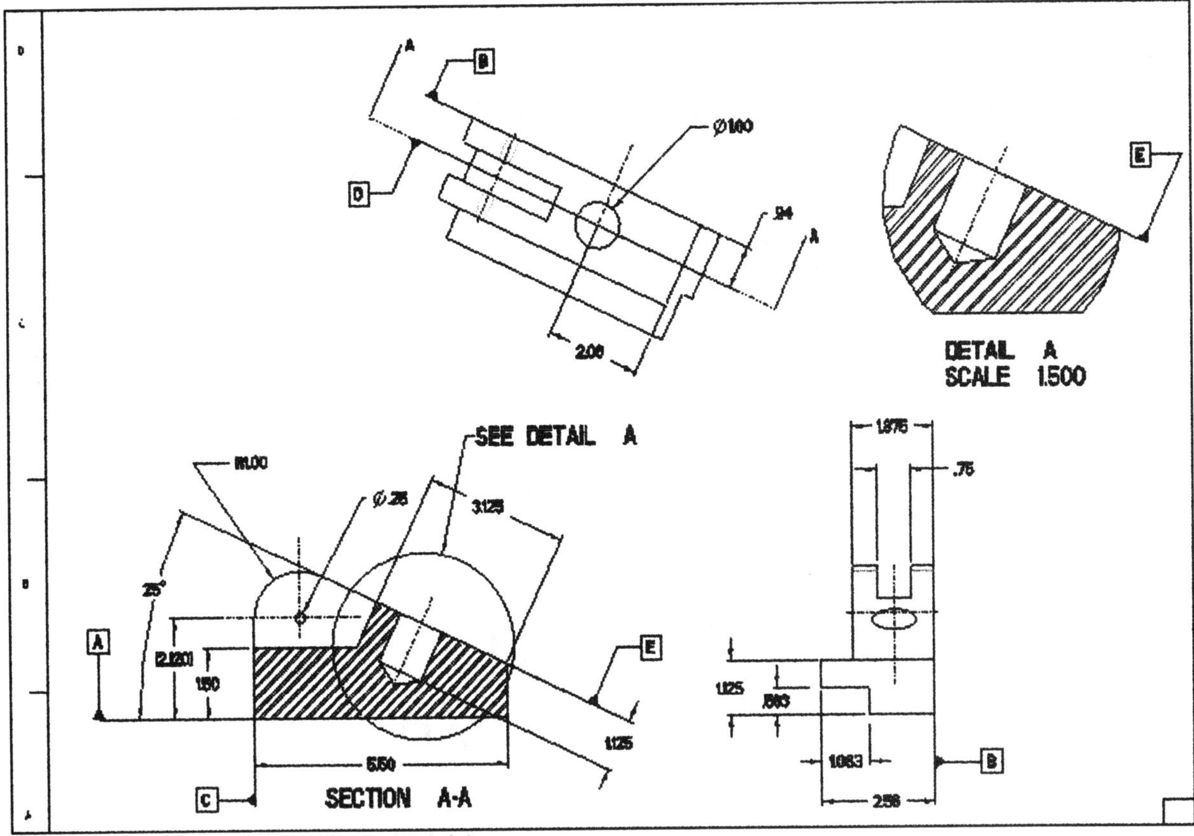

Figure 8.11(c) Changed Text Style (Height)

Change the height of the section identification lettering to **.375** [Fig. 8.11(d)] ⇒ pick on "**A**" ⇒ **RMB** ⇒ **Text Style** ⇒ Default ⇒ Height **.375** ⇒ **Apply** [Fig. 8.11(e)] ⇒ **OK** ⇒ **LMB** to deselect ⇒ ⇒ **MMB**

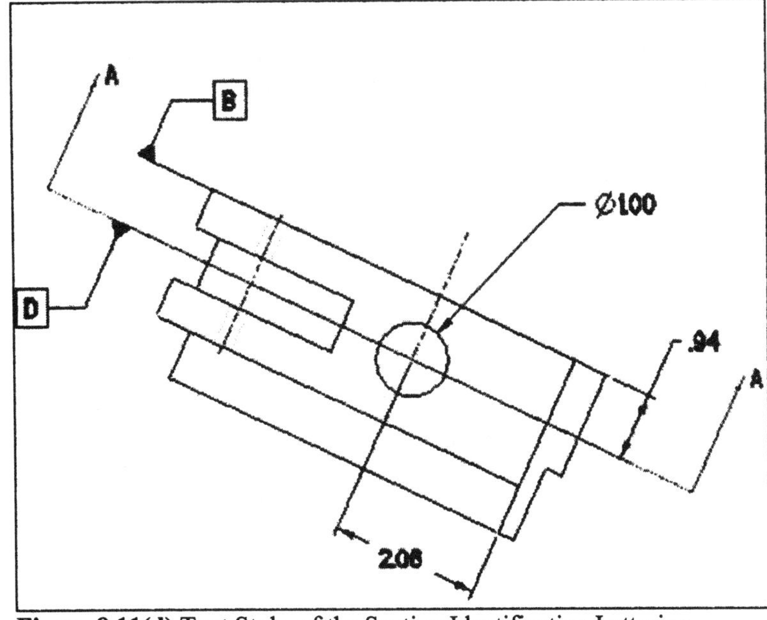

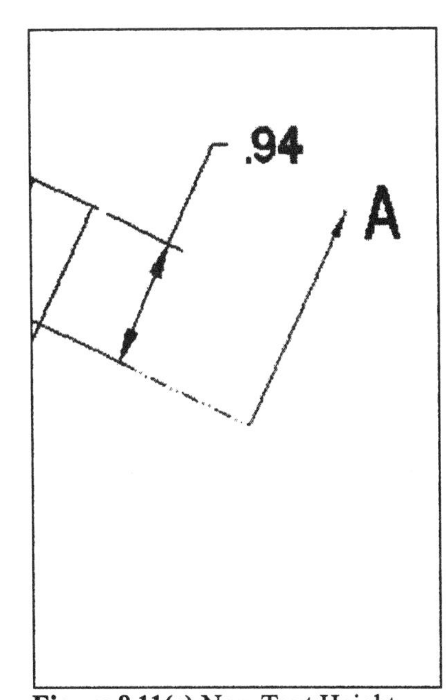

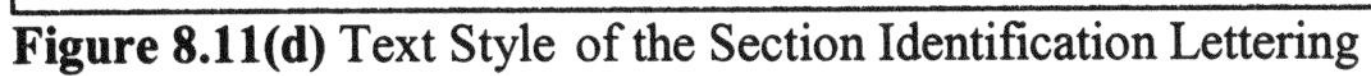
Figure 8.11(d) Text Style of the Section Identification Lettering

Figure 8.11(e) New Text Height

The Arrows for the cutting plane are now too small. These are controlled by the Drawing Options.

Click: **RMB** ⇒ **Properties** ⇒ **Drawing Options** ⇒ ***crossec_arrow_length .375***, ***crossec_arrow_width .125*** [Fig. 8.12(a)] ⇒ **Add/Change** ⇒ **Apply** ⇒ **Close** ⇒ **MMB** ⇒ [Fig. 8.12(b)]

- These options control cross sections and their arrows	|	|	|	|
crossec_arrow_length	.375	0.187500	✳	Sets the length of the arrow head on the cros
crossec_arrow_style	tail_online *	tail_online	●	Determines which end of cross-section arrow
crossec_arrow_width	.125	0.062500	✳	Sets the width of the arrow head on the cros

Figure 8.12(a) Cross Section Drawing Options

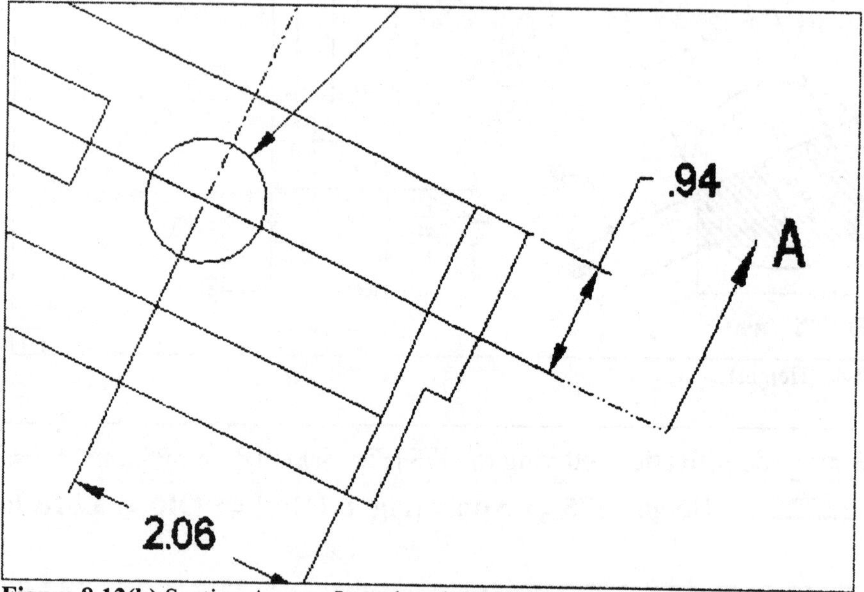

Figure 8.12(b) Section Arrows Length and Width Changed

Pick on the **1.125** dimension in the front section view [Fig. 8.13(a)] ⇒ **RMB** ⇒ **Move Item to View** ⇒ pick on view **DETAIL A** [Fig. 8.13(b)] ⇒ reposition and clip as needed [Fig. 8.13(c)] ⇒ **LMB**

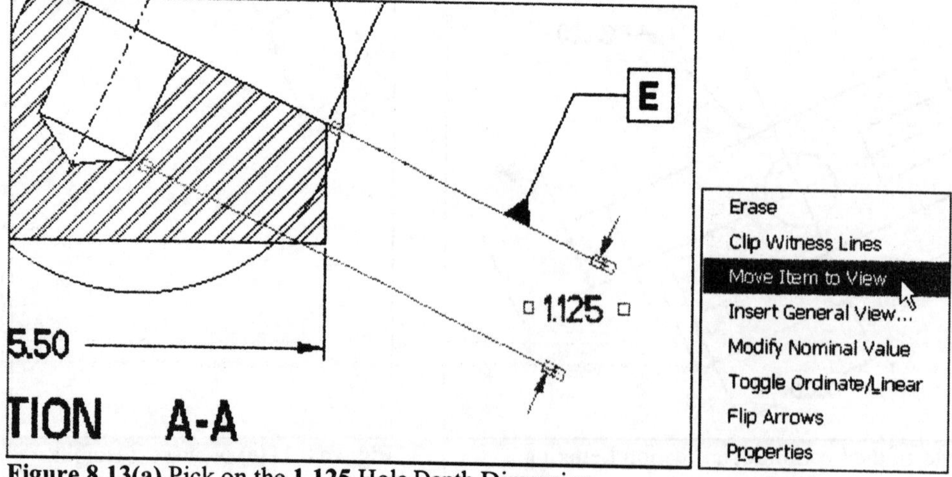

Figure 8.13(a) Pick on the **1.125** Hole Depth Dimension

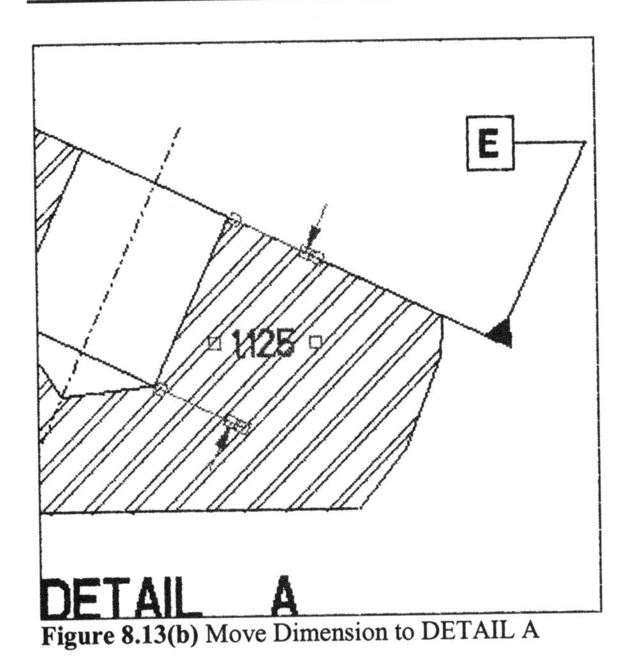

Figure 8.13(b) Move Dimension to DETAIL A

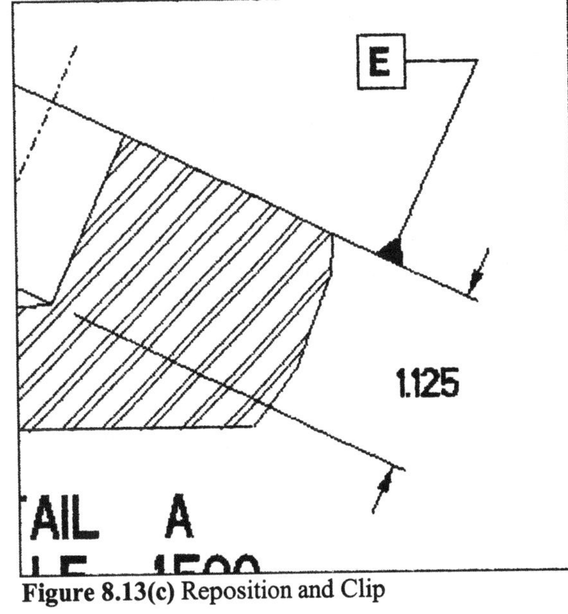

Figure 8.13(c) Reposition and Clip

Add another reference (horizontal) dimension to the small hole in the front view ⇒ keeping in mind ASME standards, cleanup the drawing as needed ⇒ ⇒ ⇒ ⇒ ⇒ **MMB** ⇒ **File** ⇒ **Delete** ⇒ **Old Versions** ⇒ **MMB** (Fig. 8.14)

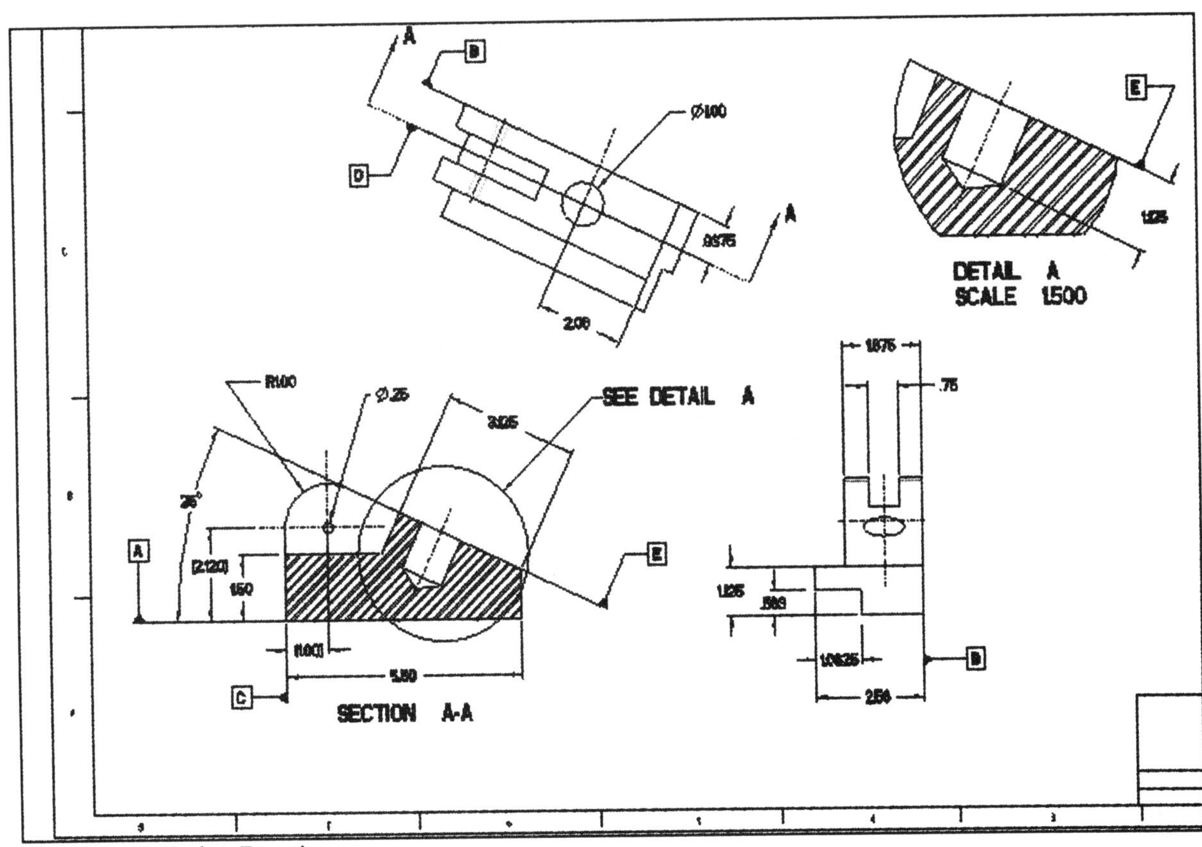

Figure 8.14 Anchor Drawing

Next, you will change the boundary of DETAIL A.

Pick on view **DETAIL A** ⇒ **RMB** (if you RMB inside the view outline, you get a pop-up list of options [Fig. 8.15(a)], whereas if you RMB outside the view outline, there are less options) [Fig. 8.15(b)] ⇒ **Properties** ⇒ click inside Spline area [Fig. 8.15(c)] ⇒ sketch the spline again in the front (SECTION A-A) view [Fig. 8.15(d)] ⇒ **Apply** [Fig. 8.15(e)] ⇒ **Close** ⇒ **LMB** [Fig. 8.15(f)] ⇒ ⇒ ⇒ ⇒ **MMB**

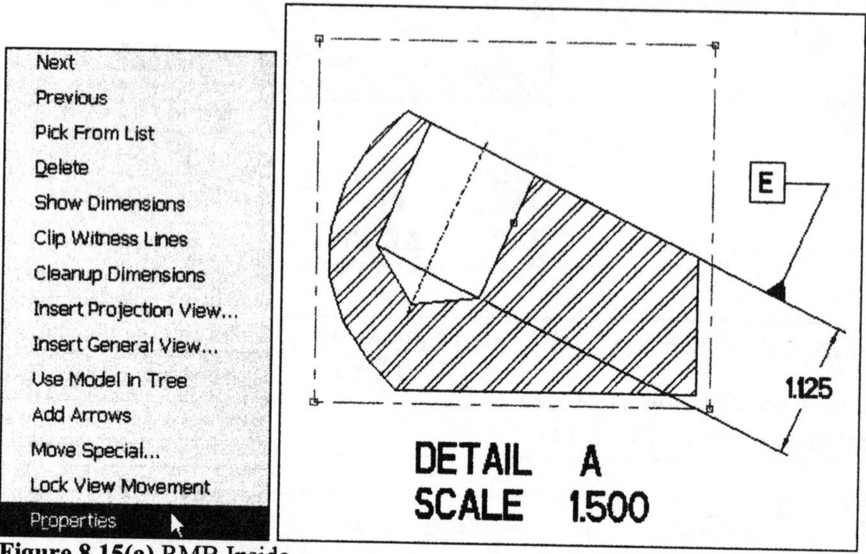

Figure 8.15(a) RMB Inside

Figure 8.15(b) RMB Outside

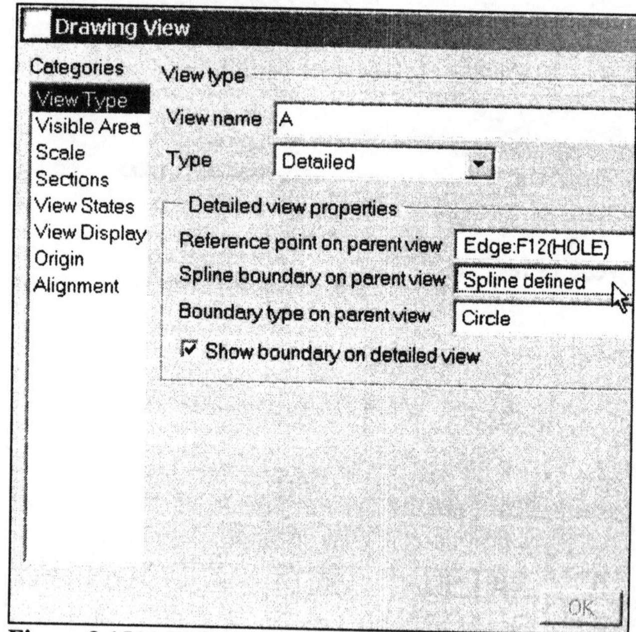

Figure 8.15(c) Pick inside Spline Collector

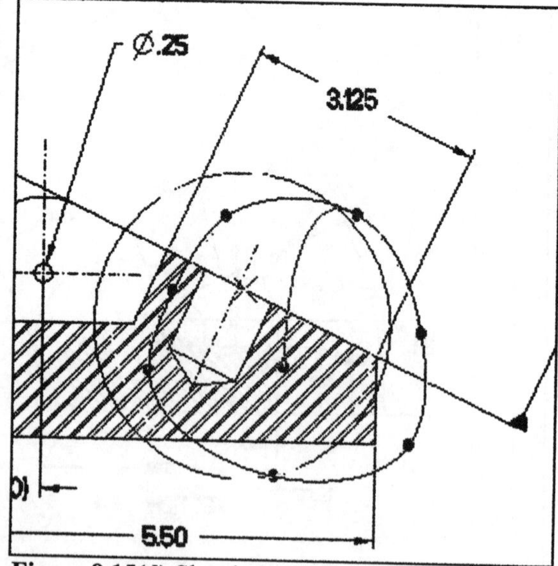

Figure 8.15(d) Sketch the Spline Again

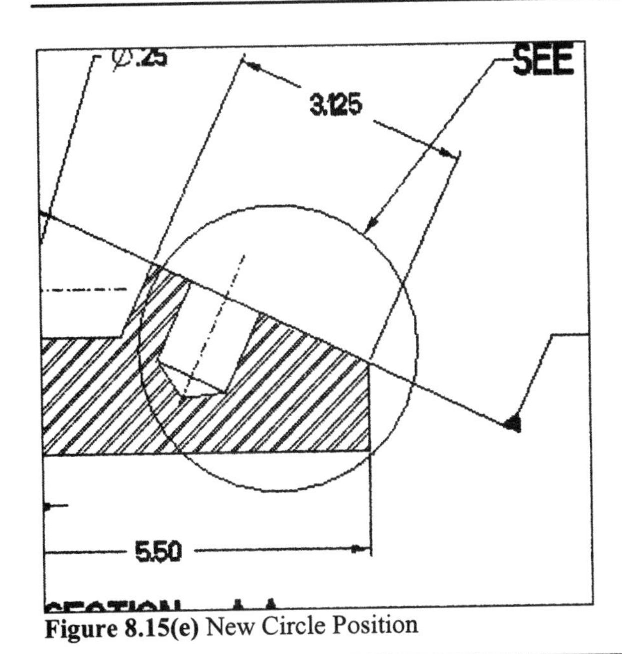

Figure 8.15(e) New Circle Position

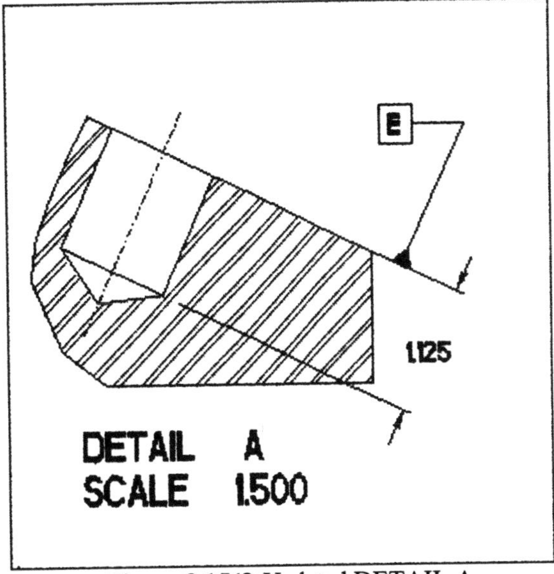

Figure 8.15(f) Updated DETAIL A

Add a dimension for the full angle of the hole's drill tip in DETAIL A ⇒ Erase the **118** degree drill tip dimension *(or the* ***59*** *degree dimension, depending on what is shown)* ⇒ **Insert** ⇒ **Dimension** ⇒ **New References** ⇒ pick both edge lines of the drill tip [Fig. 8.16(a)] ⇒ **MMB** to place the dimension [Fig. 8.16(b)] ⇒ **MMB** ⇒ reposition and clip the dimension ⇒ pick on the **118.0** dimension [Fig. 8.16(c)] ⇒ **RMB** ⇒ **Properties** ⇒ change decimal places to **0** ⇒ **Enter** ⇒ **OK** [Figs. 8.16(d-e)]

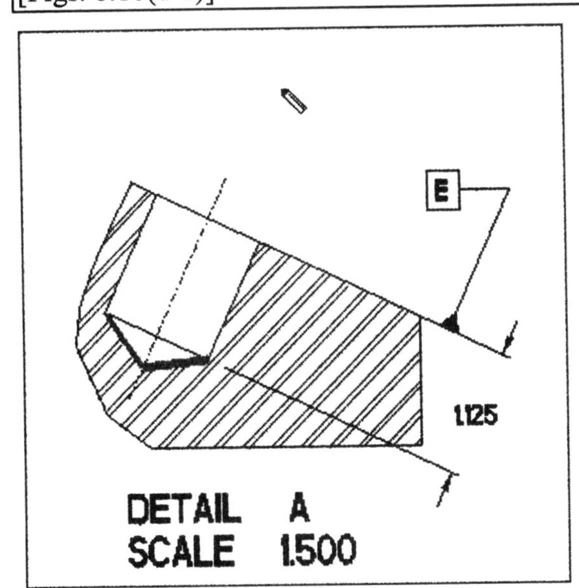

Figure 8.16(a) Drill Tip Edges

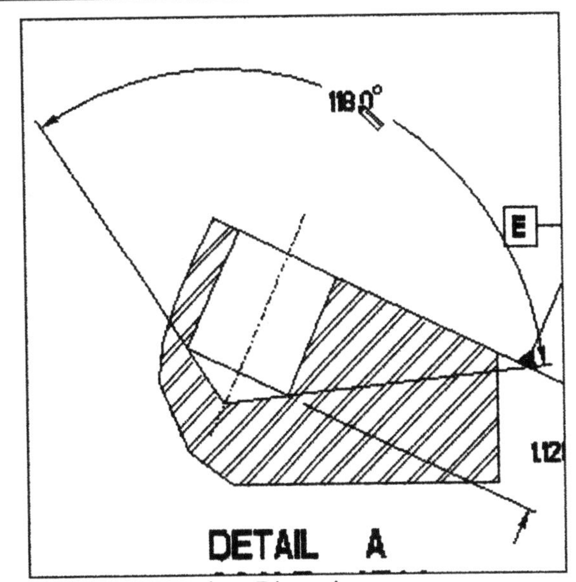

Figure 8.16(b) Place Dimension

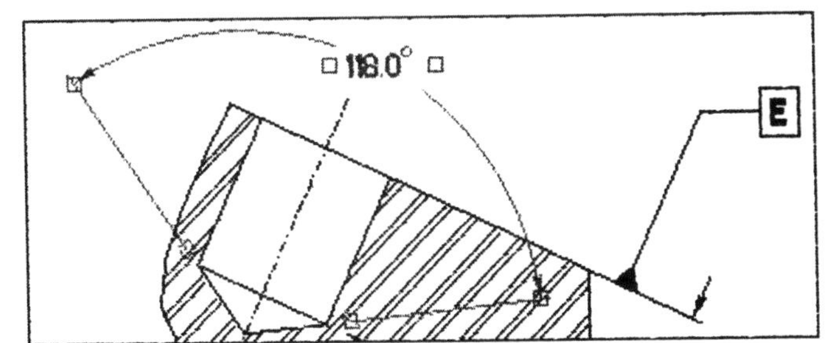

Figure 8.16(c) Reposition Dimension and Move Datum E

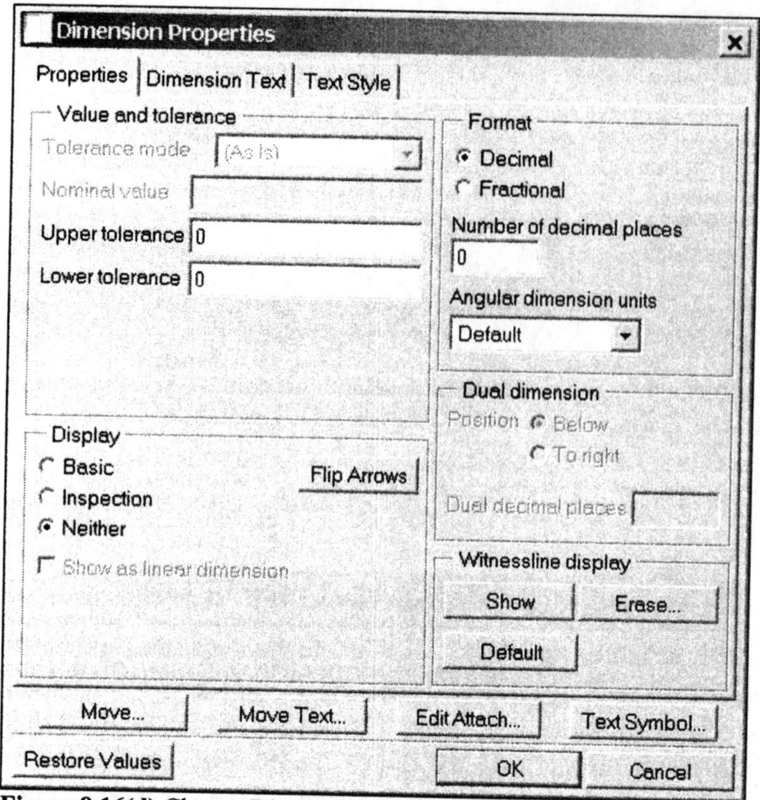

Figure 8.16(d) Change Dimension Properties of the **118** Degree Dimension

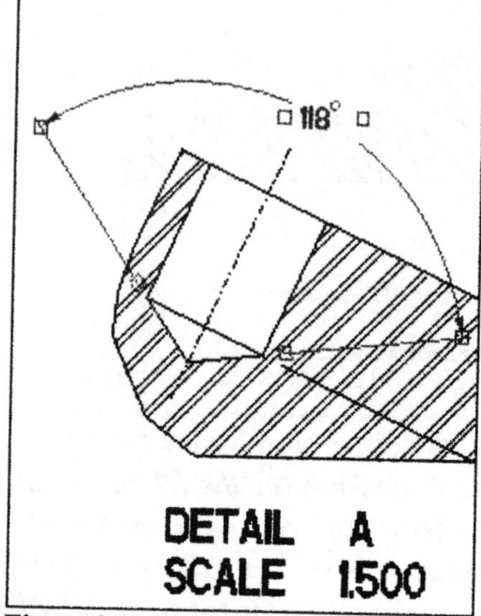

Figure 8.16(e) **118** Degrees

Pick on the **118.0** dimension in the front view ⇒ **RMB** ⇒ **Erase** ⇒ **LMB** ⇒ Change the spacing and the angle of the section lining. Pick on the hatching in SECTION A-A ⇒ **RMB** ⇒ **Properties** [Fig. 8.17(a)] ⇒ **Delete Line** ⇒ **Angle** ⇒ **30** ⇒ **MMB** [Fig. 8.17(b)] ⇒ **LMB** ⇒ ⇒ ⇒ **MMB**

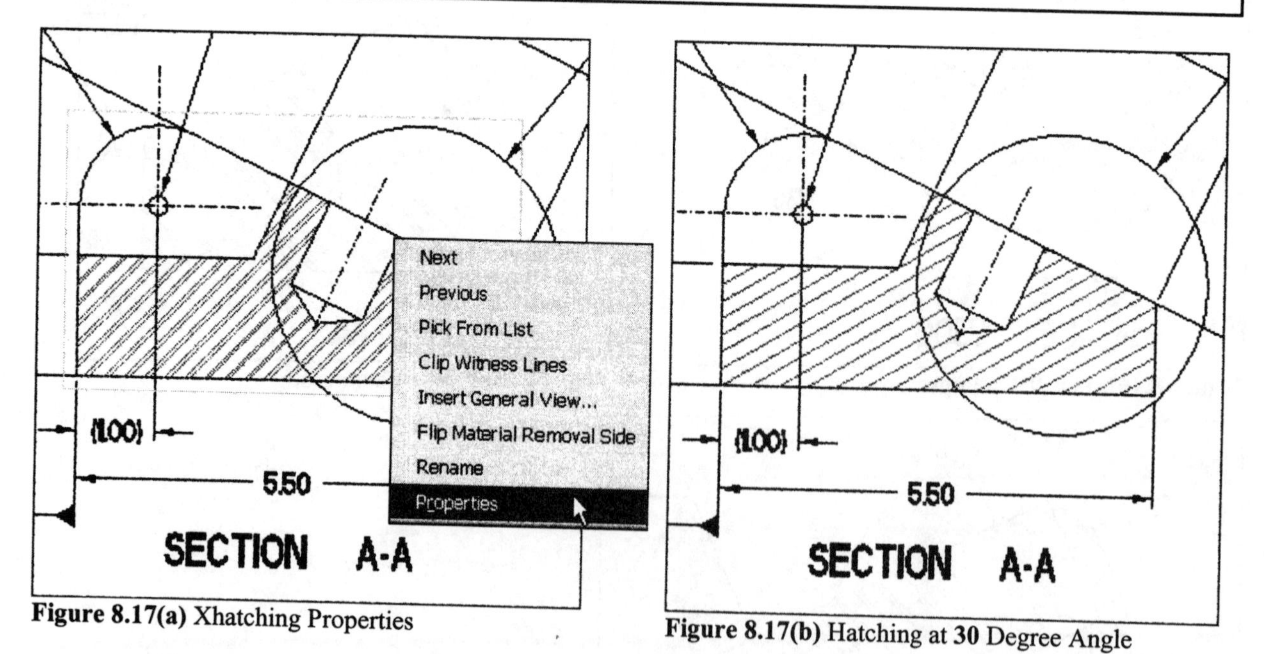

Figure 8.17(a) Xhatching Properties

Figure 8.17(b) Hatching at **30** Degree Angle

Change the spacing and the angle of the section lining in DETAIL A. Pick on the hatching in DETAIL A ⇒ **RMB** ⇒ **Properties** [Fig. 8.18(a)] ⇒ **Det Indep** ⇒ **Hatch** ⇒ **Spacing** ⇒ **Double** ⇒ **Angle** ⇒ **45** ⇒ **MMB** [Fig. 8.18(b)] ⇒ **LMB** ⇒ **RMB** ⇒ **Update Sheet** ⇒ ⇒ ⇒ **MMB**

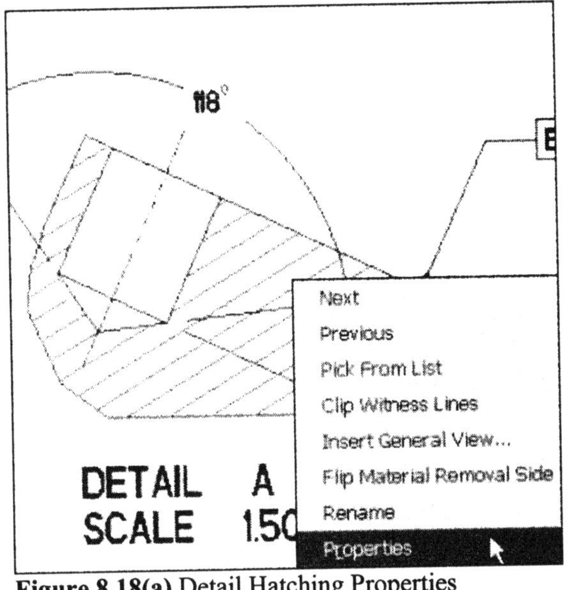

Figure 8.18(a) Detail Hatching Properties

Figure 8.18(b) Hatching at **45** Degree Angle

Change the text style used on the drawing. Click: ⇒ enclose all of the drawing text with a selection box [Fig. 8.19(a)] ⇒ **RMB** ⇒ **Text Style**

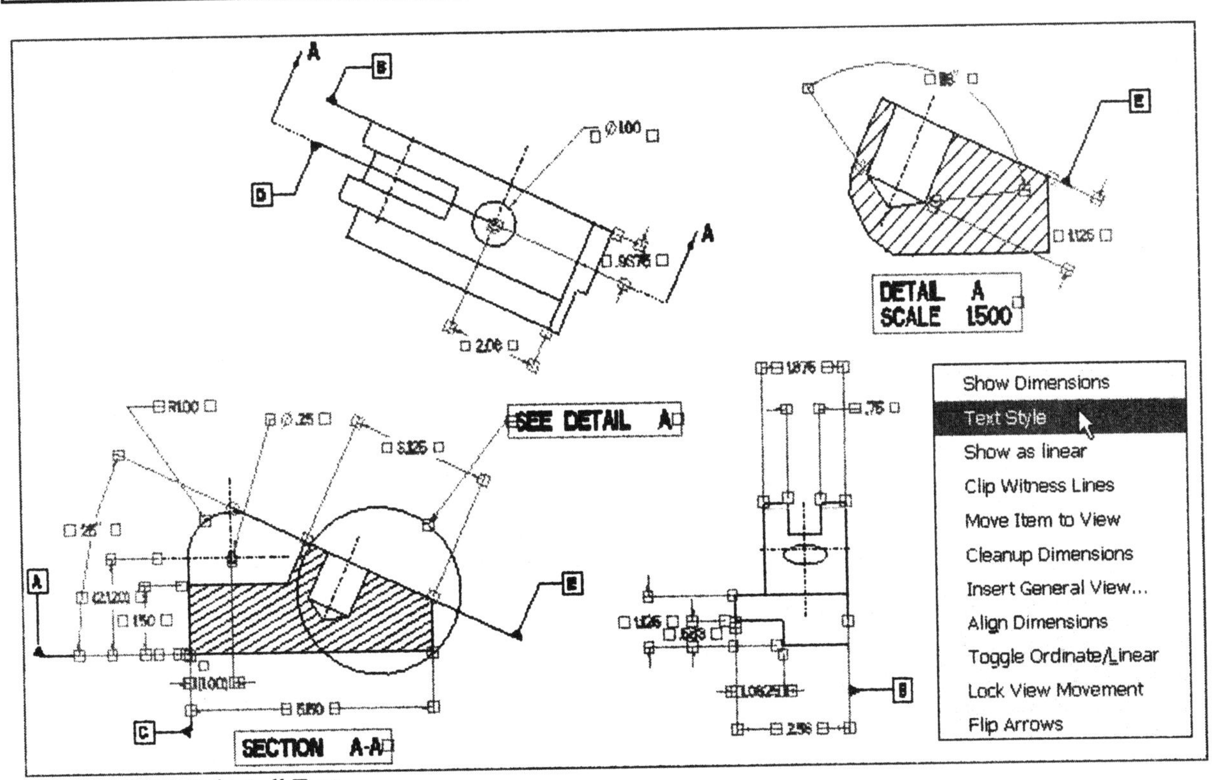

Figure 8.19(a) Select all Text

Font **Blueprint MT** [Fig. 8.19(b)] ⇒ **Apply** ⇒ **OK** ⇒ **LMB** [Fig. 8.19(c)] ⇒ [save icon] ⇒ **MMB**

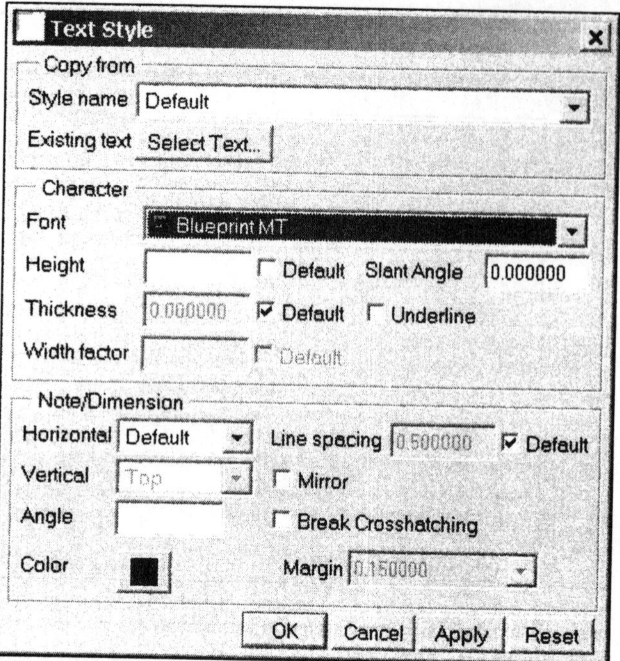

Figure 8.19(b) Text Style Dialog Box, Character- Font- Blueprint MT

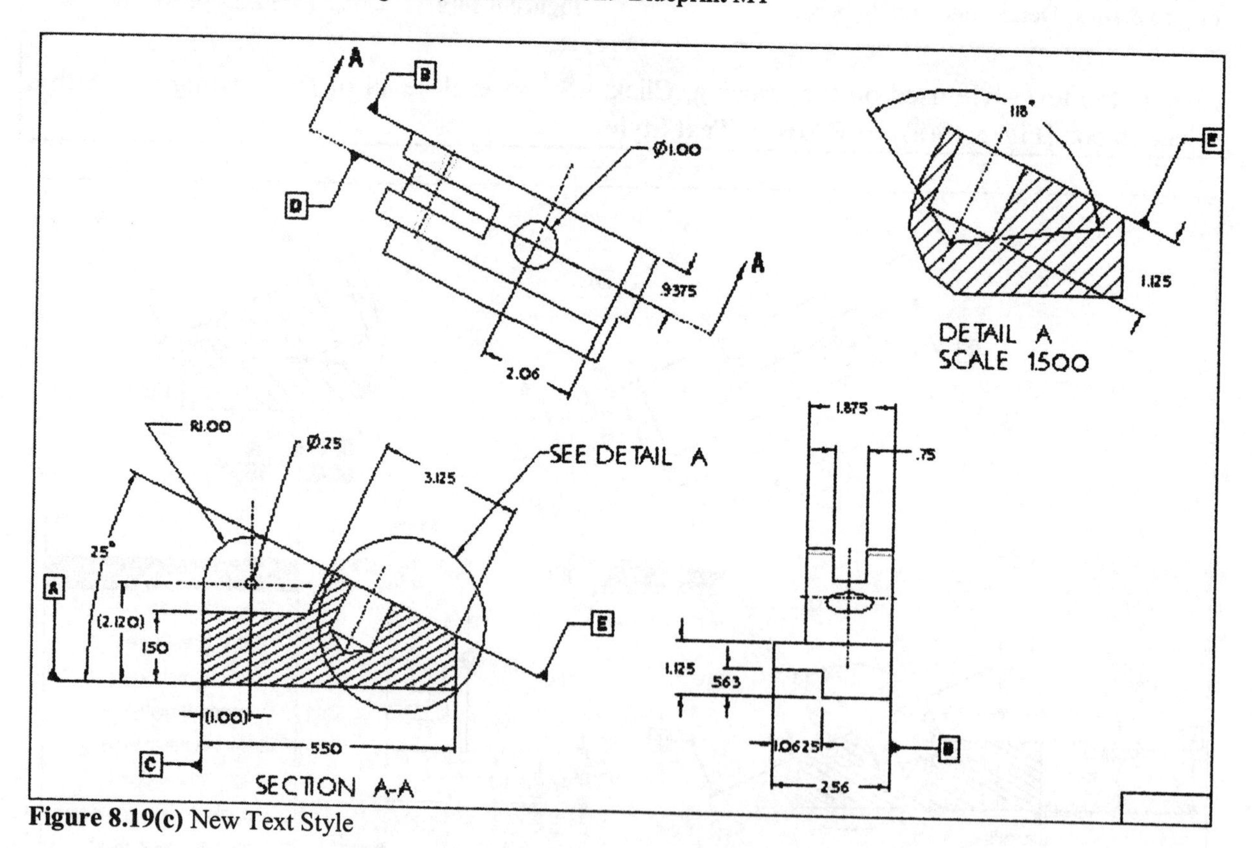

Figure 8.19(c) New Text Style

Add a geometric tolerance to the angled surface, click: **Create geometric tolerances** (*or Insert ⇒ Geometric Tolerance*) ⇒ **Angularity** ⇒ Type **Datum** [Fig. 8.20(a)] ⇒ Select Entity... ⇒ pick on datum **E** [Fig. 8.20(b)] ⇒ **Datum Refs** tab ⇒ **Primary** tab- Basic ⇒ pick on datum **A** in the front view as the Primary Reference Datum [Fig. 8.20(c)] ⇒ **Tol Value** tab ⇒ Overall Tolerance 0.005 [Fig. 8.20(d)] ⇒ **OK** ⇒ **LMB** ⇒ **RMB** ⇒ **Update Sheet** [Fig. 8.20(e)]

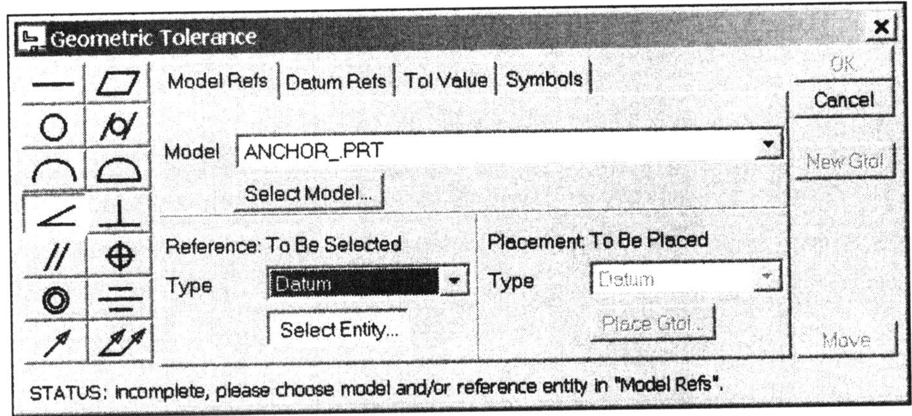

Figure 8.20(a) Angularity, Type Datum

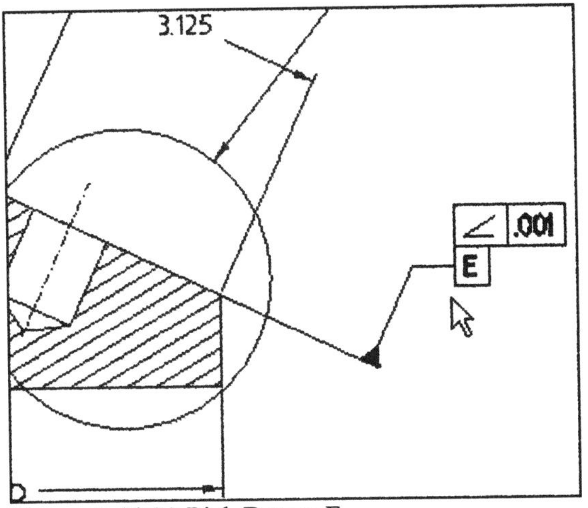

Figure 8.20(b) Pick Datum E

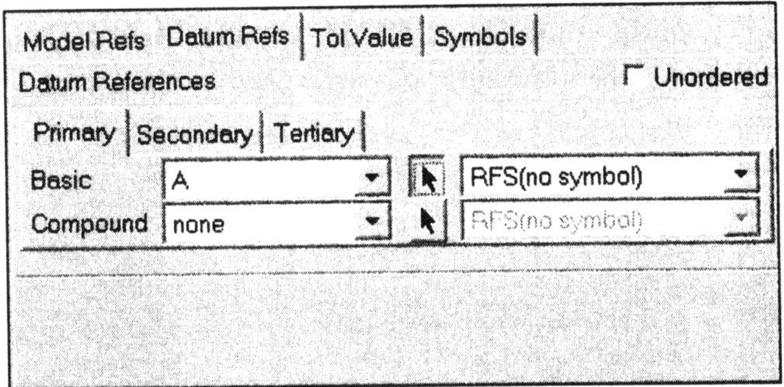

Figure 8.20(c) Primary Reference Datum A

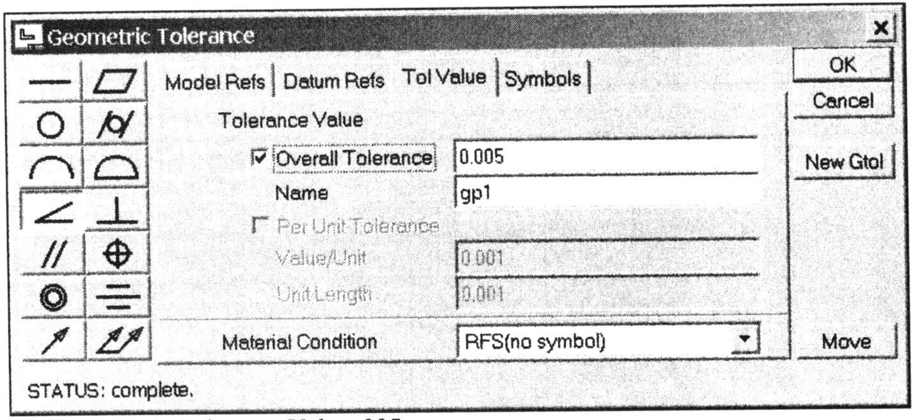

Figure 8.20(d) Tolerance Value **.005**

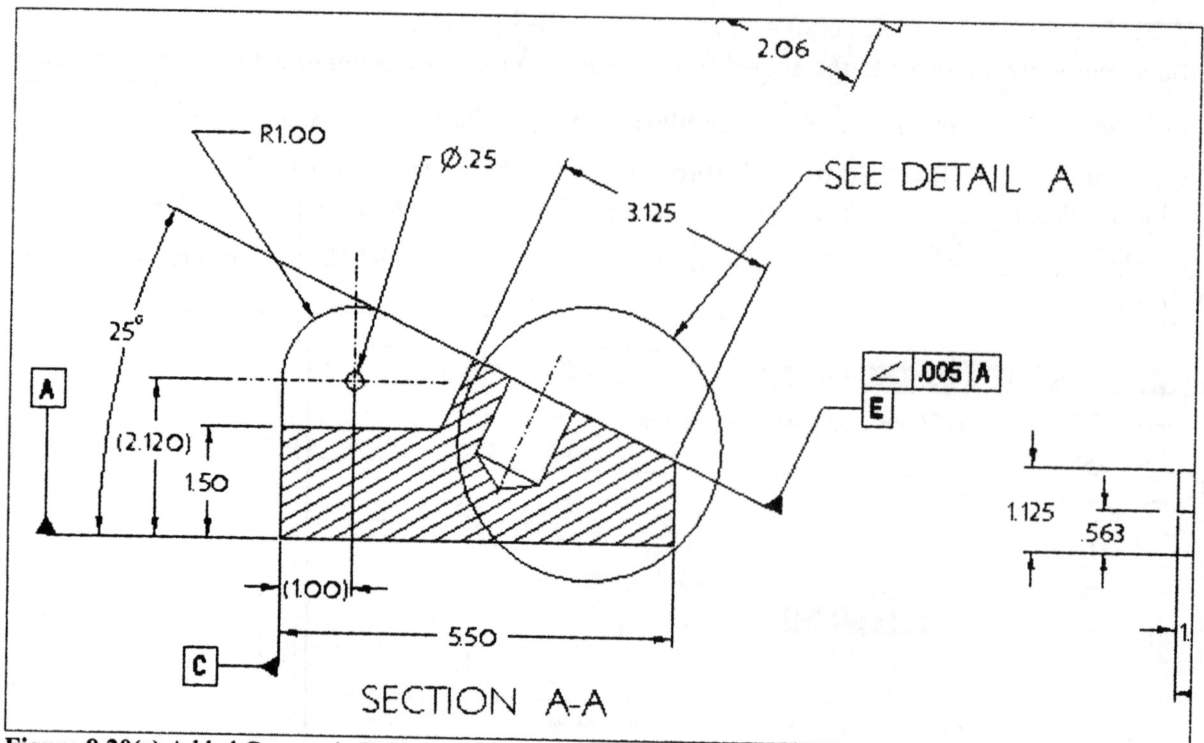

Figure 8.20(e) Added Geometric Tolerance

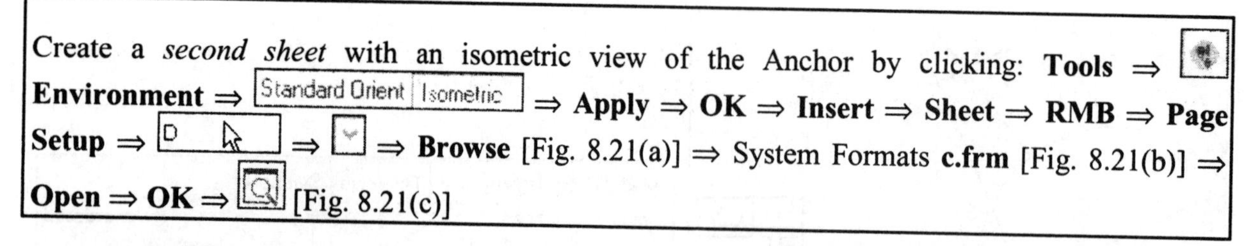

Create a *second sheet* with an isometric view of the Anchor by clicking: **Tools** ⇒ **Environment** ⇒ Standard Orient Isometric ⇒ **Apply** ⇒ **OK** ⇒ **Insert** ⇒ **Sheet** ⇒ **RMB** ⇒ **Page Setup** ⇒ D ⇒ ⇒ **Browse** [Fig. 8.21(a)] ⇒ System Formats **c.frm** [Fig. 8.21(b)] ⇒ **Open** ⇒ **OK** ⇒ [Fig. 8.21(c)]

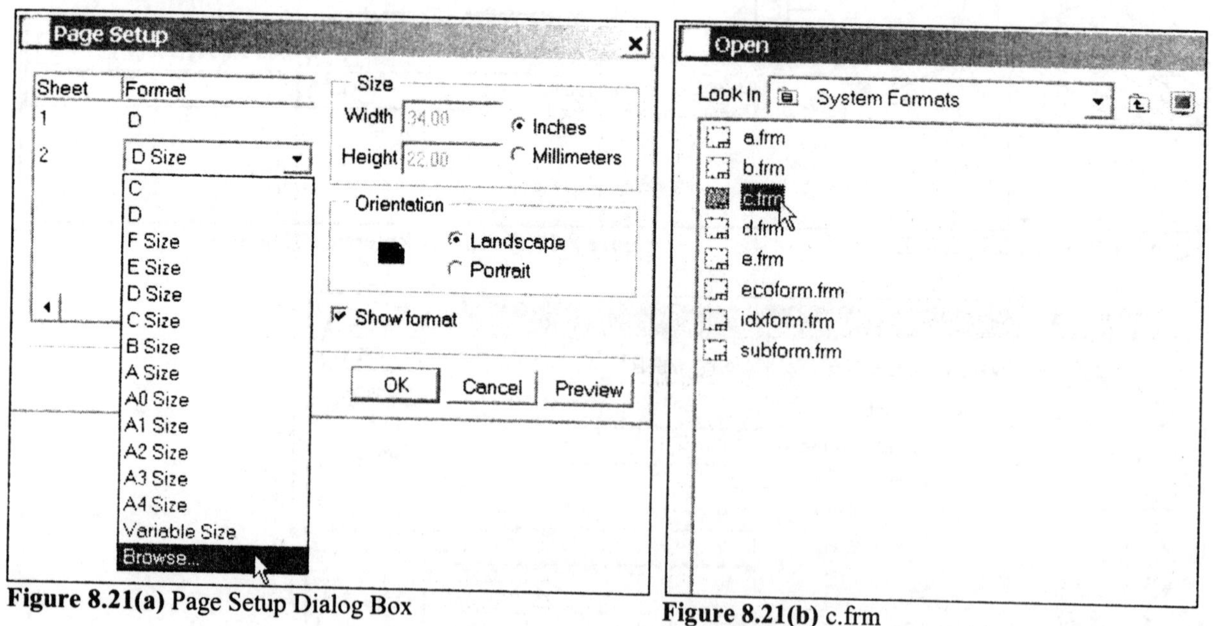

Figure 8.21(a) Page Setup Dialog Box

Figure 8.21(b) c.frm

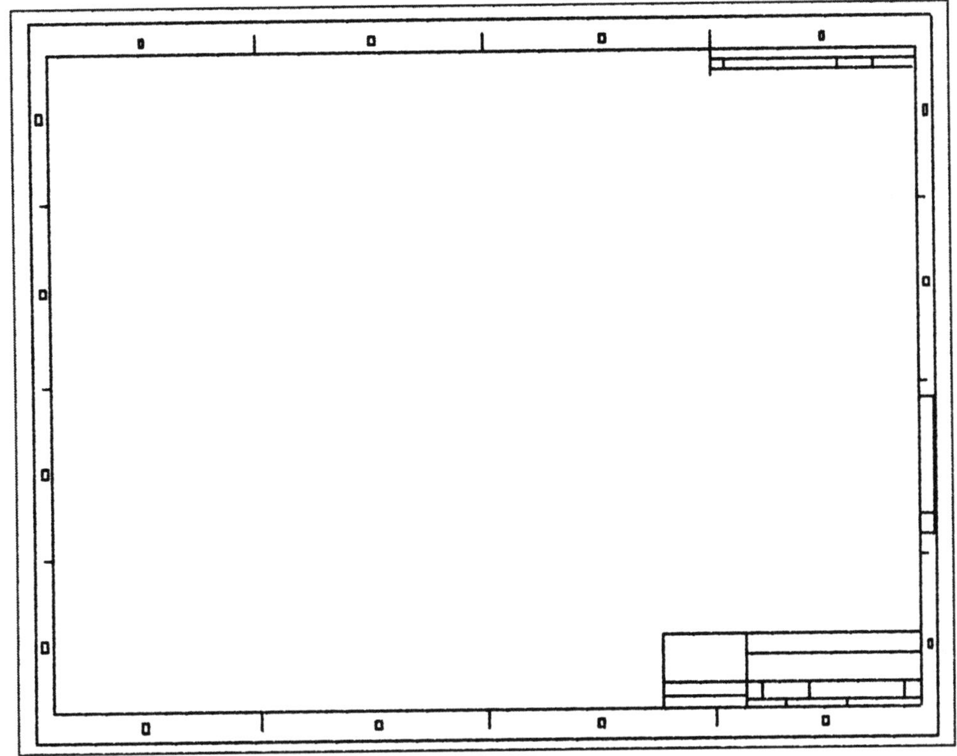

Figure 8.21(c) C Format for Page 2

Click: **Insert** ⇒ **Drawing View** ⇒ **General** ⇒ pick a center point for the view ⇒ **Scale** ⇒ •Custom Scale **1.50** ⇒ **Apply** [Fig. 8.21(d)] ⇒ **Close** ⇒ **LMB** to deselect

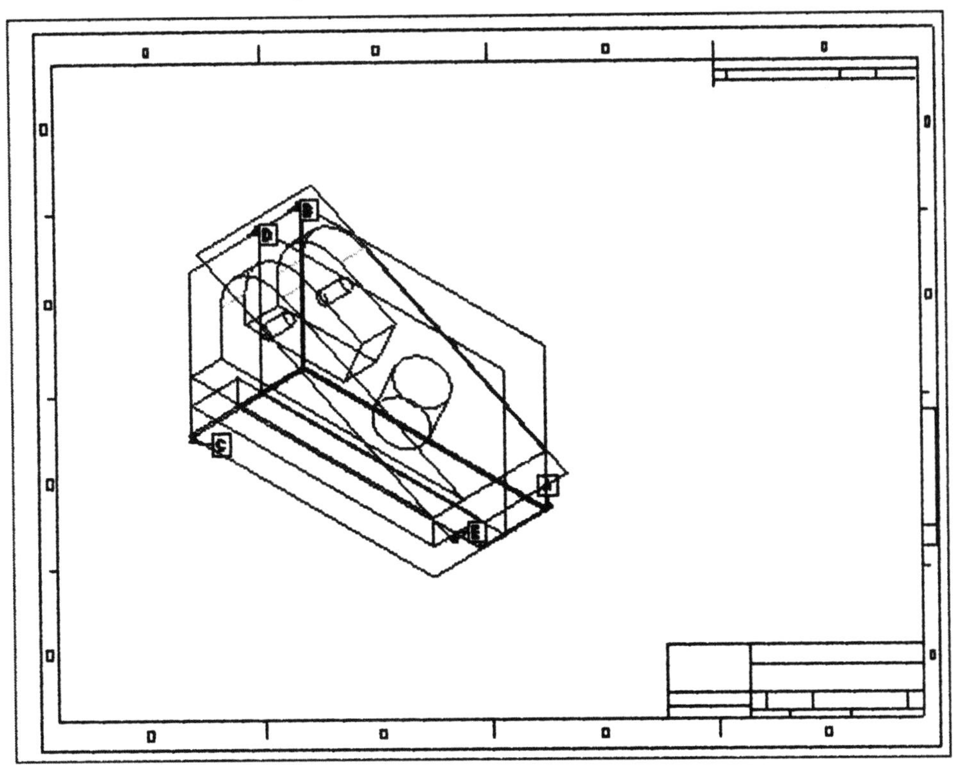

Figure 8.21(d) Pictorial View

Pick on the pictorial view ⇒ **RMB** ⇒ **Properties** ⇒ **View Display** ⇒ Display style ⇒ **No Hidden** ⇒ Tangent edges display style ⇒ **Dimmed** ⇒ **Apply** ⇒ **Close** ⇒ **LMB** [Fig. 8.21(e)] ⇒ **File** ⇒ **Save** ⇒ **MMB**

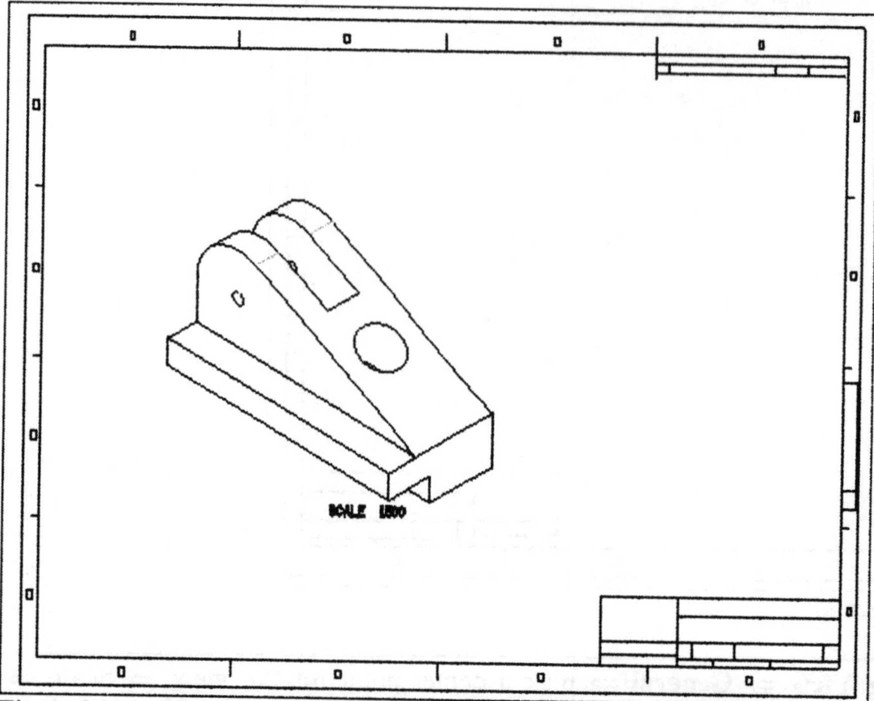

Figure 8.21(e) Sheet 2, Tangent Edges Dimmed and Hidden Lines Removed

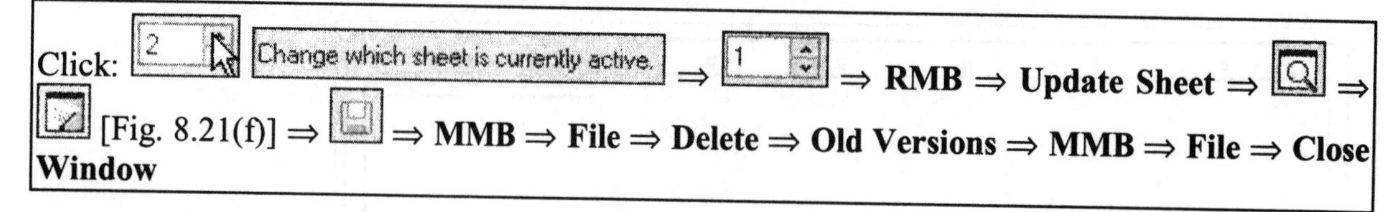

Click: [2] Change which sheet is currently active. ⇒ [1] ⇒ **RMB** ⇒ **Update Sheet** ⇒ ⇒ [Fig. 8.21(f)] ⇒ ⇒ **MMB** ⇒ **File** ⇒ **Delete** ⇒ **Old Versions** ⇒ **MMB** ⇒ **File** ⇒ **Close Window**

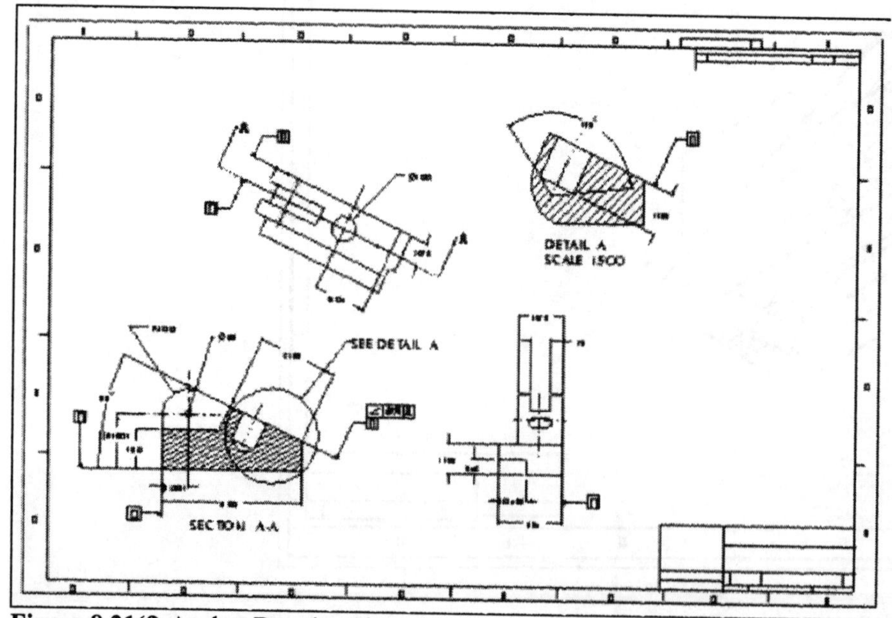

Figure 8.21(f) Anchor Drawing Sheet 1

Click: Open an existing object ⇒ ⇒ **System Formats** ⇒ **d.frm** ⇒ **Open** ⇒ **File** ⇒ **Save a Copy** ⇒ type a unique name for your format: New Name **DETAIL_FORMAT_D** ⇒ **OK** ⇒ **Window** ⇒ **Close** ⇒ ⇒ pick **detail_format_d** ⇒ **Open** ⇒ **RMB** in the graphics window ⇒ **Properties** Options dialog box opens ⇒ ***drawing_text_height .25*** ⇒ **Add/Change** ⇒ ***default_font filled*** ⇒ **Add/Change** ⇒ ***draw_arrow_style filled*** ⇒ **Add/Change** ⇒ ***draw_arrow_length .25*** ⇒ **Add/Change** ⇒ ***draw_arrow_width .08*** ⇒ **Add/Change** ● option in Status column shows Default column value active, * Status column shows pending value [Fig. 8.22(a)] ⇒ **Apply** ⇒ **Close** ⇒ **Ctrl+S** ⇒ **MMB**

The **format** will have a **.dtl** associated with it, and the **drawing** will have a different **.dtl** file associated with it. *They are separate **.dtl** files.* When you activate a drawing and then add a format, the **.dtl** for the format controls the font, etc. for the format only.

Active Drawing				
These options control text not subject to oth				
drawing_text_height	.25	0.156250	*	Sets default text height for all text in the drawing usi
text_thickness	0.000000 *	0.000000	●	Sets default text thickness for new text after regene
text_width_factor	0.800000 *	0.800000	●	Sets default ratio between the text width and text he
draft_scale	1.000000 *	1.000000	●	Determines value of draft dimensions relative to actu
These options control text and line fonts				
default_font	filled	font	*	Specifies a font index that determines default text fo
These options control leaders				
draw_arrow_length	.25	0.187500	*	Sets length of leader line arrows.
draw_arrow_style	filled	closed	*	Controls style of arrow head for all detail items involv
draw_arrow_width	.08	0.062500	*	Sets width of leader line arrows. Drives these: "draw

Figure 8.22(a) Drawing Option **.dtl** File for Format

Zoom into the title block region, and create notes for the title text and parameter text, click: **Tools** ⇒ **Environment** ⇒ ☑ Snap to Grid ⇒ **Apply** ⇒ **OK** ⇒ **View** ⇒ **Draft Grid** ⇒ **Show Grid** ⇒ **Grid Params** ⇒ **X&Y Spacing** ⇒ type **.1** ⇒ **MMB** ⇒ **MMB** ⇒ **MMB** ⇒ **Create a note** [Fig. 8.22(b)] ⇒

Make Note ⇒ pick a point for the note in the largest area of the title block ⇒ type **TOOL ENGINEERING CO.** ⇒ **Enter** ⇒ **Enter** ⇒

Make Note ⇒ ⇒ type **DRAWN** ⇒ **Enter** ⇒ **Enter** ⇒

Make Note ⇒ ⇒ type **ISSUED** ⇒ **Enter** ⇒ **Enter** ⇒

Make Note ⇒ ⇒ type **&dwg_name** ⇒ **Enter** ⇒ **Enter** ⇒

Make Note ⇒ ⇒ type **&scale** ⇒ **Enter** ⇒ **Enter** ⇒

Make Note ⇒ ⇒ type **SHEET ¤t_sheet OF &total_sheets** ⇒ **Enter** ⇒ **Enter** ⇒ **Done/Return** ⇒ **LMB** ⇒ **Tools** ⇒ **Environment** ⇒ ☐ Snap to Grid ⇒ **Apply** ⇒ **OK** ⇒ modify the text height (**.10**) and the placement of the notes as needed [Fig. 8.22(b)] ⇒ **View** ⇒ **Update** ⇒ **Current Sheet** ⇒ **View** ⇒ **Orientation** ⇒ **Refit** ⇒ **Ctrl+S** ⇒ **MMB** ⇒ **File** ⇒ **Close Window**

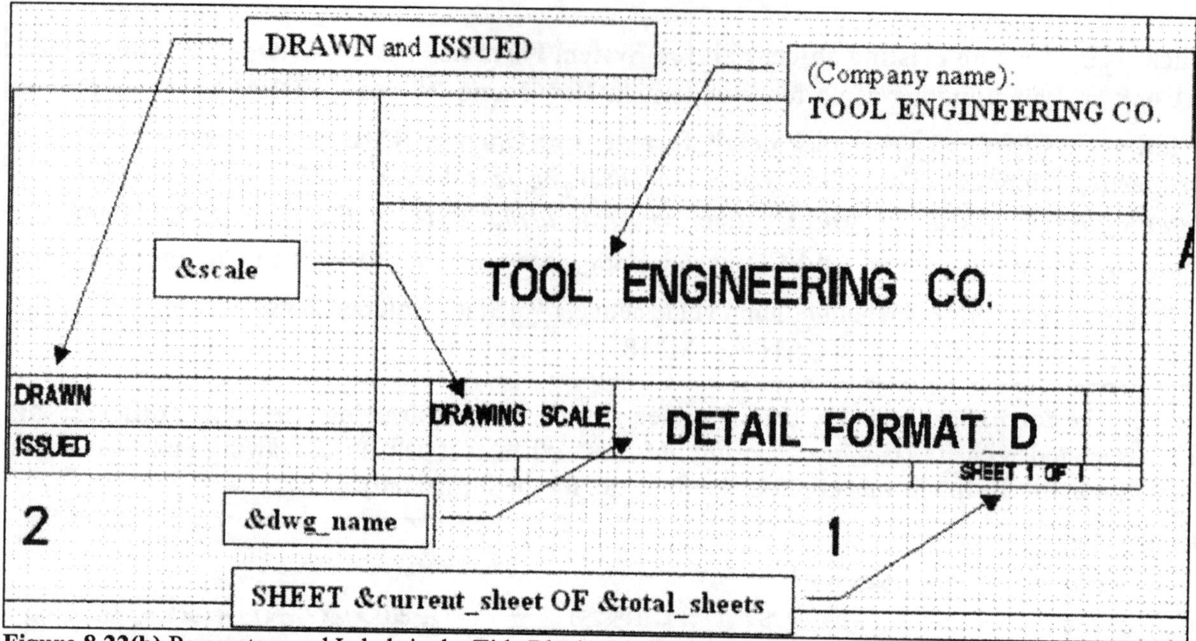

Figure 8.22(b) Parameters and Labels in the Title Block, Smaller Text is **.10** in Height

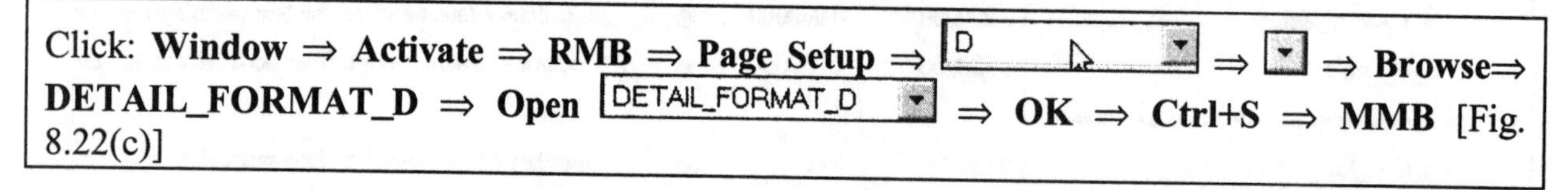

Click: **Window** ⇒ **Activate** ⇒ **RMB** ⇒ **Page Setup** ⇒ D ⇒ ⇒ **Browse**⇒ **DETAIL_FORMAT_D** ⇒ **Open** DETAIL_FORMAT_D ⇒ **OK** ⇒ **Ctrl+S** ⇒ **MMB** [Fig. 8.22(c)]

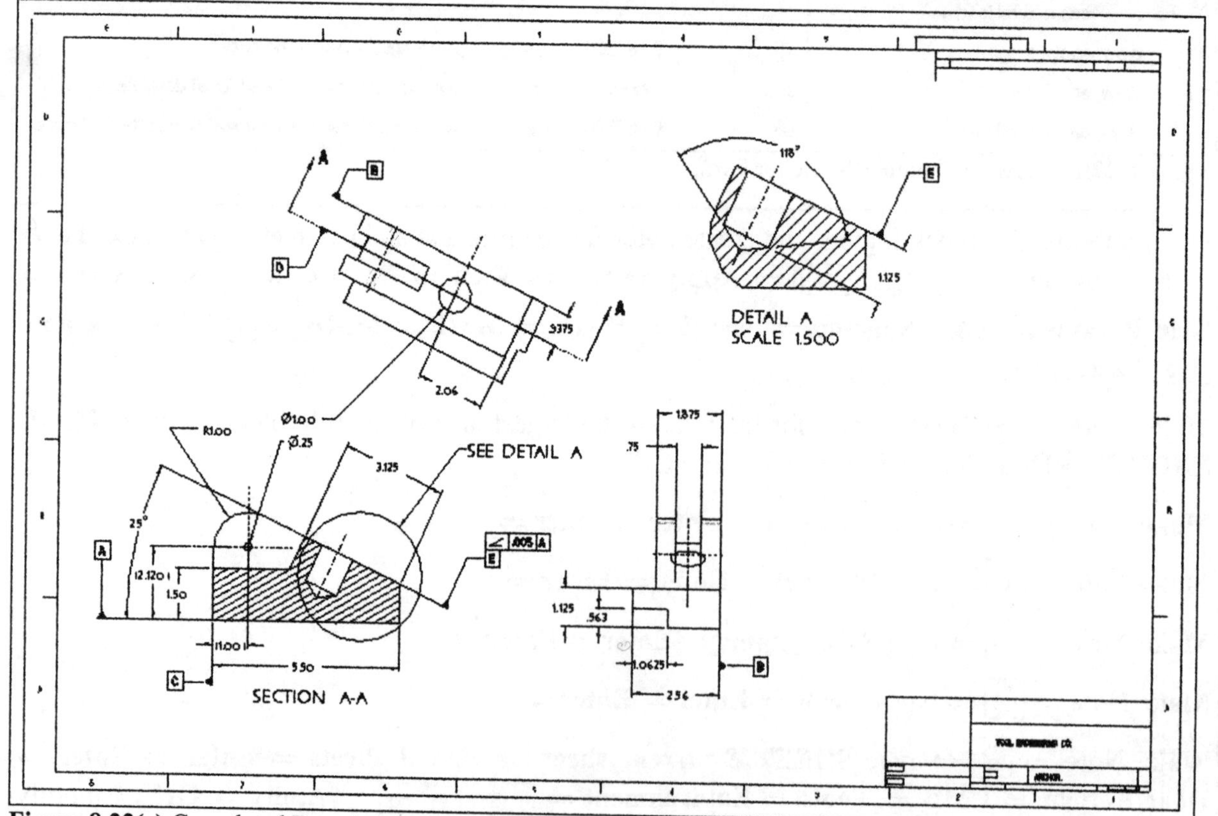

Figure 8.22(c) Completed Drawing

Detail the non-standard parts of the Clamp Assembly: Foot, Ball, Swivel, and Arm.

Lesson 9 Assembly Drawings

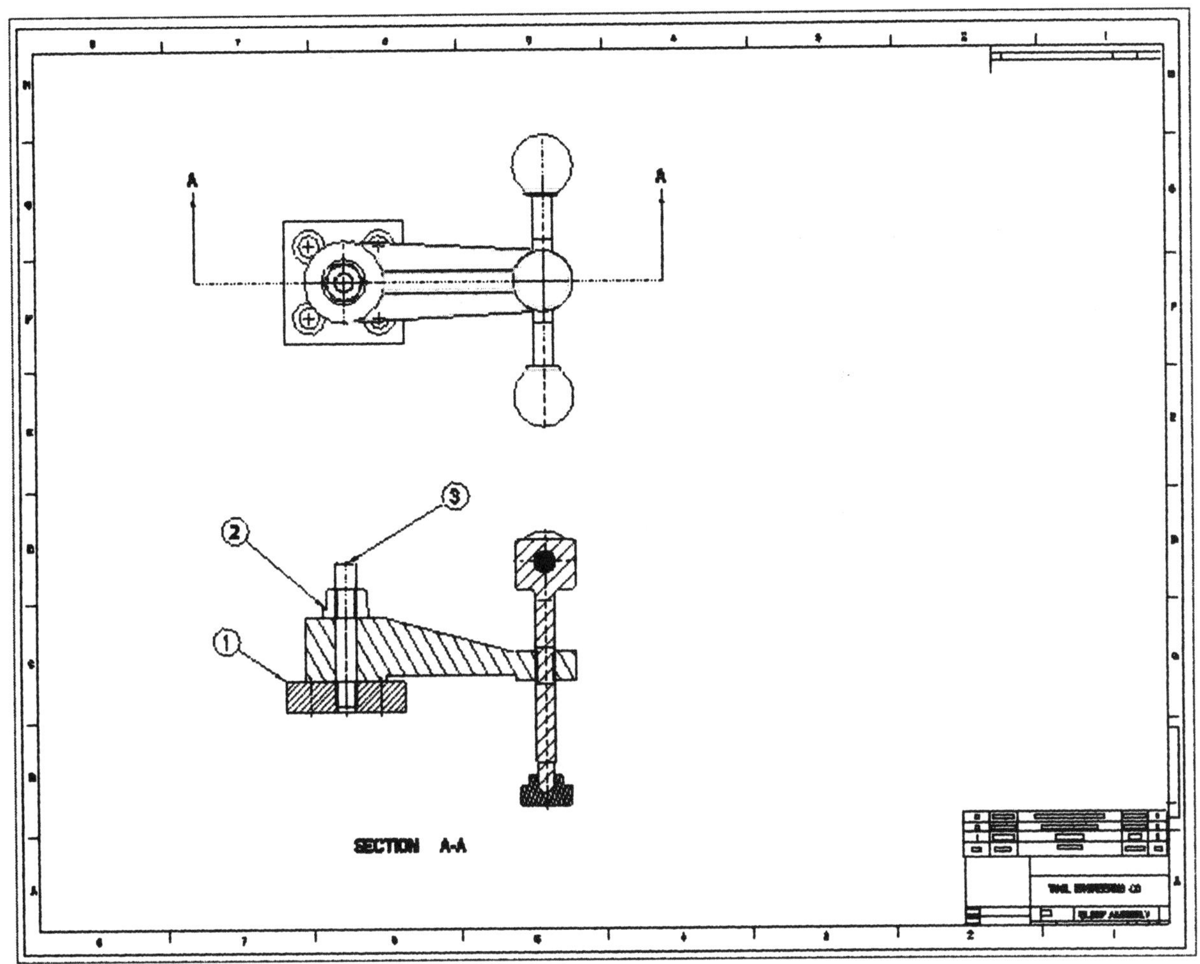

Figure 9.1(a) Clamp_Assembly Drawing

OBJECTIVES

- Create an **Assembly Drawing**
- Generate a **Parts List** from a bill of materials (**BOM**)
- **Balloon** an assembly drawing
- Create a **section assembly view** and change **component visibility**
- Add **Parameters** to parts
- Create a **Table** to generate a parts list automatically
- Use **Multiple Sheets**
- Make assembly **Drawing Sheets** with **multiple models**

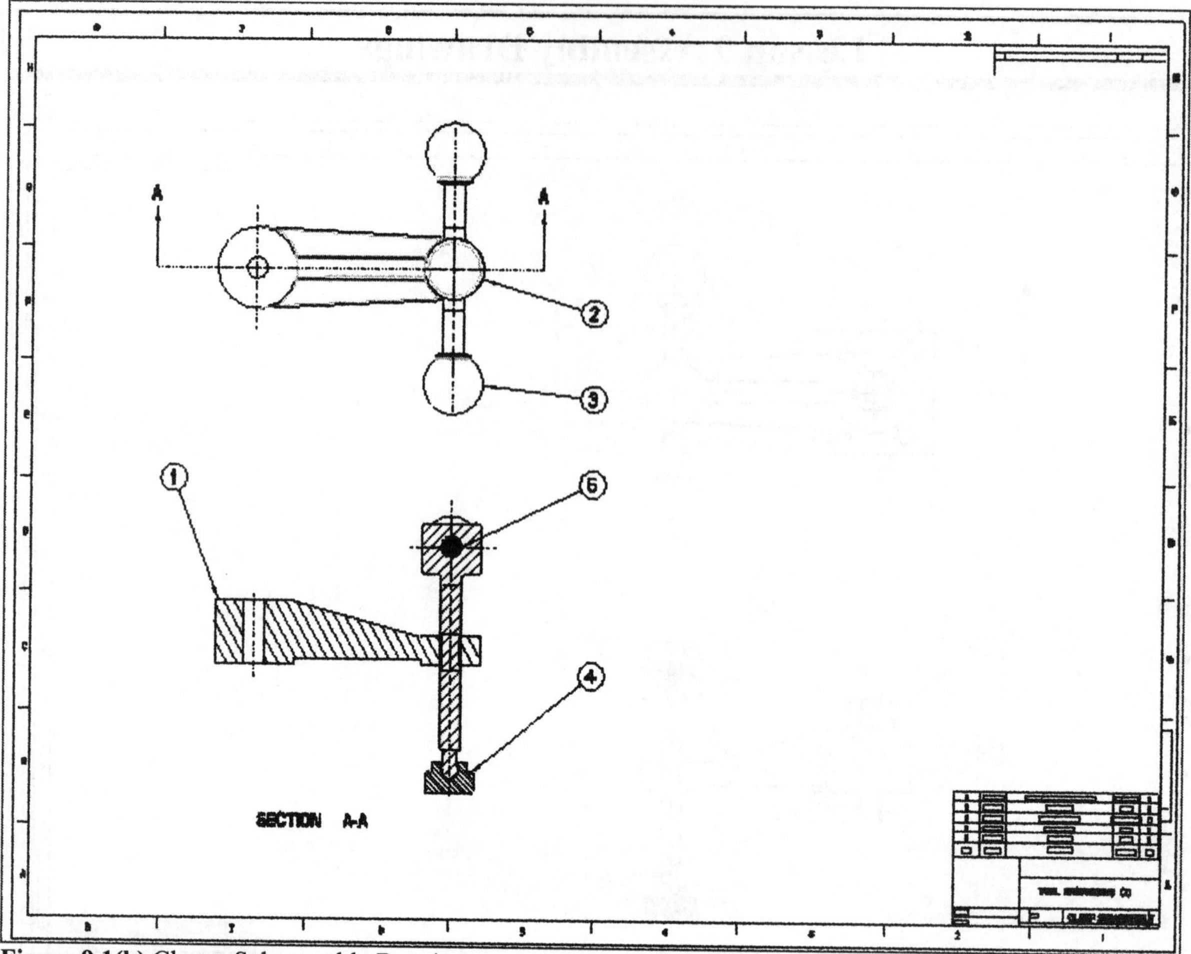

Figure 9.1(b) Clamp_Subassembly Drawing

Assembly Drawings

Pro/E incorporates a great deal of functionality into drawings of assemblies [Figs. 9.1(a-b)]. You can assign parameters to parts in the assembly that can be displayed on a *parts list* in an assembly drawing. Pro/E can also generate the item balloons for each component on standard orthographic views or on an exploded view.

In addition, a variety of specialized capabilities allow you to alter the manner in which individual components are displayed in views and in sections. The format for an assembly is usually different from the format used for detail drawings. The most significant difference is the presence of a *Parts List.*

As part of this lesson, you will create a set of assembly formats and place your standard parts list in them. A parts list is actually a *Drawing Table object* that is formatted to represent a bill of materials in a drawing. By defining *parameters* in the parts in your assembly that agree with the specific format of the parts list, you make it possible for Pro/E to add pertinent data to the assembly drawing's parts list automatically as components are added to the assembly. After the parts list and parameters have been added, Pro/E can balloon the assembly drawing automatically.

Lesson 9 STEPS

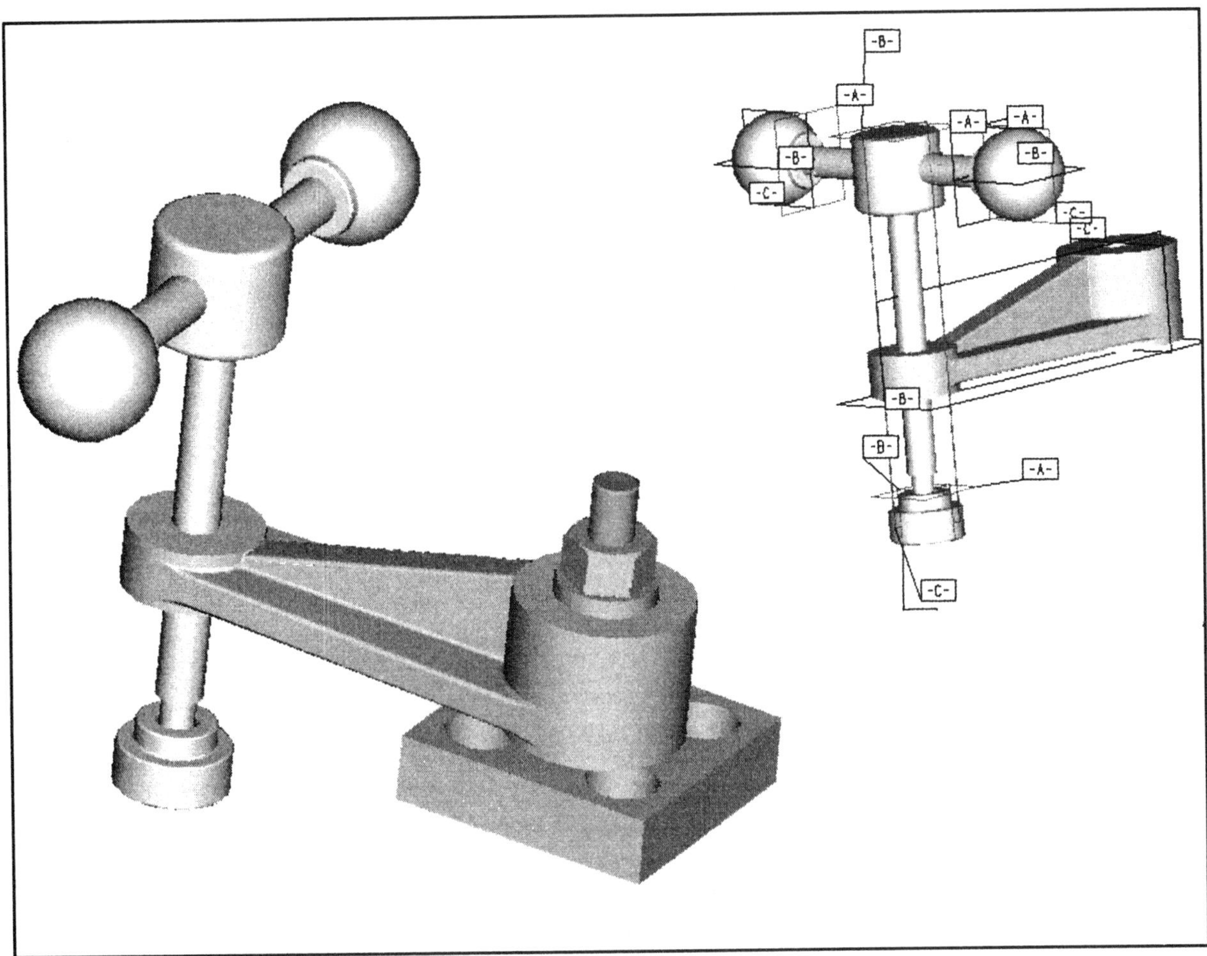

Figure 9.2 Swing Clamp Assembly and Subassembly (inset)

Swing Clamp Assembly Drawing

The format for an assembly drawing is usually different from the format used for detail drawings. The most significant difference is the presence of a parts list. We will create a standard "E" size format and place a standard parts list on it. You should create a set of assembly formats on "B", "C", "D", and "F" size sheets at your convenience.

A parts list is actually a *Drawing Table object* that is formatted to represent a bill of material (BOM) in a drawing. By defining parameters in the parts in your assembly that agree with the specific format of the parts list, you make it possible for Pro/E to add pertinent data to the assembly drawing parts list automatically, as components are added to the assembly.

After you create an "E" size format sheet with a parts list table, you will create two new drawings (each with two views) using your new assembly format. The Swing Clamp *subassembly* [Fig. 9.2(inset)] will be used in the first drawing. The second drawing will use the Swing Clamp *assembly* (Fig. 9.2). Both drawings use the "E" size format created in the first section of this lesson. The format will have a parameter-driven title block and an integral parts list.

Click: **File** ⇒ **Set Working Directory** ⇒ select working directory ⇒ **OK** ⇒ **Open an existing object** ⇒ ⇒ System Formats ⇒ e.frm ⇒ **Open** ⇒ **File** ⇒ **Save a Copy** ⇒ type a unique name for your format: New Name **ASM_FORMAT_E** ⇒ **OK** ⇒ **Window** ⇒ **Close** ⇒ ⇒ pick **asm_format_e** ⇒ **Open**

The **format** will have a **.dtl** associated with it, and the **drawing** will have a different **.dtl** file associated with it. *They are separate **.dtl** files.* When you activate a drawing and then add a format, the **.dtl** for the format controls the font, etc. for the format only. The drawing **.dtl** file that controls items on the drawing needs to be established separately in the Drawing mode.

Click: **RMB** in the graphics window ⇒ **Properties** Options dialog box opens ⇒ ***drawing_text_height .25*** ⇒ **Add/Change** ⇒ ***default_font filled*** ⇒ **Add/Change** ⇒ ***draw_arrow_style filled*** ⇒ **Add/Change** ⇒ ***draw_arrow_length .25*** ⇒ **Add/Change** ⇒ ***draw_arrow_width .08*** ⇒ **Add/Change** (Fig. 9.3) ⇒ **Apply** option in Status column has default column value active, option in Status column shows pending value for this drawing ⇒ **Close** ⇒ ⇒ ⇒ **MMB**

	Value	Default	Status	Description
Active Drawing				
- These options control text not subject to oth				
drawing_text_height	.25	0.156250	✳	Sets default text height for all text in the drawing usin
text_thickness	0.000000 *	0.000000	●	Sets default text thickness for new text after regener
text_width_factor	0.800000 *	0.800000	●	Sets default ratio between the text width and text hei
draft_scale	1.000000 *	1.000000	●	Determines value of draft dimensions relative to actu
- These options control text and line fonts				
default_font	filled	font	✳	Specifies a font index that determines default text for
- These options control leaders				
draw_arrow_length	.25	0.187500	✳	Sets length of leader line arrows.
draw_arrow_style	filled	closed	✳	Controls style of arrow head for all detail items involvi
draw_arrow_width	.08	0.062500	✳	Sets width of leader line arrows. Drives these: "draw
draw_attach_sym_height	DEFAULT *	default	●	Sets height of leader line slashes, integral signs, and
draw_attach_sym_width	DEFAULT *	default	●	Sets width of leader line slashes, integral signs, and b
draw_dot_diameter	DEFAULT *	default	●	Sets diameter of leader line dots. If set to "default," u
leader_elbow_length	0.250000 *	0.250000	●	Determines length of leader elbow (the horizontal leg
sort_method_in_region	delimited		●	Determines repeat regions sort mechanism. String_or
- Miscellaneous options				
draft_scale	1.000000 *	1.000000	●	Determines value of draft dimensions relative to actu

Figure 9.3 New Drawing Options

Zoom into the title block region, and create notes for the title text and parameter text required to display the proper information.

Click: **Tools** ⇒ **Environment** ⇒ ☑ Snap to Grid ⇒ **Apply** ⇒ **OK** ⇒ **View** ⇒ **Draft Grid** ⇒ **Show Grid** ⇒ **Grid Params** ⇒ **X&Y Spacing** ⇒ type **.1** ⇒ **MMB** ⇒ **MMB** ⇒ **MMB** ⇒ **Create a note** ⇒

Make Note ⇒ pick point for the note in the largest area of the title block (Fig. 9.4) ⇒ type **TOOL ENGINEERING CO.** ⇒ **Enter** ⇒ **Enter** ⇒

Make Note ⇒ (Fig. 9.4) ⇒ type **DRAWN** ⇒ **Enter** ⇒ **Enter** ⇒

Make Note ⇒ (Fig. 9.4) ⇒ type **ISSUED** ⇒ **Enter** ⇒ **Enter** ⇒

Make Note ⇒ (Fig. 9.4) ⇒ type **&dwg_name** ⇒ **Enter** ⇒ **Enter** ⇒

Make Note ⇒ (Fig. 9.4) ⇒ type **&scale** ⇒ **Enter** ⇒ **Enter** ⇒

Make Note ⇒ (Fig. 9.4) ⇒ type **SHEET ¤t_sheet OF &total_sheets** ⇒ **Enter** ⇒ **Enter** ⇒ **Done/Return** ⇒ **LMB** ⇒ **Tools** ⇒ **Environment** ⇒ ☐ Snap to Grid ⇒ **Apply** ⇒ **OK** ⇒ modify the text height (**.10**) and the placement of the notes so that they are positioned correctly (Fig. 9.4) ⇒ ⇒ ⇒ ⇒ ⇒ **MMB** ⇒ **File** ⇒ **Delete** ⇒ **Old Versions** ⇒ **MMB**

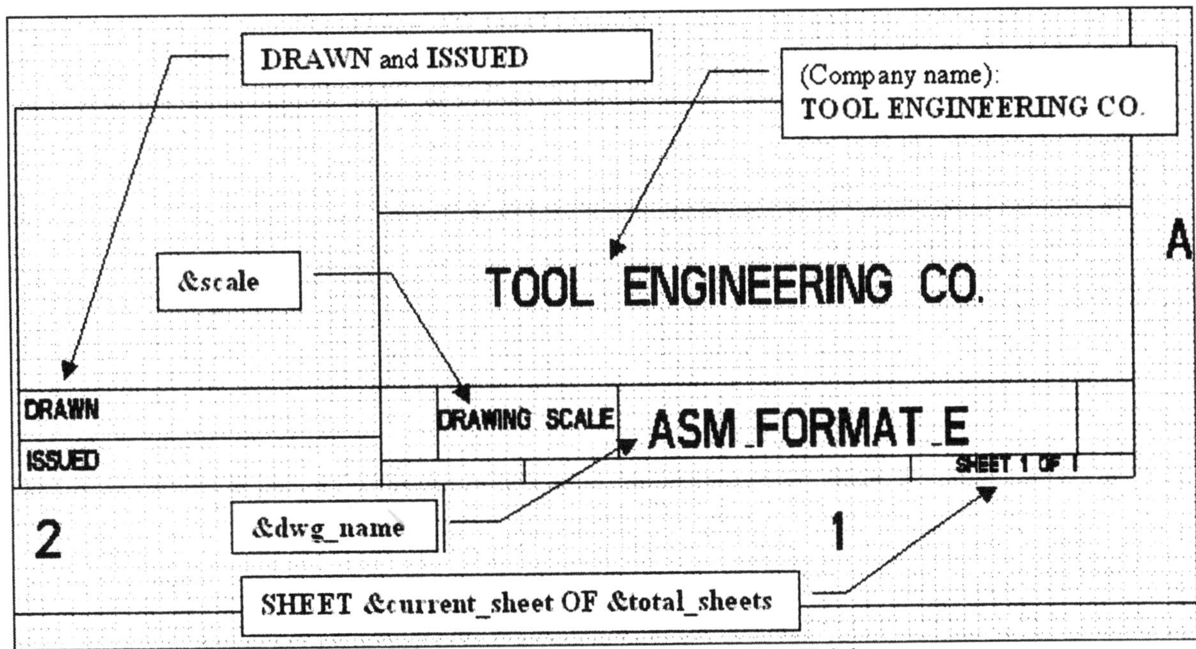

Figure 9.4 Parameters and Labels in the Title Block, Smaller Text is **.10** in Height

The parts list table can now be created and saved with this format. You can add and replace formats and still keep the table associated with the drawing. Start the parts list by creating a table.

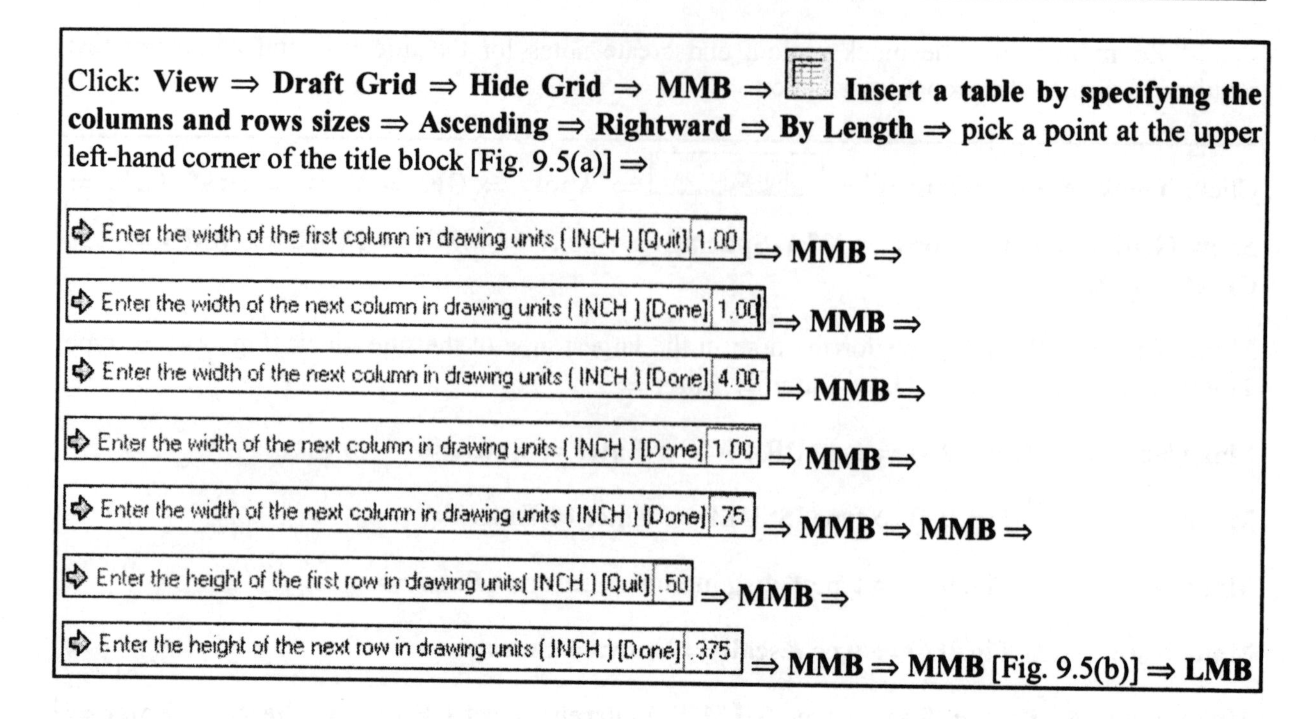

Click: **View** ⇒ **Draft Grid** ⇒ **Hide Grid** ⇒ **MMB** ⇒ **Insert a table by specifying the columns and rows sizes** ⇒ **Ascending** ⇒ **Rightward** ⇒ **By Length** ⇒ pick a point at the upper left-hand corner of the title block [Fig. 9.5(a)] ⇒

Enter the width of the first column in drawing units (INCH) [Quit] 1.00 ⇒ **MMB** ⇒

Enter the width of the next column in drawing units (INCH) [Done] 1.00 ⇒ **MMB** ⇒

Enter the width of the next column in drawing units (INCH) [Done] 4.00 ⇒ **MMB** ⇒

Enter the width of the next column in drawing units (INCH) [Done] 1.00 ⇒ **MMB** ⇒

Enter the width of the next column in drawing units (INCH) [Done] .75 ⇒ **MMB** ⇒ **MMB** ⇒

Enter the height of the first row in drawing units(INCH) [Quit] .50 ⇒ **MMB** ⇒

Enter the height of the next row in drawing units (INCH) [Done] .375 ⇒ **MMB** ⇒ **MMB** [Fig. 9.5(b)] ⇒ **LMB**

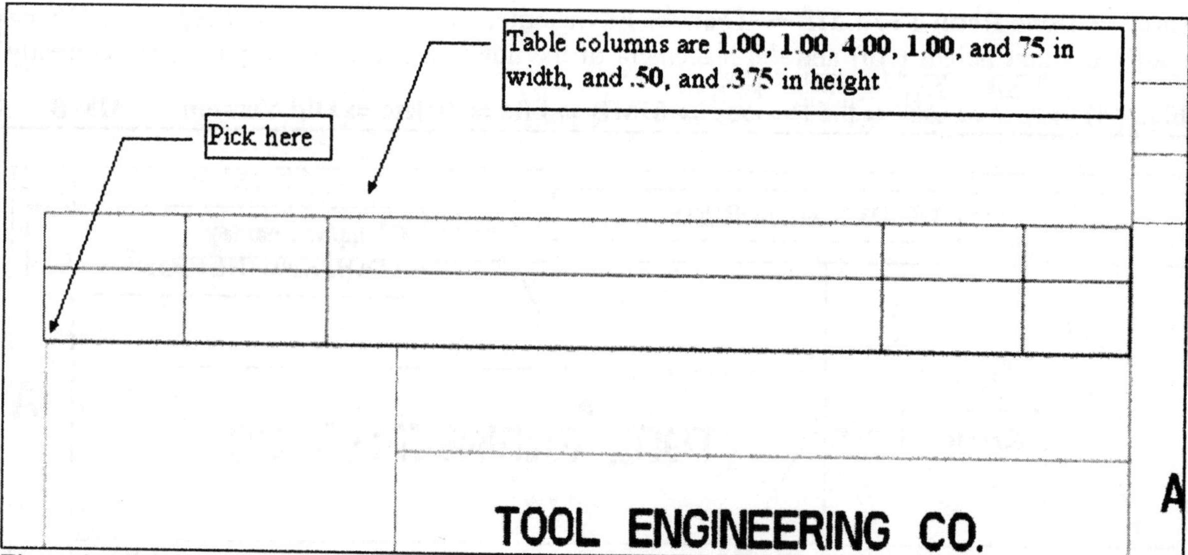

Figure 9.5(a) Inserting a Table

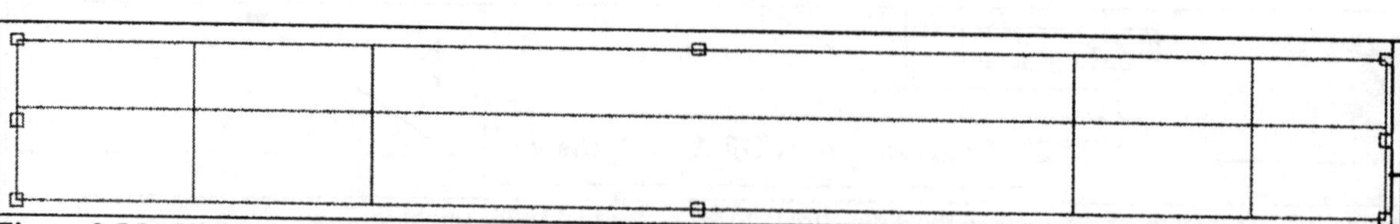

Figure 9.5(b) Highlighted Table

Click: **Table** from Menu bar ⇒ **Repeat Region** ⇒ **Add** ⇒ pick in the left block [Fig. 9.5(c)] ⇒ pick in the right block [Fig. 9.5(c)] ⇒ **Attributes** ⇒ select the Repeat Region just created ⇒ **No Duplicates** ⇒ **Recursive** ⇒ **MMB** ⇒ **MMB** ⇒ **MMB** ⇒ **Update the display of all views in the active sheet**

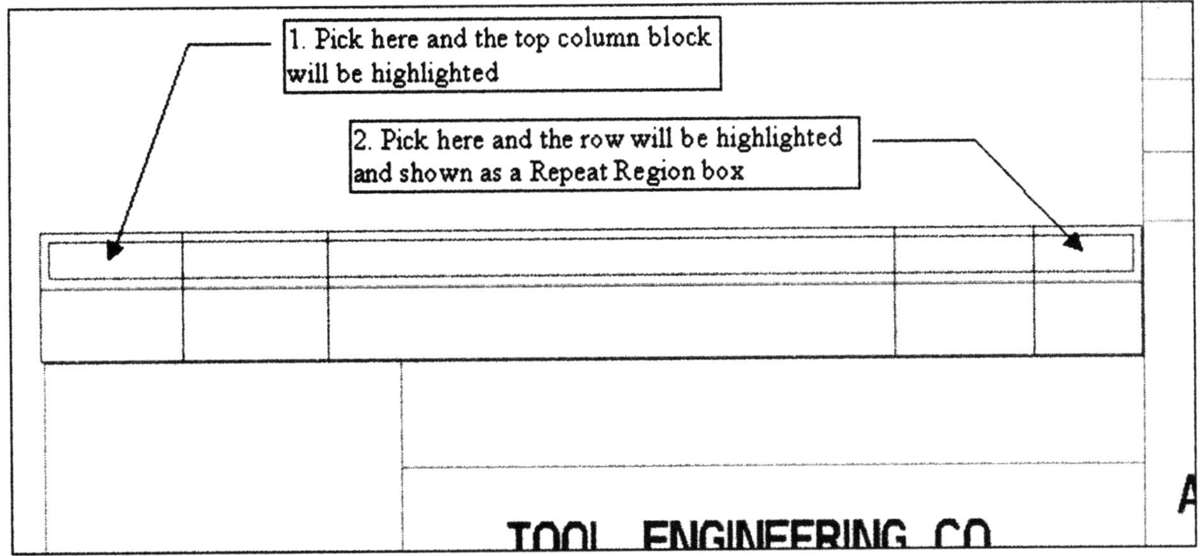

Figure 9.5(c) Repeat Region

Insert column headings using plain text: double-click on the first block [Fig. 9.6(a)] ⇒ type **ITEM** [Fig. 9.6(b)] ⇒ **OK**

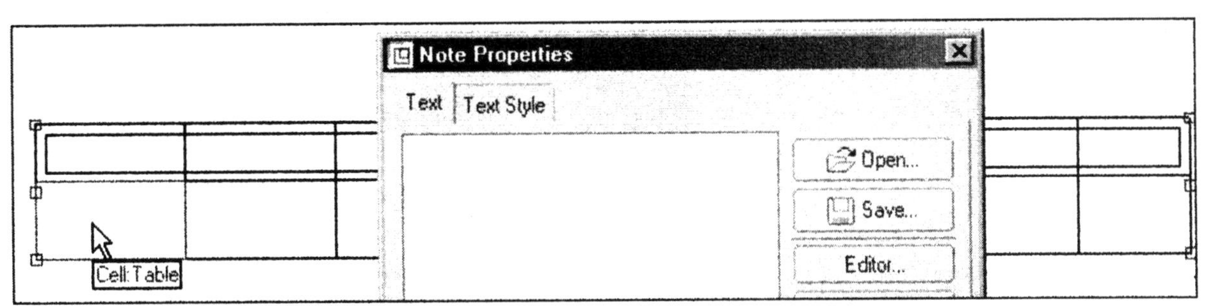

Figure 9.6(a) Select the First Block

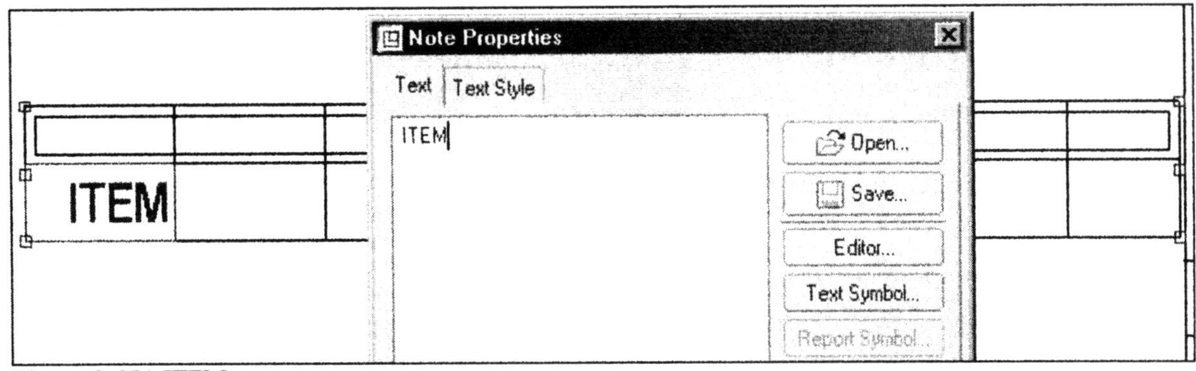

Figure 9.6(b) ITEM

The Repeat Region now needs to have some of its headings correspond to the parameters that will be created for each component model.

Double-click on the second block ⇒ type **PT NUM** [Fig. 9.6(c)] ⇒ **OK** ⇒ double-click on the third block ⇒ type **DESCRIPTION** ⇒ **OK** ⇒ double-click on the fourth block ⇒ type **MATERIAL** ⇒ **OK** ⇒ double-click on the fifth block ⇒ type **QTY** ⇒ **OK** [Fig. 9.6(d)]

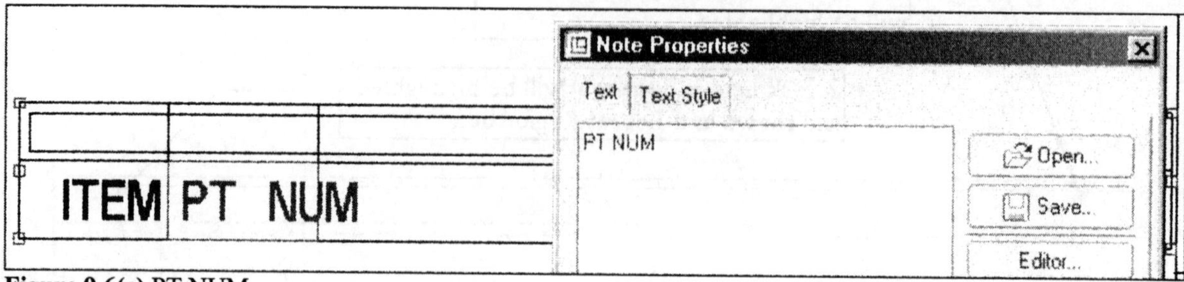

Figure 9.6(c) PT NUM

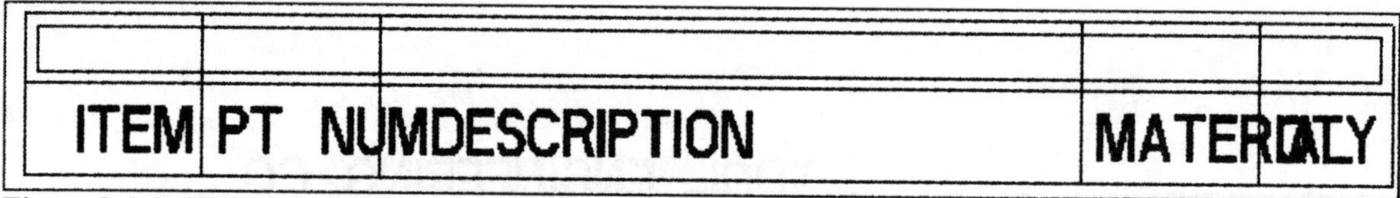

Figure 9.6(d) ITEM, PT MUM, DESCRIPTION, MATERIAL, QTY

Change the text height (**.125**), position (middle) and justification (centered) for all five titles.

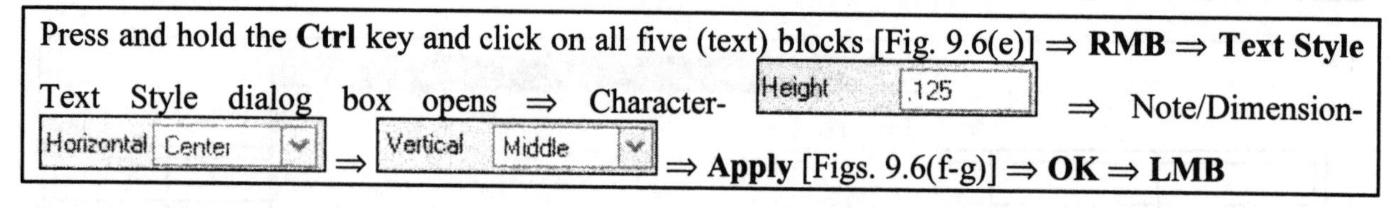

Press and hold the **Ctrl** key and click on all five (text) blocks [Fig. 9.6(e)] ⇒ **RMB** ⇒ **Text Style**
Text Style dialog box opens ⇒ Character- Height .125 ⇒ Note/Dimension- Horizontal Center ⇒ Vertical Middle ⇒ **Apply** [Figs. 9.6(f-g)] ⇒ **OK** ⇒ **LMB**

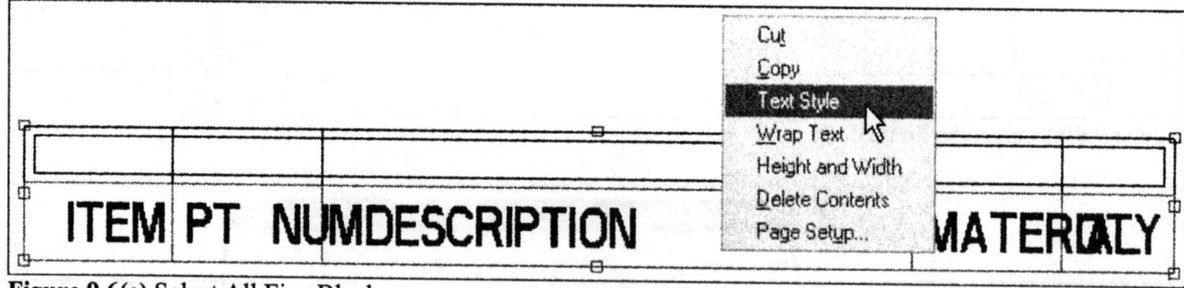

Figure 9.6(e) Select All Five Blocks

Figure 9.6(f) New Text Style

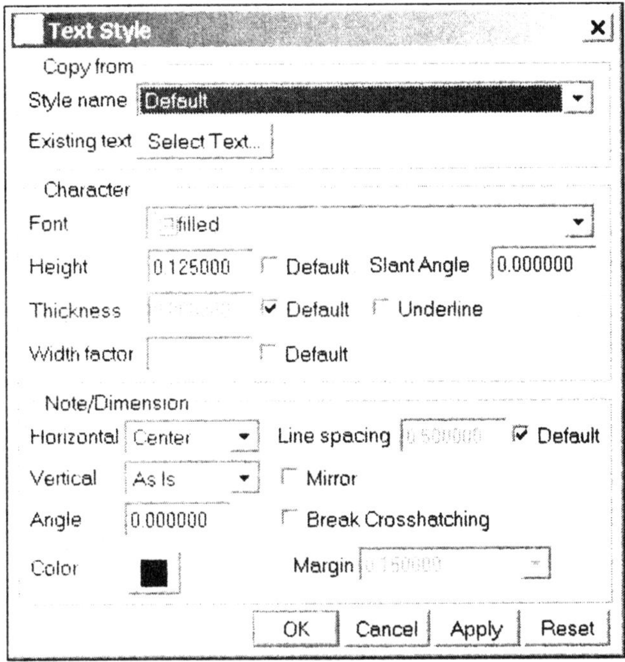

Figure 9.6(g) Text Style Dialog Box

Insert parametric text into each appropriate repeat region block.

Double-click on the first table cell of the Repeat Region [Fig. 9.7(a)] ⇒ click **rpt...** from the Report Symbol dialog box ⇒ click **index** [Figs. 9.7(b-c)]

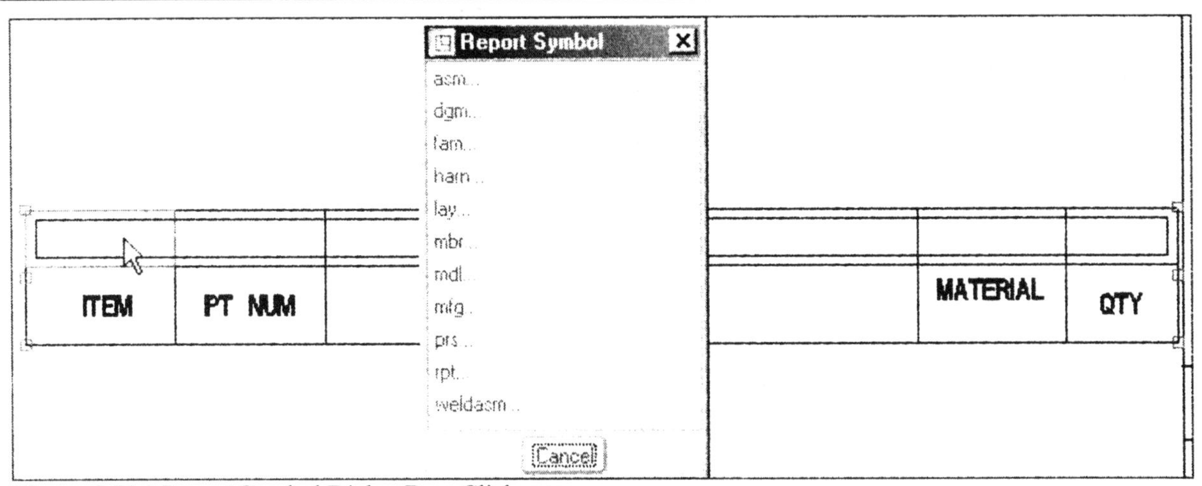

Figure 9.7(a) Report Symbol Dialog Box, Click rpt...

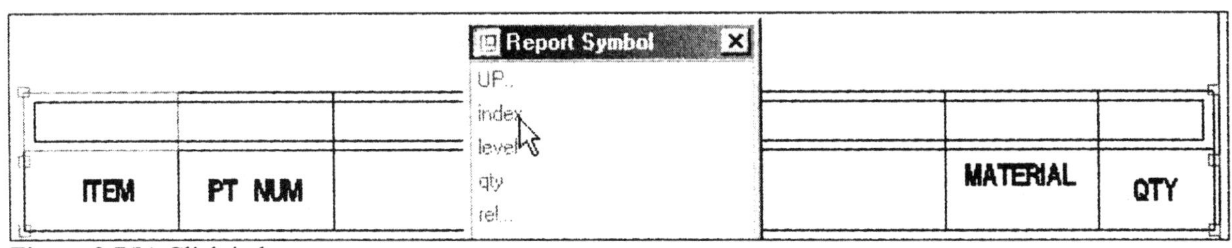

Figure 9.7(b) Click index

Figure 9.7(c) &rpt.index

Double-click on the fifth table cell of Repeat region ⇒ pick **rpt...** [Fig. 9.7(d)] ⇒ **qty** [Figs. 9.7(e-f)]

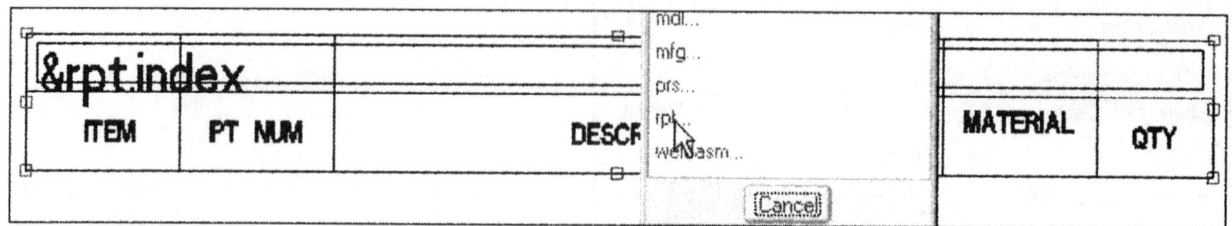

Figure 9.7(d) Pick rpt...

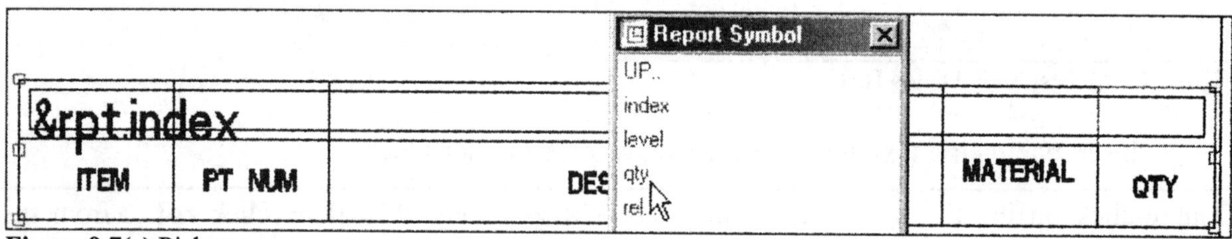

Figure 9.7(e) Pick qty

Figure 9.7(f) &rpt.qty

Double-click on the third (middle) table cell of the Repeat Region ⇒ **asm...** [Fig. 9.7(g)] ⇒ **mbr...** [Fig. 9.7(h)] ⇒ **User Defined** [Fig. 9.7(i)] ⇒ Enter symbol text: type **DSC** ⇒ **MMB** [Fig. 9.7(j)]

Figure 9.7(g) Pick asm...

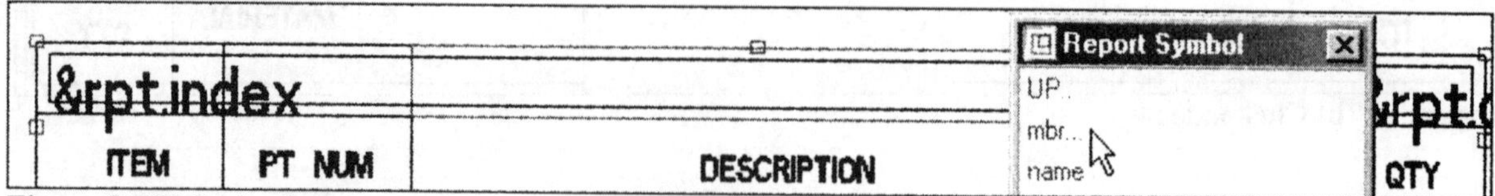

Figure 9.7(h) Pick mbr...

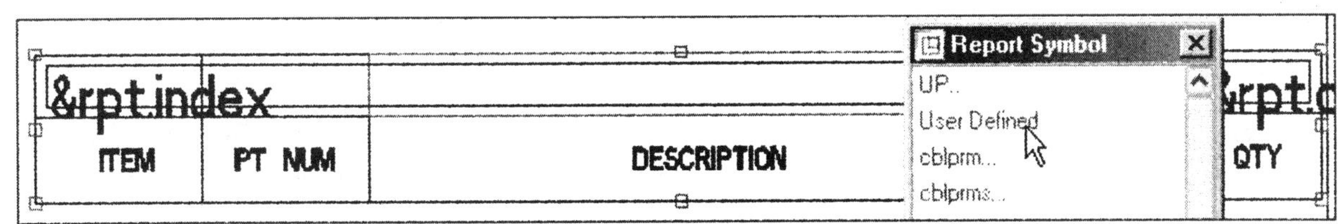

Figure 9.7(i) Pick User Defined

Figure 9.7(j) &asm.mbr.DSC

Double-click on the fourth table cell of the Repeat Region ⇒ **asm...** ⇒ **mbr...** ⇒ **User Defined** ⇒ Enter symbol text: type **MAT** ⇒ **MMB** [Fig. 9.7(k)] ⇒ double-click on the second table cell of the Repeat Region ⇒ **asm...** ⇒ **mbr...** ⇒ **User Defined** ⇒ Enter symbol text: type **PRTNO** ⇒ **MMB** [Fig. 9.7(l)] ⇒ **LMB**

Figure 9.7(k) &asm.mbr.MAT

Figure 9.7(l) &asm.mbr.PRTNO

Change the text height of the Report Symbols in the table cells of the Repeat Region to **.125**.

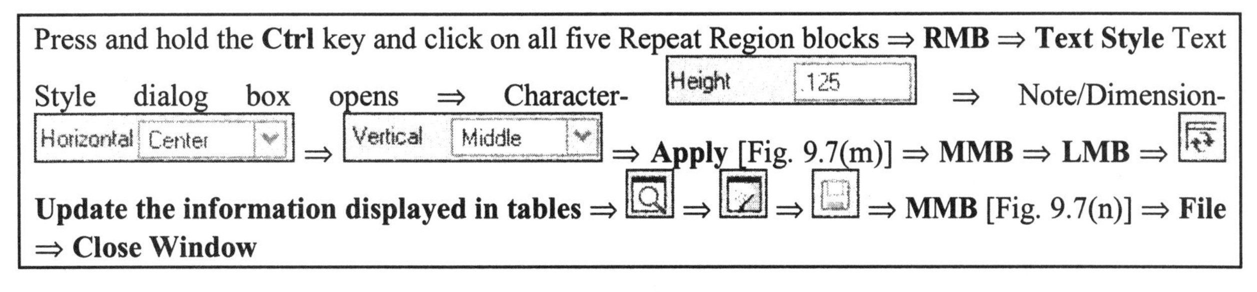
Press and hold the **Ctrl** key and click on all five Repeat Region blocks ⇒ **RMB** ⇒ **Text Style** Text Style dialog box opens ⇒ Character- Height .125 ⇒ Note/Dimension- Horizontal Center ⇒ Vertical Middle ⇒ **Apply** [Fig. 9.7(m)] ⇒ **MMB** ⇒ **LMB** ⇒ **Update the information displayed in tables** ⇒ ⇒ ⇒ ⇒ **MMB** [Fig. 9.7(n)] ⇒ **File** ⇒ **Close Window**

Figure 9.7(m) Completed BOM

&rpt.index	&asmmbr.PRTNO	&asmmbr.DSC	&asmmbr.MAT	&rpt.qty
ITEM	PT NUM	DESCRIPTION	MATERIAL	QTY

TOOL ENGINEERING CO.

DRAWN	DRAWING SCALE	ASM_FORMAT_E
ISSUED		SHEET 1 OF 1

2 1

Figure 9.7(n) Title Block and BOM Table

Adding Parts List Data

When you save your standard assembly format, the Drawing Table that represents your standard parts list is now included. You must be aware of the titles of the parameters under which the data is stored, so that you can add them properly to your parts.

As you add components to an assembly, Pro/E reads the parameters from them and updates the parts list. You can also see the same effect by adding these parameters after the drawing has been created.

Pro/E also creates Item Balloons on the first view that was placed on the drawing. To improve their appearance, you can move these balloons to other views and alter the locations where they attach.

Retrieve the clamp arm, click: **File ⇒ Open ⇒** select **clamp_arm.prt ⇒ Open** [Fig. 9.8(a)]

Figure 9.8(a) Clamp_Arm

Click: **Tools** from menu bar ⇒ **Parameters** [Fig. 9.8(b)] ⇒ **Parameters** [Fig. 9.8(c)] ⇒ **Add Parameter** from Parameters dialog box ⇒ in the Name field and type **PRTNO** [Figs. 9.8(d-e)]

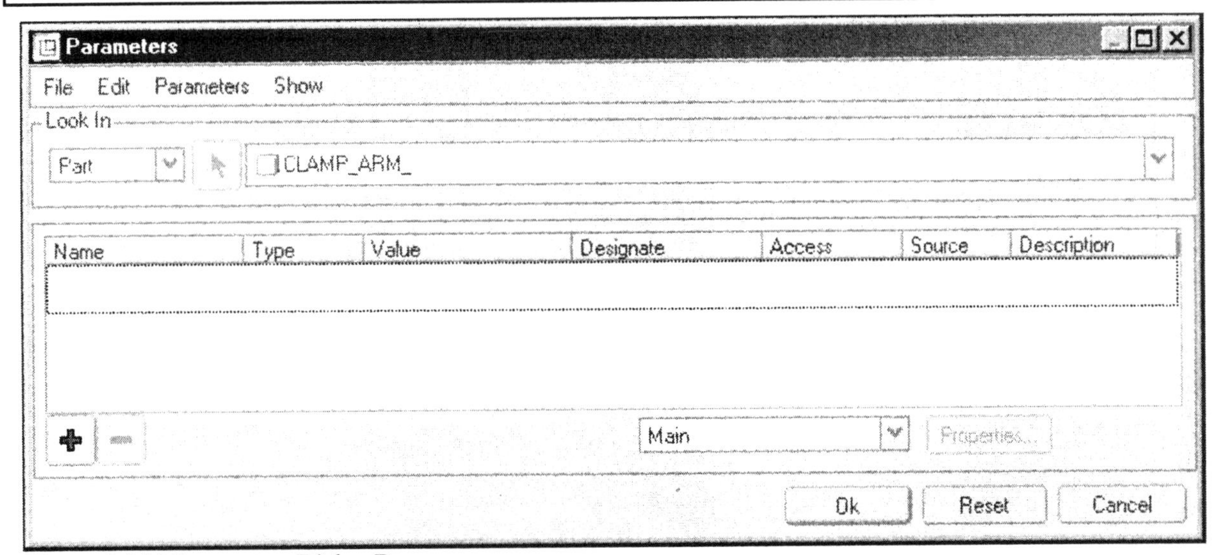

Figure 9.8(b) Parameters Dialog Box

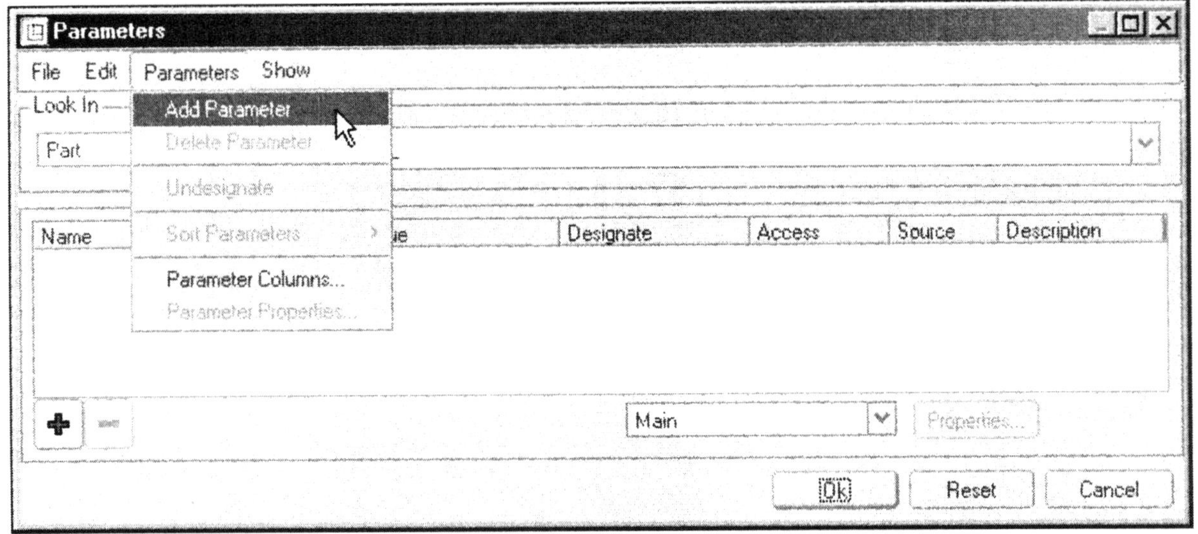

Figure 9.8(c) Add Parameter

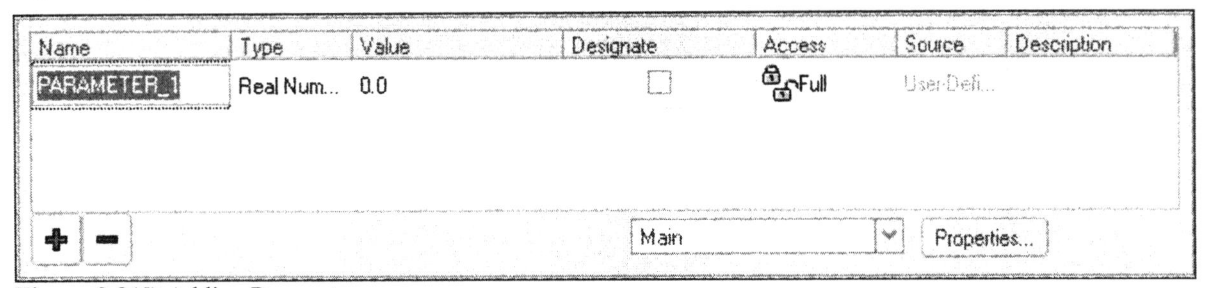

Figure 9.8(d) Adding Parameters

Name	Type	Value	Designate	Access	Source	Description
PRTNO	Real Num...	0.0		Full	User-Defi...	

Figure 9.8(e) Name PRTNO

Click in Type field [Fig. 9.8(f)] ⇒ [Fig. 9.8(g)] ⇒ String ⇒ click in Value field and type **SW101-5AR** [Fig. 9.8(h)] ⇒ click in Description field and type **part number** [Fig. 9.8(i)]

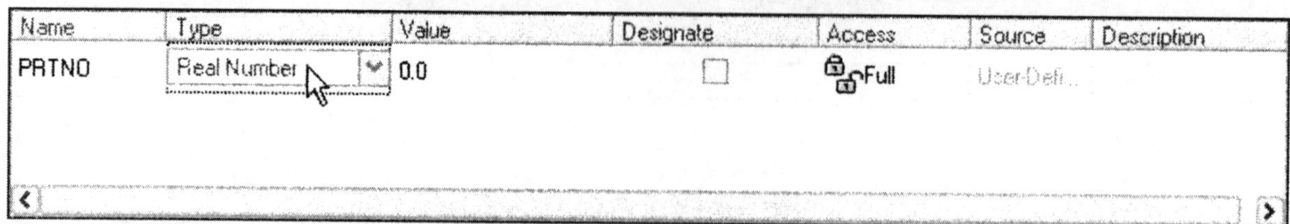

Figure 9.8(f) Click in Type Field- Real Number

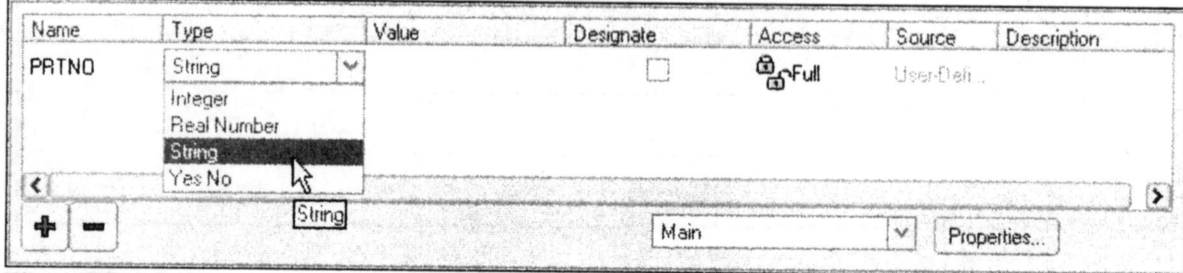

Figure 9.8(g) String

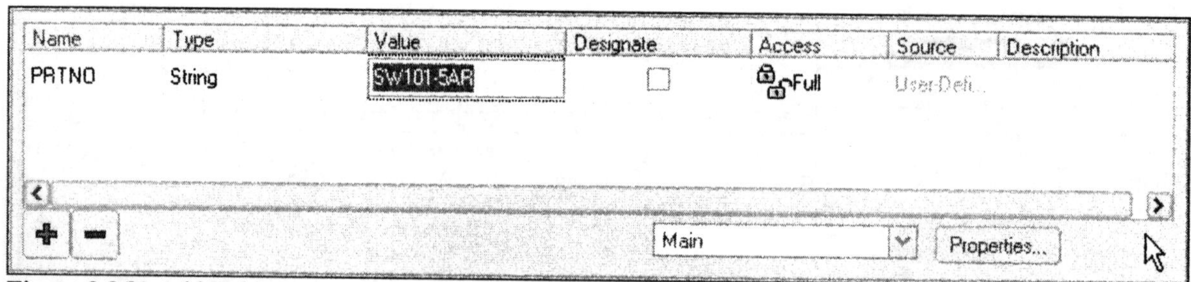

Figure 9.8(h) Add Value

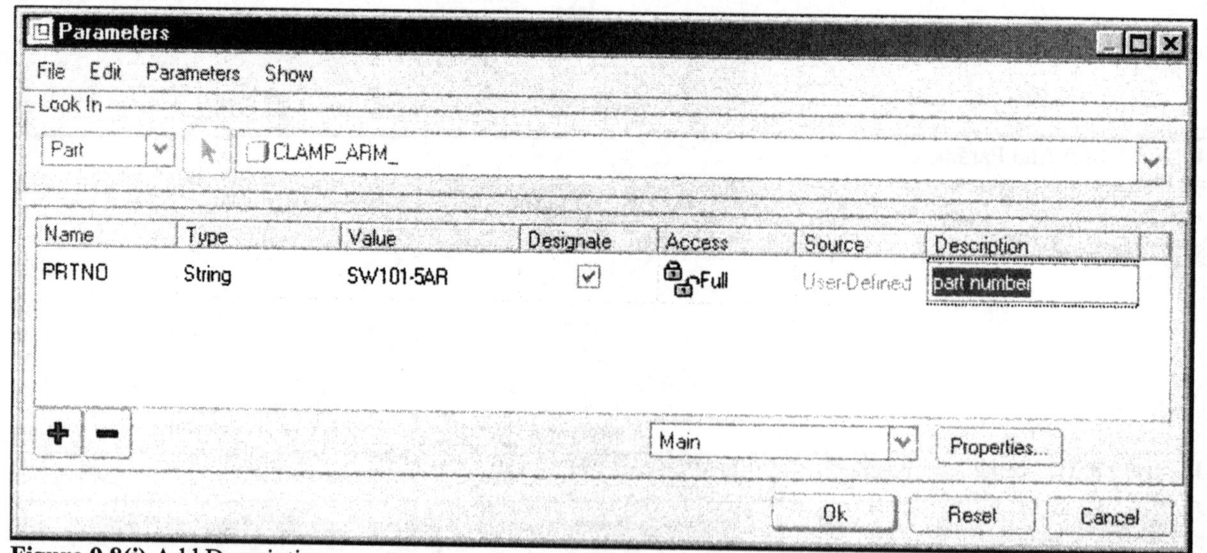

Figure 9.8(i) Add Description

Click: **+** **Add new Parameter** [Fig. 9.8(j)] ⇒ in the Name field, type **DSC** [Fig. 9.8(k)] ⇒ click in Type field ⇒ String ⇒ click in Value field and type **CLAMP ARM** ⇒ click in Description field and type **part name** [Fig. 9.8(l)] ⇒ click **+** **Add new Parameter** ⇒ in the Name field, type **MAT** ⇒ click in Type field ⇒ String ⇒ click in Value field and type **STEEL** ⇒ click in Description field and type **material** [Fig. 9.8(m)] ⇒ **Ok** ⇒ **File** ⇒ **Save** ⇒ **MMB**

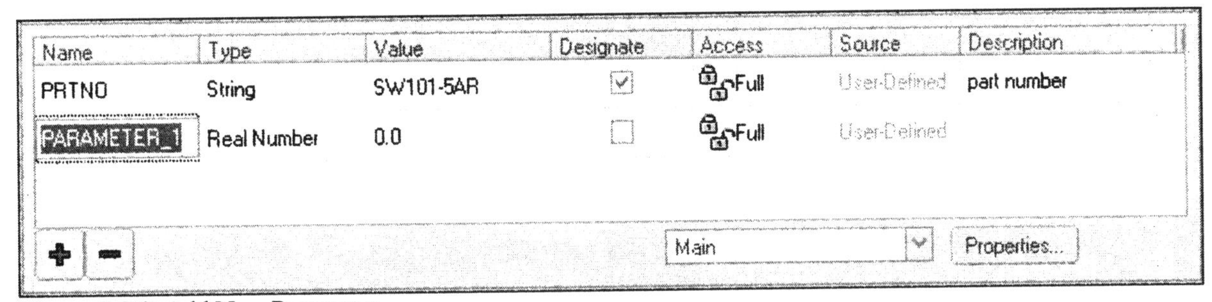

Figure 9.8(j) Add New Parameter

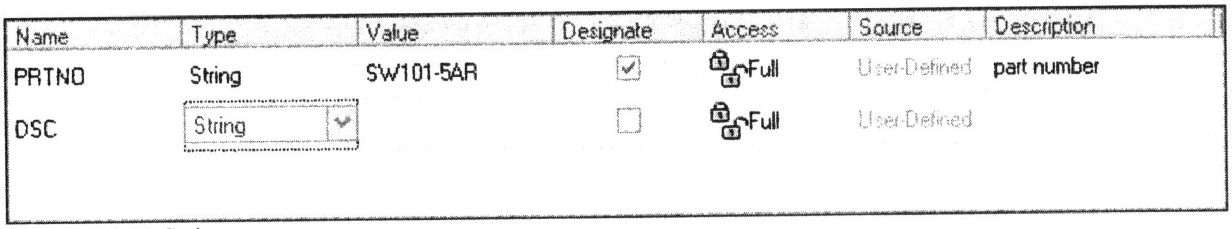

Name	Type	Value	Designate	Access	Source	Description
PRTNO	String	SW101-5AR	☑	Full	User-Defined	part number
DSC	String		☐	Full	User-Defined	

Figure 9.8(k) String

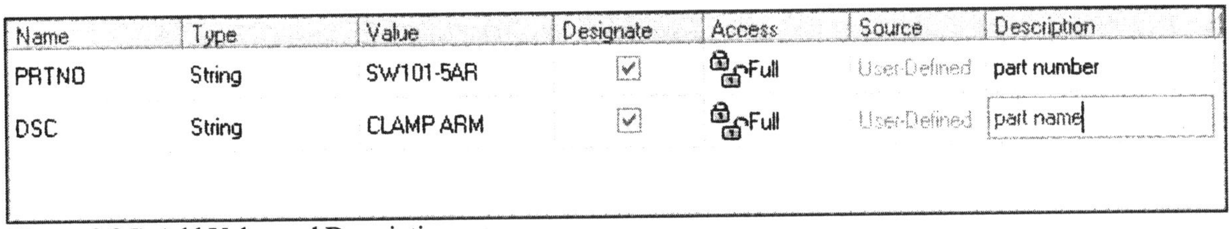

Name	Type	Value	Designate	Access	Source	Description
PRTNO	String	SW101-5AR	☑	Full	User-Defined	part number
DSC	String	CLAMP ARM	☑	Full	User-Defined	part name

Figure 9.8(l) Add Value and Description

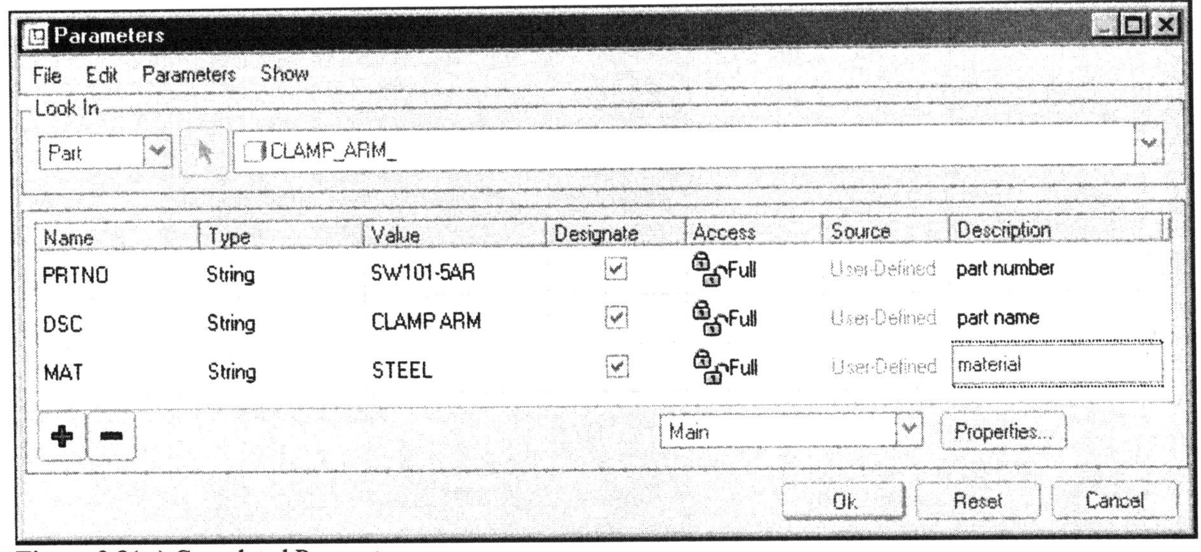

Figure 9.8(m) Completed Parameters

You can also access the parts parameters using the Relations dialog box. In the case of the Clamp_Arm, there was a relation created for controlling a features location.

Click: **Tools** ⇒ **Relations** ⇒ **Local Parameters** [Fig. 9.8(n)] ⇒ **Ok** ⇒ **File** ⇒ **Save** ⇒ **MMB** ⇒ **Window** ⇒ **Close**

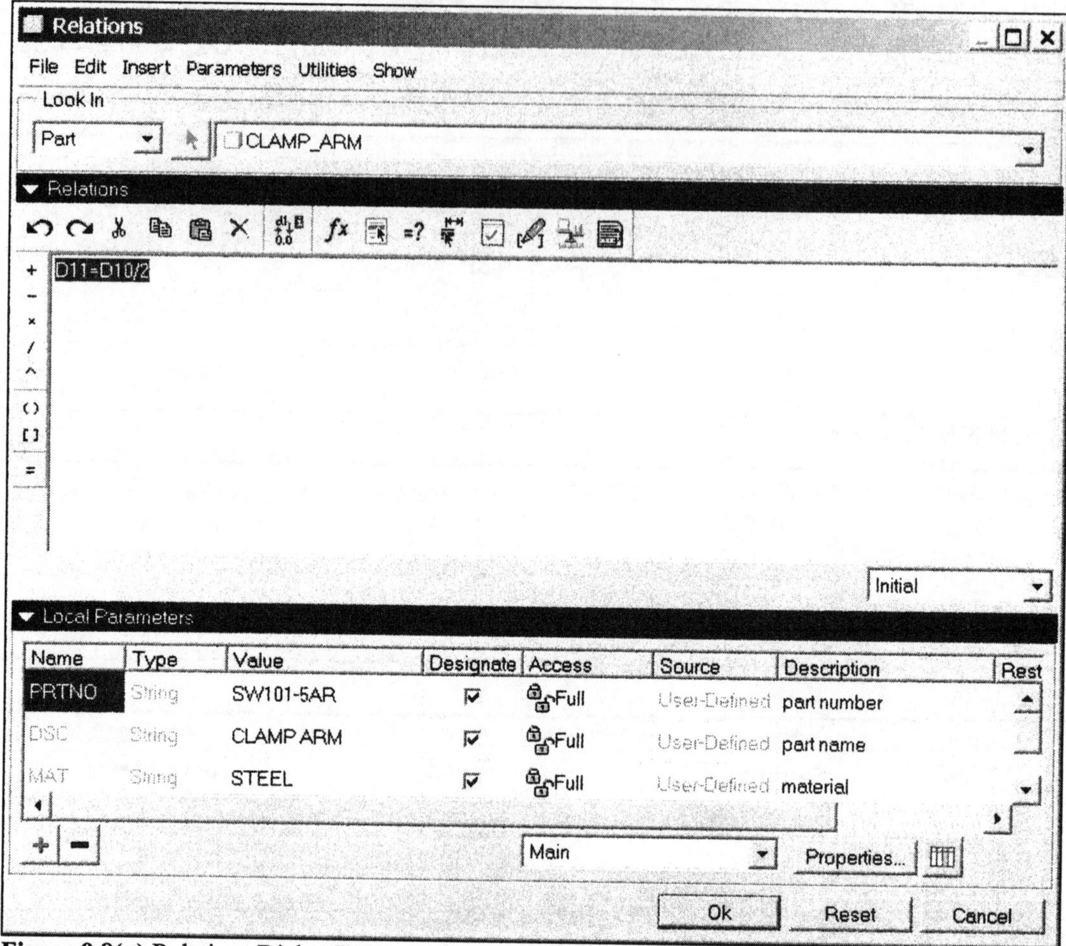

Figure 9.8(n) Relations Dialog Box

Retrieve the clamp swivel, click: **File** ⇒ **Open** ⇒ **clamp_swivel.prt** ⇒ **Open** [Fig. 9.9(a)] ⇒ **Tools** ⇒ **Parameters** ⇒ **Parameters** ⇒ **Add Parameter** ⇒ complete the parameters as shown [Fig. 9.9(b)] ⇒ **Ok** ⇒ **File** ⇒ **Save** ⇒ **MMB** ⇒ **File** ⇒ **Close Window**

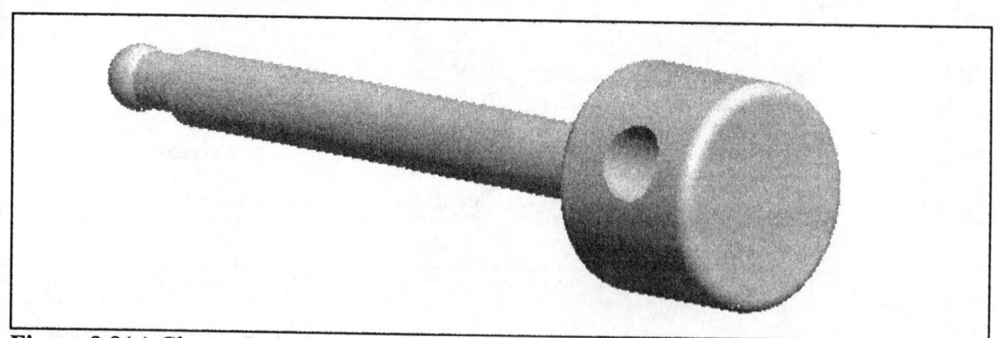

Figure 9.9(a) Clamp_Swivel

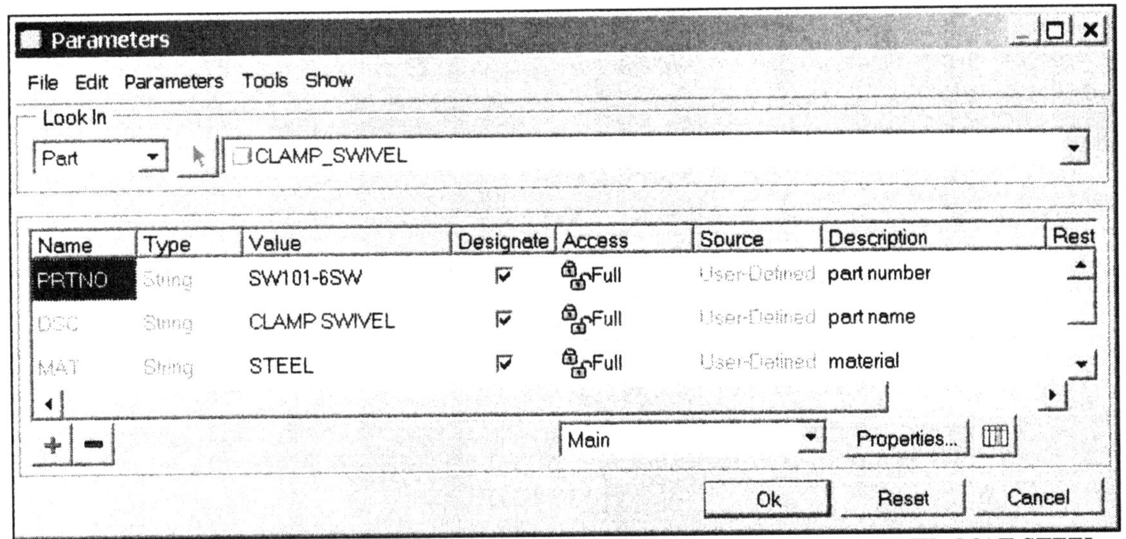

Figure 9.9(b) Clamp_Swivel Parameters: PRTNO SW101-6SW, DSC CLAMP SWIVEL, MAT STEEL

Retrieve the clamp ball, click: **File** ⇒ **Open** ⇒ **clamp_ball.prt** ⇒ **Open** [Fig. 9.10(a)] ⇒ **Tools** ⇒ **Parameters** ⇒ **Parameters** ⇒ **Add Parameter** ⇒ complete the parameters as shown [Fig. 9.10(b)] ⇒ **Ok** ⇒ **File** ⇒ **Save** ⇒ **MMB** ⇒ **File** ⇒ **Close Window**

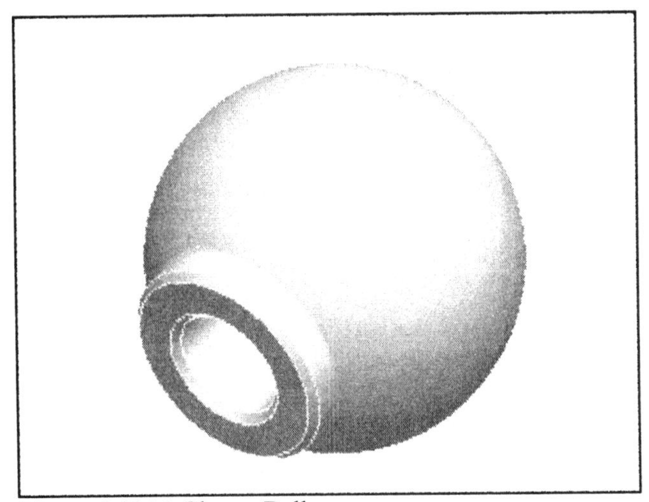
Figure 9.10(a) Clamp_Ball

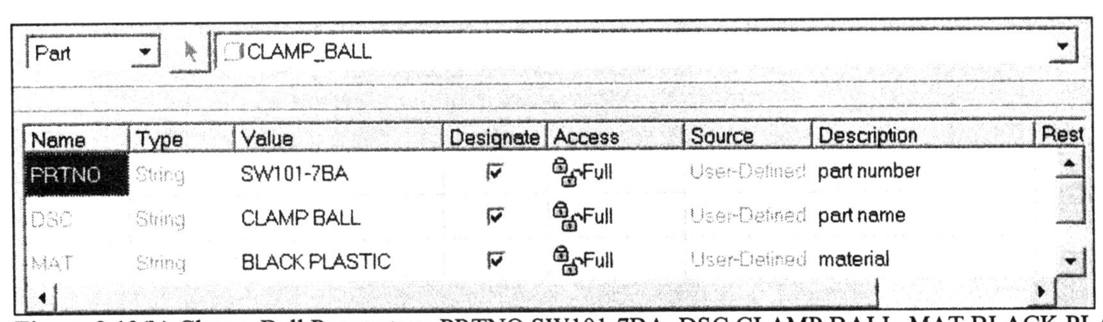

Figure 9.10(b) Clamp_Ball Parameters: PRTNO SW101-7BA, DSC CLAMP BALL, MAT BLACK PLASTIC

Parameters can be added, deleted, and modified in Part Mode, Drawing Mode, or Assembly Mode. You can also add *parameter columns* to the Model Tree in Assembly Mode and edit the parameter value. Use the following information to add parameters both to purchased components (standard parts) and to the remaining parts required for the subassembly and the assembly (Fig. 9.11). Use the Assembly Mode to input the information shown in Figures 9.12(a-e) to establish the part parameters.

Click: **File ⇒ Open ⇒ clamp_assembly.asm ⇒ Open** *(see commands on the following pages)*

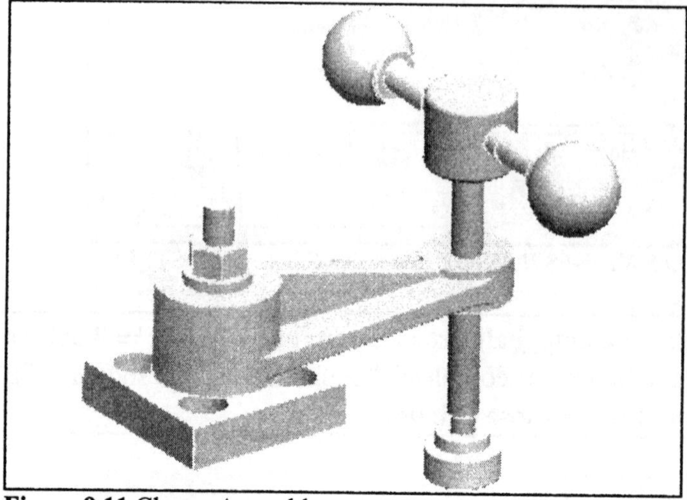

Figure 9.11 Clamp_Assembly

Component	**clamp_foot**
Part Number	**SW101-8FT**
Description	**CLAMP FOOT**
Material	**NYLON**

Figure 9.12(a) Clamp_Foot

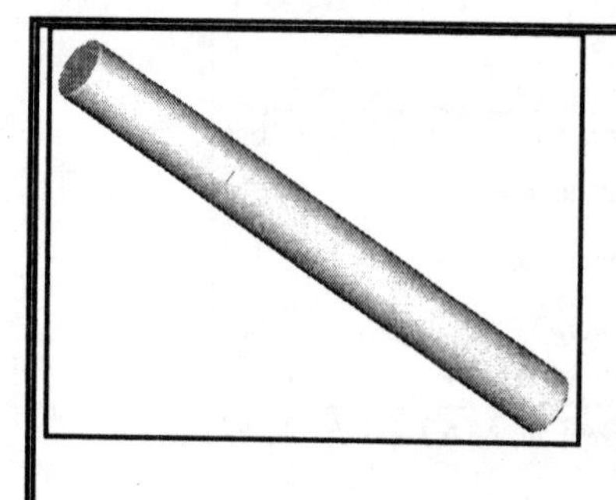

Component	**clamp_stud5**
Part Number	**SW101-9STL**
Description	**.500-13 X 5.00 DOUBLE END STUD**
Material	**PURCHASED**

Figure 9.12(b) Clamp_Stud5

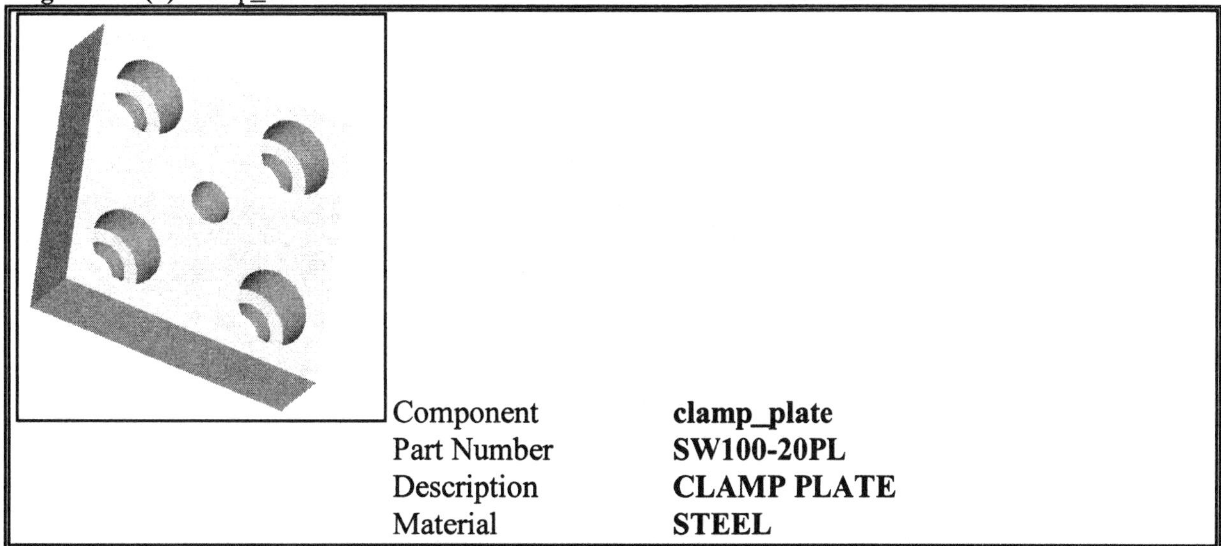

Component	**clamp_plate**
Part Number	**SW100-20PL**
Description	**CLAMP PLATE**
Material	**STEEL**

Figure 9.12(c) Clamp_Plate

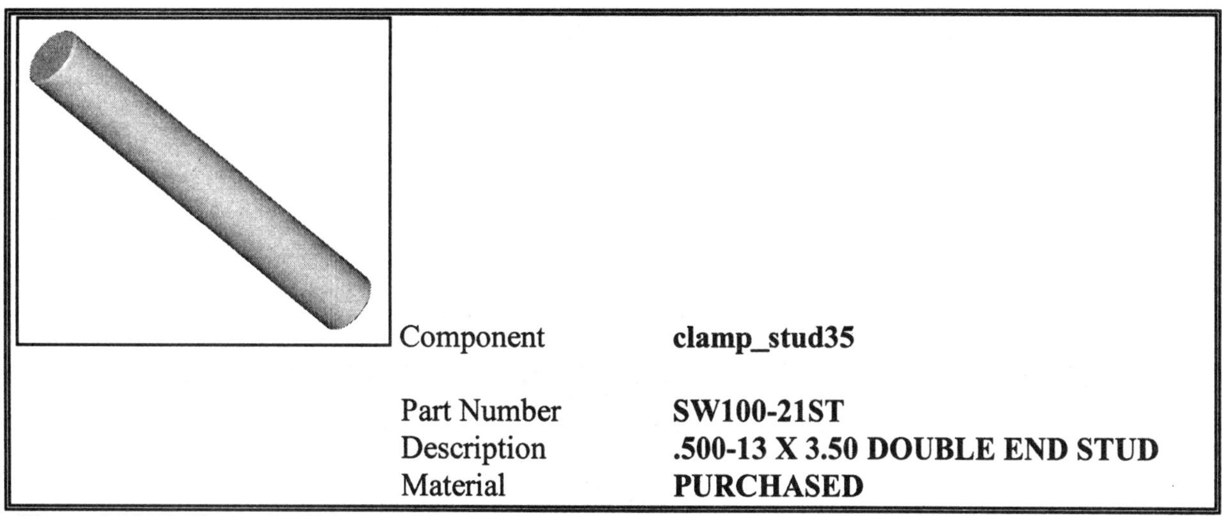

Component	**clamp_stud35**
Part Number	**SW100-21ST**
Description	**.500-13 X 3.50 DOUBLE END STUD**
Material	**PURCHASED**

Figure 9.12(d) Clamp_Stud35

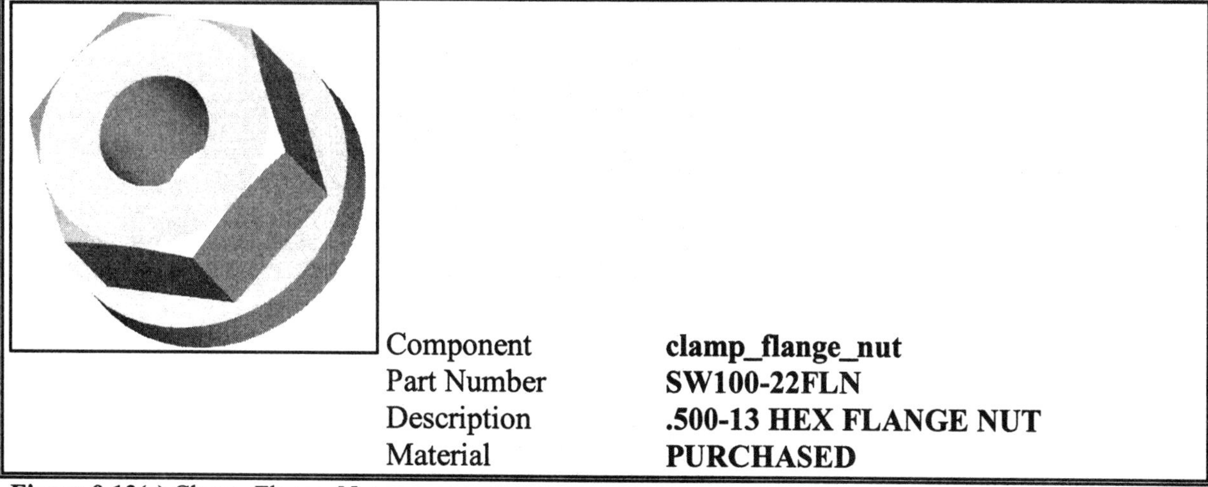

Component	**clamp_flange_nut**
Part Number	**SW100-22FLN**
Description	**.500-13 HEX FLANGE NUT**
Material	**PURCHASED**

Figure 9.12(e) Clamp_Flange_Nut

Click: **Settings** in the Model Tree ⇒ **Tree Columns** [Fig. 9.13(a)] ⇒ Type **Model Params** [Fig. 9.13(b)] ⇒ Name field, type **PRTNO** [Fig. 9.13(c)] ⇒ **Enter** ⇒ Name field, type **DSC** [Fig. 9.13(d)] ⇒ **Enter** [Fig. 9.13(e)] ⇒ Name field, type **MAT** [Fig. 9.13(f)] ⇒ **Enter** [Fig. 9.13(g)] ⇒ **Apply** ⇒ **OK**

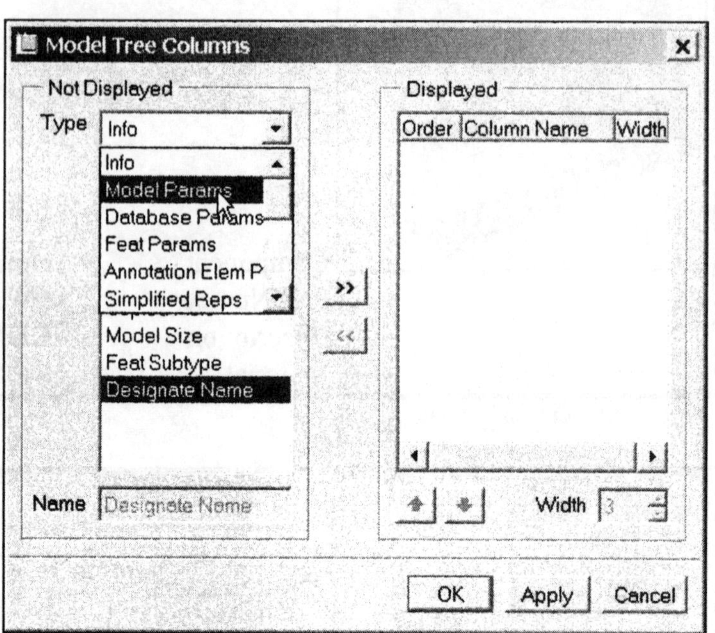

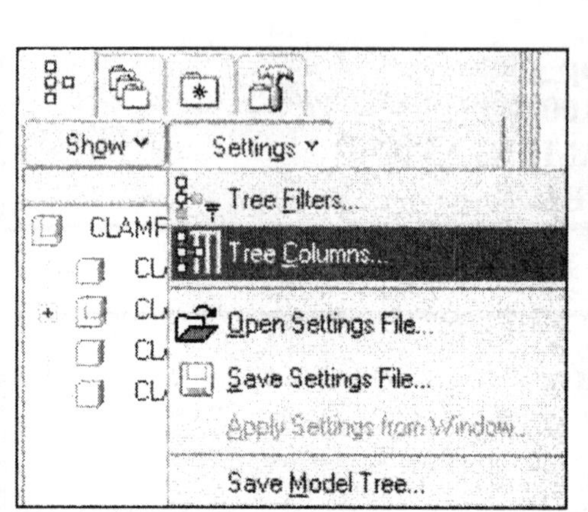

Figure 9.13(a) Tree Columns

Figure 9.13(b) Model Params

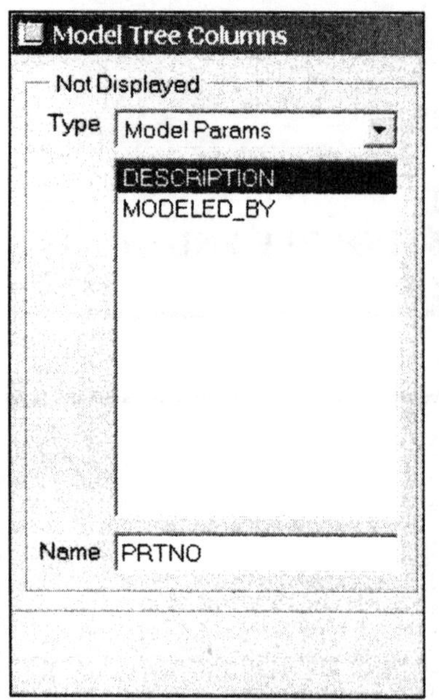

Figure 9.13(c) Name PRTNO

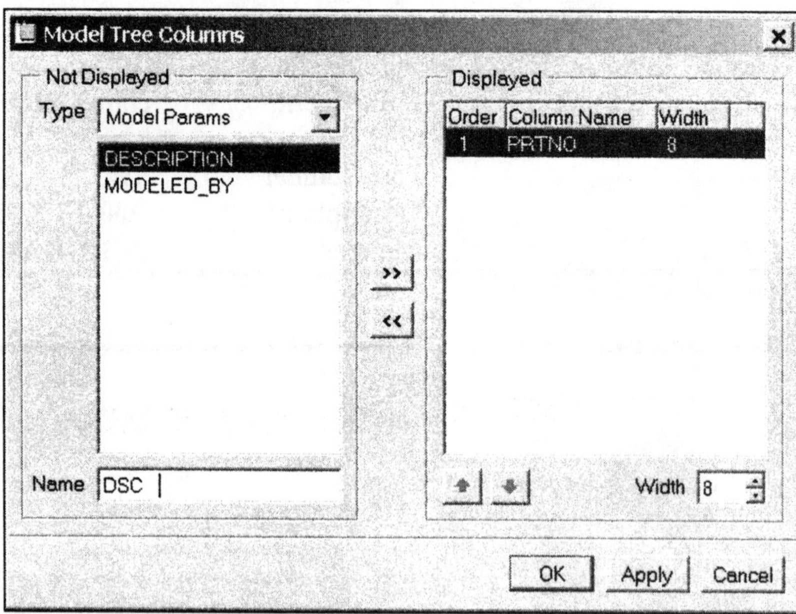

Figure 9.13(d) Name DSC

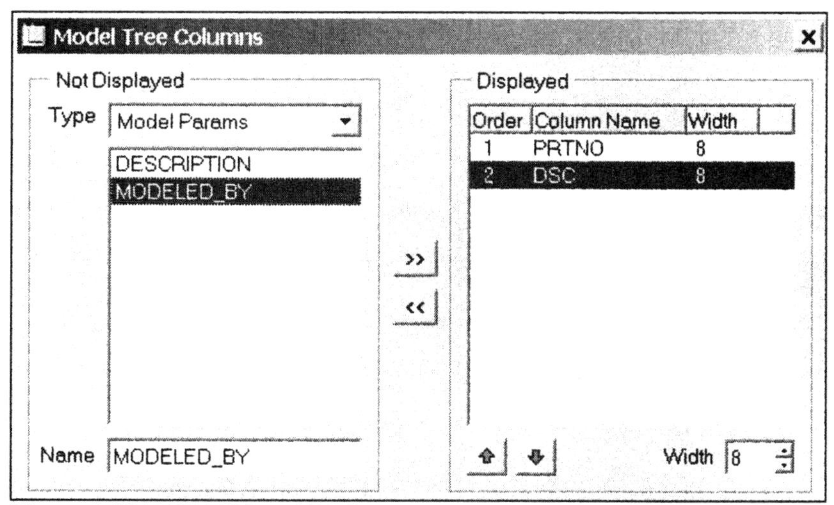

Figure 9.13(e) PRTNO and DSC

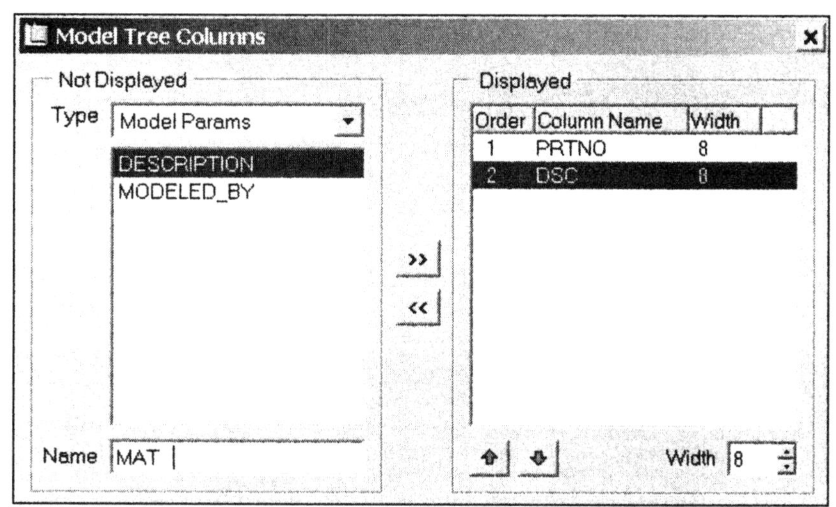

Figure 9.13(f) Name MAT

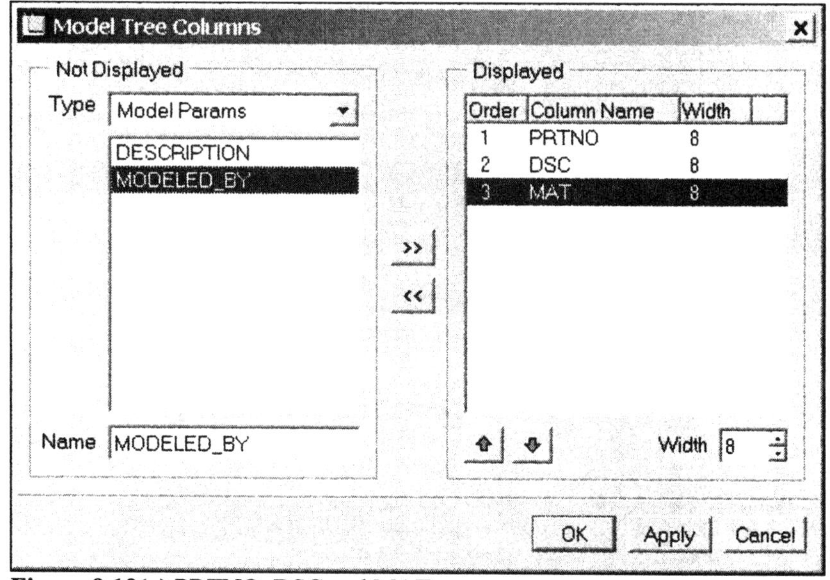

Figure 9.13(g) PRTNO, DSC and MAT

Resize the Model Tree and click on **CLAMP_PLATE.PRT** ⇒ click in **PRTNO** field [Fig. 9.14(a)] ⇒ ▾ ⇒ **String** ⇒ Value **SW100-20PL** [Fig. 9.14(b)] ⇒ **Ok** [Fig. 9.14(c)] ⇒ click in **DSC** field ⇒ ▾ ⇒ **String** ⇒ Value **CLAMP PLATE** [Fig. 9.14(d)] ⇒ **Ok** [Fig. 9.14(e)] ⇒ click in **MAT** field ⇒ ▾ ⇒ **String** ⇒ Value **STEEL** [Fig. 9.14(f)] ⇒ **Ok** ⇒ **File** ⇒ **Save** ⇒ **MMB**

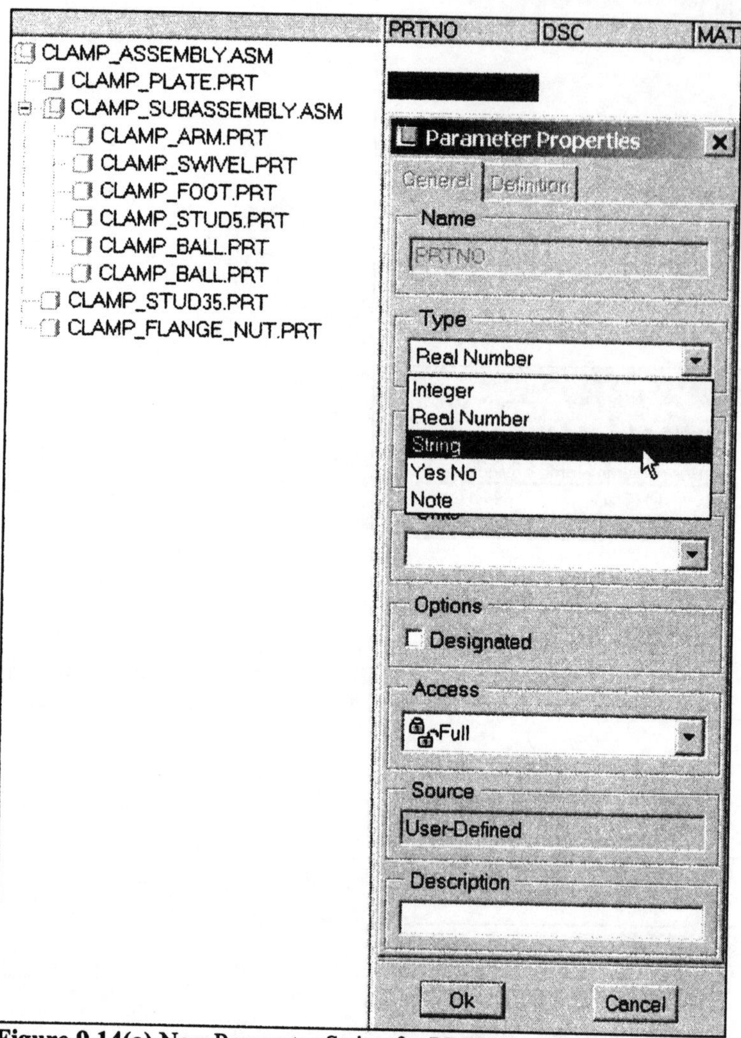

Figure 9.14(a) New Parameter String for PRTNO

Figure 9.14(b) Value SW100-20PL

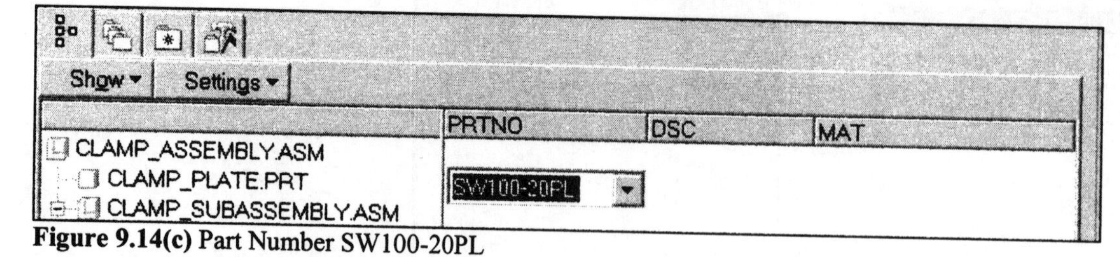

Figure 9.14(c) Part Number SW100-20PL

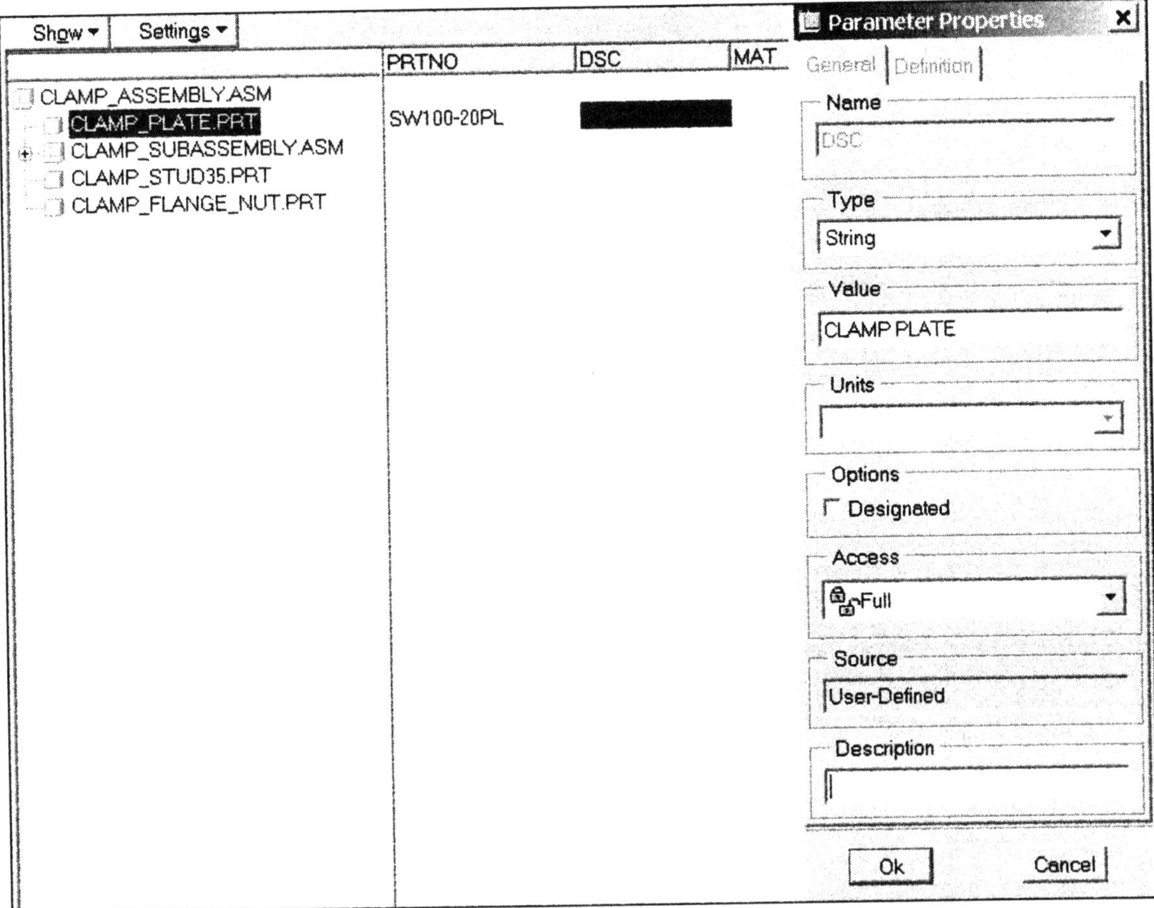

Figure 9.14(d) New Parameter String for DSC

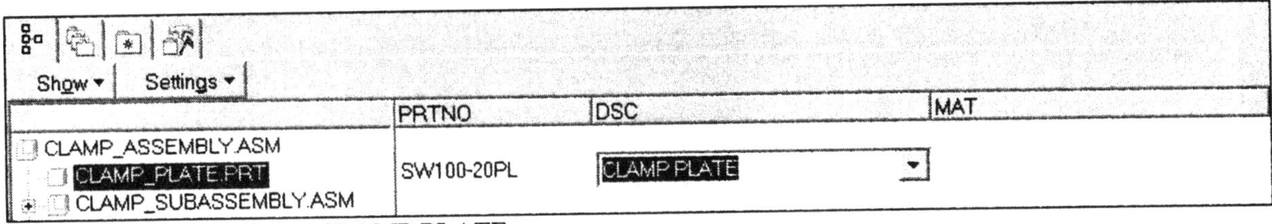

Figure 9.14(e) Description CLAMP PLATE

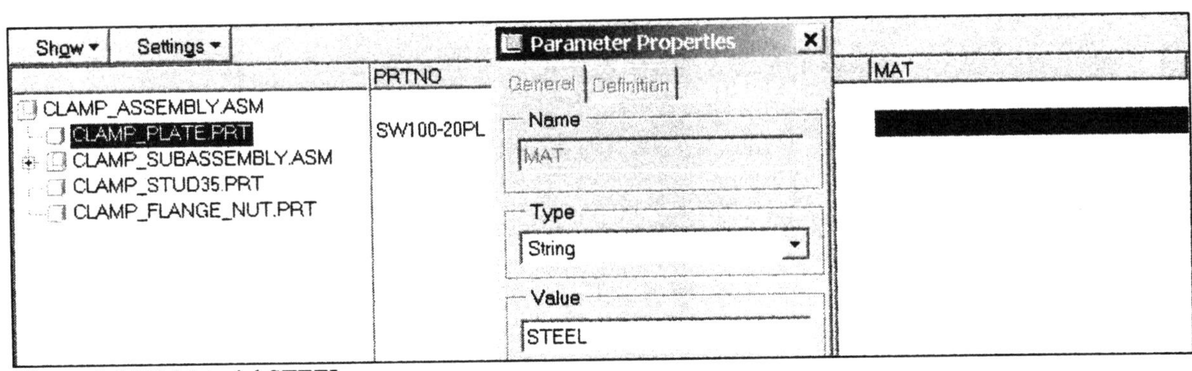

Figure 9.14(f) Material STEEL

Complete the parameters for the remaining components [Fig. 9.15(a)]

	PRTNO	DSC	MAT
CLAMP_ASSEMBLY.ASM			
CLAMP_PLATE.PRT	SW100-20PL	CLAMP PLATE	STEEL
CLAMP_SUBASSEMBLY.ASM			
CLAMP_ARM.PRT	SW101-5AR	CLAMP ARM	STEEL
CLAMP_SWIVEL.PRT	SW101-6SW	CLAMP SWIVEL	STEEL
CLAMP_FOOT.PRT	SW101-8FT	CLAMP FOOT	NYLON
CLAMP_STUD5.PRT	SW101-9STL	.500-13X5.00 DOUBLE END STUD	PURCHASED
CLAMP_BALL.PRT	SW101-7BA	SWING CLAMP BALL	BLACK PLASTIC
CLAMP_BALL.PRT	SW101-7BA	SWING CLAMP BALL	BLACK PLASTIC
CLAMP_STUD35.PRT	SW100-21ST	.500-13X3.50 DOUBLE END STUD	PURCHASED
CLAMP_FLANGE_NUT.PRT	SW100-22FLN	.500-13X HEX FLANGE NUT	PURCHASED

Figure 9.15(a) Assembly Model Tree and Parameters

Click: **Tools** ⇒ **Parameters** ⇒ ▾ ⇒ **Part** [Fig. 9.15(b)] ⇒ click on **CLAMP_FOOT.PRT** [Fig. 9.15(c)] ⇒ Designate ☑ ⇒ add Parameter Descriptions ⇒ repeat for remaining components ⇒ **Ok**

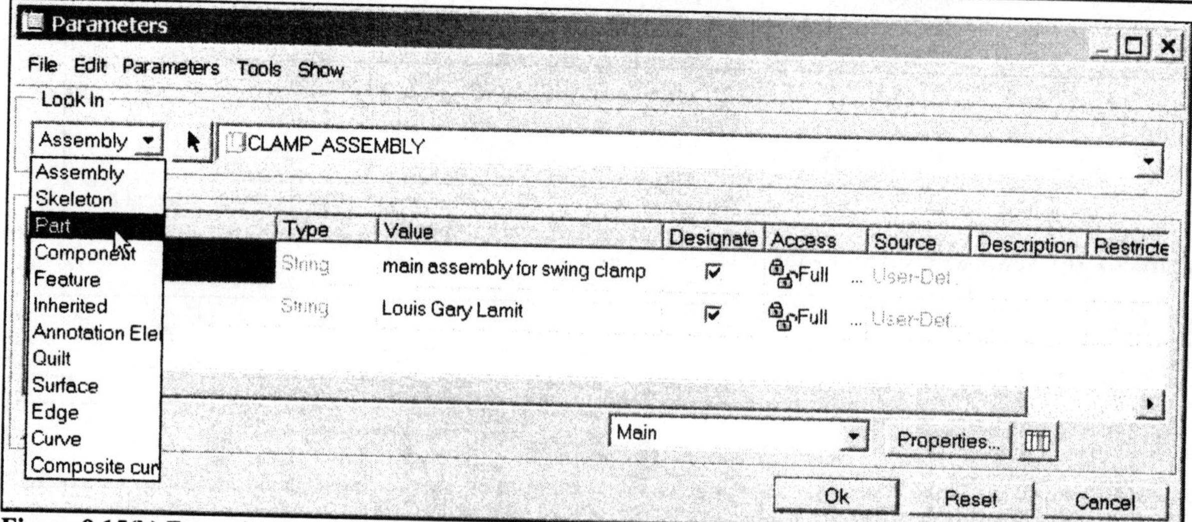

Figure 9.15(b) Extracting Part Parameters from the Assembly

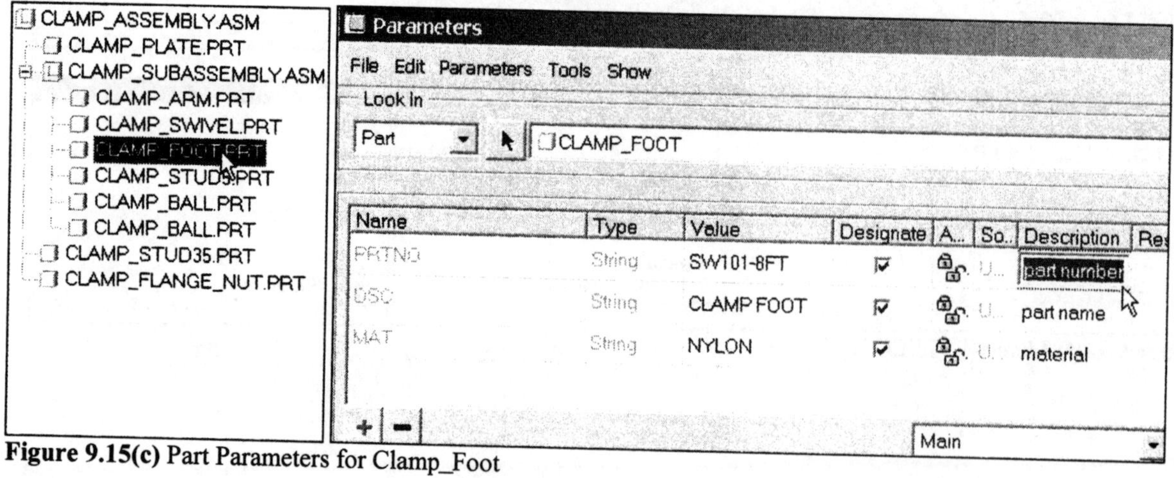

Figure 9.15(c) Part Parameters for Clamp_Foot

The Clamp_Flange_Nut, Clamp_Stud35 and the Clamp_Stud5 are standard parts that are copied from the PTC parts library; therefore they will have additional parameters [Fig. 9.16(a)] and relations [Figs. 9.16 (b-c)].

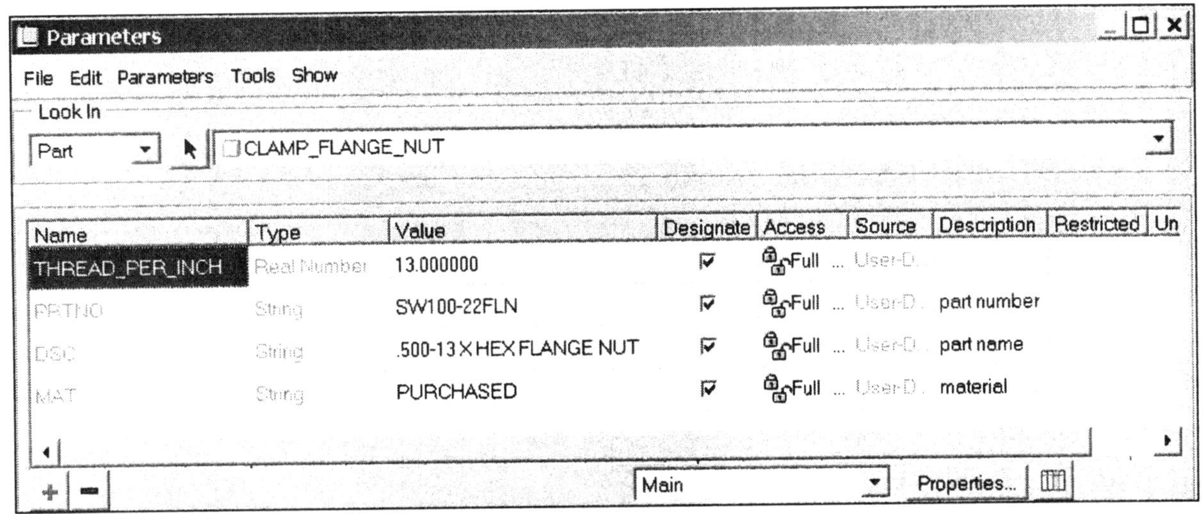

Figure 9.16(a) Clamp_Flange_Nut Parameters

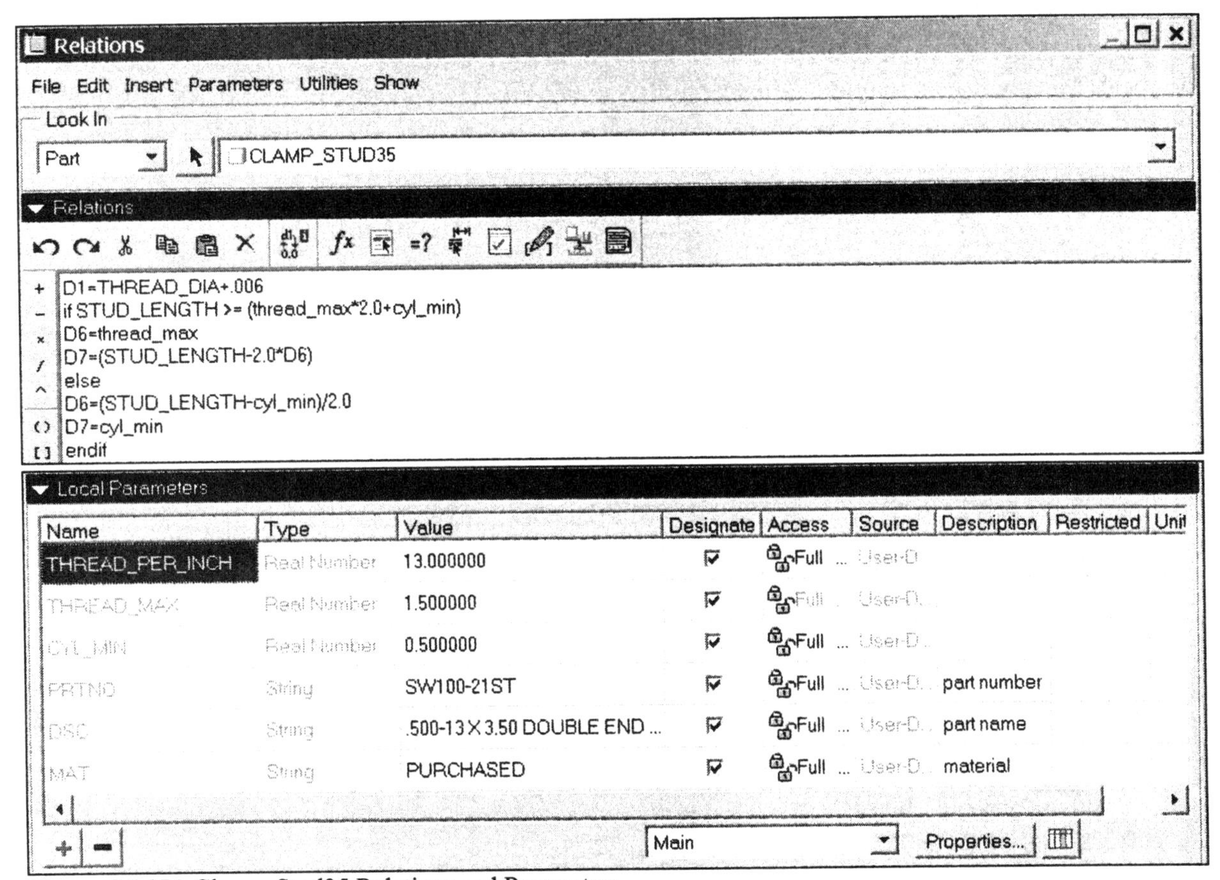

Figure 9.16(b) Clamp_Stud35 Relations and Parameters

Relations and parameters can be displayed together by clicking on the part name (here **CLAMP_STUD5**) in the CLAMP_ASSEMBLY Model Tree ⇒ **RMB** ⇒ **Open** ⇒ **Info** ⇒ **Relations and Parameters** [Fig. 9.16(c)] **File** ⇒ **Close Window**

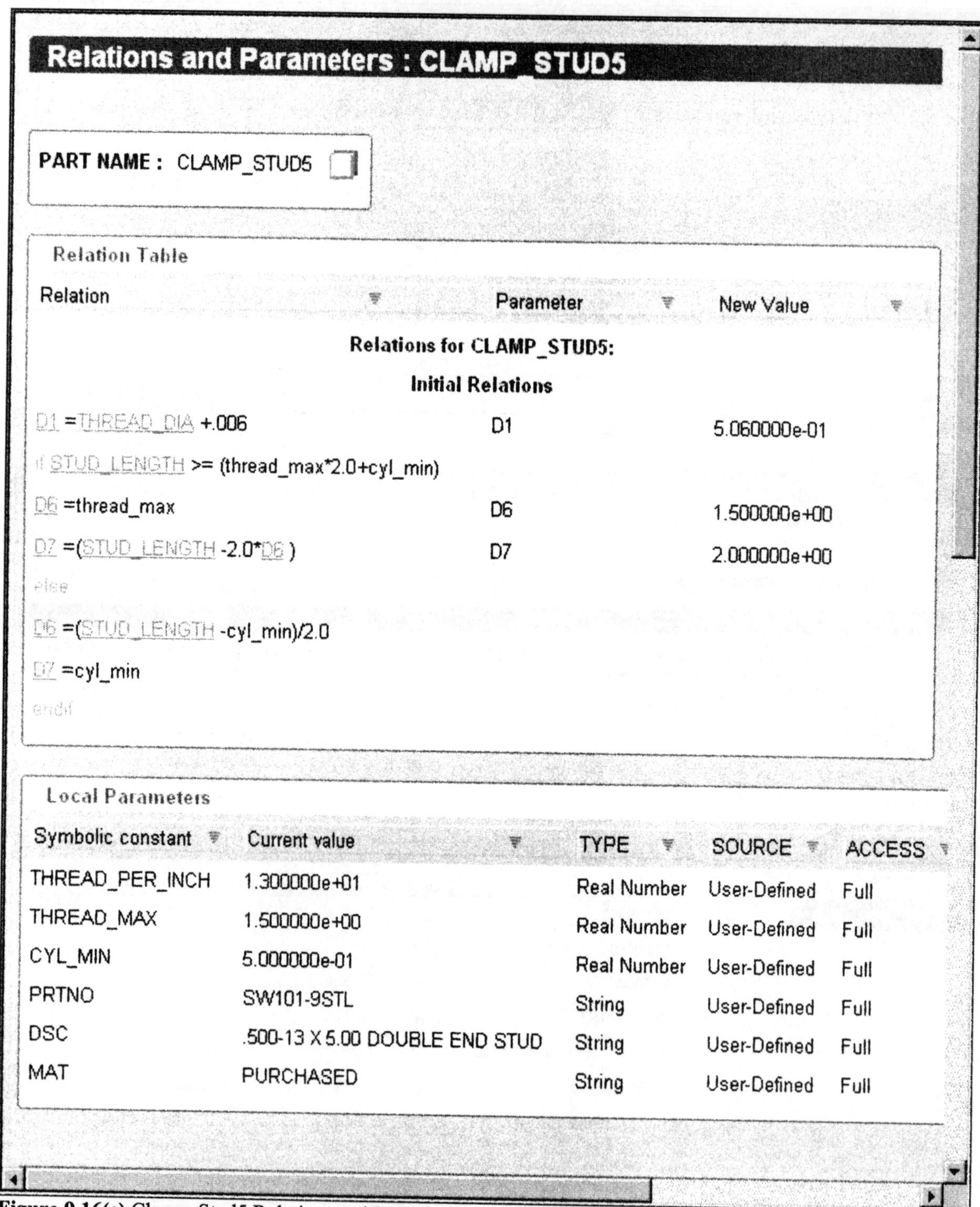

Relations and Parameters : CLAMP_STUD5

PART NAME : CLAMP_STUD5

Relation Table

Relation	Parameter	New Value
Relations for CLAMP_STUD5:		
Initial Relations		
D1 =THREAD_DIA +.006	D1	5.060000e-01
if STUD_LENGTH >= (thread_max*2.0+cyl_min)		
D6 =thread_max	D6	1.500000e+00
D7 =(STUD_LENGTH -2.0*D6)	D7	2.000000e+00
else		
D6 =(STUD_LENGTH -cyl_min)/2.0		
D7 =cyl_min		
endif		

Local Parameters

Symbolic constant	Current value	TYPE	SOURCE	ACCESS
THREAD_PER_INCH	1.300000e+01	Real Number	User-Defined	Full
THREAD_MAX	1.500000e+00	Real Number	User-Defined	Full
CYL_MIN	5.000000e-01	Real Number	User-Defined	Full
PRTNO	SW101-9STL	String	User-Defined	Full
DSC	.500-13 X 5.00 DOUBLE END STUD	String	User-Defined	Full
MAT	PURCHASED	String	User-Defined	Full

Figure 9.16(c) Clamp_Stud5 Relations and Parameters

Assembly Drawings

The parameters (and their values) have been established for each part. The assembly format with related parameters in a parts list table has been created and saved in your (format) directory. You can now create a drawing of the assembly, where the parts list will be generated automatically, and the assembly ballooned. The first assembly drawing will be of the Clamp_Subassembly [Figs. 9.17(a-b)].

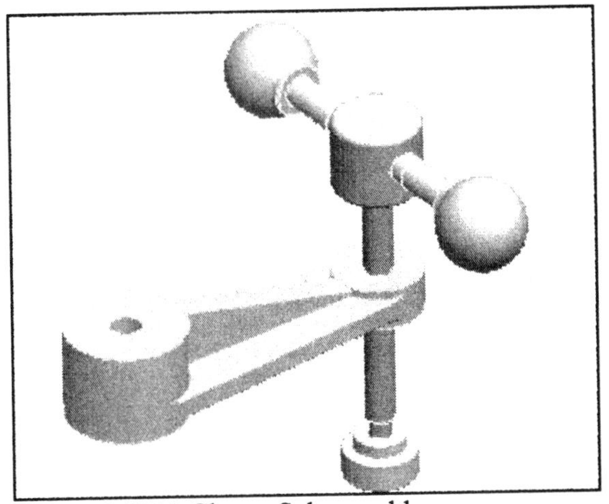

Figure 9.17(a) Clamp_Subassembly

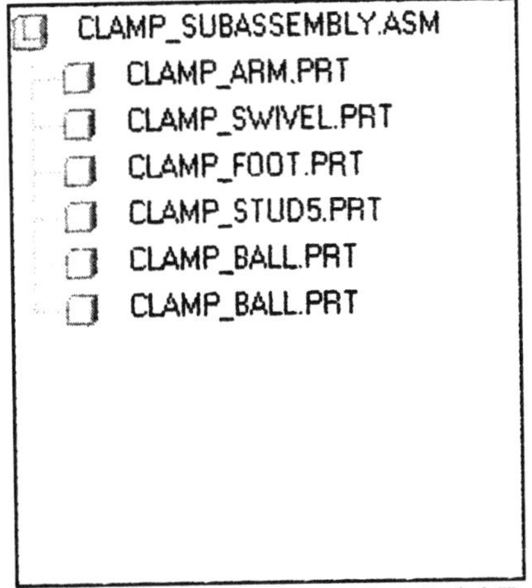

Figure 9.17(b) Subassembly Model Tree

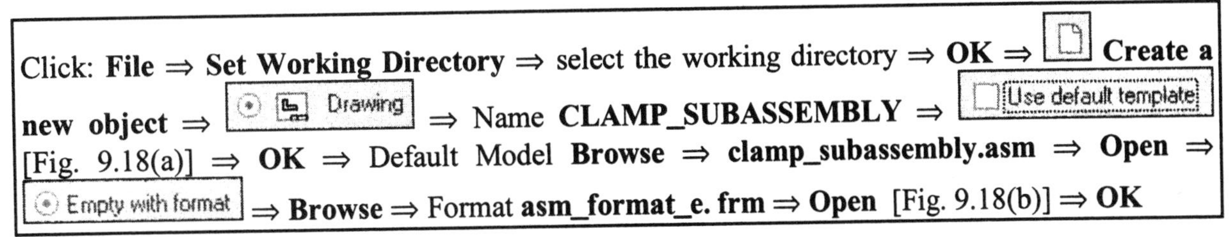

Click: **File** ⇒ **Set Working Directory** ⇒ select the working directory ⇒ **OK** ⇒ [icon] **Create a new object** ⇒ [Drawing] ⇒ Name **CLAMP_SUBASSEMBLY** ⇒ [Use default template] [Fig. 9.18(a)] ⇒ **OK** ⇒ Default Model **Browse** ⇒ **clamp_subassembly.asm** ⇒ **Open** ⇒ [Empty with format] ⇒ **Browse** ⇒ Format **asm_format_e. frm** ⇒ **Open** [Fig. 9.18(b)] ⇒ **OK**

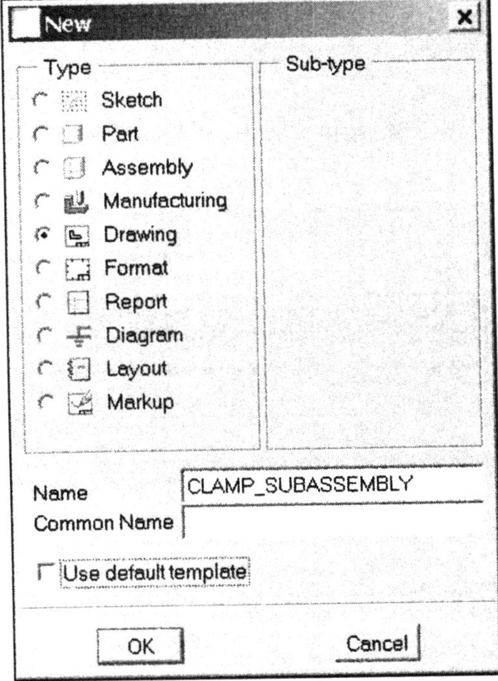

Figure 9.18(a) New Dialog Box

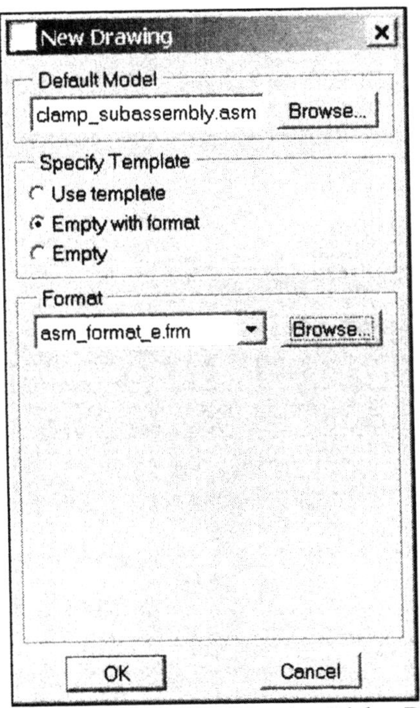

Figure 9.18(b) New Drawing Dialog Box

For the drawing to display with the correct style, modify the values of the Drawing Options.

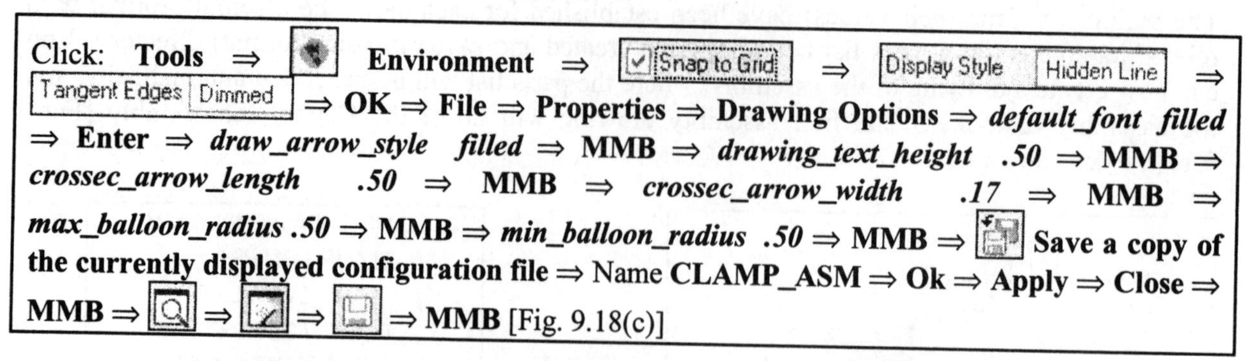

Click: **Tools** ⇒ **Environment** ⇒ Snap to Grid ⇒ Display Style Hidden Line ⇒ Tangent Edges Dimmed ⇒ **OK** ⇒ **File** ⇒ **Properties** ⇒ **Drawing Options** ⇒ ***default_font filled*** ⇒ **Enter** ⇒ ***draw_arrow_style filled*** ⇒ **MMB** ⇒ ***drawing_text_height .50*** ⇒ **MMB** ⇒ ***crossec_arrow_length .50*** ⇒ **MMB** ⇒ ***crossec_arrow_width .17*** ⇒ **MMB** ⇒ ***max_balloon_radius .50*** ⇒ **MMB** ⇒ ***min_balloon_radius .50*** ⇒ **MMB** ⇒ **Save a copy of the currently displayed configuration file** ⇒ Name **CLAMP_ASM** ⇒ **Ok** ⇒ **Apply** ⇒ **Close** ⇒ **MMB** ⇒ ⇒ ⇒ ⇒ **MMB** [Fig. 9.18(c)]

Note that the Repeat Region of the BOM [Fig. 9.18(d)] has been automatically filled in with the parameters read from the individual components.

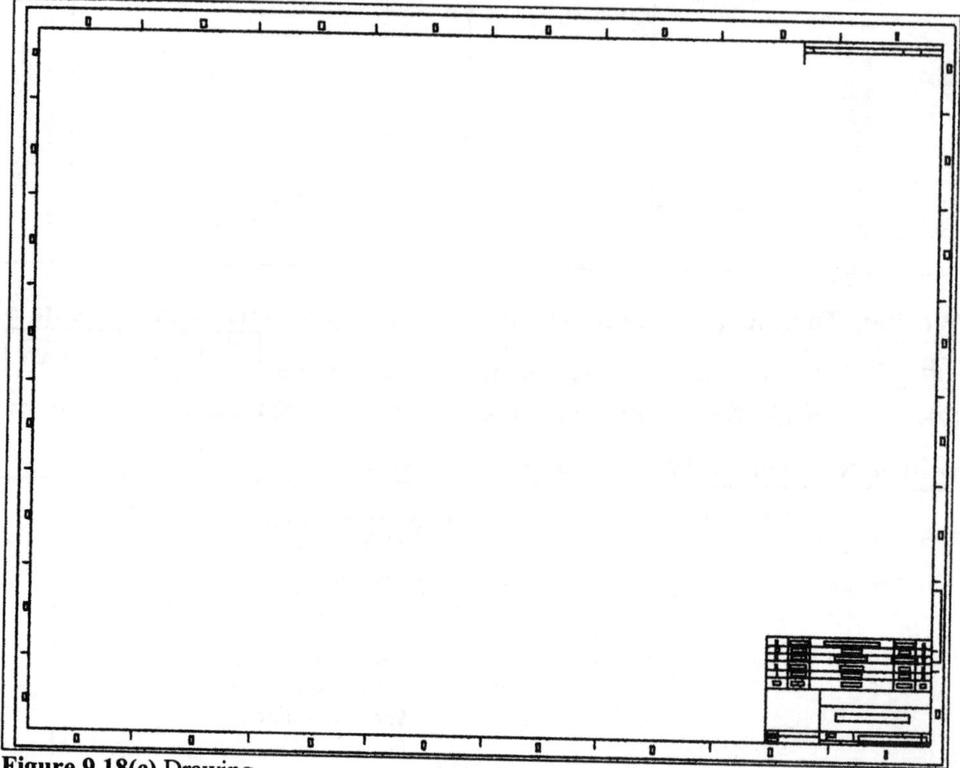

Figure 9.18(c) Drawing

5	SW101-9STL	.500-13 X 5.00 DOUBLE END STUD	PURCHASED	1
4	SW101-8FT	CLAMP FOOT	NYLON	1
3	SW101-7BA	SWING CLAMP BALL	BLACK PLASTIC	2
2	SW101-6SW	CLAMP SWIVEL	STEEL	1
1	SW101-5AR	CLAMP ARM	STEEL	1
ITEM	PT NUM	DESCRIPTION	MATERIAL	QTY

Figure 9.18(d) BOM

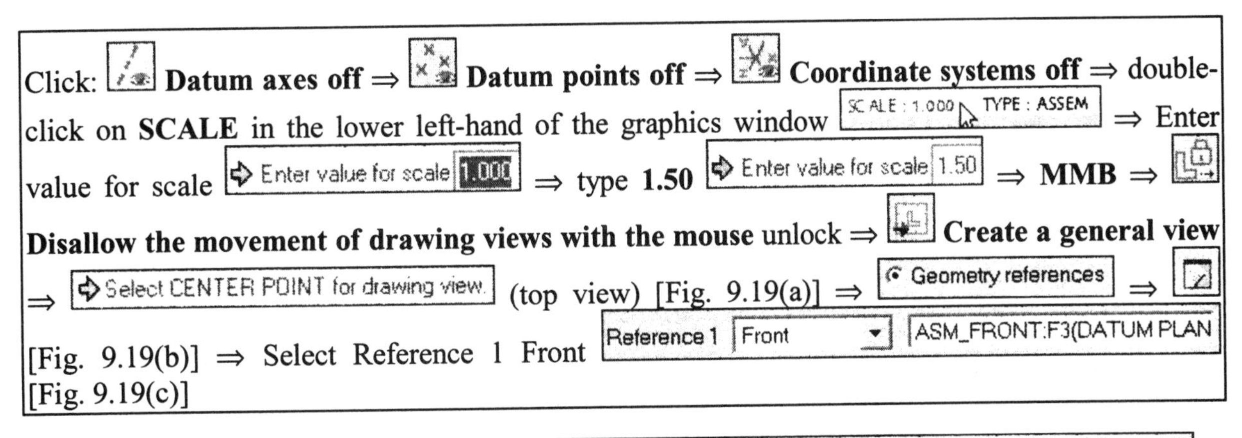

Click: **Datum axes off** ⇒ **Datum points off** ⇒ **Coordinate systems off** ⇒ double-click on **SCALE** in the lower left-hand of the graphics window SCALE : 1.000 TYPE : ASSEM ⇒ Enter value for scale Enter value for scale 1.000 ⇒ type **1.50** Enter value for scale 1.50 ⇒ **MMB** ⇒ **Disallow the movement of drawing views with the mouse** unlock ⇒ **Create a general view** ⇒ Select CENTER POINT for drawing view. (top view) [Fig. 9.19(a)] ⇒ Geometry references ⇒ [Fig. 9.19(b)] ⇒ Select Reference 1 Front Reference 1 Front ASM_FRONT:F3(DATUM PLAN [Fig. 9.19(c)]

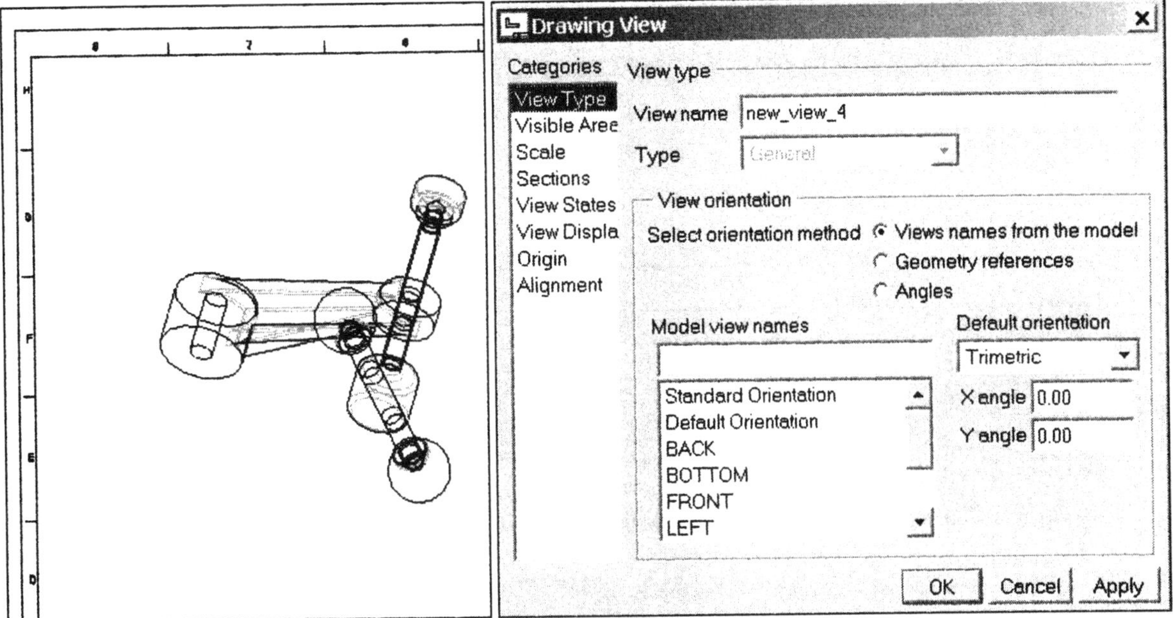

Figure 9.19(a) Select Centerpoint for First Drawing View

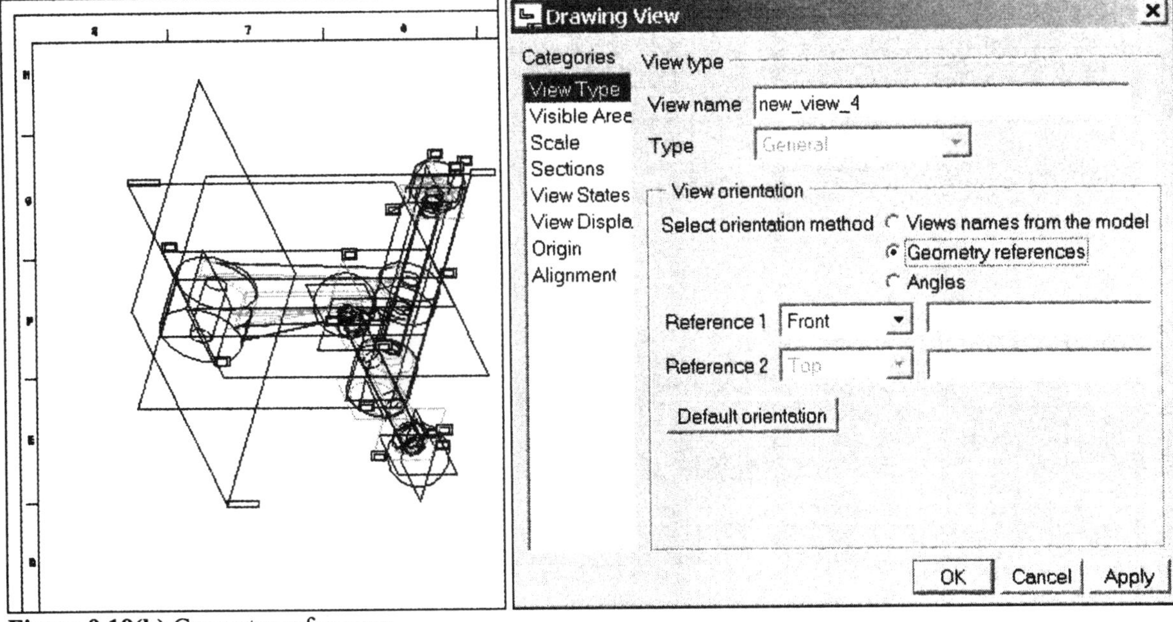

Figure 9.19(b) Geometry references

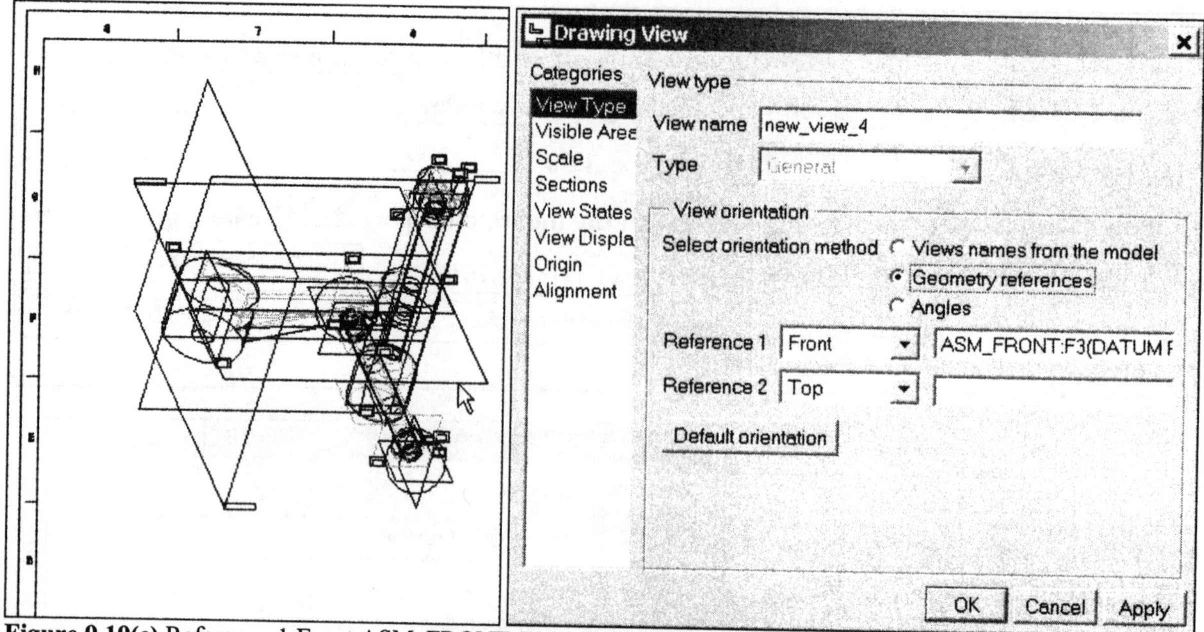

Figure 9.19(c) Reference 1 Front ASM_FRONT-F3(DATUM PLANE)

Select Reference 2 Top  [Fig. 9.19(d)] ⇒ **OK** ⇒

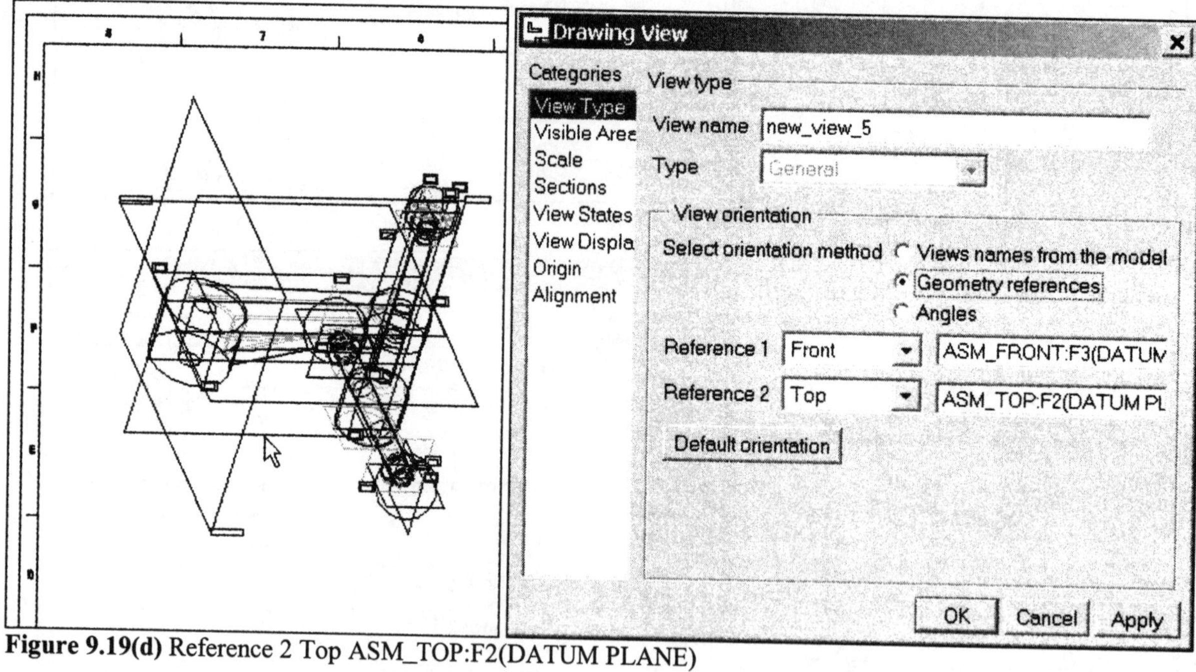

Figure 9.19(d) Reference 2 Top ASM_TOP:F2(DATUM PLANE)

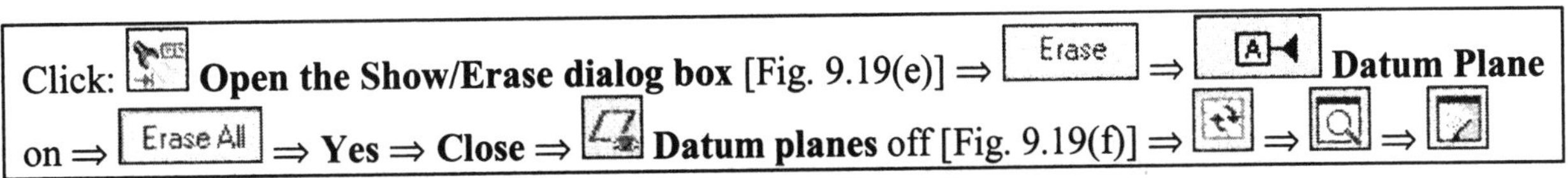

Click: **Open the Show/Erase dialog box** [Fig. 9.19(e)] ⇒ Erase ⇒ **Datum Plane** on ⇒ Erase All ⇒ **Yes** ⇒ **Close** ⇒ **Datum planes** off [Fig. 9.19(f)] ⇒ ⇒ ⇒

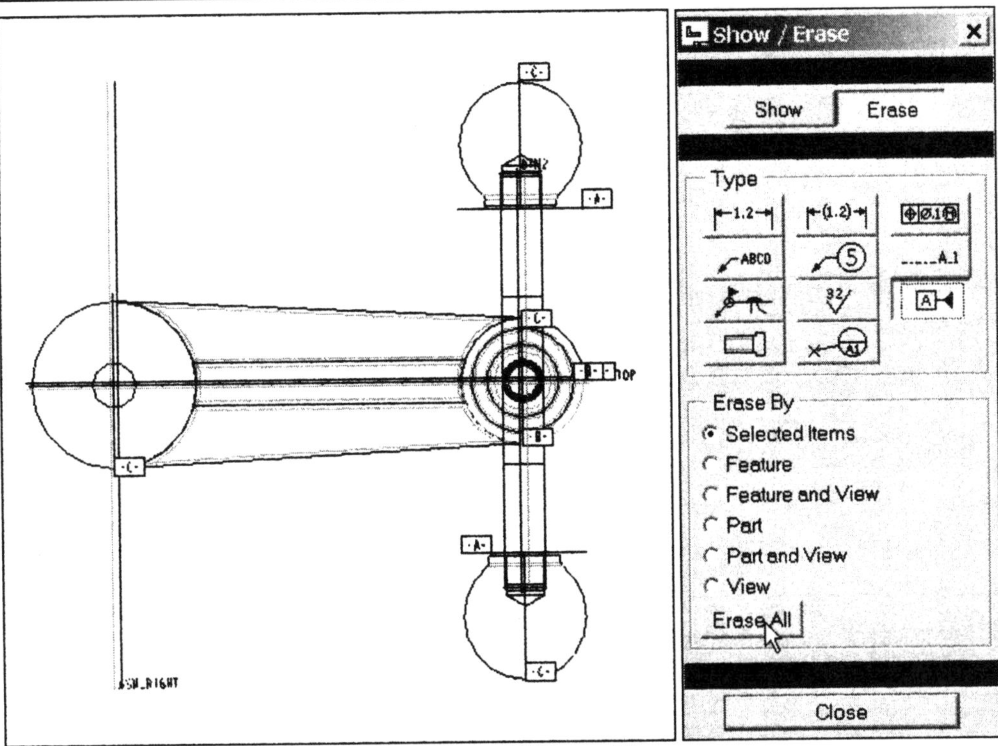

Figure 9.19(e) Show/Erase Dialog Box- Erase All Datum Planes

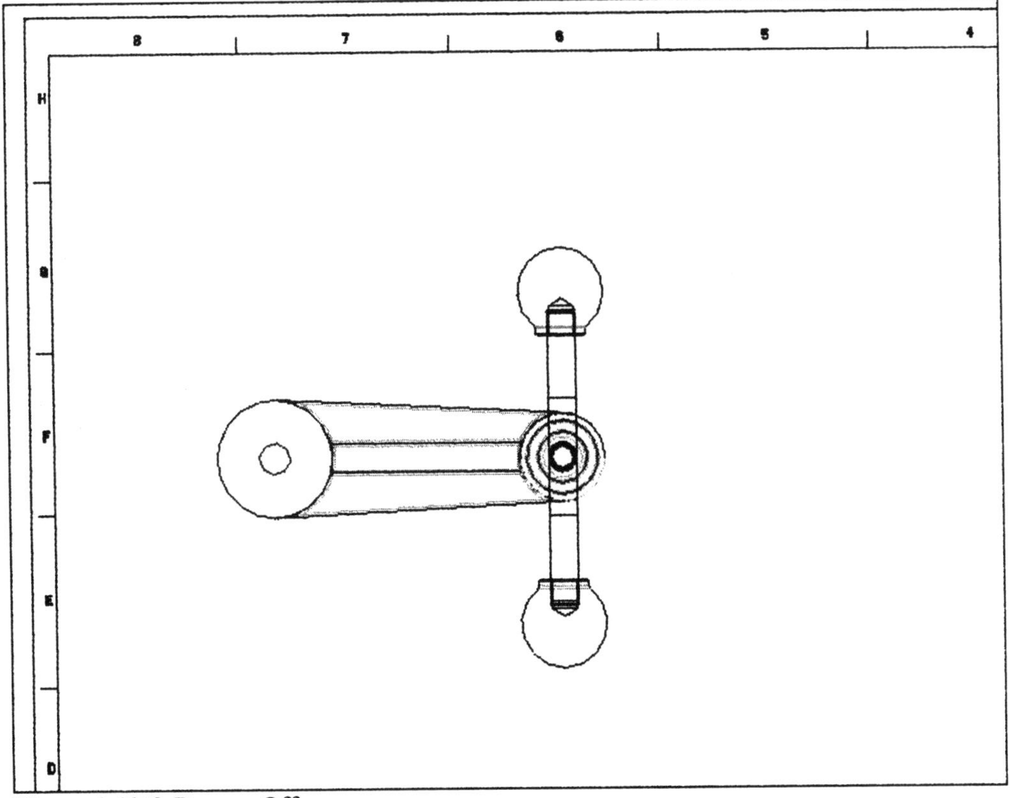

Figure 9.19(f) Datums Off

The only other view needed to show the subassembly is a front section view.

Click: [icon] ⇒ **MMB** ⇒ pick on the top view to highlight ⇒ **RMB** ⇒ **Insert Projection View** [Fig. 9.20(a)] ⇒ [Select CENTER POINT for drawing view.] select below the view previously created [Fig. 9.20(b)]

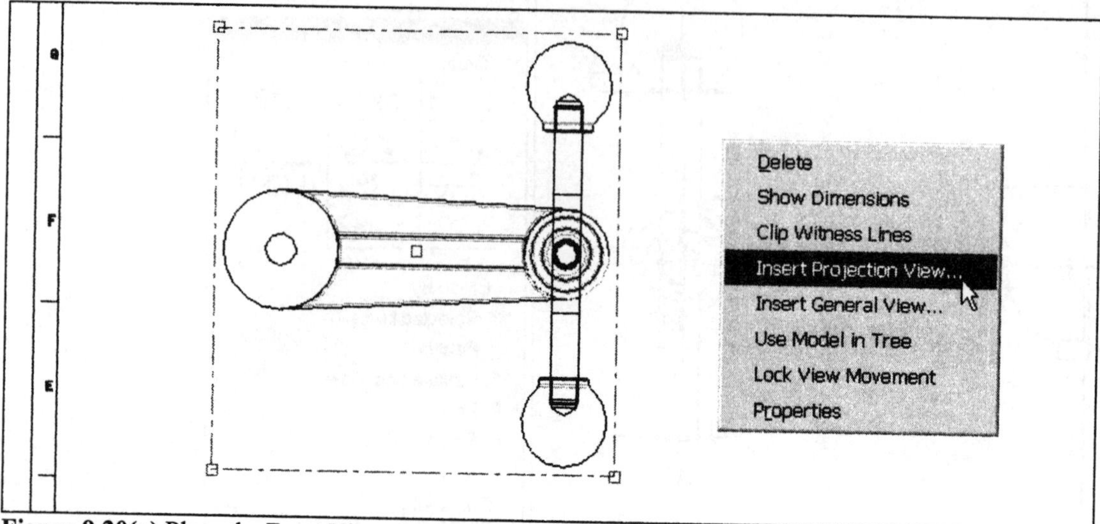

Figure 9.20(a) Place the Front View

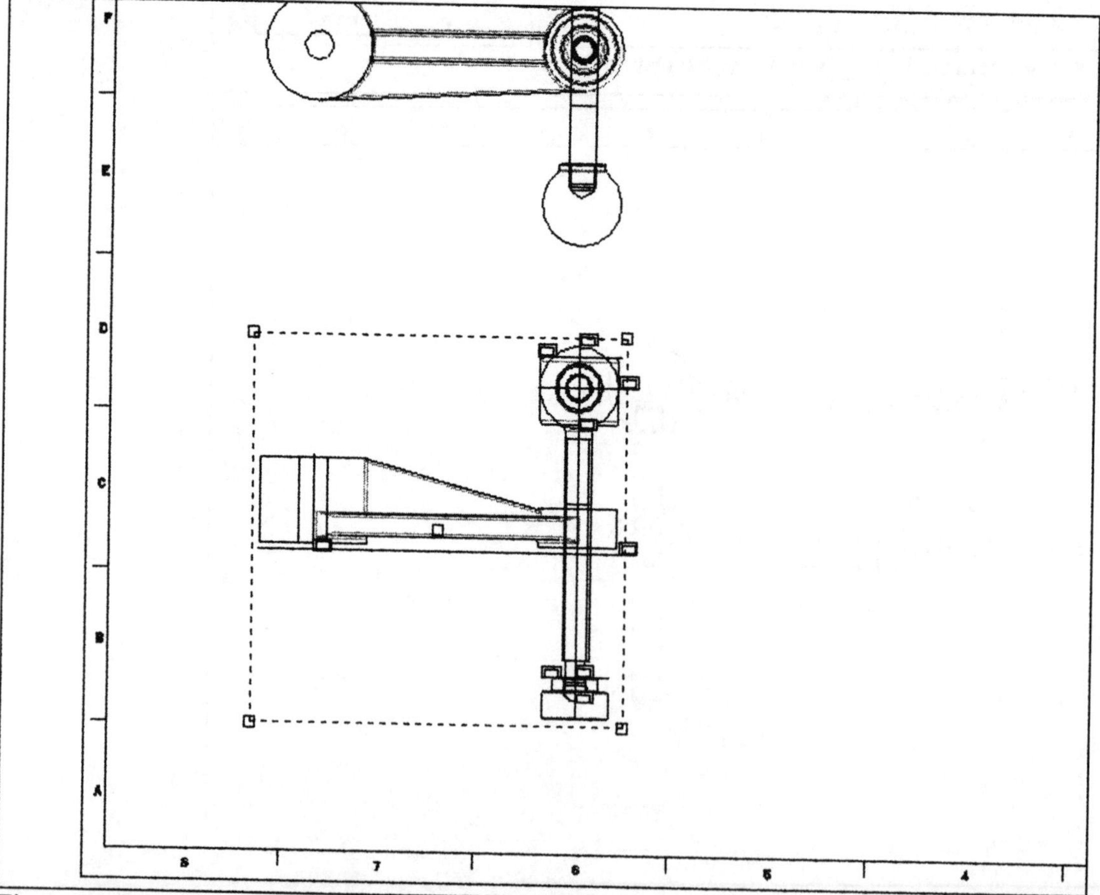

Figure 9.20(b) Select position for Front View

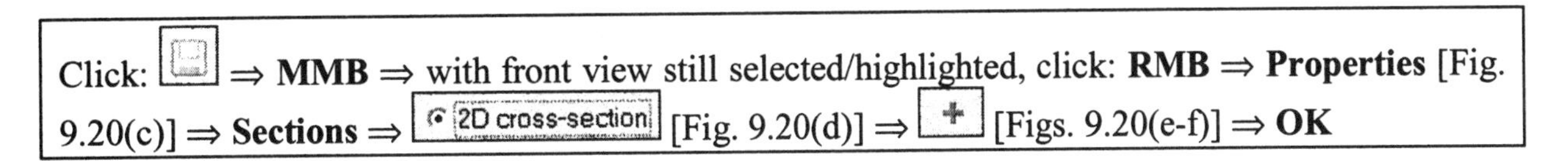

Click: [icon] ⇒ **MMB** ⇒ with front view still selected/highlighted, click: **RMB** ⇒ **Properties** [Fig. 9.20(c)] ⇒ **Sections** ⇒ [2D cross-section] [Fig. 9.20(d)] ⇒ [+] [Figs. 9.20(e-f)] ⇒ **OK**

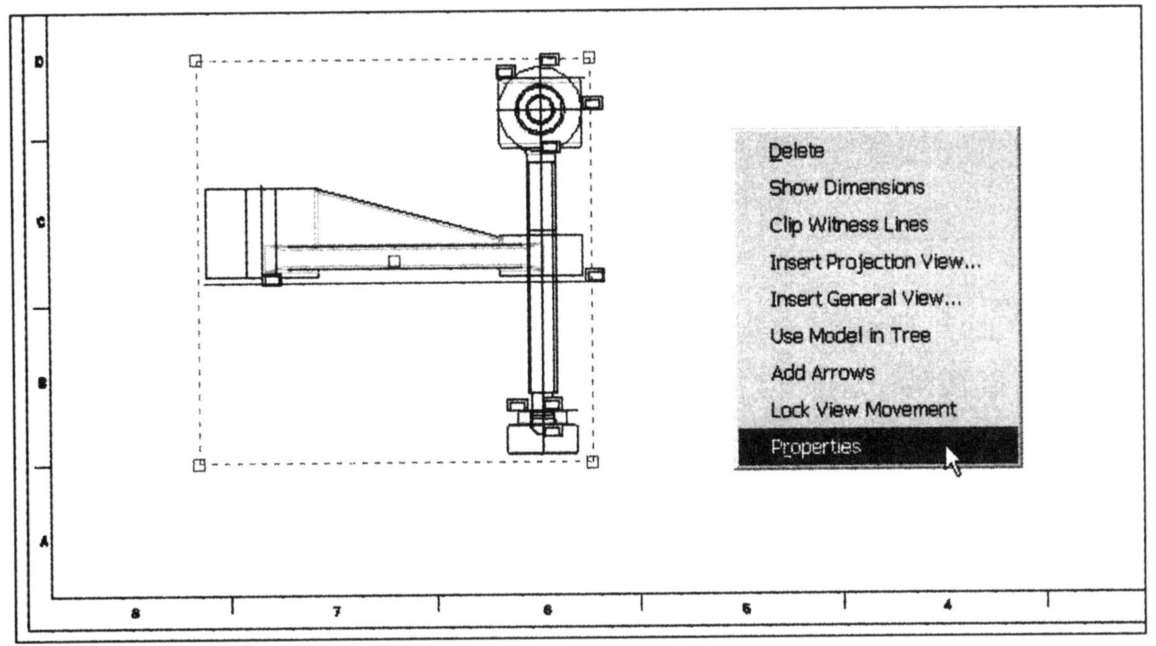

Figure 9.20(c) Properties

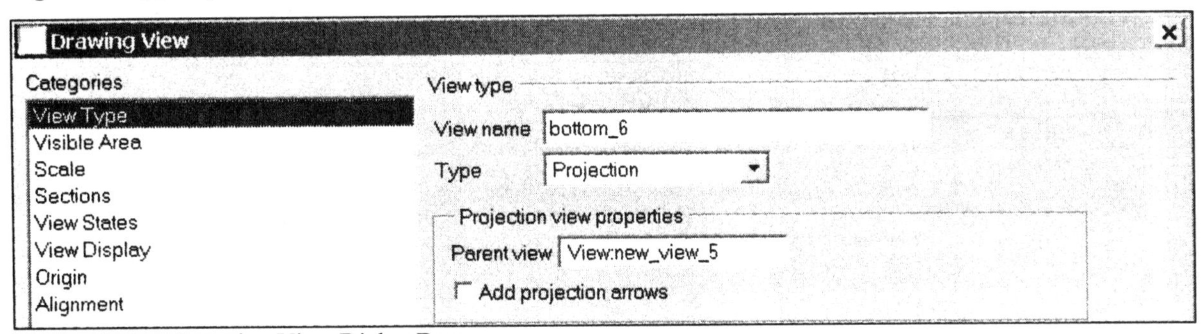

Figure 9.20(d) Drawing View Dialog Box

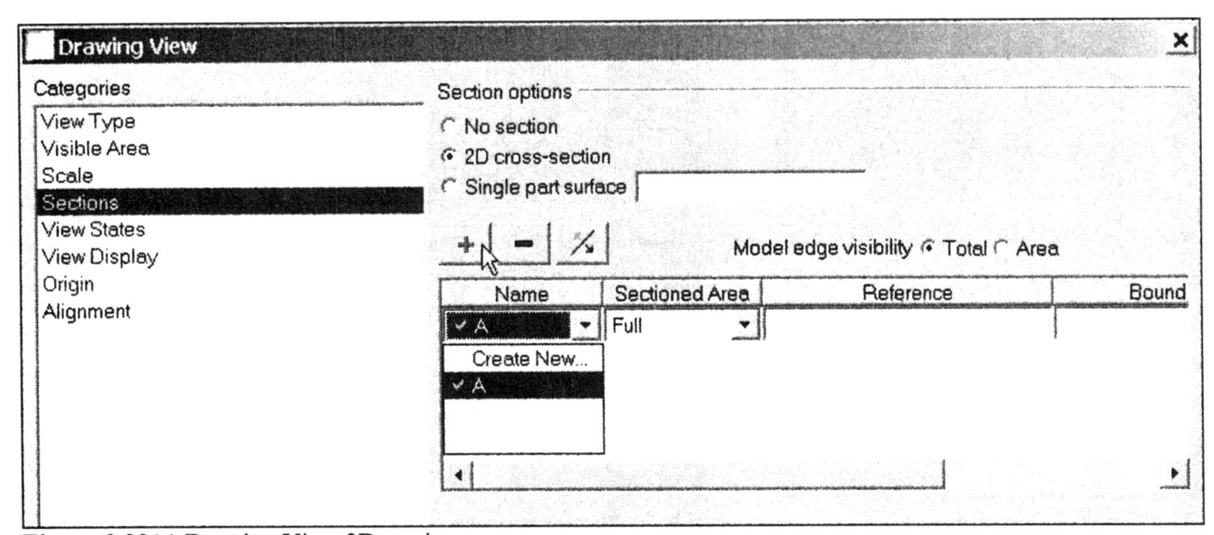

Figure 9.20(e) Drawing View 2D section

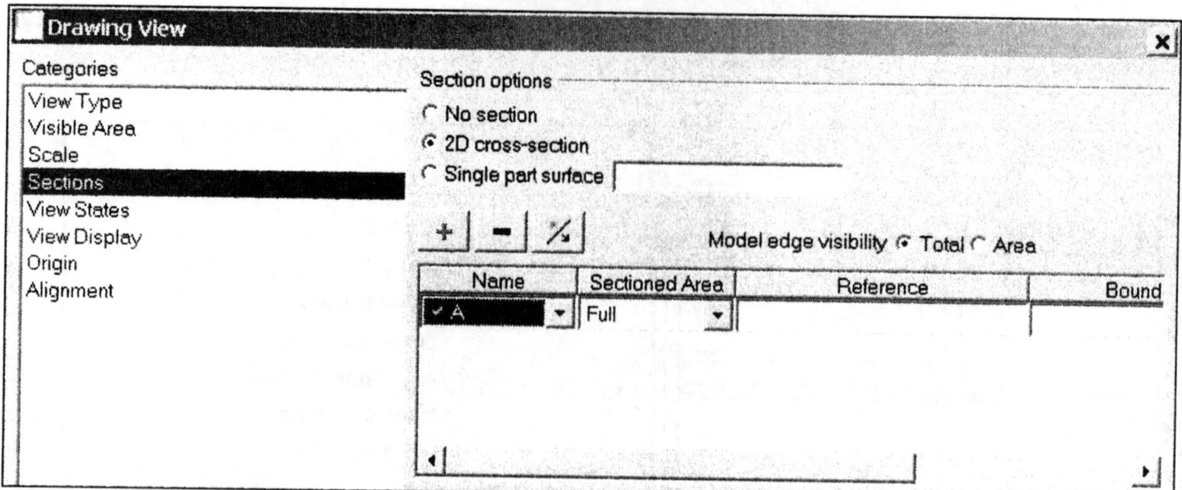

Figure 9.20(f) Section A, Full

Click **Datum planes off** ⇒ **LMB** to deselect ⇒ ⇒ ⇒ Erase ⇒ **Datum Plane** on ⇒ Erase All ⇒ **Yes** ⇒ **Close** ⇒ reposition the views as needed ⇒ **Update the display of all views in the active sheet** ⇒ ⇒ ⇒ ⇒ **MMB** [Fig. 9.20(g)] ⇒ with front view still selected/highlighted, click: **RMB** ⇒ **Add Arrows** [Fig. 9.20(h)] ⇒ pick on the top view [Fig. 9.20(i)]

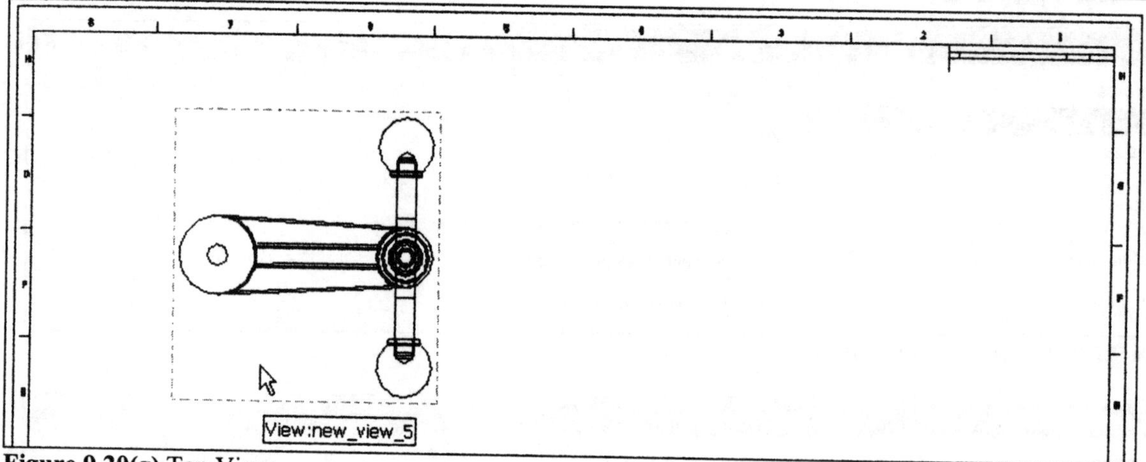

Figure 9.20(g) Top View

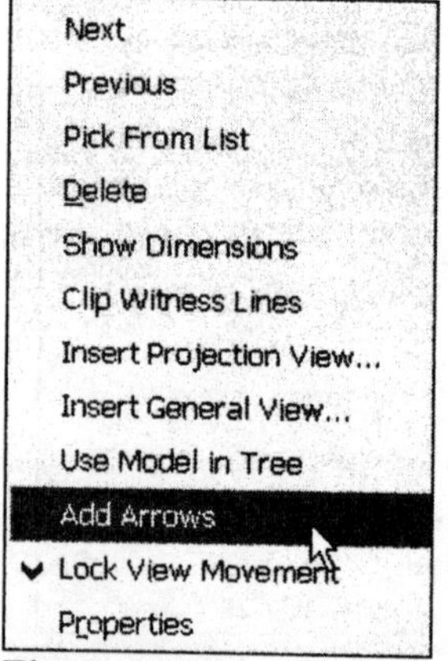

Figure 9.20(h) Add Arrows

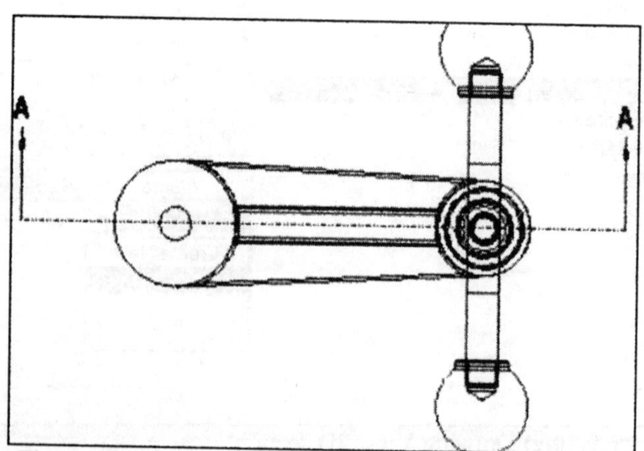

Figure 9.20(i) Top View with Section Cutting Plane

Pro/E provides tools to alter the display of the section views to comply with industry ASME standard practices. Most companies require that the crosshatching on parts in section views of assemblies be "clocked" such that parts that meet do not use the same section lining (crosshatching) spacing and angle. This makes the separation between parts more distinct. First, modify the visibility of the views to remove hidden lines and make the tangent edges dimmed. Next, show all centerlines and clip as needed.

Press and hold down the **Ctrl** key and pick on both views ⇒ click **RMB** with cursor outside of the view outlines ⇒ **Properties** ⇒ Display Style ▾ ⇒ No Hidden [Fig. 9.21(a)] ⇒ Tangent Edges display style ▾ ⇒ Dimmed [Fig. 9.21(b)] ⇒ **Apply** ⇒ **OK**

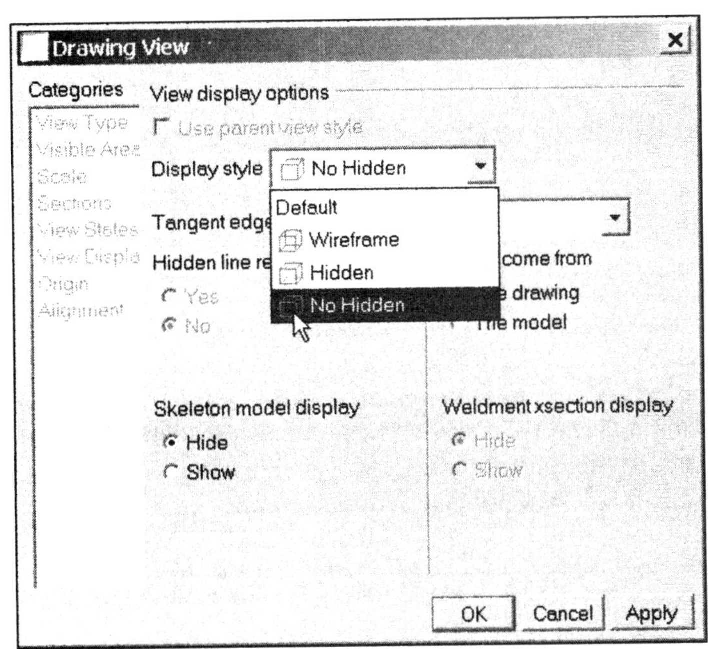

Figure 9.21(a) No Hidden

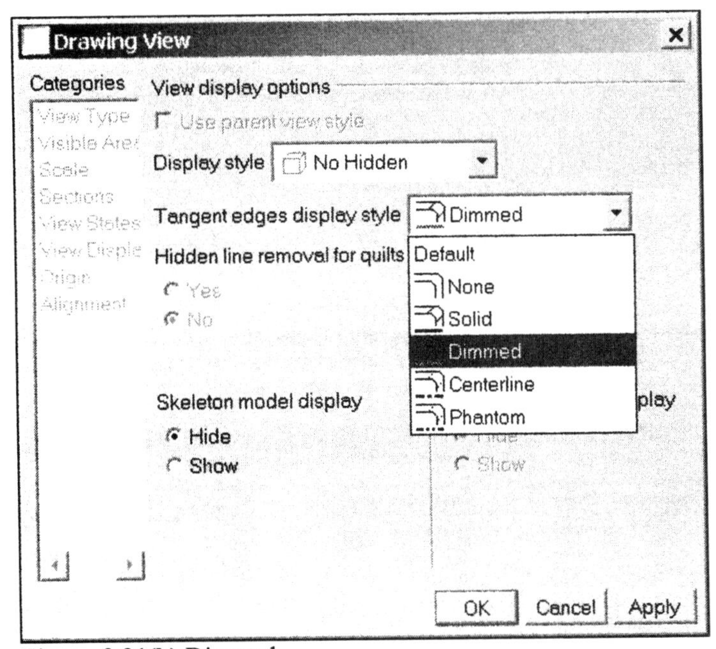

Figure 9.21(b) Dimmed

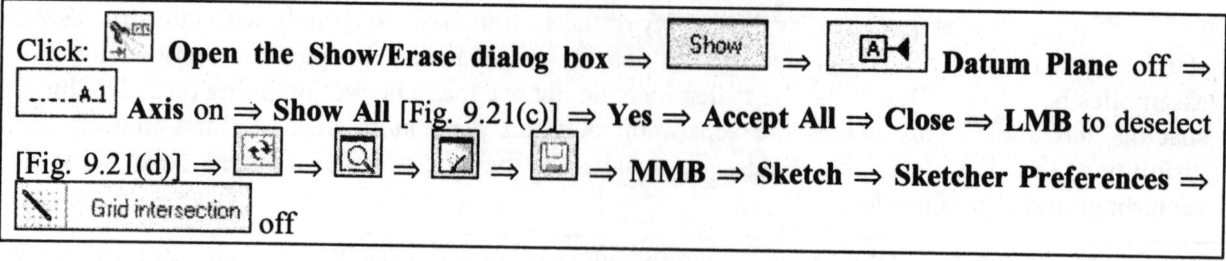

Click: **Open the Show/Erase dialog box** ⇒ Show ⇒ **Datum Plane** off ⇒ Axis on ⇒ **Show All** [Fig. 9.21(c)] ⇒ **Yes** ⇒ **Accept All** ⇒ **Close** ⇒ **LMB** to deselect [Fig. 9.21(d)] ⇒ ⇒ ⇒ ⇒ ⇒ **MMB** ⇒ **Sketch** ⇒ **Sketcher Preferences** ⇒ Grid intersection off

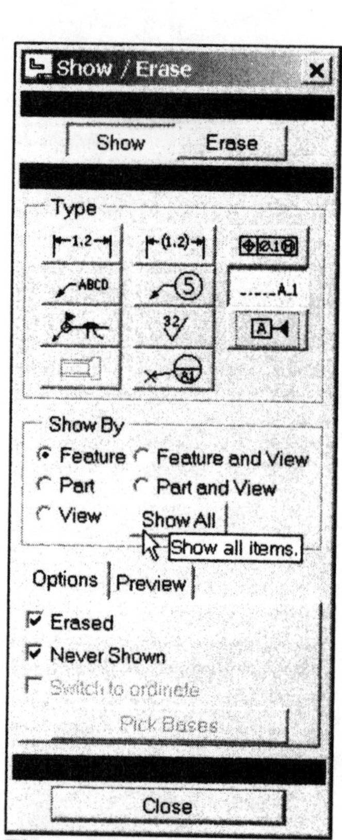

Figure 9.21(c) Show

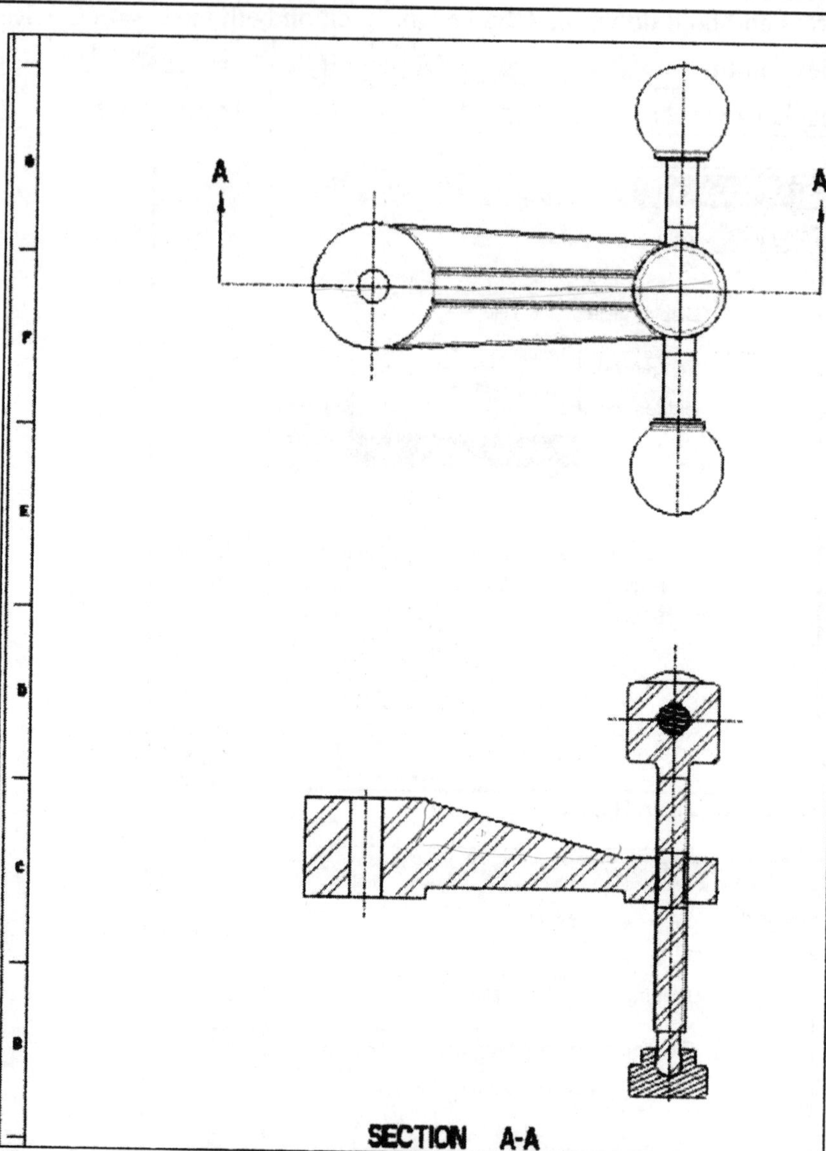

Figure 9.21(d) Hidden Lines Removed, Tangent Edges Dimmed, Centerline Lines Displayed

Double-click on the crosshatching in the front view (Clamp_Stud5 is now active) [Fig. 9.21(e)] ⇒ **Fill** [Fig. 9.21(f)] ⇒ click **Next Xsec** until Clamp_Foot is active [Fig. 9.21(g)] ⇒ **Hatch** ⇒ **Angle** ⇒ **135** [Fig. 9.21(h)]

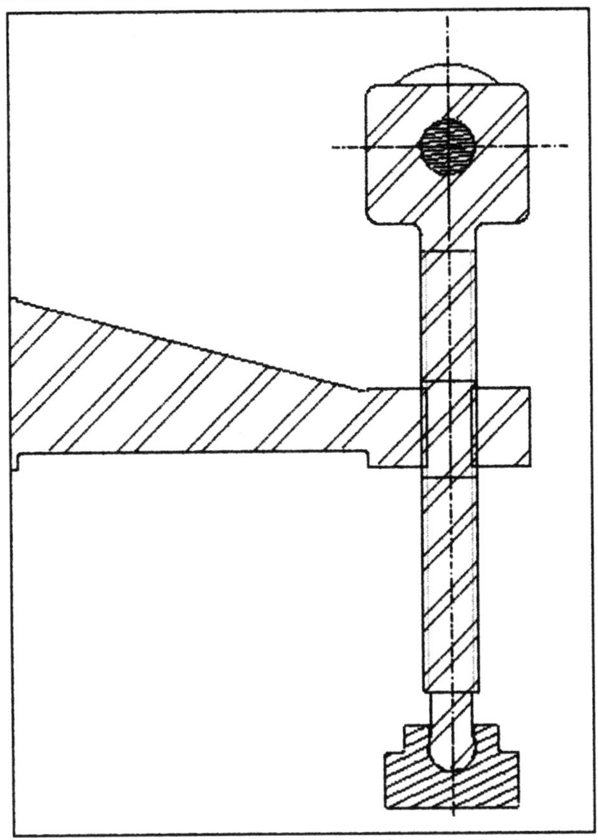

Figure 9.21(e) Clamp_Stud5 is Active

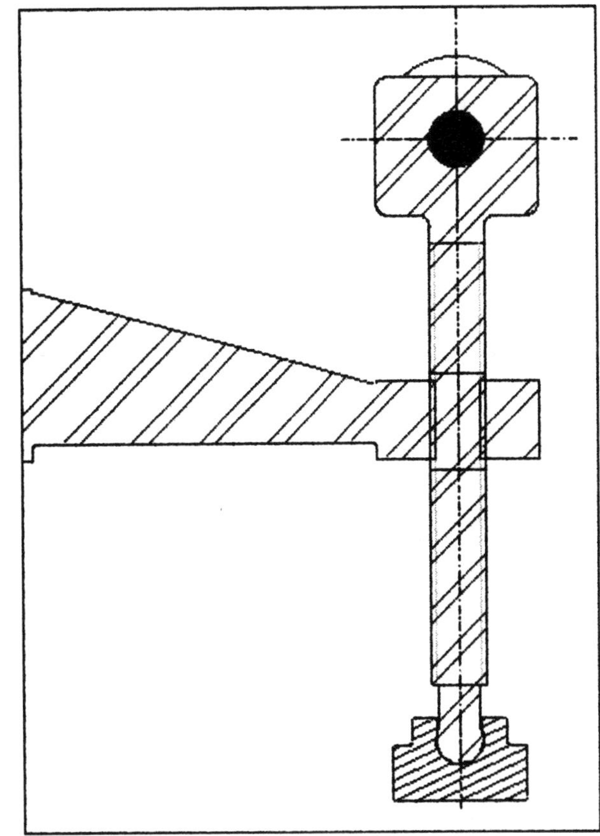

Figure 9.21(f) Clamp_Stud5 Fill Xsec

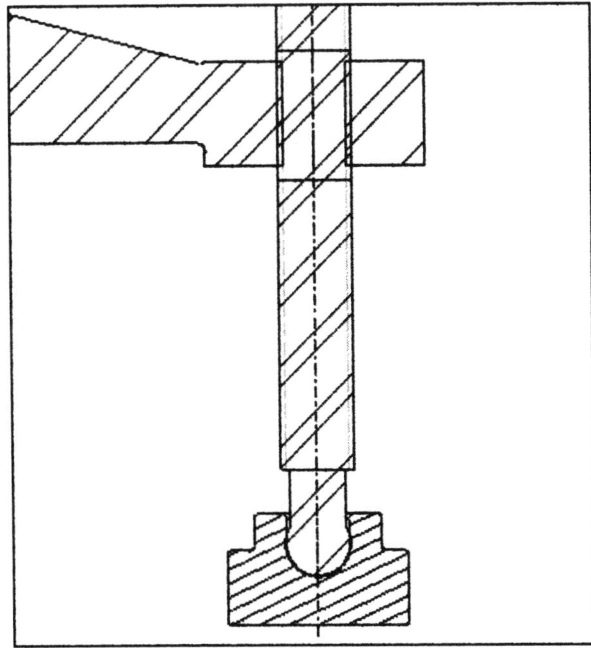

Figure 9.21(g) Clamp_Foot is Active

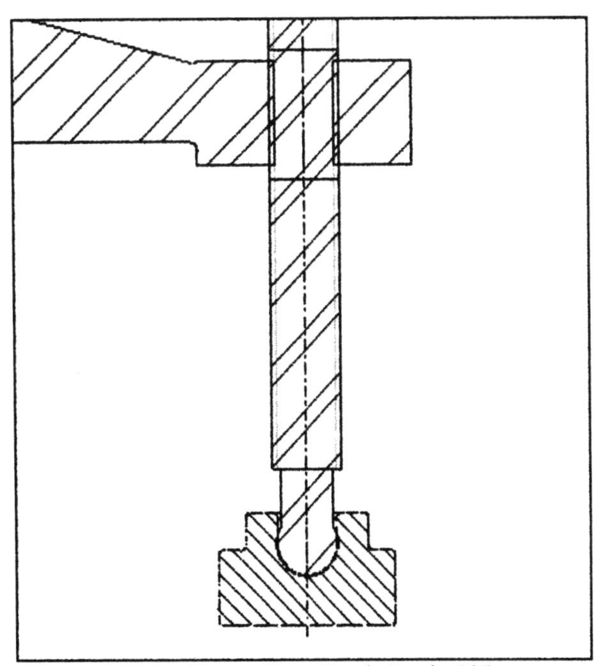

Figure 9.21(h) Clamp_Foot Hatch Angle **135**

Click: **Next Xsec** ⇒ **Next Xsec** Clamp_Arm is active ⇒ **Hatch** ⇒ **Delete line** ⇒ **Spacing** ⇒ **Half** ⇒ **Angle** ⇒ **120** [Fig. 9.21(i)] ⇒ **Prev Xsec** Clamp_Swivel is active ⇒ **Delete line** ⇒ **Spacing** ⇒ **Half** ⇒ **MMB** ⇒ **LMB** ⇒ ⇒ ⇒ [Fig. 9.21(j)] ⇒ ⇒ **MMB**

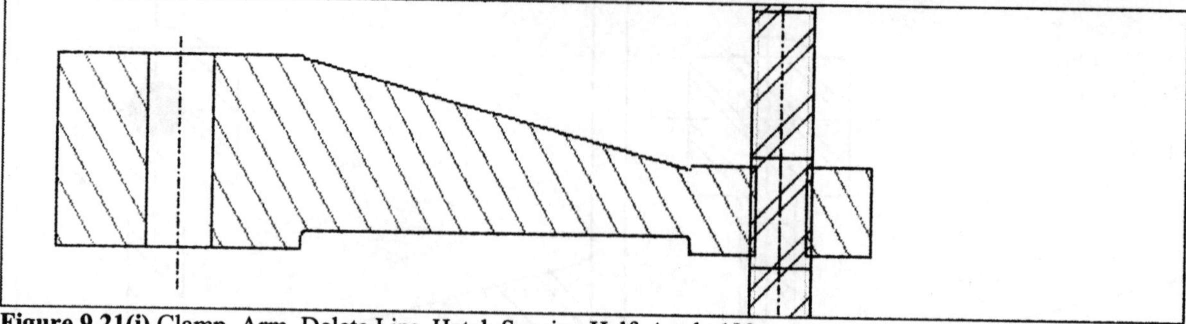

Figure 9.21(i) Clamp_Arm, Delete Line, Hatch Spacing Half, Angle **120**

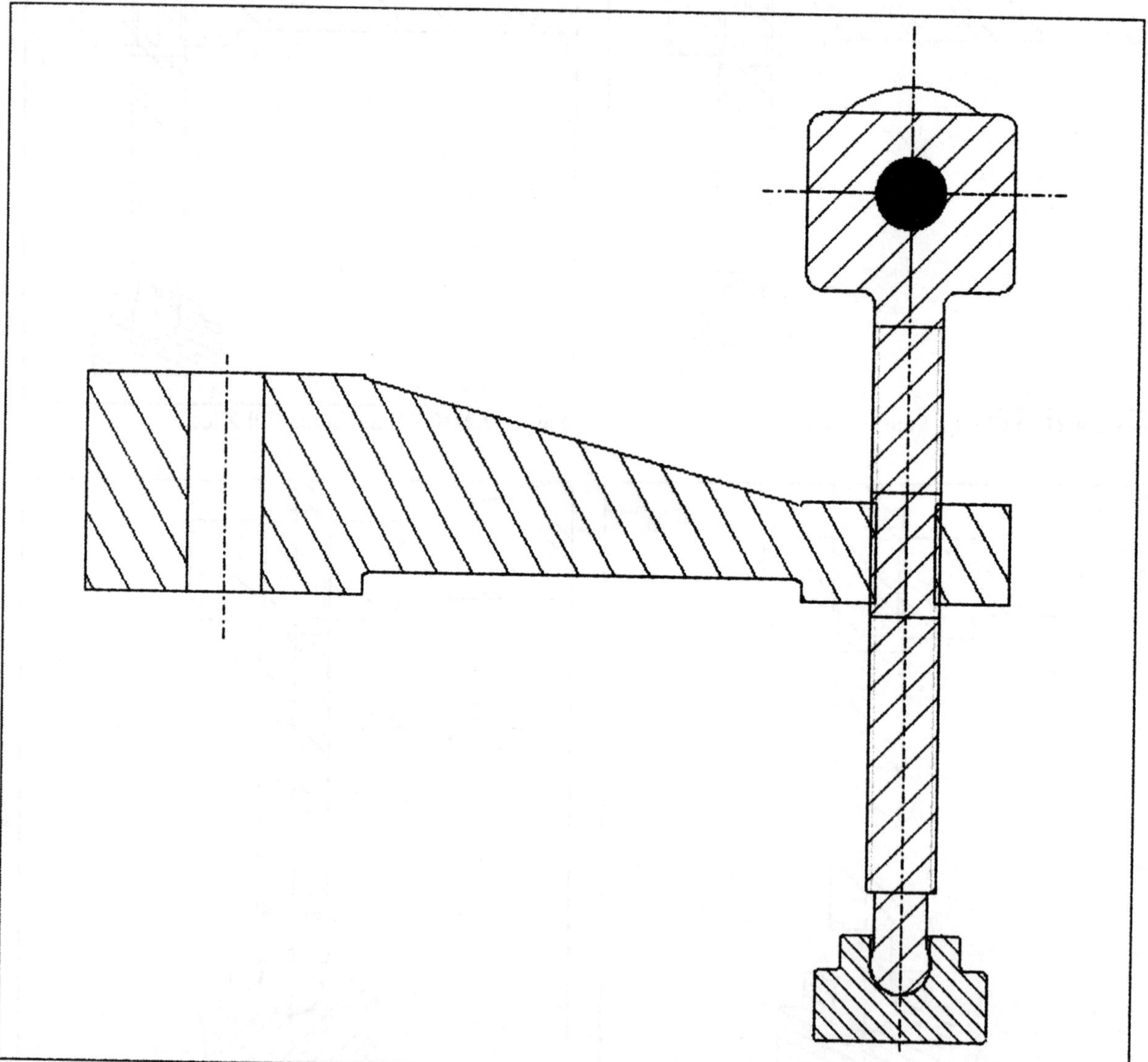

Figure 9.21(j) Completed Cross Section Hatching

To complete the subassembly drawing, you must display the balloons for each component. Show the **Item Balloons** on the drawing. Balloons are displayed in the top view, because it was the first view that was created.

Click: **Table** ⇒ **BOM Balloons** ⇒ **Set Region** ⇒ **Simple** ⇒ pick in the BOM field [Fig. 9.22(a)] ⇒ **Create Balloon** ⇒ **Show All** [Fig. 9.22(b)] ⇒ **MMB** ⇒ press and hold the **Ctrl** key and pick on the balloons for the Clamp_Arm (1), Clamp_Stud5 (5) and Clamp_Foot (4) ⇒ **RMB** ⇒ **Move Item to View** [Fig. 9.22(c)] ⇒ pick in the front view [Fig. 9.22(d)] ⇒ pick on and reposition each balloon as needed [Fig. 9.22(e)] ⇒ ⇒ ⇒ ⇒ ⇒ **MMB**

5	SW101-9STL	.500-13 X 5.00 DOUBLE END STUD	PURCHASED	1
4	SW101-8FT	CLAMP FOOT	NYLON	1
3	SW101-7BA	SWING CLAMP BALL	BLACK PLASTIC	2
2	SW101-6SW	CLAMP SWIVEL	STEEL	1
1	SW101-5AR	CLAMP ARM	STEEL	1
ITEM	PT NUM	DESCRIPTION	MATERIAL	QTY

Figure 9.22(a) Set Region

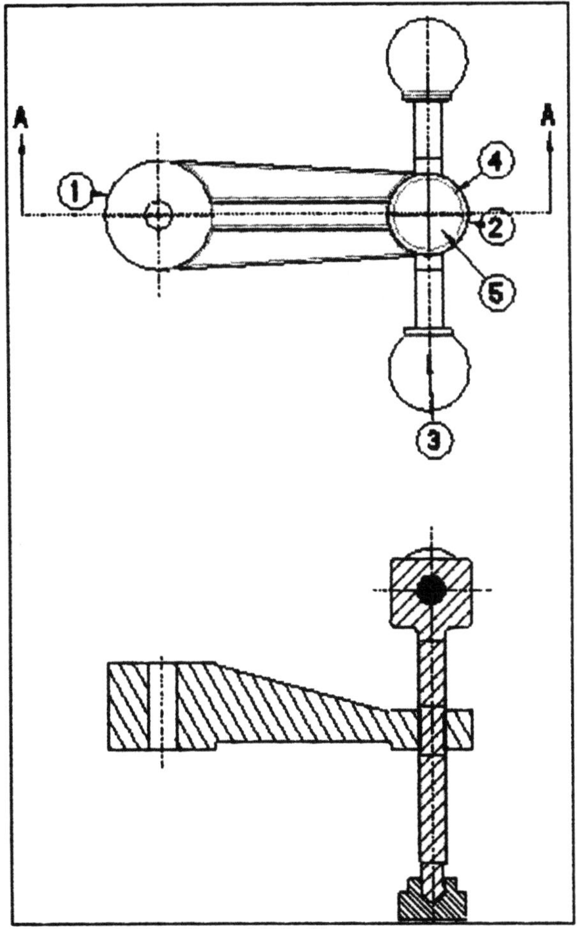

Figure 9.22(b) Show All Balloons

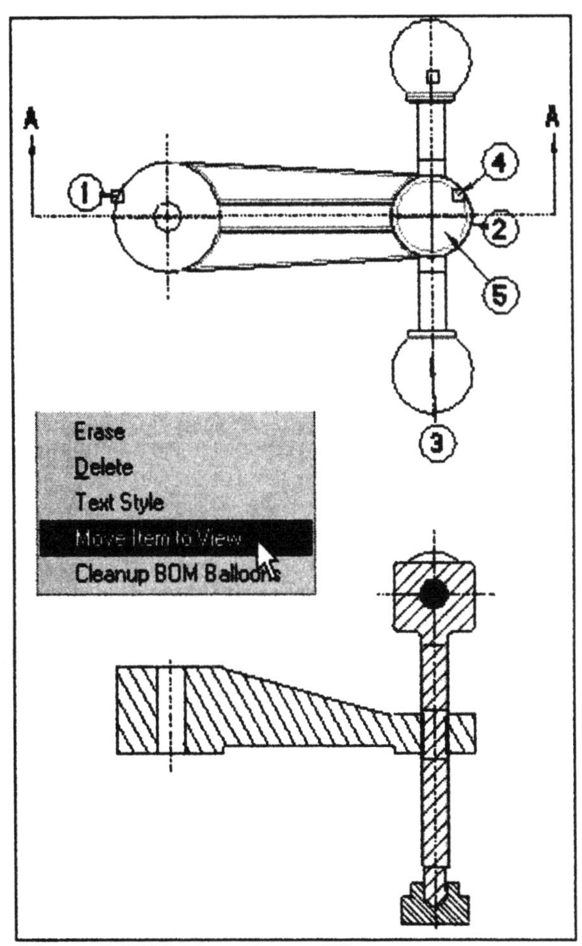

Figure 9.22(c) Move Item to View

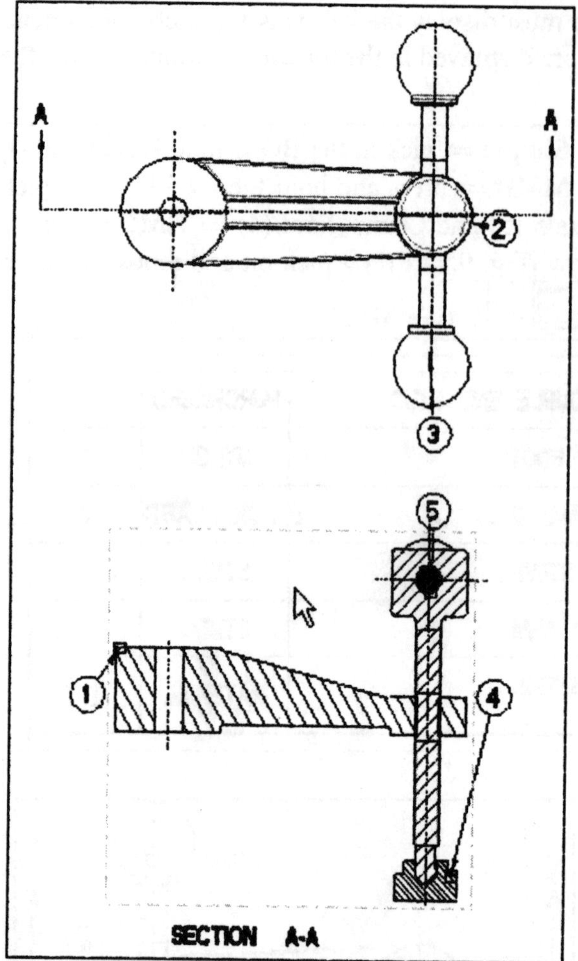

Figure 9.22(d) Balloons Moved to Front View

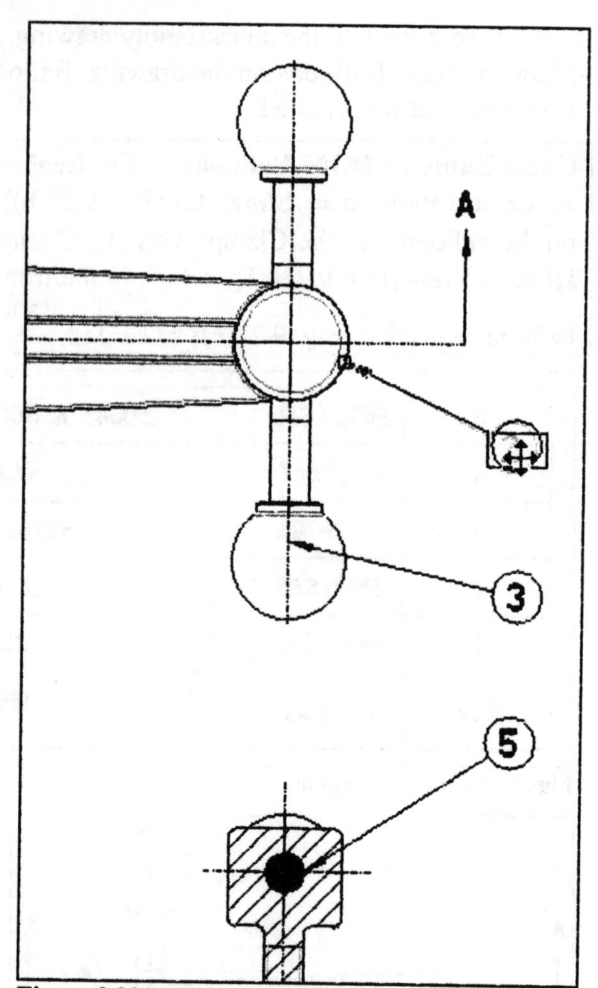

Figure 9.22(e) Reposition Balloons

Pick on balloon **3** (Clamp_Ball) ⇒ **RMB** [Fig. 9.22(f)] ⇒ **Edit Attachment** ⇒ **On Entity** ⇒ pick on the edge of the Clamp_Ball [Fig. 9.22(g)] ⇒ **MMB** ⇒ ⇒ ⇒ ⇒ ⇒ **MMB** ⇒ **File** ⇒ **Delete** ⇒ **Old Versions** ⇒ **MMB** [Fig. 9.22(h)] ⇒ **Window** ⇒ **Close**

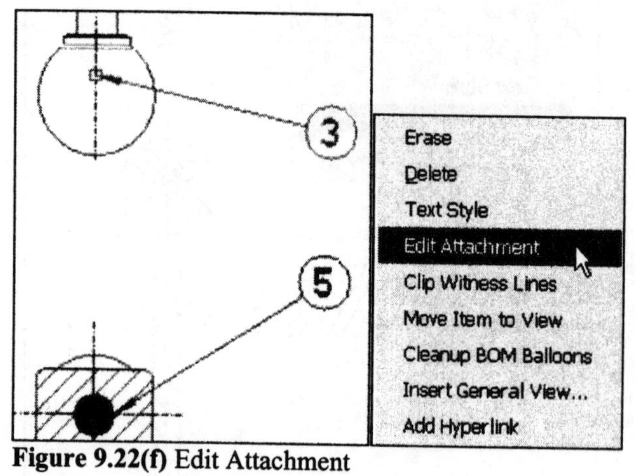

Figure 9.22(f) Edit Attachment

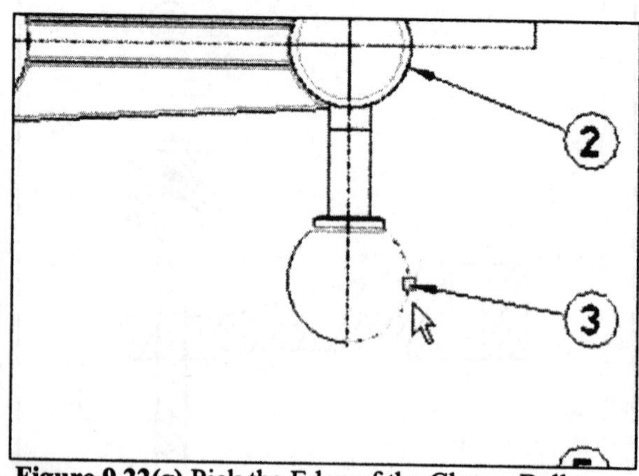

Figure 9.22(g) Pick the Edge of the Clamp_Ball

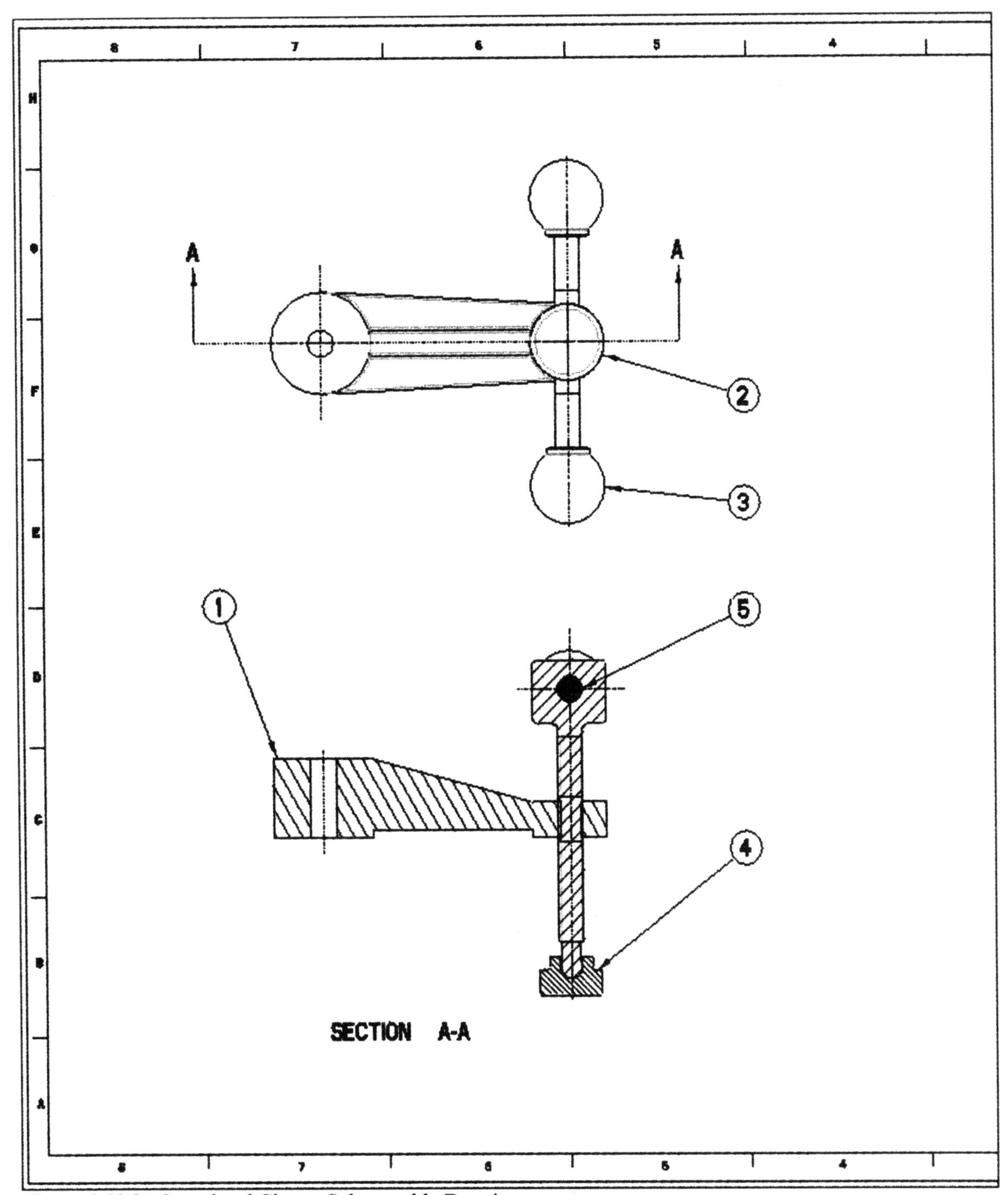

Figure 9.22(h) Completed Clamp_Subassembly Drawing

Create the drawing for the Swing Clamp Assembly. This assembly is composed of the subassembly (in the previous drawing), the plate, the short stud, and the nut. The drawing will use the same format created for the subassembly. Formats are read-only files that can be used as many times as needed.

Click: Create a new object ⇒ Drawing ⇒ Name **CLAMP_ASSEMBLY** ⇒ Use default template ⇒ **OK** ⇒ Default Model **Browse** ⇒ **clamp_assembly.asm** ⇒ **Open** ⇒ Empty with format ⇒ **Browse** ⇒ Format **asm_format_e.frm** ⇒ **Open** ⇒ **OK** ⇒ **Zoom in** on the title block [Fig. 9.23(a)] ⇒ ⇒ ⇒ **RMB** ⇒ **Properties** ⇒ **Drawing Options** ⇒ **Open a configuration file** ⇒ **clamp_asm.dtl** clamp_asm.dtl ⇒ **Open** ⇒ **Apply** ⇒ **Close** ⇒ **MMB**

ITEM	PT NUM	DESCRIPTION	MATERIAL	QTY
8	SW101-9STL	.500-13 X 5.00 DOUBLE END STUD	PURCHASED	1
7	SW101-8FT	CLAMP FOOT	NYLON	1
6	SW101-7BA	SWING CLAMP BALL	BLACK PLASTIC	2
5	SW101-6SW	CLAMP SWIVEL	STEEL	1
4	SW101-5AR	CLAMP ARM	STEEL	1
3	SW100-22FLN	.500-13 X HEX FLANGE NUT	PURCHASED	1
2	SW100-21ST	.500-13 X 3.50 DOUBLE END STUD	PURCHASED	1
1	SW100-20PL	CLAMP PLATE	STEEL	1

TOOL ENGINEERING CO.

A

DRAWN

ISSUED

1.000

CLAMP_ASSEMBLY

SHEET 1 OF 1

Figure 9.23(a) BOM and Title Block

Click: **Datum axes off** ⇒ **Datum points off** ⇒ **Coordinate systems off** ⇒ double-click on **SCALE** in the lower left-hand of the graphics window SCALE : 1.000 TYPE : ASSEM ⇒ Enter value for scale 1.50 ⇒ **MMB** ⇒ **Disallow the movement of drawing views with the mouse** unlock ⇒ **Insert a drawing view of the active model** ⇒ **MMB** ⇒ Select CENTER POINT for drawing view. top view ⇒ select Reference 1 Front **CL_ASM_FRONT** ⇒ select Reference 2 Top **CL_ASM_TOP** ⇒ **OK** ⇒ **Insert a drawing view of the active model** ⇒ **Projection** ⇒ **Full View** ⇒ **Section** ⇒ **MMB** ⇒ **Full** ⇒ **Total Xsec** ⇒ **MMB** ⇒ Select CENTER POINT for drawing view. select below the view previously created ⇒ **Create** ⇒ **Planar** ⇒ **Single** ⇒ **MMB** ⇒ type **A** ⇒ **MMB** Select or create an assembly datum ⇒ **Datum planes on** ⇒ ⇒ pick on **CL_ASM_TOP** ⇒ Pick a view for arrows where the section is perp. pick on the top view ⇒ **LMB** ⇒ **Datum planes off**

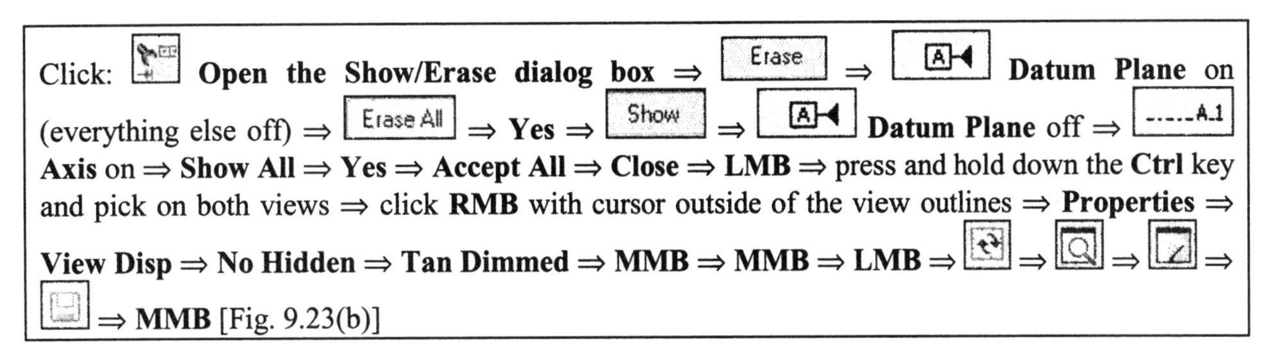

Click: Open the **Show/Erase dialog box** ⇒ Erase ⇒ **Datum Plane** on (everything else off) ⇒ Erase All ⇒ **Yes** ⇒ Show ⇒ **Datum Plane** off ⇒ **Axis** on ⇒ **Show All** ⇒ **Yes** ⇒ **Accept All** ⇒ **Close** ⇒ **LMB** ⇒ press and hold down the **Ctrl** key and pick on both views ⇒ click **RMB** with cursor outside of the view outlines ⇒ **Properties** ⇒ **View Disp** ⇒ **No Hidden** ⇒ **Tan Dimmed** ⇒ **MMB** ⇒ **MMB** ⇒ **LMB** ⇒ ⇒ ⇒ ⇒ ⇒ **MMB** [Fig. 9.23(b)]

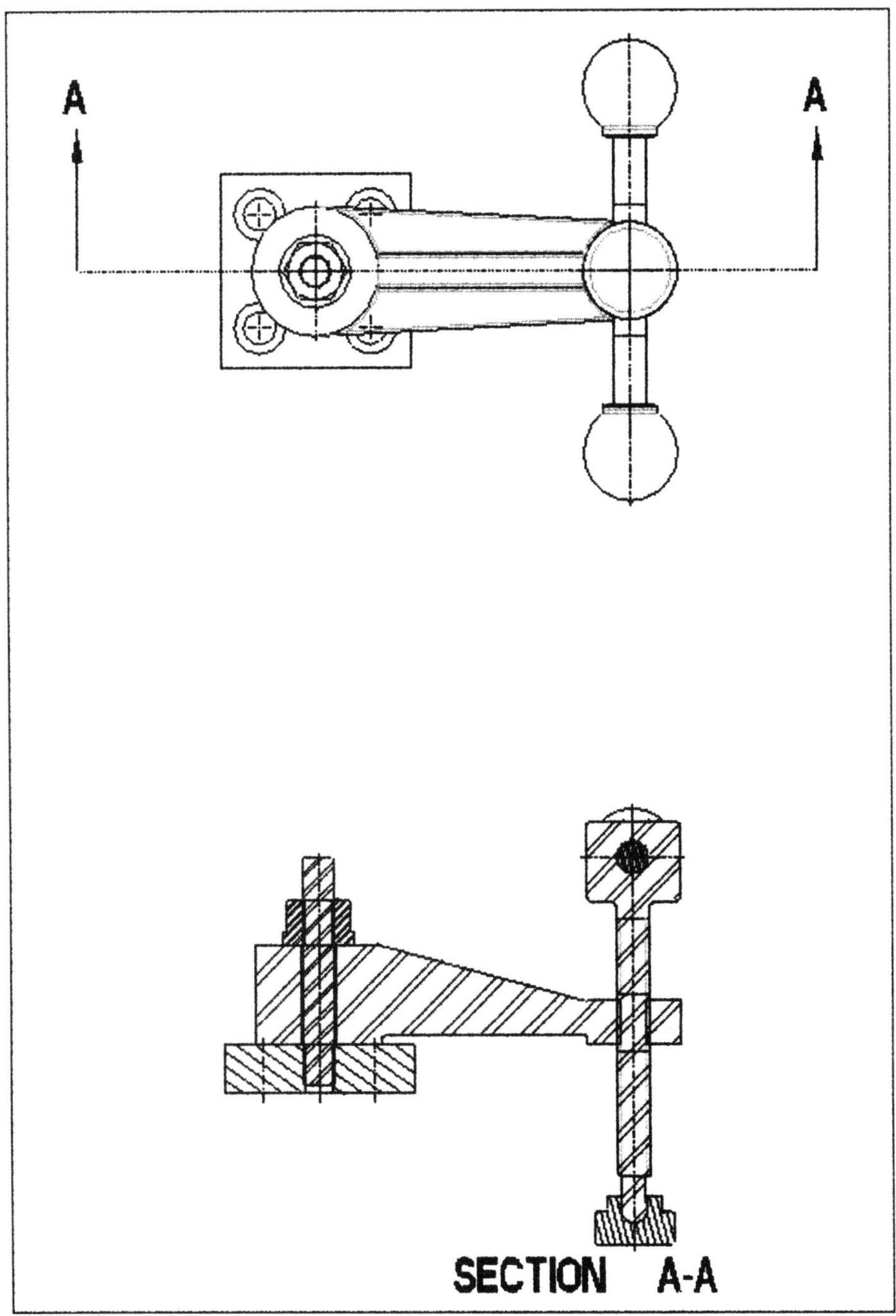

Figure 9.23(b) Assembly Drawing Views

Click: **Table** ⇒ **Repeat Region** ⇒ **Attributes** ⇒ pick in the BOM field [Fig. 9.24(a)] ⇒ **Flat** ⇒ **MMB** ⇒ **MMB** ⇒ **MMB** [Fig. 9.24(b)] ⇒ **Table** ⇒ **BOM Balloons** ⇒ **Set Region** ⇒ pick in the BOM field ⇒ **Create Balloon** ⇒ **Show All** ⇒ **MMB** [Fig. 9.24(c)]

8	SW101-9STL	.500-13 X 5.00 DOUBLE END STUD	PURCHASED	1
7	SW101-8FT	CLAMP FOOT	NYLON	1
6	SW101-7BA	SWING CLAMP BALL	BLACK PLASTIC	2
5	SW101-6SW	CLAMP SWIVEL	STEEL	1
4	SW101-5AR	CLAMP ARM	STEEL	1
3	SW100-22FLN	.500-13 X HEX FLANGE NUT	PURCHASED	1
2	SW100-21ST	.500-13 X 3.50 DOUBLE END STUD	PURCHASED	1
1	SW100-20PL	CLAMP PLATE	STEEL	1
ITEM	PT NUM	DESCRIPTION	MATERIAL	QTY

Figure 9.24(a) Showing with BOM Attribute Recursive

3	SW100-22FLN	.500-13 X HEX FLANGE NUT	PURCHASED	1
2	SW100-21ST	.500-13 X 3.50 DOUBLE END STUD	PURCHASED	1
1	SW100-20PL	CLAMP PLATE	STEEL	1
ITEM	PT NUM	DESCRIPTION	MATERIAL	QTY

TOOL ENGINEERING CO.

Figure 9.24(b) Showing with BOM Attribute Flat

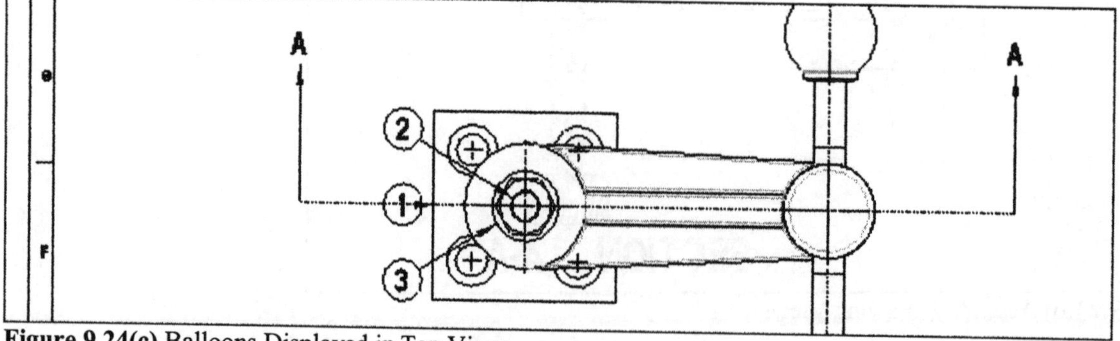

Figure 9.24(c) Balloons Displayed in Top View

While pressing the **Ctrl** key, pick on all three balloons ⇒ **RMB** ⇒ **Move Item to View** ⇒ pick in the front view [Fig. 9.24(d)] ⇒ pick on and reposition each balloon as needed ⇒ pick on balloon **2** [Fig. 9.24(e)] ⇒ **RMB** ⇒ **Edit Attachment** ⇒ **On Entity** ⇒ pick on edge [Fig. 9.24(f)] ⇒ **MMB** ⇒ **LMB**

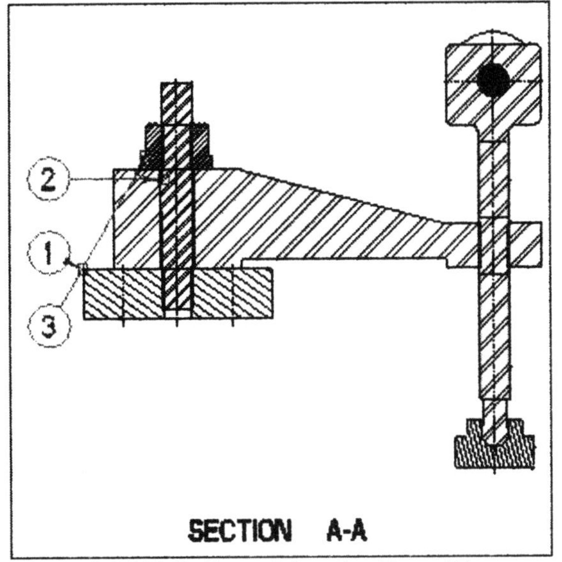

Figure 9.24(d) Balloons Moved to Front View

Figure 9.24(e) Reposition Balloons

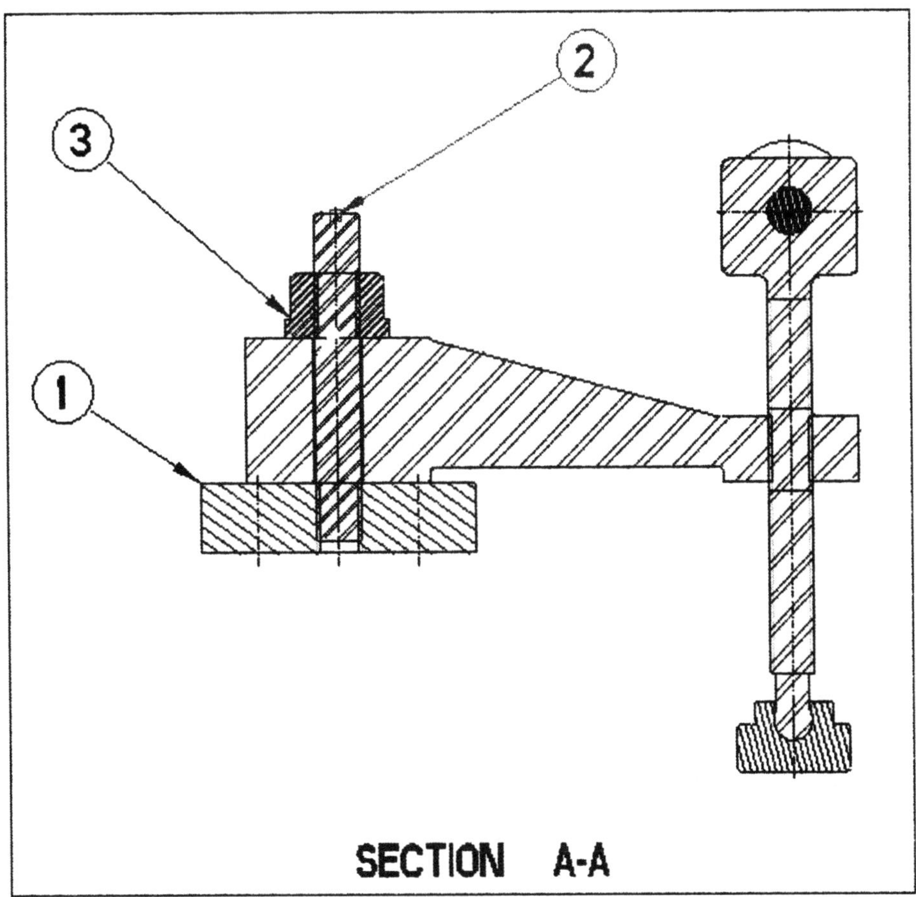

Figure 9.24(f) Attachment Changed for Balloon 2

Most companies (and as per drafting standards) require that round purchased items, such as nuts, bolts, studs, springs, and die pins be excluded from sectioning even when the section cutting plane passes through them. Remove the section lining (crosshatching) from the Clamp_Flange_Nut and the Clamp_Stud35 in the front section view.

> Double-click on the crosshatching in the front view [Fig. 9.25(a)]. The Clamp_Flange_Nut is now active [Fig. 9.25(b)]. ⇒ **Excl Comp** to eliminate Xsec of Clamp_Flange_Nut [Fig. 9.25(c)] ⇒ **Next Xsec** Clamp_Stud35 is now active ⇒ **Excl Comp** to eliminate Xsec of Clamp_Stud35 [Fig. 9.25(d)] ⇒ **MMB** ⇒ **LMB** ⇒ ⇒ ⇒ ⇒ **MMB**

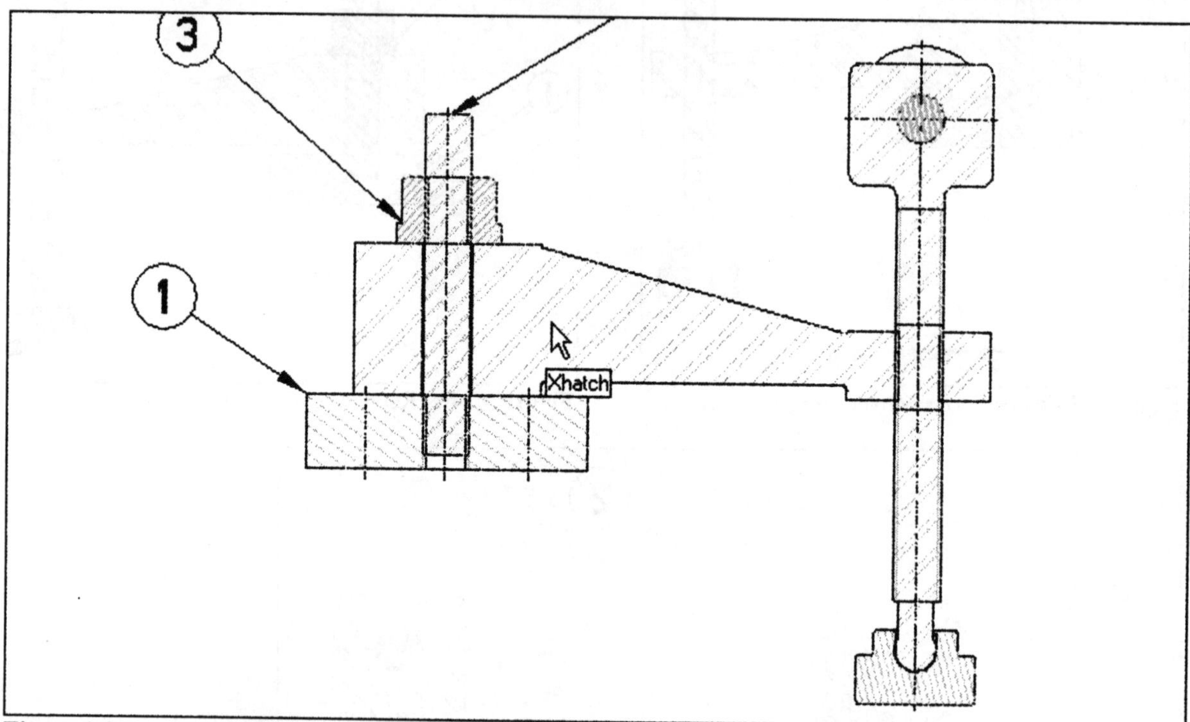

Figure 9.25(a) Double-Click on the Cross Section Lining

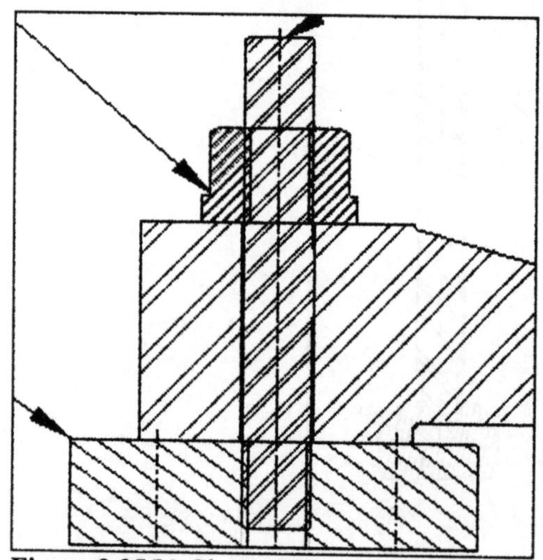

Figure 9.25(b) Clamp_Flange_Nut Active

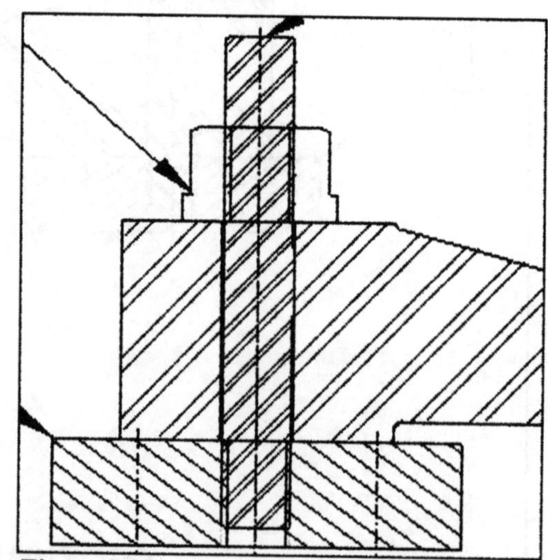

Figure 9.25(c) Section Lining Removed

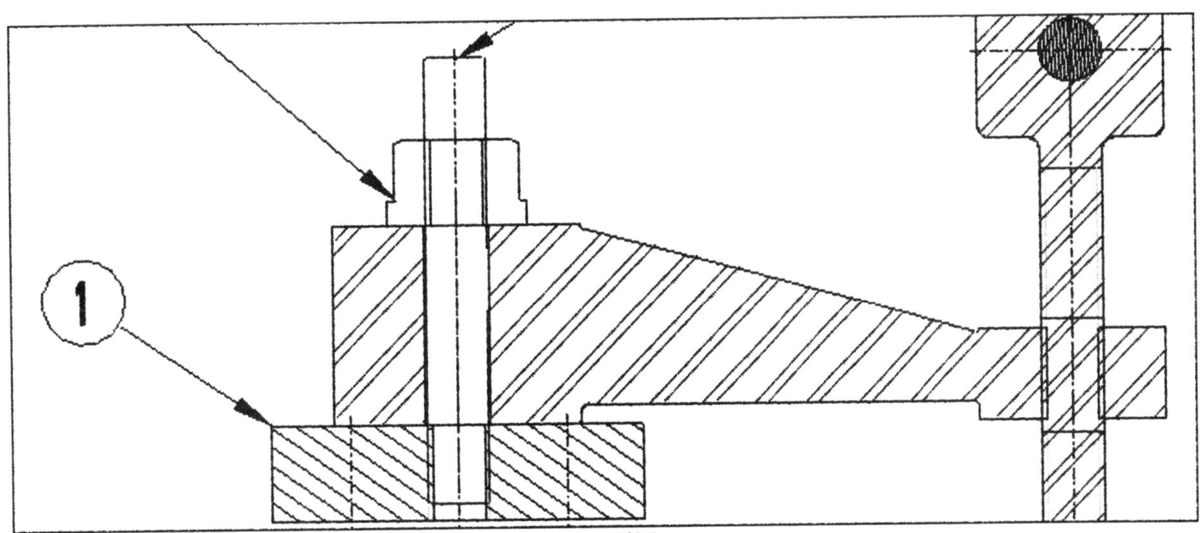

Figure 9.25(d) Section Lining Removed from Clamp_Stud35

Double-click on the crosshatching, the Clamp_Flange_Nut is now active ⇒ **Next Xsec** ⇒ **Next Xsec** Clamp_Stud5 is now active ⇒ **Fill** ⇒ **Next Xsec** ⇒ **Next Xsec** Clamp_Swivel is now active ⇒ **Delete line** ⇒ **Next Xsec** Clamp_Arm is now active ⇒ **Delete line** ⇒ **Angle** ⇒ **120** ⇒ **Next Xsec** Clamp_Plate is now active ⇒ **Angle** ⇒ **45** (Fig. 9.26) ⇒ **MMB** ⇒ **LMB** ⇒ ⇒ ⇒ ⇒ ⇒ **MMB**

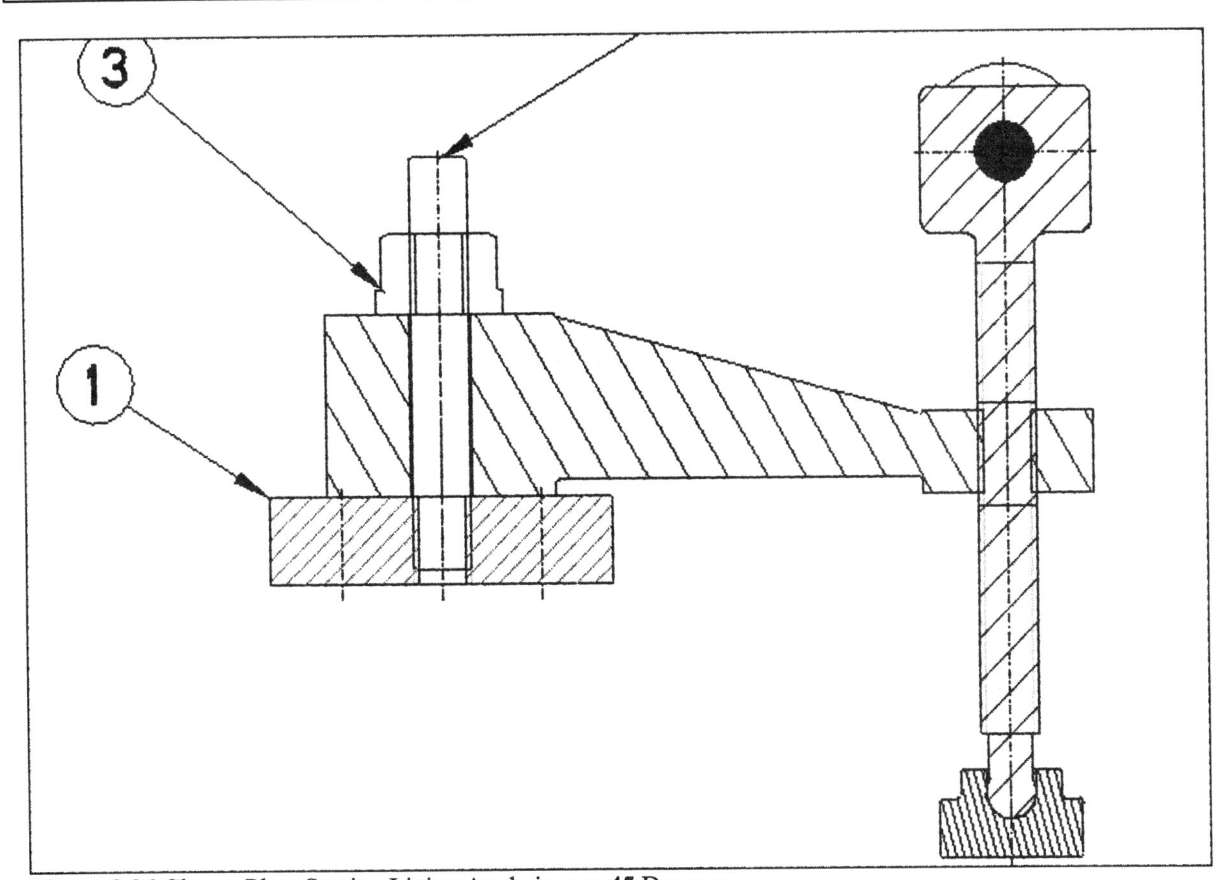

Figure 9.26 Clamp_Plate Section Lining Angle is now **45** Degrees

The numbering of the components in assemblies may need to be different from the default setting. To change the balloon numbering, you must use **Fix Index**.

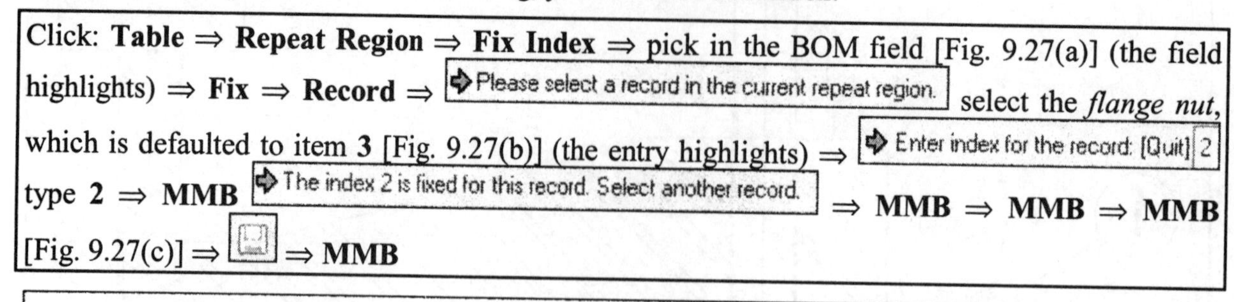

Click: **Table** ⇒ **Repeat Region** ⇒ **Fix Index** ⇒ pick in the BOM field [Fig. 9.27(a)] (the field highlights) ⇒ **Fix** ⇒ **Record** ⇒ [Please select a record in the current repeat region.] select the *flange nut*, which is defaulted to item **3** [Fig. 9.27(b)] (the entry highlights) ⇒ [Enter index for the record: [Quit] 2] type **2** ⇒ **MMB** [The index 2 is fixed for this record. Select another record.] ⇒ **MMB** ⇒ **MMB** ⇒ **MMB** [Fig. 9.27(c)] ⇒ [icon] ⇒ **MMB**

3	SW100-22FLN	.500-13 X HEX FLANGE NUT	PURCHASED	1
2	SW100-21ST	.500-13 X 3.50 DOUBLE END STUD	PURCHASED	1
1	SW100-20PL	CLAMP PLATE	STEEL	1
ITEM	PT NUM	DESCRIPTION	MATERIAL	QTY

Figure 9.27(a) Pick in the BOM Field

3	SW100-22FLN	.500-13 X HEX FLANGE NUT	PURCHASED	1
2	SW100-21ST	.500-13 X 3.50 DOUBLE END STUD	PURCHASED	1
1	SW100-20PL	CLAMP PLATE	STEEL	1
ITEM	PT NUM	DESCRIPTION	MATERIAL	QTY

Figure 9.27(b) Pick in the Clamp_Flange_Nut Table Cell

3	SW100-21ST	.500-13 X 3.50 DOUBLE END STUD	PURCHASED	1
2	SW100-22FLN	.500-13 X HEX FLANGE NUT	PURCHASED	1
1	SW100-20PL	CLAMP PLATE	STEEL	1
ITEM	PT NUM	DESCRIPTION	MATERIAL	QTY

TOOL ENGINEERING CO.

A

DRAWN

ISSUED

1500

CLAMP_ASSEMBLY

SHEET 1 OF 1

2

1

Figure 9.27(c) The Clamp_Flange_Nut is Now Listed Second

Exploded Swing Clamp Assembly Drawings

The process required to place an exploded view on a drawing is similar to adding assembly orthographic views. The BOM will display all components on this sheet. You will be required to fix the BOM sequence and manually create balloons.

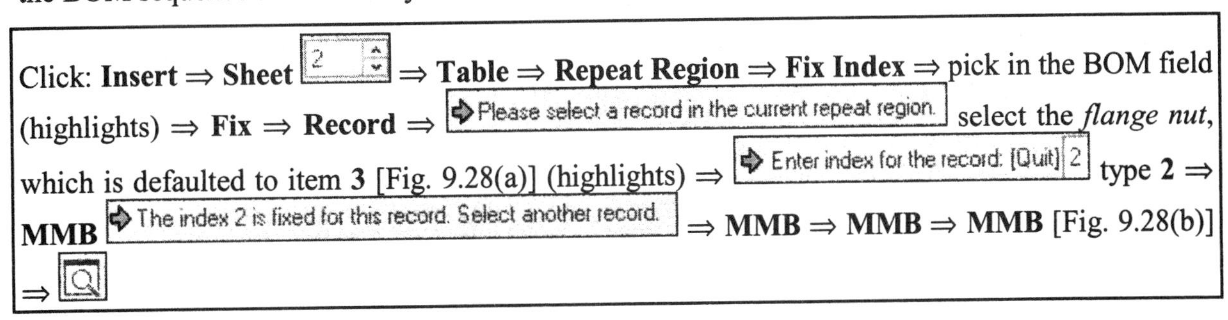

Click: **Insert** ⇒ **Sheet** 2 ⇒ **Table** ⇒ **Repeat Region** ⇒ **Fix Index** ⇒ pick in the BOM field (highlights) ⇒ **Fix** ⇒ **Record** ⇒ Please select a record in the current repeat region. select the *flange nut*, which is defaulted to item **3** [Fig. 9.28(a)] (highlights) ⇒ Enter index for the record: [Quit] 2 type **2** ⇒ **MMB** The index 2 is fixed for this record. Select another record. ⇒ **MMB** ⇒ **MMB** ⇒ **MMB** [Fig. 9.28(b)] ⇒

8	SW101-9STL	.500-13 X 5.00 DOUBLE END STUD	PURCHASED	1
7	SW101-8FT	CLAMP FOOT	NYLON	1
6	SW101-7BA	SWING CLAMP BALL	BLACK PLASTIC	2
5	SW101-6SW	CLAMP SWIVEL	STEEL	1
4	SW101-5AR	CLAMP ARM	STEEL	1
3	SW100-22FLN	.500-13 X HEX FLANGE NUT	PURCHASED	1
2	SW100-21ST	.500-13 X 3.50 DOUBLE END STUD	PURCHASED	1
1	SW100-20PL	CLAMP PLATE	STEEL	1
ITEM	PT NUM	DESCRIPTION	MATERIAL	QTY

Figure 9.28(a) Pick in the Flange Nut Table Cell on Sheet 2

8	SW101-9STL	.500-13 X 5.00 DOUBLE END STUD	PURCHASED	1
7	SW101-8FT	CLAMP FOOT	NYLON	1
6	SW101-7BA	SWING CLAMP BALL	BLACK PLASTIC	2
5	SW101-6SW	CLAMP SWIVEL	STEEL	1
4	SW101-5AR	CLAMP ARM	STEEL	1
3	SW100-21ST	.500-13 X 3.50 DOUBLE END STUD	PURCHASED	1
2	SW100-22FLN	.500-13 X HEX FLANGE NUT	PURCHASED	1
1	SW100-20PL	CLAMP PLATE	STEEL	1
ITEM	PT NUM	DESCRIPTION	MATERIAL	QTY

Figure 9.28(b) On Sheet 2 the Flange Nut is Now Listed Second

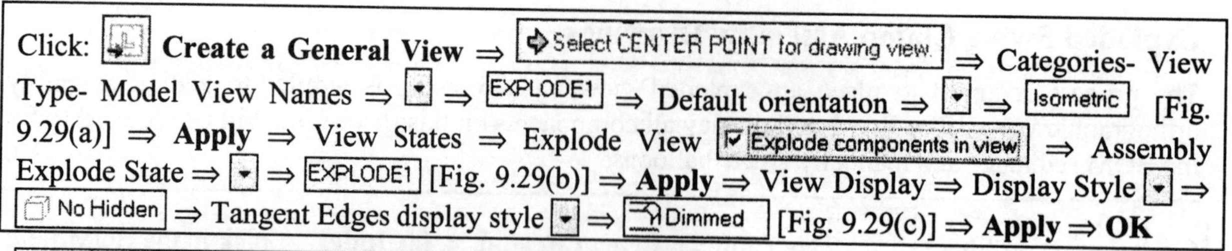

Click: Create a General View ⇒ Select CENTER POINT for drawing view. ⇒ Categories- View Type- Model View Names ⇒ ⇒ EXPLODE1 ⇒ Default orientation ⇒ ⇒ Isometric [Fig. 9.29(a)] ⇒ **Apply** ⇒ View States ⇒ Explode View Explode components in view ⇒ Assembly Explode State ⇒ ⇒ EXPLODE1 [Fig. 9.29(b)] ⇒ **Apply** ⇒ View Display ⇒ Display Style ⇒ No Hidden ⇒ Tangent Edges display style ⇒ Dimmed [Fig. 9.29(c)] ⇒ **Apply** ⇒ **OK**

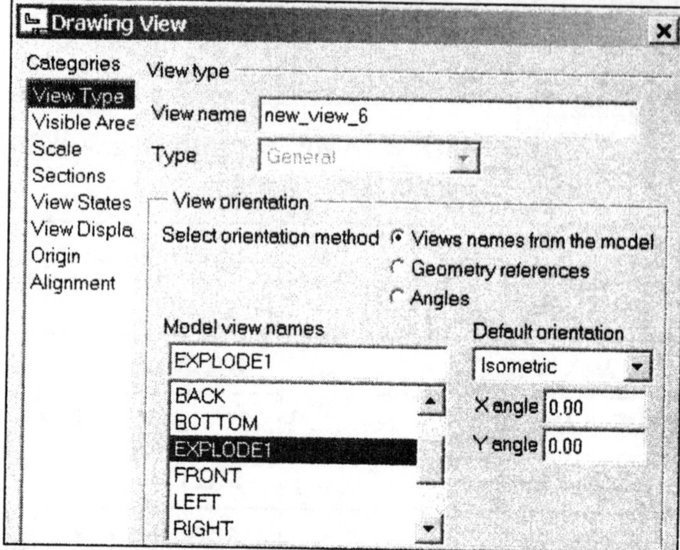

Figure 9.29(a) View State

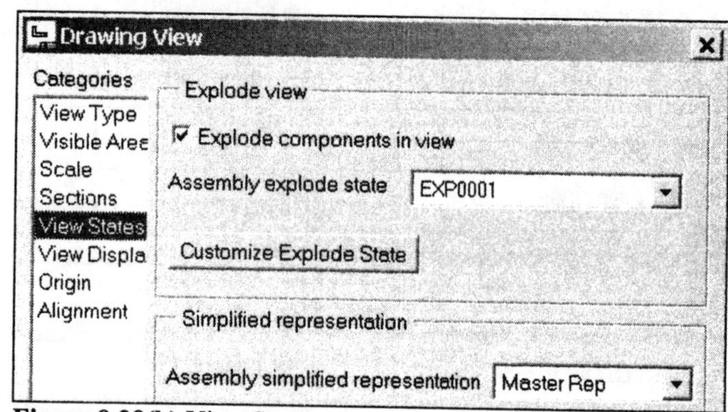

Figure 9.29(b) View States

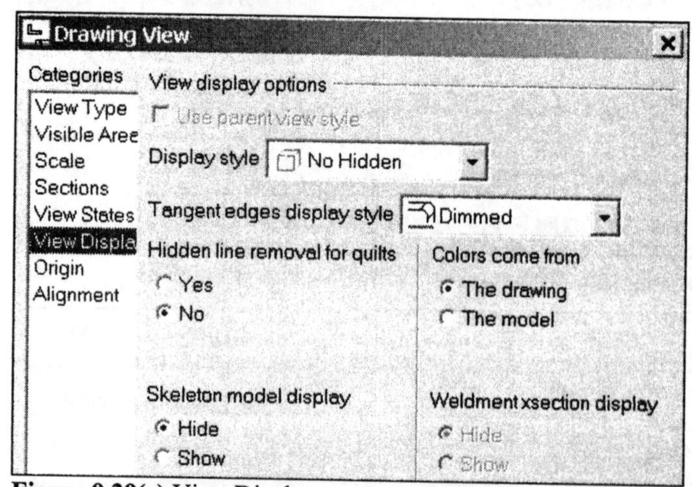

Figure 9.29(c) View Display

Click: **Disallow the movement of drawing views with the mouse** unlock ⇒ adjust and position the view as needed [Fig. 9.29(d)]

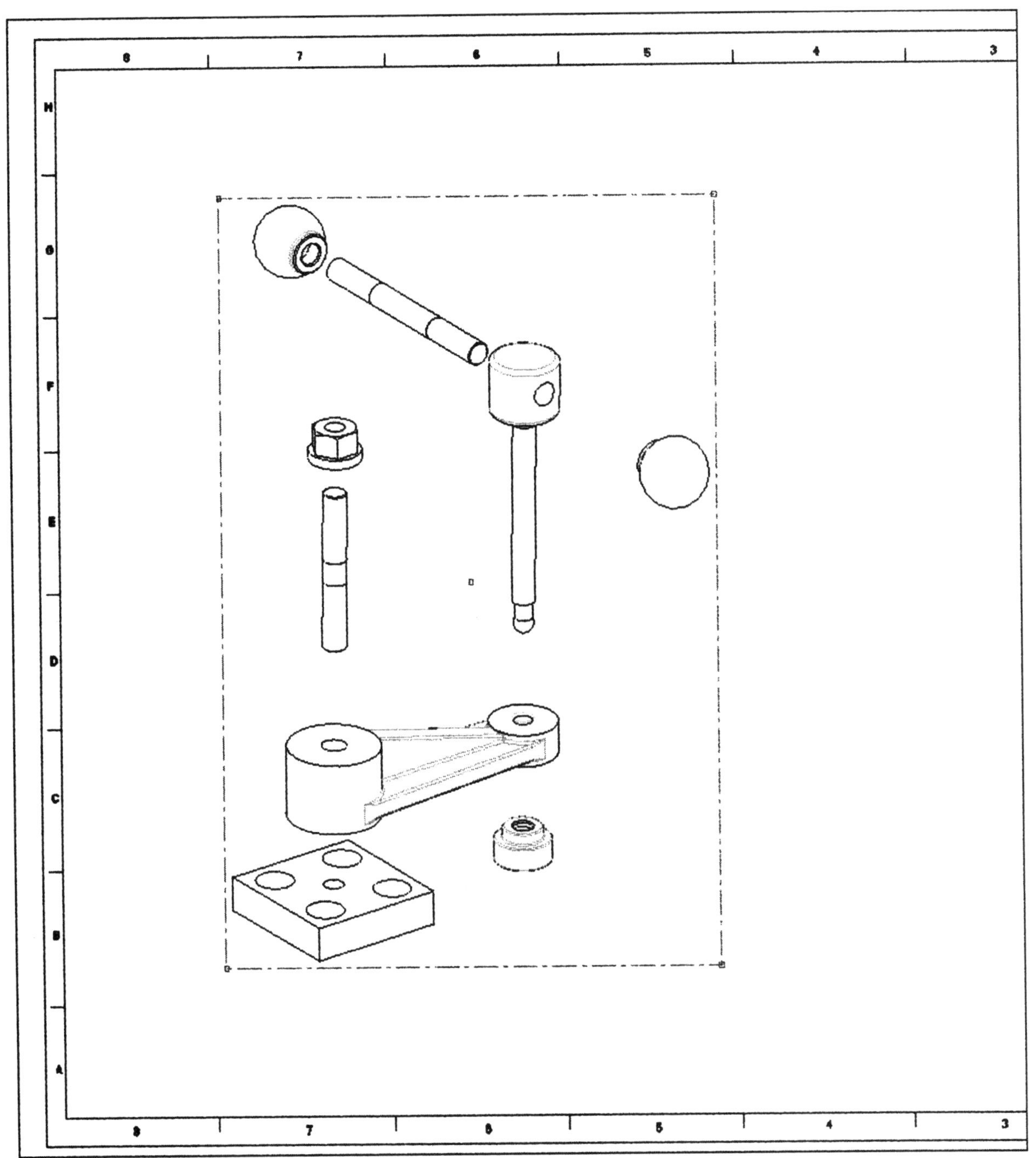

Figure 9.29(d) Repositioning the Exploded View

Click: **Open the Show/Erase dialog box ⇒** Show ⇒ -----A.1 **Axis ⇒ Show All ⇒ Yes ⇒ Accept All ⇒ Close ⇒ LMB** [Fig. 9.29(e)] ⇒ ⇒ ⇒ ⇒ ⇒ **MMB**

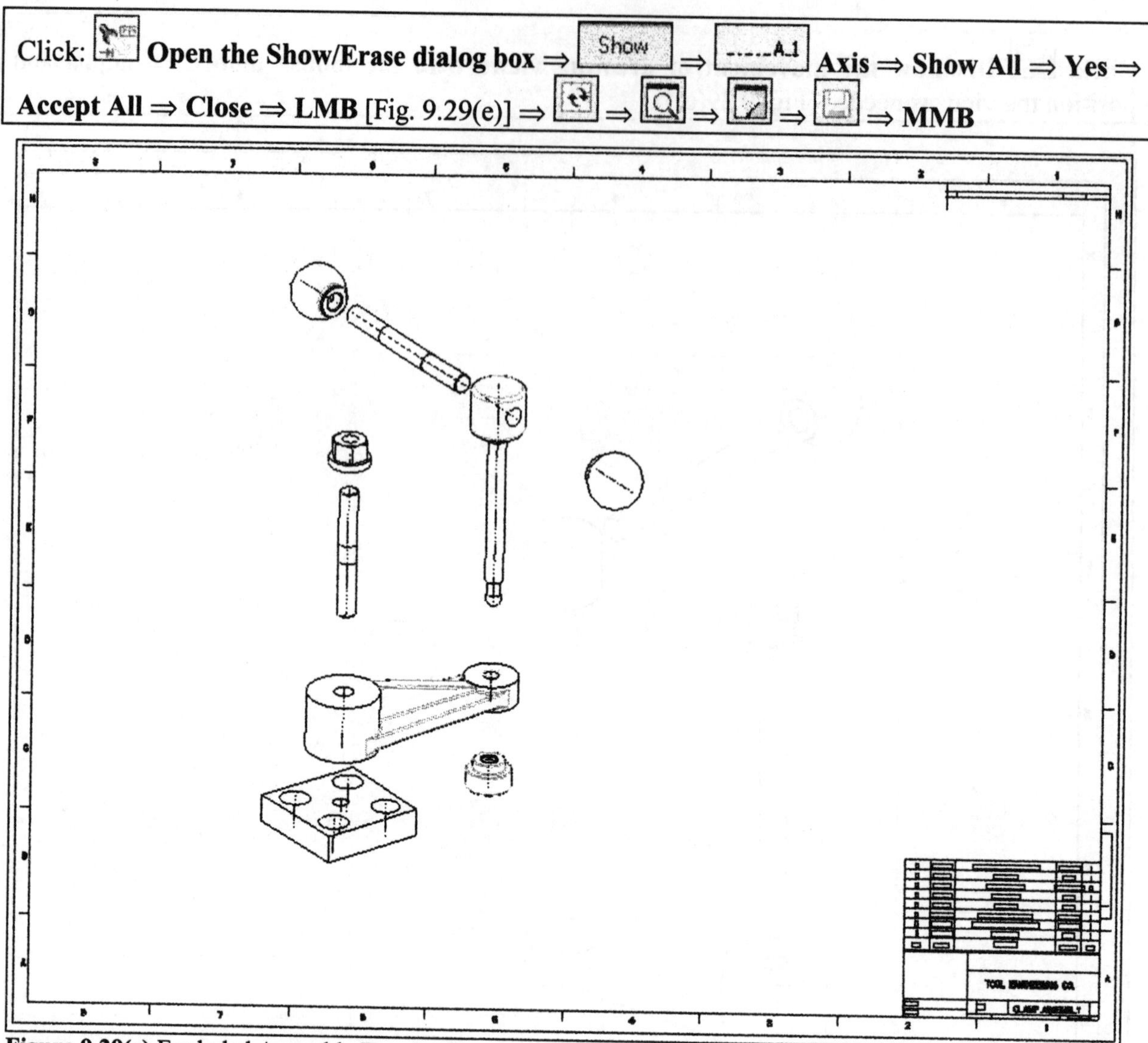

Figure 9.29(e) Exploded Assembly Drawing

Clip each centerline (axis) to extend between components that are in line [Fig. 9.29(f)]

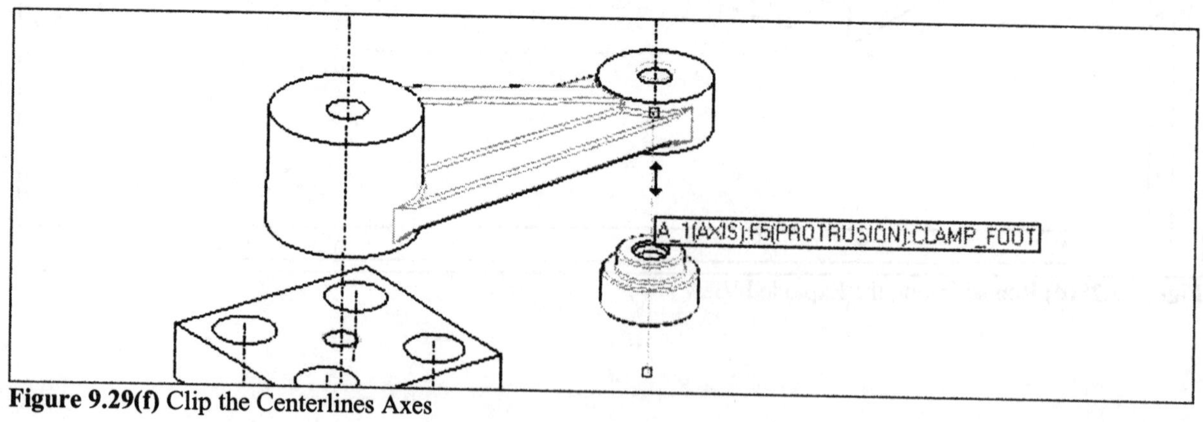

Figure 9.29(f) Clip the Centerlines Axes

To add balloons to a drawing sheet not using a parametric title block, you need to create each balloon separately. Create balloons for the components on the second sheet. The balloons added must correspond to the BOM.

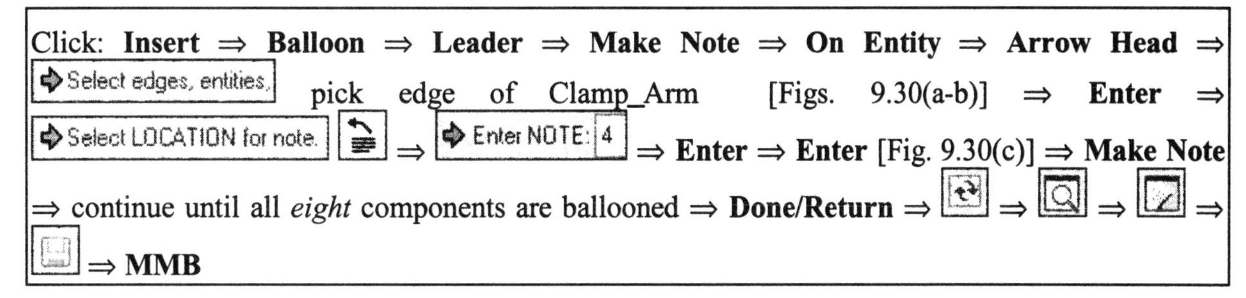

Click: **Insert** ⇒ **Balloon** ⇒ **Leader** ⇒ **Make Note** ⇒ **On Entity** ⇒ **Arrow Head** ⇒ Select edges, entities, pick edge of Clamp_Arm [Figs. 9.30(a-b)] ⇒ **Enter** ⇒ Select LOCATION for note. ⇒ Enter NOTE: 4 ⇒ **Enter** ⇒ **Enter** [Fig. 9.30(c)] ⇒ **Make Note** ⇒ continue until all *eight* components are ballooned ⇒ **Done/Return** ⇒ ⇒ ⇒ ⇒ ⇒ **MMB**

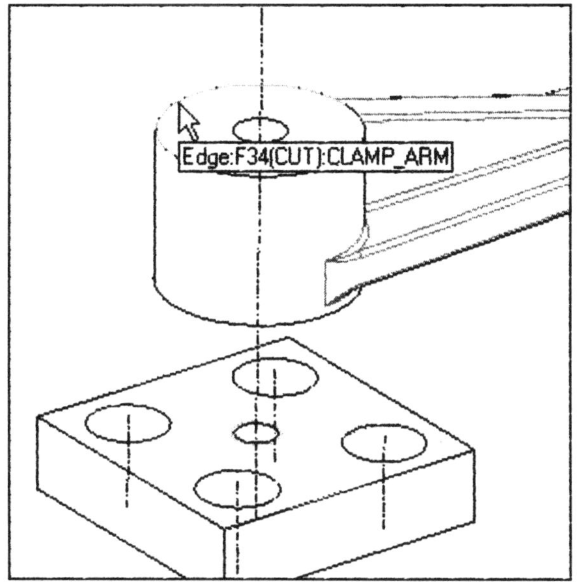

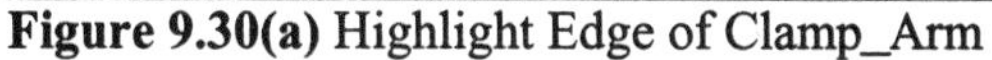
Figure 9.30(a) Highlight Edge of Clamp_Arm

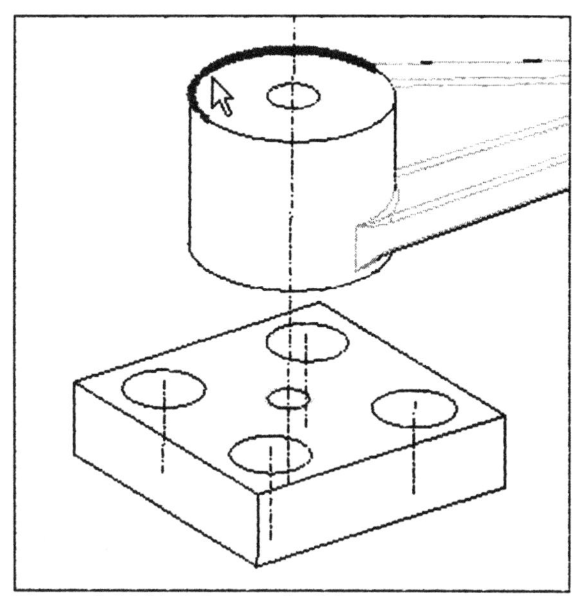

Figure 9.30(b) Select on the Edge

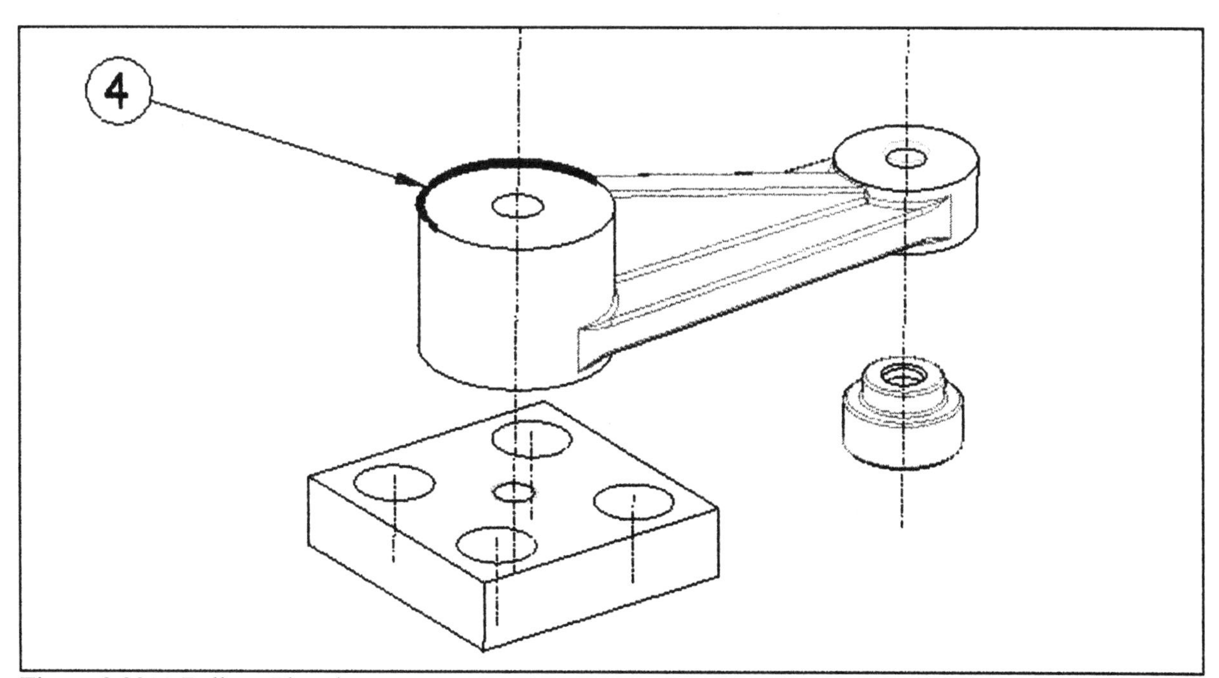

Figure 9.30(c) Balloon Placed

After the ballooning is complete, reposition the balloons and their attachment points as needed to clean up the drawing [Fig. 9.30(d)].

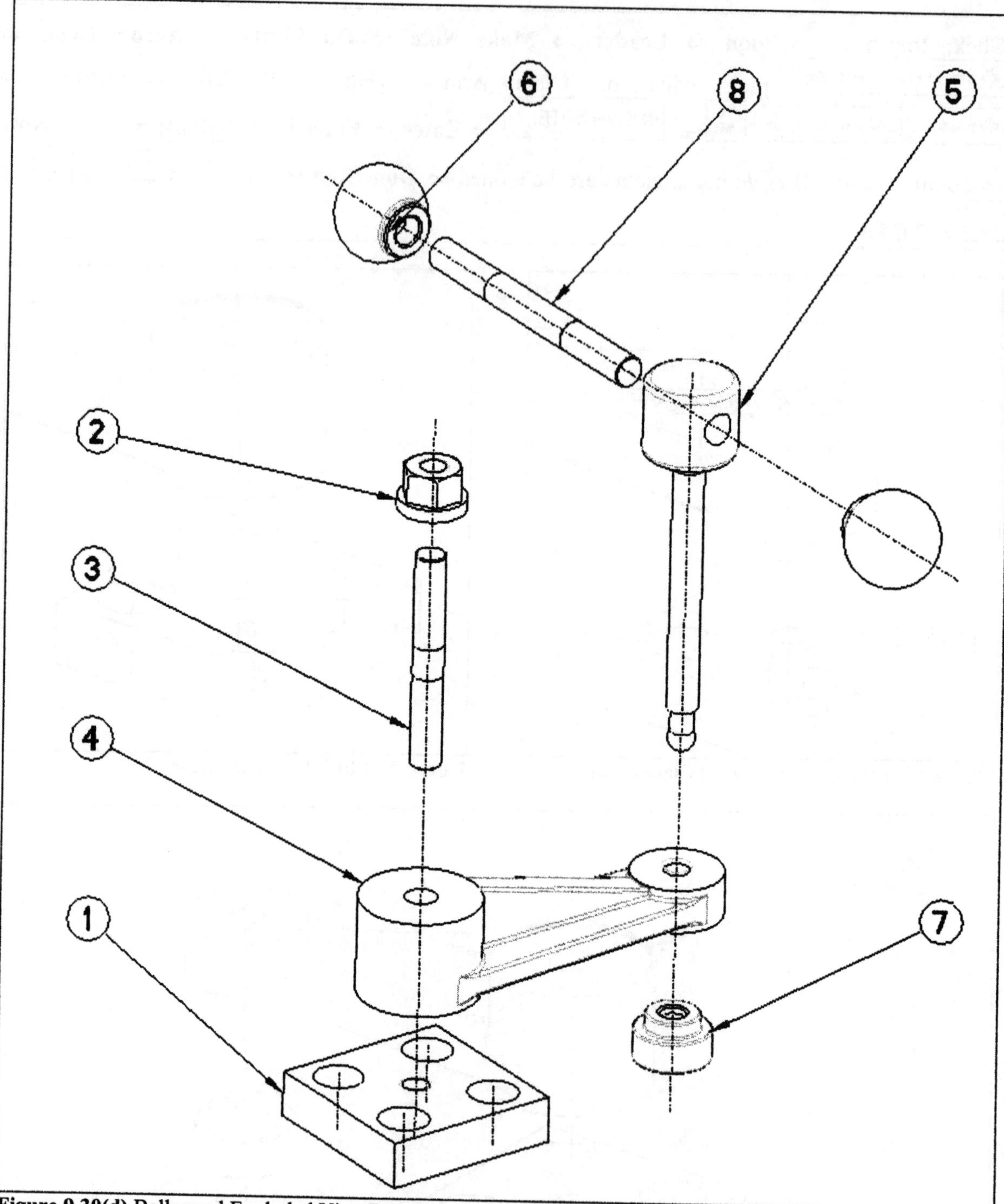

Figure 9.30(d) Ballooned Exploded View Drawing

Pick on pick balloon **6** ⇒ **RMB** [Fig. 9.30(e)] ⇒ **Edit Attachment** ⇒ **On Surface** ⇒ **Filled Dot** [Fig. 9.30(f)] ⇒ pick a place on the Clamp_Ball's surface ⇒ **MMB** ⇒ **LMB** [Fig. 9.30(g)] ⇒ ⇒ ⇒ ⇒ ⇒ **MMB** ⇒ **File** ⇒ **Delete** ⇒ **Old Versions** ⇒ **MMB** [Fig. 9.30(h)]

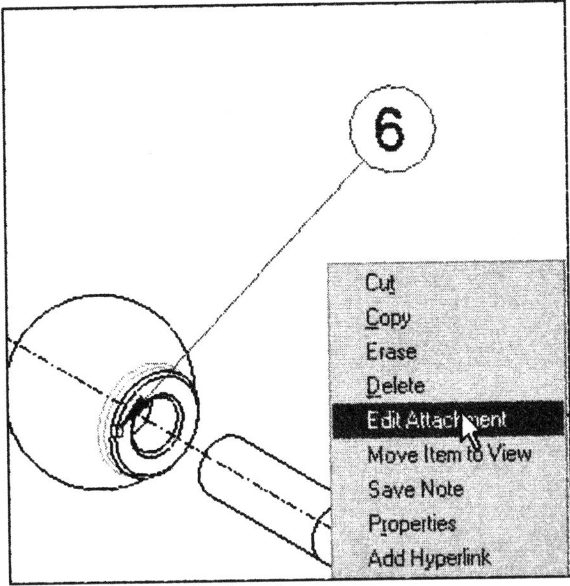

Figure 9.30(e) Edit Attachment

Figure 9.30(f) Attachment Point on Surface

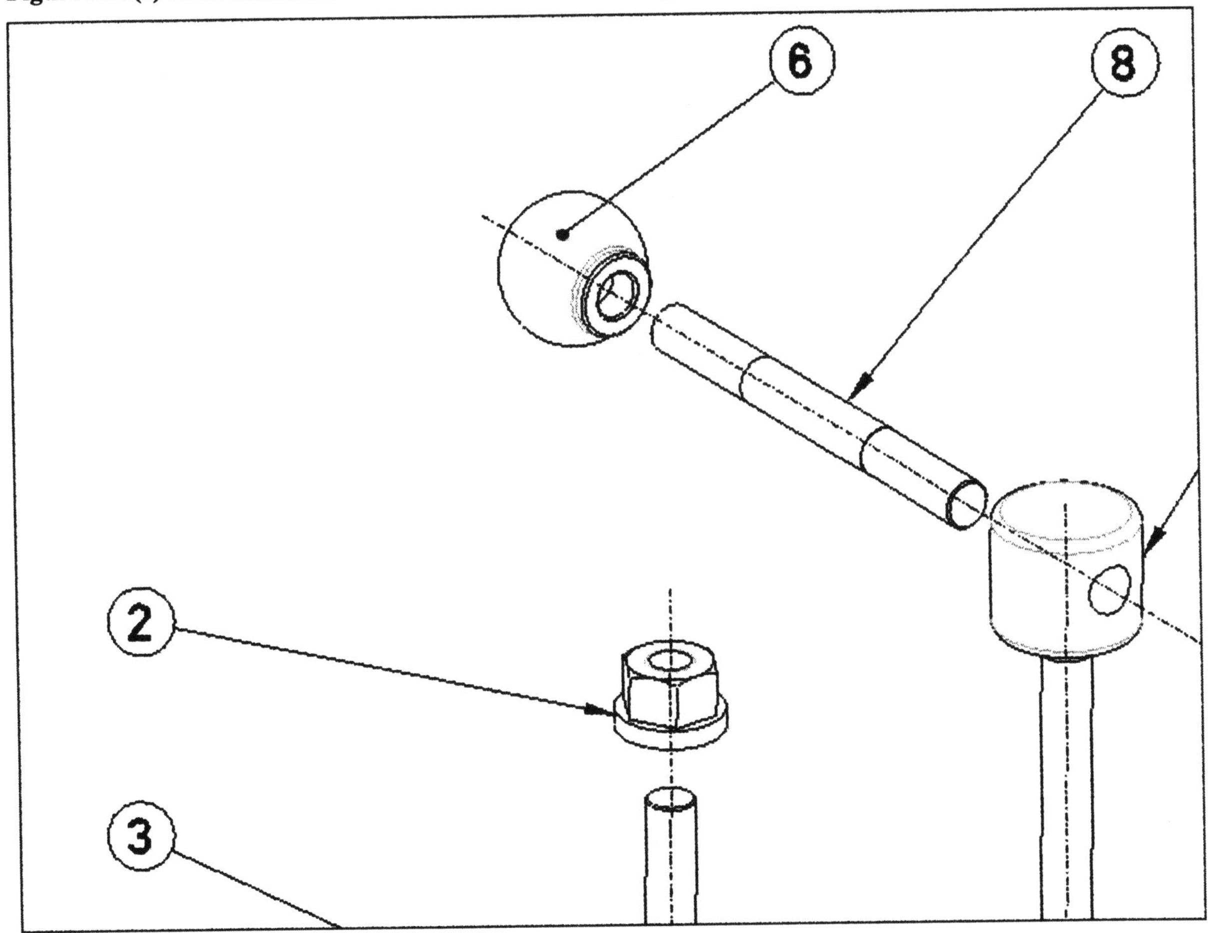

Figure 9.30(g) Filled Dot Surface Attachment

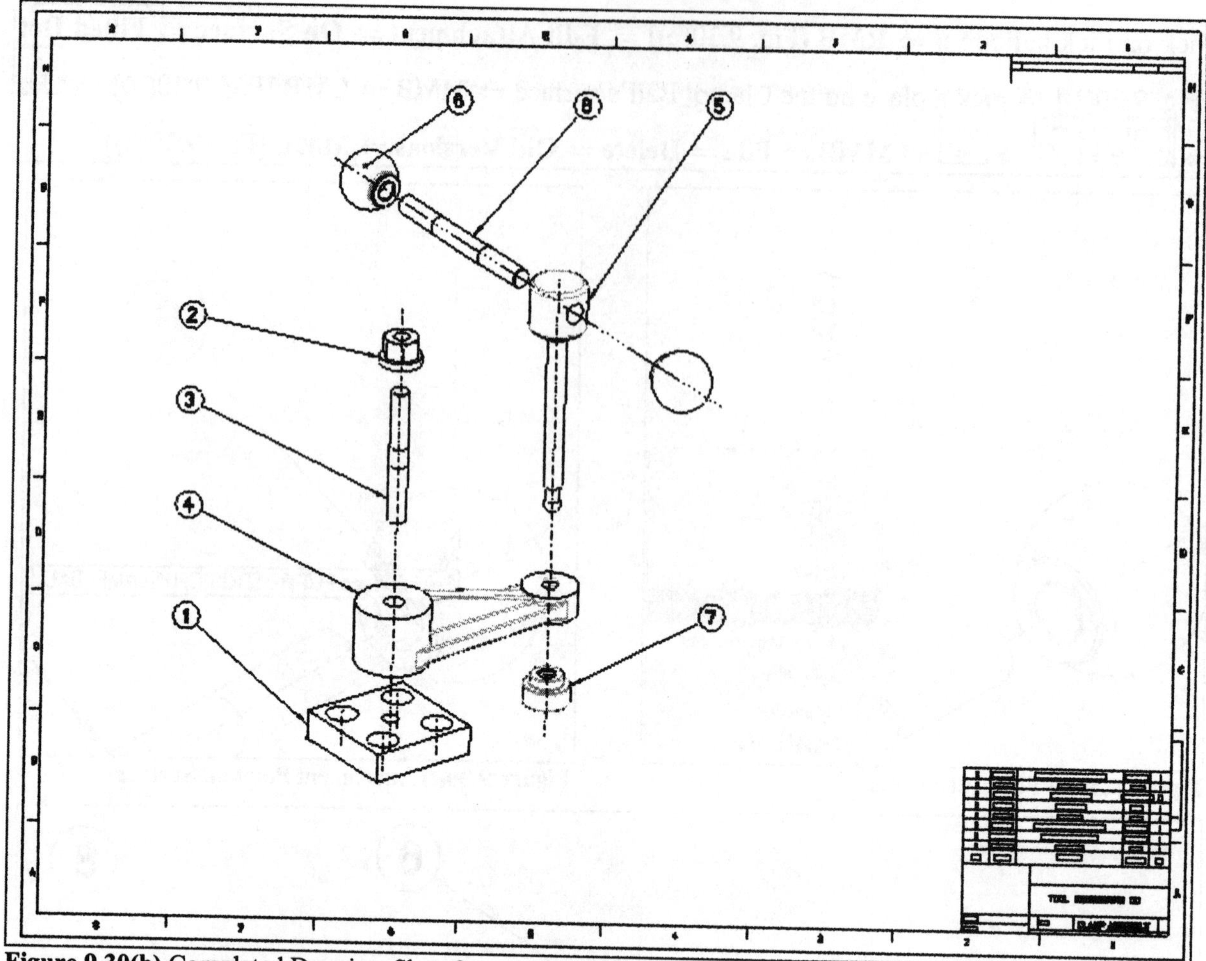

Figure 9.30(h) Completed Drawing, Sheet 2

Create a *documentation package* for the Clamp_Assembly. A complete documentation package contains all models and drawings required to manufacture the parts and assemble the components. Your instructor may change the requirements, but in general, create and plot/print the following:

- **Part Models** for all components

- **Detail Drawings** for each nonstandard component, such as the Clamp_Arm, Clamp_Swivel,
Clamp_Foot, and Clamp_Ball *(do not detail the standard parts)*

- **Assembly Drawings** using standard orthographic ballooned views

- **Exploded Subassembly Drawing** of the ballooned subassembly

- **Exploded Assembly Drawing** of the ballooned assembly

Lesson 9 is now complete. A different assembly is available at *www.cad-resources.com* ⇒ *Downloads*.

Lesson 10 More Direct Modeling

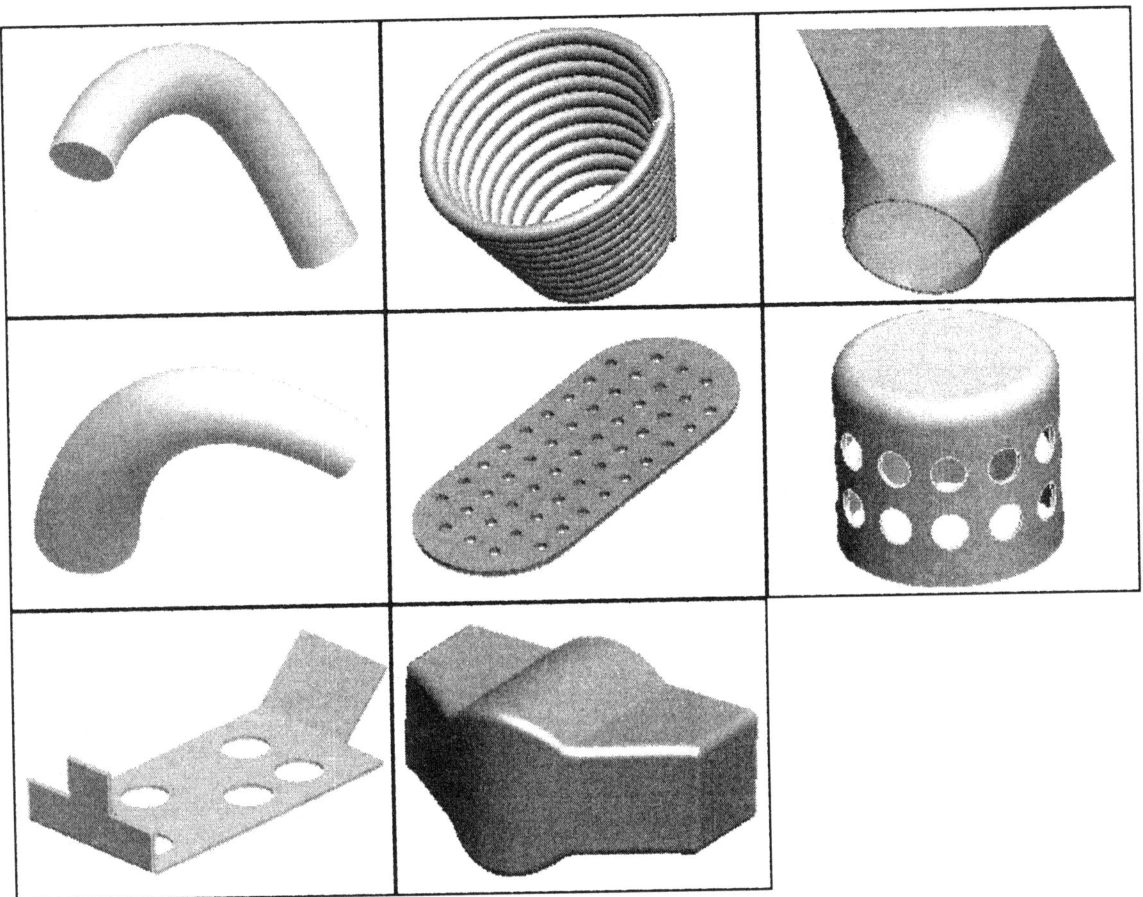

Figure 10.1 Advanced Quick Modeling Parts

OBJECTIVES

- Use the **Sweep Tool** and the **Blend Tool**
- Model a part using a **Swept Blend**
- Model a spring using the **Helical Sweep Tool**
- Create a variety of **Patterns**
- Model a **Sheetmetal** part
- Use a **Surface** to Model a Part

Other Tools

The last lesson will use a similar technique as was used in Lesson 2. In order to quickly introduce you to a few of the many modeling tools available with Pro/E, more quick modeling will be employed to expand your knowledge of Pro/E without consuming a lot of class time.

For those who wish additional information, please see ***www.cad-resources.com*** ⇒ ***Downloads***, for extra Lessons and projects.

Part Model Nine (PRT0009.PRT) (Sweep)

Click: [new icon] ⇒ ●**Part (PRT0009)** ⇒ **MMB** ⇒ **Edit** ⇒ **Setup** ⇒ **Units** [Fig. 10.2(a)] ⇒ **millimeter Newton Second** [Fig. 10.2(b)] ⇒ **Set** ⇒ ⦿ Interpret dimensions (for example 1" becomes 1mm) [Fig. 10.2(c)] ⇒ **OK** ⇒ **Close** ⇒ **Done** ⇒ **Info** ⇒ **Model** [Fig. 10.2(d)] ⇒ collapse the **Browser** using Quick Sash controls

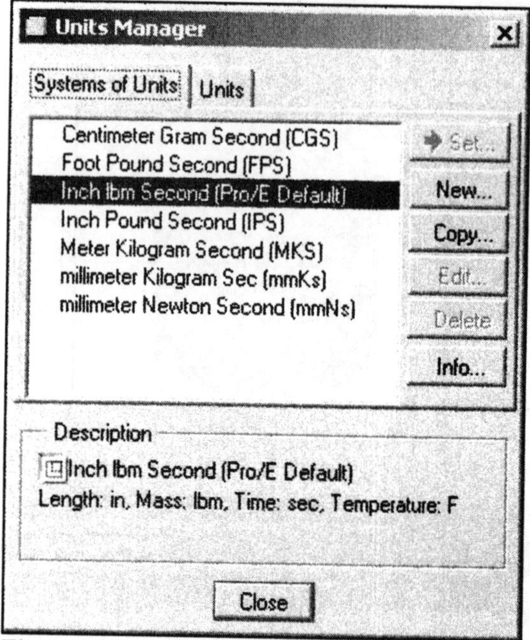

Figure 10.2(a) Units Manager Dialog Box

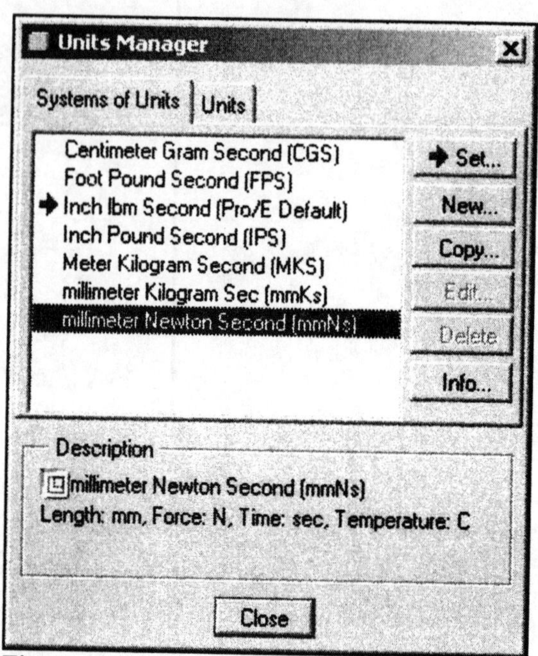

Figure 10.2(b) Select Millimeter Newton Second

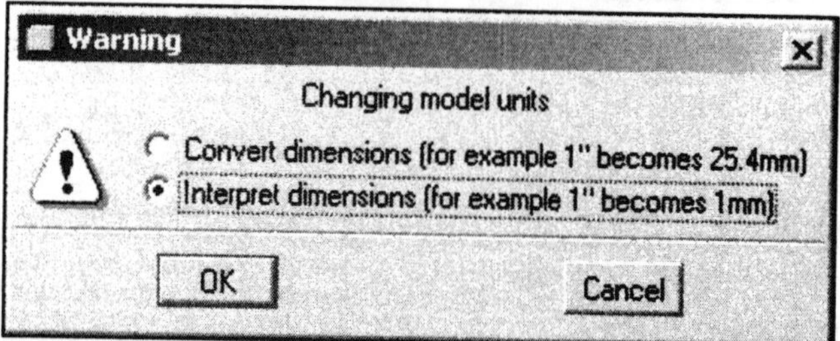

Figure 10.2(c) Interpret Dimensions

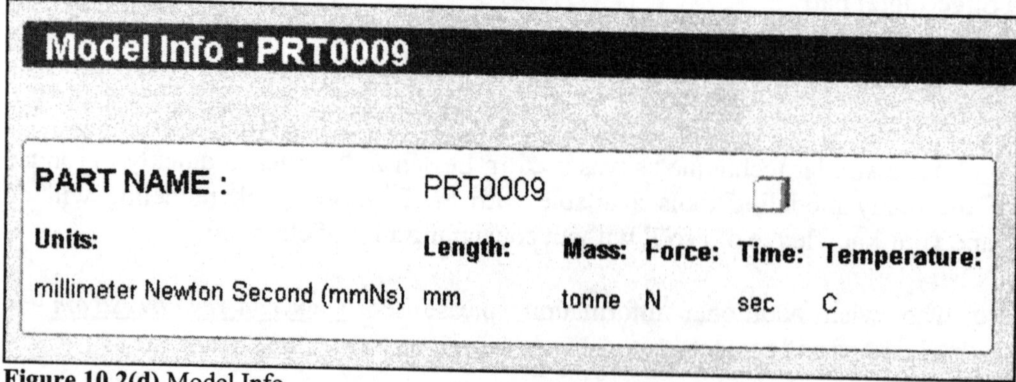

Figure 10.2(d) Model Info

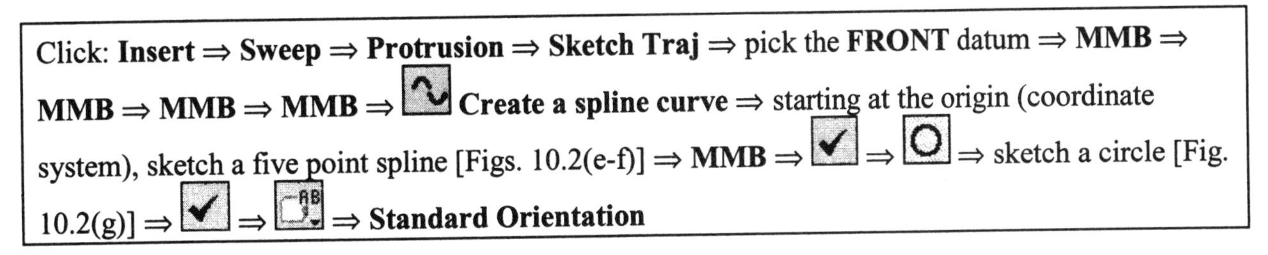

Click: **Insert** ⇒ **Sweep** ⇒ **Protrusion** ⇒ **Sketch Traj** ⇒ pick the **FRONT** datum ⇒ **MMB** ⇒ **MMB** ⇒ **MMB** ⇒ **MMB** ⇒ **Create a spline curve** ⇒ starting at the origin (coordinate system), sketch a five point spline [Figs. 10.2(e-f)] ⇒ **MMB** ⇒ ⇒ ⇒ sketch a circle [Fig. 10.2(g)] ⇒ ⇒ ⇒ **Standard Orientation**

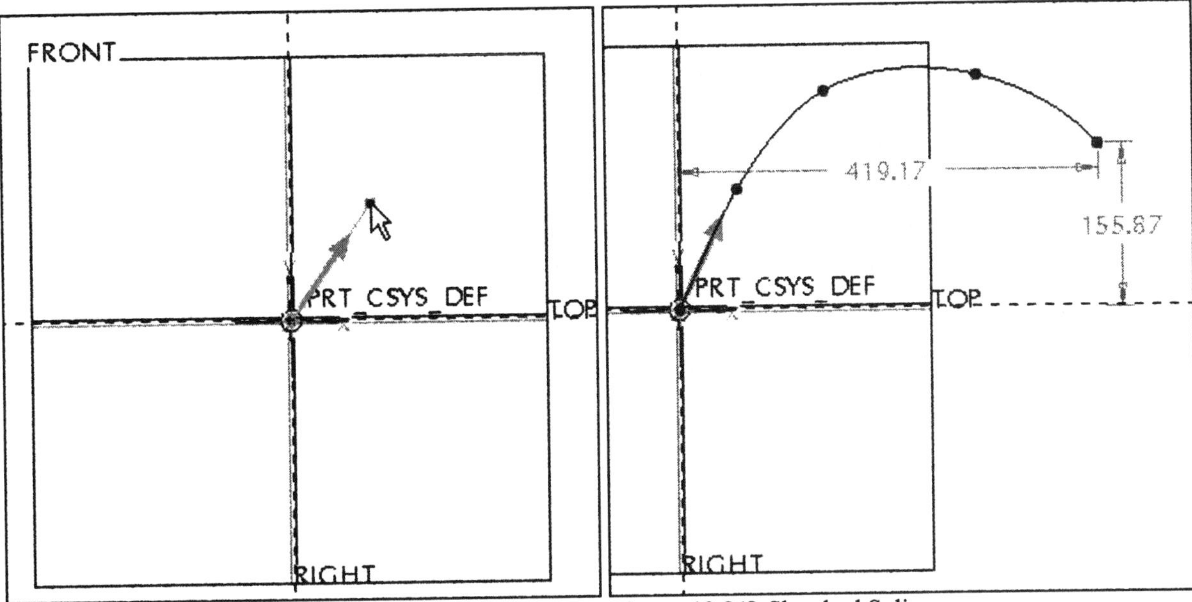

Figure 10.2(e) Start Spline

Figure 10.2(f) Sketched Spline

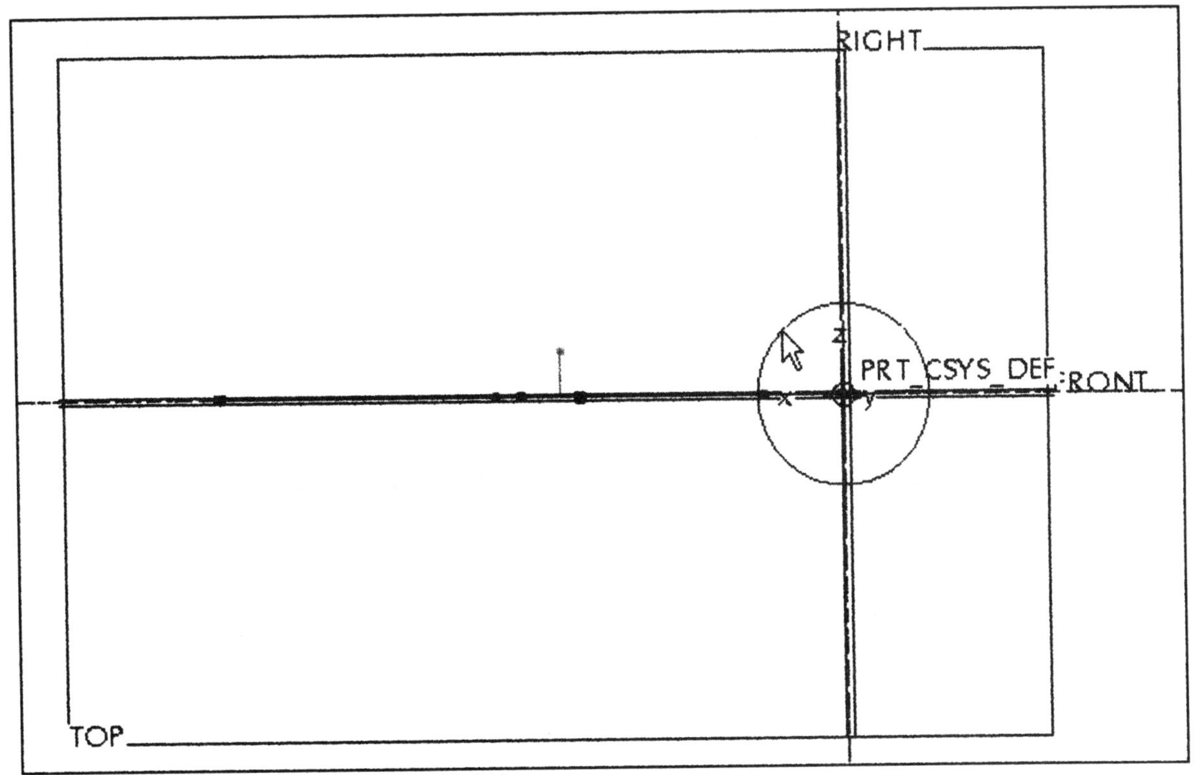

Figure 10.2(g) Sketch a Circle at the Crosshairs

Click: **Preview** [Fig. 10.2(h)] ⇒ **MMB** ⇒ **MMB** rotate the part [Fig. 10.2(i)] ⇒ **Ctrl+S** ⇒ **Enter** ⇒ **File** ⇒ **Close Window**

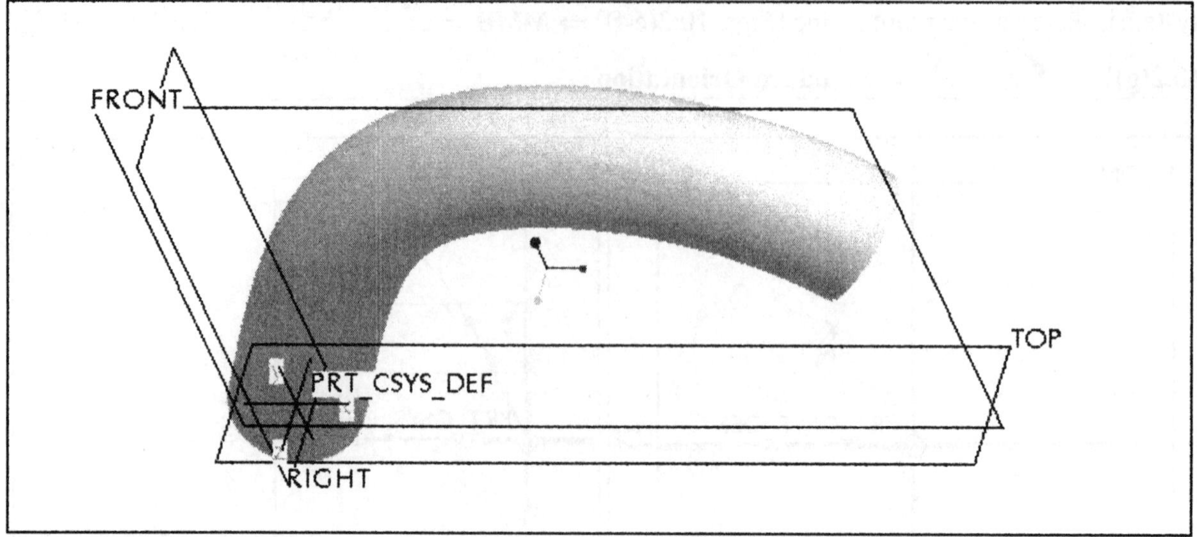

Figure 10.2(h) Preview

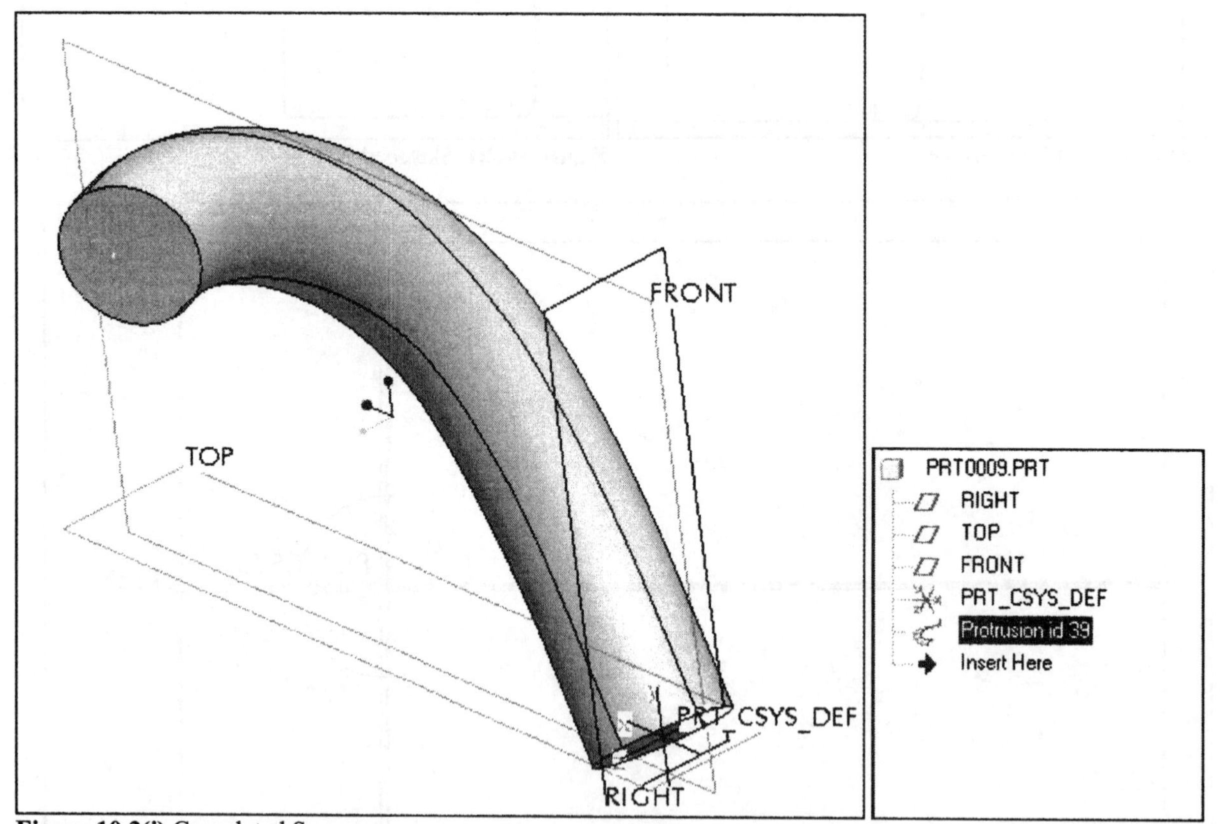

Figure 10.2(i) Completed Sweep

For those who wish additional information on sweeps, see ***www.cad-resources.com*** ⇒ ***Downloads***, for extra Lessons and projects.

Part Model Ten (PRT0010.PRT) (Helical Sweep)

Click: **File** ⇒ **New** ⇒ ●**Part (PRT0010)** ⇒ **MMB** ⇒ **Edit** ⇒ **Setup** ⇒ **Units** ⇒ **millimeter Newton Second** ⇒ **Set** ⇒ [● Interpret dimensions (for example 1" becomes 1mm)] ⇒ **OK** ⇒ **Close** ⇒ **Done** ⇒ **Insert** ⇒ **Helical Sweep** ⇒ **Protrusion** ⇒ **MMB** ⇒ pick the **FRONT** datum ⇒ **MMB** ⇒ **MMB** ⇒ **RMB** ⇒ **Centerline** ⇒ create a vertical centerline on the edge of the **RIGHT** datum ⇒ **MMB** ⇒ **RMB** ⇒ **Line** ⇒ create an angled line ⇒ **MMB** [Fig. 10.3(a)] ⇒ ✓ ⇒ **MMB** to accept default pitch value ⇒ ○ ⇒ sketch a very small circle [Figs. 10.3(b-c)] ⇒ ✓ ⇒ [icon] ⇒ **Standard Orientation** ⇒ **Preview** [Fig. 10.3(d)] ⇒ **MMB** ⇒ **LMB** to deselect *(if the spring wire intersects itself: double-click on the protrusion ⇒ double-click on the circular cross section ⇒ double-click on the diameter dimension ⇒ change the wire diameter to a smaller value ⇒ Regenerate)* ⇒ **Ctrl+S** ⇒ **MMB** ⇒ **File** ⇒ **Close Window**

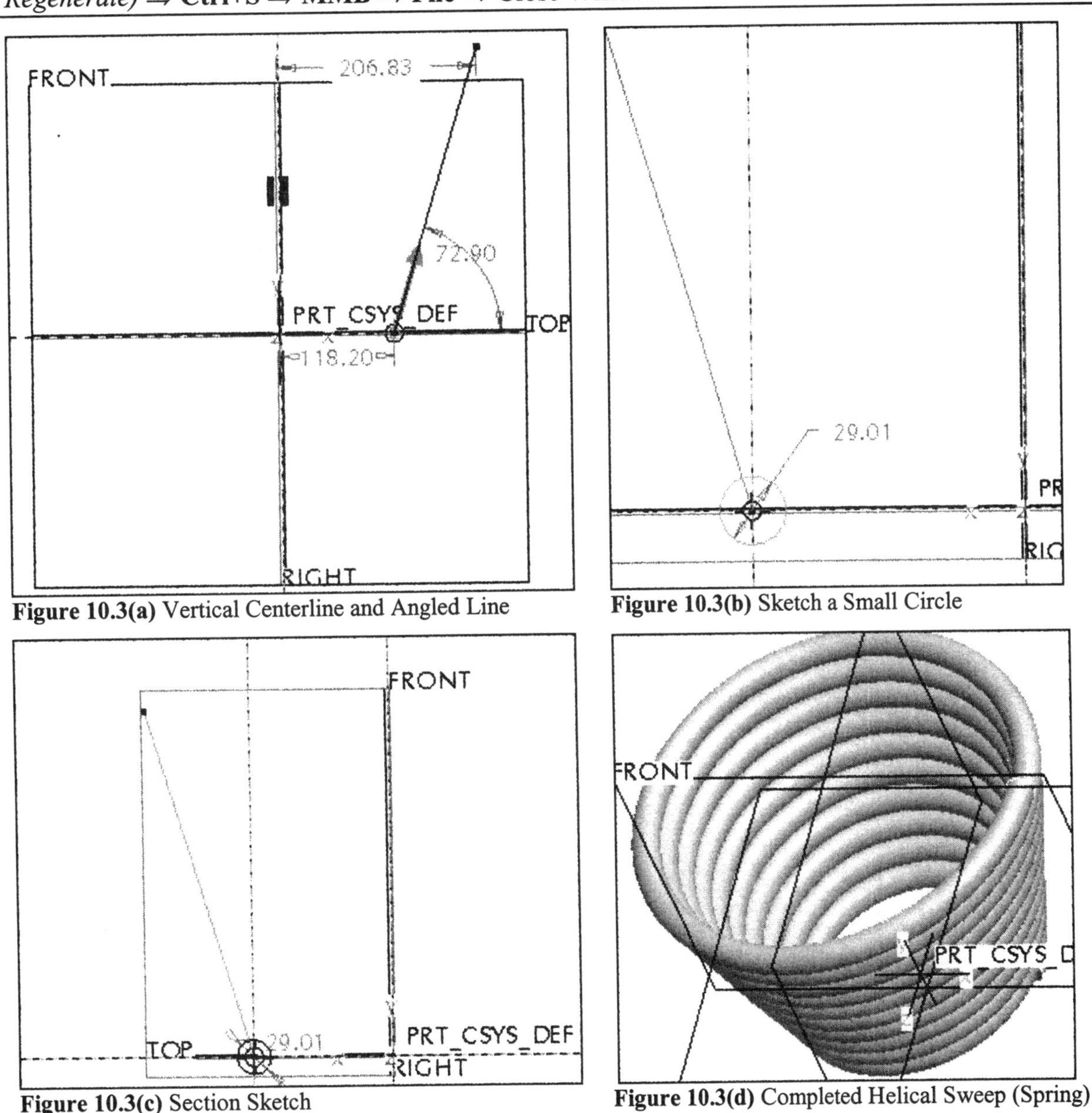

Figure 10.3(a) Vertical Centerline and Angled Line

Figure 10.3(b) Sketch a Small Circle

Figure 10.3(c) Section Sketch

Figure 10.3(d) Completed Helical Sweep (Spring)

See *www.cad-resources.com* ⇒ ***Downloads***, for extra Lessons and projects.

Part Model Eleven (PRT0011.PRT) (Blend)

Click: [icon] ⇒ ●**Part (PRT0011)** ⇒ **MMB** ⇒ **Edit** ⇒ **Setup** ⇒ **Units** ⇒ **mmNs** ⇒ **Set** ⇒ [Interpret dimensions] ⇒ **OK** ⇒ **Close** ⇒ **Done** ⇒ **Insert** ⇒ **Blend** ⇒ **Thin Protrusion** ⇒ **MMB** ⇒ **MMB** ⇒ pick the **FRONT** datum ⇒ **MMB** ⇒ **MMB** ⇒ **MMB** ⇒ **MMB** ⇒ **RMB** ⇒ **Centerline** ⇒ create vertical and horizontal centerlines ⇒ **RMB** ⇒ **Rectangle** ⇒ create a centered rectangle [Fig. 10.4(a)] ⇒ **MMB** ⇒ **RMB** ⇒ **Toggle Section** ⇒ [icon] **Create an arc by picking its center and end points** ⇒ create four quarter circle arcs *(same number of entities as the rectangle)* [Figs. 10.4(b-e)] ⇒ **MMB** ⇒ pick on the beginning point of the arrow ⇒ **RMB** ⇒ **Start Point** [Fig. 10.4(f)] *(arrows pointing in the same direction)*

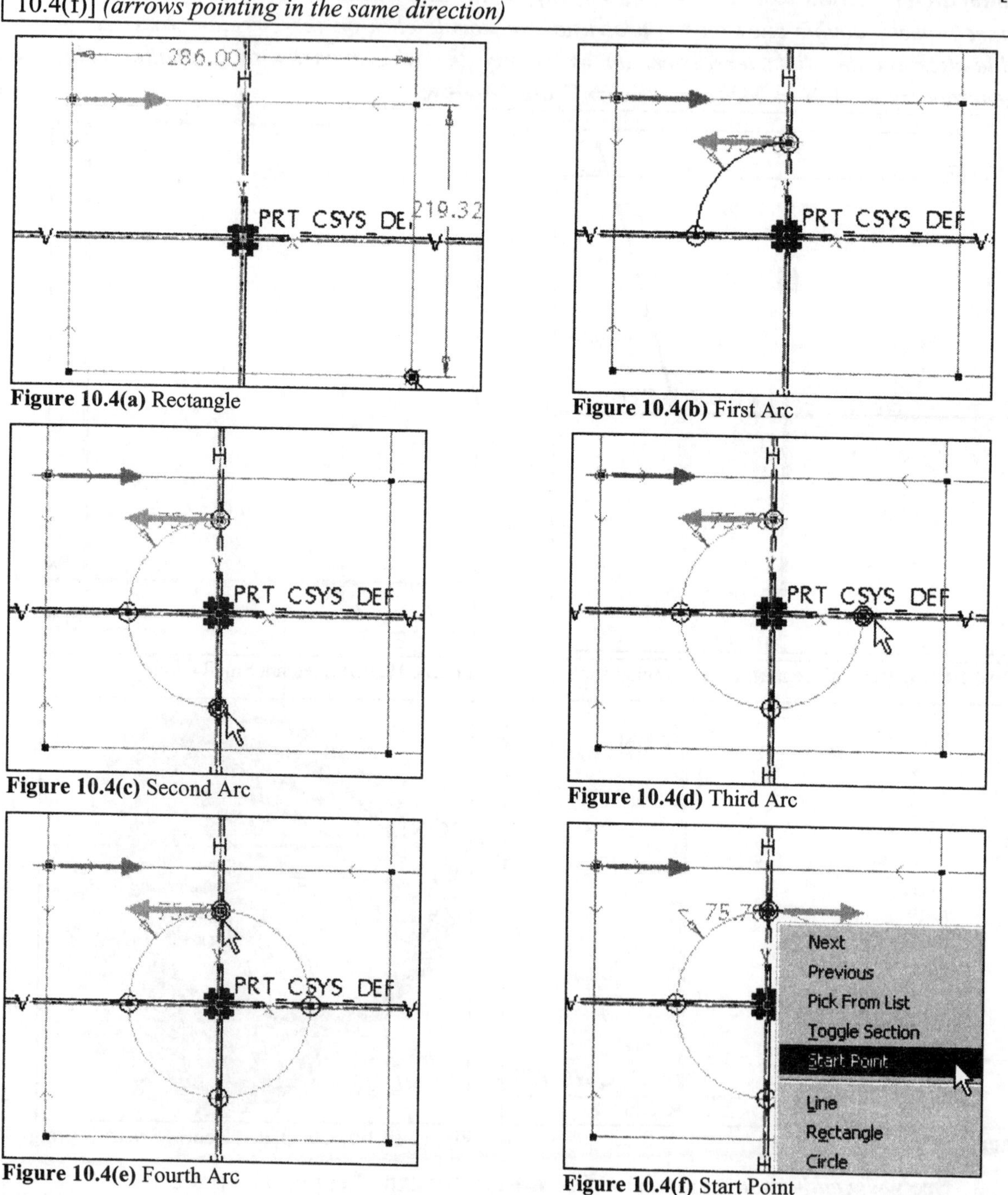

Figure 10.4(a) Rectangle

Figure 10.4(b) First Arc

Figure 10.4(c) Second Arc

Figure 10.4(d) Third Arc

Figure 10.4(e) Fourth Arc

Figure 10.4(f) Start Point

Click: ✔ ⇒ **MMB** ⇒ **MMB** ⇒ type **200** ⇒ **Enter** ⇒ ⇒ **Standard Orientation** ⇒ **Preview** [Fig. 10.4(g)] (to create a non-twisted blend, you must have the section entities of both sections inline; here the blend is twisted) ⇒ **MMB** [Fig. 10.4(h)] ⇒ **Ctrl+S** ⇒ **Enter** ⇒ **File** ⇒ **Close Window**

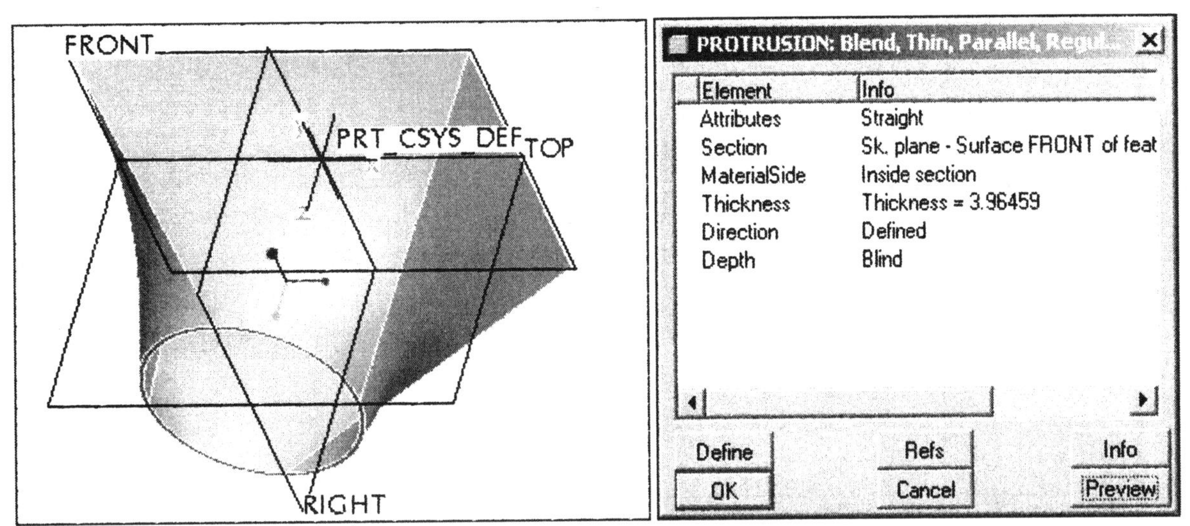

Figure 10.4(g) Preview of Blend

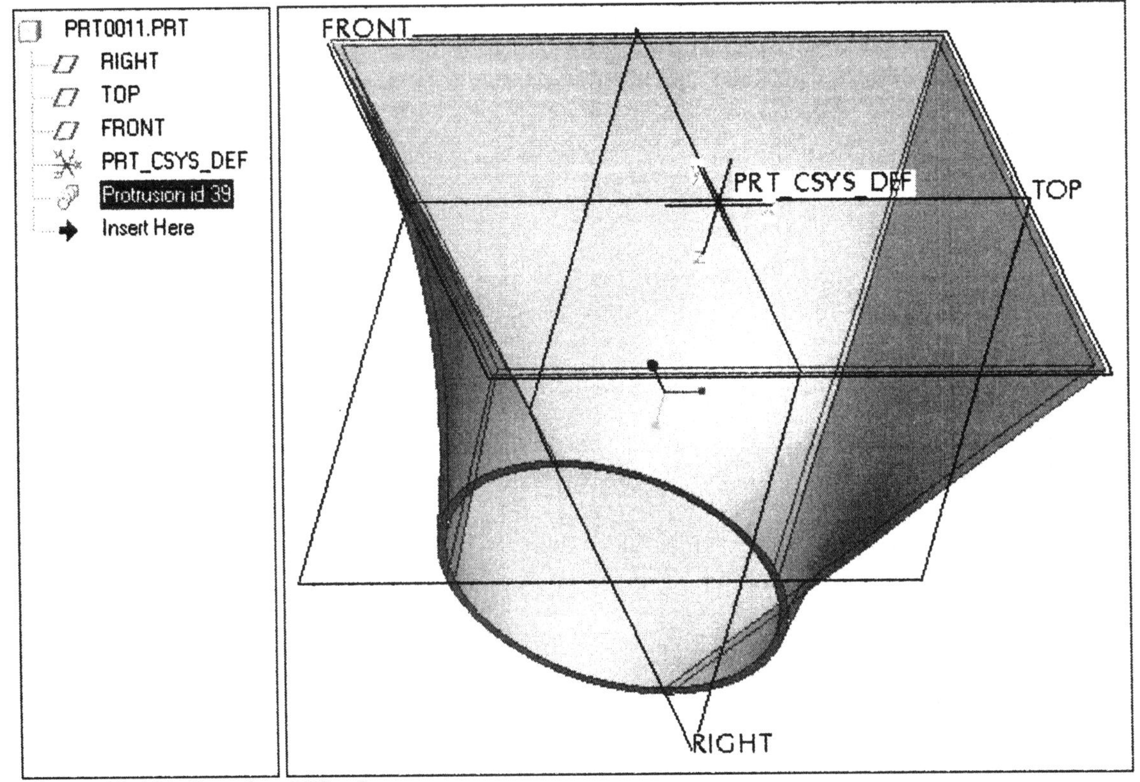

Figure 10.4(h) Completed Blend

For those who wish additional information on blends, see *__www.cad-resources.com__* ⇒ ***Downloads***, for extra Lessons and projects.

Part Model Twelve (PRT0012.PRT) (Swept Blend)

Click: [icon] ⇒ ●**Part (PRT0012)** ⇒ **MMB** ⇒ **Edit** ⇒ **Setup** ⇒ **Units** ⇒ **millimeter Newton Second** ⇒ **Set** ⇒ [Interpret dimensions (for example 1" becomes 1mm)] ⇒ **MMB** ⇒ **MMB** ⇒ **MMB** ⇒ pick the **FRONT** datum ⇒ [icon] **Sketch Tool** ⇒ **MMB** ⇒ **MMB** ⇒ **MMB** ⇒ [icon] **Create a spline curve** ⇒ starting at the origin (coordinate system), sketch a four point spline [Fig. 10.5(a)] ⇒ **MMB** ⇒ [icon] ⇒ [icon] ⇒ **Standard Orientation** [Fig. 10.5(b)] ⇒ **Insert** ⇒ **Swept Blend** ⇒ **Protrusion** ⇒ **MMB** ⇒ **Select Traj** ⇒ pick the curve from the screen⇒ **MMB** [Fig. 10.5(c)] ⇒ **MMB** ⇒ **MMB** ⇒ **MMB** [Fig. 10.5(d)]

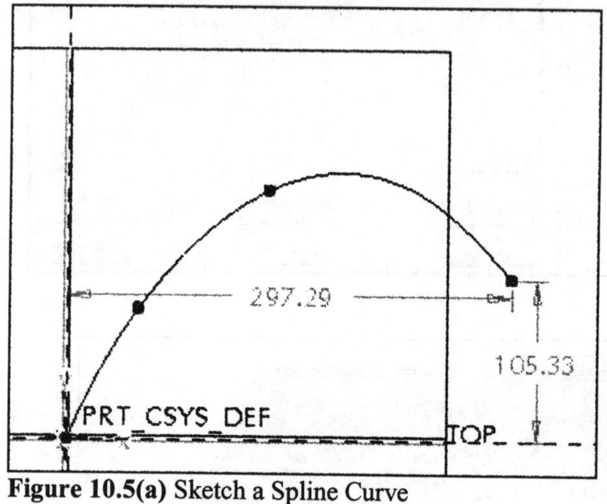

Figure 10.5(a) Sketch a Spline Curve

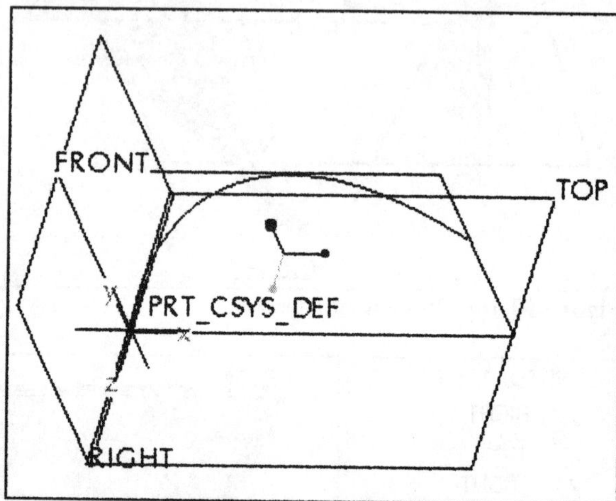

Figure 10.5(b) Completed Curve

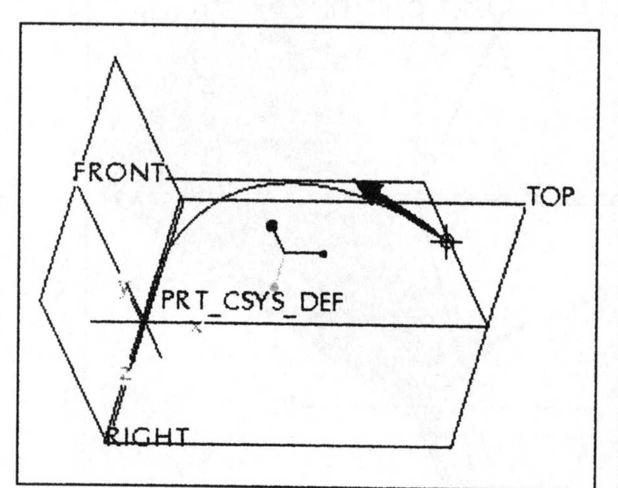

Figure 10.5(c) Selected Curve

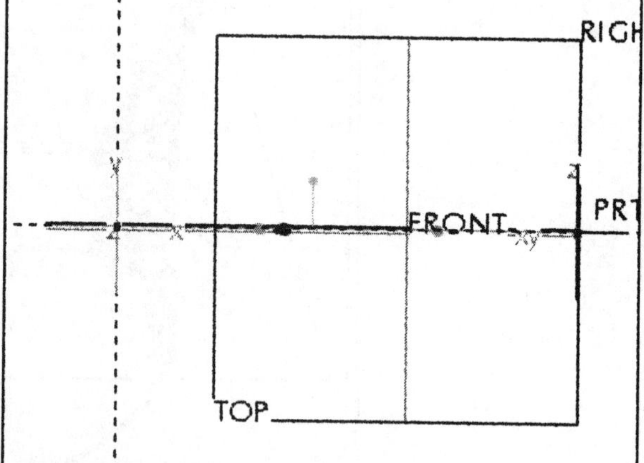

Figure 10.5(d) First Section

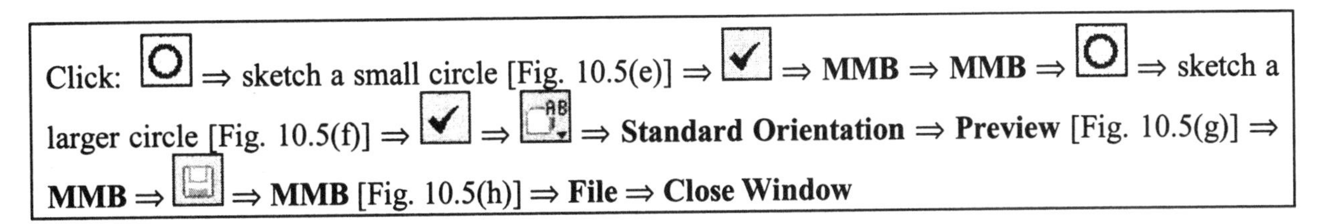

Click: [icon] ⇒ sketch a small circle [Fig. 10.5(e)] ⇒ [icon] ⇒ **MMB** ⇒ **MMB** ⇒ [icon] ⇒ sketch a larger circle [Fig. 10.5(f)] ⇒ [icon] ⇒ [icon] ⇒ **Standard Orientation** ⇒ **Preview** [Fig. 10.5(g)] ⇒ **MMB** ⇒ [icon] ⇒ **MMB** [Fig. 10.5(h)] ⇒ **File** ⇒ **Close Window**

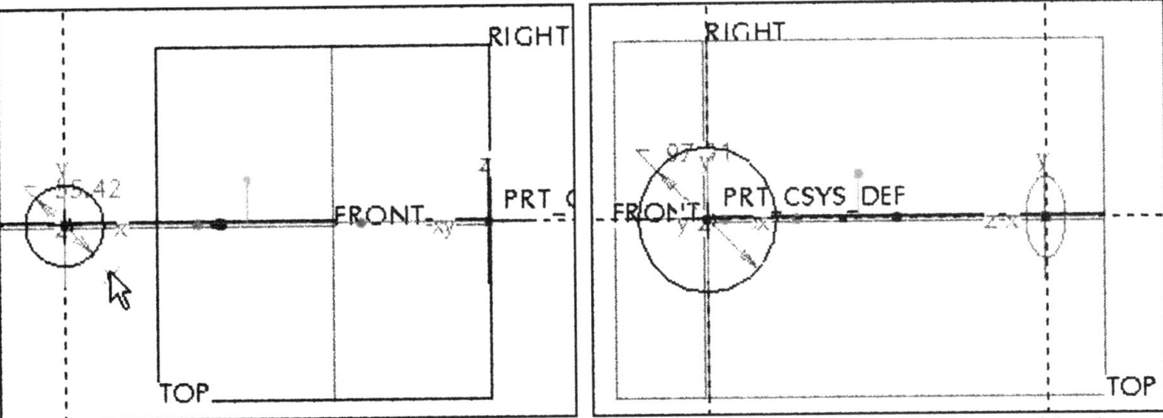

Figure 10.5(e) Small Circle for the First Section

Figure 10.5(f) Larger Circle for the Second Section

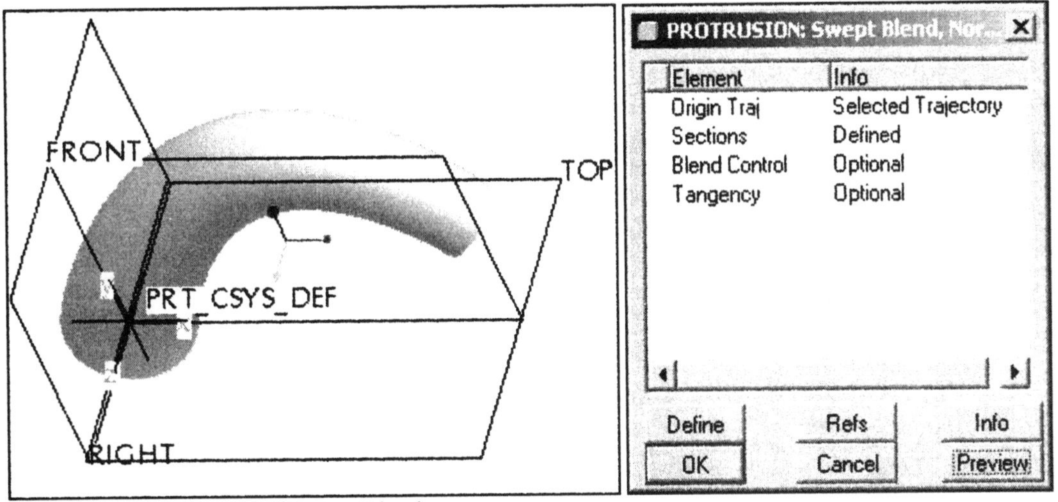

Figure 10.5(g) Preview of Swept Blend

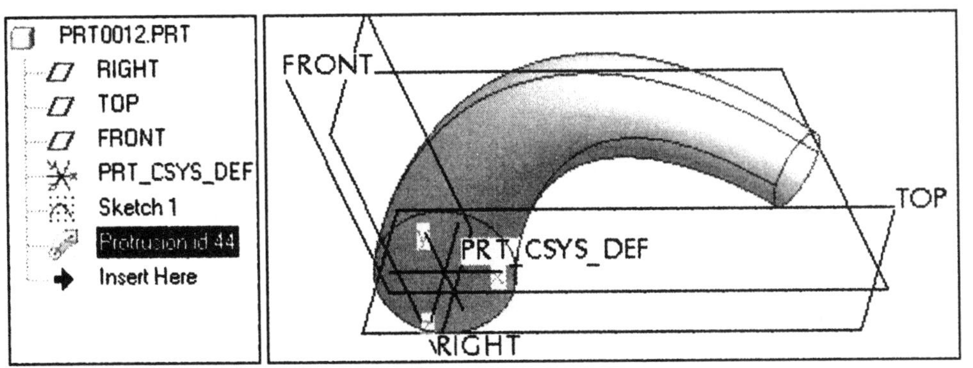

Figure 10.5(h) Completed Swept Blend

See ***www.cad-resources.com*** ⇒ ***Downloads***, for extra Lessons and projects.

Part Model Thirteen (PRT0013.PRT) (Fill Pattern)

Click: [icon] ⇒ ●**Part** **(PRT0013)** ⇒ **MMB** ⇒ **Edit** ⇒ **Setup** ⇒ **Units** ⇒ millimeter Newton Second (mmNs) ⇒ **Set** ⇒ Interpret dimensions ⇒ **OK** ⇒ **Close** ⇒ **MMB** ⇒ pick the **FRONT** datum plane ⇒ [icon] **Extrude Tool** ⇒ **RMB** ⇒ Define Internal Sketch... ⇒ **MMB** ⇒ **RMB** ⇒ **Centerline** ⇒ create vertical and horizontal centerlines ⇒ [icon] **Create an arc** ⇒ pick the center and ends ⇒ **RMB** ⇒ **Line** ⇒ create the two horizontal lines [Fig. 10.6(a)] ⇒ **MMB** ⇒ [icon] ⇒ window in the sketch ⇒ [icon] **Mirror selected entities** ⇒ pick on the vertical centerline [Fig. 10.6(b)] ⇒ [icon] ⇒ **Standard Orientation** ⇒ [icon] ⇒ adjust your part thickness to be similar to what is shown in Figure 10.6(c) ⇒ **MMB** [Fig. 10.6(c)] ⇒ [icon] ⇒ **MMB**

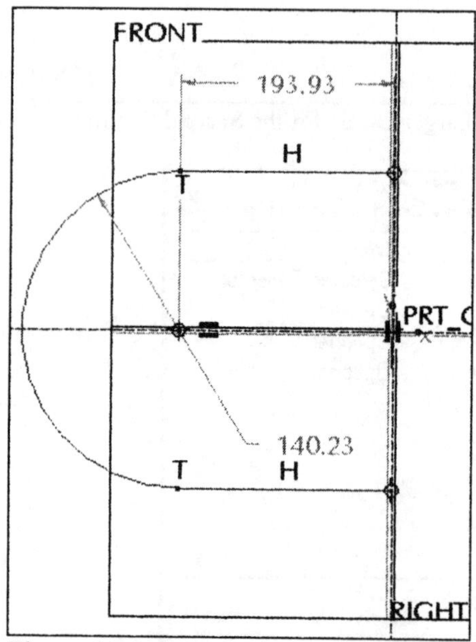

Figure 10.6(a) Arc and Lines

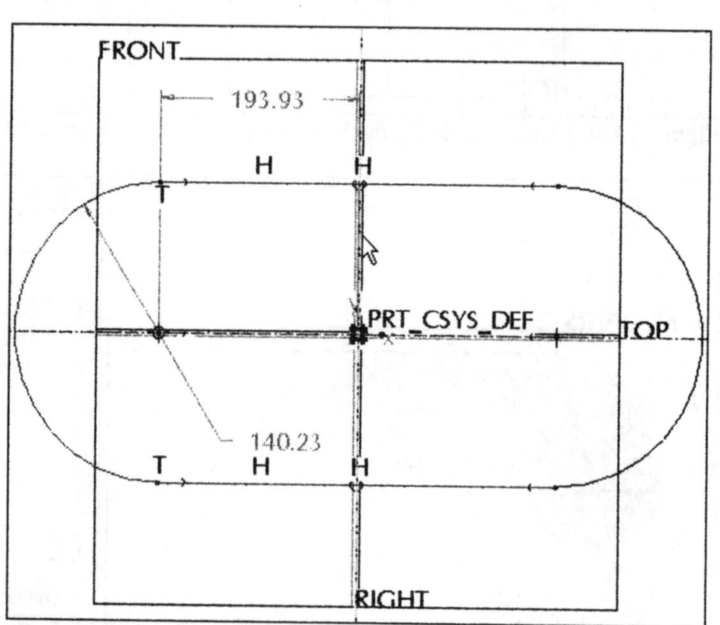

Figure 10.6(b) Mirrored Sketch

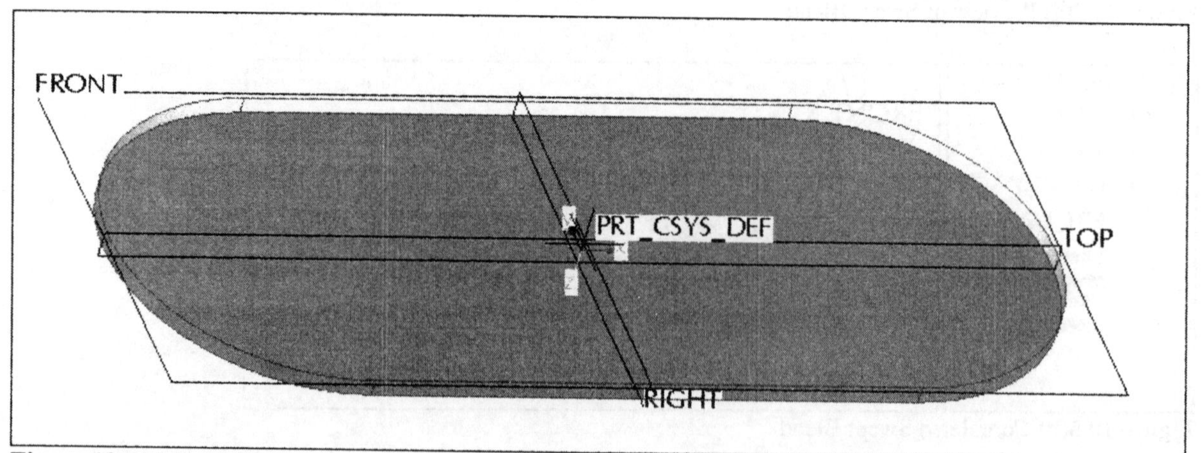

Figure 10.6(c) Extrusion

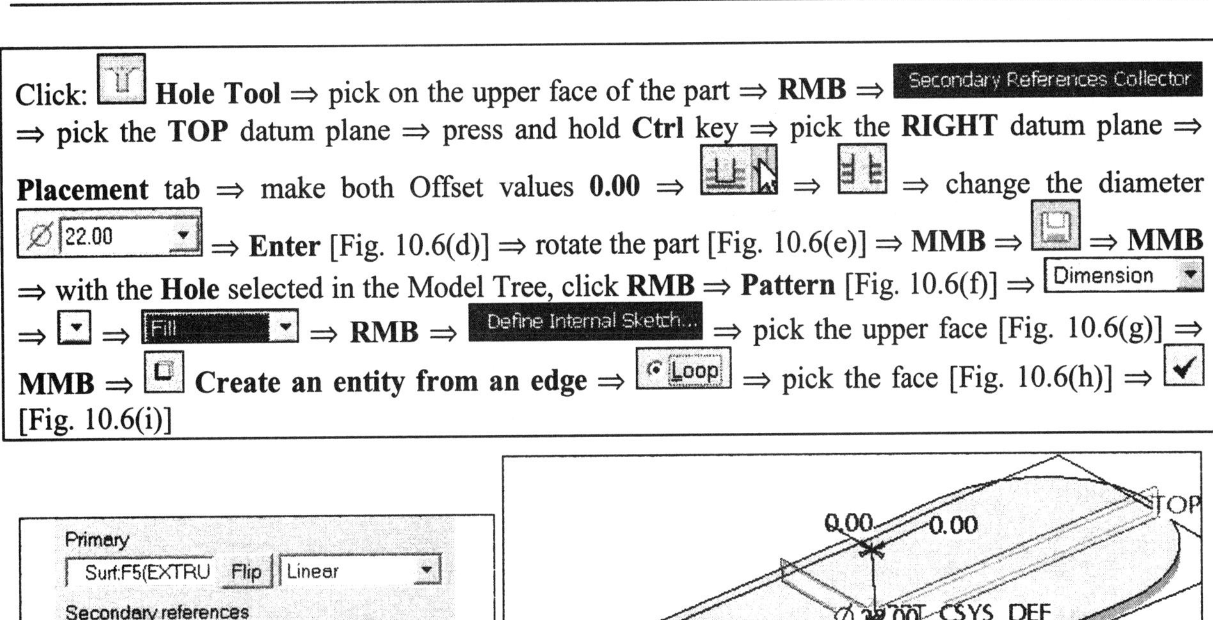

Click: [icon] **Hole Tool** ⇒ pick on the upper face of the part ⇒ **RMB** ⇒ [Secondary References Collector] ⇒ pick the **TOP** datum plane ⇒ press and hold **Ctrl** key ⇒ pick the **RIGHT** datum plane ⇒ **Placement** tab ⇒ make both Offset values **0.00** ⇒ [icon] ⇒ [icon] ⇒ change the diameter [Ø 22.00] ⇒ **Enter** [Fig. 10.6(d)] ⇒ rotate the part [Fig. 10.6(e)] ⇒ **MMB** ⇒ [icon] ⇒ **MMB** ⇒ with the **Hole** selected in the Model Tree, click **RMB** ⇒ **Pattern** [Fig. 10.6(f)] ⇒ [Dimension] ⇒ [icon] ⇒ [Fill] ⇒ **RMB** ⇒ [Define Internal Sketch...] ⇒ pick the upper face [Fig. 10.6(g)] ⇒ **MMB** ⇒ [icon] **Create an entity from an edge** ⇒ [Loop] ⇒ pick the face [Fig. 10.6(h)] ⇒ [✓] [Fig. 10.6(i)]

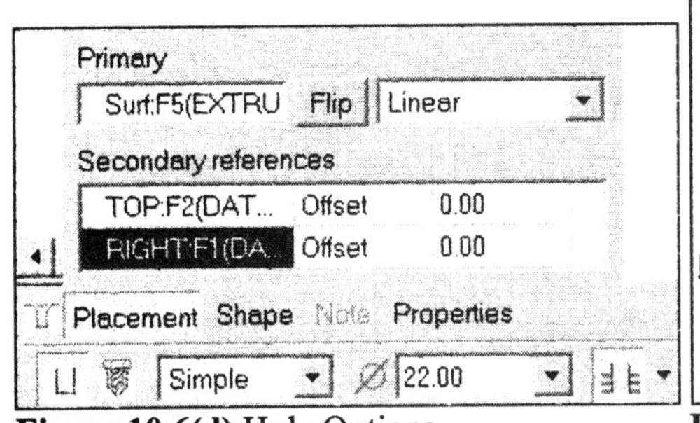

Figure 10.6(d) Hole Options

Figure 10.6(e) Hole

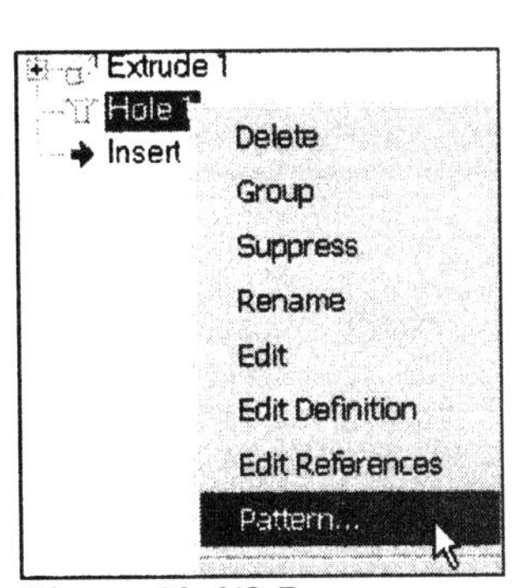

Figure 10.6(f) Pattern

Figure 10.6(g) Select the Upper Face

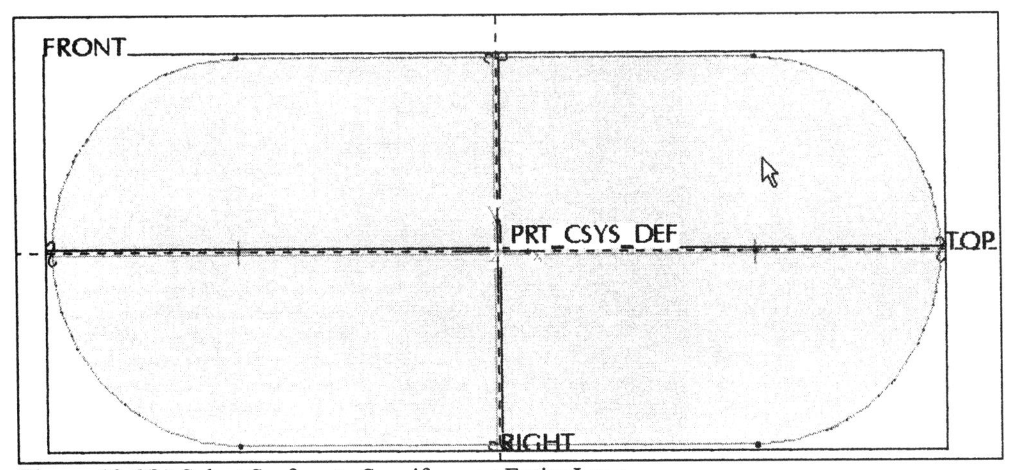

Figure 10.6(h) Select Surface to Specify as an Entity Loop

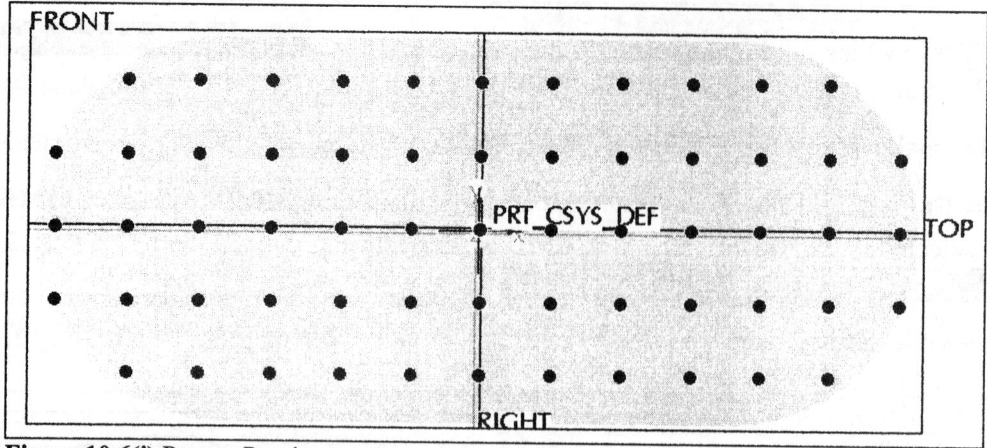

Figure 10.6(i) Pattern Preview

Pick on the unwanted copies (ten) (pick on a black dot and it changes to white) [Fig. 10.6(j)] ⇒ [icon] ⇒ **Standard Orientation** ⇒ **MMB** [Fig. 10.6(k)] ⇒ **Edit** ⇒ **Scale Model** ⇒ **.10** ⇒ **Enter** ⇒ **Yes** ⇒ [icon] ⇒ **MMB** ⇒ **File** ⇒ **Delete** ⇒ **Old Versions** ⇒ **MMB** ⇒ **File** ⇒ **Close Window**

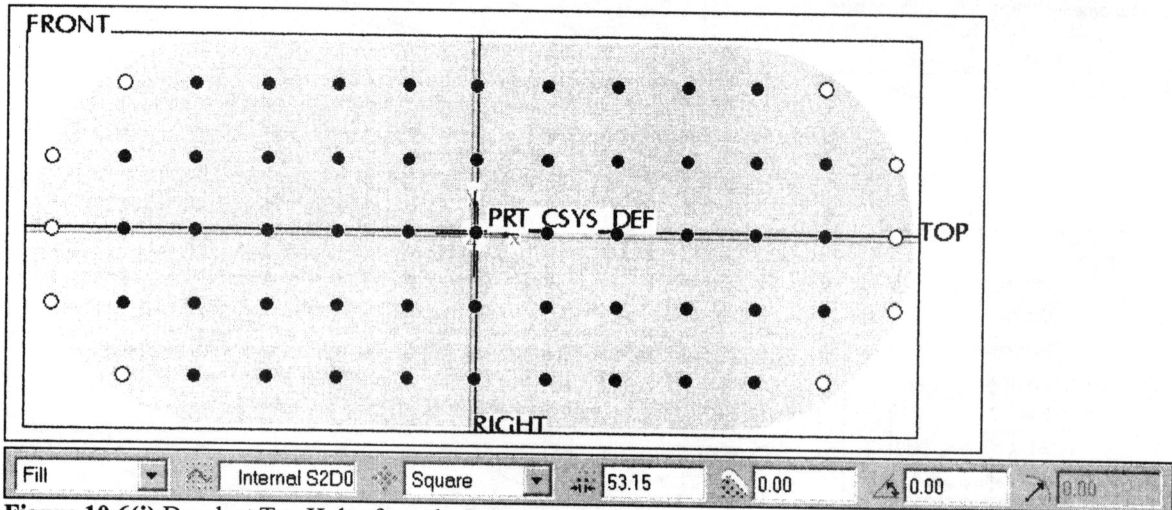

Figure 10.6(j) Deselect Ten Holes from the Pattern

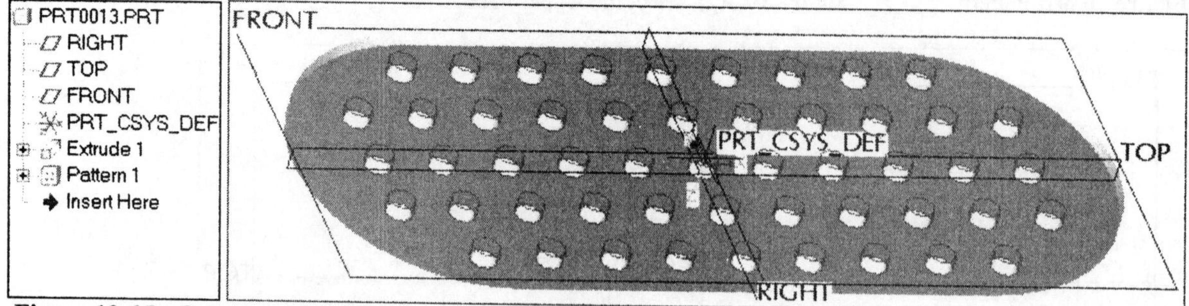

Figure 10.6(k) Completed Fill Pattern

See *www.cad-resources.com* ⇒ ***Downloads***, for extra Lessons and projects.

Part Model Fourteen (PRT0014.PRT) (Axial Pattern)

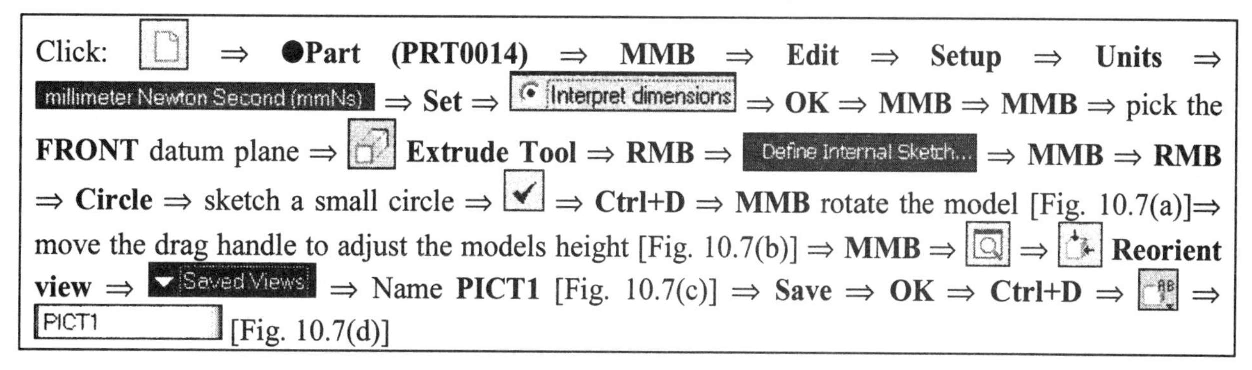

Click: ⇒ ●**Part** **(PRT0014)** ⇒ **MMB** ⇒ **Edit** ⇒ **Setup** ⇒ **Units** ⇒ millimeter Newton Second (mmNs) ⇒ **Set** ⇒ Interpret dimensions ⇒ **OK** ⇒ **MMB** ⇒ **MMB** ⇒ pick the **FRONT** datum plane ⇒ **Extrude Tool** ⇒ **RMB** ⇒ Define Internal Sketch... ⇒ **MMB** ⇒ **RMB** ⇒ **Circle** ⇒ sketch a small circle ⇒ ⇒ **Ctrl+D** ⇒ **MMB** rotate the model [Fig. 10.7(a)]⇒ move the drag handle to adjust the models height [Fig. 10.7(b)] ⇒ **MMB** ⇒ ⇒ **Reorient view** ⇒ Saved Views ⇒ Name **PICT1** [Fig. 10.7(c)] ⇒ **Save** ⇒ **OK** ⇒ **Ctrl+D** ⇒ ⇒ PICT1 [Fig. 10.7(d)]

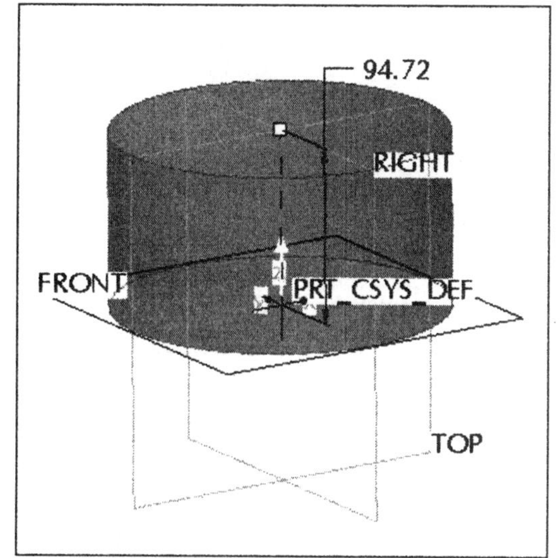

Figure 10.7(a) Circular Protrusion Preview

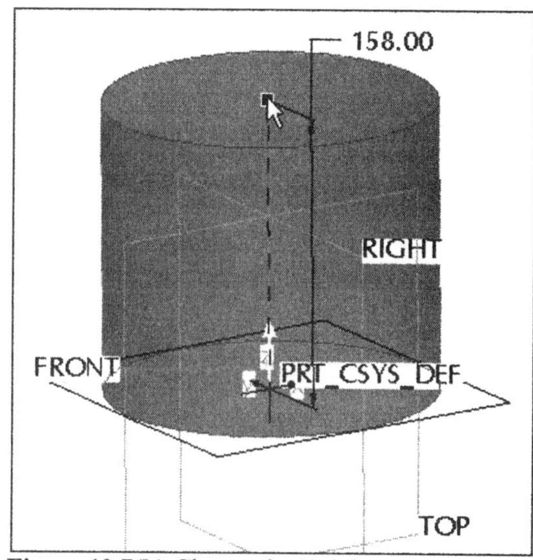

Figure 10.7(b) Change the Height

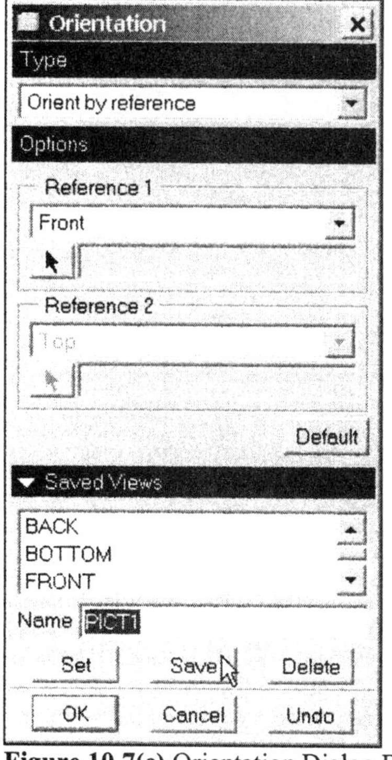

Figure 10.7(c) Orientation Dialog Box

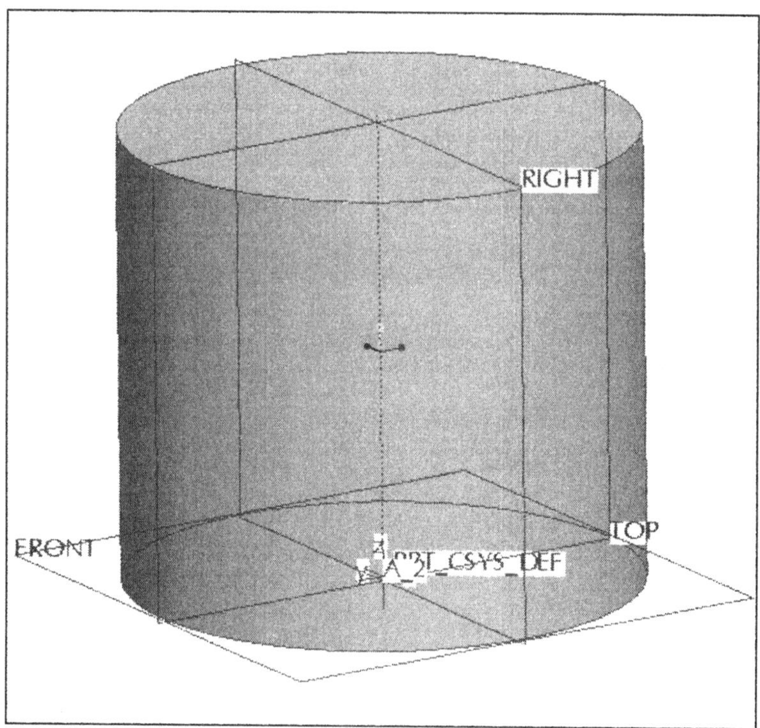

Figure 10.7(d) New Saved View

Pick on the top edge of the part ⇒ **RMB** ⇒ **Round Edges** [Fig. 10.8(a)] ⇒ **MMB** [Fig. 10.8(b)] ⇒ **Hole Tool** ⇒ pick on the vertical cylindrical face of the part ⇒ **RMB** ⇒ Secondary References Collector ⇒ pick the **RIGHT** datum plane [Fig. 10.9(a)] ⇒ press and hold **Ctrl** key ⇒ pick the **FRONT** datum plane [Fig. 10.9(b)] ⇒ **Placement** tab ⇒ edit the values [Figs. 10.9(c-d)] ⇒ ⇒ **MMB** ⇒ **Ctrl+S** ⇒ **Enter** ⇒ **LMB** to deselect

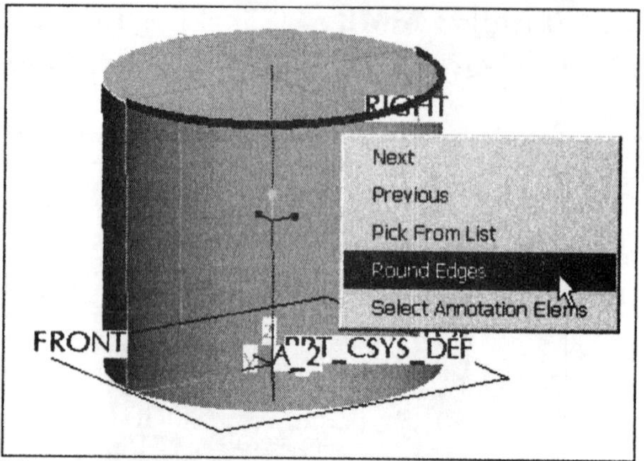

Figure 10.8(a) Round Edges

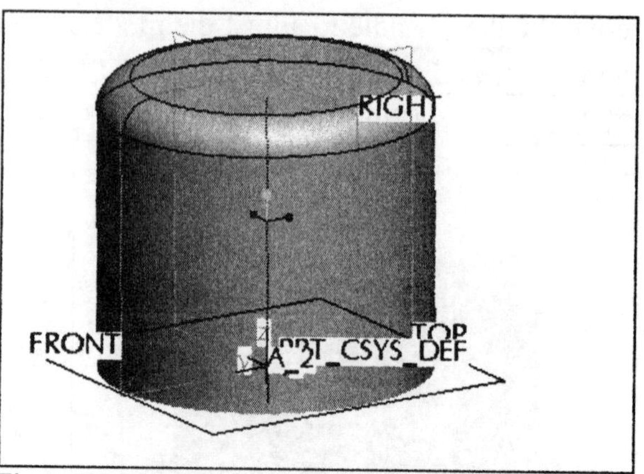

Figure 10.8(b) Completed Round

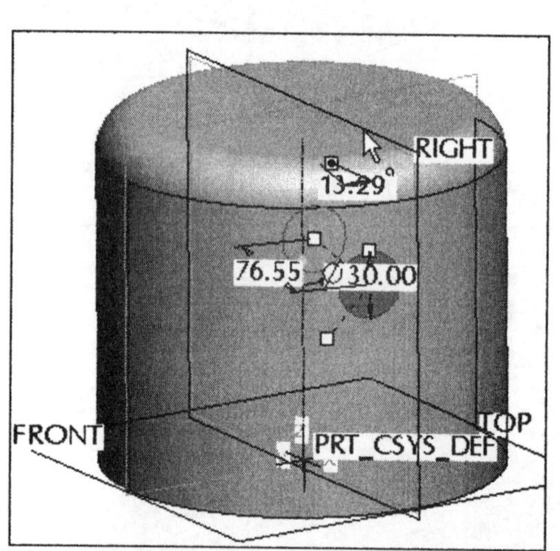

Figure 10.9(a) Right Datum Selected as Secondary Reference

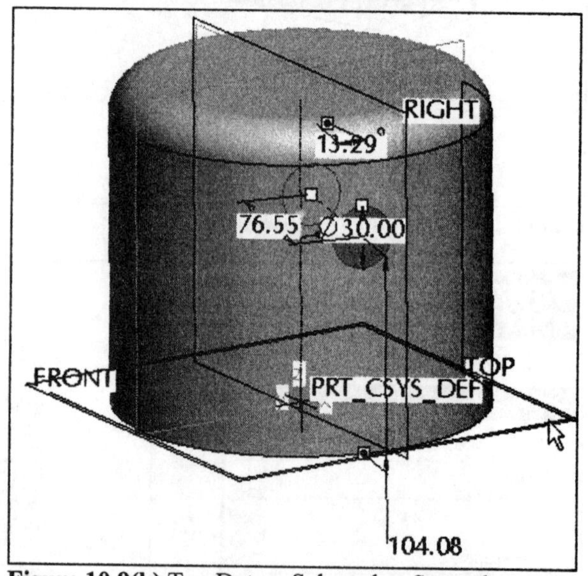

Figure 10.9(b) Top Datum Selected as Secondary Reference

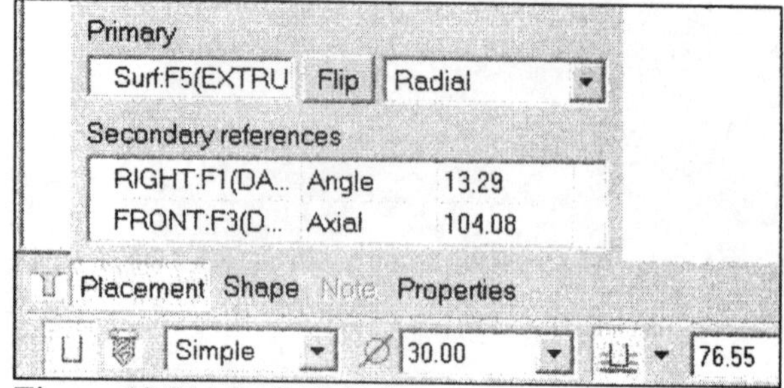

Figure 10.9(c) Default Values

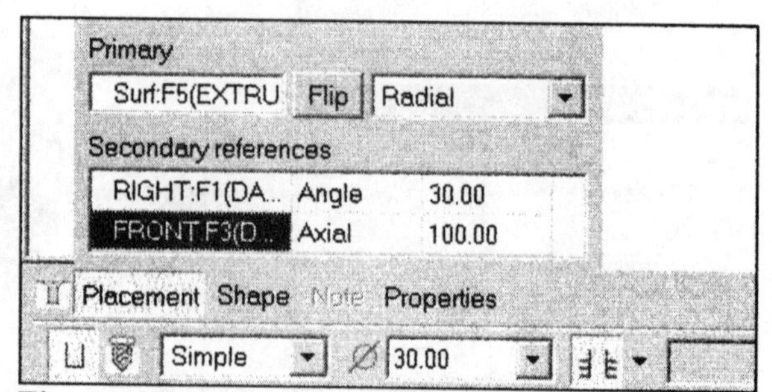

Figure 10.9(d) New Values and Depth

Pick on the hole ⇒ **Ctrl+C** ⇒ **Edit** ⇒ Paste Special ⇒ check first two options [Fig. 10.10(a)] ⇒ **OK** [Fig. 10.10(b)] ⇒ **Transformations** tab [Fig. 10.10(c)] ⇒ Direction reference- pick **Front** datum [Figs. 10.10(d-e)] ⇒ move the drag handle down [Fig. 10.10(f)] ⇒ **MMB** ⇒ **Ctrl+D** ⇒ **View** ⇒ **Orientation** ⇒ **Previous** ⇒ **LMB** to deselect ⇒ **Ctrl+S** ⇒ **Enter**

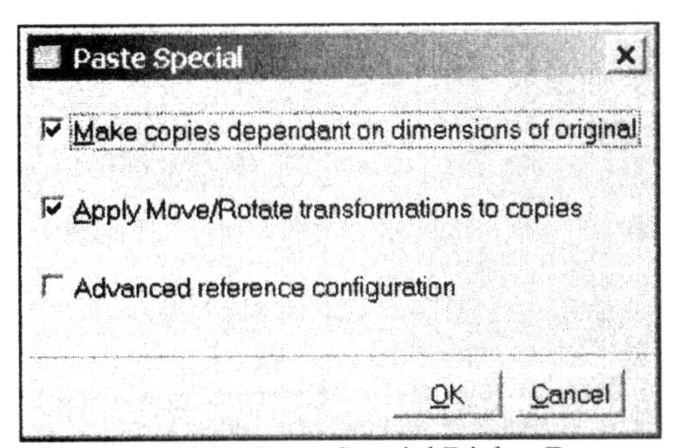

Figure 10.10(a) Paste Special Dialog Box

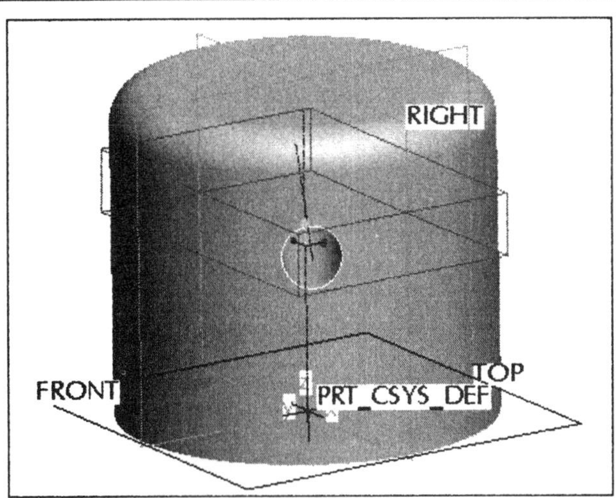

Figure 10.10(b) Paste Preview Box

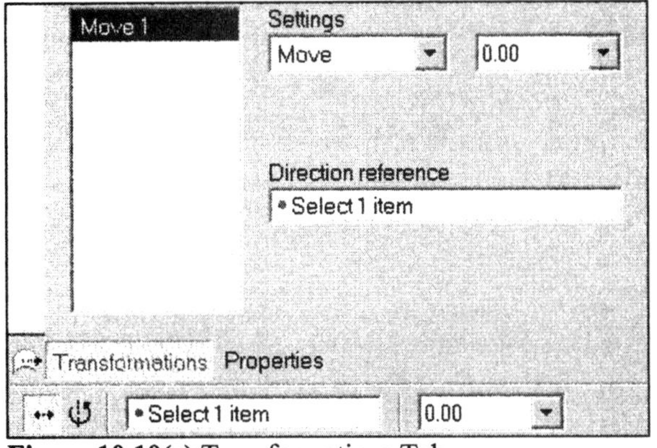

Figure 10.10(c) Transformations Tab

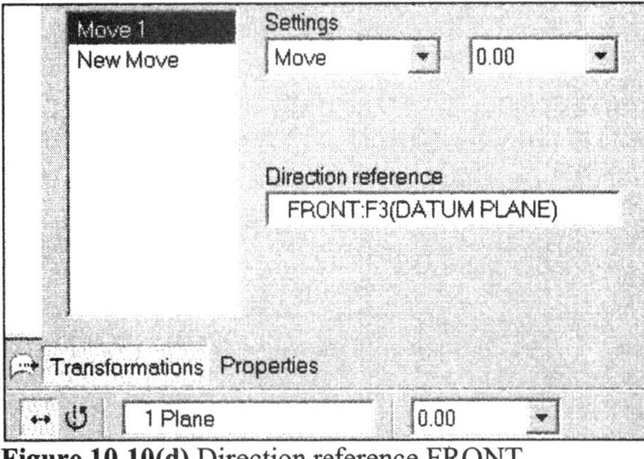

Figure 10.10(d) Direction reference FRONT

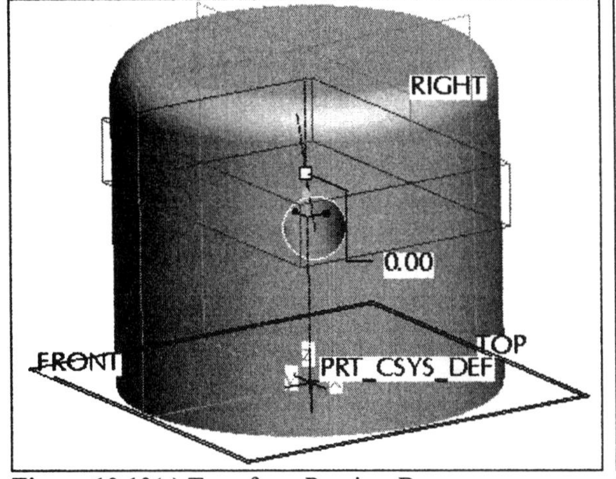

Figure 10.10(e) Transform Preview Box

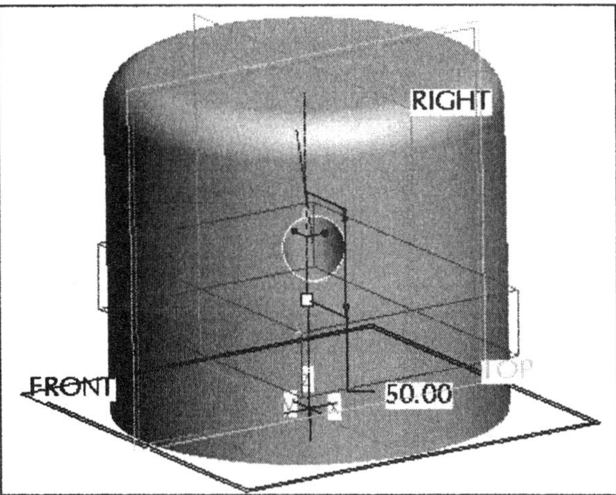

Figure 10.10(f) Drag to New Position

With the **Ctrl** key pressed, click on the **Hole** and the **Moved Copy** in the Model Tree ⇒ **RMB** ⇒ **Group** [Fig. 10.10(g)] ⇒ ⊞ ⇒ ⊞ [Fig. 10.10(h)] ⇒ **RMB** ⇒ **Pattern** [Fig. 10.10(i)] ⇒ **Dimensions** tab ⇒ **MMB** rotate the model to see the dimensions clearer ⇒ Direction 1- pick the **30°** dimension ⇒ pattern members, type **6** [1 6 1 item(s)] [Fig. 10.10(j)] ⇒ **Enter** ⇒ **MMB** [Fig. 10.10(k)] ⇒ [AB] ⇒ [PICT1] [Fig. 10.11(a)] ⇒ **LMB** to deselect ⇒ **Ctrl+S** ⇒ **Enter** [Fig. 10.11(b)]

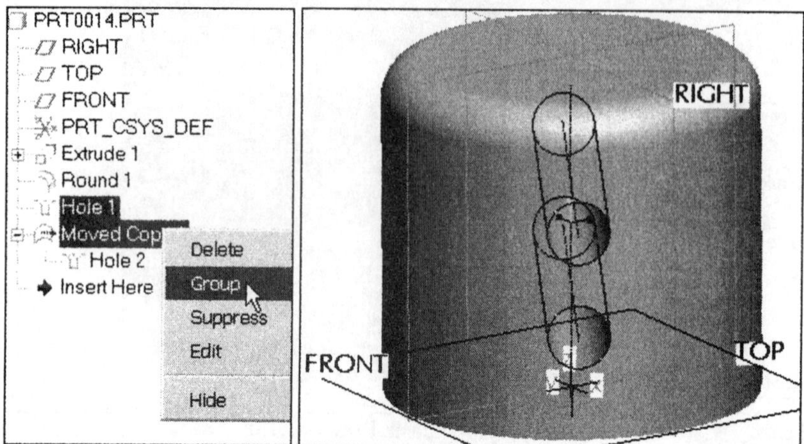

Figure 10.10(g) Grouping the Features

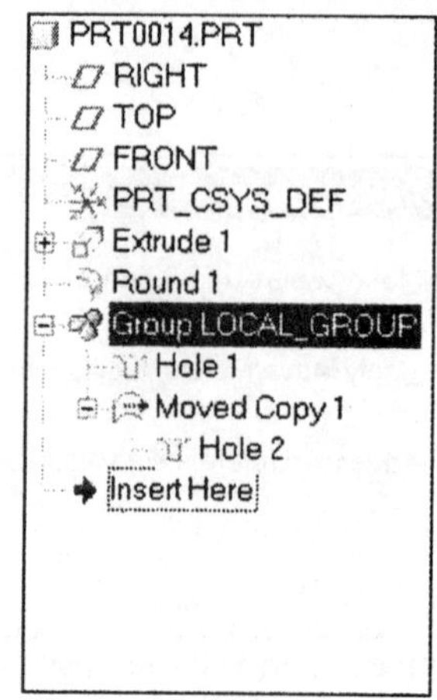

Figure 10.10(h) New Group

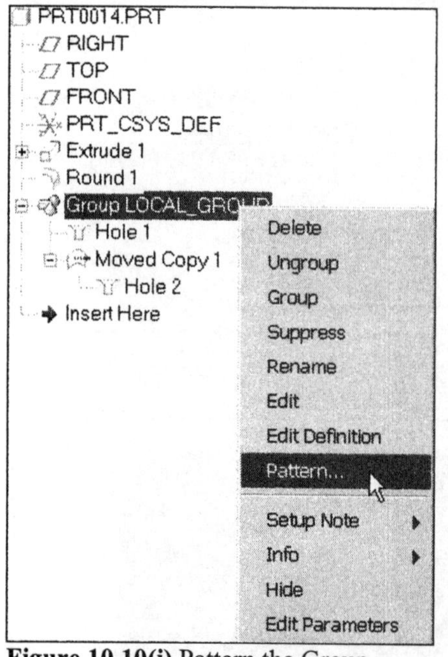

Figure 10.10(i) Pattern the Group

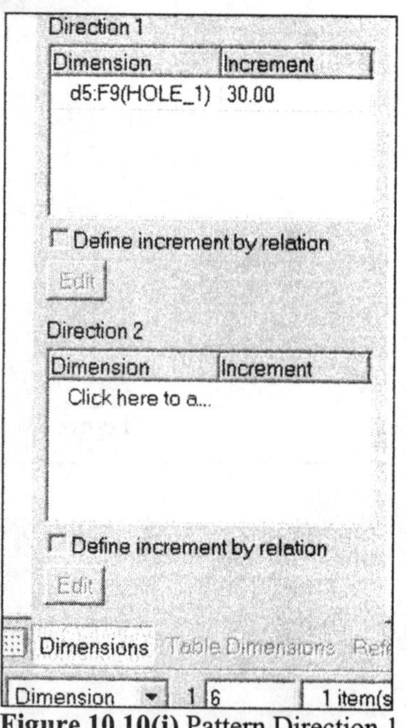

Figure 10.10(j) Pattern Direction 1

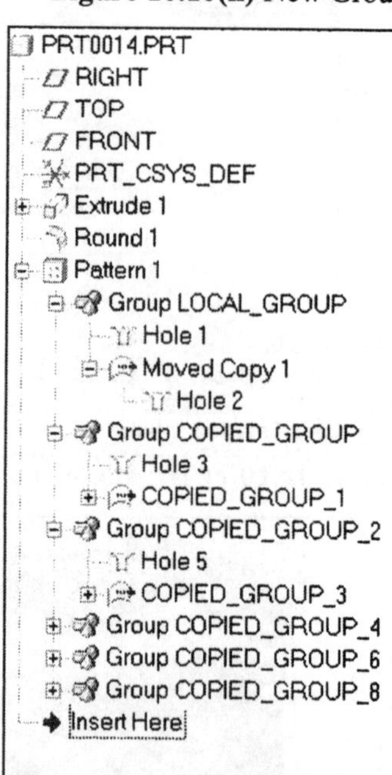

Figure 10.10(k) Pattern

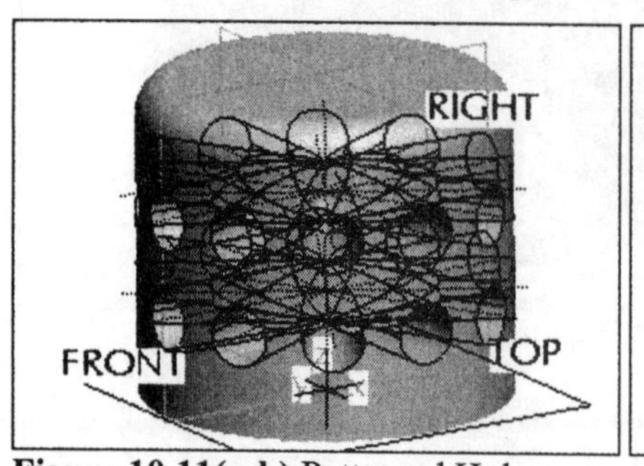

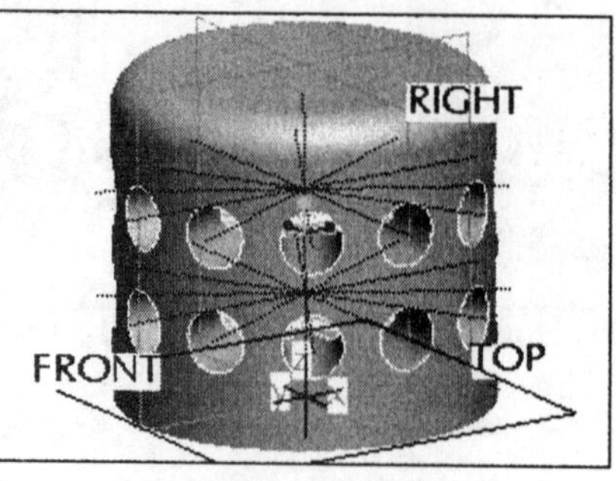

Figure 10.11(a-b) Patterned Holes

Pick on [Insert Here] in the Model Tree ⇒ drag **Insert Here** to a position before the **Pattern** [Pattern 1] and drop *(this will roll back the model to a state before the pattern was created and enter the Insert Mode)* (Fig. 10.12) ⇒ **MMB** rotate the model to see the bottom surface ⇒ pick the bottom surface [Fig. 10.13(a)] ⇒ **Shell Tool** [Fig. 10.13(b)] ⇒ [Fig. 10.13(c)] ⇒ ⇒ **MMB** ⇒ **Edit** ⇒ **Resume** ⇒ **All** ⇒ **Ctrl+S** ⇒ **Enter** [Fig. 10.13(d)]

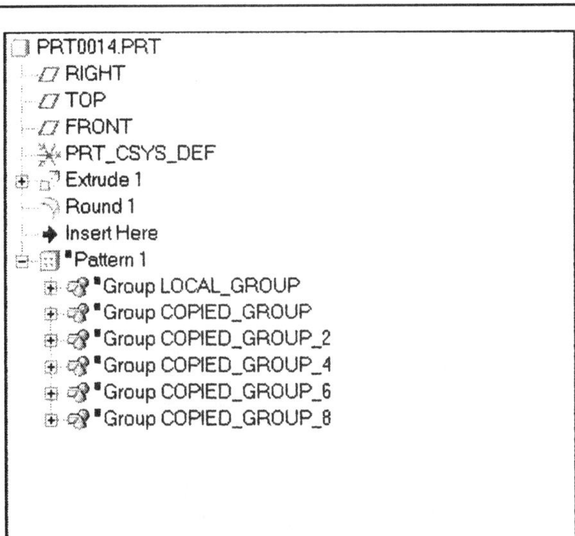

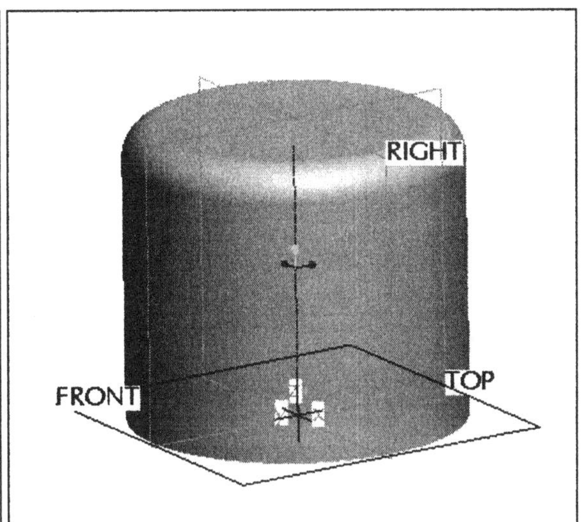

Figure 10.12 Insert Mode Activated

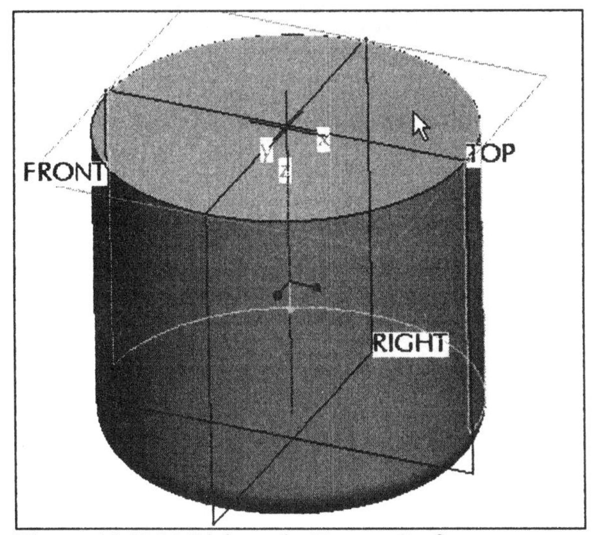

Figure 10.13(a) Pick on the Bottom Surface

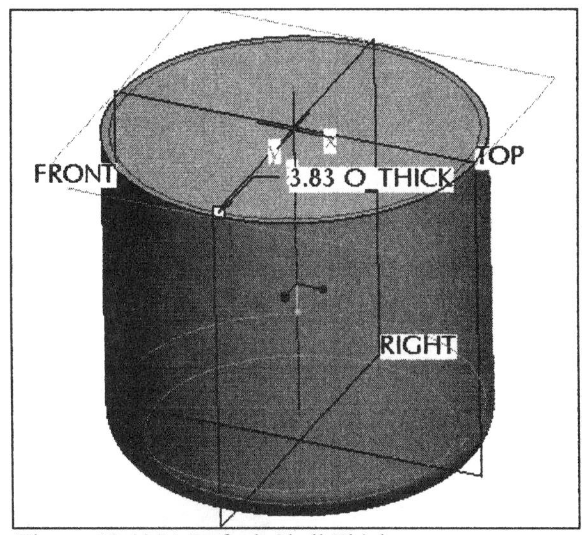

Figure 10.13(b) Default Shell Thickness

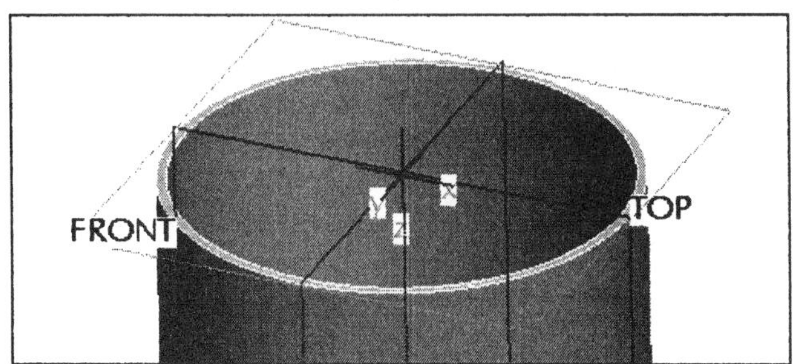

Figure 10.13(c) Shell Previewed

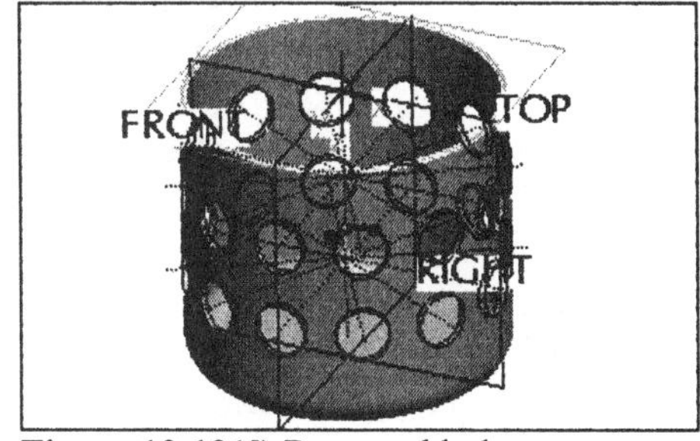

Figure 10.13(d) Resumed holes

Change the scale of the part: click: **Edit** ⇒ **Scale Model** ⇒ **.5** [Enter scale [1.0000]: .5] ⇒ **Enter** ⇒ **Yes** [Fig. 10.14(a)] ⇒ in the Model Tree, pick on **Extrude** ⇒ **Shift** key ⇒ pick on **Shell** ⇒ **RMB** ⇒ **Edit** [Fig. 10.14(b)] ⇒ [icon] ⇒ [PICT1] [Fig. 10.14(c)] ⇒ [icon] ⇒ **MMB** ⇒ **File** ⇒ **Close Window**

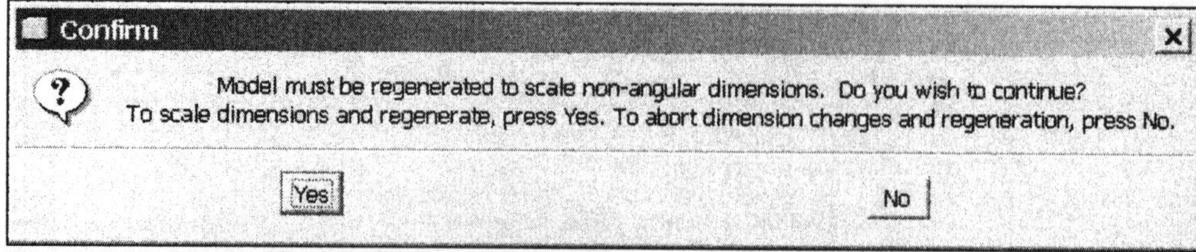

Figure 10.14(a) Confirm Dialog Box

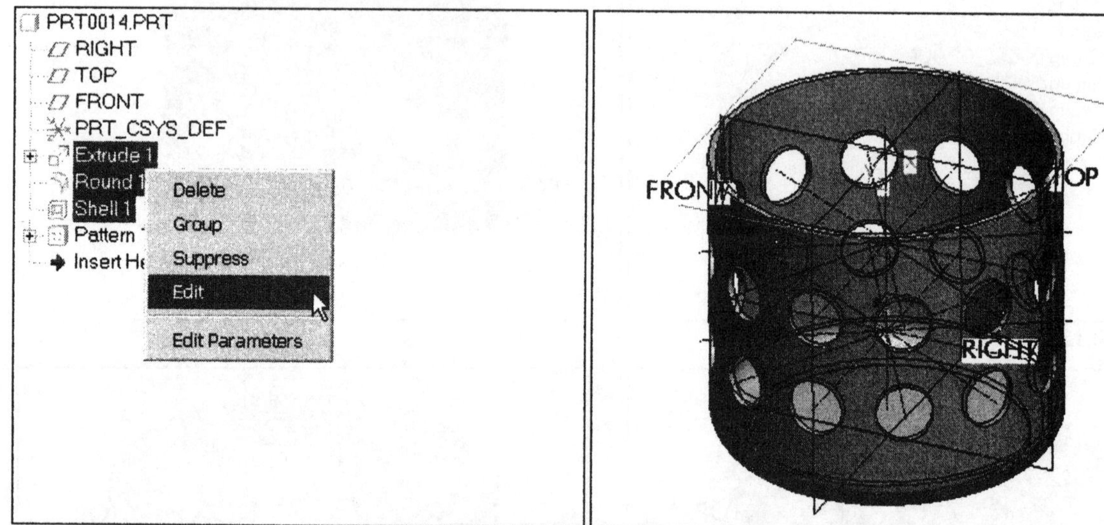

Figure 10.14(b) Edit

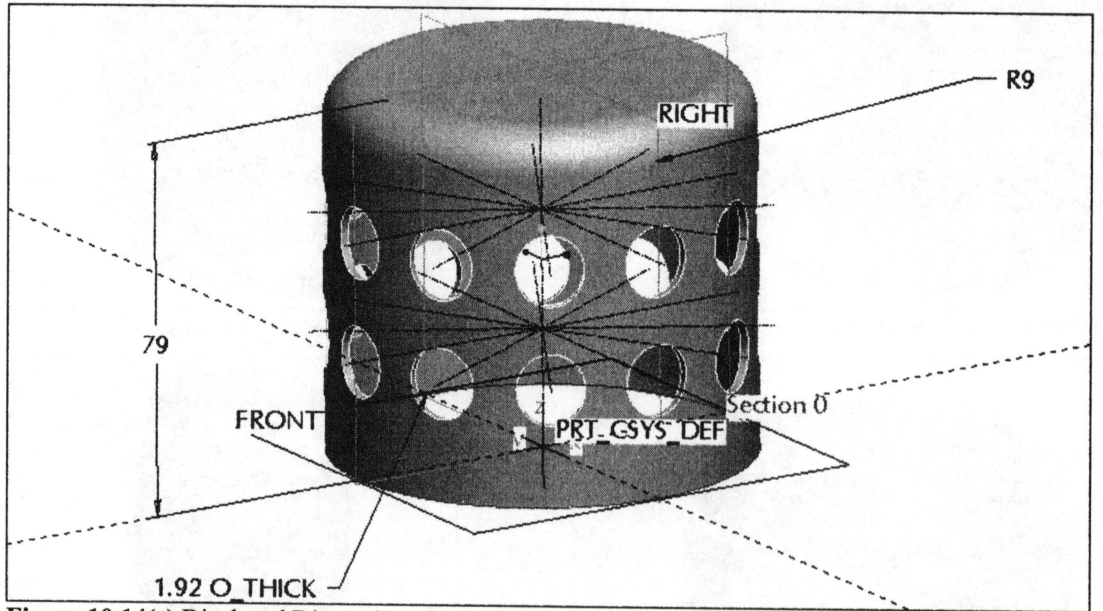

Figure 10.14(c) Displayed Dimensions (half of their original values)

See *www.cad-resources.com* ⇒ ***Downloads***, for extra Lessons and projects.

Part Model Fifteen (PRT0015.PRT) (Sheetmetal)

The Pro/E part database is different when you create parts using Pro/SHEETMETAL. All sheet metal parts are by definition thin-walled constant-thickness parts. Because of this, sheet metal parts have some unique properties that other Pro/E parts do no have. You may convert a solid part into a sheet metal part, but not a sheet metal part into a solid part. The solid must have a constant thickness.

Click: ⇒ ●**Part (PRT0015)** ⇒ ●**Sheetmetal** [Fig. 10.15(a)] ⇒ **OK** ⇒ **Edit** ⇒ **Setup** ⇒ **Units** [Fig. 10.15(b)] ⇒ **Close** ⇒ **Create Unattached Flat Wall** [Fig. 10.15(c)] ⇒ pick the **TOP** datum plane ⇒ **MMB** ⇒ **MMB** ⇒ **MMB** ⇒ **MMB** ⇒ **RMB** ⇒ **Rectangle** ⇒ sketch a rectangle [Fig. 10.15(d)] ⇒ ✓ ⇒ **MMB** ⇒ **Tools** ⇒ **Environment** ⇒ Standard Orient **Isometric** ⇒ **Apply** ⇒ **Close** ⇒ **Ctrl+D** ⇒ **Preview** ⇒ **MMB** ⇒ **Ctrl+S** ⇒ **Enter** [Fig. 10.15(e)] ⇒ **LMB** to deselect

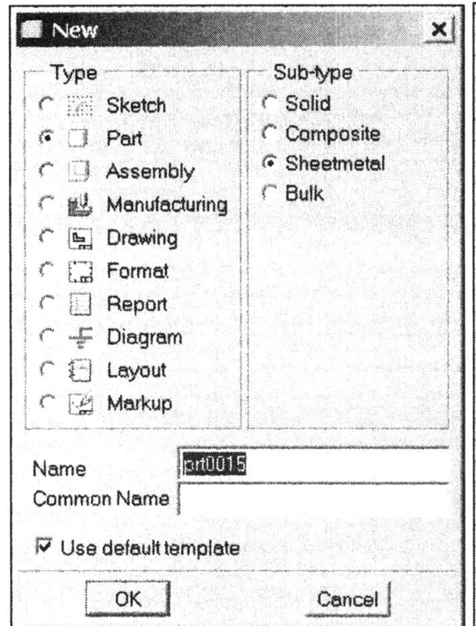

Figure 10.15(a) Sheetmetal

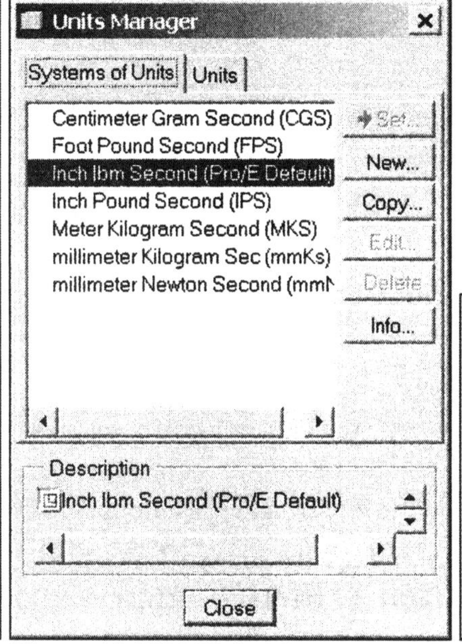

Figure 10.15(b) Units Manager Dialog Box

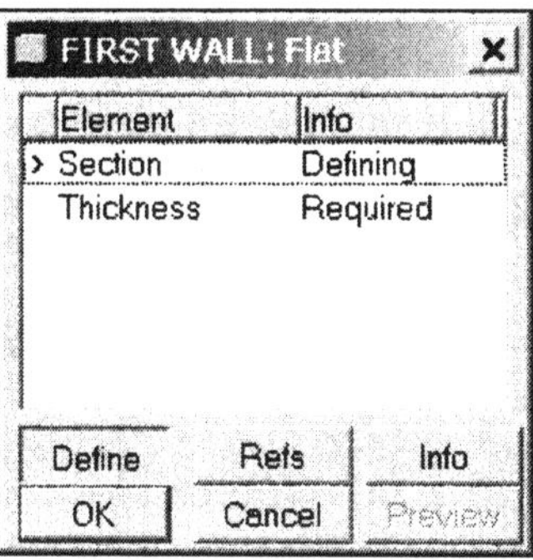

Figure 10.15(c) First Wall Dialog

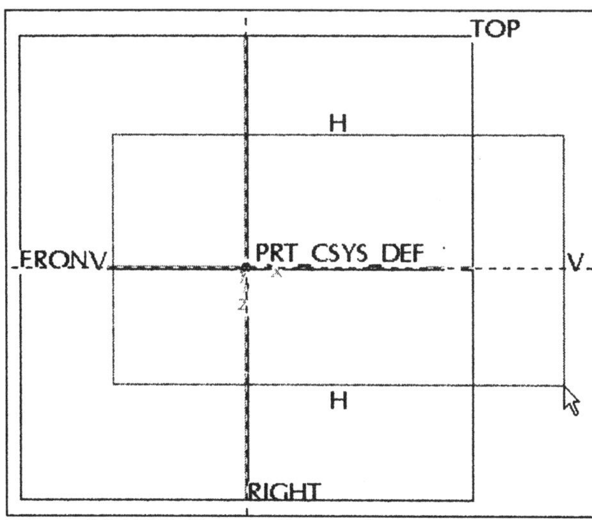

Figure 10.15(d) Sketch a Rectangle

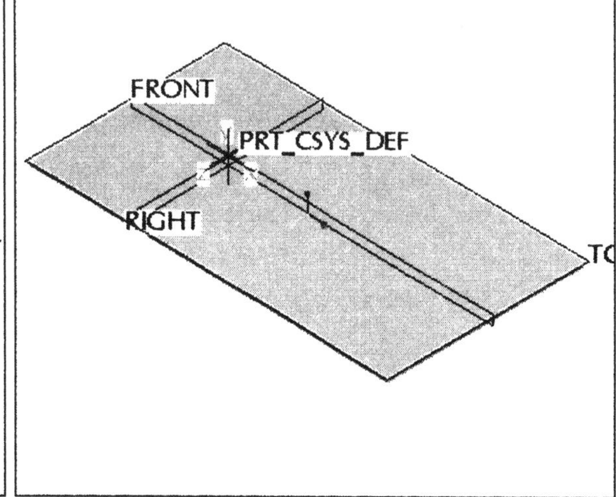

Figure 10.15(e) First Wall

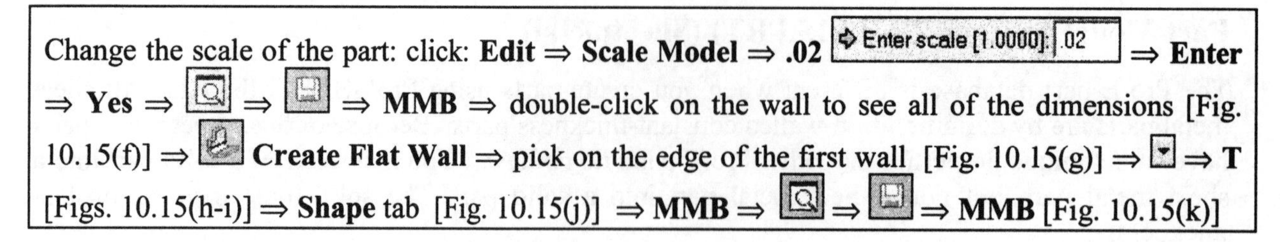

Change the scale of the part: click: **Edit** ⇒ **Scale Model** ⇒ **.02** [Enter scale [1.0000]: .02] ⇒ **Enter** ⇒ **Yes** ⇒ [icon] ⇒ [icon] ⇒ **MMB** ⇒ double-click on the wall to see all of the dimensions [Fig. 10.15(f)] ⇒ [icon] **Create Flat Wall** ⇒ pick on the edge of the first wall [Fig. 10.15(g)] ⇒ [icon] ⇒ **T** [Figs. 10.15(h-i)] ⇒ **Shape** tab [Fig. 10.15(j)] ⇒ **MMB** ⇒ [icon] ⇒ [icon] ⇒ **MMB** [Fig. 10.15(k)]

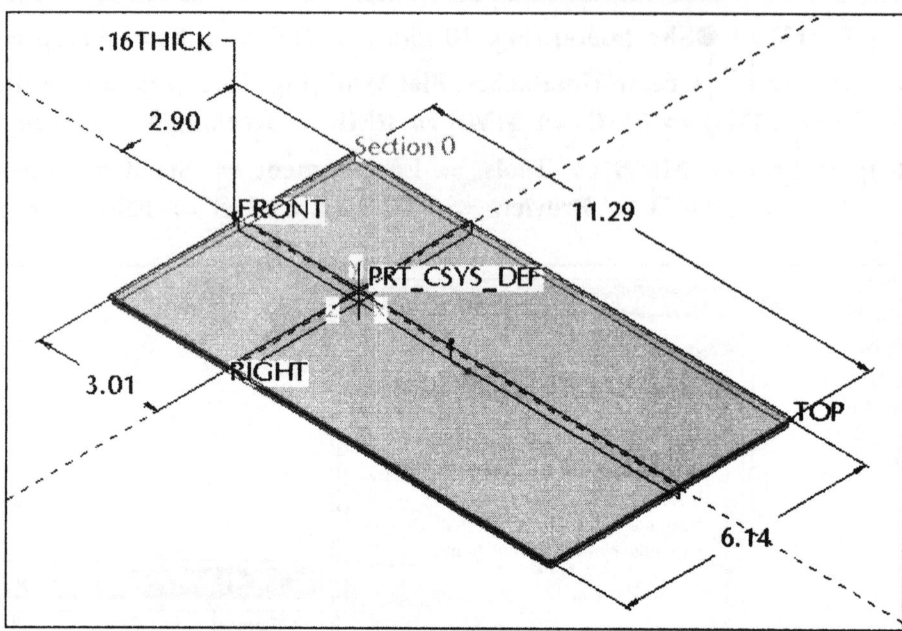

Figure 10.15(f) Walls Dimensions Displayed (yours will be different)

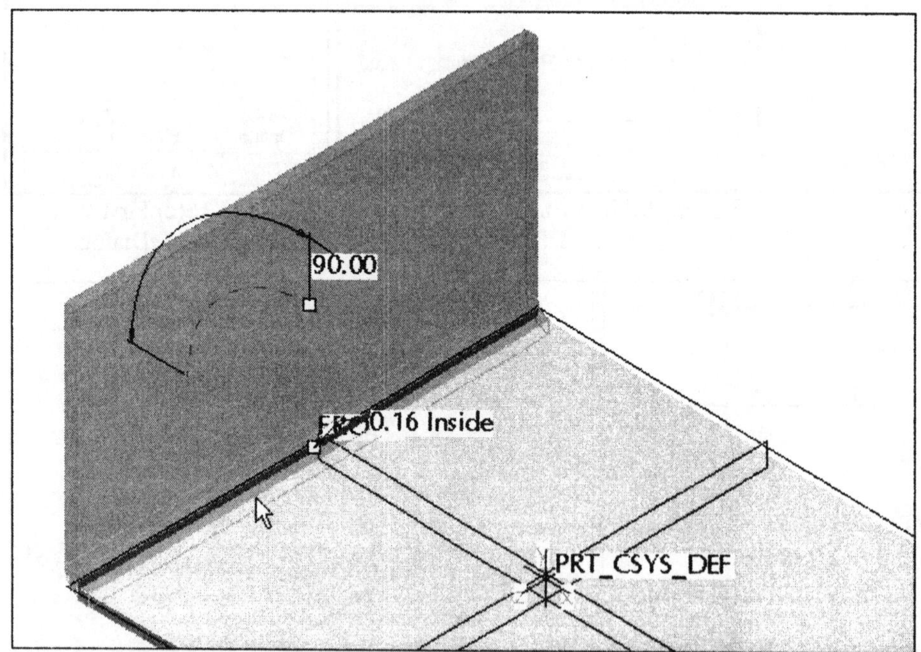

Figure 10.15(g) Pick on Edge of First Wall

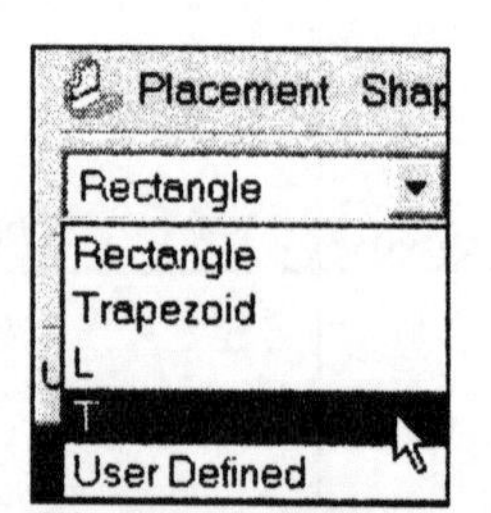

Figure 10.15(h) Select T

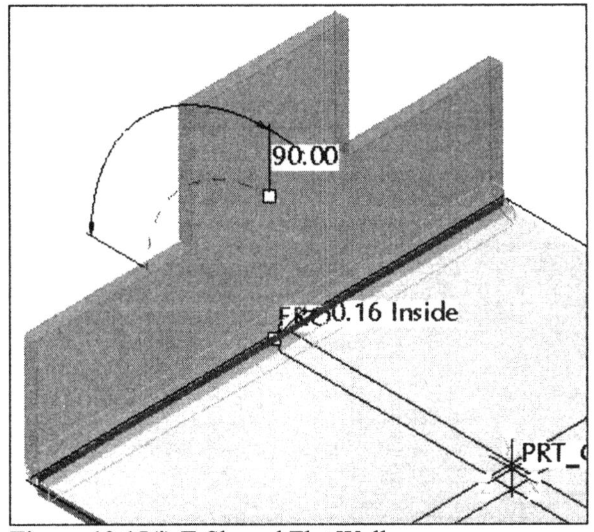

Figure 10.15(i) T-Shaped Flat Wall

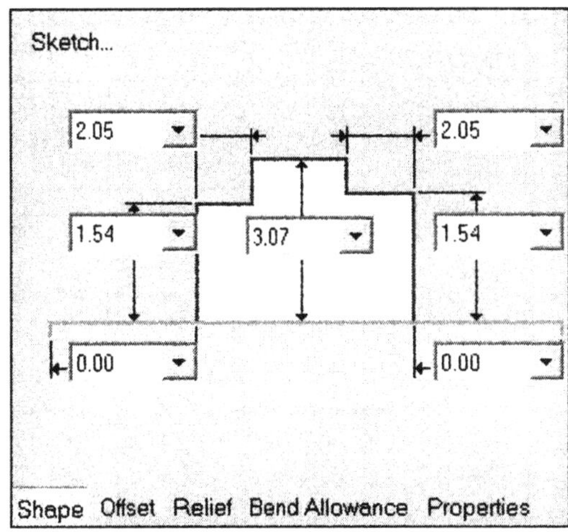

Figure 10.15(j) Shape Tab

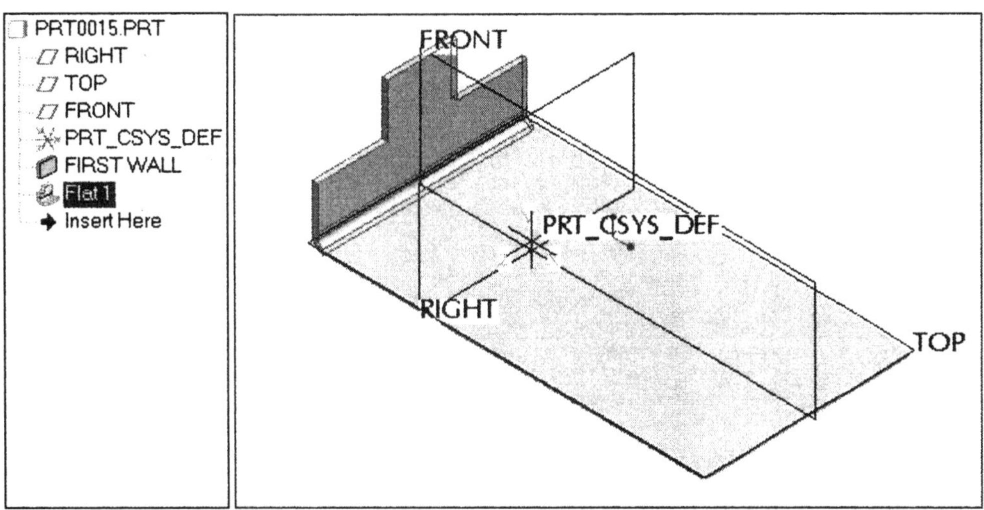

Figure 10.15(k) Flat Wall

Click:  **Create Flange Wall** ⇒ pick on the opposite edge of the first wall [Fig. 10.15(l)]

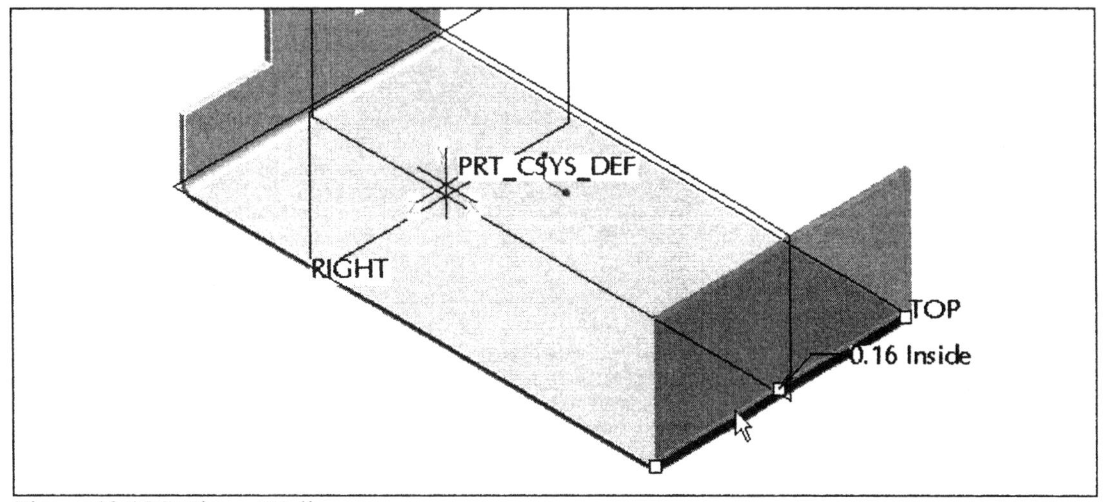

Figure 10.15(l) Flange Wall

Click: **Profile** tab [Fig. 10.15(m)] ⇒ **45** ⇒ **MMB** rotate model [Fig. 10.15(n)] ⇒ move a drag handle to shorten the wall [Fig. 10.15(o)] ⇒ **MMB** ⇒ **MMB** rotate model [Fig. 10.15(p)] ⇒ **Ctrl+S** ⇒ **Enter**

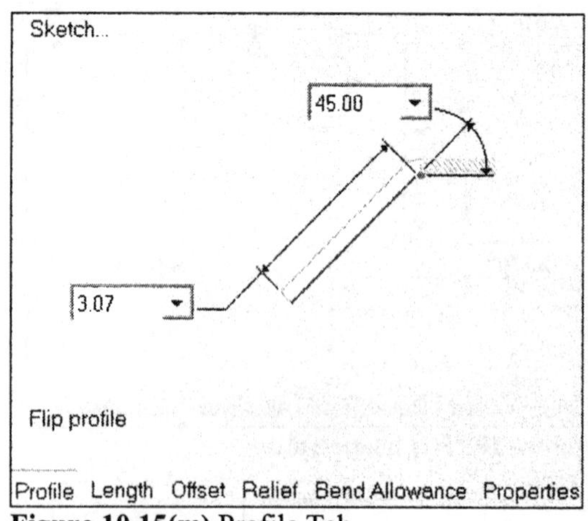

Figure 10.15(m) Profile Tab

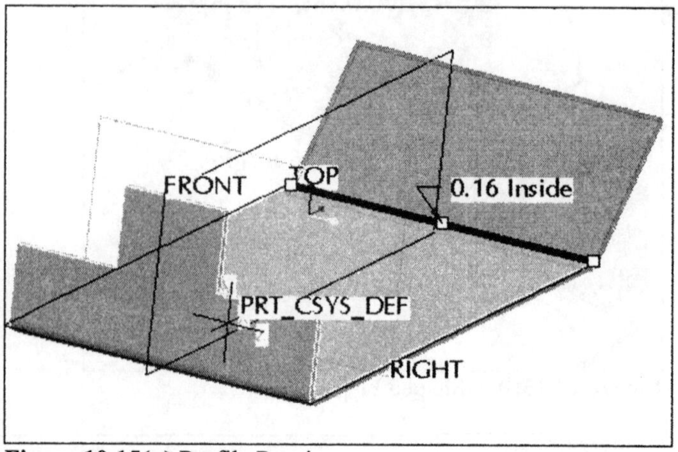

Figure 10.15(n) Profile Preview

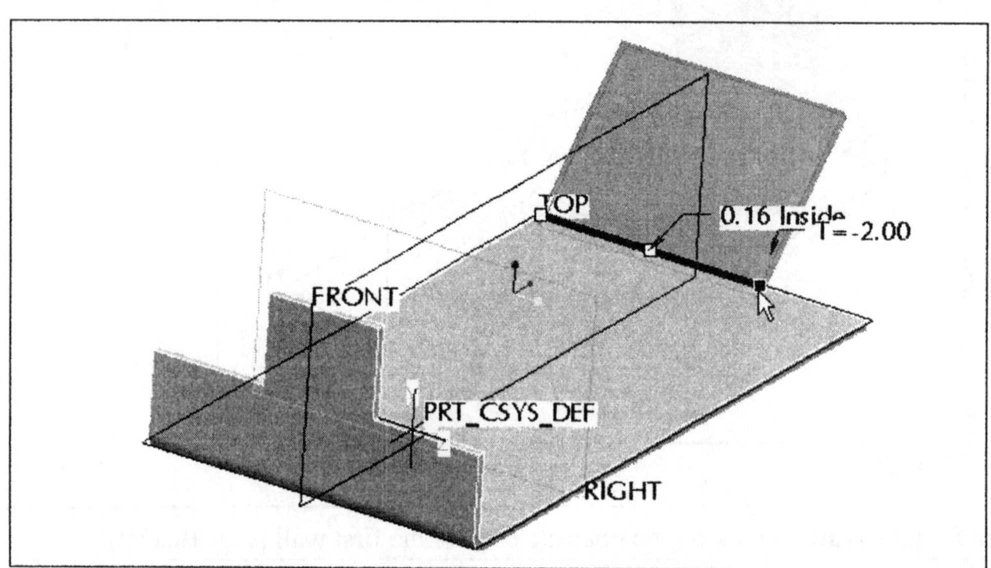

Figure 10.15(o) Move a Drag Handle

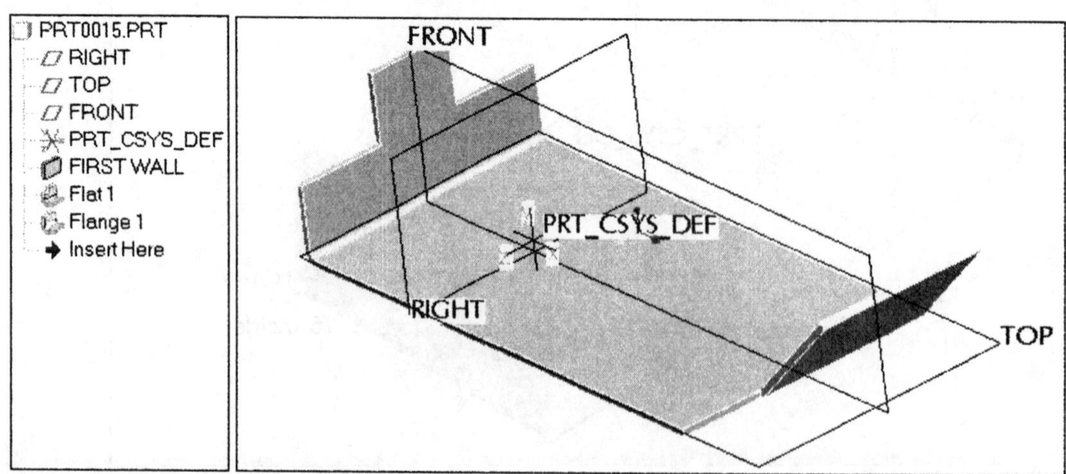

Figure 10.15(p) Completed Flange Wall

Click: **Insert** ⇒ Hole... ⇒ pick on the face of the part [Fig. 10.15(q)] ⇒ **RMB** ⇒ Secondary References Collector ⇒ pick the front edge [Fig. 10.15(r)] ⇒ press and hold **Ctrl** key ⇒ pick on the adjacent edge [Fig. 10.15(s)] ⇒ adjust dimensions ⇒ [Fig. 10.15(t)] ⇒ **MMB** [Fig. 10.15(u)]

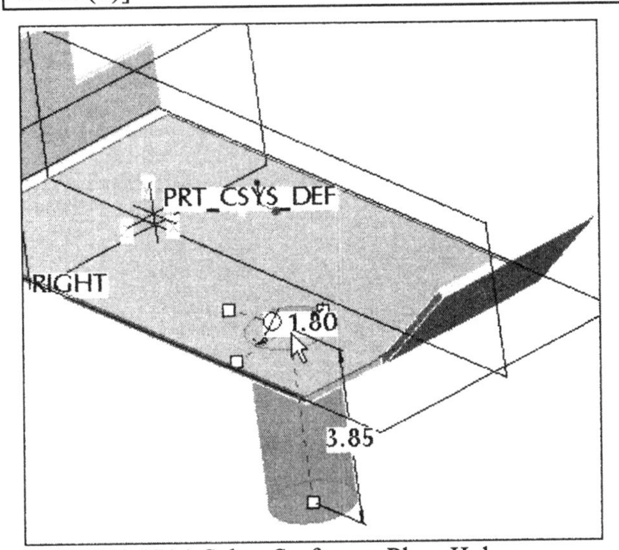

Figure 10.15(q) Select Surface to Place Hole

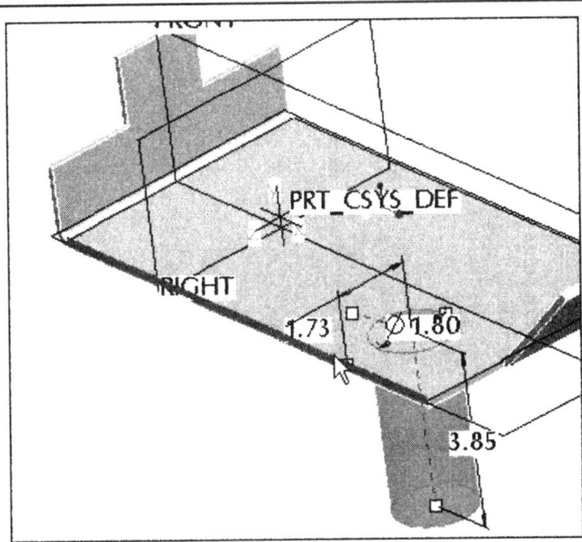

Figure 10.15(r) First Edge Reference

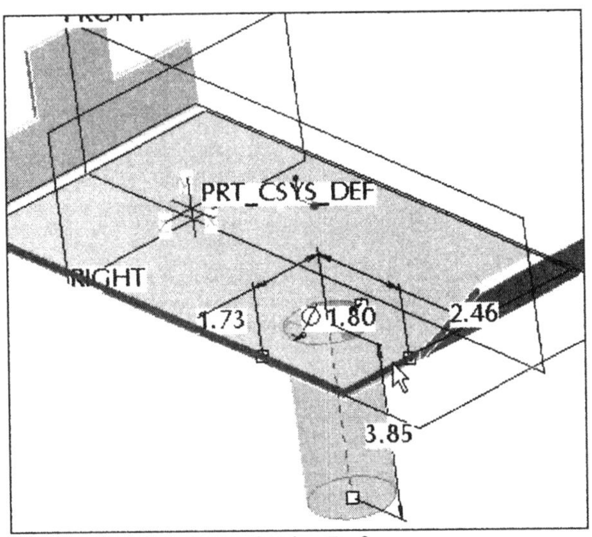

Figure 10.15(s) Second Edge Reference

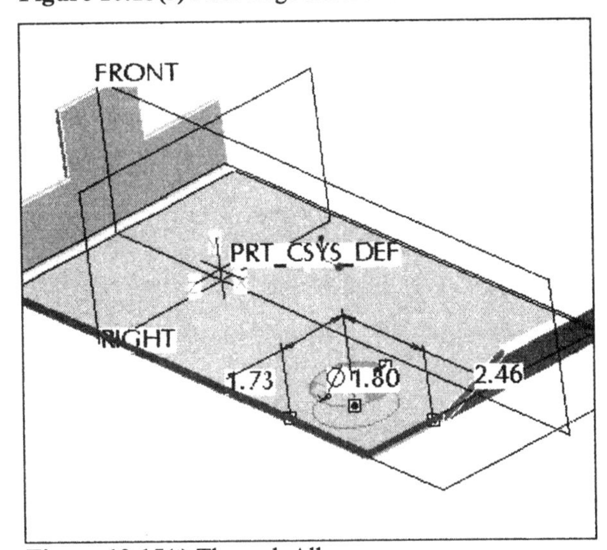

Figure 10.15(t) Through All

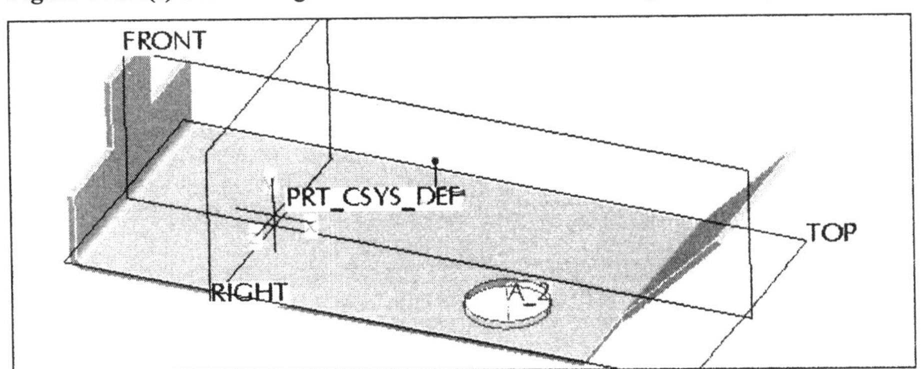

Figure 10.15(u) Completed Hole

With the **Hole** selected in the Model Tree, click **RMB** ⇒ **Pattern** ⇒ **Ctrl+MMB** zoom in ⇒ Dimension ⇒ ⇒ Direction ⇒ pick on the same first edge used to locate the hole [Fig. 10.15(v)] ⇒ **Flip the first direction** ⇒ type **4** as the number of holes in this direction ⇒ **Enter** ⇒ move the drag handle to space the holes [Fig. 10.15(w)] ⇒ 2 Click here to add ⇒ 2 Select 1 item ⇒ pick on the same second edge used to locate the hole [Fig. 10.15(x)] ⇒ move the drag handle [Fig. 10.15(y)]

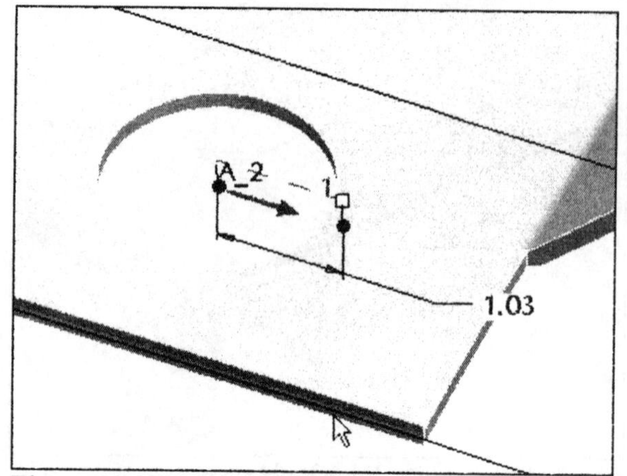

Figure 10.15(v) First Direction

Figure 10.15(w) First Direction Reversed

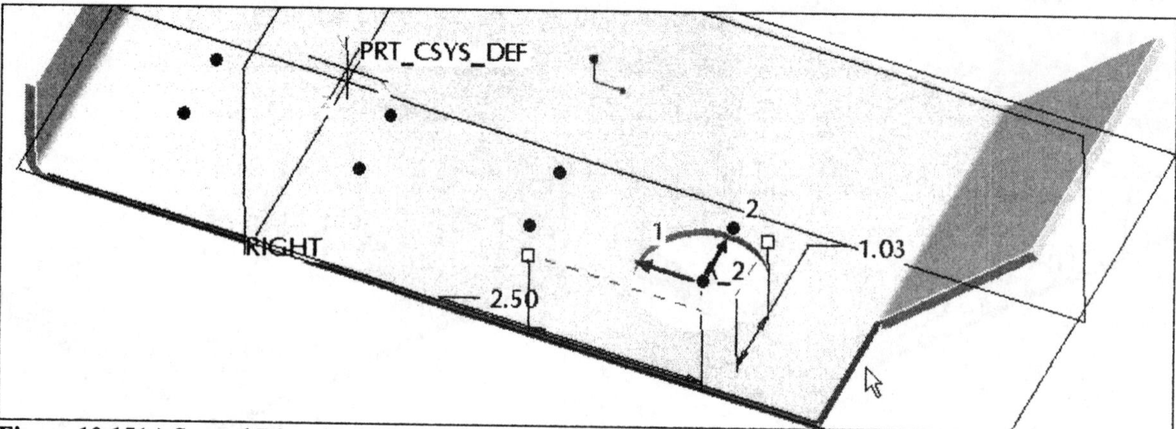

Figure 10.15(x) Second Direction

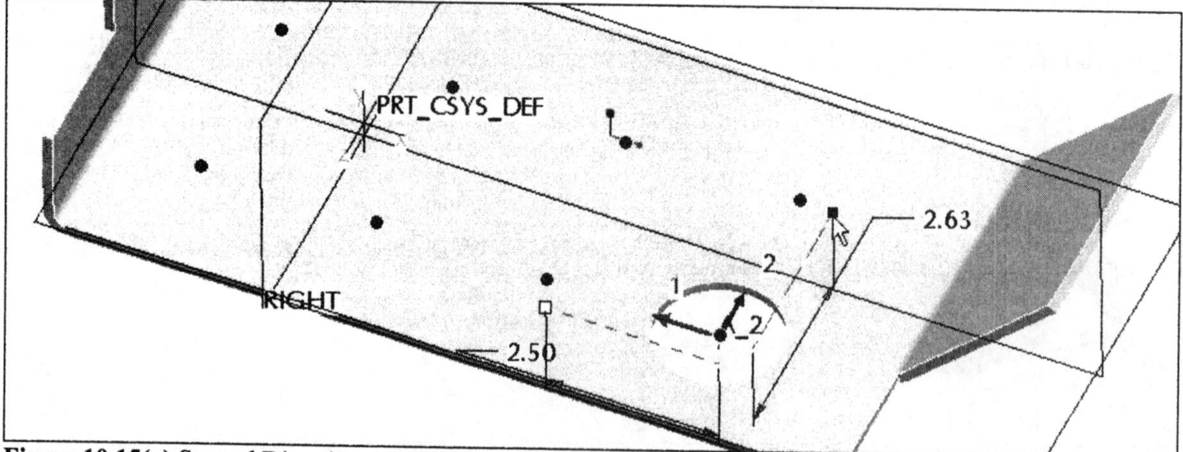

Figure 10.15(y) Second Direction Adjusted

Pick on the two pattern hole preview dots (black) to remove them from the pattern (dots turn white) [Fig. 10.15(z)] ⇒ **MMB** (Fig. 10.16(a) ⇒ **LMB** to deselect

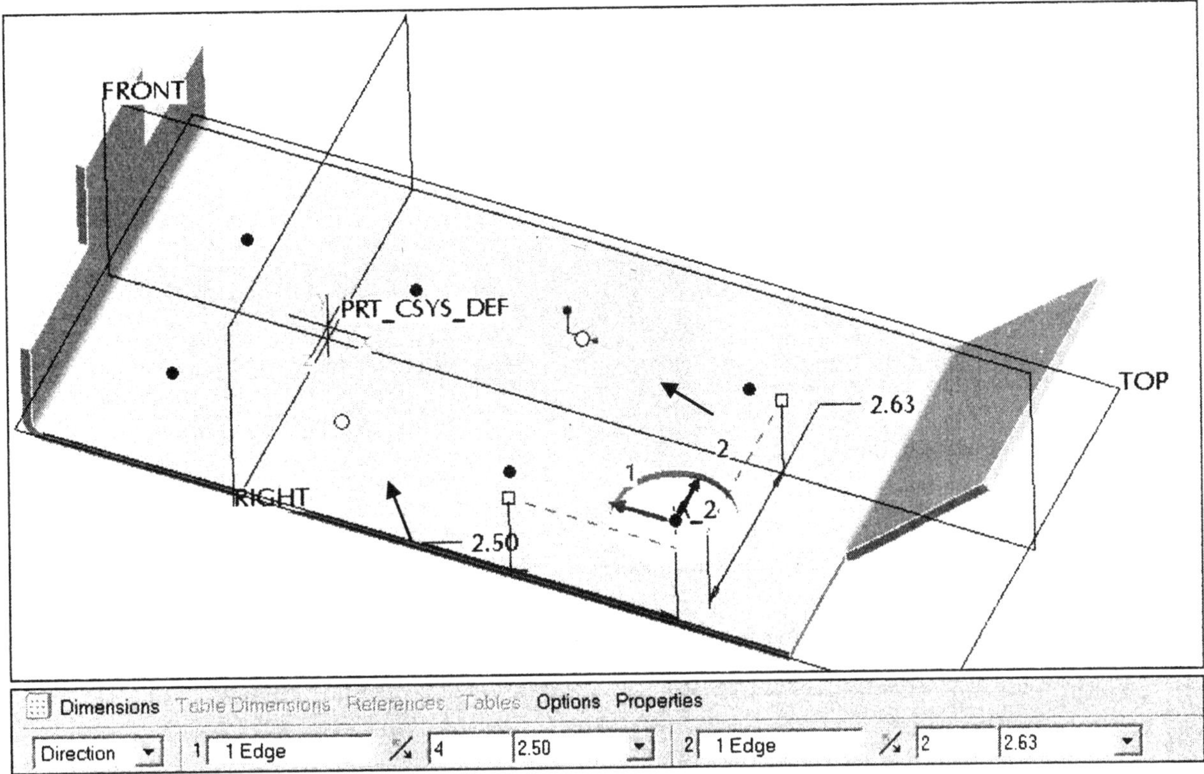

Figure 10.15(z) Remove Two Holes from the Pattern

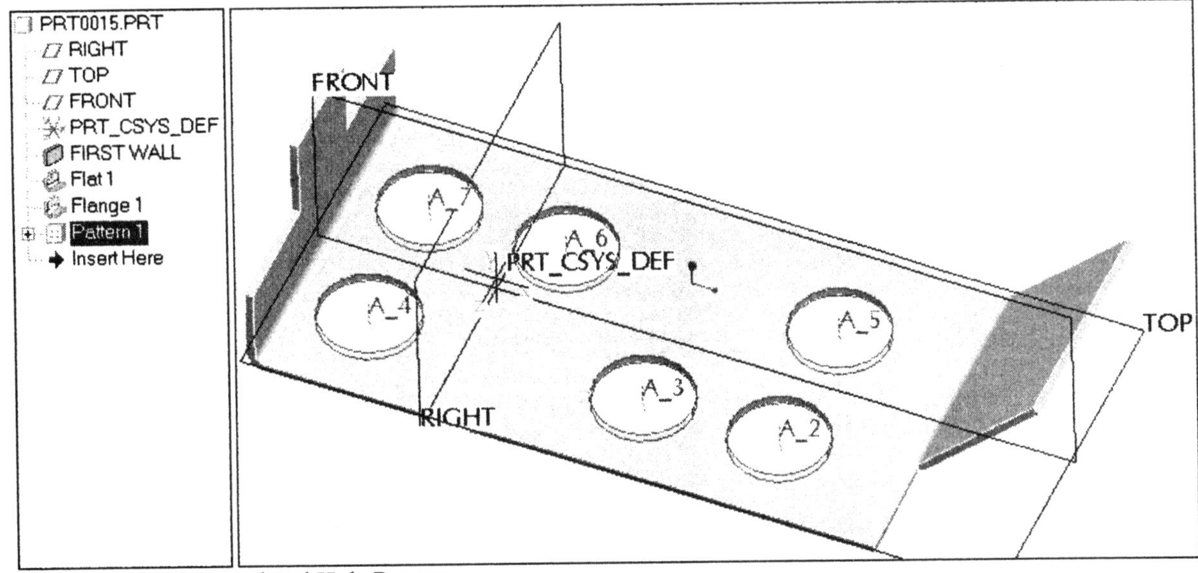

Figure 10.16(a) Completed Hole Pattern

Click: [icon] ⇒ **Ctrl+D** ⇒ [icon] **Create Flat Pattern** ⇒ pick on the upper surface [Figs. 10.16(b-c)] ⇒ [icon] ⇒ **Ctrl+S** ⇒ **Enter** ⇒ **Tools** ⇒ **Environment** ⇒ Standard Orient **Trimetric** ⇒ **Apply** ⇒ **Close** ⇒ **File** ⇒ **Close Window**

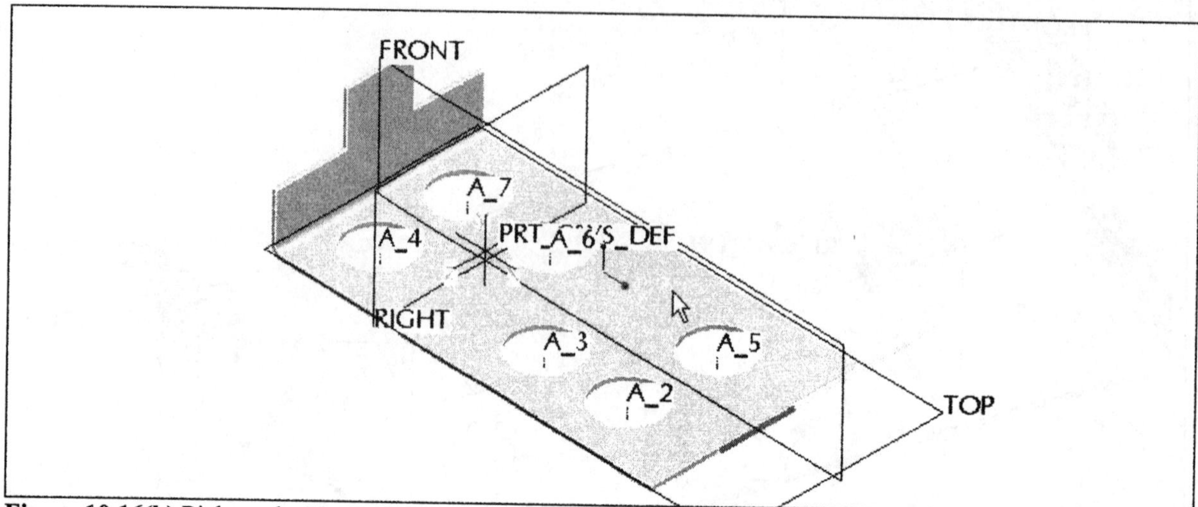

Figure 10.16(b) Pick on the Upper Surface

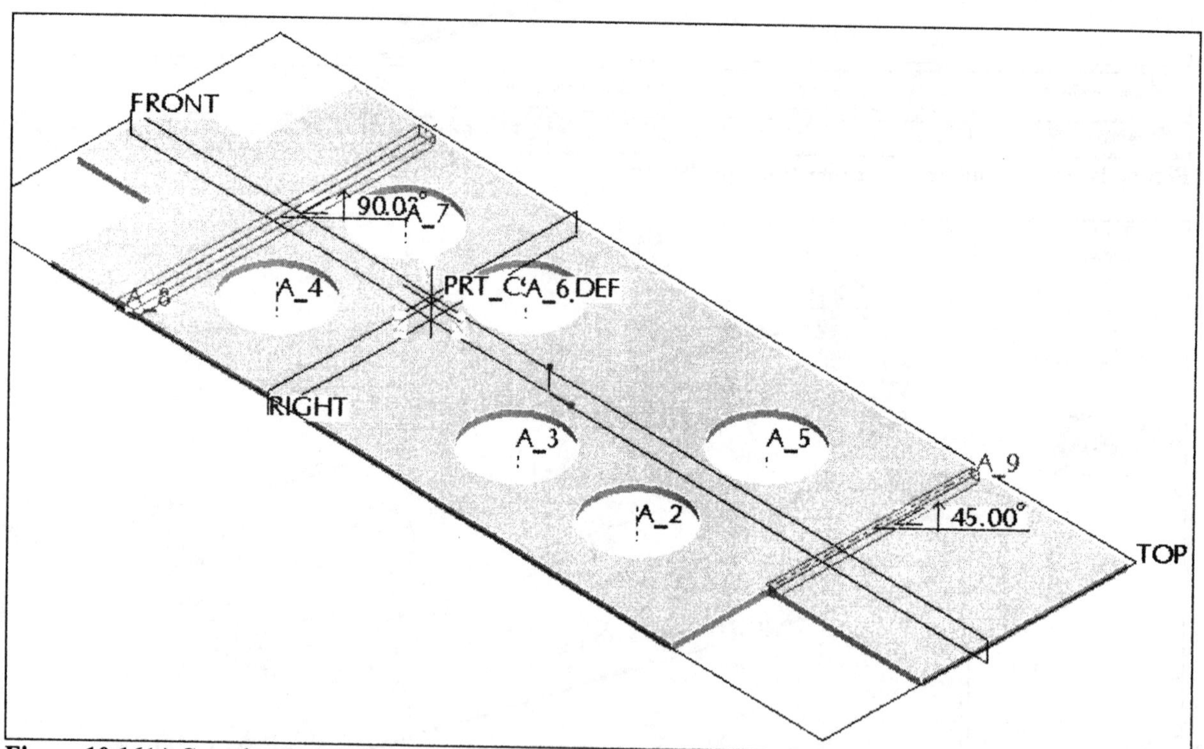

Figure 10.16(c) Completed Flat Pattern

See ***www.cad-resources.com*** ⇒ ***Downloads***, for extra Lessons and projects.

Part Model Sixteen (PRT0016.PRT) (Surface Offset)

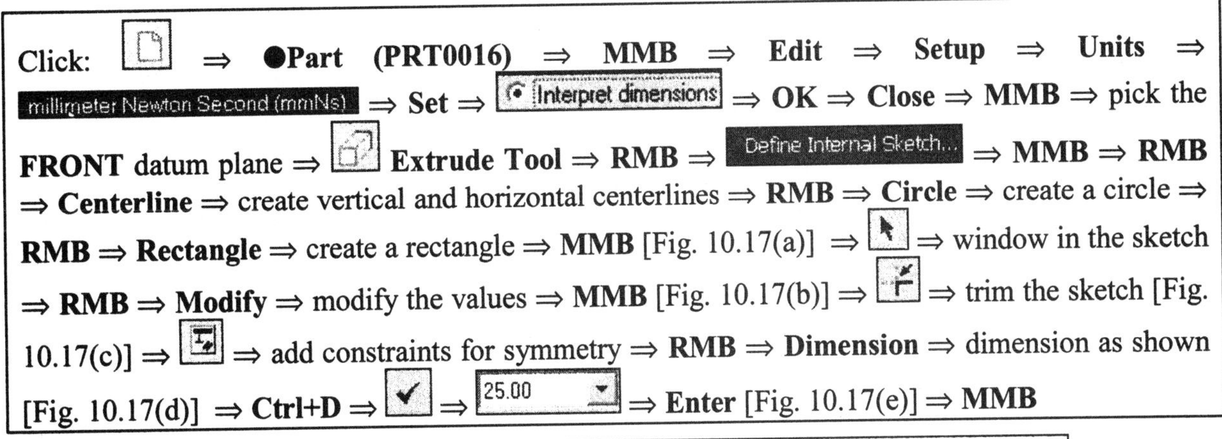

Click: ⇒ **●Part** **(PRT0016)** ⇒ **MMB** ⇒ **Edit** ⇒ **Setup** ⇒ **Units** ⇒ millimeter Newton Second (mmNs) ⇒ **Set** ⇒ Interpret dimensions ⇒ **OK** ⇒ **Close** ⇒ **MMB** ⇒ pick the **FRONT** datum plane ⇒ **Extrude Tool** ⇒ **RMB** ⇒ Define Internal Sketch... ⇒ **MMB** ⇒ **RMB** ⇒ **Centerline** ⇒ create vertical and horizontal centerlines ⇒ **RMB** ⇒ **Circle** ⇒ create a circle ⇒ **RMB** ⇒ **Rectangle** ⇒ create a rectangle ⇒ **MMB** [Fig. 10.17(a)] ⇒ ⇒ window in the sketch ⇒ **RMB** ⇒ **Modify** ⇒ modify the values ⇒ **MMB** [Fig. 10.17(b)] ⇒ ⇒ trim the sketch [Fig. 10.17(c)] ⇒ ⇒ add constraints for symmetry ⇒ **RMB** ⇒ **Dimension** ⇒ dimension as shown [Fig. 10.17(d)] ⇒ **Ctrl+D** ⇒ ⇒ 25.00 ⇒ **Enter** [Fig. 10.17(e)] ⇒ **MMB**

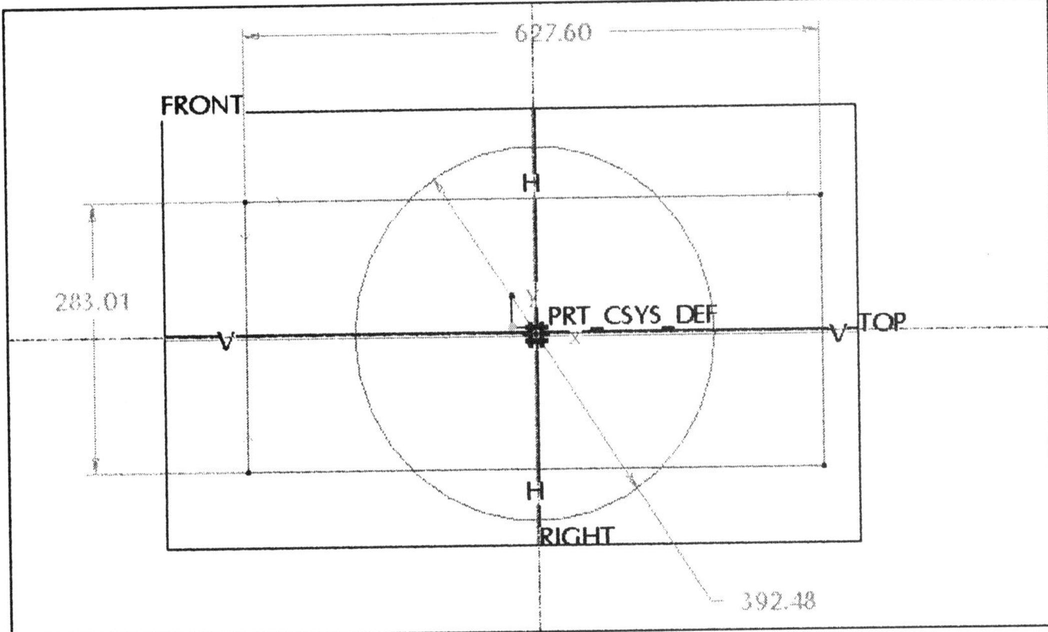

Figure 10.17(a) Sketch the Circle and Rectangle

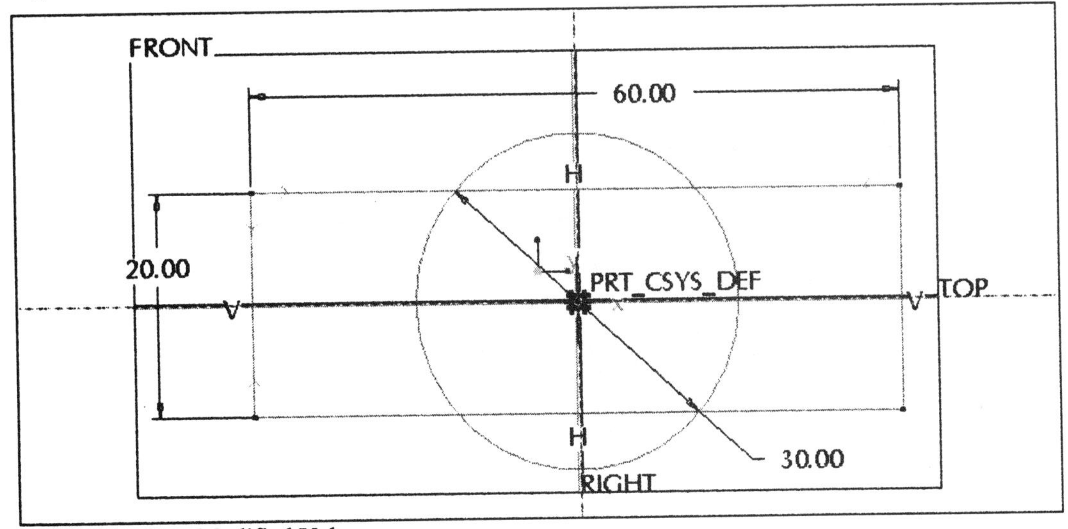

Figure 10.17(b) Modified Values

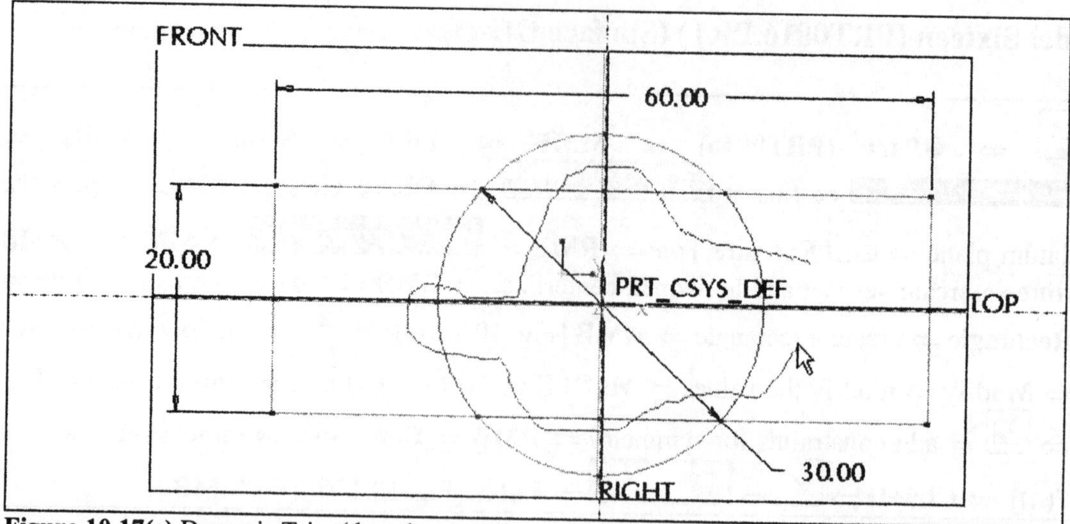

Figure 10.17(c) Dynamic Trim (drag the cursor over the sketch entities you want deleted)

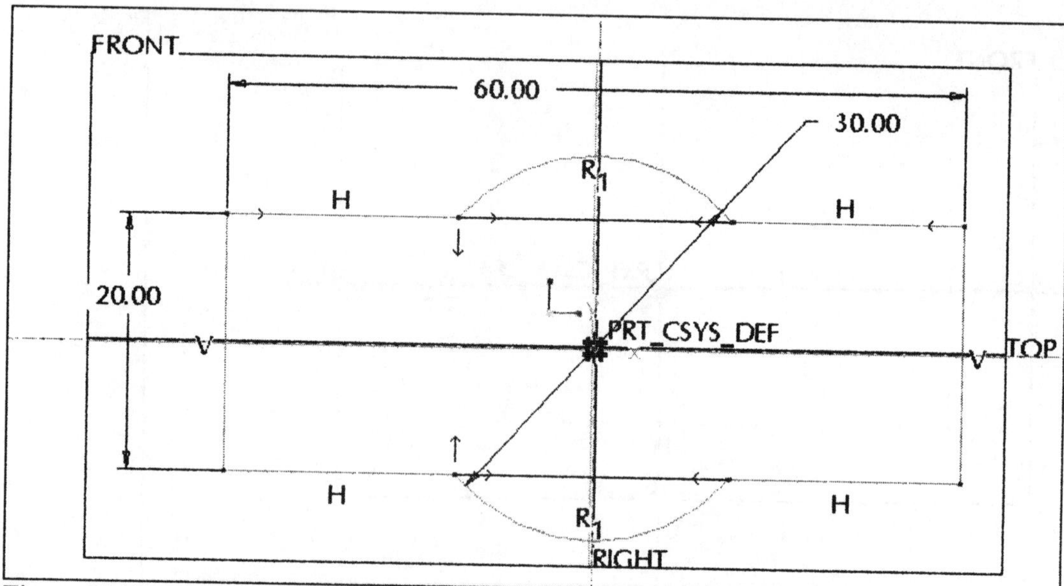

Figure 10.17(d) Add Constraints and Dimensions

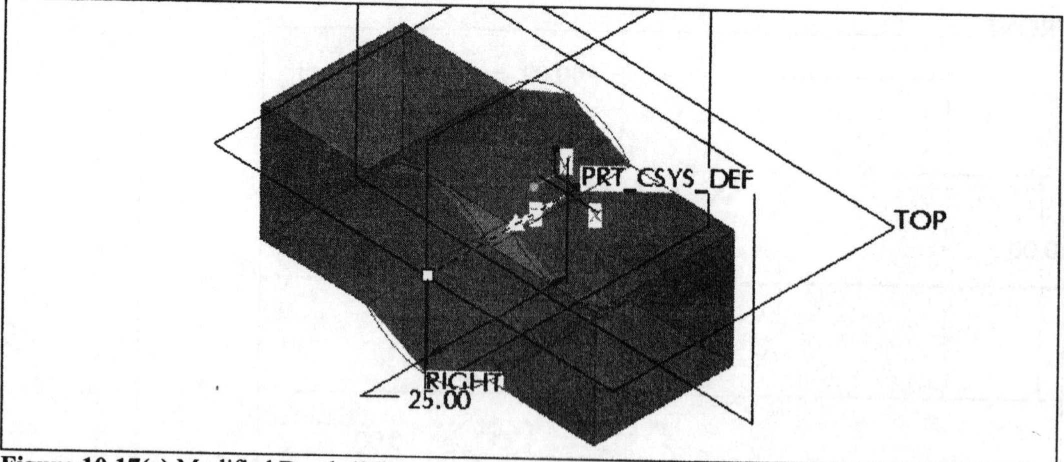

Figure 10.17(e) Modified Depth (**25**)

Pick an edge, press **Ctrl** key and then pick on the remaining seven edges [Fig. 10.17(f)] ⇒ **RMB** ⇒ **Round Edges** ⇒ 2.00 ⇒ **Enter** ⇒ **MMB** ⇒ pick on the **TOP** datum plane ⇒ **Extrude Tool** ⇒ **Extrude as surface** ⇒ ⇒ **40** depth ⇒ **Enter** ⇒ **Placement** tab ⇒ **Define** [Fig. 10.17(g)] ⇒ pick on the arrow to reverse the sketch view direction [Fig. 10.17(h)]

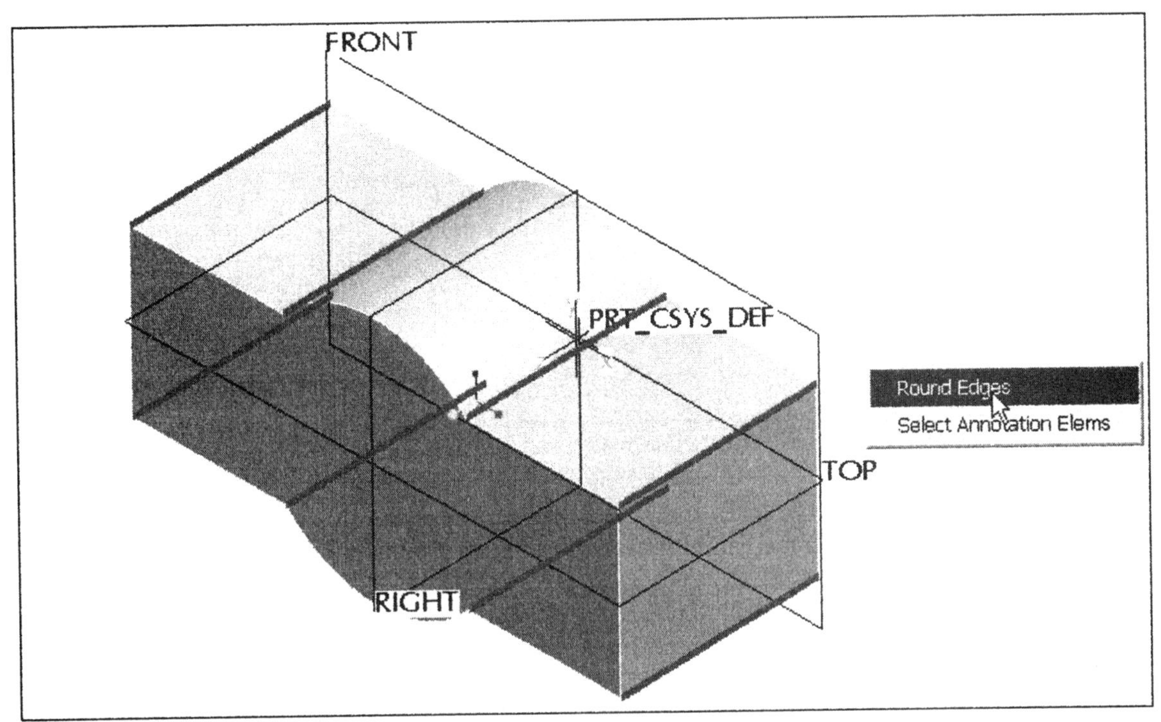

Figure 10.17(f) Round Edges

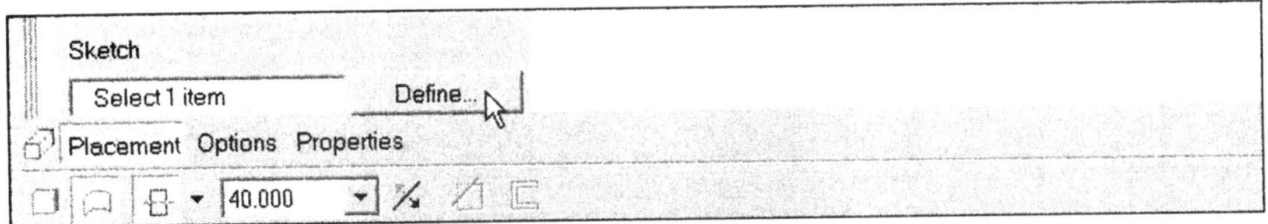

Figure 10.17(g) Surface Extrude Dashboard Options and Values

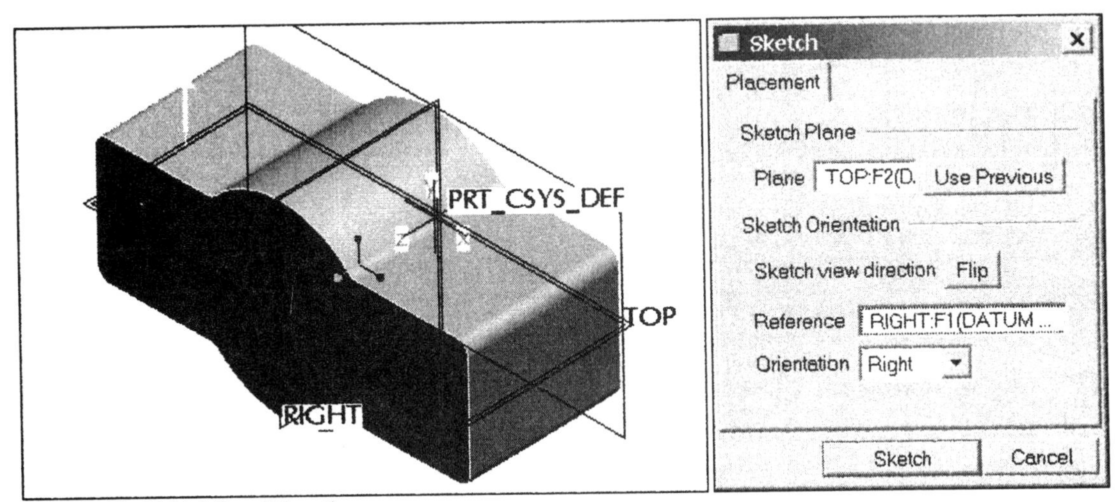

Figure 10.17(h) View Direction

Click: **MMB** ⇒ **RMB** ⇒ **Centerline** ⇒ create a vertical centerline ⇒ **Create a spline curve** ⇒ sketch a curve [Fig. 10.17(i)] ⇒ **MMB** ⇒ make all points symmetrical and modify the dimensions [Fig. 10.17(j)] ⇒ ⇒ **Ctrl+D** [Fig. 10.17(k)] ⇒ **MMB** ⇒ **LMB** to deselect

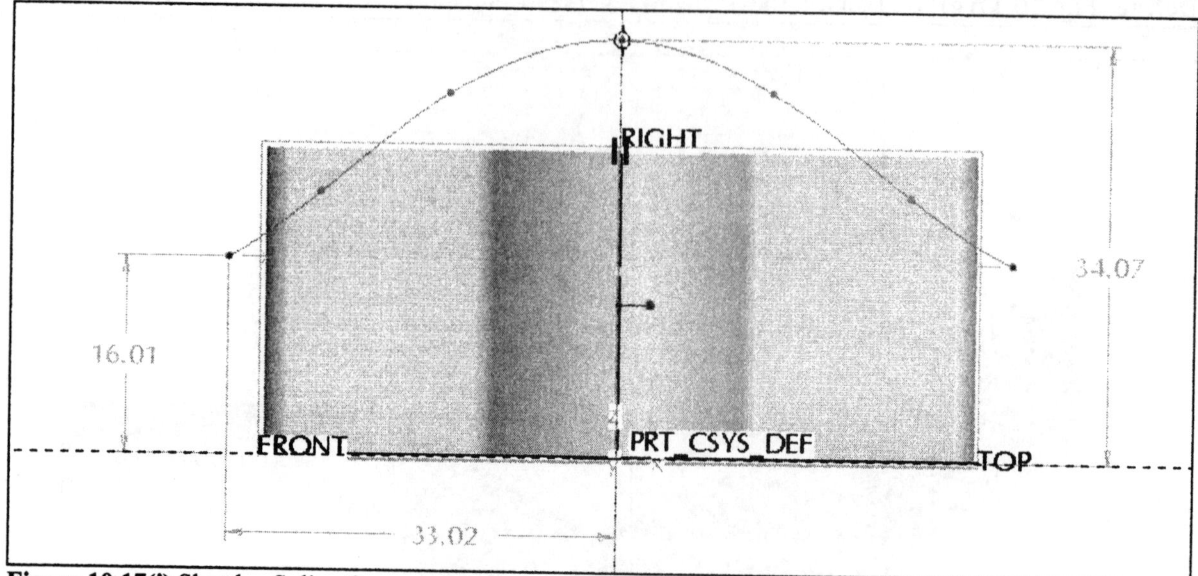

Figure 10.17(i) Sketch a Spline Curve

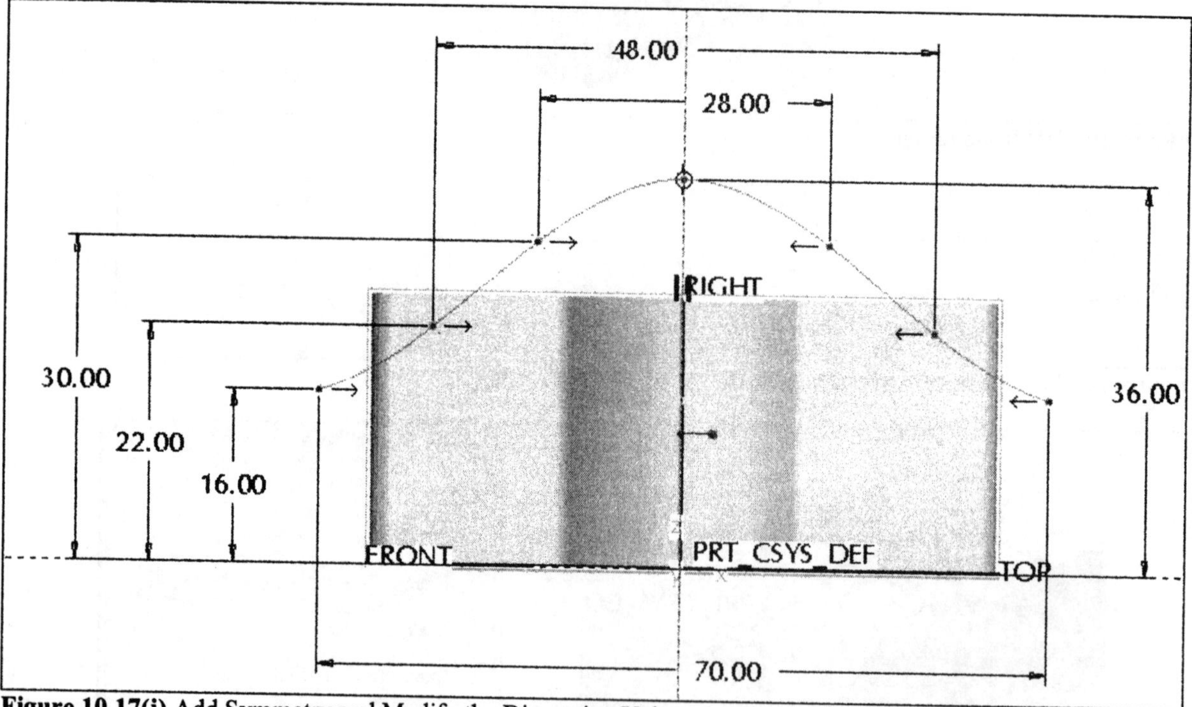

Figure 10.17(j) Add Symmetry and Modify the Dimension Values

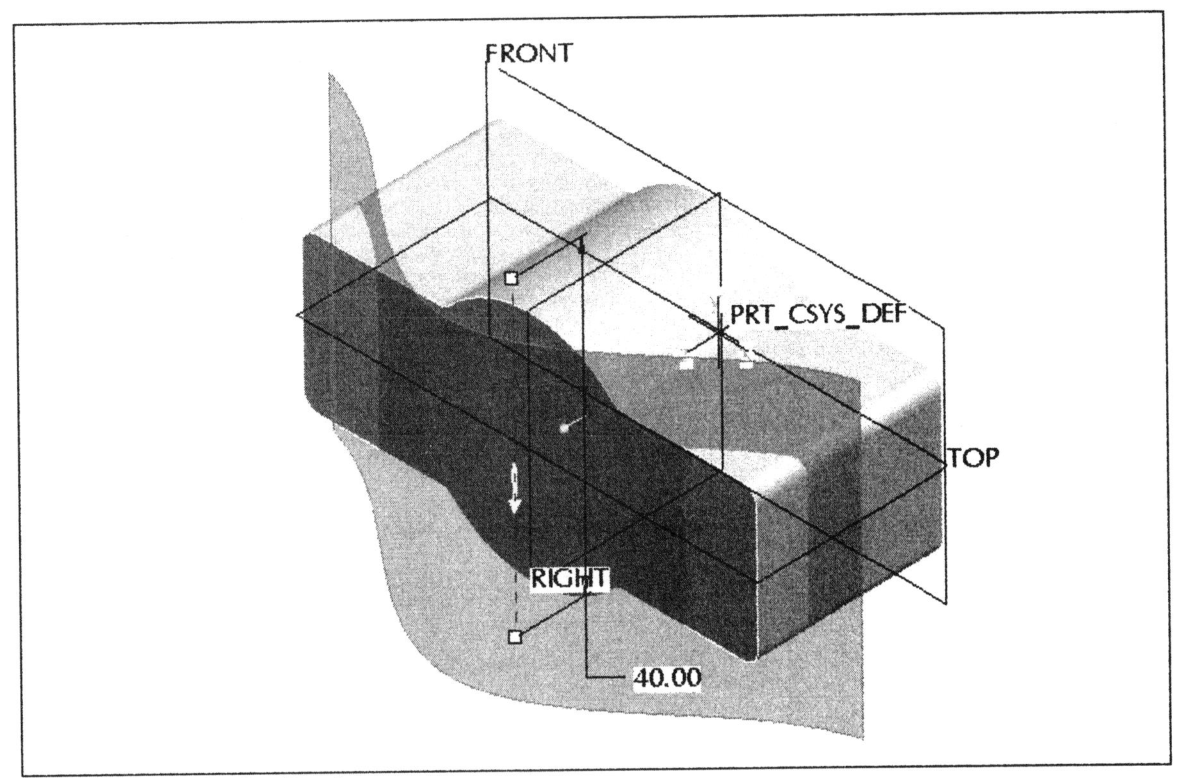

Figure 10.17(k) Surface Preview

Pick on the front surface of the protrusion until it highlights in pink [Fig. 10.17(l)] ⇒ [Fig. 10.17(m)] ⇒ **Edit** ⇒ **Offset** ⇒ [Fig. 10.17(n)] ⇒ **Replace Surface Feature** ⇒ **References** tab [Fig. 10.17(o)] ⇒ ⇒ pick on the extruded surface [Fig. 10.17(p)] ⇒ **MMB** ⇒ **Ctrl+S** ⇒ **MMB** [Fig. 10.17(q)] ⇒ **LMB** to deselect

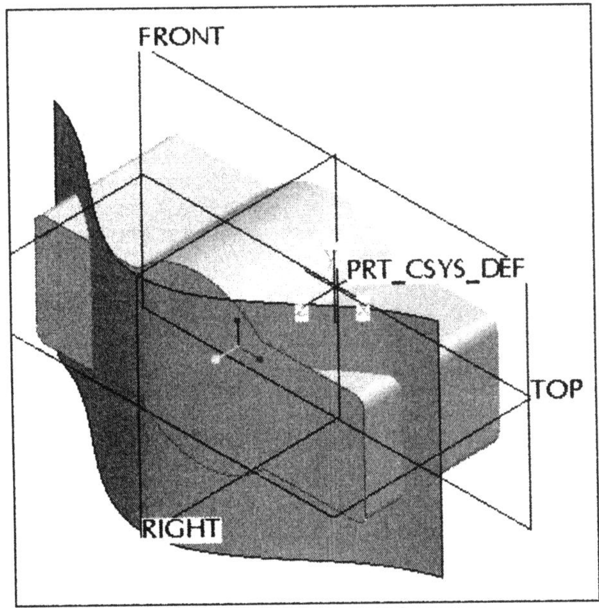

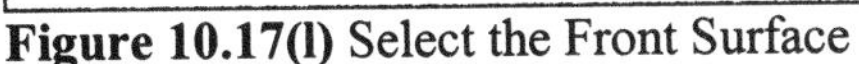

Figure 10.17(l) Select the Front Surface

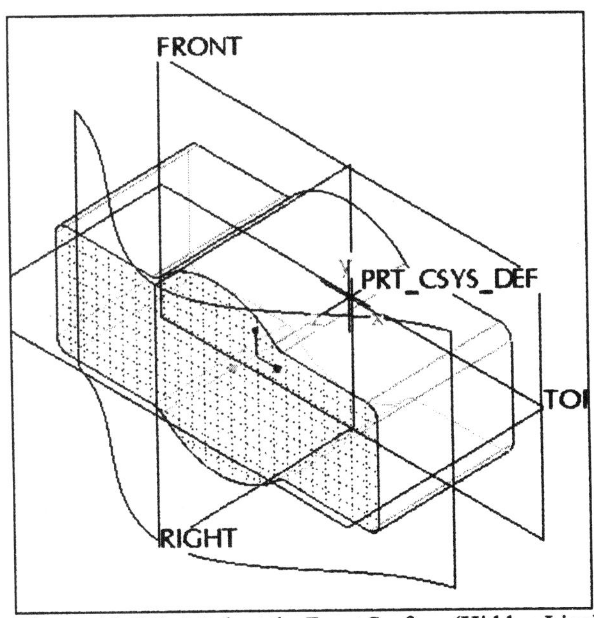

Figure 10.17(m) Select the Front Surface (Hidden Line)

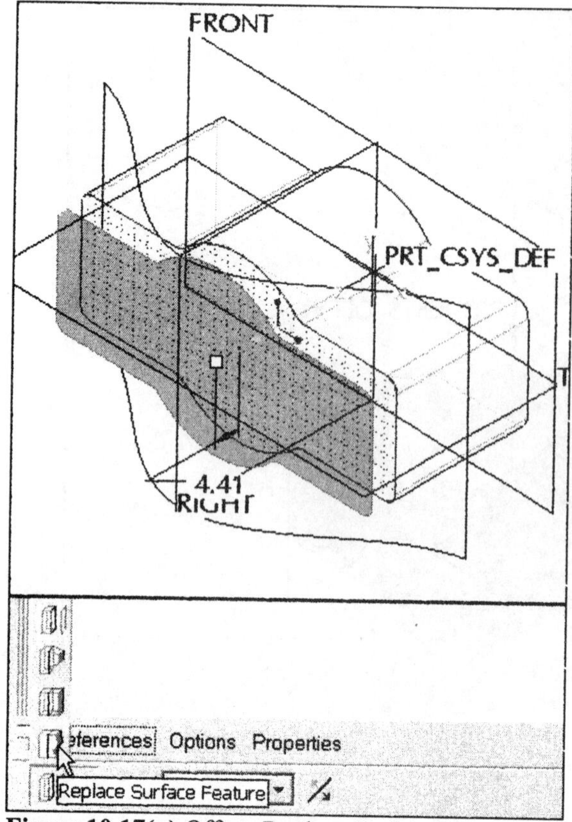

Figure 10.17(n) Offset: Replace Surface Feature

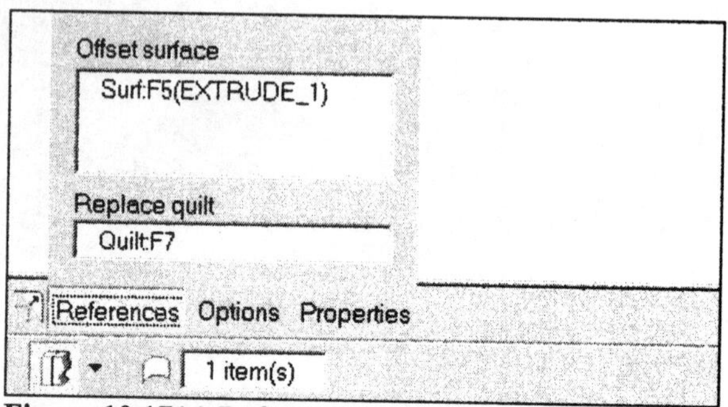

Figure 10.17(o) References Tab

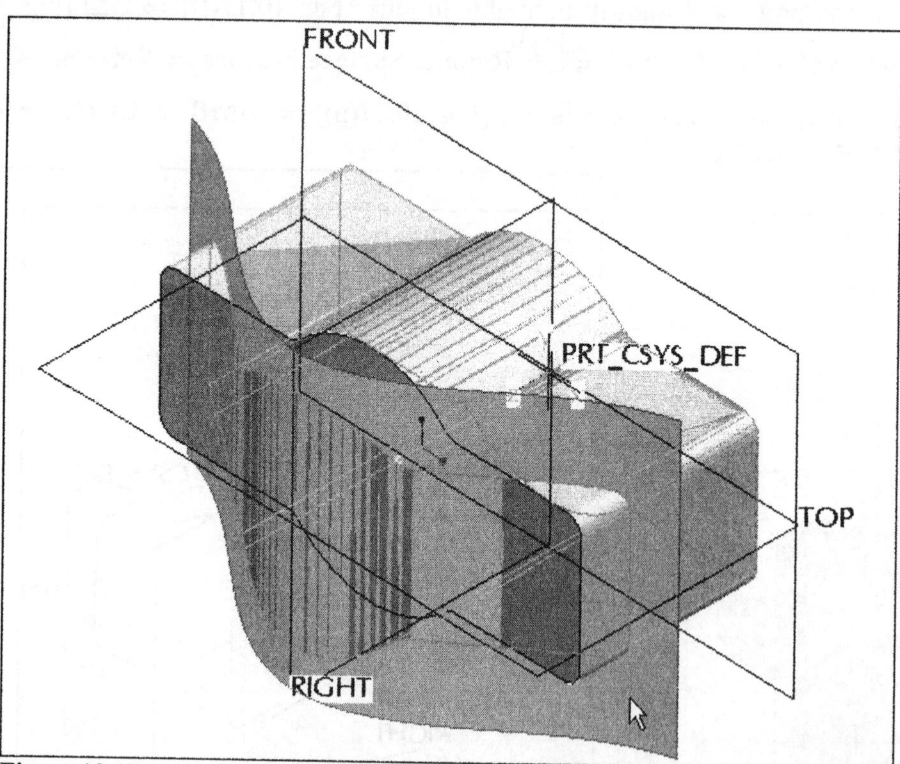

Figure 10.17(p) Pick on the Extruded Surface

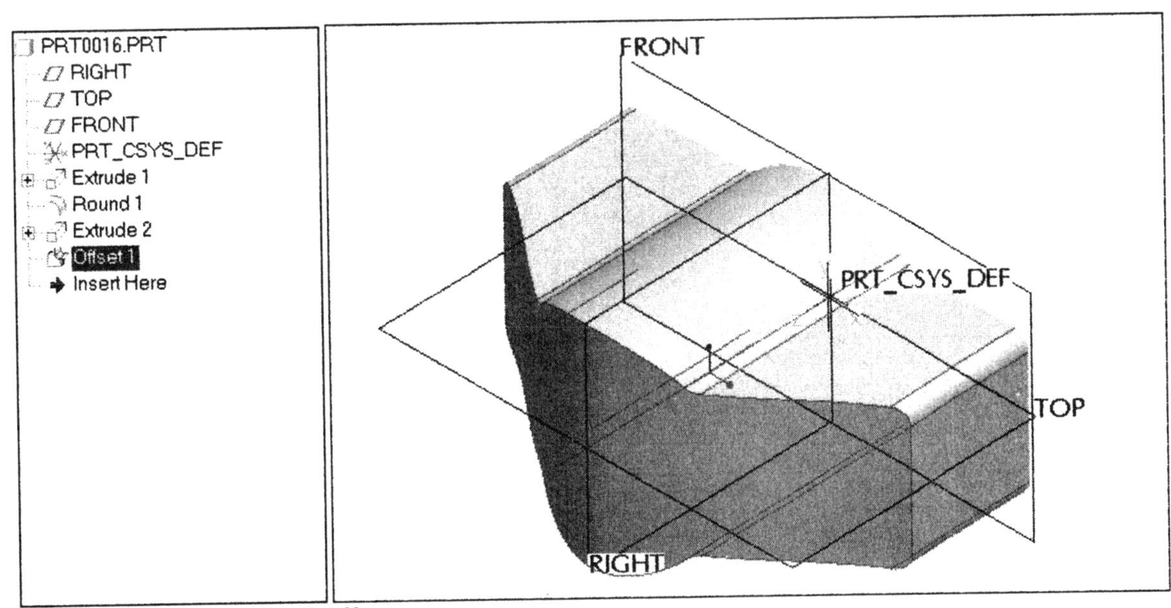

Figure 10.17(q) Completed Offset

Pick on an edge [Fig. 10.17(r)] ⇒ press and hold **Shift** key ⇒ pick on the front surface [Fig. 10.17(s)] ⇒ **RMB** ⇒ **Round Edges** ⇒ 2.00 ⇒ **Enter** ⇒ **MMB** ⇒ **View** ⇒ **Shade** [Fig. 10.17(t)] ⇒ **LMB**

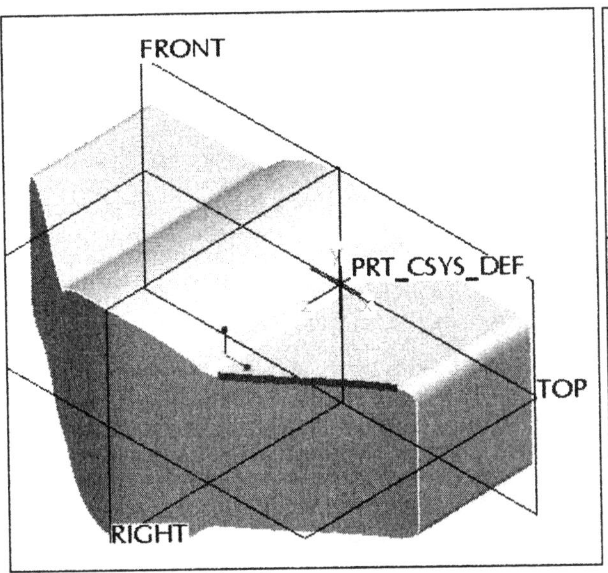

Figure 10.17(r) Pick Edge

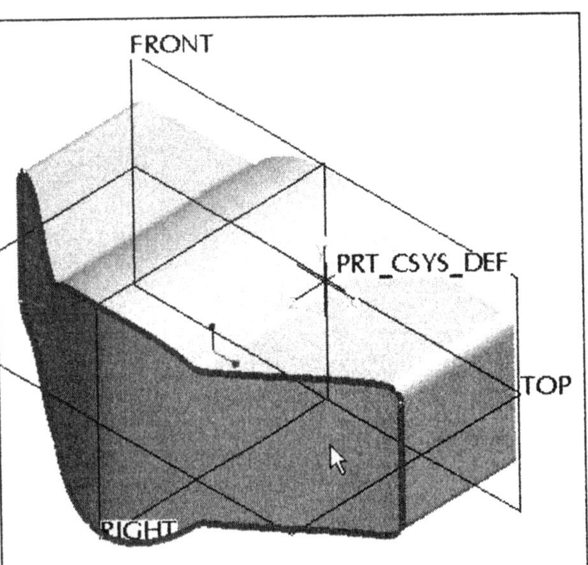

Figure 10.17(s) Shift and Pick the Front Surface

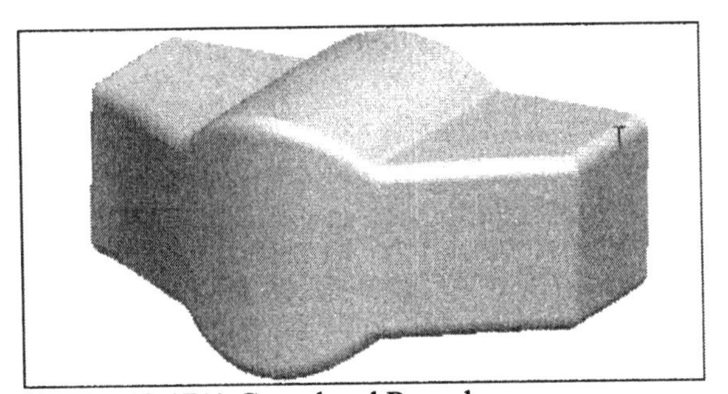
Figure 10.17(t) Completed Rounds

Click: **Ctrl+S** ⇒ **Enter** ⇒ **Ctrl+R** ⇒ **MMB** rotate the part ⇒ pick on the flat bottom surface until it is pink [Fig. 10.17(u)] ⇒ **Shell Tool** ⇒ Thickness 1 ⇒ **Enter** ⇒ ⇒ **MMB** ⇒ **MMB** [Fig. 10.17(v)] ⇒ **Ctrl+D** ⇒ **LMB** ⇒ **Ctrl+S** ⇒ **OK**

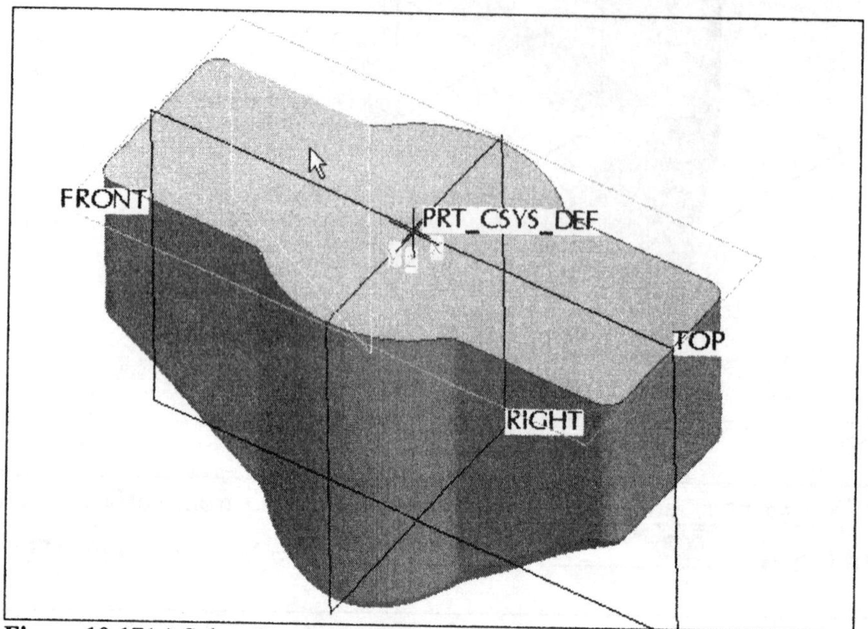

Figure 10.17(u) Select Bottom Surface

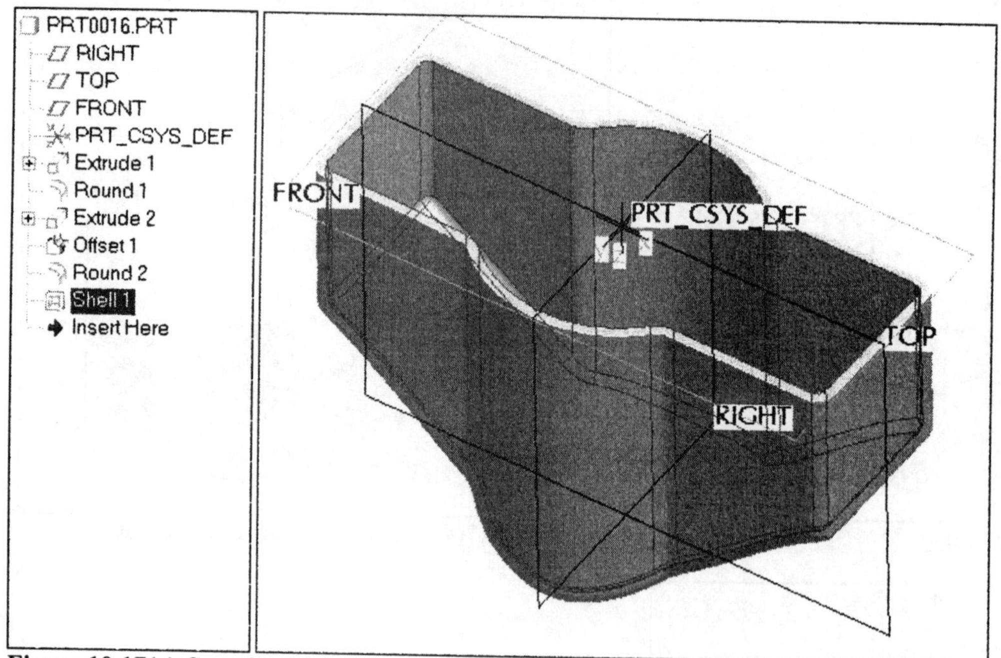

Figure 10.17(v) Completed Part

For those who wish additional information on surfaces, see ***www.cad-resources.com*** ⇒ ***Downloads***, for extra free projects.

Appendix

Customizing the User Interface (UI)

Customize your user interface to increase your efficiency in modeling. You can customize the Pro/E user interface, according to your needs or the needs of your group or company, to include the following:

- Create keyboard macros, called *mapkeys*, and add them to the menus and toolbars
- Add or remove existing toolbars
- Add split buttons to the toolbars (split buttons contain multiple closely-related commands and save toolbar space by hiding all but the first active command button)
- Move or remove commands from the menus or toolbars
- Change the location of the message area
- Add options to the Menu Manager
- Blank (make unavailable) options in the Menu Manager
- Set default command choices for Menu Manager menus

With a Pro/E file active, click: **Tools** ⇒ **Customize Screen** ⇒ Customize dialog box opens with the Commands tab active (Fig. A.1) ⇒ Categories, click: **View** (Fig. A.2) ⇒ click on **Display object in standard orientation** ⇒ drag and drop in the top Tool chest, Tool bar ⇒ Categories, click: **Model Display** ⇒ **colors on/off** ⇒ drag and drop (Fig. A.3)

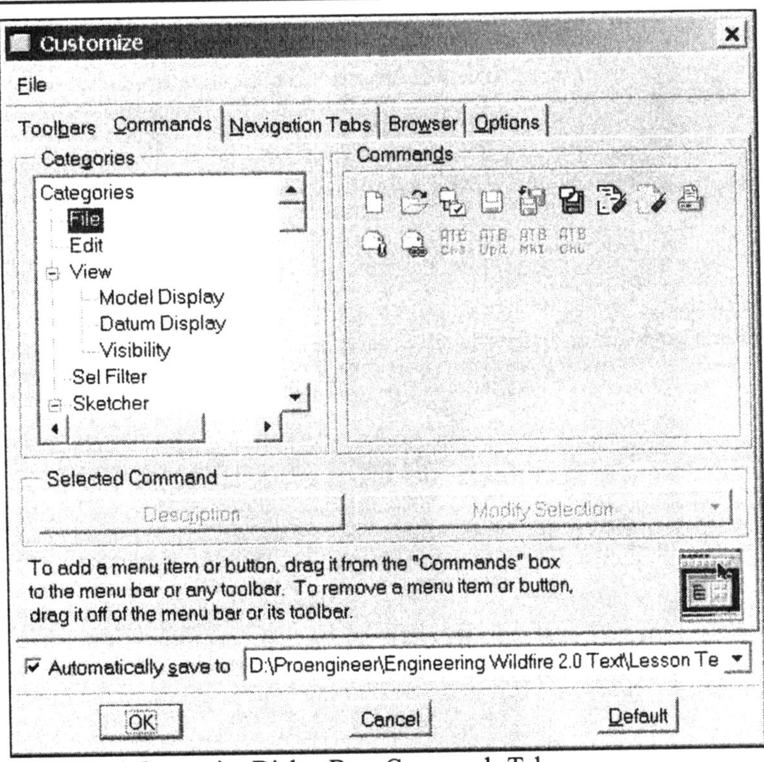

Figure A.1 Customize Dialog Box, Commands Tab

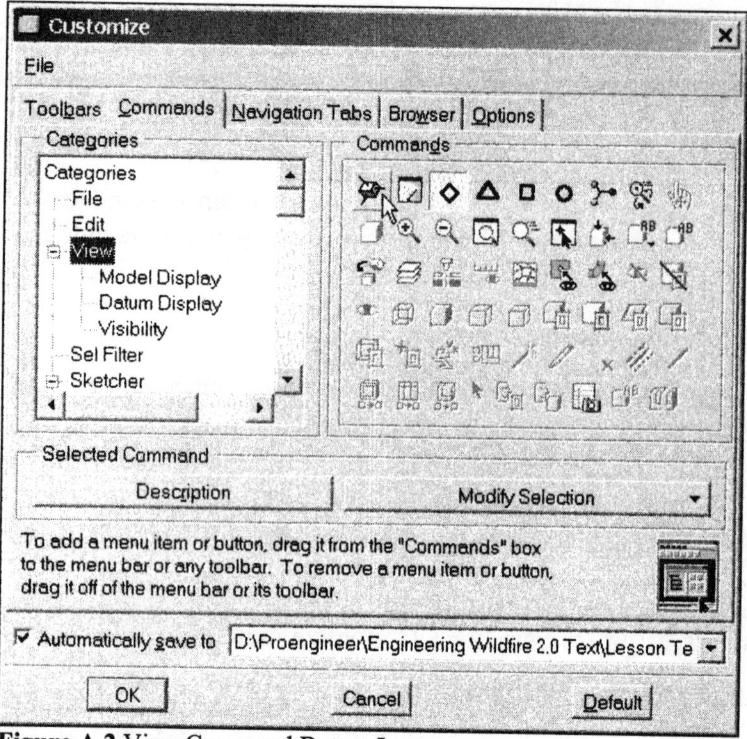

Figure A.2 View Command Button Icons

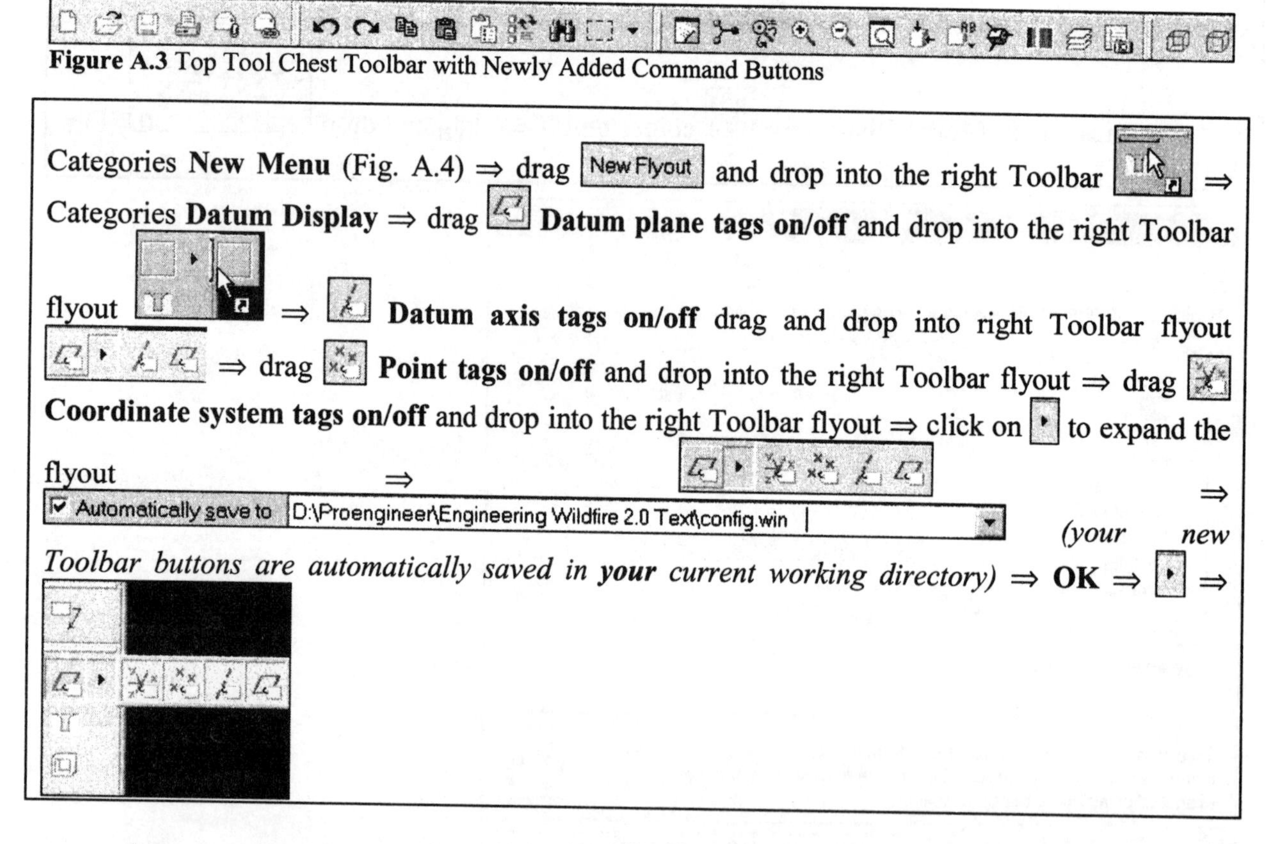

Figure A.3 Top Tool Chest Toolbar with Newly Added Command Buttons

Categories **New Menu** (Fig. A.4) ⇒ drag New Flyout and drop into the right Toolbar ⇒ Categories **Datum Display** ⇒ drag **Datum plane tags on/off** and drop into the right Toolbar flyout ⇒ **Datum axis tags on/off** drag and drop into right Toolbar flyout ⇒ drag **Point tags on/off** and drop into the right Toolbar flyout ⇒ drag **Coordinate system tags on/off** and drop into the right Toolbar flyout ⇒ click on to expand the flyout ⇒ ⇒ Automatically save to D:\Proengineer\Engineering Wildfire 2.0 Text\config.win *(your new Toolbar buttons are automatically saved in **your** current working directory)* ⇒ **OK** ⇒ ⇒

You can recall saved settings by clicking: *Tools ⇒ Customize Screen ⇒ File ⇒ Open Settings ⇒ select the file ⇒ Open ⇒ OK.* Buttons can be removed from the Toolbar using the exact same method, except, drag the buttons away from the Toolbar and release the mouse button.

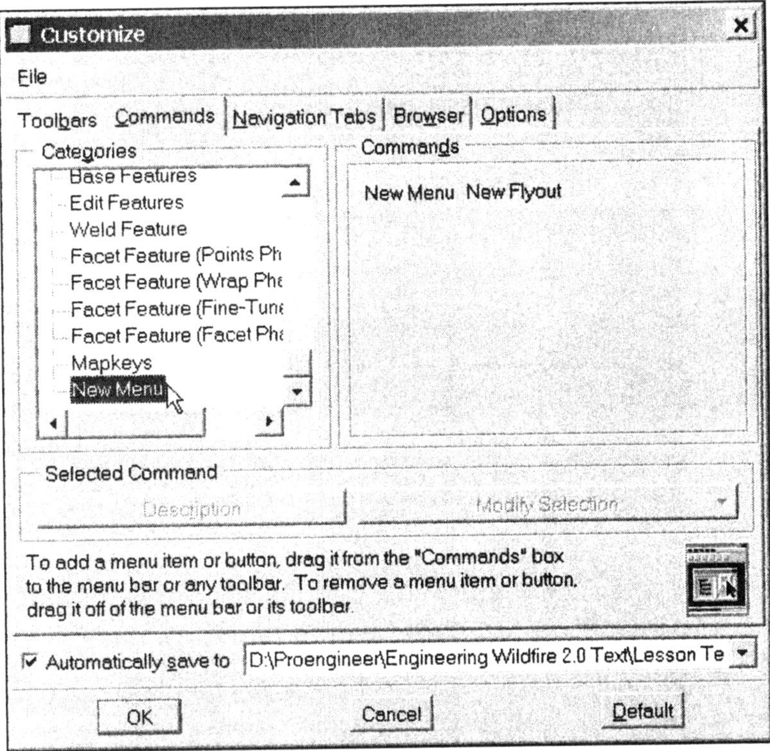

Figure A.4 New Menu Category

Next, add a Toolbar to the left side of the Navigator Window, click: **Tools** ⇒ **Customize Screen** ⇒ click **Toolbars** tab ⇒ Tools (Fig. A.5) ⇒ **Left** (Fig. A.6) ⇒ **OK**

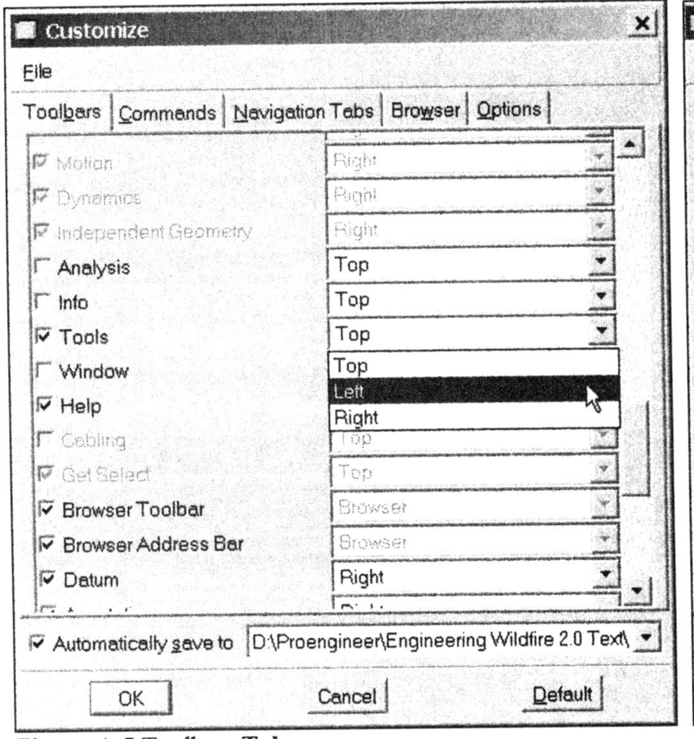

Figure A.5 Toolbars **Tab**

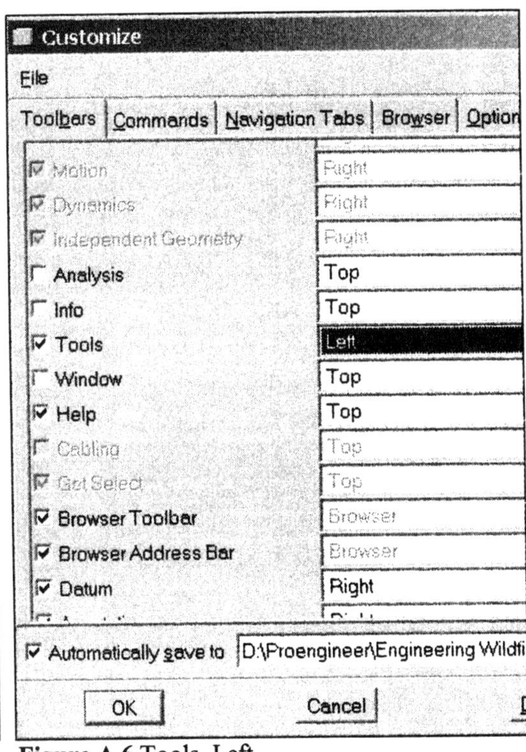

Figure A.6 Tools, Left

The Tools toolbar (Fig. A.7) includes: **Set various environment options**, **Run trail or training file**, **Create macros**, and **Select hosts for distributed computing**.

Figure A.7 Newly Added Command Buttons

The Navigation Tabs tab provides options for controlling the location of the Navigator (left or right), its width setting, and its placement in relation to the Model Tree settings. Click: **Tools ⇒ Customize Screen** ⇒ click: **Navigation Tabs** tab and explore the settings (Fig. A.8)

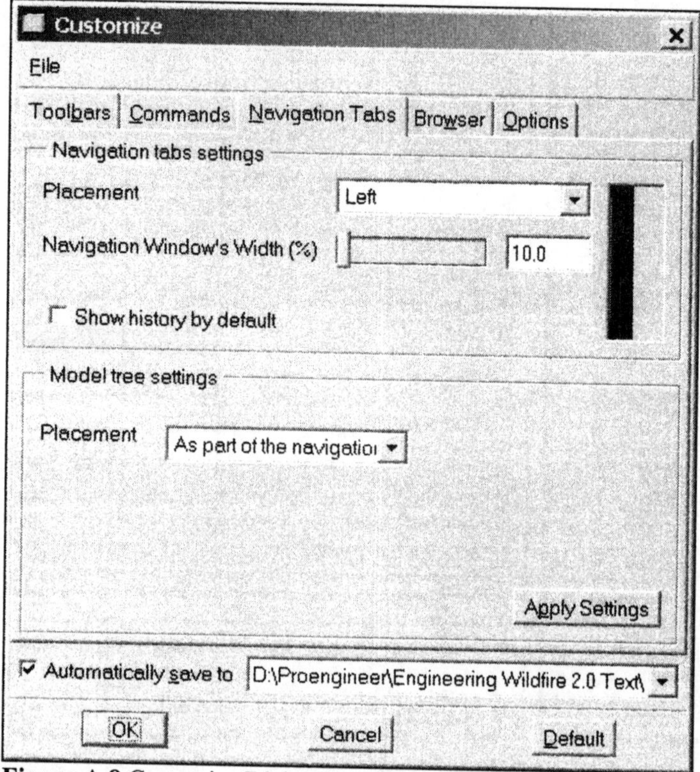

Figure A.8 Customize Dialog Box, Navigation Tabs

The Browser tab is used to control its window width and animation option. Click **Browser** tab and explore the options (Fig. A.9) ⇒ The Options tab provides settings to locate the Dashboard, Secondary Window size, and Menu display. Click: **Options** tab and explore its capabilities (Fig. A.10) ⇒ **OK**

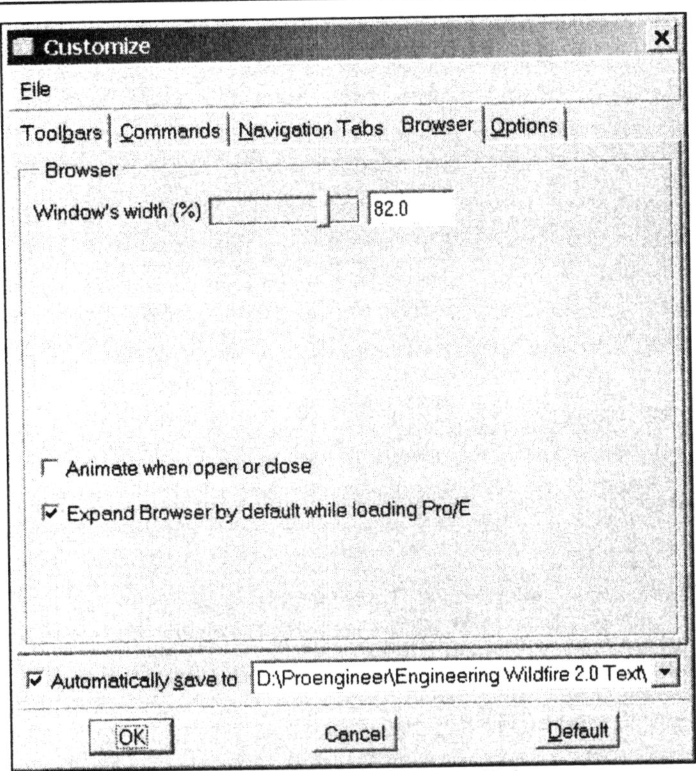

Figure A.9 Customize Dialog Box, Browser Tab

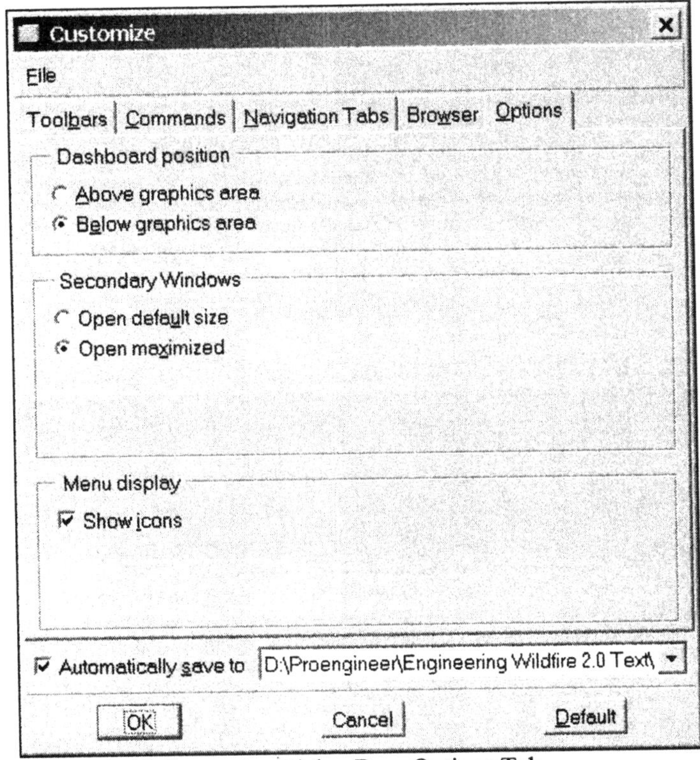

Figure A.10 Customize Dialog Box, Options Tab

Mapkeys

In Pro/E, a **Mapkey** is a macro that maps frequently used command sequences to certain keyboard keys or sets of keys. Mapkeys are saved in the configuration file, and are identified with the option *mapkey*, followed by the identifier and then the macro. You can define a unique key or combination of keys which, when pressed, executes the mapkey macro (for example, **F6** on your keyboard). You can create a mapkey for virtually any task you perform frequently within Pro/E.

By adding mapkeys to your toolbar or menu bar, you can use mapkeys with a single mouse click or menu command and thus streamline your workflow in a visible way.

To create a mapkey, you can use the configuration file option *mapkey*, or, on the Pro/E menu bar, click Tools ⇒ Mapkeys, and then in the Mapkeys dialog box, you click New and record your mapkey in the Record Mapkey dialog box. Pro/E records your mapkey as you step through the sequence of keystrokes or command executions to define it. After you define the mapkey, Pro/E creates a corresponding icon and places it in the Customize dialog box under the Mapkeys category. To open the Customize dialog box, click: Tools ⇒ Customize Screen. On the Commands tabbed page, select the Mapkeys category. You can then drag the visible mapkey icon onto the Pro/E main toolbar. You can also create a label for the new mapkey.

You can also nest one mapkey within another, so that one mapkey initiates another. To do so, you include the mapkey name in the sequence of commands of the new mapkey you are defining.

Mapkeys include the ability to do the following:

- Pause for user interaction.
- Handle message window input more flexibly.
- Run operating system scripts and commands. The Record Mapkey dialog box contains the OS Script tabbed page, whose options allow you to run OS commands instead of Pro/E commands.

When you define a mapkey, Pro/E automatically records a pause when you make screen selections, so that you can make new selections while the mapkey is running. In addition, you can record a pause (at any place in the mapkey) along with a user-specified dialog prompt, which will appear at the corresponding point while the mapkey is running.

If you create a new mapkey that contains actions that open and make selections from dialog boxes, then when you run the mapkey, it does not pause for user input when it opens the dialog box. To set the mapkey to pause for user input when opening dialog boxes, you must select *Pause for keyboard input* on the Pro/E tab in the Record Mapkey dialog box before you create the new mapkey.

Mapkeys Dialog Box

You use the Mapkeys dialog box (Fig. A.11), to define new mapkeys, modify, and delete existing mapkeys, run a mapkey chosen from the list, and save mapkeys to a configuration file. To open the Mapkeys dialog box, click Tools ⇒ Mapkeys. The following defines each command option on the dialog box:

- **New** Allows you to define a new mapkey and opens the Record Mapkey dialog box
- **Modify** Allows you to modify the selected mapkey
- **Run** Allows you to run the selected mapkey
- **Delete** Allows you to delete the selected mapkey
- **Save** Allows you to save the selected mapkey to a configuration file
- **Changed** Allows you to save only the mapkeys changed in the current session
- **All** Save all the mapkeys

In the Record Mapkey dialog box (Fig. A.12), you can type the key sequence that is to be used to execute the mapkey in the Key Sequence text box. To use a function key, precede its name with a dollar sign ($). For example, to map to **F7**, type **$F7**. Type the Name and Description of the mapkey in the appropriate text boxes. On the Pro/E tab, specify how Pro/E will handle the prompts when running the mapkey by selecting one of the following commands:

- **Record keyboard input** (default selection) Record the keyboard input for prompts when defining the mapkey, and use it when running the macro
- **Accept system defaults** Accept the system defaults when running the macro
- **Pause for keyboard input** Pause for user input when running the macro

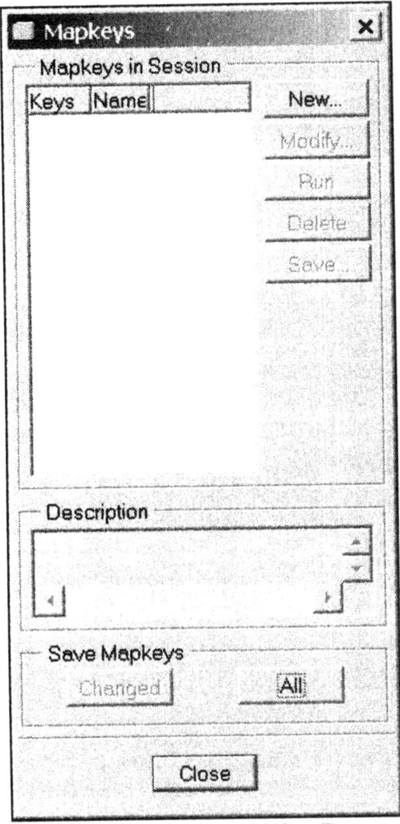

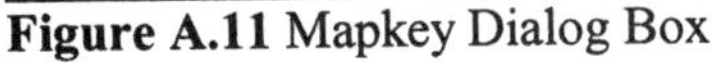

Figure A.11 Mapkey Dialog Box

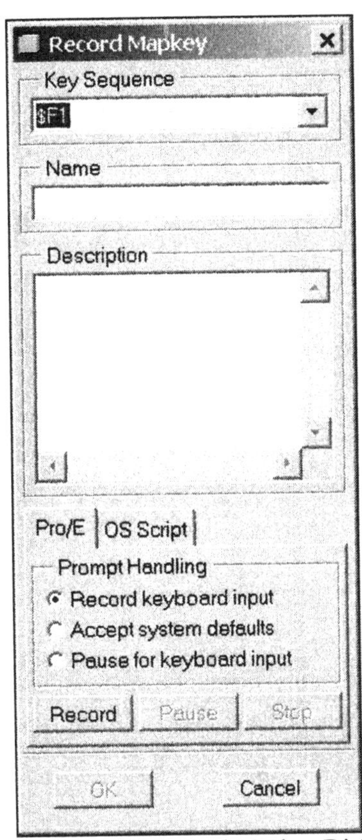

Figure A.12 Record Mapkey Dialog Box

Use Record to start recording the macro by selecting menu commands in the appropriate order. Use Pause to indicate where to pause while running the mapkey. Type the prompt in the Resume Prompt dialog box. Use Resume and continue recording the mapkey. When you run the macro, Pro/E will pause, display the prompt you typed, and give you the options to Resume running the macro or Cancel. Use Stop when you are finished recording the macro.

After you define the mapkey, a corresponding button appears in the Customize dialog box. You can then drag the mapkey onto the toolbar just like the Pro/E- supplied buttons.

Mapkeys include the ability to pause for user interaction, handle message window input flexibly, and run operating system commands.

Create a mapkey, click: **Tools** ⇒ **Mapkeys** Mapkeys dialog box opens ⇒ **New** Record Mapkey dialog box opens ⇒ Key Sequence- **$F5** ⇒ Name **COLORS** ⇒ Description **Opens Appearance Editor and starts New Color Definition** (Fig. A.13) ⇒ **Record** ⇒ **View** from menu bar ⇒ **Color and Appearance** ⇒ [+] **Add new appearance** ⇒ Color: **Color** button [Color] ⇒ **Stop** ⇒ **OK** ⇒ [All] **Save all mapkeys to config file** ⇒ [Name my_config.pro] ⇒ **Ok** ⇒ **Close** ⇒ **Close** Color Editor ⇒ **Close** Appearance Editor ⇒ **Tools** ⇒ **Customize Screen** ⇒ **Commands** tab ⇒ **Mapkeys** from the Categories list ⇒ click on the new mapkey **COLORS** ⇒ press the **RMB** (Fig. A.14) ⇒ click **Choose Button Image** ⇒ select [◇] (Fig. A.15) ⇒ **Modify Selection** (Fig. A.16) ⇒ **Edit Button Image** Button Editor Opens (Fig. A.17) ⇒ click on a color block and edit the picture as you wish (Fig. A.18) ⇒ **OK**

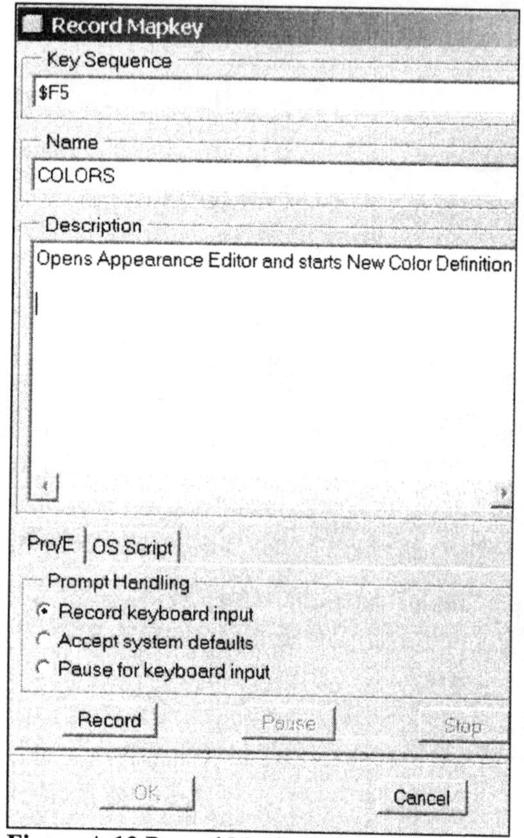

Figure A.13 Record Mapkey Dialog Box

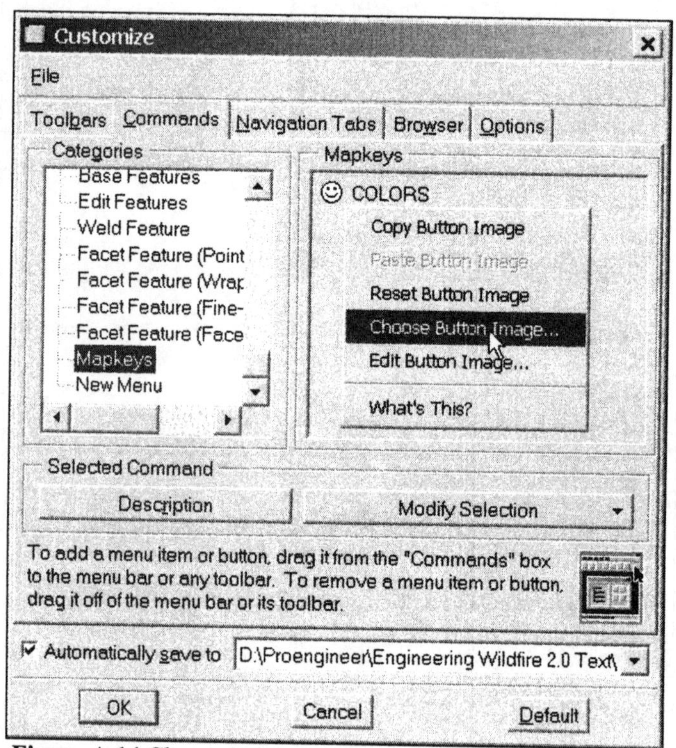

Figure A.14 Choose Button Image

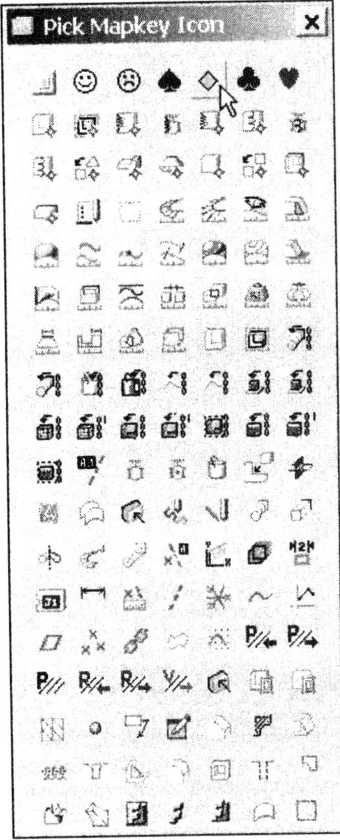

Figure A.15 Select an Icon

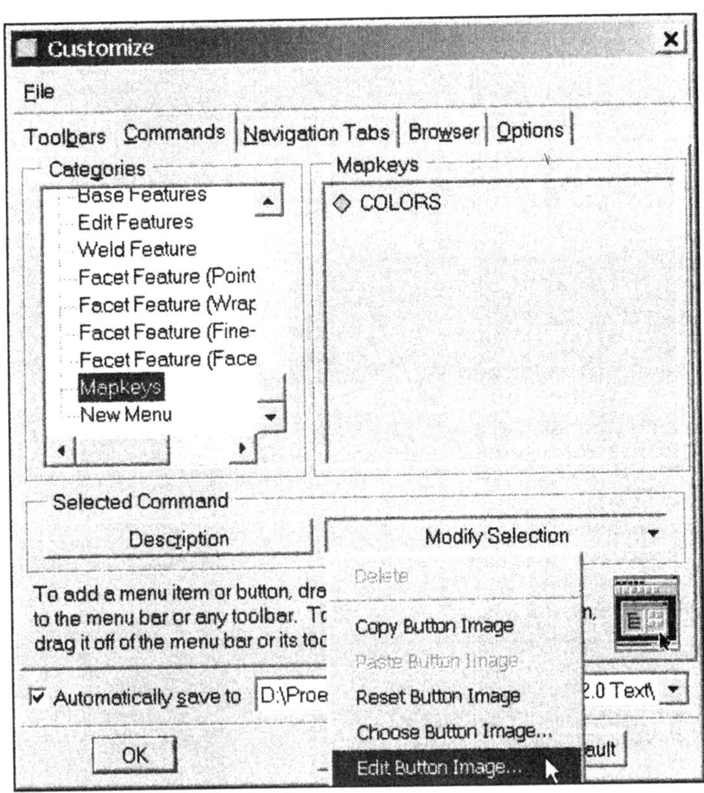

Figure A.16 Edit Button Image

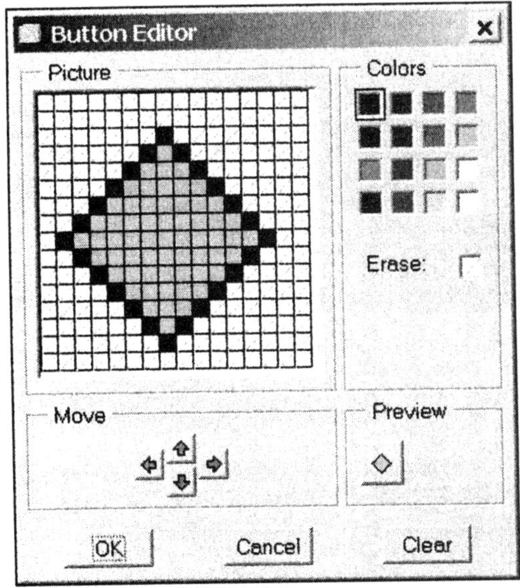

Figure A.17 Button Editor

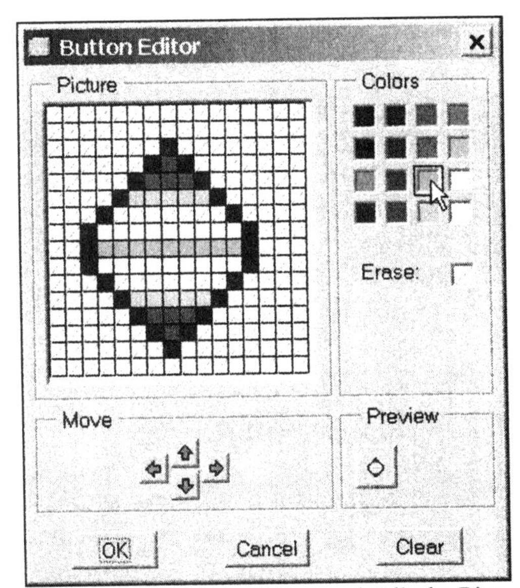

Figure A.18 Pick on Colors and Edit the Picture

Pick the new mapkey: COLORS (Fig. A.19) ⇒ drag to the Toolbar and drop (Fig.A.20) ⇒ **OK** ⇒ test the new button, click: **Opens Appearance Editor and starts New Color Definition** Color Editor dialog box opens (Fig. A.21) ⇒ **Close** ⇒ **Close** ⇒ ⇒ **MMB**

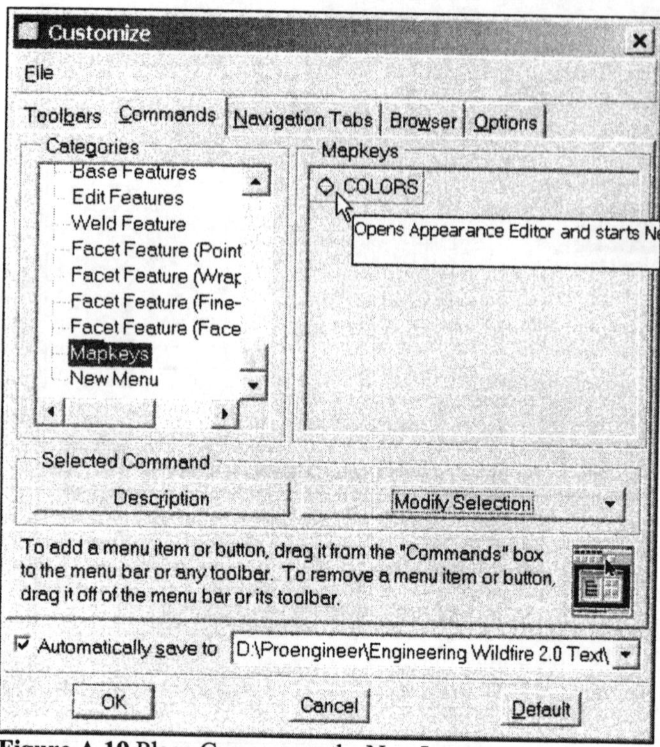

Figure A.19 Place Cursor over the New Icon to see the Description

Figure A.20 Drag and Drop in Toolbar

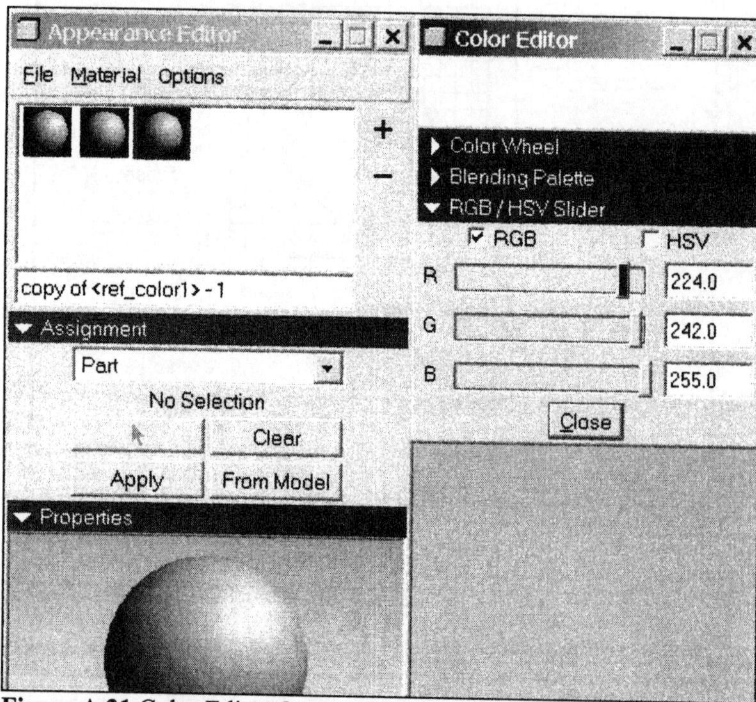

Figure A.21 Color Editor Opens

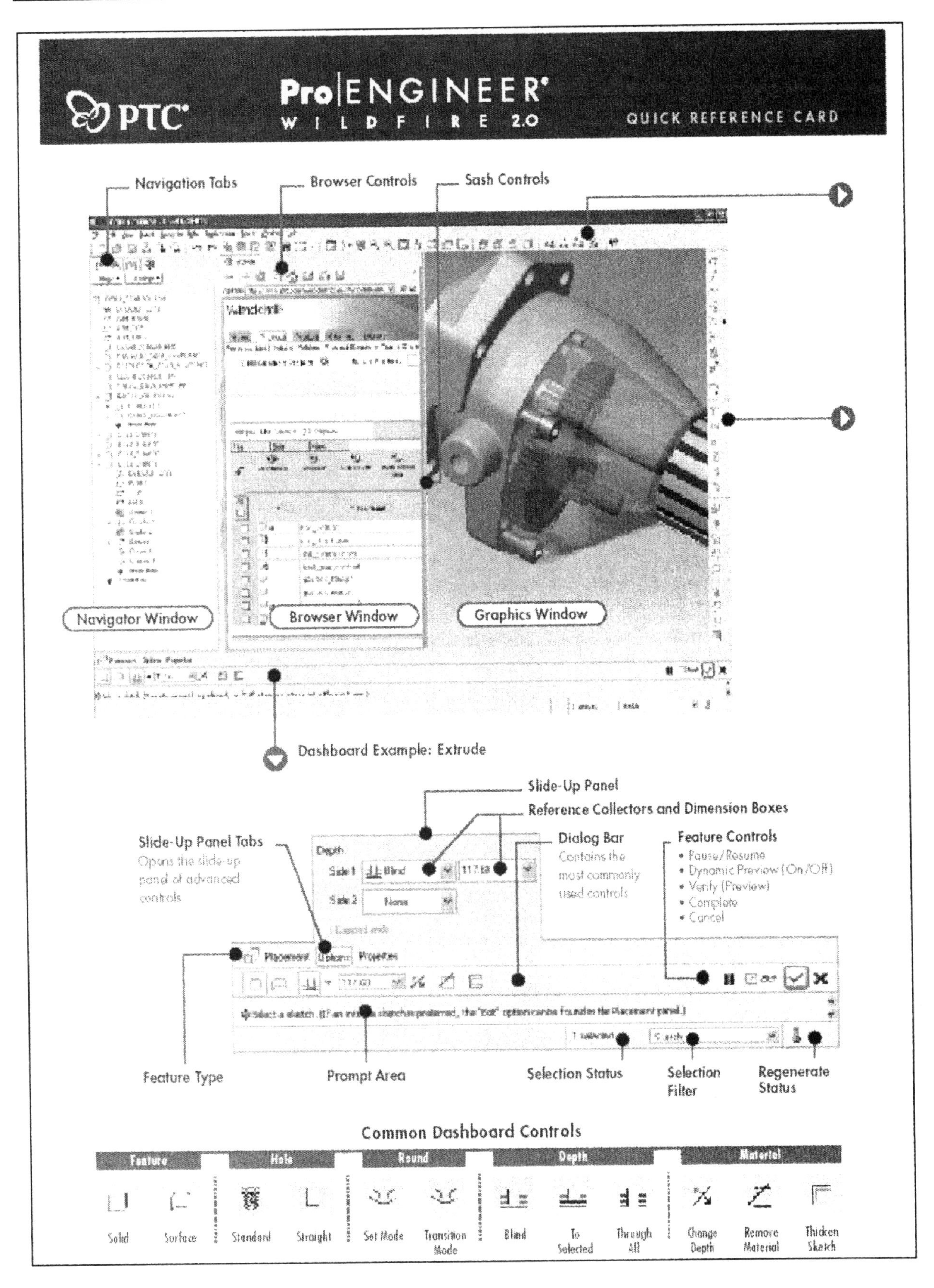

See www.ptc.com for PDF download of Quick Reference Cards.

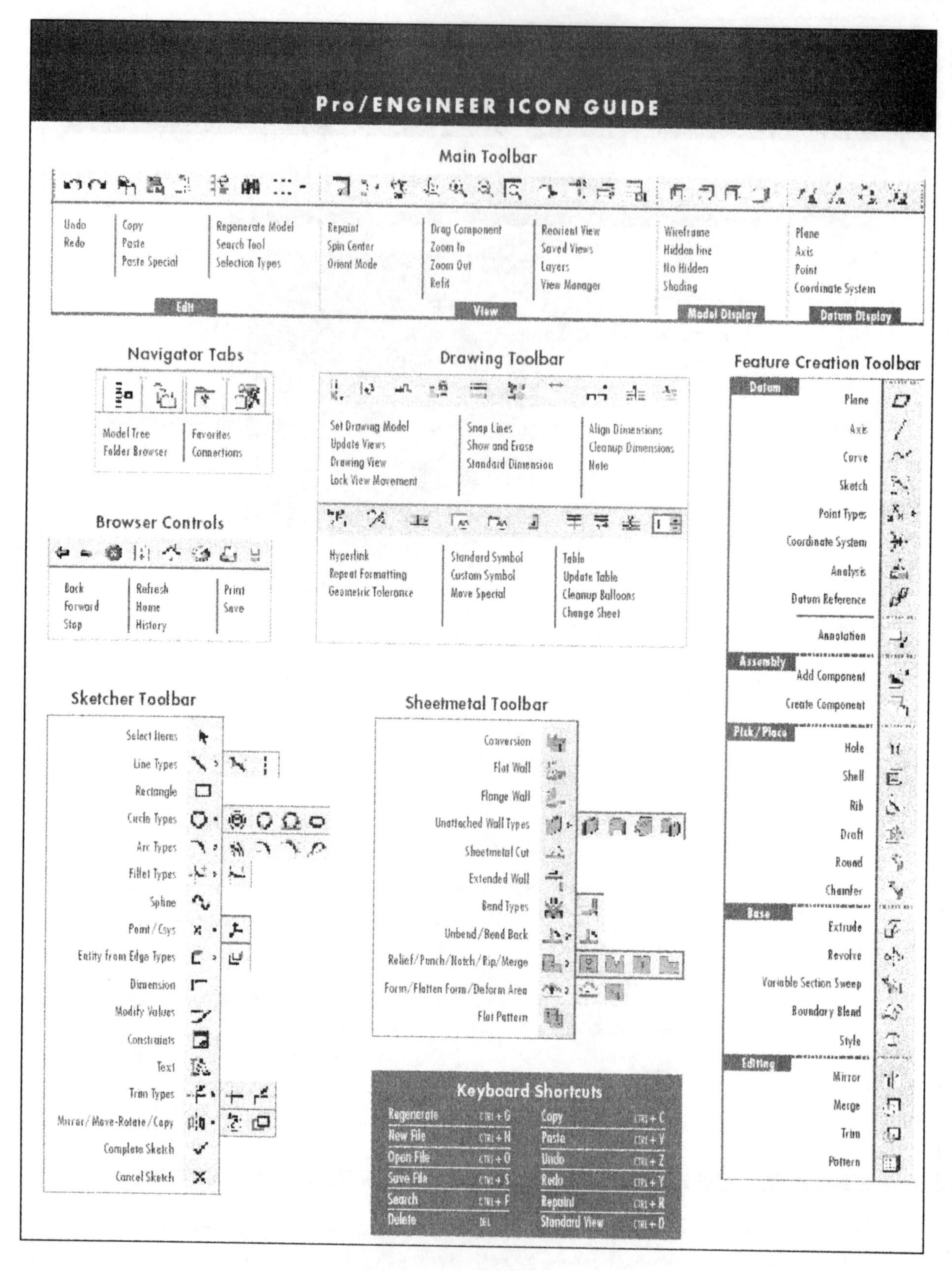

See www.ptc.com for PDF download of Quick Reference Cards.

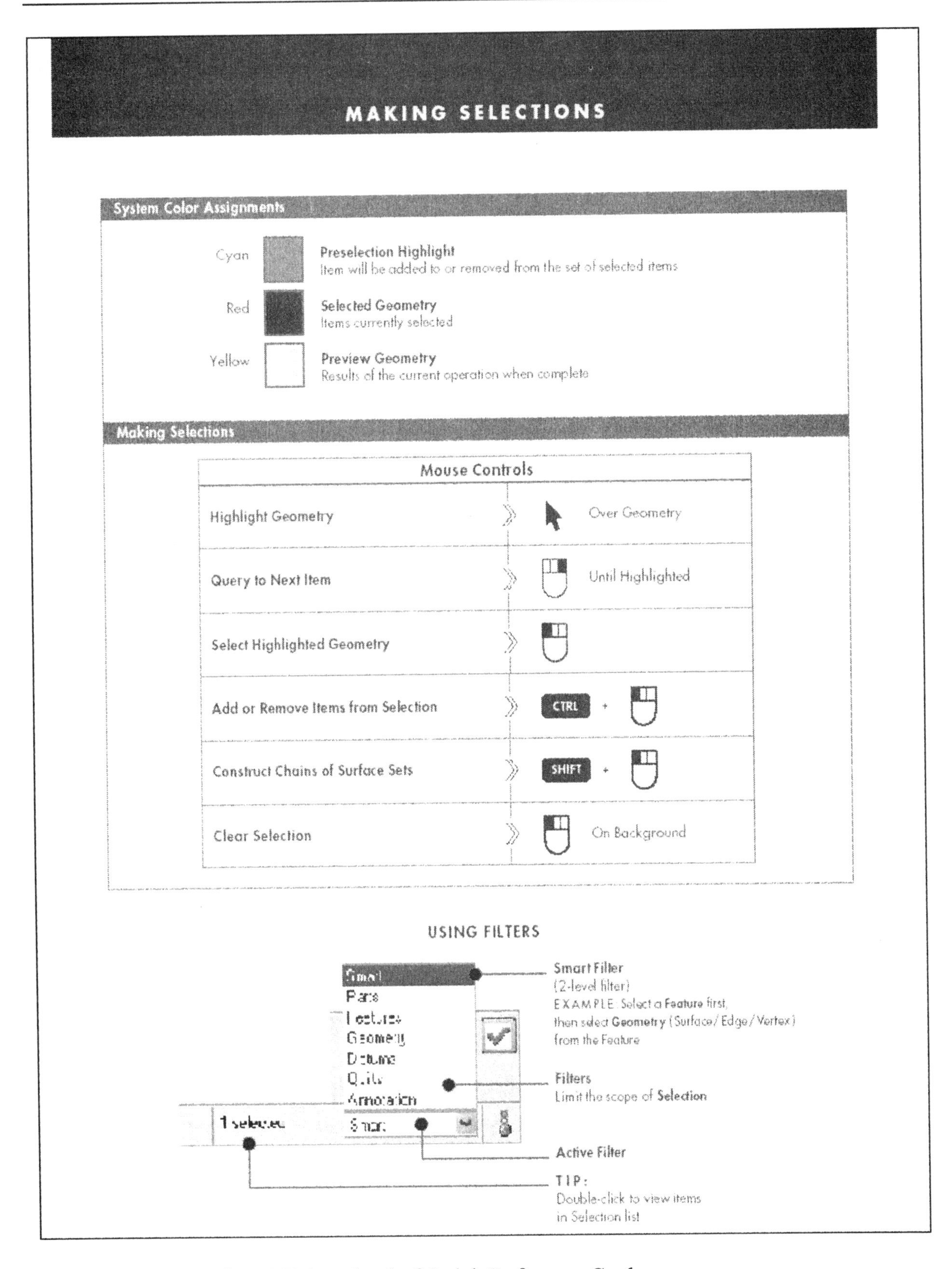

See www.ptc.com for PDF download of Quick Reference Cards.

ADVANCED SELECTION: Chain and Surface Set Construction

DEFINITIONS

General Definitions

Chain
A collection of adjacent edges and curves that share common endpoints. Chains can be open-ended or closed-loop, but they are always defined by two ends.

Surface Set
A collection of surface patches from solids or quilts. The patches do not need to be adjacent.

Methods of Construction

Individual
Constructed by selecting individual entities (edges, curves, or surface patches) one at a time. This is also called the One-by-One method

Rule-Based
Constructed by first selecting an anchor entity (edge, curve, or surface patch), and then automatically selecting its neighbors (a range of additional edges, curves, or surface patches) based on a rule. This is also called the Anchor/Neighbor method.

CONSTRUCTING CHAINS

Individual Chains

One-by-One
To select adjacent edges one at a time along a continuous path:

1 Select an edge
2 Hold down SHIFT
3 Select the edge again
4 Select adjacent edges
5 Release SHIFT

Rule-Based Chains

Tangent
To select all the edges that are tangent to an anchor edge:

1 Select an edge
2 Hold down SHIFT
3 Highlight **Tangent** chain (Query may be required)
4 Select tangent chain
5 Release SHIFT

Boundary
To select the outermost boundary edges of a quilt:

1 Select a one-sided edge of a quilt
2 Hold down SHIFT
3 Highlight **Boundary** chain (Query may be required)
4 Select boundary chain
5 Release SHIFT

Surface Loop
To select a loop of edges on a surface patch:

1 Select an edge
2 Hold down SHIFT
3 Highlight **Surface** chain (Query may be required)
4 Select surface loop
5 Release SHIFT

From-To
To select a range of edges from a surface patch or a quilt:

1 Select the **From** edge
2 Hold down SHIFT

3 Query to highlight the desired **From-To** chain
4 Select From-To chain
5 Release SHIFT

Multiple Chains

1 Construct initial chain
2 Hold down CTRL
3 Select an edge for new chain
4 Release CTRL
5 Hold down SHIFT
6 Complete new chain from selected edge

See www.ptc.com for PDF download of Quick Reference Cards.

CONSTRUCTING SURFACE SETS

Individual Surface Sets

Single Surfaces

To select multiple surface patches from solids or quilts one at a time:

1 Select a surface patch

2 Hold down CTRL

3 Select additional patches (Query may be required)

4 Release CTRL

Rule-Based Surface Sets

Solid Surfaces

To select all the surface patches of solid geometry in a model:

1 Select a surface patch on solid geometry

2 Right-click and select **Solid Surfaces**

Quilt Surfaces

To select all the surface patches of a quilt

1 Select a surface feature

2 Select the corresponding quilt

Loop Surfaces

To select all the surface patches that are adjacent to the edges of a surface patch:

1 Select a surface patch

2 Hold down SHIFT

3 Place the pointer over an edge of the patch to highlight the **Loop Surfaces**

4 Select the Loop Surfaces (The initial surface patch is de-selected)

5 Release SHIFT

Seed and Boundary Surfaces

To select all surface patches, from a **Seed** surface patch up to a set of **Boundary** surface patches:

1 Select the **Seed** surface patch

2 Hold down SHIFT

3 Select one or more surface patches to be used as boundaries

4 Release SHIFT (All surfaces from the Seed up to the Boundaries are selected)

Excluding Surface Patches from Surface Sets

To exclude surface patches during or after construction of a surface set:

1 Construct a surface set

2 Hold down CTRL

3 Highlight a patch from the surface set

4 Select the patch to de-select it

5 Release CTRL

CONSTRUCTING CHAINS AND SURFACE SETS USING DIALOG BOXES

To explicitly construct and edit Chains and Surface Sets, click **Details** next to a collector:

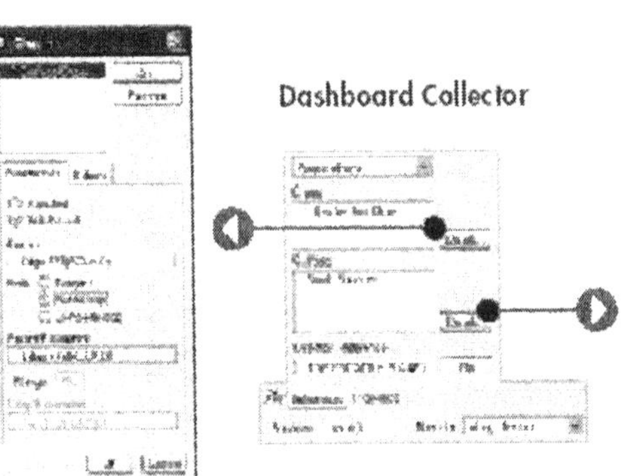

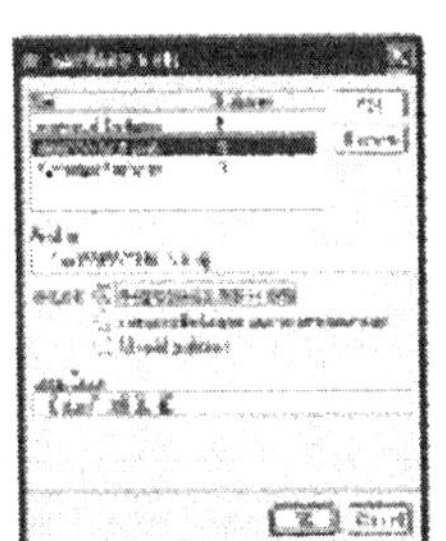

See www.ptc.com for PDF download of Quick Reference Cards.

ORIENTING THE MODEL

DYNAMIC VIEWING

3D Mode Hold down the key and button. Drag the mouse	
SPIN	
PAN	SHIFT +
ZOOM	CTRL +
TURN	CTRL +

2D Mode	
PAN	
ZOOM	CTRL +

2D and 3D Modes Hold down the key and roll the mouse wheel	
ZOOM	
FINE ZOOM	SHIFT +
COARSE ZOOM	CTRL +

Using the Spin Center

Click the icon in the Main Toolbar to enable the Spin Center.

- Enabled – The model spins about the location of the spin center
- Disabled – The model spins about the location of the mouse pointer

Using Orient Mode

Click the icon in the Main Toolbar to enable Orient mode.

- Provides enhanced Spin/Pan/Zoom Control
- Disables selection and highlighting
- Right-click to access additional orient options
- Use the shortcut: CTRL + SHIFT + Middle-click

Using Component Drag Mode in an Assembly

Click the icon in the Main Toolbar to enable Component Drag mode.

- Allows movement of components based on their kinematic constraints or connections
- Click a location on a component, move the mouse, click again to stop motion.
- Middle-click to disable Component Drag mode

COMPONENT PLACEMENT CONTROLS

Allows reorientation of components during placement

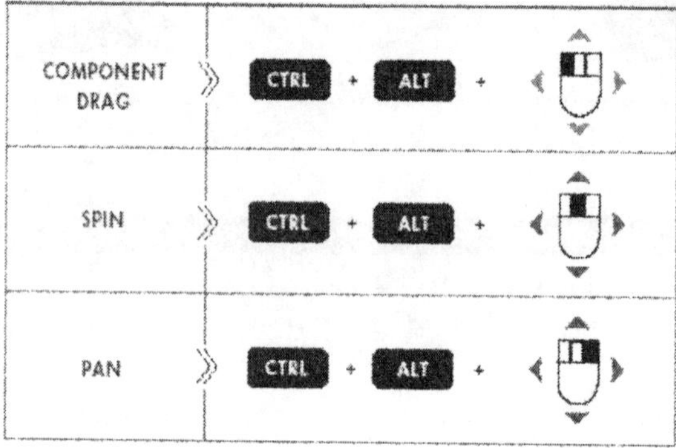

COMPONENT DRAG	CTRL + ALT +
SPIN	CTRL + ALT +
PAN	CTRL + ALT +

Object Mode

Provides enhanced Spin/Pan/Zoom Control:

1 Enable Orient mode
2 Right-click to enable Orient Object mode
3 Use Dynamic Viewing controls to orient the component
4 Right-click and select Exit Orient mode

DOC-RC60388-EN-340

See www.ptc.com for PDF download of Quick Reference Cards.

Index

E

F

G

H

I

L